Aquatic Ecosystems

Concern about future supplies of fresh water to society, to meet the full range of human needs, now comes very high on the priority list of global societal issues. An overarching issue that the book tries to address is whether global climate change is a dominant driver of change in the structure and function of natural aquatic ecosystems, or whether direct human population growth and accelerated consumption are playing an equal or greater role. This book divides the aquatic realm into 21 ecosystems, from those on land (both saline and fresh water) to those of the open and deep oceans. It draws on the understanding of leading ecologists to summarize the state and likely condition by the year 2025 of each of the ecosystems. Written for academic researchers and professionals, the aim is to put the climate change debate into a broader context as a basis for conservation science.

NICHOLAS V. C. POLUNIN is Professor of Marine Environmental Science at the School of Marine Science and Technology in Newcastle. He is also President of the Foundation for Environmental Conservation.

Aquatic Ecosystems
Trends and Global Prospects

Edited by
Nicholas V. C. Polunin
Foundation for Environmental Conservation
and
Newcastle University, UK

 CAMBRIDGE
UNIVERSITY PRESS

CAMBRIDGE UNIVERSITY PRESS
Cambridge, New York, Melbourne, Madrid, Cape Town, Singapore, São Paulo, Delhi

Cambridge University Press
The Edinburgh Building, Cambridge CB2 8RU, UK

Published in the United States of America by Cambridge University Press, New York

www.cambridge.org
Information on this title: www.cambridge.org/9780521833271

First published 2008

Printed in the United Kingdom at the University Press, Cambridge

A catalogue record for this publication is available from the British Library

Library of Congress Cataloguing in Publication data
Aquatic ecosystems : trends and global prospects / edited by Nicholas V. C. Polunin.
    p. cm.
   "Foundation for Environmental Conservation."
   Includes bibliographical references and index.
   ISBN 978-0-521-83327-1 (hardback)   1. Aquatic ecology.   I. Polunin, Nicholas.
   II. Foundation for Environmental Conservation.   III. Title.
   QH541.5.W3A679 2008
   577.6—dc22   2008012213

ISBN 978-0-521-83327-1 hardback

Contents

Contributors

PAUL ADAM
School of Biological, Earth and Environmental Sciences
University of New South Wales
Sydney, NSW 2052
Australia

NIKOLAY V. ALADIN
Zoological Institute of the Russian Academy of Sciences
Laboratory of Brackish Water Hydrobiology
St Petersburg
190034 Russia

JAVIER ALCOCER
Tropical Limnology Research Project (PILT)
UIICSE, FES Iztacala
Department of Biology
Universidad National Autónoma de México (UNAM)
54090 Mexico

BARBARA E. BEDFORD
Department of Natural Resources
121 Fernow Hall
Cornell University
Ithaca, NY 14853
USA

ALFRED M. BEETON
Scientist Emeritus
NOAA Great Lakes Enviromental Research
Laboratory (GLERL)
2205 Commonwealth Blvd.,
Ann Arbor, MI 48105
USA

MARK D. BERTNESS
Department of Ecology and Evolutionary Biology
Brown University
Providence, RI 02912
USA

JENS BORUM
Department of Biology

Freshwater Biological Laboratory
University of Copenhagen
51 Helsingørsgade
3400 Hillerød
Denmark

GEORGE M. BRANCH
Department of Zoology
University of Cape Town, Private Bag X3
Rondebosch 7701
Cape Town
South Africa

CHARLES M. BREEN
Center for Environment and Development
Institute of Natural Resources
University of Natal, Private Bag X01
Scotsville 3209
Pietermaritzburg
South Africa

ANDREW S. BRIERLEY
Gatty Marine Laboratory
University of St Andrews
Fife, KY16 8LB
UK

MARK M. BRINSON
Biology Department
East Carolina University
Greenville, NC 27858
USA

CHRISTER BRÖNMARK
Department of Limnology
University of Lund
221 00 Lund
Sweden

ALEXANDER C. BROWN
† (1931–2005), South African marine biologist

who authored 200 scientific publications including
several books, was a devoted teacher,
head of the Zoology Department at the University
of Cape Town, Director of the Centre for
Marine Studies and President of the Royal
Society of South Africa

ROBERT W. BUDDEMEIER
Kansas Geological Survey
University of Kansas
1930 Constant Avenue
Lawrence, KS 66045
USA

STUART E. BUNN
Australian Rivers Institute
Centre for Riverine Landscapes
Griffith University
Brisbane, QLD 4222
Australia

RODRIGO H. BUSTAMANTE
CSIRO Marine and Atmospheric Research
PO Box 120
Cleveland, QLD 4163
Australia

JUAN CARLOS CASTILLA
Centro de Estudios Avanzados en Ecología &
Biodiversidad
Facultad de Ciencias Biológicas
Pontificia Universidad Católica de Chile
Alameda 340, C.P. 6513677
Casilla 114-D, Santiago
Chile

ANDREW CLARKE
British Antarctic Survey
National Environment Research Council
High Cross, Madingley Road
Cambridge, CB3 0ET
UK

ALAN P. COVICH
Institute of Ecology
University of Georgia
Athens, GA 30602
USA

TASMAN P. CROWE
Department of Zoology
University College Dublin

Belfield, Dublin 4
Ireland

DAVID C. CULVER
Department of Biology
Hurst Hall 110A
American University
4400 Massachusetts Avenue
Washington, DC 20016
USA

DAN L. DANIELOPOL
Institute of Limnology
Austrian Academy of Sciences
Mondseestrasse 9
5310 Mondsee
Austria

ANTHONY J. DAVY
Centre for Ecology, Evolution and Conservation
(CEEC)
School of Biological Sciences
University of East Anglia
Norwich, NR4 7TJ
UK

PAUL K. DAYTON
Scripps Institution of Oceanography
University of California–San Diego
9500 Gilman Drive
La Jolla, CA 92093
USA

RICHARD S. DODD
Department of Environmental Science Policy and
Management
College of Natural Resources
321 Mulford Hall
University of California–Berkeley
Berkeley, CA 94720
USA

CARLOS M. DUARTE
Instituto Mediterráneo de Estudios Avanzados (IMEDEA)
Natural Resources Department
Miquel Marqués 21
07190 Esporles (Mallorca)
Spain

ROBERT ENGELMAN
Worldwatch Institute, Vice President of Programs
1776 Massachusetts Ave NW

Washington, DC 20036–1904
USA

C. MAX FINLAYSON
Global Research Division
Global Change
Water and Environment Group
International Water Management Institute (IWMI)
127 Sunil Mawatha
Pelwatte
Battaramulla
Sri Lanka

GLEN GEORGE
Institute of Geography and Earth Sciences
University of Wales–Aberystwyth
Old College
King Street
Aberystwyth, SY23 2AX
UK

JANINE GIBERT
University Claude Bernard Lyon 1, UMR/CNRS 5023
Groundwater Hydrobiology and Ecology Laboratory
Bât. Forel 403, 43 Boulevard du 11 Novembre 1918
69622 Villeurbanne Cedex
France

ADRIAN G. GLOVER
Zoology Department
The Natural History Museum
Cromwell Rd
London, SW7 5BD
UK

BRIJ GOPAL
School of Environmental Sciences
Jawaharlal Nehru University
New Delhi 110067
India

CHRISTOPHER GORDON
Volta Basin Research Project and Centre for
African Wetlands
University of Ghana
PO Box 209
Legon
Accra
Ghana

NICHOLAS A. J. GRAHAM
School of Marine Science and Technology

Newcastle University
Newcastle upon Tyne, NE1 7RU
UK

CHRISTIAN GRIEBLER
Institute of Groundwater Ecology
GSF–National Research Center for Environment
and Health
Ingolstädter Landstrasse 1
85764 Neuherberg
Germany

AMARA GUNATILAKA
Water Resources Management
Verbundplan Ltd, Engineers and Consultants
Parkring 12
1010 Vienna
Austria

STEPHEN J. HALL
WorldFish Center
PO Box 500 GPO
Penang, 10670
Malaysia

LARS-ANDERS HANSSON
Department of Limnology
University of Lund
22100 Lund
Sweden

COLIN M. HARRIS
Environmental Research and Assessment Ltd.
219c Huntingdon Road
Cambridge, CB3 0DL
UK

STEPHEN J. HAWKINS
Marine Biological Association of the UK
The Laboratory
Citadel Hill
Plymouth, PL1 2PB
UK

ROBERT E. HECKY
Department of Biology
University of Waterloo
200 University Avenue West
Waterloo, Ontario N2L 3G1
Canada

ALAN G. HILDREW
School of Biological and Chemical Sciences
Queen Mary and Westfield College (University of
London)
London, E1 4NS
UK

ALISTAIR J. HOBDAY
CSIRO Division of Marine Research
PO Box 120
Cleveland, QLD 4163
Australia
and
PO Box 1538, School of Zoology
Life Science Building 3/41
University of Tasmania
Hobart, Tasmania 7000
Australia

OVE HOEGH-GULDBERG
The Centre for Marine Studies
The University of Queensland
Brisbane, QLD 4072
Australia

VENUGOPALAN ITTEKKOT
Centre for Tropical Marine Ecology
Fahrenheitstrasse 6
28359 Bremen
Germany

NANCY L. JACKSON
Graduate Program in Environmental Policy Studies
Department of Chemistry and Environmental Science
New Jersey Institute of Technology (NJIT)
Newark, NJ 07102
USA

ROBERT JELLISON
Marine Science Institute
University of California–Sauta Barbara
Santa Barbara, CA 93106
USA

ERIK JEPPESEN
National Environmental Research Institute
Department of Freshwater Ecology
Vejlsøvej 25, PO Box 314
8600 Silkeborg
Denmark

GEOFFREY P. JONES
School of Marine Biology and Aquaculture
James Cook University
Townsville, QLD 4811
Australia

WOLFGANG J. JUNK
Max Planck Institut für Evolutions Biologie
Postfach 165
24302 Plön
Germany

MICHAEL J. KENNISH
Institute of Marine and Coastal Sciences
Rutgers University
71 Dudley Road
New Brunswick, NJ 08901
USA

ANTHONY KOSLOW
CSIRO Marine and Atmospheric Research
Floreat Laboratories
Private Bag 5, Wembley
Perth, WA 6014
Australia

OLIVIA LANGMEAD
School of Earth, Ocean and Environmental Sciences
University of Plymouth
A517, Portland Square
Drake Circus
Plymouth, PL4 8AA
UK

LISA A. LEVIN
Integrative Oceanography Division
Scripps Institution of Oceanography
9500 Gilman Drive
La Jolla, CA 92093
USA

ROBERT J. LIVINGSTON
Center for Aquatic Research and Resource Management
Department of Biological Science
Florida State University
Tallahassee, FL 32306
USA

DAN LUBIN
Center for Atmospheric Sciences
Scripps Institution of Oceanography

University of California–San Diego
9500 Gilman Drive
La Jolla, CA 92093
USA

NILS MALMER
Department of Ecology, Section Plant Ecology and
Systematics
Ecology Building, Lund University
22362 Lund
Sweden

BJÖRN MALMQVIST
Department of Ecology and Environmental Science
Umeå University
90187 Umeå
Sweden

TIM R. MCCLANAHAN
The Wildlife Conservation Society
Kibaki Flats no. 12
Bamburi
Kenyatta Beach
PO Box 99470
Mombasa 80107
Kenya

ANTON MCLACHLAN
College of Agricultural and Marine Sciences
Sultan Qaboos University
Oman

BETH MIDDLETON
National Wetlands Research Center
US Geological Survey
700 Cajundome Boulevard
Lafayette, LA 70506
USA

PETER D. MOORE
Department of Life Sciences
King's College London, Waterloo Campus
Franklin Wilkins Building
150 Stamford Street
London, SE1 9NN
UK

BRIAN R. MOSS
School of Biological Sciences
University of Liverpool
Liverpool, L69 3BX
UK

ANNETTE MÜHLIG-HOFMANN
Foundation for Environmental Conservation
Fellendsweg 33
28279 Bremen
Germany

ROBERT J. NAIMAN
School of Aquatic and Fishery Sciences
University of Washington, Box 355020
Seattle, WA 98195
USA

KARL F. NORDSTROM
Institute of Marine and Coastal Sciences
Rutgers University
71 Dudley Road
New Brunswick, NJ 08901
USA

JOS NOTENBOOM
Netherlands Environment Assessment Agency
(MNP)
PO Box 303
3720 AH Bilthoven
The Netherlands

JIN E. ONG
Centre for Marine and Coastal Studies
Universiti Sains Malaysia
11800 Penang
Malaysia

DANIEL PAULY
Fisheries Centre, 2204 Main Mall
University of British Columbia
Vancouver, BC V6T 1Z4
Canada

JOHN K. PINNEGAR
Centre for Environment, Fisheries and Aquaculture
Science (CEFAS)
Lowestoft Laboratory
Pakefield Road
Lowestoft, NR33 0HT
UK

NICHOLAS V. C. POLUNIN
School of Marine Science and Technology
Newcastle University
Newcastle upon Tyne, NE1 7RU
UK

RONALD G. PRINN
Department of Earth, Atmosphere, and Planetary Sciences
Massachusetts Institute of Technology 54-1312
77 Massachusetts Ave
Cambridge, MA 02139
USA

MICHAEL C. F. PROCTOR
School of Biosciences
University of Exeter
The Geoffrey Pope Building
Stocker Road
Exeter, EX4 4QD
UK

GERRY P. QUINN
School of Life and Environmental Sciences
Faculty of Science and Technology
Deakin University
Warrnambool Campus, PO Box 423
Princes Highway
Warrnambool, Victoria 3280
Australia

DAVE RAFFAELLI
Environment Department
University of York
Heslington
York, YO10 5DD
UK

FEREIDOUN RASSOULZADEGAN
Centre National de la Recherche Scientifique (CNRS)
Université Pierre et Marie Curie (Paris VI)
Laboratoire d'Océanographie de Villefranche
UMR 7093, Station Zoologique,
06234 Villefranche-sur-Mer Cédex
France

KARSTEN REISE
Biosciences/Coastal Ecology
Alfred Wegener Institute for Polar and Marine
Research (AWI)
Wadden Sea Station Sylt
Hafenstrasse 43
25992 List/Sylt
Germany

CHRISTOPHER T. ROBINSON
Department of Aquatic Ecology
Swiss Federal Institute of Aquatic Science and
Technology (EAWAG/ETH)
Ueberlandstrasse 133, Postfach 611
8600 Duebendorf
Switzerland

STUART I. ROGERS
Centre for Environment, Fisheries and Aquaculture
Science (CEFAS)
Lowestoft Laboratory
Pakefield Road
Lowestoft, NR33 OHT
UK

SIMON D. RUNDLE
School of Biological Sciences
Davy Building, Drake Circus
University of Plymouth
Plymouth, PL4 8AA
UK

PAUL SAMMARCO
Louisiana Universities Marine Consortium (LUMCON)
Defelice Center
8124 Highway 56
Chauvin, LA 70344
USA

DOUGLAS J. SHERMAN
Department of Geography
Texas A&M University
College Station, TX 77843
USA

FREDERICK T. SHORT
Department of Natural Resources
Jackson Estuarine Laboratory
University of New Hampshire
85 Adams Point Road
Durham, NH 03824
USA

BORIS SKET
Faculty of Chemistry and Chemical Technology
University of Ljubljana
Aškerčeva 5
1000 Ljubljana
Slovenia

CRAIG R. SMITH
Department of Oceanography
Marine Sciences Building

University of Hawaii at Manoa
1000 Pope Road
Honolulu, HI 96822
USA

RAYMOND C. SMITH
Institute of Computational Earth System Science (ICESS)
6812 Ellison Hall
University of California–Santa Barbara
Santa Barbara, CA 93106
USA

JACK A. STANFORD
Flathead Lake Biological Station
University of Montana
311 Bio Station Lane
Polson, MT 59860
USA

JOHN H. STEELE
Marine Policy Center, MS 41
Woods Hole Oceanographic Institution
Woods Hole, MA 02543
USA

ROBERT S. STENECK
School of Marine Sciences,
Darling Marine Center,
University of Maine,
Walpole, ME 04573
USA

KENTON M. STEWART
Department of Biological Sciences
University at Buffalo
State University of New York
119a Hochstetter Hall, North Campus
Buffalo, NY 14260
USA

T. FREDE THINGSTAD
Department of Microbiology
University of Bergen
Jahnebakken 5
5020 Bergen
Norway

RICHARD C. THOMPSON
School of Biological Sciences
Davy Building, Drake Circus

University of Plymouth
Plymouth, PL4 8AA
UK

SIMON F. THRUSH
National Institute of Water and Atmospheric Research (NIWA)
PO Box 11-115
Hamilton
New Zealand

BRIAN TIMMS
Faculty of Science and Information Technology
School of Environmental and Life Sciences
Social Sciences Building
University of Newcastle
Callaghan, NSW 2308
Australia

KLEMENT TOCKNER
Leibniz-Institute of Freshwater Ecology and Inland Fisheries (IGB)
Müggelseedamm 310
12587 Berlin
Germany

COLIN R. TOWNSEND
Department of Zoology
University of Otago
340 Great King Street
PO Box 56
Dunedin
New Zealand

PAUL A. TYLER
School of Ocean and Earth Science
National Oceanography Centre
University of Southampton
European Way
Southampton, SO14 3ZH
UK

JOS T. A. VERHOEVEN
Section of Landscape Ecology
Department of Geobiology
University of Utrecht
Sorbonnelaan 16
3584 CA Utrecht
The Netherlands

PETER G. VERITY
Skidaway Institute of Oceanography
10 Ocean Science Circle
Savannah, GA 31411
USA

DIANA I. WALKER
School of Plant Biology / Faculty of Natural and
Agricultural Sciences
The University of Western Australia
35 Stirling Highway
Crawley, WA 6009
Australia

WILLIAM D. WILLIAMS
†(1936–2002), Australian limnologist,
worked tirelessly toward a vision of conservation and
the wise use of natural resources. His passionate interest
in salt lakes resulted in a prodigious number of
scientific publications and continues to inspire a
generation of salt lake limnologists.

JOY B. ZEDLER
Department of Botany, Genbotany
College of Letters and Science
University of Wisconsin–Madison
302 Birge Hall, 430 Lincoln Drive
Madison, WI 53706
USA

DIRK ZELLER
Fisheries Centre, 2204 Main Mall
University of British Columbia
329–2202 Main Mall
Vancouver, BC V6T 1Z4
Canada

Preface

Concern about future supplies of fresh water to society, to meet the full range of human needs, now comes very high on the priority list of global societal issues. Water cuts across both ecological and human systems in a way that no other resource other than energy does. It spans from the highest watersheds to the deepest ocean, it has constrained the development of civilizations and continues to limit densities of biota including those of the human species. Water links all the Earth's ecosystems and sustains agriculture, households and industries, in fact most forms of human creativity including art. The future availability of water lies at a moving interface between climate systems, human population size and natural ecosystems, and it begs many questions about how and on what basis conflicting demands of people and nature can be lastingly resolved.

This book has its origins in the Foundation for Environmental Conservation and the International Conferences on Environmental Future which my father (Nicholas Polunin Senior) initiated, the first of the latter pre-dating the original 1972 United Nations Conference on the Human Environment by a year. The specific idea was to bring together many of the best-qualified ecologists to consider at a global level the present and possible future states of all the Earth's aquatic ecosystems. The book is dedicated to Nicholas Polunin Senior and Helen Polunin, who together made it possible, and owes much also to those aquatic ecologists and many others around the world who contributed across the many stages of its development.

Nicholas V. C. Polunin
Newcastle University

Acknowledgements

This book would not have been possible without the funding of the Foundation for Environmental Conservation, through its Board (J. McNeely, C. Martin, A. McCammon) and ultimately N. Polunin Senior. EAWAG staff A. Zehnder, J. Ward and C. Rapin and assistants E. Keller, J. Wolynska, J. Rüegg, A. Breitenstein and S. Aebischer made the 5th ICEF meeting possible. N. Graham, A. Mühlig-Hofmann, C. Sweeting, C. Wabnitz also logistically supported the 5th ICEF, and special thanks are due to A. Kingsford, A. Bentley, A. Mühlig-Hofmann and G. Wilson for editorial input. N. Graham conducted the expert survey. Scientific illustrator and artist T. Gunther produced the black-and-white illustrations for each chapter.

Aside from the chapter authors, substantial input to the book was also given by the following individuals: T. Anderson, N. Andrew, M. Angel, K. Arnebrant, M. Attrill, R. Babcock, P. Balabanis, A. Baltanás, K. Banse, C. Bégin, J. Bennett, Z. Billinghurst, D. Bilton, J. Borum, T. Buijse, J. Callaway, B. Carney, G. Chapman, M. Chauhan, J. Cole, E. Cooksey, B. Cowen, J. Dolan, J. Dreher, K. Dunton, I. Everson, J. Fourqurean, D. Fujita, S. Gaines, D. Galat, B. Gardner, N. Hairston Jr, G. Harris, K. Helmle, J. Hill, U. Humpesch, W. F. Humphreys, J. Jackson, S. Jenkins, C. Job, C. Johnson, G. Kendrick, S. Kimmance, S. King, J. Kleypas, D. Krause-Jensen, P. Kremer, C. Langdon, A. Leland, J. Lopez, S. Madon, S. Maki, M. Marhuenda, S. Matouch, M. May, G. Meyer, G. Nihous, P. Ojeda, S. Ormerod, C. Otto, V. Patrick, D. Peterson, R. Psenner, K. Reid, M. Risk, J. Roff, B. Scheibling, R. Schmidt, C. Short, J. Simons, P. Steinberg, P. Swart, H. Thiel, P. Tyler, T. Underwood, J. Vasquez, J. Vavrinec and S. Wilson.

The following groups and programmes significantly supported work which underpinned individual chapters: the Antarctic Peninsula 2101 group-members of the British Antarctic Survey; the Australian–Spain cooperation programme funded by the Spanish Plan Nacional de I+D; the Austrian Fonds zur Förderung der wissenschaftlichen Forschung; the Countryside Council for Wales; the Department of Energy USA; English Nature; the Environment Agency and Scottish Executive; the European Commission LIFE programme; the International Association for the promotion of co-operation with scientists from the New Independent States of the former Soviet Union (INTAS); the International Lake Environment Committee Foundation (ILEC); Joint Nature Conservation Council; the MarClim project supported by The Crown Estate; the Foundation for Strategic Environmental Research (MISTRA, Sweden) VASTRA programme; the National Center for Ecological Analysis and Synthesis; the US National Science Foundation; the National Science Research Council; the National Undersea Research Program's National Research Center at the University of Connecticut at Avery Point; NERC Grant in Aid funded Marine Biological Association Fellowship programme; NOAA's Sea Grant program to the University of Maine; National Research Foundation and Andrew Mellon Foundation grants; the Pew Foundation for Marine Conservation; the Russian Fund for Basic Research; the David the Lucile Packard Foundation; Scottish Natural Heritage; the State of Jersey, USA; the Swedish Environmental Protection Agency; the Swedish Research Council for Forestry and Agriculture; Swiss Federal Institute for Environmental Science and Technology; the UK Department for the Environment, Food and Rural Affairs; the University of New Hampshire Jackson Estuarine Laboratory; and the World Wide Fund for Nature.

1 · Introduction: Climate, people, fisheries and aquatic ecosystems

ROBERT ENGELMAN, DANIEL PAULY, DIRK ZELLER, RONALD G. PRINN,
JOHN K. PINNEGAR AND NICHOLAS V. C. POLUNIN

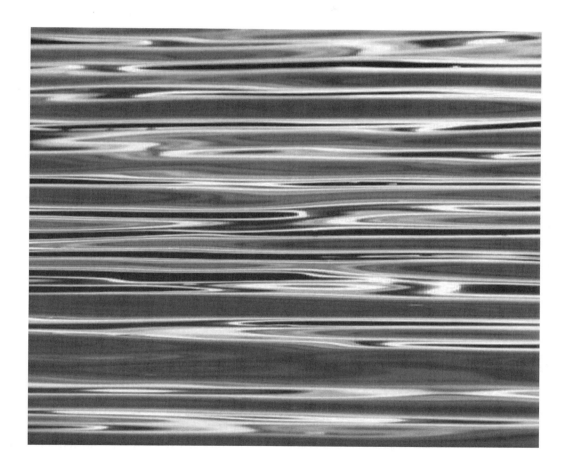

INTRODUCTION TO THE BOOK

Evidence of human damage to natural resources and the environment is long-standing even in the sea, but some 50 years ago awareness of human degradation of natural environments around the globe grew substantially. This concern was expressed above all in the creation of protected areas, organizations and agencies focused on nature conservation. The extent to which wilderness areas everywhere were contaminated or otherwise influenced by human agency was beginning to be generally recognized, and some people predicted these impacts would only grow in future. But few predicted the extent and number of global environmental changes that are now occurring. The limits to growth of the world's economies, individually or

Aquatic Ecosystems, ed. N. V. C. Polunin. Published by Cambridge University Press. © Foundation for Environmental Conservation 2008.

1

collectively, were little questioned in ecological contexts. Uncertainties surrounding how economics and ecology might relate to each other were little explored, and questions such as 'At what points will economies become constrained by the decline and loss of natural ecological goods and services?' persist to this day.

Water is increasingly seen as a constraint on sustainable human development and focus of potential human conflict. Nevertheless, it has been rare for the provision of this good, and related ecological services such as fish production and waste disposal, to be viewed holistically in the context of the natural environment. Also, there have been no comprehensive reviews of status and trends encompassing all aquatic environments (Clark *et al.* 2006), from the fresh and terrestrial saline to those of the deepest seas. Certain fresh waters and saline lakes have been radically altered by humans, but the extent and nature of these impacts vary among ecosystem types and geographical locations. The seas and oceans retain a greater complement of resource-development frontiers than does the land, for example in mining and fisheries (e.g. Berkes *et al.* 2006); their wilderness value remains substantial. Clearly, the extent of human disturbance is increasing across environment types and locations, yet there is little comprehensive scientific appraisal of the magnitude and nature of this permeation.

It is important at this time to consider what is happening ecologically to the world's aquatic environments, and where possible project these forward to an appropriate time horizon at an appropriate ecological scale. The terminology surrounding ecological units is large and application can be problematic especially where, as here, freshwater and marine environments are to be bridged. Thus the term 'biome' is discerning of terrestrial environments but not with respect to those of the sea, while 'ecoregions' (e.g. Bailey 1998) omit the continental shelves. It was important in the design of this book to focus on a number of natural environments that would be comprehensive yet feasible in terms of relevant data and expertise. One aim of this chapter is to consider alternative comprehensive categorizations of aquatic environments as a basis for this book.

Natural processes and ecological services rely on notions of system functioning that have long been integral to the concept of the 'ecosystem'. Other recognized types of ecological unit such as the biotope exist, but the term 'ecosystem' is the most relevant, and most widely used and recognized, and is thus the unit of choice for this book. What was also needed in the design of this book was a time horizon sufficiently far in the future to provoke thoughtful and imaginative projections across all the ecosystems, but within a time-frame for which some detailed scenarios such as of human population growth and global ambient temperatures existed. The year 2025 was chosen as an appropriate compromise between those two constraints. This book aims to assemble the views of expert ecologists on the status of all ecosystems across the aquatic realm, review contemporary changes and their drivers, and consider likely outcomes at the 2025 time horizon. Major drivers considered are climate change and the direct impacts of human population growth and economic development.

Climate is usefully defined as the average of the weather experienced over a 10–20-year period. Temperature and rainfall changes are typical measures of change that can be expressed at local, regional, national or global scales. Global warming or cooling can be driven by any imbalance between the solar energy the Earth receives from the Sun and the energy it radiates back to space as invisible infrared light. The greenhouse effect is a warming influence caused by the presence of gases and clouds in the atmosphere, which are very efficient absorbers and radiators of this infrared light. The greenhouse effect is opposed by substances at the Earth's surface (such as snow and desert sand) and in the atmosphere (such as clouds and white aerosols) that efficiently reflect sunlight back into space (albedo) and are thus a cooling influence. Global climate changes and their possible consequences are and will remain a major area of debate within and among most nations. International negotiations in the Framework Convention on Climate Change and its Kyoto Protocol are contentious in particular because the components of the problem are many, the science is uncertain and the whole issue has a major bearing on other global problems such as poverty, inequality and economic development. Another aim of this introductory chapter is to consider uncertainties in climate predictions and review warming, sea-level rise and related trends as they might affect the world's aquatic environments.

The human population is approaching the 7 billion mark, increasing by *c.*1.2% per year. Mapping and photography of the Earth's night lights (Weier 2000) show how unevenly human beings are settled (Deichmann *et al.* 2001). Especially dense populations characterize northern South Asia, East Asia, Europe, eastern North America and south-western Africa. More than half of the world's population lives within 60 km of the coast, and this figure could reach 75% by the year 2020 (Roberts & Hawkins 1999). In addition, human economic wealth and output are often concentrated close to major rivers or estuaries, especially in the temperate regions of the northern hemisphere (Cohen 1997; Nicholls & Small 2002). Various human interactions with critical natural resources and the environment illustrate that the human population over the past several decades has reached levels that significantly modify aquatic ecosystems. While some of this change is economically and socially beneficial, some is ecologically detrimental (e.g. biodiversity loss; see below). A third aim of this chapter is to introduce a human backdrop for assessing global changes in aquatic ecosystems; this will be done in particular by reviewing some of the consequences of human population growth for water, water supply and its natural sources.

Fisheries have been recognized as a major driver of change in the aquatic realm. They constitute a major impact at the interfaces among human population growth, economic development and environmental changes, both anthropogenic and natural. Modern industrial fisheries resulted from the economic and technological development of Europe, North America and Japan over a century ago. Frontiers in the development of global fishery resources persist in the use of Southern Ocean and deep-sea or reef resources. What are the current trends in global fisheries, and how do these relate to what is happening in the aquatic ecosystems involved? The intention in this chapter is also to introduce marine fisheries as an important example of human intrusion into the oceans and seas; this is followed by an overview of global trends in aquatic biodiversity and specific extinctions.

COMPLEXITIES OF CLIMATE CHANGE AND ITS CONSEQUENCES

Global climate has remained relatively stable (temperature changes of $<1\,°C$ over a century) during the last 10 000 years. However, society now faces potentially rapid changes because human activities have altered the atmosphere's composition and changed the Earth's radiation balance (IPCC [Intergovernmental Panel on Climate Change] 1996).

The most important greenhouse gas is water vapour, which typically remains for a week or so in the atmosphere. Concerns about global warming, however, revolve around much longer-lived greenhouse gases, especially carbon dioxide but also other long-lived gases (methane, nitrous oxide, chlorofluorocarbons), concentrations of which have increased substantially since c.1750. Carbon dioxide (CO_2) has risen by about 30%, methane (CH_4) by more than

100% and nitrous oxide (N_2O) by about 15%. These gases are now at higher concentrations than at any time in the past 160 000 years (IPCC 1996). The combustion of fossil fuels and, to a lesser extent, changes in land use account for anthropogenic CO_2 emissions. Agriculture is responsible for nearly 50% of human-generated CH_4 emissions and about 70% of anthropogenic N_2O emissions.

When the concentration of a particular greenhouse gas increases, it tends to lower the flow of infrared energy to space and increases the flow of infrared energy down toward the surface. The Earth is then receiving more energy than it radiates to space. This 'radiative forcing' warms the surface and the lower atmosphere; however, the rate of surface warming is slowed by uptake of heat by the world's oceans. The greenhouse effect as quantified by this radiative forcing is real and the physics relatively well understood. What is more uncertain, and the cause of much of the scientific debate, is the response of the global system that determines climate to this radiative forcing. Feedbacks in the system can either amplify or dampen the response in ways that are still only partially understood.

The IPCC was jointly established by the World Meteorological Organization and the United Nations Environment Programme (UNEP) in 1988, in order to (1) assess available scientific information on climate change, (2) assess the environmental and socioeconomic impacts of climate change, and (3) formulate response strategies. The Integrated Global System Model (IGSM) is one means that the IPCC has at its disposal to analyse carefully the scientific and economic implications of proposed mitigation policies (Fig. 1.1). The IGSM consists of a set of coupled submodels of economic development and associated emissions, natural biogeochemical cycles, climate and natural ecosystems (Prinn et al. 1999; MIT [Massachusetts Institute of Technology] 2003). It attempts to include each of the major areas in the natural and social sciences that are amenable to quantitative analysis and are relevant to the issue of climate change (Schneider 1992; Prinn & Hartley 1992; IPCC 2001a, b, c). A major challenge inherent in global climate modelling is to decide what is important and what is unimportant. In the IGSM, the coupled atmospheric chemistry and climate model (Fig. 1.1) is driven by a combination of anthropogenic and natural emissions. The essential components of this model are chemistry, atmospheric circulation and ocean circulation, each of which, by itself, can require enormous computer resources. The atmospheric chemistry is modelled in sufficient detail to capture its sensitivity to climate and different mixes of emissions, and to address the

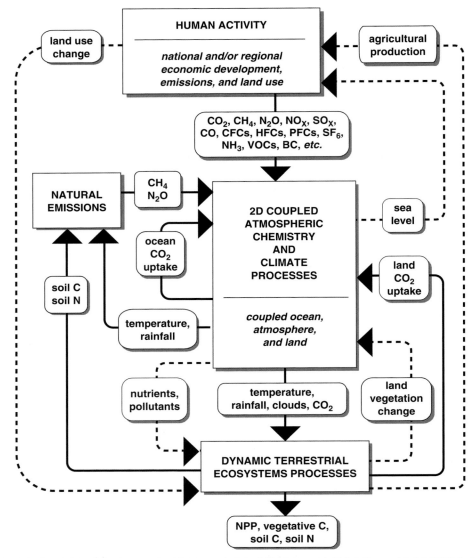

Fig. 1.1. Schematic illustration of the framework and processes of the MIT Integrated Global System Model (IGSM). Feedbacks between the component models that are currently included or proposed for inclusion in the next generation are shown as solid and dashed lines respectively (Prinn *et al.* 1999; MIT 2003).

relationship among policies proposed for control of emissions related to air pollution, aerosols and greenhouse gases (Wang *et al.* 1998). But linking complex models together leads to many challenges, illustrated by the failure of coupled ocean–atmosphere models (including IGSM) to accurately simulate current climate without arbitrary adjustments. The IGSM coupled chemistry–climate model outputs

drive a terrestrial ecosystems model that predicts vegetation changes, land CO_2 fluxes and soil composition (Xiao *et al.* 1998), and these feed back to the climate model, chemistry model and natural emissions model. The effects of changes in land cover and surface albedo on climate, and the effects of changes in climate and ecosystems on agriculture and anthropogenic emissions are also included.

Hundreds of runs of the IGSM have been used to quantitatively assess uncertainty in climate projections (Webster *et al.* 2002, 2003). One study shows a global median surface temperature rise from 1990 to 2100 of 2.4 °C, with a 95% confidence interval of 1.0–4.9 °C (Webster *et al.* 2003). For comparison, IPCC (2001*a*) reported a range for the global mean surface temperature rise by 2100 of 1.4–5.8 °C, but did not provide likelihood estimates for this key finding although it did do so for others. The implications of the projected level of climate warming for the aquatic realm are unclear, but will vary with type of ecosystem. For example, small surface freshwater bodies will typically vary readily with climate, while in the open ocean, ambient temperature will lag that of the atmosphere by decades or more.

When the probability distribution functions for the mean global surface temperature and sea-level increases between 1990 and 2100 are compared between no–policy and applied-policy scenarios (designed to simulate strict regulations leading to stabilization of atmospheric CO_2 concentrations at about twice pre-industrial levels), a median of only 1.6 °C and 95% range of only 0.8–3.2 °C is forecast (Fig. 1.2). There is a 50% chance of warming exceeding 2.4 °C in the no–policy case and a 1-in-7 chance in the policy case. Implementing the policy therefore lowers the probability of large amounts of warming to a substantial degree.

To better appreciate the risks of the no–policy scenario, it is important to examine the latitudinal distribution of the projected warming. In common with other climate models, the computed temperature increases in polar regions are much greater than those in equatorial regions (no–policy case: Fig. 1.3). Polar regions contain vulnerable ecosystems with large carbon storage (e.g. Chapter 8), and the Greenland and Antarctic ice sheets with large water storage (e.g. Chapter 21). Release of some of this stored carbon and water is of significant concern. A 1-in-40 chance of warming by 8–12 °C or greater in polar regions for the no-policy case (Fig. 1.3) is particularly worrisome. The policy scenario lowers the polar warming in the 1-in-40 calculation to 5–7 °C (Webster *et al.* 2003).

Similar significant reductions in the probability of large and risky amounts of sea-level rise due to the hypothetical policy are also evident (Fig. 1.2). Emissions reductions are predicted to lower the chance of exceeding an extreme climate outcome but not to eliminate the risk entirely, and analysis of the reduction in probability is an important policy consideration. Future climate assessments would better serve the policy process by including formal analysis of

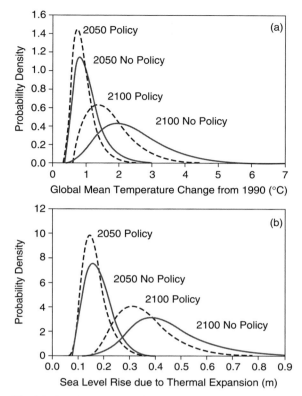

Fig. 1.2. Probability density function (PDF) for the change in global mean: (a) surface temperature and (b) sea-level rise from 1990 to 2100 estimated as a best fit to 250 simulations using latin hypercube sampling from input PDFs for uncertain variables. The solid line shows the PDF resulting from no explicit emissions restrictions and the dashed line is the PDF under hypothetical emissions policy leading to steady levels of atmospheric CO_2 of about twice pre-industrial values (Webster *et al.* 2003). The IPCC (2001*a*, *b*, *c*) upper estimate is beyond the 95% confidence limit. Based on this distribution there is a 12% chance that the temperature change in 2100 would be less than the IPCC lower estimate.

uncertainty for key projections, with an explicit description of the methods used (Allen *et al.* 2001; Reilly *et al.* 2001).

Sea-level rise

Global mean sea level has risen by 10–25 cm over the last 100 years, and although there has been no detectable change in the rate of sea-level rise over the course of the century, it would appear that this has been significantly higher than the rate averaged over the last several thousand years (IPCC

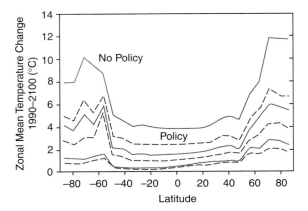

Fig. 1.3. Zonal mean temperature change in surface warming by latitude band between 1990 and 2100 in the case assuming no explicit policy in Fig. 1.2. There is a 1-in-40 chance of being above or below the upper and lower curves and a 1-in-2 chance of being above or below the middle curve respectively (Webster *et al.* 2003).

2001*a*, *b*). It is likely that the rise in sea level has been largely due to the increase in global temperature over the last 100 years. The possible factors include thermal expansion of the oceans and melting of glaciers, ice caps and ice sheets. Changes in surface water and groundwater storage (related to changes in human activities and changes to drainage basins) may also have affected sea level.

Model simulations have predicted 2–7 cm of the reported sea-level rise over the last 100 years (10–25 cm) and attributed it to thermal expansion of the oceans, yet many of the world's mountain glaciers have retreated substantially over this time period and may have accounted for 2–5 cm of the observed rise. However, most of the non-oceanic water on Earth resides in great ice sheets, namely those of Antarctica and Greenland, and most of their volume lies on land above sea level. Loss of only a small fraction of this volume could have a significant effect on sea level (Warrick *et al.* 1996). In Antarctica, discharges of enormous icebergs and recent break-ups of Antarctic Peninsula ice shelves have focused public attention on the possibility of 'collapse' of this ice reservoir within the next century, with major potential impacts on sea level.

According to the IPCC (2001*a*, *b*) 'best estimate' climate change scenario, global sea level is projected to rise by 20 cm by the year 2050 (within a range of uncertainty of 7–39 cm) and by 49 cm by 2100 (within a range of uncertainty of 20–86 cm); more than half of this is attributed to oceanic thermal expansion.

Lakes, streams and wetlands

Inland aquatic ecosystems are expected to be greatly influenced by climate change through altered water temperatures, flow regimes and water levels (e.g. Kaczmarek *et al.* 1996; Öquist *et al.* 1996). Water-level declines may be severe in lakes and streams in dry evaporative drainages and in basins with small catchments, whilst the distribution of wetlands is likely to shift with changes in temperature and precipitation (IPCC 2001*a*; Chapters 2–10). Wetlands temporarily store runoff water, thereby reducing floodwater peaks and protecting downstream areas, consequently a reduction of wetland area due to climate change could severely hamper flood-control efforts in some regions (e.g. Kaczmarek *et al.* 1996; Öquist *et al.* 1996; Chapters 9 and 10).

Coastal systems and small islands

Coastal systems are economically and ecologically important. Climate warming, sea-level rise and increases in storms and storm-surge frequencies may result in erosion of shores and associated habitats, altered tidal ranges in rivers and bays, changes in sediment and nutrient transport, and increased coastal flooding (see Bijlsma *et al.* 1996). Some coastal ecosystems are particularly at risk, namely saltmarshes, mangroves, coral reefs and atolls, and river deltas (Chapters 11–13 and 16). Changes in these ecosystems are likely to negatively affect tourism, fisheries and biodiversity (IPCC 2001*b*). Many small island countries could lose a significant part of their land area with a sea-level rise of 0.5–1 m, the Maldives for instance having average altitudes of only 1–1.5 m (Bijlsma *et al.* 1996).

Oceans

Climate change could lead to altered ocean circulation (e.g. weakening of the Atlantic thermohaline circulation) and wave climates, and reductions in sea-ice cover (IPCC 2001*a*, *b*, *c*; ACIA [Arctic Climate Impact Assessment] 2004). As a result, nutrient availability, biological productivity, the structure and functions of marine ecosystems, and heat and carbon storage capacity may be affected with important feedbacks to the global climate system (see Ittekkot *et al.* 1996). These changes would have implications for coastal regions, fisheries, tourism and recreation, transport and offshore structures.

Most CO_2 released into the atmosphere as a result of burning fossil fuels will eventually be absorbed by the ocean. As the amount of CO_2 in the atmosphere rises, more of the gas reacts with seawater to produce bicarbonate and hydrogen ions that increase the acidity of the surface layer. Ocean pH was around 8.3 after the last ice age and 8.2 before CO_2 emissions took off in the industrial era. Ocean pH is now 8.1 and if atmospheric CO_2 exceeds 1900 ppm by the year 2300, pH at the ocean surface could fall to 7.4 (Caldeira & Wickett 2003). There is limited understanding of the effect increased acidity might have on marine biota, but coral reefs, calcareous plankton and other organisms, skeletons or shells of which contain calcium carbonate, may be substantially affected.

PEOPLE AND WATER

Perhaps the best-documented human interaction with aquatic ecosystems involves the use of fresh water (Engelman & LeRoy 1993; Engelman et al. 2000). Precious little of the world's water is salt-free, and only a small proportion of this fresh water is accessible to human beings. Water use has increased for many centuries, but in many major watersheds (the key geographic unit of interest for freshwater use), only in the past few centuries has the scale reached the point where natural variations in water supply have begun to collide with growing human use. This is most evident in western Asia and eastern and southern Africa, but to varying degrees such problems of availability are also present elsewhere. Humans already use more than half of the renewable fresh water that is readily accessible (Postel et al. 1996), and human population growth is currently the dominant factor in the increase of water withdrawals worldwide.

The world's urban population is currently growing at four times the rate of the rural population. Between 1990 and 2025, the number of people living in urban areas is projected to double to more than 5 billion out of a total 8 billion (Table 1.1). An estimated 90% of the increase will occur in developing countries and at the current pace, over 60 million people are added to urban populations each year, placing great strain on local governments to provide even the most basic services. Of all urban inhabitants in developing countries, 25–50% live in impoverished slums and squatter settlements, with little or no access to adequate water, sanitation or refuse collection (UN [United Nations] 1997).

Human water demand for agricultural irrigation (an increasing proportion of food production comes from irrigated cropland), industrial and household uses almost invariably takes precedence over environmental needs. In

Table 1.1. *World population by year and UN projection variant*

Year	Low-variant projection	Medium-variant projection	High-variant projection
2010	6 843 645 000	6 906 558 000	6 967 407 000
2025	7 568 539 000	8 010 509 000	8 450 822 000
2050	7 791 545 000	9 191 287 000	10 756 366 000

Source: UNPD (2006).

all but the most remote and best-protected areas, there is constant pressure on water resources. By hydrological benchmarks of water stress (c.1000–1700 m^3 per person per year) and scarcity (<1000 m^3 per person per year) (Falkenmark & Widstrand 1992; Engelman & LeRoy 1993), the numbers of human beings living in these two high-risk categories is growing much faster than population growth generally. One-third of humanity lives in countries experiencing moderate to high water stress (WRI [World Resources Institute] 1998), and a billion people could face severe water shortages by 2025 (Potts 2000).

Apart from fisheries, which are considered below, several other interactions between human population and the environment are especially salient for aquatic ecosystems. In terms of food security, much is made of agricultural intensification as the antidote to forest loss. When more food can be produced on the same land, there is less need to convert relatively wild land to farmland. This can, of course, benefit wetlands, which historically have been lost to farmland more frequently than to settlement, although this ratio is changing as coasts and river valleys become more densely populated (Chapters 9–13).

Water is the greatest transportation network for chemical wastes of all types. Humanity now exceeds nature as a fixer and producer of nitrogen compounds (Smil 1997), and studies of rivers have shown that nitrogen loads are highly correlated with human population density along river banks (Cole et al. 1993). Intensive farming can also exacerbate soil erosion if farmers do not have the resources to manage their soil properly, and the world's rivers offload much of the resulting silt onto the continental shelves and beyond. Conversion of forest to farmland is another source of soil erosion, also closely correlated with population density and growth.

Introduction of alien species is an indirect result of human population growth and economic and technological development, because human beings uniquely face few barriers to their movement over the Earth's surface. From Nile perch to zebra mussels, the list of introduced aquatic species altering aquatic ecosystems is lengthening. In addition, marine life is at growing risk from a range of diseases, the spread of which is being hastened by global warming, global transport and pollution (Harvell *et al.* 1999, 2002). Well-documented cases include crab–eater seals in Antarctica infected with distemper by sled-dogs, sardines in Australia infected with herpes virus caught from imported frozen pilchards, and sea-fans in the Caribbean killed by a soil-borne fungus.

The further interaction between human populations and economics offers an additional obstacle to the conservation or restoration of aquatic ecosystems. When ecosystems are degraded, a common action of last resort is to attempt some sort of protection, where human use and access are restricted. But the process of setting land and/or fishing space aside becomes more difficult as population density increases. People may bid up the price of land to the point where conservation efforts become financially impossible (Cincotta & Engelman 2000). Land conservation organizations have increasingly focused on wetlands, coasts and other aquatic ecosystems, knowing they are in a race with time to buy key parcels of land or convince governments to protect them. Too rarely, however, do conservationists acknowledge that continued population growth limits their prospects for long-term success, simply because land will become too valuable to serve environmental rather than economic interests.

Market forces and global economics can also greatly influence the need for and development of port and shipping facilities. Historically, estuaries and saltmarshes have provided cheap sources of land for development. As commodities continue to be traded on global markets and ships become ever bigger, the demand for coastal land for development will increase and much of it will become degraded (Pinnegar *et al.* 2006).

While these interactions between human population and aquatic ecosystems paint a bleak picture, there are also reasons for hope. One of the most positive trends is that the human population is seen by most demographers as unlikely to double again. One way to explain this is that women all over the world are having fewer children than ever before, and seemingly want to have even fewer in the future. Average family size has shrunk from five children per woman in the early 1960s to a bit more than 2.5 children per woman today. In more than two-fifths of the

world's countries, couples are only just replacing themselves in the population, or they are having fewer than the *c.*2.1 children needed for net replacement (UNPD [United Nations Population Division] 2002). Contrary to public impressions, this does not mean that these populations are now stable or declining. The large proportion of people of childbearing age in populations that up until recently were growing fairly rapidly guarantees that more births than deaths will occur for essentially an average human lifetime after replacement fertility is reached. Also, high levels of migration mean that most nations experiencing low fertility rates will continue to grow (International Organization for Migration 2000; UNPD 2002).

GLOBAL TRENDS: THE CASE OF GLOBAL MARINE FISHERIES

Given the means and opportunity, fishers like hunters before them ultimately deplete the resources that they target (e.g. Clovis of North America: Alroy 2001). Understanding this analogy is important, as it provides a framework for understanding the severe depletion of smaller and mid-size mammals in Africa ('bushmeat': Bowen-Jones 1998) and large fishes in the world's oceans. The recent marine fish biomass declines (e.g. Christensen *et al.* 2003 for the North Atlantic) are primarily attributable, not to scientific incompetence in monitoring (Malakoff 2002), nor shifts in distribution (Bigot 2002), nor regime shifts (Steele 1998), but to overfishing (Jackson *et al.* 2001).

When fishing starts in a new area, the large fishes go first, as they are relatively easier to catch than small fishes (e.g. with harpoons or lines) and tend to provide a better return on investment (Pauly *et al.* 2002). Large fish, with their low natural mortalities and relatively high age at first maturity (Pauly 1980; Froese & Binohlan 2000), cannot sustain substantial fishing pressure and rapidly decline (Denney *et al.* 2002), forcing the fishers either to move on to smaller fishes, and/or to other, previously unexploited areas. As larger fish tend to have higher trophic levels than smaller fish – indeed, the latter are usually the prey of the former – this process, now called 'fishing down marine food webs', leads to declining trends in the mean trophic level of fisheries landings (Pauly *et al.* 1998; though see Caddy *et al.* 1998).

The 'fishing down' process has occurred around the world (Table 1.2) and the broad pattern is that fishing targets a succession of species until the residual species mix ceases to support a fishing economy. This implies a

Table 1.2. *Occurrence of 'fishing down' using local/detailed data sets, following the original presentation of this phenomenon by Pauly* et al. *(1998), based on the global FAO catch dataset*

Country/area[a]	Years	Decline	Source and remarks
Iceland	1900–1999	1918–1999	Valtsson and Pauly (2003), based on comprehensive catch database of Valtsson (2001)
Celtic Sea	1945–1998	1946–2000	Pinnegar *et al.* (2002), based on trophic levels estimated from stable isotopes of nitrogen
Gulf of Thailand	1963–1982; 1963–1997	1965–1982; 1965–1997	Christensen (1998); Pauly and Chuenpagdee (2003)
Eastern Canada	1950–1997	1957–1997	Pauly *et al.* (2001), based on data submitted to FAO by the Canadian Department of Fisheries and Oceans
Western Canada	1873–1996	1910–1996	Pauly *et al.* (2001), based on comprehensive dataset assembled by S. Wallace
Cuban EEZ	1960–1995	1960–1995	Pauly *et al.* (1998) and Baisre (2000)
East Coast, USA	1950–2000	all	R. Chuenpagdee *et al.* unpublished data; emphasis on Chesapeake Bay
Chinese EEZ	1950–1998	1970–1998	Pang and Pauly (2001)
West Central Atlantic	1950–2000; 1950–2000	1950–2000; 1950–2000	D. Pauly and M. L. Palomares unpublished data, based on FAO data (Area 41), disaggregated into USA (North) and other countries (South)
World, tuna and billfishes	1950–2000	1950–2000	D. Pauly and M. L. Palomares unpublished data, based on FAO data (ISSCAAP [International Standard Statistical Classification of Aquatic Animals and Plants] group 36 only)
World, all fishes	1950–2000	1950–2000	Pauly and Watson (2003), based on spatially disaggregated data

[a] EEZ, Exclusive Economic Zone.

transition from fisheries targeting large fish (e.g. northern cod) to those targeting smaller fishes (e.g. capelin) or invertebrates (e.g. northern shrimp and snow crab). In the north-east Atlantic, fish species that mature later, grow more slowly and have lower rates of potential population increase, have exhibited greater long-term declines in abundance than closely related species which are smaller, grow faster and are more fecund (Jennings *et al.* 1998, 1999; Dulvy *et al.* 2000). In sharks, rays and skates, large size combined with low fecundity makes them particularly vulnerable to overexploitation; many populations have been rapidly declining (Dulvy *et al.* 2000) and are likely to be rendered extinct in the coming decades (Dulvy *et al.* 2003). In some cases, depletion of keystone species may have important indirect effects on community and habitat structure (e.g. Dulvy *et al.* 2004).

Exports have become a major issue in fisheries, with marine products being amongst the most heavily traded commodities (Pauly *et al.* 2002). The general trend is for developed countries to increasingly to compensate for the shortfall of products from traditional fishing grounds in their exclusive economic zones (EEZs) by fishing in developing countries (Pauly *et al.* 2005). Given the debts that most developing countries have run up with respect to international lenders, this implies that marine resources, and their underlying ecosystems, suffer from increased pressure to make up shortfalls. Examples include the countries of West Africa, whose dependence on financial support from the European Union forces them to sign fisheries agreements providing access for European fishing fleets under terms that appear unfair to these countries (Kaczynski & Fluharty 2002). In Argentina, demersal

Table 1.3. *Marine fish that are unlikely to survive, given continuation of present fisheries trends*

Major features	Representative groups
Large- to moderate-sized, predaceous, territorial reef fishes and rockfishes with late age at maturity, very low natural mortality rates and low recruitment rates versus adult stock size	Snappers, sea basses, emperors, rockfishes, sea breams
Large- to moderate-sized shelf dwelling, soft bottom predators susceptible to bottom trawling	Cods, flounders, soles, rockfishes, croakers, skates
Large- to moderate-sized schooling midwater fishes susceptible to midwater trawling	Hakes, rockfishes, armorheads, rougheyes
Large- to moderate-sized shelf dwelling, schooling, pelagic fishes	Bonitos, sierras, capelin, eulachon, salmon, sharks
Any species with exceptionally high monetary value	Bluefin tuna, red snappers, halibuts, medicinal fishes, aquarium fishes, groupers, salmon, red mullets, billfishes

Source: Adapted from Parrish (1995, 1998) and Pauly (2000).

resources which in the early 1980s were among the few that were both large and underexploited have now collapsed under the pressure of both national and international fleets, licensed for their ability to generate foreign exchange (Sánchez 2002).

Given present trends and pressures, present target species are not expected to survive and small non-palatable non-schooling species will be those that will fare best (Table 1.3). In extreme cases, finfish may be progressively replaced by consumer species such as jellyfish, which have few predators and may impede any finfish recovery (e.g. northern Benguela, Namibia: Lynam *et al.* 2006).

The fishing industry on its own is incapable of reversing the fishing down of food webs. There are other social forces, which can and will play an increasing role in the international debates on fisheries. Foremost is the community of non-governmental organizations devoted to maintaining or re-establishing 'healthy' marine ecosystems, and striving for ecosystem-based fisheries management. Public debate about fisheries was unheard of 20 years ago, and many conservation biologists feel they need to debunk those who deny the need for action (e.g. Lomborg 2001). Global catches are declining (Watson & Pauly 2001; Pauly *et al.* 2002) but catches can and usually do remain high when stocks collapse. Thus the cod off eastern Canada yielded good catches until the fishery had to be closed because there were literally no fish left (Myers *et al.* 1997).

Some suggest that aquaculture could help compensate for overfishing, and a quarter of human fish consumption now derives from aquaculture. However, as currently practised, aquaculture also causes environmental damage, raising questions about how to meet food demands and preserve environmental quality (WRI 1998). Aquaculture in fact exists in two fundamentally different forms. One is devoted to the farming of bivalves (e.g. oysters, mussels) and/or freshwater fish (e.g. carp, tilapia) and relies mainly on plant matter to generate a net addition to the fish food supply available to consumers. This is based predominantly in developing countries (mainly in China, but also in countries such as the Philippines and Bangladesh), and supplies cheap animal protein where it is needed (New 2002). By contrast, the other form of aquaculture involves the farming of carnivorous fish such as salmon or sea bass, and increasingly, the fattening of wild-caught bluefin tuna. In nature, salmon, sea bass and bluefin tuna have high trophic levels, hence it is impossible to feed them only on vegetable matter. This implies that as this form of aquaculture increases, there will be fewer cheap fish (e.g. sardine, herring, mackerel and anchovies) available for humans to buy and eat. It is thus not adding to global fish supply, and instead increases the pressure on wild fish stocks (Naylor *et al.* 2000).

This second type of aquaculture predominates, and it has led to massive imports by developed countries of meal from fishes caught and ground up in developing countries, exacerbating fishing pressures in these regions. Coastal pollution and diseases emanating from the uneaten food and faeces of these marine feed-lot operations are also seen as a major constraint to the development of the industry (New 2002).

Fisheries need to be reinvented as providers of a healthy complement to grain-based diets. Nor can they remain subject to a free-for-all among distant-water fleets; they can however become a regular source of income for communities whose members act within natural constraints (Pitcher 2001). Such reinvented fisheries will be smaller in size, and they will hopefully rely more on fish biomass being exported from marine protected areas closed to fishing.

Humans consume around 86 million tonnes of fish per year, nearly 15.7 kg per person, which is more than twice the 1950 level. Inhabitants of the economically less developed countries (including China) will likely increase their total consumption of food fish to the 2025 time horizon by as much as 34 million tonnes, whereas consumption will remain relatively static elsewhere (Delgado *et al.* 2003). Given current population trends and the global market for fishery products, the question remains as to whether fishing pressure can ever be reduced and aquatic ecosystems restored.

GLOBAL TRENDS IN AQUATIC BIODIVERSITY AND EXTINCTION

With the exception of seven countries, all nations of the world are now signatories to the UN Convention on Biological Diversity, introduced in 1992 at the Rio 'Earth Summit'. The Convention committed nations 'to achieve by 2010 a significant reduction of the current rate of biodiversity loss at the global, regional and national level as a contribution to poverty alleviation and to the benefit of all life on earth'. However, both marine and freshwater ecosystems continue to be degraded and species lost all around the world.

Freshwater systems occupy only 0.8% of Earth's surface (McAllister *et al.* 1997), but they are rich in species and vital as habitat. Perhaps 12% of all animal species live in fresh water (Abramovitz 1996). Due to their limited area, freshwater ecosystems contain only about 2.4% of all Earth's plant and animal species (Reaka-Kudla 1997); however, on a hectare-for-hectare basis, they are richer in species than the more extensive terrestrial and marine ecosystems (Table 1.4).

The alteration and damming of river systems for industrial and domestic use, irrigation and hydroelectric power have fragmented more than half of the world's large river systems. Some 83% of their total annual flow is affected

Table 1.4. *Species richness by ecosystem*

Ecosystem	Per cent of Earth's total habitat	Per cent of known species[a]	Relative species richness[b]
Freshwater	0.8	2.4	3
Terrestrial	28.4	77.5	2.7
Marine	70.8	14.7	0.2

[a] Species do not sum to 100% because 5.3% of known symbiotic species are excluded.
[b] Relative species richness is the ratio of the per cent of known species and the per cent of area occupied by the ecosystem.
Source: McAllister *et al.* (1997).

(52% moderately, 31% severely), with Europe's river flow being the most regulated and Australasia's the least (WWF [World Wide Fund for Nature] 2006). While many factors can simultaneously contribute to freshwater fish extinctions, habitat alteration and introduction of non-native species have been the major causes of species losses (Harrison & Stiassny 1999). Habitat alteration has contributed to 71% of extinctions, non-native species (which can compete with or feed on native species) to 54%, overfishing to 29% and pollution to 26% (Harrison & Stiassny 1999).

Dramatic declines in amphibian populations, including population crashes and mass localized extinction, have been noted since the 1980s from locations all over the world, and amphibian declines are thus perceived as one of the most critical threats to global biodiversity. A number of causes are believed to be involved, including habitat destruction and modification, overexploitation, pollution, pesticide use, introduced species, climate change, increased ultraviolet-B radiation (UV-B) and diseases. However, many of the causes of amphibian declines are still poorly understood, and amphibian declines are currently a topic of much ongoing research.

In an attempt to provide a quantitative overview of the long-term changing health of the planet, in 1998 the WWF produced a 'Living Planet Report', which has since been updated annually. The stated aim of this report was to answer the question: how fast is nature disappearing from the Earth? It introduced several useful concepts, including the Living Planet Index (LPI), the aggregate of the Forest Ecosystems Index, Freshwater Ecosystems Index (FEI) and Marine Ecosystems Index (MEI) (Fig. 1.4).

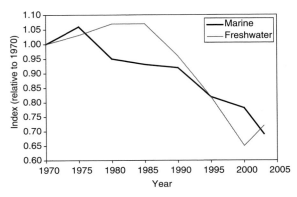

Fig. 1.4. Freshwater and marine Living Planet Indices (FEI [Freshwater Ecosystems Index] and MEI [Marine Ecosystems Index] respectively) from 1970 to 2003 (WWF 2006).

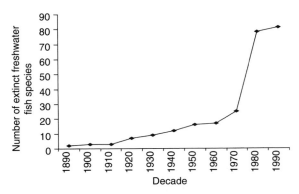

Fig. 1.5. Cumulative sum of known fish species extinctions by decade (WCMC 1998).

The FEI is based on the population trends of 344 freshwater vertebrate species, of which 387 are temperate and 51 tropical. These species include every mammal, bird, reptile, amphibian and fish for which time series population data are available, and they now encompass 11 mammals, 153 birds, 17 reptiles, 69 amphibians and 94 fishes (WWF 2006). The FEI points to freshwater species having on average declined by about 28% since 1970 (Fig. 1.4), the decline being more marked in temperate than tropical systems. Studies around the world have pointed to declines in freshwater animal populations, notable among which are amphibians (e.g. Wyman 1990; Wake 1991; Pounds & Crump 1994; Houlahan *et al.* 2000). However, large declines have also occurred in many freshwater fish (e.g. Moyle & Leidy 1992; Stiassny 1996; WCMC [World Conservation Monitoring Centre] 1998) and bivalve populations (e.g. Bogan 1993; Ricciardi *et al.* 1998). Some 81 fish species are recorded to have become extinct during the past century, and a further 11 are extinct in the wild but remain as captive populations (Fig. 1.5).

A major proportion of known extinctions (notably, 50 species of cichlid) have resulted from the ecological effects of the deliberate introduction of the Nile perch *Lates niloticus* into Lake Victoria in the mid twentieth century (WCMC 1998). In North America alone, some 123 freshwater animal species have been recorded as extinct since 1900, and hundreds of additional species of fishes, molluscs, crayfishes and amphibians are imperilled (Ricciardi & Rasmussen 1999). In total, 734 species of fish are classified as threatened, of which 84% are freshwater species (IUCN [World Conservation Union] 1996).

The MEI is based on population trends of 274 marine species, including 48 marine mammals, 112 seabirds, 7 reptiles (turtles) and 107 fishes. The MEI has declined by about 27% since 1970, a slightly slower decline than is evident in the FEI (WWF 2006) but nonetheless substantial (Fig. 1.4). Relatively stable trends are evident in the Pacific and in the Arctic/Atlantic Oceans, in comparison with dramatic declines in the Indian/Southeast Asian and Southern Oceans (Table 1.5). However, overall increases in the populations of seabirds and some mammal species in the Atlantic and Pacific Oceans since 1970 mask declines in many fish species, especially those of economic importance such as cod and tuna, as well as turtles and other species that are caught as by-catch.

There is a public misconception that there is little likelihood of a marine species ever becoming extinct (Roberts & Hawkins 1999). However, this has happened in the past and is set to continue and possibly increase in frequency in the future, as habitats are changed and exploitation continues. Notable among large vulnerable species are the sea mammals, populations of which were decimated by whaling and hunting up until the second half of the twentieth century. Steller's sea cow *Hydrodamalis gigas* was hunted to extinction in 1767 and the Caribbean monk seal *Monachus tropicalis* in 1952 (Day 1990), while species such as the Mediterranean monk seal *Monachus monachus* and Gulf porpoise *Phocoena sinus* remain critically endangered. Several species of fish and molluscs have also been lost, and many remain critically endangered (Roberts & Hawkins 1999).

Among fisheries globally, 75% have been judged to be fully fished, overfished or depleted (FAO [Food and

Agriculture Organization of the United Nations] 2004). In addition to the effects of exploitation, the biodiversity of many coral reef and coastal marine species may also be influenced by habitat loss (Friedlander & Parrish 1998; Wilson *et al.* 2006). Of 58 marine fish population extirpations, exploitation was the primary cause of decline in 40 cases, the remainder being attributable to habitat loss, although the effects of habitat loss and exploitation on abundance are not always readily separable (Dulvy *et al.* 2003). Given ongoing increases in fishing effort and globally declining catches (Watson & Pauly 2001; Pauly 2002), and in the absence of a global network of relatively large marine protected areas (Russ & Zeller 2003), it is expected that large marine fishes, both pelagic and demersal, could be fished to extinction in the next few decades (e.g. see Sadovy & Cheung 2003; Dulvy *et al.* 2003). The fisheries targeting these species will obviously go bankrupt in the process, although they may continue to last for a while if propped up by sufficient subsidies.

HOW MANY ECOSYSTEMS?

The number of distinct ecosystems recognized differs markedly and this is particularly so for the aquatic realm. It has often been a matter of personal taste as to how many are deemed to exist and on what criteria they are separated. Ecosystems blend into one another and sharp boundaries are a convenience for cartographers rather than a reality of nature. There are many potential ways of dividing up the world's aquatic systems including by depth, geography, climatic or hydrological conditions, and primary productivity, or the nature of the biological communities. Which of these criteria are the most useful will depend on the purpose of the classification. At one end of the spectrum authors have recognized only two aquatic ecosystems, namely marine and freshwater systems, whilst other texts describe as many as 57 distinct systems in the marine environment alone (Longhurst *et al.* 1995).

Marine systems

The oceans cover about 71% of the Earth's surface and reach depths of more than 10 000 m. They extend from regions where precipitation exceeds evaporation to where the opposite is true. Perhaps the most general possible division consists of five marine 'biomes' separated on the basis of depth, temperature and nutrient status. These include: warm continental shelves, cold continental shelves,

warm open oceans, cold open oceans and upwelling waters (Bradshaw 1977). More detailed treatments generally take their lead from work in terrestrial biomes and divide aquatic systems into distinct types of biological community such as coral reefs, saltmarshes or mangroves. Thus, based on other work (Whittaker 1975; Friday & Ingram 1985; Barnes & Hughes 1988; Archibold 1995), Chapman and Reiss (1998) suggested 10 distinct marine systems and five freshwater systems. Other schemes reflect less the types of assemblage present and more the geographic location or climatological/hydrological conditions.

An approach adopted by IUCN, several UN agencies and the US National Oceanographic and Atmospheric Administration (NOAA), is that of Large Marine Ecosystems (LMEs). The 64 recognized LMEs are relatively large geographical units, characterized by distinct bathymetry, hydrography, productivity and trophically dependent populations (Sherman & Alexander 1986; Sherman *et al.* 2005). Another possible scheme for dividing up the world's aquatic systems is that used by the FAO. Twenty-seven major fishing areas are recognized for statistical purposes, and consist of eight major inland areas, each covering the waters of one of the eight continents, and 19 major marine areas covering the waters of the Atlantic, Indian, Pacific and Southern Oceans, and adjacent seas. The latter are generally divided along lines of latitude and longitude and range in size from 3 million km^2 (Mediterranean and Black Sea) to 49 million km^2 (Pacific, Eastern Central) (FAO 1994).

A further possible scheme for dividing the world's marine systems is that of the UNEP Regional Seas Programme initiated in 1974 as an approach to the control of marine pollution and resource management. The Programme has action plans for 13 regions and involves the participation of more than 140 coastal states and territories. Thus far, however, there are large parts of the world's oceans that are not covered by this scheme, and the open ocean is a major source for concern (Pauly *et al.* 2003; Chapters 20–22).

Freshwater environment

In nearly all existing schemes to describe the world's aquatic systems, there are fewer types of freshwater than marine systems. At one extreme, the scheme of Baily (1998) included no freshwater system *per se*, but instead all rivers, streams, lakes and ponds were included as part of the particular ecoregion in which they lay, as defined by climatic conditions and altitude. A similar approach was adopted by Bradshaw (1977). The most common distinction is that between running waters (e.g. streams and rivers) and still waters (e.g. lakes and ponds) (e.g. Chapman & Reiss 1998). Three other types of freshwater systems may be highlighted, namely cool temperate bogs, tropical freshwater swamp forests (e.g. those of the Amazon basin) and temperate freshwater swamp forests (e.g. the Everglades) (Chapman & Reiss 1998); these have also been recognized in the *Ecosystems of the World* book series (Goodall 1977–2000), albeit viewed as terrestrial rather than freshwater systems (Gore 1983; Lugo *et al.* 1990).

Discussion

All of the schemes mentioned have both advantages and disadvantages. For example, the FAO system is comprehensive in that all the world's oceans and freshwater systems are included, but the divisions used are not always biologically appropriate. Thus, many of the species found in the north-east Atlantic (e.g. cod and haddock) are also found in the north-west Atlantic and Arctic Sea, whilst inland boundaries such as those between Asia, the former USSR and Europe are not clearly defined in terms of their catchments, where certain rivers (e.g. Danube and Amur) start in one continent but flow into another.

For this book, the most appropriate scheme was one based on a categorization of ecosystems largely by particular biological communities (e.g. coral reefs, rivers), although this system has problems of its own. For example, is an apparently coherent community of organisms sufficiently similar throughout the world to be grouped? For instance, are coral reefs on the Great Barrier Reef adequately similar to those in the Caribbean to be labelled together? For the purposes of this book, the answer to this question is 'yes', since coral reefs around the world generally face similar types of threat including global warming, eutrophication and overfishing, and might thus be expected to respond in a broadly similar manner (Chapter 16). Likewise, rivers across the world face threats such as canalization, channelization, eutrophication, pollution and excessive water extraction for irrigation (Chapter 2). Scientists in general are more accustomed to writing global reviews relating to a single type of community, rather than reviewing issues relating to all types of community at a particular geographic location or region.

Starting with the list of 15 aquatic systems and their definitions in Chapman and Reiss (1998), a number of factors were important in determining which of these groupings were to be retained, which further subdivided and which joined into a single category. Chapman and Reiss (1998) recognized many types of shallow community underlain by soft substrata (e.g. marine mudflats, sandy beaches, temperate saltmarshes, mangroves), but did not include seagrass beds, which are ecologically distinctive and comparatively well studied (e.g. McRoy & Helfferich 1977). In addition, 'marine mudflats' and 'sandy beaches' tend to encompass only shallow or intertidal communities and do not seem to include soft-sediment systems at greater depths on continental shelves, many of which support substantial trawl fisheries (Jennings & Kaiser 1998). A better solution for describing the myriad soft-bottom categories was to include 'sandy beaches', 'salt-marsh/mudflats' and 'subtidal soft-bottom' communities.

For marine systems underlain by hard substrata, Chapman and Reiss (1998) recognized 'marine rocky shore' 'continental shelf benthos' and 'coral reef'. Both the terms 'marine rocky shore' and 'continental shelf benthos' are somewhat vague, the former being used to encompass rocky intertidal communities, and the latter to include subtidal communities such as kelp beds which may reach 50 m depth (Chapman & Reiss 1998). Therefore, clearer terminology for these widely researched ecosystems was required ('rocky intertidal': Chapter 14).

With respect to freshwater systems, the large inland seas such as the North American and African Great Lakes, and the Aral and Caspian Seas, are here recognized as being distinct from smaller bodies of standing water, since many of these have their own characteristic biotas and are affected by somewhat different threats, where factors such as eutrophication may represent a more important challenge. One type of aquatic system which has been neglected from nearly all global syntheses, but which has been the subject of considerable scientific interest (e.g. Wilkens *et al.* 2000), is that of subterranean water courses. Finally, given their importance to the origins and modern sustenance of human societies, and in spite of overlaps with other ecosystems, it was felt useful to include flood plains as a separate ecosystem.

GOALS OF THIS BOOK

While much emphasis is being put on climate change, its anthropogenic drivers and its likely impacts on ecosystems and society, the direct impacts of human population growth and industrialization have long had, are having, and will continue to have substantial consequences for the extent, structure and functioning of aquatic ecosystems. It is broadly known that many salty and freshwater ecosystems on land have already been particularly affected by human agency, and those of the open oceans, especially the deep sea, little altered. However there have been no comprehensive attempts to compare expert opinions of the status and likely trends across all aquatic ecosystems at a global level. This book originated in a series of reviews published in the journal *Environmental Conservation* which were further progressed by expert groups at the 5th International Conference on Environmental Future at the Eidgenössische Technische Hochschule (ETH, Zürich, Switzerland, March 2003) and brought to fruition thereafter. The book will systematically explore current and likely future states of aquatic ecosystems, spanning from running waters and small water bodies to the massive marine systems around the poles, and in the open oceans. The overall questions to which answers were sought included the following: What are the key threats to the status of each of the aquatic ecosystems? To what extent might each of these ecosystems change between now and the 2025 time horizon? How do the patterns of threat and

likely change vary among these ecosystems and what drives these variations? What information does this overview yield for the conservation science and management of the world's aquatic ecosystems? Because the 21 ecosystems selected for treatment are classifiable into seven categories (flowing waters, still waters, freshwater wetlands, coastal wetlands, rocky shores, soft shores, and vast marine systems), the opportunity was provided not only to make comparisons between particular ecosystems but also to compare among the different ecosystem types. The book accordingly has seven principal parts, each with its own preamble, and in the final Chapter 23 aims to synthesize the information on trends and prognoses, and draw broad conclusions about the threats and trajectories of the world's aquatic ecosystems, their overall structures and ecological functions. The IPCC process has helped to show how building scientific consensus and effectively addressing audiences outside that of science is a long iterative process. The global environmental assessment of aquatic ecosystems that this book represents is one scientifically based step towards sharing sound information and ensuring this ultimately contributes to the development of appropriate international policies and actions.

Part I
Flowing waters

One per cent of the world's fresh water is held in flowing-water, or lotic, ecosystems and provides essential environmental goods and services to human societies. The rivers and streams that most people think of in this context are indeed major sources of water, energy, biological resources and waste assimilation (Chapter 2). Groundwater aquifers are often neglected in terms of their environmental services, yet they contain 30% of the global freshwater reserves, are major providers of potable and irrigation water, prevent land subsidence and erosion, control floods through absorption of runoff, and improve water quality through mechanical and biochemical water purification (Chapter 3). Flood plains are not usually considered a separate ecosystem but they are among the most biologically productive and diverse ecosystems on Earth, and in the developing world, large human populations directly benefit from them (Chapter 4). Their seminal role in the origins and history of civilizations and as foci of current impact are such that they deserve special consideration. As societal nurseries, flood plains have borne the brunt of much human demand for resources – through water abstraction, netting, water-course diversion and damming – and for avoidance of major flooding events – through riparian protection and other infrastructure.

Being exploited for domestic and irrigation water supply, electricity generation, food and waste disposal, lotic ecosystems are known in many cases to have been heavily impacted by humans. These flowing waters permeate what in many cases appear at least to be very resilient ecosystems. Many of the resident species are for example adapted to floods, requiring high flows and spates to maintain habitats and essential ecosystem functions. These species and the underlying food webs may be threatened by a variety of human flow-calming measures such as damming, while the implied physical dynamism also offers some hope for resilience of the system at least to some types of environmental change. However, attempts at a global overview of lotic ecosystem status are few, and the perceived resilience belies many instances of ecosystem sensitivity.

2 · Prospects for streams and rivers: an ecological perspective

BJÖRN MALMQVIST, SIMON D. RUNDLE, ALAN P. COVICH, ALAN G. HILDREW,
CHRISTOPHER T. ROBINSON AND COLIN R. TOWNSEND

INTRODUCTION

Although only a fraction of 1% of the world's fresh water is found in streams and rivers at any instant, running waters are a very valuable human resource: many of the world's rivers have acted as magnets for human settlement and there are now very few river catchments that are unaffected by people in some way. Rivers contain a rich and varied biota, much of which remains undescribed, particularly in the tropics. Yet, understanding of the functional role of biodiversity in running waters is rather rudimentary (Covich 1996; Jonsson *et al.* 2001), and it is difficult to gauge the importance of ecosystem services attributable to stream biota. However, with the global human population predicted to increase by approximately one-third by 2025 (UN [United Nations] 2003), the pressure on lotic systems will certainly increase.

Aquatic Ecosystems, ed. N. V. C. Polunin. Published by Cambridge University Press. © Foundation for Environmental Conservation 2008.

Running waters encompass a wide range of habitats, spanning a continuum from small mountain springs to immense lowland rivers. The susceptibility of lotic systems to anthropogenic activities is exacerbated by their linear and unidirectional nature; almost any activity within a river catchment has the potential to cause ramifying environmental change and any pollutant entering a river is likely to exert effects for a large distance downstream (Fig. 2.1). The nature of threats to running waters also differs from region to region. Pressure on water resources is extremely high and rising in many economically less developed countries, because of human population growth and the understandable desire to provide more people with safe water (e.g. Chapters 3, 5 and 6). Water stress may also rise in developed countries for reasons of climate change and socioeconomics. The destruction of running-water habitats has been so extensive in the past in some developed countries that there is now a need to protect what is left, or to restore degraded systems. In countries where industrial development has been slower, the destructive processes are still growing in magnitude, and present an immediate threat (Dudgeon 1999).

Malmqvist and Rundle (2002) reviewed current impacts and future threats to running water ecosystems (Table 2.1). The focus here is on the implications of key threats to the

year 2025. The examples given illustrate the diversity of impacts on running waters, the complex issues that surround management, and the variety of scales at which human activities influence the ecological integrity of rivers and their catchments. The scales at which stressors operate should help guide efforts at amelioration and restoration.

THREATS TO STREAMS AND RIVERS: SCALES OF CAUSE AND EFFECT

That river systems constitute physical hierarchies in the landscape is a familiar ecological concept (Frissell *et al.* 1986; Hildrew & Giller 1994; Townsend 1996; Poole 2002). Forcing factors at the largest scale, including geology, relief and climate, constrain fluvial processes and structures, such as local bedform and flow patterns, at levels lower in the hierarchy. It is an attractive proposition that such large-scale factors also force the local ecological communities that are most easily observed (Rundle *et al.* 2000).

Some stressors of running-water ecosystems act at scales greater than those of catchments and may be global, including climate warming, acidic airborne pollution and increases in UV irradiance (Fig. 2.1). These issues can only truly be addressed by global cooperation, though local action can ameliorate their effects to some extent (see Hildrew & Ormerod 1995). Many other impacts on river systems are caused by more local economic activities, particularly through land-use change and associated, often diffuse, nutrient and sediment additions. Such impacts are only global in the sense that they are being repeated throughout the world and on a massive scale. Stressors such as water abstraction, polluting effluents, species additions and river regulation also fall into this 'local but widespread' category. Of course, individual 'point-source' perturbations have local effects and are easier to manage than the truly global or diffuse impacts.

Although threats to running waters vary with the spatial scale considered, the magnitude of their effects can also vary along longitudinal river profiles. For example, the effects of global climate change might be most pronounced on small tributaries that are more prone to fluctuations in discharge and temperature; at the same time, the effects of drought and saline intrusion are likely to be most pronounced in lowland sections of rivers (Fig. 2.1). For stressors such as nutrient additions, it is more likely that effects will be concentrated in middle and lowland river segments where human settlement and agricultural development are more pronounced (Fig. 2.1).

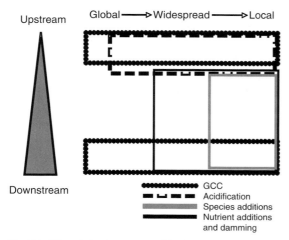

Fig. 2.1. The relative importance of five key future threats to river ecosystems. Rectangles illustrate where these factors are likely to have the most impact along the length of rivers and whether this impact is likely to be truly global (local in occurrence but repeated throughout the globe) or local. GCC, global climate change.

Table 2.1. *Anthropogenic factors and their abiotic and biotic effects in running–water ecosystems*

Ultimate forcing factor	Sub-factor	Proximate cause	Abiotic alteration	Biotic effects
Ecosystem destruction		Urban and agricultural expansion, water abstraction	Complete ecosystem loss	Species and population extinction
Habitat alteration	Hydrology	Damming, channelization, water abstraction, deforestation, water transfer schemes	Loss of natural flow periodicity, increased risk of drought, severing of upstream–downstream linkages	Altered habitat conditions, reduced dispersal
	Siltation	Deforestation, agricultural development	Reduced substratum complexity	Altered habitat conditions
	Alteration of riparian corridor	Urbanization, channelization, agriculture	Altered energy inputs (organic matter/light), altered in-stream marginal habitat	Altered trophic dynamics, altered habitat conditions
Water chemistry	Acidification	Industrial emissions (SO_2 and NO_X), exhaust emissions (NO_X)	Reduced pH, increased Al^{3+}	Direct physiological effects, indirect (food chain) effects
	Nutrient addition	Agriculture/deforestation, industry, sewage works/landfill, atmospheric emissions of NO_X	Increased N and P	Increased primary production, algal blooms
	Toxic metals	Mining, industrial gaseous emissions, landfill/sewage works	Many trace metals (e.g. Cu, Hg, Zn, Al, Pb, Cd)	Direct toxic effects
	Organochlorine toxins	Industry (atmospheric and water emissions), landfill/sewage works, waste incineration, agriculture	PCBs, organochlorine pesticides (e.g. DDT, dieldrin)	Toxic effects through biomagnification
	Organic pollution	Urbanization, sewage works, agriculture	Reduced O_2, increased suspended solids	Reduced habitat availability
	Endocrine disruptors	Industry, agriculture, waste incineration	Organohalogens (e.g. dioxins, furans, PCBs), pesticides (e.g. DDT, dieldrin), pharmaceuticals (oestrogens)	Interference with naturally produced hormones
Species removal and addition		Fisheries, aquaculture/aquarium trade, sport fishing, horticulture (riparian plants)	Invasive species	Increased/reduced competition, altered energy inputs (riparian) and ecosystem dynamics

Source: After Malmqvist and Rundle (2002).

As well as operating at different spatial scales, threats to running waters may also vary temporally. As ecosystems, rivers have proved remarkably resilient to pulsed perturbations, and they depend on episodic flow disturbances to maintain their biotic and physical diversity. They usually recover well from isolated pulses of point-source pollutants or one-off physical disturbances, particularly if the full physical complexity of the river hierarchy is intact, since there are then a wide variety of refugia in which the biota can survive and from which they can recolonize (see for example Milner 1994). Other temporal characteristics of anthropogenic disturbance are more problematic. Pulsed perturbations with a rapid return time, such as frequent bouts of pollution or sediment input, as is often the case in intensely managed landscapes, may not allow system recovery. Most damaging of all are persistent or 'press' perturbations, where there are long-term changes to average conditions. Ecological resilience may then be lost and system changes occur, though even then recovery is possible if the conditions can be restored sufficiently and there are nearby sources of potential colonist species.

GLOBAL CLIMATE CHANGE: DROUGHT AND FLOOD DISTURBANCES

The complex consequences for riverine systems of increasing concentrations of greenhouse gases (Coutant 1981; Firth & Fisher 1992; Covich *et al.* 1997; Grimm *et al.* 1997; Puckridge *et al.* 1998; Yarnell 1998; Meyer *et al.* 1999; Wood & Petts 1999; Murdoch *et al.* 2000; Schindler 2001; Poff *et al.* 2002), and El Niño–Southern Oscillation (ENSO) and North Atlantic Oscillation (NAO) climate dynamics, are increasingly being recognized (Dahm & Molles 1991; Piechota *et al.* 1997; Cayan *et al.* 1999; McCabe & Dettinger 1999; Bradley & Ormerod 2001; Cluis & Laberge 2002). Shifts in mean annual temperature and precipitation remain difficult to forecast, however, especially at the seasonal and regional level. Current global circulation models (GCMs) predict some increase in mean annual global temperature and changes in global precipitation over the next century (Felzer & Heard 1999; Xu 1999; Chapter 1). These directional changes are expected to alter regional patterns of evaporation and precipitation and thereby alter regional hydrology, especially in the frequency and intensity of floods and droughts. Statistical downscaling models are being linked to GCMs and runoff models to generate regional and basin-level predictions of climate and precipitation runoff (Xu 1999; Busuioc *et al.* 2001; Hellström *et al.* 2001). This approach appears to be more accurate than others used previously, because GCMs do not include effects of surface features and land-surface processes that influence runoff in mountainous basins.

Although climate change models vary widely in their regional predictions, extreme hydrological events (i.e. floods and droughts) are likely to increase and alter the functioning of riverine ecosystems. In arid regions, especially those with fast-growing human populations, the relationships among the interannual and seasonal distribution of precipitation, land use and rate of evaporation are critically important for river basins. In many other regions, the predicted increase in extreme climatic events may result in long-term, large-scale hydrological and geomorphological responses that will greatly modify the ecology of entire drainage basins, through landslides, sediment loading and the infilling of pools that then dry up during prolonged droughts. Normally, the recovery of aquatic communities after infrequent and brief floods and droughts (i.e. infrequent pulse perturbation) (IPCC [Intergovernmental Panel on Climate Change] 2001*a*) is relatively rapid (Smock *et al.* 1994; Feminella 1996; Holmes 1999; Magoulick 2000; Ledger & Hildrew 2001). The rate of recovery will probably be slower or fundamentally different if drought frequency, duration and intensity increase (frequent pulse perturbation) or if the physical system is degraded, for example if substratum interstices are occluded by sediment (press perturbation).

Researchers attempting to evaluate the effects of future climate change on hydrology and precipitation confront similar challenges to those who seek to evaluate the effects of other perturbations. However, the increase both in freshwater use per caput and in human population density in river basins is rapidly increasing the demands for fresh water. These accelerating demands will continue to compete with the need to maintain natural flow regimes or ensure the minimum-flow requirements of native species. Many biodiversity 'hotspots' around the globe contain unique riverine freshwater and riparian species that are under cumulative stresses (Allan & Flecker 1993; Covich *et al.* 1999; Malmqvist & Rundle 2002). Changes in precipitation, even within the natural range of variability, will further increase environmental stress because of the increased human demand for high-quality fresh water. Moreover, extremely low flows will exacerbate the effects of pollution from sewage outfalls and industrial toxins.

Natural droughts and floods maintain a complex spatiotemporal mosaic of aquatic habitats and natural flow variability that is essential to sustain these dynamic habitats (Boulton *et al.* 2000; Malmqvist 2002; Robinson *et al.* 2002). However, the combination of natural flow variability and increased human demand for fresh water (Covich *et al.* 1997; Grimm *et al.* 1997), along with a possible trend towards much longer and more severe droughts, could reduce the abundance of many species in many regional riverine communities beyond the threshold for recovery. Thus, the cumulative effect of regional drought on riverine ecosystem functions results from a combination of natural and societal driving variables.

The frequency and intensity of climatic fluctuations often generate societal responses to minimize future vulnerability, such as the construction of dams and storage reservoirs and the implementation of regional water-transfer projects. These engineering solutions create more persistent alterations of natural hydrological regimes well past the end of natural drought events (see below). Furthermore, as human populations grow and transform drainage basins, the extensive construction of impervious surfaces such as roofs and roads causes rapid runoff and flash floods. The lack of infiltration to recharge groundwater later results in decreased stream flows during periods of low precipitation (Chapter 3). Consequently, major rivers around the world (such as the Yellow River, Ganges, Indus, Nile, Colorado and Rio Grande) are overexploited and their waters often no longer reach coastal zones.

ACID RAIN AND RIVER RECOVERY: IMPORTANT MANAGEMENT LESSONS

While global climate change presents a future challenge to river managers, in the developed world at least, acidification may now have seen its day as a large-scale environ-mental threat to rivers. The recognition of acidification as a widespread environmental problem in the 1970s led to national and international agreements to reduce sulphur emissions. As a result, there has been a significant decline in both emissions and deposition in North America and Europe (Cambell & Lee 1996; Driscoll *et al.* 2001). In the north-eastern and upper mid-western USA, for example, sulphur deposition decreased by 29% and 35%, respect-ively, between 1980 and 1995 (Stoddard *et al.* 1999). In some instances, stream pH responded positively to these reductions. In the Hubbard Brook Experimental Forest, for example, stream water and precipitation pH have increased (Driscoll *et al.* 2001) and the acid neutralizing capacity of streams has also increased in much of Europe (Fig. 2.2). However, due to a strong historical decline in the base-cation concentration of catchments in other regions such as the Adirondack Mountains (USA) (see also Lawrence *et al.* 1999), no such recovery has been apparent. Such a lag in the recovery of pH following emission reductions may be common in catchments where soils have a high storage capacity for sulphate (Alewell *et al.* 2000).

Despite the evidence of chemical recovery of acidified streams in many instances, there are few data on concomi-tant changes in biota; those that do exist suggest variable responses (for example Soulsby *et al.* 1995; Lancaster *et al.* 1996; Monteith *et al.* 2005). Potential response of biota to changes in pH can be inferred from streams recovering from liming. In limed streams in mid-Wales (UK), recovery of invertebrates has lagged behind chemical changes, and recolonizing acid-sensitive species have not persisted, pos-sibly because of recurrent acid episodes (Bradley & Ormerod 2002). Chemically recovering streams in the UK's Acid Waters Monitoring Network have seen distinct, though only modest and patchy, biological recovery. So far

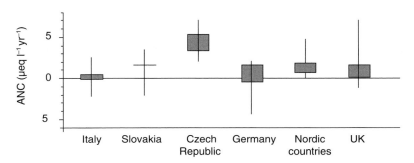

Fig. 2.2. Ranges in trend slopes of acid neutralizing capacity (ANC) in 56 lakes and streams within different European regions in the last one to two decades of the twentieth century. Boxes show interquartile range of slope estimates. (Redrawn from Evans *et al.* 2001.)

this includes increases in predatory invertebrates, rather than a wholesale replacement of acid-tolerant by acid-sensitive species. This has led to speculations that acidification takes aquatic communities through a regime change, reversal of which may be resisted and perhaps delayed by the new community (Ledger & Hildrew 2005; Monteith *et al.* 2005).

Although the benefits of a reduction in acidifying emissions in the developed world have not yet been fully realized, there are positive lessons to be gleaned from the management approaches to the problem and the ecological responses observed. In particular, the successful international implementation of large-scale emission reductions informed by monitoring programmes such as the critical-loads approach (Cresser 2000) should lend much hope. Such knowledge should be used to positive effect when dealing with other atmospheric pollutants or with the same kinds of acidifying emissions which may become a problem for fresh waters in developing countries (Rodhe & Herrera 1988).

FUTURE DESTRUCTION OF LOTIC HABITATS: DAMMING AND FRAGMENTATION

One of the most obvious ways in which a river system can be degraded is when human activities cause it to dry up completely or cause substantial damage to its physical structure, in effect creating a completely different habitat (Chapter 4). Running-water habitats are being increasingly subjected to such pressures, mostly due to the rapidly expanding human population (Malmqvist & Rundle 2002) and the massive increase in water use (Postel 2001).

The physical nature of rivers has undoubtedly been affected most profoundly by the creation of dams (e.g. Chapters 4 and 6). There are currently more than 45 000 large dams in the world (WCD [World Commission on Dams] 2000) and very few of the world's major rivers have not been impounded in some way. Hence, damming is a form of press perturbation repeated around the globe that both fragments and destroys lotic habitats. River fragmentation due to damming has important consequences, hindering migratory mammals, fish and crustaceans, and blocking the dispersal of plants (Dynesius & Nilsson 1994). This interruption of connectivity can lead to reduced genetic variation and increased extinction risks. Dammed rivers also lose sections with fast-flowing habitats and the organisms that require such habitats. It

is often high-profile, economically important species such as the Atlantic salmon (*Salmo salar*) that are particularly vulnerable to the conversion of rivers with steep gradients to a step-like profile with cumulative impacts. Moreover, the destruction of upstream juvenile habitats means that the construction of fish passages is often not worthwhile.

There are currently some very large dam projects; the biggest enterprise presently, the Three Gorges Dam on the Yangtze, is due to be completed by 2010. Planning is also at an advanced stage for large dams on the Mekong and large Indian rivers such as the Narmada. Apart from huge socioeconomic impacts, the ecological consequences of these projects will be great, but hard to predict due to poor knowledge about the complex ecosystems involved (Dudgeon 1999, 2000*a*). There is currently a tendency, however, towards damming of not only very large, but also small rivers (for example in Japan: Morita & Yamamoto 2002), largely because medium-sized rivers are already maximally exploited. The development of new dams stands substantially to change running-water biota in many parts of the world, although elsewhere there is a trend towards breaching dams that do not fulfil environmental, economic and safety requirements (WCD 2000; Stanley & Doyle 2003). An increasing number of dams on streams and small rivers, while having little local impact, may produce a substantial cumulative regional effect.

Fragmentation of stream systems through processes such as damming puts specific ecological demands on the capacity of organisms to disperse and survive at small effective population sizes. Even in pristine rivers, populations are not always as well connected as they may seem. In fact, particular species are rarely distributed throughout the river continuum, but are restricted to specific segments that are thus patchily distributed throughout the landscape. It is therefore plausible that the river hierarchy imposes a metapopulation structure on freshwater organisms in the landscape (see Berendonk & Bonsall 2002). Individual patches can thus be lost, or the population in that patch driven extinct by damming or other impacts such as pollution events or drought constituting frequent pulse or press perturbations. 'Patch loss' can thus be a consequence of true global climate change, for instance through drying, or of local impacts such as land drainage, which may be viewed as 'apparent climate change'. A loss of suitable habitat patches will reduce local population density of dependent species and is likely thereby to reduce overall geographical range, leading ultimately to extinction even where apparently suitable patches remain or are conserved (see Hanski 1998; Lawton 2000). Essentially, the likelihood of successful recolonization of suitable habitat patches declines as they become rare in the landscape. We predict that such species extinctions will become acute as a consequence of climatic drying or habitat loss (e.g. Chapters 4 and 6).

Direct assessments of the adults of aquatic insects, particularly of species with poor powers of flight (Townsend *et al.* 2003), do not suggest frequent widespread dispersal away from the stream corridor or an influence of catchment land use on dispersal potential (Petersen *et al.* 1999, 2004; Briers *et al.* 2002; Wilcock *et al.* 2003). Rather, the stream corridor itself seems the main 'highway' for dispersal, adding further to the requirement for its sensitive management. Recent evidence, however, suggests that inter-patch dispersal may be more frequent. This includes whole-stream labelling with stable isotopes (Briers *et al.* 2004) and the use of molecular markers. For instance, genetic differentiation between pairs of local populations of a patchily distributed caddisfly in the UK was negligible up to distances of about 20 km, suggesting frequent adult dispersal, but differentiation increased at greater distances (Wilcock *et al.* 2001, 2003). This suggests a regional structure in the landscape, in which dispersal across inhospitable areas is possible if patches of suitable habitat remain and can be used as stepping stones (Kelly 2002). Under such a scenario, fragmentation of the landscape by real or 'apparent' climate change could have genetic and demographic consequences for lotic species.

CHEMICAL POLLUTANTS PRESENT GLOBAL MANAGEMENT CHALLENGES

Nutrient concentrations have increased substantially in rivers throughout the world and, globally, fewer than 10% of rivers can be classified as pristine in terms of their nitrate status (i.e. <0.1 mg NO_3-N l^{-1}: Heathwaite *et al.* 1996). Hence, nutrient additions are a classic example of a local impact on rivers repeated throughout the globe on a massive scale (e.g. Chapters 4, 5 and 6). Although the management of point-source nutrient additions can be achieved comparatively easily, diffuse sources present more of a management challenge. Fertilizer additions for agricultural purposes are by far the biggest single driver of nutrient loadings, and global fertilizer additions have been predicted to increase by 145% over 1990 levels by 2050 (Kroeze & Seitzinger 1998), representing an increase in the global

average fertilizer use from 15 to 21 kg per caput per year. These projected increases vary dramatically between regions, with the largest absolute increases in countries bordering the western Pacific (China) and the Indian Ocean, although substantial increases are also predicted throughout Asia, Africa and Latin America (Kroeze & Seitzinger 1998). Legislative measures, such as the designation of nitrate-vulnerable areas in Europe, are a good example of how developed countries have tackled damaging agricultural nutrient additions. The creation of buffer strips and the education of farmers to employ methods that are less likely to lead to erosion and nutrient release are other local management options that have been employed (Sweeney *et al.* 2004; Chapter 4). Such approaches require substantial finances, however, and it is difficult to see how dramatically increased loads to rivers can be avoided without financial assistance in developing areas of the world, where the need for high food production will be at a premium. It is also important that investigations of the ecological effects of nutrients in rivers are increased because knowledge of the implications of eutrophication for running waters is substantially less than that of standing fresh waters (Chapter 6).

In addition to nutrients, there are many other chemical pollutants of running waters, including past legacies from the industrial age and present-day threats from the continued release of pesticides, petro-hydrocarbons, organic solvents, heavy metals and sewage-borne pathogens (Paul & Meyer 2001). Substantial numbers of previously overlooked chemical pollutants, such as pharmaceuticals (Daughton & Ternes 1999; Kolpin *et al.* 2002) and endocrine disruptors (Kolpin *et al.* 2002), have also been highlighted by recent analytical advances. For example, over 60% of the pesticides used in the USA are now known to have endocrine-disrupting properties (Dudgeon 2000*a*) and the implications of such chemicals may be severe, considering the evidence of their effects on the development of running-water biota such as fish (see Noaksson *et al.* 2003).

Inputs of heavy metals into running waters have also continued to increase, but now tend to be production- rather than consumption-related (Dudgeon 2000*a*). Their supply is strongly associated with sediment/floodplain dynamics (Chapter 4), especially in urbanized settings (Foster & Charlesworth 1996). Not only is the quantity of heavy metals important, but their speciation must be understood to appreciate their potential environmental effects. Mercury pollution from coal-fired power plants and mining has increased in recent decades and the bioaccumulation of methylmercury will probably continue to increase in consumers at the top of food webs such as river otters (Gutleb *et al.* 1998; Taastrom & Jacobsen 1999).

The complexities of managing chemical pollution are increasing as more compounds enter aquatic ecosystems (Halling-Sørensen *et al.* 1998). Complex mixtures may cause synergies, antagonistic toxicities and genotoxic activities (Kolpin *et al.* 2002). The persistence of various chemicals may be related to degradation processes for pulse releases, or continued exposure that allows particular agents to persist for much longer periods. Exposure routes of many chemical agents can also be complex (Halling-Sørensen *et al.* 1998) and derive from point and non-point sources. Some chemicals cause multixenobiotic and antibiotic resistance in bacteria, fungi and viruses, all of which have high dispersal abilities. Medical substances are released through both human and veterinary sources via direct disposal in sewage or through treatment of animals in field settings. The evolution of microbial resistance to antibiotics widely used in farming and medicine is creating new threats to human health, as well as potentially altering natural ecosystem function (Wiggins *et al.* 1999; Whitlock *et al.* 2002).

Hence, chemical pollutants represent a global management challenge that is relevant to developed and developing countries alike (Baer & Pringle 2000). Perhaps the greatest challenge lies in forecasting which chemicals are likely to be of most importance in the future and the degree to which their impact will be manifest. Three different methods are currently used to reduce uncertainty and improve forecasting: empirical regression models, Bayesian modelling and adaptive management (Nilsson *et al.* 2003). All three have been criticized for various reasons, but can be improved as knowledge is acquired. For example, uncertainty decreases as the amount of evidence and the level of agreement or consensus increase (Moss & Schneider 1997). Other strategies include modifying the scale or precision of analysis, or incorporating hybrid approaches and new research (Clark *et al.* 2001; Benda *et al.* 2002). Certainty in forecasting can also be improved by developing models for whole river systems (an individualistic approach), improved determination of exposure routes that incorporates the life histories of organisms and by proactive behaviour through more effective application of what is known already (Dudgeon 2000*a*).

INVASIVE SPECIES: PROBLEMS, RISKS AND REMEDIES

There has been much concern over the increasing number of invasions by exotic species and the global homogenization of biota (Vitousek *et al.* 1996; e.g. Chapters 4, 5 and 6). Some argue that invasive species constitute a great threat to natural ecosystems and to the economy (for example Schmitz & Simberloff 1997; Pimentel *et al.* 2000), whereas others take the view that exotic species generally increase diversity and only in exceptional cases trigger negative consequences (Gido & Brown 1999; Rosenzweig 2001). There is no doubt that several invaders in running waters are causing major problems. In the USA, for example, exotic plants such as the Eurasian milfoil (*Myriophyllum spicatum*) and *Hydrilla* clog waterways, and riparian invaders such as *Tamarix* cause water loss through their high transpiration rates (Brock 1994; Dahm *et al.* 2002). In some regions, such as southern Africa where water stress is high, increased water loss in river catchments due to invasive plants has significant economic and human implications (Van Wilgen *et al.* 2001). Animal invaders also significantly depress or replace populations of native species and affect ecological processes in running waters (e.g. Townsend 2003). Various species of fish, mussels and crustaceans are the most successful invaders, with the zebra mussel (*Dreissena polymorpha*) in North America being perhaps the best-known example (see Malmqvist & Rundle 2002).

Hitherto, the main focus has been on harmful invasives, and not exotic species in general (Kolar & Lodge 2001). Most exotic species do not, however, have dramatic negative effects (Williamson & Fitter 1996). In those cases when they do, it is beneficial if the mechanisms underpinning their success can be identified (Simon & Townsend 2003). Invasions may be more likely, for example, when the invader is a top predator, when the invaded community has low diversity or is affected by human or natural disturbances (Moyle 1999). The number of individuals or propagules introduced and the number of introductions are also important (Kolar & Lodge 2001). Generally, however, predicting invasion success is intrinsically difficult (Williamson 1999).

Theoretical models are being used increasingly to develop strategies for the control of invaders. In particular, such models can be used to predict the spread of biological control agents in relation to that of the pest, and they often consider invasion as a two-stage process of (1) establishment and (2) local dispersal and replacement of native species, requiring different control approaches (Allendorf & Lundquist 2003) (Fig. 2.3). The rate of spread depends as much on population growth rate as on the dispersal abilities of both host and agent, and a non-linear spatial

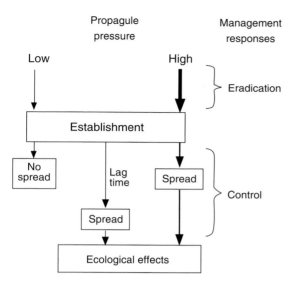

Fig. 2.3. The two phases of invasion: (1) arrival and establishment of the non-indigenous species in a new area, and (2) its spread and replacement of native species. Also shown are the associated management responses. (Redrawn from Allendorf & Lundquist 2003.)

interaction between them may determine the success of the biological control agent (Fagan *et al.* 2002). Alternatively, high dispersal rates in a control agent species might be negatively related to its local impact.

The existence and probable importance of long-distance dispersal limit the value of the models presently available. Moreover, the introduction of species as control agents may not itself be safe. Hitherto, it seems that the successful control of invasive species has primarily been achieved by mechanical and chemical methods (Simberloff 2003). To meet the increasing problem of invasive species in running waters, further research is needed, in addition to improved legislation and education. One way forward may be to compare the traits of invasive and native species, and those of successful invaders with species that fail, to detect general features (Kolar & Lodge 2001).

WILL THE LOSS OF BIODIVERSITY AFFECT STREAM AND RIVER ECOSYSTEM FUNCTION?

High biodiversity is generally considered to be a positive quality of ecosystems. For instance, species provide harvestable food and may harbour yet undiscovered

substances that can be of use in the future. High diversity may also make ecosystems more stable in the face of disturbances (Tilman *et al.* 1996; Naeem 1998; Naeem *et al.* 2000) and deliver important 'ecosystem services' (Hector *et al.* 1999).

Following the growing concern about the accelerating rate of biodiversity loss from natural ecosystems, scientists have begun to ask what consequences this loss might have for ecosystem function (see Schultze & Mooney 1993; Loreau *et al.* 2002). The question of the effects of species loss is complex and has been addressed through theoretical and observational approaches. Perhaps most interesting, however, are those manipulative experimental studies in which the number of species, or other components of biodiversity (such as species evenness, composition and interactions) and spatial and temporal variation in these qualities (Symstad *et al.* 2003), are altered. Hitherto the vast majority of such experiments have focused on ecosystem processes in grasslands and soils (Tilman *et al.* 1996; Hector *et al.* 1999; Wardle 2002).

Many species in natural systems are functionally important in their habitats, including those living in streams and rivers (for example Wallace & Webster 1996). However, only a few studies on the effects of biodiversity on ecosystem processes have been conducted in aquatic ecosystems. These have primarily addressed the effects of complementary resource use by component species on ecosystem processes, the mechanisms underlying these effects and the influence of the ordering of species in the assembly or disassembly of communities (McGrady-Steed *et al.* 1997; Jonsson & Malmqvist 2000, 2003*a, b*; Cardinale *et al.* 2002; Jonsson *et al.* 2002; Fukami & Morin 2003; Larned *et al.* 2003). For stream environments, the loss of consumers that play highly significant roles in the transformation of organic matter into animal tissue may remove important links between primary resources and other trophic levels. Organic material may accumulate and fish production dwindle in the absence of herbivorous and detritus-feeding invertebrates (Jonsson *et al.* 2001). Experimental studies also suggest that the extinction of individual detritivore species may affect the rate of litter breakdown through the loss of positive interactions among species (Jonsson & Malmqvist 2000, 2003*a*). However, Ledger and Hildrew (2005) suggested that there was a good deal of functional redundancy among stream invertebrates, such that algal grazing was supported in acid streams by generalist 'detritivores', even though most specialized grazers (mayflies and gastropods) were lost.

Stream biota may also play an important role as bio-turbators. Experimental manipulations of the diversity of bioturbating species in marine sediments had significant effects on nutrient release (Emmerson *et al.* 2001). In soft-bottomed reaches of rivers, mussels, worms, larval midges and meiofaunal taxa burrow in the substratum with as yet unknown diversity effects on the rates of processes related to their mixing activities. Indeed, the functional significance of certain groups of stream biota, such as meiofauna and chironomid larvae, is still poorly understood (Hakenkamp & Morin 2000; Robertson *et al.* 2000).

Covich *et al.* (2004) reviewed 18 empirical studies from aquatic environments addressing the effects of various components of biodiversity on ecological processes, such as elemental cycling, decomposition or productivity. Effects of changed species diversity varied strongly in magnitude and direction in a context-dependent way. Experimental studies on the relationship between biodiversity and ecosystem processes have mainly been conducted in microcosms, i.e. at small spatial and temporal scales (Loreau *et al.* 2001; Symstad *et al.* 2003). Moreover, previous studies have rarely considered the great variability in species abundances (Ruesink & Srivastava 2001). Other aspects of biodiversity, such as evenness, have been studied only in terrestrial plant systems, and then often with conflicting results (Wilsey & Potvin 2000; Polley *et al.* 2003). However, decomposition in stream ecosystems can be influenced both by species rich-ness and dominance, and the magnitude of this influence may vary seasonally and spatially (Dangles & Malmqvist 2004). Considerable development is expected in this field by integrating theoretical and experimental work with large-scale field investigations, to provide forecasts with higher realism and applicability to the urgent problem of global species loss. At this point, the presence of biodiversity effects is widely recognized, although their indirect conse-quences and importance in relation to other factors in streams and rivers require increased study.

Two types of information are necessary for predicting future effects of biodiversity loss on ecosystem processes. These are (1) accurate estimates of which species will be lost and which will remain, and (2) precise quantitative data on the link between diversity and ecosystem function. At pre-sent, extinction order may be predicted based on the known sensitivities of various taxa (Jonsson *et al.* 2002), but the resulting assemblage may be more-or-less functionally dis-similar (Jonsson & Malmqvist 2003*b*; Heemsbergen *et al.* 2004). Although recent research shows that diversity has significant effects on ecological processes, these data are primarily qualitative and from small-scale experiments. Scaling up such results to make meaningful predictions for natural ecosystems is a major challenge in biodiversity research (Loreau *et al.* 2001).

FUTURE MANAGEMENT OF STREAMS AND RIVERS: THE VITAL ROLE OF ECOLOGY

The examples presented in this chapter serve to illustrate the range of temporal and spatial scales over which environmental impacts operate in running waters. If management of the threats posed by these impacts is to be improved, it is essential that not only overall under-standing of riverine ecology (particularly in tropical and subtropical regions) be improved but that links between impacts, patterns and processes occurring at different scales be better comprehended.

Some of the environmental impacts on running waters are part of ongoing, large-scale disturbances, which means that the possibilities for protection or restoration are limited and that impacts will be repeated throughout the globe. The issue of global climatic change is a prime example where immediate political decisions informed by ecologists as well as forecasters and modellers are essential. Climate modelling can indicate where the effects of global warming are likely to be most pronounced and ecologists can, on the basis of such models, make informed predictions of consequences for populations, communities and ecosystems. To date, predictions have mostly been based on the erroneous assumption that a suitable climate at a site would cause species requiring that specific climate to become established with no regard for ecological attributes of those species, such as their dispersal ability, population dynamics or inter-actions with other taxa (Travis 2003). Hence, predictive models need to be developed that take climate change and species' ecologies into account. In a perspective spanning the next 20 years, the impacts of global warming on rivers is likely to act alongside rather than override those of pollution and habitat destruction (Malmqvist & Rundle 2002). In a 100-year perspective this impact is likely to become relatively more critical, and therefore it is urgent to act forcefully and early to ameliorate potentially wide-ranging effects on biodiversity and human societies.

Habitat loss, at all scales, is another major issue (e.g. Chapters 4, 5 and 13) where prediction and management need to be underpinned by improved ecological know-ledge. There is no doubt that major losses of habitat in riverine landscapes will be seen over the next few decades, as human population pressure intensifies. Most current ecological theory and experience indicate that habitat loss equates with biodiversity loss and furthermore with reduced stability and resilience (Brooks *et al.* 2002) and impaired ecological services (Tilman *et al.* 1996). Here, there is a clear role for metapopulation and dispersal theory for predicting the consequences of habitat loss for species. Such knowledge will also be highly relevant for making predictions of system recovery following disturbance by chemical pollutants. Small-scale disturbances from pulse pollution events may also generate regional to global problems by being repeated on a widespread basis. There are likely to be marked differences in pollution effects between economically developed countries, where trends in pollutant discharges have levelled off or are decreasing, and developing countries where they are increasing (e.g. Chapters 5, 7 and 10). The reduction in acid rain in Europe and the USA at least provides a positive example of a chemical pollutant of rivers that has been reduced through large-scale, international cooperation between scientists and decision-makers. It would be satisfying if the emerging threats from increasing nitrogen and, ultimately CO_2 emissions, could be handled with equal success.

Global homogenization of biota following the spread of exotic species is a growing threat, the final extent and impact of which is uncertain. Ecologists are trying to understand the unifying factors behind successful invasions and how to prevent establishment of invasives and control already established species. The emerging discipline of invasion ecology is likely to play an increasing role in the development of strategies to cope with non-indigenous species. Clearly, research into the functional role of biodiversity and on species interactions generally should inform the management of invasive species.

In conclusion, ecologically informed modelling at species, population, community and ecosystem levels should incorporate the issue of scale. The unequal distribution of future threats throughout the regions of the world calls for creative solutions and, although streams and rivers will be affected most in dry and/or densely populated regions, global cooperation will be required. Ultimately, changes in management practices and public awareness will hopefully improve running-water ecosystems in economically developed countries, and underpin conservation strategies in developing countries. This hope can only be realized through relevant implementation based on sound ecology.

3 · Groundwater ecosystems: human impacts and future management

JANINE GIBERT, DAVID C. CULVER, DAN L. DANIELOPOL, CHRISTIAN GRIEBLER,
AMARA GUNATILAKA, JOS NOTENBOOM AND BORIS SKET

INTRODUCTION

Groundwater plays an important role in the sustainability of many Earth ecosystems and a crucial role in human life and socioeconomic developments. In the past, it has offered insurance against drought by providing 'buffer storage', a vital factor for human survival; groundwater is a globally important valuable renewable resource. In many parts of the world, mostly during the last 50 years, groundwater has been depleted and degraded (Morris *et al.* 2003).

Groundwater is not only important in supporting human welfare, it is also the basis for the life of diverse organisms existing below the Earth's surface. The complex relationships between groundwater and subsurface organisms generate dynamic ecological systems, reviewed in detail by Danielopol *et al.* (2003).

During the last 10–15 years, groundwater problems have been increasingly analysed within an integrated hydrological and ecological framework (Gibert *et al.* 1994;

Aquatic Ecosystems, ed. N. V. C. Polunin. Published by Cambridge University Press. © Foundation for Environmental Conservation 2008.

Griebler *et al.* 2001; Sophocleous 2003), and the importance of ecohydrology has been stressed repeatedly (Klijn & Witte 1999; Rodríguez-Itrube 2000, 2003; Sophocleous 2002). Building on other reviews of the topic (Tóth 1999; Sophocleous 2002, 2003; Morris *et al.* 2003), the aim here is to highlight how groundwater will need to be carefully managed throughout the twenty-first century if its use and importance for the functioning of many ecosystems are to be sustained.

This chapter presents a brief overview of subterranean aquatic systems, with emphasis on three topics: (1) present-day negative human impacts on the subterranean hydrosphere, (2) policy/management measures that integrate socioeconomic and ecological arguments, and (3) likely developments up to the 2025 time horizon.

Groundwater ecosystems

Groundwater is an important phase of the hydrological cycle (Fig. 3.1). Most of the flow of perennial streams originates from subsurface water and most of the Earth's liquid fresh water is not found in rivers and lakes, but is stored underground in aquifers. The volume of fresh groundwater comprises an extensive subsurface aquatic domain of 30% of the total freshwater reserves, less than 1% of which are constituted by surface waters such as lakes, marshes, wetlands and rivers (Table 3.1). Nearly 2 billion people in the world today depend on aquifers for drinking water, and 40% of the world's food produced by agriculture is irrigated largely from groundwater sources (Morris *et al.* 2003). These aquifers also provide a valuable base flow, supplying water to rivers during periods of no rainfall (Chapter 2).

Subterranean aquatic ecosystems are open systems through which energy flows and matter are processed in various ways. Such systems are built from three major components: the geological substratum, groundwater itself and living organisms, each playing a distinct role. The geological substratum can be either consolidated rocks through which water sometimes carves large solution channels and flows very rapidly (for example karstic aquifers), or unconsolidated rocks (sand, gravel and pebbles) through which water moves much more slowly (porous aquifers) (Fig. 3.2). Permeable sediments play the same role in the subsurface as open reservoirs on the surface. Groundwater can be stored or further transmitted to other

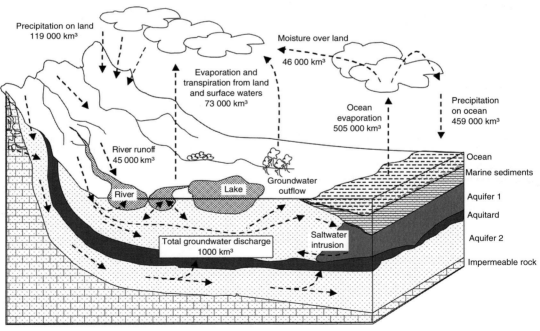

Fig. 3.1. Interactions among groundwater surface water systems, rivers and their watersheds, lakes and oceans, including values for annual fluxes of the water volume in the hydrological cycle. (Adapted from Danielopol *et al.* 2003 and Morris *et al.* 2003.)

Table 3.1. *Distribution of water on the Earth's surface*

Major stocks of water	Volume (10^3 km^3)[a]	Total hydrosphere (%)	Fresh water (%)
Salt water			
Oceans	1 338 000	96.54	
Saline/brackish water	12 870	0.93	
Salt water lakes	85.4	0.006	
Inland waters			
Glaciers	24 064	1.74	68.70
Permafrost ice	300	0.022	0.86
Fresh groundwater	10 530	0.76	30.06
Freshwater lakes	91	0.007	0.26
Marshes, wetlands	11.5	0.001	0.03
Rivers	2.12	0.000 2	0.006
Soil moisture	16.5	0.001	0.05
Atmospheric water	12.9	0.001	0.04
Living organisms	1.12	0.000 1	0.003
Total water	1 386 000	100	
Total fresh water	35 029.2	2.53	100

[a] According to Shiklomanov and Rodda (2003, Table 1.8) the total volume of groundwater inclusive of both gravity and capillary water, is $23 400 \times 10^6$ km^3, representing 1.7% of the total hydrosphere. Not included is 2×10^6 km^3 of deep underground water. Fresh groundwater in the deep Antarctic is estimated to be 1×10^6 km^3.

Source: Based on data from Shiklomanov and Rodda (2003) and UNEP [United Nations Environment Programme] (2002).

geological formations. The renewal capacity of the aquifers within a given geological formation or a geographical area is an important criterion for water management decisions.

The groundwater domain is inhabited by diverse micro-, meio- and macroorganisms (Wilkens *et al.* 2000; Chapelle 2001) (Fig. 3.2). Due to lack of light, groundwater ecosystems are dominated by secondary producers and the microbial community is mainly heterotrophic. Primary producers, represented by chemoautotrophic bacteria and archaea, seem to play a minor role in most groundwater ecosystems. However, Gold (1999) suggested that chemoautotrophic iron bacteria can produce enough biomass to clog the interstitial space. Microorganisms have invaded subsurface systems to depths of several thousand metres where elevated temperature limits life (Ghiorse 1997). Most of the living microorganisms are associated with the sediment surfaces where they form microcolonies and biofilms (Griebler *et al.* 2002). The biomass of groundwater-dwelling prokaryotes in the unconsolidated subsurface domain accounts for about 6–40% of the Earth's total microbial biomass. Protozoans (flagellates, amoebae and ciliates) and fungi seem restricted in their distribution in groundwater habitats located close to the vadose (unsaturated) zone or to the soil. In aquifers, microorganisms are responsible for major turnover of energy and matter. They play an important role in weathering and formation of minerals, and they store significant quantities of carbon, nitrogen and phosphorus in their biomass. Moreover, they contribute to the development in the subsurface of microhabitats that are chemically distinguished by their redox reactions. In the context of strong human impacts on the environment, the high purification potential of groundwater ecosystems is of increasing interest and importance. This purification service is mainly provided by microorganisms. Protozoa are also suggested to participate in biodegradation activities of microbial communities, resulting in stimulation (Kinner *et al.* 1998) or reduction (Kota *et al.* 1999) of actual bacterial degradation rates via grazing on bacteria and viruses.

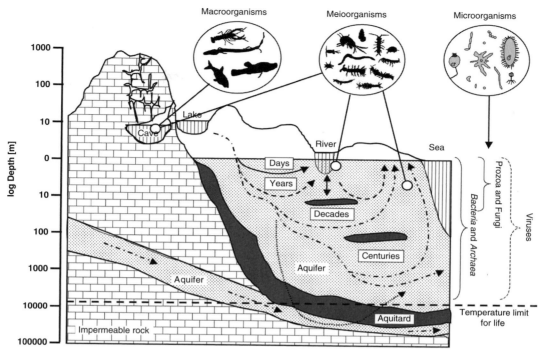

Fig. 3.2. Karstic and porous aquifers: diagrammatic representation of water flow direction, water residence time and of major types of groundwater organisms. (Adapted from Danielopol *et al.* 2003.)

Groundwater ecosystems harbour an impressive number of animal species that are exclusively known from subterranean waters (referred to as stygobionts), such as crustacean species belonging to Bathynellacea, Thermosbaenacea, Remipedia, Mictacea and Spaeleogriphacea (Sket 1999). As an adaptation to energy-poor environments, stygobiouts typically show life histories involving delayed maturity, greater longevity, smaller clutch size, larger eggs and lower percentages of mature ovigerous females. In many systems, population densities are low, food chains are simple, trophic levels are few and feeding behaviour is unspecialized. The ecological characteristics of subterranean animal assemblages offer information on the functional state of groundwater ecosystems and/or on the degree of connectivity, especially between above (epigean) and below (hypogean) surface ecosystems (Gibert 2001a).

Groundwater ecosystems vary in size and structural complexity. Small-scale systems (on the scale of metres to hundreds of metres), such as those existing in alluvial sediments along rivers, are potentially understandable in terms of their functional properties (Danielopol *et al.* 2000). Large

aquifers (on the scale of kilometres) encompass alluvial plains or whole river sectors within watersheds. They form regional or basin groundwater systems that are especially important for water management planning and sustainability (Sophocleous 1998). Karstic systems form large complex ecosystems, sometimes spanning kilometres of galleries and fissures below the Earth's surface (Gibert *et al.* 2000).

Relationships between groundwater and surface ecosystems

Hydrology controls the composition and functioning of aquatic and terrestrial subsurface ecosystems. Here the interactions between groundwater and surface water play a dominant role in shaping the structure of subterranean species assemblages. The interactions are governed by climate, landform, geology and several biotic factors.

The mechanisms of groundwater–surface water interactions affect the closely interrelated recharge and discharge processes. Groundwater moves along complex flow paths that form hierarchical nested structures in space, forming flow systems which depend on the topography of a region and give rise to local, intermediate and regional flow systems (Fig. 3.2). These flow systems are generally characterized by regions of recharge, throughflow and discharge operating as conveyer belts which effectively interact with their ambient environment for mobilization, transport and accumulation. These interactions produce *in situ* environmental effects (Tóth 1999; Sophocleous 2002, 2003).

Shallow subterranean ecosystems are directly connected to surface aquatic and terrestrial systems like rivers, lakes and wetlands (Figs. 3.1 and 3.2). Therefore to obtain a better understanding of the complexity of the groundwater domain, ecologists and hydrologists extend their cooperative studies to composite systems, such as river–aquifer systems (Stanford & Ward 1993; Brunke & Gonser 1997) or complex wetland areas along large rivers that include both terrestrial and aquatic components (Chapters 4, 9 and 10). Hence the interdisciplinary approach that ecologists and hydrologists now favour involves treating groundwater ecosystems within a framework of ecohydrology, which integrates mutual interaction between the hydrological cycle and ecosystems (Klijn & Witte 1999; Hunt & Wilcox 2003).

HUMAN IMPACTS

Human activities have various types of impact on the state of groundwater ecosystems; these are generally the result of synergistic actions and have negative effects at various space and timescales (Danielopol *et al.* 2003). Only the main impacts are reviewed here.

Agricultural demand

Irrigated agriculture accounted for 80% of global water consumption and for 86% of water consumption in the less-developed countries in 1995 (Rosegrant *et al.* 2002). The highest percentage of irrigated land lies in Asia (67%), followed by North America (12.6%) and Europe (11%) (Vörösmarty *et al.* 2000). The USA uses *c.*63% of its groundwater withdrawal for irrigation and *c.*3% for live-stock watering (US EPA [Environmental Protection Agency] 2000). Tilman *et al.* (2001) predicted a 30% increase in the world's irrigated land between the year 2000 (when it stood at 250 million hectares) and 2020.

The transfer of technologies from economically developed countries in order to reduce water scarcity in the less-developed countries, and aggressive marketing strategies that have increased sales of groundwater, have led to depletion of groundwater reserves in many countries; at present, water tables are falling on every inhabited continent. It is estimated that 160 billion m^3 of groundwater are being over-pumped in India, China, the Middle East, North Africa and the USA (Brown 2002). Such overexploitation has resulted in continuous drops in water tables of 1 m per year or more in India, China (Shah *et al.* 2000; Yang & Zehnder 2001) and Indonesia.

Urban, domestic and industrial demand

In 1990, 43% of the world human population was urban, and in less-developed countries 35% of the population was urban (UN [United Nations] 1997). In these countries in 2000, there were 18 mega-cities (with populations >10 million people), which drain groundwater resources and fill streams that cross them with domestic and industrial wastewaters (Lundquist 1998). The additional urban population in the less-developed countries is likely to be 1.8 billion people in 2020. This will also increase pressure on domestic and drinking water supplies, sanitation and food supply. In many urban areas of the world, the continual extraction of domestic and drinking water from aquifers will rapidly lead to groundwater resource depletion. For example, in Bangkok, the water table has fallen by 25 m since 1958 and continues to drop at a rate of 3 m per year, and seawater intrusion into the aquifer is occurring at the rate of 450 m per year (Shah *et al.* 2000).

There is also a high demand for water from manufacturing and service industries (cooling aggregates, cleaning and manufacture of products). The quantities of water consumed strongly depend on the respective industrial sector and its activities, and can often be extremely high (e.g. thermoelectricity generation and mining). Industrial water demand in Europe accounts for just over half of the total water withdrawal.

Pollution from agriculture

Along with the development of crop production during the last 40–50 years, huge quantities of fertilizers (phosphates and nitrates) and pesticides, and intensive irrigation practices have been applied.

Nitrates represent the major source of groundwater pollution all over the world, leading to unwanted eutrophication. In the year 2000, *c.*87 billion tonnes of nitrogenous compounds were used for crop cultivation worldwide and a 60% increase in use of these compounds is expected by 2020 (Tilman *et al.* 2001); there is thus a foreseeable rise in groundwater pollution. In Europe, where riverbank filtrates are a primary drinking water source, 5–10% of these aquifers exceed the EU (European Union) nitrate limits (25 mg l^{-1}) and, in intensively cultivated areas, the NO$_3$-N concentration is increasing by 1–2 mg N per litre per year (Dudley 1990).

The production of pesticides is expected to increase in volume by *c.*74% by 2020 (Tilman *et al.* 2001). Herbicides and insecticides are strong inhibitors of metabolic activity in both plants and animals. Some of these compounds, such as parathion and lindane, are potential carcinogens. Excessive application of fertilizers and/or pesticides over wide field areas also introduces toxic heavy metals like mercury, cadmium and selenium into groundwater, and it may increase the salt concentration (US EPA 2000).

Organic pollution of groundwater is occurring also from animal feedlots and irrigation practices. Animal feedlots produce large amounts of liquid waste from manure, and pose public-health risks through the possible microbial contamination of drinking water. A large number of newly emerging contaminants in groundwater, such as steroids and antibiotics, are used in animal feeding operations in large quantities; these may end up in wastewater and ultimately find their way into aquifers. These cattle feed additives are also in the manure that is often spread directly on land; this is an additional pathway of groundwater contamination.

Pollution from urban and industrial waste

The increased quantity of wastewater in urban areas combined with the leakage of sewage systems allows the introduction of large quantities of dissolved organic matter, heavy metals, salts and pathogens into groundwater (US EPA 2000). Petroleum products are complex mixtures of various weakly or highly soluble polycyclic organic compounds. One of the most common forms of contamination of groundwater is through the seepage of oil products from fuel storage tanks. There are about 400 000 confirmed leaks from underground petroleum tanks in the USA and every second petrol station in the UK stands on contaminated soil and groundwater (Sampat 2000).

Additionally, many accidental spills in and around urban regions are responsible for organic pollution of high-quality groundwater areas. Landfills for solid waste disposal over time leak organic pollutants, nutrients like nitrogen and phosphorous salts, heavy metals and various other salts that change the total salinity of the groundwater. Industry often needs a large amount of cooling water which in many cases is extracted from subsurface aquifers and, once used, is reinfiltrated into the subsoil, generating a source of chemical and microbial pollution.

In the past, burial has been used as a means of waste disposal, and many subsurface sites are therefore now highly contaminated. Over 100 000 contaminated groundwater sites containing toxic metals are estimated to exist in Europe. In the USA, about 35 million litres of solvents, heavy metals and radioactive material are injected annually into deep groundwater via wells, below the deepest source of drinking water (Sampat 2000).

The high number of cars, especially in developed countries and in and around urban centres generally, represents a source of pollution for the subsurface environment from fossil fuels that can accidentally spill into the groundwater. The fuel additive methyl tertiary-buthyl ether (MTBE) is an example. About 20 Mt of MTBE are used annually in the world, 60% of them in the USA and 15% in Europe. This compound has high water solubility and, once introduced by diffuse seepage or by point accidental spills, can contaminate large volumes of groundwater, producing a negative effect on taste and odour of drinking water even at concentrations of 2–50 μg l^{-1} (Schmidt *et al.* 2001).

Injection of highly toxic chemicals or radiochemicals by industry into deep aquifers can, in the long term, also be dangerous for drinking-water reserves and indirectly for human health, because these sites allow diffuse leakage and dispersion of contaminated water into shallow aquifers used for human needs. Injection of organic waste into aquifers could also favour development of thick bacterial biofilms and clogging of porous systems. Chemical acid waste injected into karstic systems can dissolve rock (Danielopol *et al.* 2003).

Land use and forest mismanagement

During the last 50 years, land use has been greatly modified. The major impacts of deforestation, increase in cultivated land area, selection of water-demanding crops, increase in impervious coverings of urban areas, stream regulation and embankment building have isolated aquifers from surface ecosystems. In all these cases, the volume of groundwater recharged from surface runoff has declined with time. For example, in the arid western regions of the USA, the degradation of vegetation has led to a reduction in groundwater recharge, up to 60% of the total runoff now flowing at the soil surface. The widespread clearing of native vegetation in Australia mobilized naturally occurring salts that accumulated in superficial earth layers (Peck & Williamson 1987; Salama *et al.* 1993). The forest can be considered a water reservoir that gradually releases water to the underlying aquifer. Conversely forest may extract a high quantity of groundwater. The conversion of land in Brazil from savannah to *Eucalyptus* forest has reduced the subsurface water recharge by nearly 60% (Le Maitre *et al.* 1999). Forest also plays an important role in nutrient purification and protection of groundwater quality (Haycock *et al.* 1997).

Deforestation and intensive land use for agriculture along streams and rivers frequently induce soil erosion and clogging of shallow groundwater habitats with fine sediment, leading to a reduction in subterranean biodiversity (see Hahn 2002).

Climate change

The present global warming is now recognized to be promoted by anthropogenic greenhouse gases (IPCC [Intergovernmental Panel on Climate Change] 2001*a*). Shallow aquifers loaded with high concentrations of organic matter from anthropogenic activities (such as in areas of organic-waste disposal) by denitrification release nitrous oxides (Ronen *et al.* 1988) that have a potential greenhouse effect about 275 times greater than that of CO_2 (IPCC 2001*a*). The concentration of these gases in contaminated aquifers is up to 1000-fold higher than expected in equilibrium with the atmosphere.

The warming of the Earth's atmosphere may have significant implications for the state of groundwater ecosystems. During the last two decades, the Earth's average surface temperature has increased by about 0.5 °C and is expected to rise similarly up to 2025 (IPCC 2001*a*), but not uniformly all over the world. This also applies to precipitation.

These climatic changes will have negative impacts on the groundwater reserves of many aquifers. In the semi-arid and arid zones, the recharge of aquifers will be reduced through decrease in runoff. In the case of areas with high net humidity, runoff water will not optimally recharge the depleted aquifers because of its fast flow. In the northern hemisphere, the recharge of aquifers will not be able to

proceed through slow snow melting and longer periods of infiltration into the soil (IPCC 2001*a*). Moreover, climate changes over shorter periods can be expected to influence rainfall and temperature patterns in many parts of the world, as does the El Niño–Southern Oscillation (ENSO) (IPCC 2001*a*), which affects precipitation in arid and semi-arid regions in mid-latitudes of North and South America, Asia and Australia. As the ENSO increases in frequency and intensity, there will be a greater strain on groundwater recharge rates.

POLICY AND MANAGEMENT SUGGESTIONS

The need for sustainable groundwater management

Sustainability (*sensu* Lubchenco *et al.* 1991) of groundwater ecosystems concerns the maintenance of their ecological functioning and/or potential human exploitation. It also means that ecosystems are able to integrate disturbances and to perpetuate renewable resources. However, in some countries, more water is being consumed by humans than renewed by nature, and many aquifers are polluted and can no longer be used for drinking water. As population grows, more countries and regions will reach an unsustainable situation (Tables 3.2 and 3.3). A water crisis is likely to arise in the twenty-first century (see references in Danielopol *et al.* 2003), being only partly eased by the implementation of long-term environmental policies. These policies consist mainly of reducing water withdrawal from aquifers, increasing their water recharge and controlling groundwater contamination.

An approach to sustainable groundwater systems involves managed depletion and subsequent natural or artificial recharge of aquifers. Especially in tropical areas, the controlled depletion of aquifers before the arrival of the monsoon, during which much runoff water is lost at the surface, enables more water to be used for a longer time (Ambroggi 1977).

Although shown to be a flawed concept (Sophocleous 1998), the 'safe yield' is still used as a basis for groundwater management policies. Groundwater management should not be reduced to the classic approach of sustainable yield; the volume of groundwater that can be extracted annually from a groundwater basin should not cause adverse effects on the subterranean system. Both hydrologists and ecologists should view the problem of

Table 3.2. *Groundwater contribution to drinking water use by region and population size*

Area	Drinking water from groundwater (%)	Population (10^6)
Europe	75	200–500
North America (USA)	51	135
Asia–Pacific	32	1000–1200
South America	29	150
Australia	15	3

Source: Sampat (2000), modified by Danielopol *et al.* (2003).

Table 3.3. *The human population (at continental and global scales) and the volume of usable water supply for 1985 and projections for 2025*

Area	Population (millions)		Usable water supply (km^3 yr^{-1})		Supply/demand 2025 (%)	
	1985	2025	1985	2025	Scenario 2[a]	Scenario 3[b]
World	4830	8010	39 300	37 100	50	61
Africa	543	1440	4520	4100	73	92
Asia	2930	4800	13 700	13 300	60	66
Australia + Oceania	22	33	714	692	30	44
Europe	667	682	2770	2790	30	31
North America	395	601	5890	5870	23	28
South America	267	454	11 700	10 400	93	121

[a] Scenario 2: reflects only the rising demand to 2025.
[b] Scenario 3: considers both climate and development effects (Vörösmarty *et al.* 2000, Table 4).
Source: Based on Vörösmarty *et al.* (2000), modified from Danielopol *et al.* (2003).

groundwater management through a holistic framework where groundwater management is coupled with the protection of both subterranean and surface aquatic and/or terrestrial ecosystems.

A key point regarding the sustainable use of groundwater is the maintenance of the integrity of structure and functioning of groundwater ecosystems (Fig. 3.3). Subterranean ecosystems should not be considered only as a water resource, but also as a whole system, in which organisms play an important ecological role. In this perspective, the quality of the environment must be protected at surface and subsurface levels and a sound management strategy for the use of groundwater resources applied (Sophocleous 1998, 2002; Notenboom 2001; see also Tables 3.4 and 3.5).

Effect of different human attitudes on groundwater use

ECONOMIC ASPECTS

Like other natural resources, such as oil or coal, groundwater has a quantifiable value, is tradable and represents a source of political power (Gleick 1993a). Compared to other Earth resources, the market value of high-quality water is too low and should be increased following classic economic criteria (OECD [Organization for Economic Cooperation and Development] Working Committees 2000). The human appropriation of renewable freshwater reserves lies at the centre of the water crisis. Both of the following socioeconomic scenarios can improve this situation.

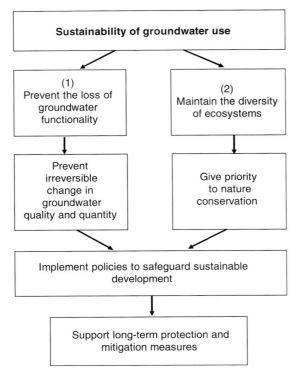

Fig. 3.3. A model for development and management of sustainable use of groundwater. (Adapted from Notenboom 2001 and Danielopol *et al.* 2003.)

Table 3.4. *Global water use by sector and total in 1990 (Gleick 1996; Seckler et al. 1998) or 1996 (Alcamo et al. 1997), and projections for 2025*

Estimation	Water use (km^3 yr^{-1}) Agriculture	Industry	Domestic	Total water use (km^3 yr^{-1})
Gleick (1996)				
1990	2680	973	300	3953
2025	2930	1000	340	4270
Alcamo *et al.* (1997)				
1996	2022	728	296	3046
2025	2022	2235	621	4580
Seckler *et al.* (1998)				
1990	2084	823		2907
2025		1193		4569
2025 HIE	2432	1193		3625

Source: HIE, high irrigation efficiency (note that Seckler *et al.* 1998 present the volume of the industrial and domestic water use together).

(1) Full-cost pricing for water supply. At present, water pricing does not take into account additional costs related to the safe maintenance of well-functioning ecosystems or the infrastructure necessary for water supply and sanitation. This suggestion to develop an economic market for well-maintained water resources can be expected to have negative consequences exactly in those parts of the world with high dense human populations where people with very low incomes will not be able to purchase the water needed (Shah *et al.* 2000). Barlow and Clarke (2002) consider that groundwater does not belong to any country, private person or corporation, should therefore not be sold as are other resources, and should be viewed as part of the Earth's heritage. As such it should be shared by all the organisms of the planet.

(2) Reduce or eliminate environmentally perverse subsidies. It has been proposed to reduce or suppress incentives which sustain economic development but are detrimental to the water supply and sustainable ecosystems (OECD Working Committees 2000). Moreover, financial support for fertilizer and pesticide use should be reduced in the future, because their excessive use will continue to pollute groundwater.

CULTURAL ATTITUDES

Traditional human societies have repeatedly adapted to chronic water scarcity with low and efficient water consumption. Many springs have been protected against environmental pollution or overexploitation of water for centuries. The opposite has occurred in areas where there has been economic progress (Gleick 1993a). Yet poorer countries which respect water value suffer from water scarcity, whereas richer industrial countries overexploit and contaminate large available freshwater reserves.

TECHNOLOGICAL ASPECTS

With technical progress over the last 50 years, groundwater extraction and distribution methods have improved and their price has declined, giving easier and greater access to drinking water (Shah *et al.* 2000). Perhaps the biggest revolution in water-resource management has been the small cheap diesel or electric pump that gives

Table 3.5. *Possible activities to alleviate the pressure on groundwater systems by 2025*

1. Create new incentives for artificial recharge activities in both semi-arid and monsoon areas. For monsoonal regions, development of strategies to replenish the pre-monsoon drawdown from enhanced recharging from rainwater. Storage development. Drought prevention through underground storage
2. Ameliorate the water distribution, e.g. reduce consumption, increase the use of recycled water
3. Reduce the volume of wastewater
4. Decrease the tourism pressure in groundwater vulnerable areas
5. Reduce the amount of fertilizers and toxic substances use, e.g. pesticides
6. Use new technologies for irrigation to avoid waterlogging or intense evapotranspiration and salinization problems
7. For water-stressed areas import water in form of grain (or vegetables and fruit) from water-rich regions
8. Improve the waste treatment techniques
9. Better control of waste disposal, e.g. for long-term spread of highly toxic pollutants reduce the volume of injected material; select suitable areas for dumping sites, use improved technologies for landfill construction
10. Educate people to reduce the volume of wastes
11. Find remedial treatments for contaminated areas and brownfield sites that endanger aquifers
12. Avoid mining in areas important for drinking-water purposes or in landscapes with high value for biological conservation
13. Define protection zones for groundwater aquifers using holistic approaches, e.g. aquifer basin management
14. Implement sustainable groundwater management, to balance water withdrawal and recharge
15. Avoid constructions along rivers which in the long term diminish the quality of the groundwater
16. Control the surface water quality and protect the zones of surface infiltration

Source: From Danielopol *et al.* (2003).

farmers the means to invest in self-management of groundwater irrigation. In Pakistan, private investment in groundwater development through tube wells (360 000 wells in 1993 alone) has driven growth. In India, almost half of all irrigated areas depend wholly or partly on groundwater. In China, more than 2 million pumps irrigate some 9 million hectares. In the USA, one of the world's largest aquifers, the Ogallala, has been developed through privately financed wells feeding sprinkler systems. While groundwater irrigation has substantially contributed to world food production and provided farmers with a dependable source of water, it has also led to massive overuse and falling groundwater tables (WWC [World Water Council] 2003*a*).

ECOLOGICAL ENGINEERING
Ecological engineering of groundwater systems has rapidly expanded in the recent past (see Dreher & Gunatilaka 2001; Gunatilaka & Dreher 2001). For example, natural attenuation was implemented in the USA in the early 1990s as a biologically efficient method and, in some cases, as the only feasible means of contaminant remediation (Hazen 1997). In Europe, however, natural attenuation has

yet to be authorized as a sanitation strategy. A recent German network of projects should lead to the integration of the natural-attenuation concept into the guidelines and recommendations for environmental and remediation policies (see Danielopol *et al.* 2003).

Improved water-use efficiency

In all countries throughout the world, there has been an increase in water development and withdrawals, not only in building dams and reservoirs, but also in capturing springs or pumping groundwater. The technological means of groundwater extraction and distribution have become widely available through their low costs (Shah *et al.* 2000). However, the systems supplying agricultural, industrial and urban water are often highly ineffective and wasteful. Irrigation, which uses up to 80% of the annual freshwater supply that is appropriated for human use, bears the greatest responsibility. During the past two decades, irrigated areas have expanded quite rapidly; for example, in Egypt, irrigated areas account for 100% of the total cropland (Yang & Zehnder 2002). Town water supply systems often lose water through leaky taps and connections, 50% or more losses being commonly recorded (WWC 2003*a*). In order to reduce the volume of groundwater withdrawals, some countries are using new technologies in water distribution, such as root-zone irrigation of crops, on a wider scale (WWC 2000). Savings could make large amounts of water available for other purposes.

Different water-users can adapt to waters of different quality. Groundwater, the preferred source of drinking water for most people in the world, should be prioritized for human consumption. With controls and checks, treated urban effluents could be used in agriculture and industry. Optimizing the various uses together is the first step to achieving the most efficient water distribution possible.

Because groundwater resources are limited, other hydrological aspects of aquifers need to be considered. For example, throughout the world, submarine springs might potentially be found along all the limestone coasts, including in Australia, Vietnam, Saudi Arabia, Mexico, Senegal, Florida and around the Mediterranean. In many countries, water extraction from submarine springs might represent a valuable supply that could provide several small towns with drinking water (G. de Marsily, personal communication 2003).

It is expected that increasing competition and conflict among alternative uses will lead to a transfer of agricultural water to the sectors where the value of water use is higher, for example to those of industry, municipalities or tourism activities. The trend has been evident in many less-developed countries (Yang & Zehnder 2001).

Alternatively, projections of water demand in different regions throughout the world indicate spiralling future problems. For water-scarce countries, lack of water constrains food production. Importing food can effectively reduce the countries' water demand; water shortage may locally be compensated for by food imports. The southern Mediterranean countries and China are typical examples. Compensation for physical water scarcity by 'virtual water imports' of cereal grain or non-cereals (sugar and vegetable oils) can be included in current regional and national agricultural strategies (Yang & Zehnder 2002).

Different perceptions of the needs and effects of water use in different countries have meant that while some problems are identified, others are disregarded. There is need for increasing collaboration and partnership at all levels, based on political commitment to and wider societal awareness of the importance of water security and management of water resources (Wagner *et al.* 2002; see also Fig. 3.4).

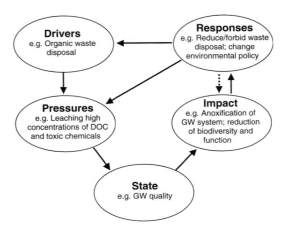

Fig. 3.4. The DPSIR (drivers, pressures, state, impacts, responses) framework (arrows indicate direction of actions: continuous arrow, the best remedial/response action; dashed arrow, possible action for remediation, more expensive to implement as compared to the best action). GW, groundwater, DOC, dissolved organic carbon. (From Danielopol *et al.* 2003.)

Integrated water management

It has become obvious that water resources management has to be integrated to promote development and sustainable use of land and water resources, and boost fair economic and social progress. Integrated management of water resources must put human populations back at the heart of the management process. How we intensify watershed and groundwater management, and thus groundwater quantity and quality, should both be considered when making decisions that affect dependent ecosystems. Equally important is the need to consider surface water, soils, vegetation and land-use planning issues when making decisions about groundwater. In watersheds in the USA, Japan, Switzerland and Brazil, effective governance of water management has to focus on a watershed scale and must integrate all relevant aspects such as hydrology, ecology, urban water management, agriculture and waste management (Wagner *et al.* 2002).

With the new policies on groundwater protection and water pollution control (1994) and 'Clean and safe water for the 21st century' (2000), the USA has been focusing legislation on a watershed scale (R. B. Levin *et al.* 2002). In its Water Framework Directive, the European Union is trying for the first time to apply such an approach (EU 2001). The Directive describes sustainable bodies of groundwater in terms of maintaining the ecological quality of surface waters and terrestrial ecosystems, and providing water for abstraction.

However, integrated management of water resources is probably inadequate when watersheds have common aquifers. The Indus Delta Water Consortium in Pakistan has a specific objective to manage the quantity and quality of shallow groundwater. The Beauce aquifer in France is also an example of shared groundwater management (WWC 2003a).

Fully integrated water management will be difficult to achieve, however, because the fragmentation of institutions in the water sector is a serious obstacle. Thus the people, organizations, laws and regulations for water supply and sanitation for population use often have very little to do with those applicable to water used for irrigation, hydropower or tourism.

Reduction of poverty

Lack of safe sources of water for drinking and basic hygiene is a major cause of human poverty around the world; one person out of five does not have access to safe and affordable drinking water, and half of the world population does not have access to sanitation. Each year at least 3–4 million people die of waterborne diseases and more than 2 million children die of diarrhoea (WHO [World Health Organization] 1996).

Gleick (1999) stated that access to a minimum of 30–50 l of high-quality water per day per person (roughly 11–18 m^3 per year) at an affordable price should be a human right. Governments have the responsibility to ensure this basic requirement is met for all (WWC 2002). This is now enforced by the general Comment of the United Nations Committee on Cultural and Social Rights. It means that the 145 countries which have ratified the International Covenant on Economic, Social and Cultural Rights appendix to the UN Human Rights Document (UN 2002) must progressively ensure that everyone has access to safe and secure drinking water and sanitation facilities, equitably and without discrimination (Dasgupta 2001). However, this does not imply that drinking water has to be supplied free of charge. A common argument for a low water price is that the poor cannot afford a real market price. In many places, because of lack of taps in their houses or neighbourhoods, the poor have to buy water from street vendors. For example, in Jakarta (Indonesia), the ratio between prices charged by vendors and those of public utilities ranges between 4 and 400. There are also new initiatives to reduce

water consumption: in Mexico City, replacement of leaky house toilets with more efficient toilets has allowed enough water for the needs of more than 200 000 new residents to be saved (Brown 2002).

Installations and services for basic water supply and sanitation in poor countries can be economically feasible if locally adapted technologies requiring small investments are implemented step-by-step with the participation of all stakeholders and under the supervision of local authorities.

But many questions arise. Do the less-developed countries really have the capacity to radically improve their water management? Are the donors' activities in the less-developed countries adapted to the national needs and priorities? Does the expertise generated by the international activities reach the people living in less-developed countries that really need it? Opinions vary, but very often, countries have not as much capacity as they should have, the donor activities are not always appropriate to the setting and thus the efficiency of the international system needs to be questioned.

Given the high demands for reducing poverty, efforts should be concentrated on countries in which poverty and diseases related to water are particularly acute and in which there is a promising potential for improvement. The WWC (2003a) suggests that (1) donor networks should be created and expert groups from different countries should share their experiences, (2) training should be offered to professionals involved in water management and (3) expert groups should work with specialized institutes in less-developed countries. Countries such as Brazil, China and India have expertise and training kits that they have used to improve the skills of their water personnel, and this kind of expertise should be disseminated in other countries.

Protection of groundwater systems and establishment of conservation areas

The need to protect groundwater resources increases every day with the growth of the world population and the misuse of aquifers. Protection of groundwater resources as a sustainable source of drinking water requires the protection of natural ecosystems. As in other heterogeneous aquatic ecosystems, the protection of groundwater ecosystems has to be based on ecological risk assessment (US EPA 1992). There are at present some tools and methods giving a basis for regulatory actions and development of strategic plans (Danielopol et al. 2003). However, ecological risk assessment and management at the watershed level require (1) strong scientific and technological partnership between the various disciplines, (2) the development of up- and downscaling procedures for different model types, (3) development of particular toxicity-screening procedures and (4) critical evaluation and interpretation approaches for the data, including probability risk assessments, quantitative uncertainties analysis and other innovative tools.

During the last few years, it has become obvious that full clean-up of polluted groundwater sites is hardly feasible because of the high costs and technical difficulties. It is also evident that remediation of polluted groundwater sites is much more expensive than the implementation of preventive protection programmes. Thus in the USA, the costs of cleaning up a municipal well contaminated by petroleum were more than 30 times greater than the costs of a protection plan for the surrounding area (US EPA 2000). Since it is impossible to establish full protection of all groundwater systems all over the world, some hydrologists have proposed the creation of 'hydrological nature reserves' in order to safeguard pristine systems from contamination (De Marsily 1992). Groundwater remediation could then be deployed in addition for those sites that pose the largest risks in terms of pollution of drinking-water wells and dispersion of pollution towards these reserves.

For the establishment of 'hydrological nature reserves', groundwater biodiversity (for example taxonomic richness and endemism) should be another important criterion, as in Europe, in addition to aquifer vulnerability and human activities (Gibert 2001*b*).

Ecological aspects have been considered in the protection of surface ecosystems, but this is not the case for groundwater. For example, in the EU Water Framework Directive (EU 2001), a new planning system was established for 2000–2027 for the protection, improvement and sustainable use of rivers, lakes, estuaries, coastal waters and groundwater. For surface waters, the status objectives that must be achieved by 2015 are good ecological and chemical status. Groundwater has not yet attained a good ecological status, only a good chemical status; biotic factors within groundwater systems are yet to be taken into account. However, groundwater systems, exactly as surface water systems, are subject to hydrological (hydrogeological), chemical, biological and physical/dynamic processes, which together define their ecological status (see Gibert *et al.* 1994; Griebler *et al.* 2001).

PROJECTIONS TO 2025

Lower water tables and land subsidence

The total human appropriation of fresh water has increased dramatically over the past century, and is expected to grow by at least 8% by 2025 (Gleick 1996). It is estimated that about half of accessible surface freshwater runoff is appropriated for human use and that this will rise to 70% by 2025 (Postel *et al.* 1996). While surface waters, particularly rivers and streams (Chapter 2), are in many regions the most easily developed water source, there are many regions of the world where the major source of water is groundwater rather than surface water. These include the Arabian Peninsula, much of the western USA, and many karst regions. In many parts of the developing and developed world, there were significant reductions in the water table in the last half of the twentieth century. The water table is lowered by increased groundwater withdrawal which, in turn, has several causes. As we have seen, these include (1) overuse of surface waters and contamination of surface waters forcing increased groundwater use, (2) increased population density with a concomitant increase in domestic water needs, (3) increased per caput domestic water consumption, (4) increased irrigation as agricultural productivity increases and (5) increased water use for industry, including power-plant cooling. There is every reason to think these trends of increased water use will continue over the 2025 time horizon, although there should be a levelling off of demand towards the end of this period. In industrialized countries overall per caput water use has actually levelled off and declined (Gleick *et al.* 2002). The causes of this are complex, but seem to have resulted mostly from increased efficiencies of manufacturing as a result of increased costs of water. Domestic water use, which is heavily dependent on groundwater, is but a small fraction of total water use. Total water use, including both agricultural and industrial uses, could also show declines as efficiency increases and waste declines. Counteracting these trends is the increased use of groundwater as a domestic water source. The nearly worldwide use of bottled spring water is one highly visible sign of this trend. For example, in Europe, people aware of pollution problems do not rely on the quality of tap water.

In many regions, increased groundwater extraction has resulted in land subsidence. A spectacular example of this is the state of Florida in the USA, where hundreds of new sinkholes form every year as a result of a lowered water table and the structural instabilities that follow the draining of solution cavities. Sinkholes can form from anthropogenic action in a variety of ways, including collapse and subsidence, and they have important economic consequences (Beck & Herring 2002). In the relatively well-watered eastern USA, much of the lowering of the water table is the result of increased numbers of individual wells for domestic use. In regions where groundwater is used for irrigation, the economic consequences of land subsidence can be severe. There are now entire conferences devoted to sinkhole formation and the engineering solutions to the problem (Beck & Herring 2002). The economic costs of land subsidence may provide a counterforce to the increased dependence on groundwater.

Biodiversity loss

There are estimated to be 50 000 obligate subterranean species (Culver & Holsinger 1992), about one-third of which are aquatic. The majority of these species, probably 80–90% of the total, are undescribed, and nowhere is this truer than for the aquatic fauna of small cavities, namely interstitial habitats. In contrast, the obligate aquatic fauna of karst is among the best-known invertebrate faunas. Two features of the obligate groundwater fauna make it especially vulnerable. Firstly, the range of nearly all species is highly restricted. Over one-quarter of the obligate groundwater species in the USA is known from a single site and nearly half is known from a single county (D. C. Culver & H. Hobbs III, unpublished data 2000). For many species, local environmental degradation can result in species extinctions. Secondly, groundwater-dwelling species are outcompeted by surface animals, even with only moderate increases in organic pollution which enriches the food supply (Sket 1977). In addition, these animals generally have long reproductive cycles and genetically isolated populations, making them potentially sensitive to environmental changes (Notenboom et al. 1994). There are very few data from which to estimate biodiversity loss, but Elliott (2000) suggested that four out of the 400 obligate groundwater-dwelling species in the USA became extinct in the second half of the twentieth century as a result of sealing and damming springs. Local extirpations of obligate cave fauna may result from spills, illegal dumping and other activities (see Crunkilton 1985; Culver et al. 1992). These trends are likely to continue up to the year 2025. However, widespread extirpation of groundwater fauna seems unlikely because, except in karst, groundwater

moves slowly, thus localizing the problem, and legislation such as the USA Endangered Species Act and the EU Habitat Directive, and the increased publicity that threatened species receive from organizations such as the World Conservation Union, have resulted in increased protection of rare species. What is more likely is that increased range restriction and fragmentation will occur.

Restriction of functionality (water uses become reduced in diversity)

Groundwater becomes less useful as it becomes contaminated, until finally it cannot be used at all. The increased use of fertilizers and pesticides on a worldwide basis portends an international decline in groundwater quality, and automobile use also exacts a heavy environmental price. While improved technology in the industralized countries may slow the growth in new contaminated sites, the number of contaminated sites in the less-developed world is likely to continue to rise. For example, access to advanced and/or cheap well technology for groundwater abstraction in Asian countries has increased water use for both crop production and drinking necessities, but, in many places, arsenic salts were extracted and became a source of disease for humans (Nickson et al. 1998; Smith et al. 2002).

An important factor in projecting groundwater quality is its unique spatiotemporal aspect. Except in karst areas, where contaminants can spread through underground conduits, the movement of contaminant plumes is very slow, often only a few metres per year. The breakdown and shrinkage of the contaminant plume can take decades or centuries under 'natural attenuation' or decades under more aggressive measures (Herman et al. 2001). Thus, the landscape is likely to become increasingly fragmented into areas of essentially clean groundwater intercalated with areas of heavily contaminated groundwater.

Salinization

It is estimated that salinization affects productivity on about 22 Mha of land and has less severe impacts on another 55 Mha; thus 23% of the total irrigated land in the USA is affected by salinity (Morris et al. 2003).

Coastal areas are especially affected as a result of excessive drawdown of groundwater, as are arid areas with the evaporation and subsequent entry into groundwater of irrigation waters. With human populations concentrated along the margins of the oceans, this problem is certain to grow. For example, the Jakarta (Indonesia) water supply and disposal systems were designed for 500 000 people, but today service more than 9 million people; 54% of these depend on groundwater. The city suffers continuous water shortages and less than 25% of the population has direct access to water-supply systems. The water level in what was previously an artesian aquifer is now generally below sea level and, in some places, 30 m below. Saltwater intrusion and pollution have largely ruined this as a source of drinking water (WWC 2003a). Salinization will require the abandonment of agricultural areas, or the use of new water sources.

Changes in surface ecosystems

Changes in groundwater quantity and/or quality have significant hydrological and ecological effects on surface waters, significantly damage terrestrial ecosystems, can mean a failure to achieve the objectives of protected area or river-basin management planning and can compromise the restoration of a surface-water body. Excessive groundwater withdrawal, as well as dry periods, can lead to reduced connectivity with surface aquatic ecosystems. Since groundwater can contribute to the base flow to surface-water systems, wetlands may dry out, losing both their important capacity to filter nutrients and pollutants and their water storage with slow release maintaining stream flows.

Physicochemical conditions of many aquatic and terrestrial habitats at the Earth's surface are determined by the upwelling of quality, mostly well-buffered lithotrophic groundwater. This holds not only for springs, spring brooks and wetlands, but also for vegetation types dependent on groundwater recharge in the root zone. For example, basiphilous open species-rich vegetation types of young dune slacks have declined throughout Europe in recent years, and have largely been replaced by often acidophilous tall marsh and scrub vegetation. This succession appears to be accelerated by a decrease in the discharge of calcareous groundwater from sandy ridges or small dune hummocks. Acidifying effects of rainfall were reflected in the chemical composition of the groundwater below small dune hummocks that acts directly on the plant community dynamics (Silva *et al.* 1997).

There may be important consequences for terrestrial ecosystems. For example, in locally humid areas in the Mediterranean region, such as around Madrid (Spain), excessive withdrawal of groundwater in discharge areas during the summer (for example El Pardo) has transformed the system to a recharge type with negative consequences for the vegetation and fauna; several species of phreatophiles have disappeared, and there has been an increase in parasitic attacks and diseases in various plants (González Bernáldez *et al.* 1985).

Moreover, groundwater-influenced areas at the surface are of significant nature value. The attainment or maintenance of favourable conditions in nature reserves, for example, often depends on proper integrated water management, including maintenance of surface–GW interactions, and infiltration and groundwater recharge areas (Fig. 3.4; Table 3.5).

CONCLUSIONS

Relative to 2000, the global average annual per caput availability of renewable water resources may by 2025 fall from 6600 m^3 to 4800 m^3 (Zehnder *et al.* 2003). Tensions between increasing needs (such as demography, domestic consumption and agriculture) and progressively more exploited resources continue to grow. Climatic changes and the present global warming might be the cause of food disasters, plaguing millions of people with water deprivation. With 100 million more people to feed every year, solutions are difficult to find.

Shiklomanov's (1998) projections assume a large increase in water for agriculture and more high-quality water for human consumption. For these projections to be realized, groundwater depletion will have to continue apace. Recharging and storing water in aquifers might be a potent solution, but given the overdrawing of groundwater in India, China, North Africa, the USA and elsewhere, it is definitely not the best management solution. Increasing needs and demand for better water quality will not be met without far-reaching projects and actions for better-integrated management. Management programmes for the protection of groundwater have to incorporate both socioeconomic and environmental targets. To harvest sufficient groundwater of good quality, management in the future should unify resource sustainability and groundwater ecosystem integrity.

The challenge is impressive. Huge investments are required to halt groundwater deterioration and over-exploitation, take compensatory measures and implement the types of management policies previously highlighted. At the same time, there needs to be an emphasis on fundamental changes in the relationship between humans and the environment, maintenance of ecosystem functionality, protecting biodiversity and providing drinking water to everybody.

4 · Flood plains: critically threatened ecosystems

KLEMENT TOCKNER, STUART E. BUNN, CHRISTOPHER GORDON,
ROBERT J. NAIMAN, GERRY P. QUINN AND JACK A. STANFORD

INTRODUCTION

Riparian zones, river-marginal wetland environments and flood plains are key landscape elements with a high diversity of natural functions and services. They are dynamic systems that are shaped by repeated erosion and deposition of sediment, inundation during rising water levels, and complex groundwater–surface water exchange processes (Chapter 3). This dynamic nature makes flood plains among the most biologically productive and diverse ecosystems on Earth (Junk *et al.* 1989; Gregory *et al.* 1991; Naiman & Décamps 1997; Tockner & Stanford 2002; Naiman *et al.* 2005). Flood plains are also of great cultural and economic importance; most early civilizations arose in fertile flood plains and throughout history people have learned to cultivate and use their rich resources. Flood plains have also served as focal points for urban development and exploitation of their natural functions.

Aquatic Ecosystems, ed. N. V. C. Polunin. Published by Cambridge University Press. © Foundation for Environmental Conservation 2008.

Awareness has been growing during the past decade of the global significance of freshwater biodiversity, of the array of factors that have (or will have) threatened it and the extent to which real damage has already been done (Abramovitz 1996; McAllister *et al.* 2000). Flood plains in particular have been highly degraded throughout the world by river and flow management and by land-use pressures. Nowadays, they are among the most endangered ecosystems worldwide (Olson & Dinerstein 1998). Accordingly, flood plains deserve increased attention for their inherent biodiversity, for the goods and services provided to human societies and for their aesthetic and cultural appeal.

This chapter builds upon a recent comprehensive environmental review of riverine flood plains (Tockner & Stanford 2002). It starts with a short summary of the distribution and extent of the world's flood plains, followed by an overview of their primary economic, ecological and cultural values. The multifaceted threats that make flood plains one of the most endangered landscape elements worldwide are then discussed. Finally, future trends in floodplain exploitation in developing and developed countries are contrasted, and the ecological consequences of rapid alteration of floodplain ecosystems forecasted. Since flood plains, wetlands and river systems are not always clearly distinguished, general examples of rivers or floodplain wetlands are included in the present review when appropriate. The flood-inundated wetlands and lakes of the humid tropics are treated in more detail later (Chapter 10).

FLOOD PLAIN DISTRIBUTION

Flood plains are defined as 'areas of low lying land that are subject to inundation by lateral overflow water from rivers or lakes with which they are associated' (Junk & Welcomme 1990) (Fig. 4.1). This definition includes fringing flood plains of lakes and rivers, internal deltas and the deltaic flood plains of estuaries. Brinson (1990) and Brinson and Malvaréz (2002; Chapter 9) proposed a hydrogeomorphic classification of wetlands based on the (1) geomorphic setting, (2) water source and (3) hydrodynamics. Considering these components, riverine flood plains are located on low-gradient alluvial 'shelves' and water originates primarily from lateral overspill of rivers, although other water sources (groundwater, precipitation) may significantly contribute to inundation (see for example Mertes 2000).

Although accurate global assessments of wetlands are elusive due to the complexity and highly dynamic nature of wetlands and floodplain ecosystems (Finlayson & Davidson

Fig. 4.1. (a) Tagliamento River, Italy, a semi-natural braided floodplain corridor in Central Europe (photograph: K. Tockner). (b) Flood plains along the Nyaborongo/Akagera River, Rwanda, Africa, colonized by dense stands of *Papyrus cyperus* (photograph: K. Tockner). (*c*) Cooper Creek in the Lake Eyre Basin, Australia, at the height of a flood in March 2000, which covered approximately 14 000 km² (photograph: R. Ashdown).

1999), wetlands cover nearly 10% of the Earth's land surface and of this 15% are flood plains (Ramsar [Ramsar Secretariat] & IUCN [World Conservation Union] 1999). Total global floodplain area estimates range from 0.8×10^6 km²

Table 4.1. *Global extent of flood plains that are functionally largely intact, and major human threats*

Region	Major river systems	Floodplain area (km²)[a]	Major impacts
Africa	Nile (Sudd), Congo basin, Niger	310 000	Hydrological change
Europe	Danube and Volga deltas	40 000	Embankment, drainage
North America (USA)	50% in Alaska, Mississippi	240 000	Embankment, hydrological change, drainage
South America	Primarily Amazonian basin	1 100 000	Urbanization
South-East Asia (including China)	Mekong, Irrawaddy	400 000	Urbanization, hydrological change
Australasia	Fly River, Paroo River and Cooper Creek	150 000	Hydrological change
Russia/Central Asia	Lena, Mongolia	unknown	Climate change
Total		2 240 000	

[a] Conservative estimation of area based on Tockner and Stanford (2002) and Table 4.2. A clear separation of individual wetland types is often difficult (e.g. flood plains, marshes and swamps).

(Aselmann & Crutzen 1989), to 1.65×10^6 km² (flood plains and swamps: Costanza *et al.* 1997) and 2.2×10^6 km² (flood plains along rivers and lakes: Ramsar & IUCN 1999) (Tables 4.1 and 4.2).

Based on data from 145 major river corridors around the world, the extent of human impacts on riparian zones is apparent (Fig. 4.2a). For example, 11% of the riparian area of African rivers (mean population density 24 people km^{-2}) is intensively cultivated, compared to 46% for North American rivers (mean population density, excluding northern Canada and Alaska, 24 people km^{-2}) and 79% for European rivers (mean population density 75 people km^{-2}). The most impacted riparian corridors with respect to land use are found in Europe and in the densely populated areas of Asia (catchments with population densities >200 people km^{-2}) (Fig. 4.2a). There, 60–99% of the entire riparian corridor has been transformed into cropland and/or is urbanized, the latter particularly so in Europe, where the Seine River (France) shows the highest impact of all rivers investigated. Along Asian and African river corridors, there is a highly significant linear relationship between human population density and the rate of land transformation. Along European and American corridors the relationship is logarithmic (Tockner & Stanford 2002). The analysis considers only those areas with intensive agricultural development, excluding mosaics of cropland and natural vegetation. Therefore, a more traditional and perhaps a

more sustainable use of riparian and floodplain areas can explain the low levels of land use for African and many Asian river corridors. This does not imply, however, that flood plains with little intensive agricultural development or low population densities have not been markedly impacted in these regions (Chapter 3). For example, construction of impoundments, water harvesting and floodplain development (for example on-farm storages and levees) in Australia have reduced flooding of many wetlands, and changes in aquatic plant communities have occurred together with declines in abundances of waterbirds, fish and invertebrates. As a direct result of river regulation, and deliberate filling and draining, most of the wetlands in south-eastern Australia have been destroyed. About 90% of the floodplain wetlands in the Murray–Darling Basin, 75% of the coastal wetlands of New South Wales and 75% of Swan Coastal Plain wetlands in south-western Australia have been lost (Bunn *et al.* 1997).

Today, the largest remaining flood plains are in South America where about 20% of tropical lowlands are flooded annually (Junk 2002). In Africa, many large flood plains are still relatively untouched (for example Sudd, Congo Basin). However, they are disappearing or are being transformed at an accelerating rate as a result of water-management activities, in particular by large-scale irrigation schemes and the ongoing construction of dams (Gordon 2003). In the USA, flood plains originally covered 7% or

Table 4.2. *Distribution and extent of selected fringing riverine flood plains (including a few rain-fed flood plains)*

Drainage system/geographical area	Area (km^2)[a]	Major flood plains/comment/reference
Africa[b]		
Zaire/Congo system	70 000	Middle Congo depression, Kamulondo, Malagarasi
Niger/Benue system	38 900	Niger central delta, Benue River
Nile system	93 000	Sudd, Kagera basin
Zambezi system	19 000	Kafue flats, Barotse plain, Liuwa plain
Western systems	19 000	Flood plains along the Senegal (excluding delta), Volta and Ouémé
South-east systems	100	Pongolo flood plain
Eastern systems	8 600	Kilombero, Rufiji, Tana River
Chad system[c]	63 000	Chari and Lagone River system
Gash River	3 000	Inner delta in Sudan (primarily woodland and savannah; Kirkby & O'Keefe 1998)
Tana Delta	670	Endangered by upstream dams (Hughes 1990)
Europe		
Switzerland	200	A total of 234 flood plains of national importance (Forum Biodiversität Schweiz 2004)
The Netherlands	498	Area regularly flooded by rivers (primarily meadows; Yon & Tendron 1981)
Danube National Park, Austria	93	The last remaining semi-natural flood plain along the Upper Danube (Tockner *et al.* 2000)
Tagliamento	150	The last morphologically intact river corridor in the Alps (Tockner *et al.* 2003)
Lonjsko polje (Save River), Croatia	507	One of largest and best-preserved flood plains in Europe (Spanjol *et al.* 1999)
Kopacki rit, Croatia	177	Semi-natural flood plain at the intersection of the Danube and the Drava (Spanjol *et al.* 1999)
Upper Rhine	70[d]	Originally, flood plains covered 1000 km^2 (Carbiener & Schnitzler 1990)
French Rhône (fringing flood plain)	70	Mostly functionally extinct flood plains. Former extent 830 km^2 (Bravard 1987)
Rhône Delta, Carmargue	750	Former delta of 1644 km^2 (Bravard 1987)
Guadiana River, Spain	450	Floodplain marshes in the Doñana National and Nature Parks (Benayas *et al.* 1999)
Danube main stem (including delta)	17 400	Mostly disconnected, therefore functionally extinct
Danube Delta	5 800	Danube Delta Biosphere Reserve, *c.*50% of this area belongs to the 'Danube Delta' (RIZA 2000)
Danube islands, Bulgaria	107	75 islands in the main stem (Bulgarian Ministry of Agriculture and Forests, unpublished report 2001)
Dnieper River Delta, Ukraine	*c.*500	Hydrologic impact by unpredictable flood releases from upstream dams, pollution (Timchenko *et al.* 2000)
Elbe River, Germany	840	Original extent 6170 km^2 (Helms *et al.* 2002)
Volga Delta	18 000	Largest European delta (Czaya 1981)
Tisza, Hungary, Ukraine and Romania	1 800	Remaining area represents only 4.7% of former flood plains (Haraszthy 2001)

Table 4.2. (cont.)

Drainage system/geographical area	Area (km²)[a]	Major flood plains/comment/reference
Poland	820	Originally, floodplain forests covered 27 800 km² (Sienkiewicz et al. 2001)
European part of Russia	9 000	Approximate estimation (Shatalov 2001)
North America		
Ogeechee	150	Subtropical river in south-east USA (Benke et al. 2000)
Kissimmee (Lower basin)	180	Disconnected at present, will be partly restored (see Warne et al. 2000)
Altamaha and Tone Rivers	400	Mertes (2000)
Upper Mackenzie River	60 000	A complex of marshes, fens and flood plains (Fremlin 1974)
Mackenzie Delta	13 000	Including 24 000 lakes (Marsh et al. 1999)
Lower Missouri River flood plain	7 700	At present mainly agricultural land (D. Galat, personal communication 2003)
Mississippi River flood plain	20 000	Remaining bottomland hardwood forests out of formerly 85 000 m² (Llewellyn et al. 1996)
Rocky Mountain states, USA	c.4 000	Abernethy and Turner (1987)
Washington and Oregon	12 500	Abernethy and Turner (1987)
Alaska	120 000	c.50% of present floodplain area in USA (Mitsch & Gosselink 2000)
Pánuco River, Mexico	1 400	Inundated area excluding permanent waters (Hudson & Colditz 2003)
South America		
Amazon River	890 000	Large and small river flood plains combined (Aselman & Crutzen 1989; Sippel et al. 1998)
Orinoco Delta	30 000	Complex of flood plains, marshes and swamps (Groombridge 1992)
Orinoco fringing flood plain	7 000	Hamilton and Lewis (1990)
Pantanal	130 000	The 'largest' single wetland complex on Earth (Hamilton et al. 1996)
Parana	20 000	Fringing flood plains (Welcomme 1979)
Magdalena	20 000	Deltaic flood plain (Welcomme 1979)
Flooding pampa grassland, Argentina	90 000	80% of the area still covered by natural grasslands, fed by rain (Perelman et al. 2001)
Asia		
China	80 000	Riverine and partly lacustrine wetlands in the Yangtze and Yellow-Huaihe Basin (Lu 1995)
Lena and Yana Deltas	38 700	Largest northern delta complex
Mekong, Kampuchea	11 000	Inundated forests along the Mekong and the shores of Le Grand Lac (Pantulu 1986)
Irrawaddy, Burma	31 000	Welcomme (1979)
Indonesia	119 500	Primarily in Kalimantan and Irian Jaya (Lehmusluoto et al. 1999)
Bangladesh	98 000	Primarily cultivated, including 28 000 km² of rice fields (Welcomme 1979)
Ganges and Brahmaputra, India	23 000	Flood-prone area, heavily cultivated (FAO, unpublished report 2001a)
Yellow River, China	120 000	Including parts of the Hai and Huai Rivers (DHI Water & Environment, unpublished report 2001)
Tigris and Euphrates	20 000	Mesopotanic region; cultivated flood plain (Al Hamed 1966); about 7600 km² has disappeared since 1973 (UNEP, unpublished report 2001a)

Table 4.2. (cont.)

Drainage system/geographical area	Area (km²)[a]	Major flood plains/comment/reference
Australasia		
Fly river, Papua New Guinea	45 000	Swales *et al.* (1999)
Kakadu National Park, Australia	260	13% of the park area (Gill *et al.* 2000)
Lower Balonne flood plain, Australia	2 460	Sims & Thoms (2002)
Chowilla anabranch, Australia	200	Largest remaining natural floodplain forest along the lower Murray river (Jolly 1996)
Cooper Creek and Paroo river, Australia	106 000	Endorheic flood plains in Central Australia (Kingsford *et al.* 1998)

[a] Area of flood plain is shown during the season of maximum inundation (extended from Tockner & Stanford 2002). The reported areas differ substantially between reference sources, for example Upper Nile Swamps (Sudd): >30 000 km² (Mitsch & Gosselink 2000), 50 000 km² (Groombridge 1992), >90 000 km² (Howard-Williams & Thompson 1985). Central Niger Delta: 30 000 km² (Howard-Williams & Thompson 1985), 320 000 km² (Mitsch & Gosselink 2000). Middle Congo depression: 70 000 km² (Howard-Williams & Thompson 1985), 200 000 km² (Mitsch & Gosselink 2000). Several major 'flood plains' listed are composed of different wetland types including swamps, wet grasslands or shallow lakes (for example Kagera valley, Central Congo basin, Sudd, Orinoco delta).

[b] Data compiled by Howard-Williams & Thompson (1985) and Thompson (1996). Flood plains in Mozambique or Angola are not included because not enough information was available at the date of compilation (Howard-Williams & Thompson 1985; Thompson 1996).

[c] Dadnadji and van Wetten (1993) report that wetlands along the Lagone and Chari cover totally 78 000 km² including inundated flats, riverine flood plains, marshes and smaller lakes.

[d] Plantations cover 50% of the remaining flood plains.

about 700 000 km² of all land (Kusler & Larson 1993), but nowadays, about 50% of the unmodified flood plains occur in Alaska. Along the Mississippi, 90% of the flood plain (former area 123 000 km²: Sparks *et al.* 1998) is leveed, and has therefore become 'functionally' extinct, meaning that the basic attributes that sustain the flood plain such as regular flooding or morphological dynamics are gone. In Australia, the largest relatively intact flood plains can be found along Cooper Creek and other rivers of the Lake Eyre Basin in Central Australia (catchment: 1.14×10^6 km²) where, during floods, up to 106 000 km² of flood plains become inundated (Kingsford *et al.* 1998). The Lower Balonne River in eastern Australia also features a very large flood plain, with approximately 2500 km² of flood plain regularly inundated (Sims & Thoms 2002). Despite these examples, river regulation throughout much of south-east Australia has altered the connection between flood plains and their parent rivers, resulting in loss of 18–90% of floodplain wetlands, according to type and region in the Murray–Darling Basin (MDBMC [Murray–Darling Basin Ministerial Council] 1995). The current situation for European flood plains is critical (see Wenger *et al.* 1990; Klimo & Hager 2001; Hughes 2003), with 95% of the original floodplain area converted to other uses (WWF [Worldwide Fund for Nature] & EU [European Union] 2001) (Table 4.2). In 45 European countries, 88% of alluvial forests have disappeared from their potential range (UNEP–WCMC [United Nations Environment Programme–World Conservation Monitoring Centre] 2004). Many of the remaining European flood plains are far from pristine and have lost most of their natural functions. For example, of the former 26 000 km² of floodplain area along the Danube and its major tributaries, about 20 000 km² are isolated by levees (Busnita 1967; Nachtnebel 2000). However, major flood events have highlighted the vast extent of plains still subject to flooding (Table 4.3).

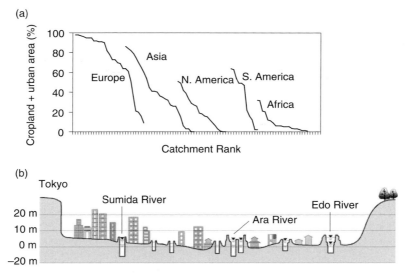

Fig. 4.2. (a) Relative proportions (%) of riparian zones (2 km along both sides of the river) along all major river corridors that have been transformed into urban or agricultural areas (total: 145 river corridors). River corridors are ranked for each continent from highest to lowest transformation rate (Revenga *et al.* 1998; Tockner & Stanford 2002). (b) Cross-section of the Tokyo 'flood plain'. The flood water level is much higher than the surrounding ground level. (After Seki 1994.)

ECOLOGICAL VALUES OF FLOOD PLAINS

Flood plains develop in all geographic regions and at different locations along river corridors (see Tockner *et al.* 2000; Junk & Wantzen 2004). Flood plains are also topographically unique ecosystems occupying the lowest position in the landscape; they tend to integrate upstream catchment-scale processes. There exist some basic principles that drive the ecology of river–floodplain ecosystems (Bunn & Arthington 2002). The flow regime determines the magnitudes of ecological processes and the temporal variability in floodplain communities and ecosystem processes. Fluvial dynamics, including the expansion/contraction of surface waters ('flood and flow pulses'), is also the driving force that sustains connectivity in flood plains and the river channel (Junk *et al.* 1989; Petts 1990; Tockner *et al.* 2000; Ward *et al.* 2002). Hydrologic connectivity, a key process in riverine flood plains, refers to water-mediated transfer of energy, matter and organisms within or among elements of riverine corridors (Ward *et al.* 1999; Pringle 2003). For example, diversity of birds and fish is significantly higher in floodplain waters with a natural hydrological regime com-pared to those with a regulated regime (Ward *et al.* 1999). Even small changes in the relative contribution of individual water sources may drastically alter species composition and diversity. For example, local groundwater upwelling (Chapter 3) is often associated with a higher standing crop of algae, higher zoobenthos biomass, faster growth rates of cottonwood trees and a higher species richness of woody and herbaceous plants (Harner & Stanford 2003). Despite its overwhelming importance in flood plains, hydrology is often given only cursory attention in restoration and mitigation projects (e.g. Bedford 1996; see also Chapter 3).

Flood plains form a complex mosaic of landforms, communities and environments (Naiman *et al.* 1993). The exceptionally high biological diversity of floodplain ecosystems results from a number of attributes such as soil moisture availability, structural complexity, microclimate characteristics and nutrient enrichments. Indeed, far more species of plants and animals occur on flood plains than in any other landscape unit in most regions of the world. In the Pacific coastal ecoregion of the USA, for example, approximately 29% of wildlife species ranging from 12%

Table 4.3. *The maximum area flooded (km^2) and the economic damage caused by recent flood disasters, compared to the 1995 Kobe earthquake*

River, geographical area	Year of event	Area flooded (km^2)	Economic loss, casualties	Source
Odra River, Poland and Germany	1997	6000	US$ 2 billion	Swiss Re (1998)
Mississippi, USA	1993	60 000	US$ 18 billion	Swiss Re (1998)
Queensland and New South Wales, Australia	1990	220 000	na	Swiss Re (1998)
Rhine River, Germany and the Netherlands	1995	na	US$ 3 billion	Swiss Re (1998)
Yangtze River, PR China	1998	640 000	£12.82 billion, 4000 people died	Zong and Chen (2000)
Elbe River, Germany and Czech Republic	2002	na	US$ 12 billion	Becker and Grünewald (2003)
Red River, USA and Canada	1997	2000	US$ 0.22 billion, 28 000 people evacuated	Simonovic and Carson (2003)
Kobe earthquake, Japan	1995		US$ 100 billion, 5500 people died	Swiss Re (1998)

Note: na, not available.

of mammals to 60% of amphibians found in riparian forests are riparian obligates (Kelsey & West 1998). Although less than 1% of the landscape of the western USA supports riparian vegetation, this vegetation provides habitat for more species of breeding birds than any other vegetation association (Knopf & Samson 1994). Similarly, the dense paperbark and lowland monsoon forests which fringe the larger rivers of Kakadu National Park in northern Australia are an important habitat for birds, despite occupying a small part of the landscape (Woinarski *et al.* 1988; Morton & Brennan 1991). In the Amazon basin, approximately 1000 flood-adapted tree species occur in flood plains. In a small floodplain segment upstream of Manaus (Mamirauá Reserve, Brazil), approximately 500 tree species have been identified, about 80% of which are flood–plain–specific (Wittmann 2002). The occurrence of a distinct flooding gradient seems to maintain high landscape heterogeneity and therefore biodiversity on Amazonian flood plains (Ferreira & Stohlgren 1999). Despite these important attributes, flood plains remain one of the least investigated landscape elements in terms of their contribution to regional biodiversity (Sabo *et al.* 2005).

Flood plains are among the most productive landscapes on Earth, owing to continual enrichment by import and retention of nutrient-rich sediments from the headwaters and from lateral sources, and they are usually more productive than the parent river and adjacent uplands (for example E. Balian & R. J. Naiman, unpublished data 2004). There is a positive correlation between fish catch and the maximum inundated floodplain area in African rivers, with fish yield being most influenced by the flood state in previous years (Welcomme 1975, 1979). The so called 'flood-pulse advantage' (*sensu* Bayley 1995) was recognized by the ancient Egyptians, since tax levies were based on the extent of the annual flood of the Nile.

Globally, flood plains are key strategic natural resources and they will continue to play a pivotal future role as focal nodes for biodiversity and bioproduction. Due to their distinct position along a river corridor, they are influenced by changes in the river as well as alterations at the catchment/landscape scale, resulting in a dynamic ecosystem under threat from all sides.

ECONOMIC VALUE OF FLOOD PLAINS

Flood plains offer a remarkably diverse array of natural services and functions. They provide biodiversity, flood retention, a nutrient sink, opportunities for pollution control, groundwater recharge (Chapter 3), carbon sequestration, timber and food production, organic matter production and export, recreational facilities and aesthetic value. Some of these are complementary service functions and the benefits are experienced simultaneously, while others are in mutual conflict, so that exploiting one benefit implies loss of another (Barbier & Thompson 1998; Acharya 2000). It is difficult to place quantitative values on some services without undertaking in-depth empirical socioeconomic research (Turner 2000), and the valuation of different functions and services of flood plains is likely to be context and location specific.

The services provided by flood plains are estimated to be worth US$3920 $\times 10^9$ per year worldwide, assuming a total floodplain area of 2×10^6 km^2 with a value of US$19 580 per ha per year (Costanza *et al.* 1997). In total, flood plains contribute >25% of all terrestrial ecosystem services, although they cover only 1.4% of the land surface area (for discussions see Aselmann & Crutzen 1989; Mitsch & Gosselink 2000; Tockner & Stanford 2002; Table 4.2).

The major services provided by flood plains include flood regulation (37% of their total value), water supply (39%) and waste treatment (9%). The value of floodplain land in Illinois (USA) is estimated as high as US$7500 per ha per year, with 86% of the value based on regional floodwater storage (Sheaffer *et al.* 2002). Nitrogen removal, an important floodplain service, ranges from 0.5 to 2.6 kg N per ha per day (Tockner *et al.* 1999). Flood plains along the Danube are valued at €384 per ha per year for recreation and nutrient removal (Andréasson-Gren & Groth 1995). Similarly, the nitrogen reduction capacity of Estonian coastal and floodplain wetlands is worth €510 per ha per year. The Danube Delta Biosphere Reserve area of 5800 km^2 yields 5000–10 000 t per year of fish, equivalent to US$6.3 million annually; 15 000 people within the delta and approximately 160 000 people from adjacent regions depend at least partly on this fishery resource (Navodaru *et al.* 2001). The economic value of selected African wetlands, mainly flood plains, is in the range of US$67–1900 per ha per year (Schuijt 2002). The aggregated value of agricultural, fishing and fuel wood benefits of a Sahelian flood plain is US$34–51 per ha per year (Barbier & Thompson 1998). The natural value of the flood plain would be even higher if other important benefits such as the role in pastoral grazing and recharging groundwater were included. Agricultural benefits of a planned irrigation project would, however, be only in the range of US$20–31 per ha per year. Firewood, recession agriculture, fishing and pastoralism generate US$32 per 1000 m^3 of flood water, compared to a value of US$0.15 per 1000 m^3 for irrigation. In the Inner Delta of the Niger River,

over 550 000 people with about 2 million sheep and goats use the flood plain for post-flood dry-season grazing (Dugan 1990). There are many other examples of how local communities make use of the productivity and services provided by flood plains. The benefits of natural flood plains are clear and multifaceted, especially in dry regions.

HAZARDS AND HUMAN BENEFITS OF FLOOD PLAINS

The double-edged face of floods

Instead of allowing rivers to fan out and take advantage of the natural flood-control function of flood plains, many governments have spent large amounts of money to force rivers into tight channels, which has encouraged human use of flood plains throughout the developed world (Fig. 4.2). Moreover, reclamation of flood plains has often led to a massive increase in peak flow in downstream sections. Nowadays, about half of Europe's population and *c.*50% of Japanese people live on flood plains (Statistics Bureau of Japan 2003). Despite engineering efforts to control rivers and protect life and property, flood losses continue to increase in most parts of the developed world. Floods are among the costliest natural disasters worldwide, in particular where human encroachment is intense (for example Burby 2002; Table 4.3). According to the Emergency Disaster Database (EM-DAT 2004), in the period from 1900 to 2004, floods affected 2.9 billion people, with 2.9 million killed and >130 million made homeless worldwide. In Japan, the average annual flood damage costs are US$1.53 billion (Seki 1994). In the USA, with 6 million buildings located within the boundaries of a 100-year-old flood plain, flood losses are widespread, have increased dramatically over the last few decades (the average damage cost is US$115 million per week) and will continue to do so over the next decades (Congressional Natural Hazard Caucus Work Group 2001). Industrial complexes in floodplain areas are also potential sources of contamination. For example, during the recession phase of the Elbe floods in Central Europe in 2002, threats from dioxin, mercury and other contaminants arose from flooding of sewage plants and release from river and floodplain beds.

Whereas the increasing exposure of property to flood risks is likely to be responsible for the major increase in flood damage over the next decades (Mitchell 2003), the invasion of flood plains by humans in search of new land for farming and homes is one of the most important drivers of flood losses in economically developing countries.

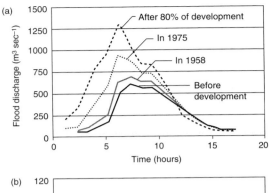

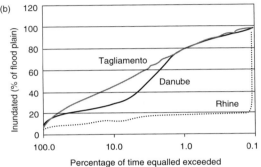

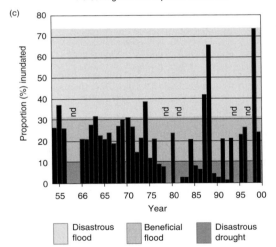

Fig. 4.3. (a) The impact of floodplain development (primarily urbanization) on flood discharge in the Tsurumi River (Kanto plain, Japan) (after Sakaguchi 1986). (b) Inundation–duration curve for a natural (Tagliamento, Italy), a regulated (Danube, Austria) and a channelized (Rhine, the Netherlands) river–floodplain system. (Data from Rhine: T. Buijse, personal communication 2003). (c) Historic flood events in Bangladesh. The relative proportion (%) of the country that has been inundated during the annual flood (1954–1999) (after Mirza 2003). Eighty per cent corresponds to an inundated area of 110 000 km². Beneficial floods, disastrous floods and droughts are distinguished. nd, no data.

In Bangladesh, flood plains occupy 80% of the country and annual flooding (locally known as *borsha*) is a part of peasant life to which people are resilient, since one-fifth of the country is regularly flooded (Mirza 2002, 2003) (Fig. 4.3c). Regular flooding is viewed as beneficial; it creates and maintains the high fertility of soils and supports the world's most densely populated country (average population density 1209 persons km⁻²: Haque & Zaman 1993; World Bank 2002).

Increasing population density has increased the vulnerability of human people to flooding. This increasing risk has led to the development of multiple large-scale management plans including the construction of large dams in upstream regions (for example Nepal), high embankments and bypass channels. All of these plans, however, ignore the multiple benefits of the natural flood regime.

The cultural value of flood plains

The multiple resources provided by flood plains have favoured the development of ancient cultures along the Nile, the Euphrates and the Indus. Rice cultivation started in China about 7000 years ago (Boulé 1994) and continues to be the nutritional basis for much of the human population worldwide. In the flood plain of the Amazon River, pre-Columbian human population density was several times higher than in adjacent uplands (Junk & Wantzen 2004).

Flood plains are inhabited by a variety of indigenous human societies that are well adapted to the conditions and have developed a rich local culture. Many of these peoples are already listed as threatened and there is a great risk that the ongoing environmental impact on large river–floodplain complexes will also increase the future risk of their extinction (Box 4.1). Examples include the Lotzi ('Water People') in the Borotse flood plain (Zambezi River basin, southern Africa), the Tonga people in the Kafue flood plain (Zambezi River basin, southern Africa) or the Ogoni in the Niger delta (Nigeria), the last threatened by oil exploitation in addition to changing hydrology. The Nuer, Dinka and Shiluk in the Sudd area (Nile) are threatened by civil war and by the construction of the Jonglai Canal.

MAJOR DRIVERS IN FLOODPLAIN CHANGE

Flood plains are among the most altered landscapes worldwide and they continue to disappear at an alarming

Box 4.1. Outcast of Eden

'At the height of the floods great tracts of the desert adjoining the marshes are covered by sheets of open water that vary each year in size but can extend for a distance of more than two hundred miles from the outskirts of Basra almost to Kut. As the floods recede most of this inundated land reverts into desert' (Thesiger 1964). At the edge of the Mesopotamian wetlands, human history in Iraq began. By the fifth millennium BC the ancient Madan (Marsh Arabs) built reed houses, made boats and harpooned and netted fish. This was considered to be the true Garden of Eden. The Madan are considered the oldest people worldwide, with direct ethnic links back to the Sumerians and Babylonians.

The Mesopotamian marshlands were one of the world's greatest wetlands, covering an estimated area of 20 000 km^2, and an important centre of biodiversity, playing a vital role in the intercontinental migration of birds. By the 1990s,

more than 90% of the former wetlands had been transformed into bare land and salt crusts. The creation of large reservoirs in the upstream sections (for example the Southeastern Anatolia Project [GAP], URL http://www.gap.gov.tr) and, in particular, the active drainage of the wetlands themselves led to the collapse of the entire ecosystem complex (Gleick et al. 2001; UNEP [United Nations Environment Programme] 2001a). Draining the wetlands was done primarily to punish the Madan after they started a rebellion against the Saddam Hussein regime following Gulf War I; most of the wetland area disappeared between 1992 and 1994 (Munro & Touron 1997).

The destruction of these Mesopotamian wetlands is one of the world's great environmental disasters, affecting the ecology, and social, cultural and economic base of the Madan. Hundreds of thousands of Madan are internally displaced within Iraq, while c.40 000 seek refuge in nearby Iran (UNEP 2001a). The impact on biodiversity has included the loss of several endemic vertebrates (UNEP 2001a).

rate, since the 'reclamation' rate is much higher than for most other landscape types (Vitousek et al. 1997b; Olson & Dinerstein 1998; Revenga et al. 2000). The decline of floodplain diversity is much greater than that of terrestrial systems (Stein 2001), which is attributable to habitat alteration, pollution, competition for water, invasive species and overharvesting (Abramovitz 1996). Among these factors, land transformation including alteration of flow regime is the single most important cause of species extinction (Vitousek et al. 1997b). Habitat degradation and loss contribute to the endangerment of 85% of the threatened species in the USA (Wilcove et al. 1998). Drivers such as climate change, species invasion and habitat degradation also cause irreversible changes, at least on ecological timescales from decades to centuries.

For river–floodplain ecosystems, expected impacts vary latitudinally. In tropical ecosystems, land use is expected to have the greatest effect, with climate change being minimal. In temperate systems, both land-use change and invasion of non-native species can be expected equally to affect biodiversity, and in high latitude/altitude systems climate change is by far the most dominant driver, although region-specific differences exist (Sala et al. 2000). Anthropogenic threats to river–floodplain biodiversity will differ between economically developed and developing countries (Table 4.4).

Human population growth

Increase in human population will be by far the most important driver in future biodiversity change in the developing countries. The current annual human population increase of 80 million will probably remain constant until 2015 (Fischer & Heilig 1997). The total population growth from 6 billion in 2000 to an expected 8 billion in 2025 will be generated almost exclusively in developing countries, primarily in Asia, although average annual growth rates will be highest in Africa (Chapter 1). The regions that will experience the largest difficulties in meeting future demand for land resources and water, or alternatively will have to cope with increased dependency on external supplies, include western and south-central Asia, and northern Africa. A large stress on resources is to be expected also in many countries in the remainder of Africa (Tinker 1997). The strong population momentum indicates the next 20–30 years to be the most critical for achieving food security. Since remaining natural flood plains generally promise high food production rates, these are expected to be among the first ecosystems to disappear. High human population growth rate, fast economic development and limited financial resources for conservation will lead to a disproportional impact on global biodiversity in developing countries (Table 4.4).

Table 4.4. *Different states and future trends in developing and developed countries*

	Developing countries	Developed countries[a]
Population growth 2000–25 (%)	2–3	±0
Human population in 2025 (billions)	6.4	1.6
Water abstraction (% increase 2000–25)	+50	+18
Present water use for agriculture (% of total withdrawal)	82	30
Urbanization (% in 2025)	50	75
Increase in hydropower production (% 1995–2010)	+100	+10
Mega-biodiverse countries	9	1
Global hotspot areas	20	5
Major future impacts	Urbanization, pollution, abstraction, climate change	Morphological modification, pollution, climate change
Major measures required	Conservation, pollution control	Restoration

[a] Mainly Member States of the OECD (Organization for Economic Cooperation and Development).
Source: UN (1999); Cosgrove and Rijsberman (2000); Meyers *et al.* (2000); IUCN (2000); UNEP (2001*b*, 2003).

Climate change

More intensive precipitation events over many areas, increased summer drying over most mid-latitude continental interiors and associated risks of drought, and higher minimum temperatures are likely to result from climate change (IPCC 2001*b*). These could increase the risk of abrupt and non-linear changes in flood plains, affecting their function, biodiversity and productivity. A warming of the air by 3–4 °C is predicted to eliminate 85% of all remaining wetlands (UNEP & WCMC 2004). Projected effects for flood plains include lower water levels during the growing season and higher mean water temperatures. The return period of extreme precipitation events could decrease almost everywhere. Climate simulations have suggested that global warming induced by greenhouse gases would lead to a moistening of the atmosphere (7% per degree warming) and an intensification of the hydrological cycle. Numerical modelling carried out in Europe for the fall season predicts a substantial shift towards more frequent large-precipitation events, in particular in southern Europe and in the Alpine regions, since these areas are expected to be particularly receptive to the 'moisture effect' (Frei *et al.* 1998). For North Carolina, an increase in precipitation of 15% (5–30%) is predicted by 2100, which will shift floodplain boundaries and make recent human developments in floodplain areas especially vulnerable to flood damages. Similar effects are predicted

for the Pacific Northwest (USA). In 2070, for example, floods that now have 20-year return periods are projected to occur twice as often (Zwiers & Kharin 1998). For the eastern USA, a 10% decline in annual runoff could result in a nearly 50% increase in intermittent streams (Poff 1992). In the Fraser River (British Columbia, Canada), annual floods may occur on average 24 days earlier in 2070, summer temperature may increase by 1.9 °C, and the potential exposure of salmon (*Oncorhynchus* spp.) to water temperatures above 20 °C, limiting their spawning success, could increase by a factor of 10 (Morrison *et al.* 2002). Palaeoecological evidence suggests that droughts are likely to occur more frequently in the future in Australia, where anthropogenic effects are likely to exacerbate the trend (Humphries & Baldwin 2003).

Flow modification

The alteration of flow regimes is the most serious and continuing threat to the ecological integrity of river–floodplain ecosystems (Nilsson & Berggren 2000; Bunn & Arthington 2002; Naiman *et al.* 2002) by reducing flood peaks, flooding frequency and duration, and changing the nature of dry periods (McMahon & Finlayson 2003). These flow changes reduce lateral connectivity between the flood plain and parent river, disrupt sediment transport and reduce channel-forming flows. Inundation–duration curves used to assess the hydroecological integrity of these ecosystems indicate an almost linear relationship between water level and inundated area in natural flood plains (Benke *et al.* 2000; Van der Nat *et al.* 2002) (Fig. 4.3b). In regulated rivers such as the Austrian Danube, floodplain inundation is short and the increase in inundation area is abrupt. Along channelled rivers, flood plains only receive surface flooding from the river during major flood events (see Table 4.3). Large-scale floodplain development and water-resource development often lead to major decreases in the reactive floodplain area. For example, the Barmah-Millewa Forest on the upper Murray River (Australia) is the largest remaining river red gum (*Eucalyptus camaldulensis*) forest in Australia and is in part listed under the Ramsar Convention. Regulation of river flows in the upper catchment has largely alienated the flood plain from the river, and flooding of the forest has been reduced from 80% to 35% of years (Close 1990). Reduced flooding frequency has changed the community composition, growth and regeneration of the floodplain vegetation (Chesterfield 1986; Bren 1988). A similar fate has befallen the Macquarie Marshes in the upper Murray–Darling Basin. These major floodplain wetlands at the end of the Macquarie River cover about 130 000 ha during large floods. Water diversion from the Macquarie River has greatly reduced the area of floodplain inundation and the Macquarie Marshes have contracted to 40–50% of their original size (Kingsford & Thomas 1995). The abundance and species richness of waterbirds in the northern part of the Macquarie Marshes have declined as a consequence and major changes in the composition of floodplain vegetation have also occurred (Kingsford 2000).

Changes in catchment hydrology, caused by clearing of native vegetation for agriculture, have also led to a marked increase in dryland salinity in some Australian flood plains (Box 4.2).

Species invasion and pollution

Species invasion is one of the most important causes of the overall decline in aquatic biodiversity. The higher percentage of exotic plants and animals in flood plains compared to uplands demonstrates the vulnerability of the riparian zone to invasion (Pysek & Prach 1993). Despite great differences in climate, species richness and land-use history, 20–30% of species in riparian corridors of Europe, the Pacific Northwest (USA) and South Africa are invasive (Hood & Naiman 2000; Tabacchi & Planty-Tabacchi

2000). Fourteen of the top 18 environmentally noxious weeds in Australia occur in flood plains and wetlands (Humphries *et al.* 1991; Bunn *et al.* 1997), and species such as *Mimosa pigra* and *Urochloa mutica* have had major impacts in the largely undeveloped floodplain river systems of the Australian wet–dry tropics (Douglas *et al.* 1998). Although invasive species are frequently responsible for the decline of native species, the introduction of exotic species may increase overall diversity, although the impacts of exotic species may vary at different spatiotemporal scales (Sax & Gaines 2003).

Pollution is still a major issue in both the economically developed (for example Van Dijk *et al.* 1994) and developing countries. In developing countries, an estimated 90% of wastewater is discharged directly into rivers without treatment (Johnson *et al.* 2001). For example, in China, 80% of the 50 000 km of major rivers are too polluted to sustain fisheries and fish have been completely eliminated from 5% (FAO [Food and Agricultural Organization of the United Nations] 1999). In Poland, 75% of the water in the Vistula, which retains many semi-natural flood plains, is unsuitable even for industrial use (Oleksyn & Reich 1994). High nutrient concentrations of the parent river are also a major obstacle to restoring flood plains along many rivers such as the Danube and Rhine (Buijse *et al.* 2002).

FORECASTING ENVIRONMENTAL FUTURE OF FLOOD PLAINS

Quantitative forecasting of the ecological state of flood plains over the next decades, in the face of increasing human population density, climate change and changing land use, is a difficult goal (Nilsson *et al.* 2003). For example, accurate methods to forecast land use changes over a period of 20 years or more are not available, especially not at spatially explicit scales. A logical approach to coping with uncertainty in predicting the environmental future is to develop alternative scenarios since they focus attention on the unusual, the uncertain and the surprising when making decisions (Sala *et al.* 2000; Bennett *et al.* 2003).

Long-term ecological data in freshwater and wetland habitats are rare, and this makes it difficult to determine whether observed changes are part of long-term environmental trends or responses to specific human actions. Qualitative information on population trends (increasing, stable, decreasing) of more than 200 freshwater, wetland and

---

Box 4.2   Dryland salinity issues for flood plains

Salinity is a natural feature of the Australian landscape, particularly in the drier regions of the south-west and south-east, the combination of several factors making dryland salinity a significant problem to both terrestrial and aquatic ecosystems (see Chapters 2, 3, 10). Native Australian vegetation has evolved to take full advantage of available water, and little movement of rainwater into the root zone typically occurs (for example 1–5 mm per year) in dry lands. The low topographical relief, poor drainage and older, less permeable soils are conducive to accumulation of salt. These conditions have meant that salts are not flushed from the landscape, reaching levels of over 10 000 t per ha in some regions (Hatton & Salama 1999).

Clearing of catchments for dryland cropping and grazing has replaced deep-rooted perennial native vegetation with shallow-rooted annual species, allowing more rainfall to penetrate the deeper soils, causing water tables to rise. Wherever the groundwater contains salt or intercepts salt stored in the landscape, salt is mobilized to the land surface and ultimately into streams, rivers and wetlands. In the drier parts of this region (rainfall <900 mm per year), catchments with as little as 10% of the native vegetation removed have streams with marginal salinities (Schofield *et al.* 1989). The problem has been compounded by irrigation (e.g. in parts of the Murray–Darling Basin in south-east Australia), where the water use has hastened rises in groundwater levels, and increasing regulation of rivers by dams and weirs and diversion of water have reduced the natural flushing.

Dryland salinity currently affects 1.8 million ha in Australia and may rise to 15 million ha in the next 50 years (PMSEIC [Prime Minister's Science, Engineering and Innovation Council] 1998). The economic costs are currently estimated at A$700 million in lost land and A$130 million per year in lost production. Salinity is degrading rural towns and infrastructure, as seen in crumbling building foundations and roads. In the lower Murray River in South Australia, salinity is projected to rise to a level close to the World Health Organization limit of 800 EC units for desirable drinking over the next 50 years. In some northern parts of the Murray–Darling Basin, river salinity will rise to levels that will seriously constrain the use of river water for irrigation.

Biological effects of secondary salinization include losses of remnant vegetation, riparian vegetation, wetlands and aquatic biota. About 25% of the Murray River flood plain in south Australia is currently affected by salt, and this could rise to 30–50% within 50 years (MDBMC 1999). The Ramsar-listed Chowilla Wetlands are under serious threat and without intervention more than half of the 20 000 ha will be lost (MDBMC 1999). About 80% of the length of streams and rivers in south-western Australia are degraded by salinity and half of the waterbird species have disappeared from wetlands that were once fresh or brackish (PMSEIC 1998).

The area of salinized land in Australia is likely to at least triple in the short term and further losses of biodiversity and ecosystem services can be expected. Extensive catchment revegetation to control recharge would be required to reverse these trends but, even then, response times would be slow (Hatton & Salama 1999). Engineering solutions (for example pumping) can have local effects on water levels if maintained over long timescales; however, there are obvious impacts associated with off-site disposal of saline groundwater. Investment in restoration may be better placed elsewhere or at least focused on protecting key assets (Hatton & Salama 1999).

---

water margin vertebrate species indicates most species are declining in population size (WWF 1999). Quantitative data are available for only 70 species. An index generated from these data shows a decline of around 50% from the 1970 baseline, which is a much greater decline than for terrestrial or marine biodiversity (Tockner & Stanford 2002).

The time lag between habitat loss and species disappearance means that measured extinction rates have considerable inertia. The effect of the main drivers on an ecosystem is also expected to be non-linear, including major time lags between impact and response. The response of a biological community to its modified habitat condition can take a century or more, especially for organisms with long life cycles (like riparian trees), and is a potential source of misleading interpretations. This also implies that present-day activities may not yield demonstrable changes for several years or even for decades. Floodplain examples of lag times in the range of decades to centuries between onset of an impact and biodiversity response include riparian flood control (Poiani *et al.* 2000), and invasion of riparian ecosystems by exotic plants (Pysek & Prach 1993). The results of these studies set the timescale over which humanity must take conservation actions in impacted terrestrial and aquatic ecosystems (years to several decades).

Land-use change is the major environmental driver of the condition of freshwater ecosystems. At the global scale, the land area significantly impacted by human activities may increase from 15–20% to 50–90% within 50 years, and pressure on flood plains will be among the highest (UNEP 2001b). Flood plains are often the focus of urban or agricultural development. The Caspian region, the Middle East, parts of Argentina and Chile, south-east Africa and South-East Asia have been identified as the most critical regions of water resource vulnerability with respect to climate change and land use within the next 20 years (Kabat et al. 2002). In South-East Asia, the combined effects of climate change, human population growth and economic development will increase the pressure on freshwater biodiversity probably more than in any other part of the world. In particular, aquatic organisms will be affected by water pollution, flow regulation, habitat degradation and exotic species invasion (Dudgeon 2002). None of these threats is unique to South-East Asia, but the rate and extent of environmental change in Asia is exceptional. In Asia, more than 5000 km^2 of wetlands are lost every year because of agricultural development, urbanization and dam construction (McAllister et al. 2000). If this trend continues, which is probable because of rapid economic and demographic development, by 2025 an additional 105 000 km^2 of wetlands will be converted to other uses. In particular, India and China belong to the rapidly developing countries where an increasing demand for water from a growing population and rapid economic development will lead to substantial water shortages. In 2025, India, Pakistan and China are expected to join the list of those countries where per caput water availability will fall below the critical threshold of c.1500 m^3 per year (Zehnder et al. 2003). The increasing water demand for agriculture and industry will reinforce the current pressure on flood plains. It is predicted that China will be able to feed its population in the future only if all rivers are dammed and no fresh water reaches the sea except during major floods. The ecological consequences for river–floodplain ecosystems are expected to be disastrous.

In the Ganges–Brahmaputra–Meghna Basin, an increase in seasonal and interannual variability, as already experienced during the past 20 years, can be expected (Fig. 4.3c). India is also planning to invest €100 billion within the next 15 years to build 32 large dams for transferring water from large rivers such as the Ganges to water-scarce areas via c.10 000 km of new canals. The irrigation of 350 000 km^2 more land should allow a doubling of present Indian food production within 50 years (Anon. 2003). Freshwater ecosystems will obviously be affected.

FLOODPLAIN MANAGEMENT REQUIREMENTS

Inventory and indicators

In spite of the Convention on Biological Diversity (1992), there is still no programme to document changes in biodiversity in natural habitats and the dearth of information on trends at the global scale is therefore not surprising (but see UNEP 2001b). For example, there exist no global estimates of rates of change in the extent of flood plains or in their conditions (Jenkins et al. 2003). The estimated percentage of wetland area in Africa ranges between 1% and 16% as a result of problems of wetland definition and limited scientific knowledge (Schuijt 2002). Even for vertebrates, available data are inadequate for discerning reliable population size trends for flood plains. There is a critical need for better and more up-to-date environmental data to determine trends. Flood plains need to be inventoried in a way that identifies the level of anthropogenic impact, for example: (1) natural flood plains (2) semi-natural flood plains (with a sustainable use), (3) flood plains with a high degree of reversibility (see Amoros & Petts 1993), (4) flood plains with a very low degree of reversibility and (5) degraded flood plains. A list of the mostly natural areas can provide a guide to opportunities for effective conservation. These are also those areas where the widest range of biodiversity might be conserved with a minimum of conflict, and might be seen as seeds of wilderness (Sanderson et al. 2002). Indeed, in Europe, North America and Japan, there are few remaining dynamic flood plains (see Fig. 4.1a) where large-scale natural disturbance events still occur.

The selection of indicators of environmental conditions is crucial for the interpretation of environmental changes. These indicators need to be integrated over different spatial and temporal scales. For example, changes in biodiversity and ecosystem processes are expected to differ among different scales. A major decline in biodiversity at one scale can be contrasted by an increase in biodiversity at another scale, as demonstrated in recent empirical studies (Sax & Gaines 2003).

Cost–benefit calculations

The flood plains of large rivers can no longer be regarded as wastelands suitable for draining and agricultural development. One move towards resolving the difficulties now besetting the world's rivers and flood plains is an ecosystem-based approach to management, which includes careful cost–benefit calculations. For example, cost–benefit analyses are available for the Elbe flood plains (Germany), which may serve as a model for comparable management projects. This project considers, beside other measures, the reintegration of 150 km^2 of former flood plains. Although these areas will only slightly mitigate future flood damage, there will be major ecological benefits. Even under very conservative assumptions, the benefits outweigh the costs and for each €1 invested, a return of €2.5 is expected via increased ecosystem services such as higher nutrient removal rates. The Elbe River clearly shows that restoration of flood plains is economically beneficial to society (Meyerhoff & Dehnhardt 2002).

Environmental flow requirements and management

There has been a major move toward the evaluation of river flow regimes in relation to the needs of natural ecosystems (both in streams and on flood plains) as legitimate users of fresh water, as well as those from agriculture, industry and domestic water supply (King & Louw 1998; Arthington & Pusey 2003) (Fig. 4.4). Both high flows and low flows may be managed (in terms of the timing, frequency, magnitude and duration of flow regimes) to encourage sustainable river–floodplain ecosystems (Lambs & Muller 2002; Hughes & Rood 2003). Unfortunately, the amount of water allocated to rivers through environmental flows is rarely enough to replace the small to medium floods that regulation and abstraction have affected. In these situations, water is best targeted on key floodplain resources, such as Ramsar wetlands. There are exceptions where environmental flows can benefit whole flood plains. For example, in Cameroon, a flood plain is being brought back to life through the Waza–Lagone rehabilitation scheme. Water released through newly constructed openings in the main river levee has enabled restoration of about 60% of the dam-affected flood plain (Bergkamp *et al.* 2000).

Institutional framework and floodplain restoration

The twenty-first century has frequently been designated as the century of nature restoration, at least in the developed world. However, major difficulties exist in the implementation of restoration schemes. Firstly, there is a perceived conflict between the goals of flood management and ecological conservation or restoration, which partly reflects the different values and paradigms within organizations and corresponding institutions (e.g. Pahl-Wostl 2002). Secondly, there are conflicts that arise because of the transboundary nature of many major river ecosystems, with the attendant

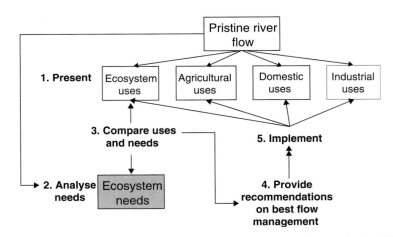

Fig. 4.4. Adaptive management to allocate water for ecosystems. (After C. Nilsson, personal communication 2004.)

problems of continuity of floodplain management policy and practice between upstream and downstream riparian states with different environmental priorities. Thirdly, a tension exists between water management, land-use planning and other aspects of environmental management.

Viable floodplain management can only be achieved when property owners, developers, environmental activists and governmental officials at all levels work together (Naiman 1992). It must be recognized that floodplain management is a public–private venture (flood control vs. land ownership). Incentives are suitable for floodplain management because they consider private as well as public benefit (Sheaffer et al. 2002). A positive example is the recent initiation of a new approach to flood mitigation and river restoration by the US Army Corps of Engineers. In a pilot phase, the Corps pays 65% of the costs of resuming properties on flood plains, tearing down dams and dykes, and relocating property owners. The local communities finance the remaining 35%. Initial results demonstrate a high interest by communities and affected people in the property buyouts on the USA's flood plains (IUCN 2000).

CONCLUSIONS

Flood plains are unique and dynamic ecosystems that link rivers with their catchments. They are highly productive environments, supporting a diverse biota, but are also intensively used by humans for agricultural and urban development, resulting in loss of biodiversity and ecological functioning. The priority for flood plains is to conserve those that are still intact and to attempt to rehabilitate those that are degraded. In both cases, protecting or restoring key components of the natural flow regime is essential, while maintaining sustainable use of floodplain resources by local communities, particularly in developing countries. Finding this compromise between conservation and resource use requires a greater understanding of the role of flow relative to other stressors in driving ecological processes in flood plains. Floodplain management and restoration strategies must also take into account climate change models that predict significant changes to flow regimes in most of the world's rivers, especially in temperate and arid regions.

Part II
Still waters

Lakes may contain just 0.01% of the Earth's water but they have inspired poets, painters, philosophers, musicians, scientists and environmentalists. The English Lake District helped nurture the poets W. Wordsworth and S. T. Coleridge, the painter J. M. W. Turner, and the environmental and social thinking of J. Ruskin. Karelia in Finland fostered the ringing tones of J. Sibelius. H. D. Thoreau's *Walden Pond* is an icon of American environmentalism; and the lacustrine landscapes of New England and Wisconsin nurtured the lively writings of the ecologist G. E. Hutchinson and conservationist A. Leopold. Lakes are a prominent feature in travel brochures, culturally encompassing tranquility, rest and refreshment.

The water demand for agricultural, domestic and industrial purposes that has risen with human population increase and economic development is going to continue to grow. By the year 2025, North Africa, the Middle East, South Africa and northern China will face very severe water shortages due to human demand exceeding natural supply. Lakes are going to be especially important and vulnerable, and the water allocation to society will typically continue to be met at the expense of the natural environment. With water scarcity, other issues such as water-based diseases can be expected to become more prominent. Despite differences in morphometry, physical dynamics, chemistry and biota, small and large water bodies, whether saline or freshwater, are subject to similar threats and will often respond similarly to human uses and abuses, albeit at different scales. The cut-off between 'large' and 'small' lakes is arbitrary but size is an important determinant of environmental vulnerability. Ponds and small lakes (here <500 km^2 in area) are relatively vulnerable to water abstraction, pollution, habitat loss and resource depletion (Chapter 5). Because of their greater volume, large lakes (>500 km^2) are relatively less susceptible to environmental perturbations; for this and other reasons, including geographical isolation, some of these, such as Lake Baikal, are important biodiversity foci (Chapter 6). Saline lakes are not always regarded as a separate ecosystem; however they span from the smallest to the greatest lakes (Caspian Sea: volume 78 200 km^3, area 374 000 km^2) and have many distinctive features that warrant individual treatment (Chapter 7).

5 · The future of small lakes and ponds

BRIAN MOSS, CHRISTER BRÖNMARK, GLEN GEORGE, LARS-ANDERS HANSSON
AND ERIK JEPPESEN

INTRODUCTION

Small lakes have changed immensely in response to the pressures beginning, only a few thousand years ago, with the rise of settled agriculture and consequent urbanization. Human population increase and potent engineering have had big impacts over the last two centuries. Lakes are not isolated features of the landscape. They cannot be considered separately from the rivers, streams and groundwater that feed them (e.g. Chapters 2 and 3), nor from the catchment areas through and over which this water, altered chemically and biologically, passes to them. Nor even can they be isolated from the rain that falls on these catchments, bringing with it dissolved gases and pollutants, sometimes from great distances.

Because of these connections, small lakes are especially good barometers of these pervasive pressures. Water is

Aquatic Ecosystems, ed. N. V. C. Polunin. Published by Cambridge University Press. © Foundation for Environmental Conservation 2008.

Box 5.1   Case study: Lake Naivasha, Kenya

While the larger lakes of the East African rift valleys are famed for their endemic fish faunas, Lake Naivasha lacks endemic fish. Naivasha has dried up at least once in the past 10 000 years, and was once part of a much larger lake (Worthington & Worthington 1933).

Reduced 3000 years ago to a windblown flat (Richardson & Richardson 1972) that would have killed any previous fauna, Naivasha was probably recolonized by animals flying in or entering in river floods. Many cycles of complete or partial drying out have occurred since then, and there are probably no native fish left. The Kenyan Highlands have been a favoured area for human colonization. Since the nineteenth century, European farmers have introduced several cichlid fish from elsewhere in East Africa to provide bases for food fisheries, American bass (*Micropterus salmoides*) for recreation, and several small species to control malaria-bearing mosquitoes. Occasionally, native fish have re-entered from the inflow river system but have not survived in the face of competition from the robust introduced species. A mammal, the coypu (*Myocastor coypus*), also escaped from fur farms into the papyrus wetlands around the lake, but has not survived. However, the Louisiana red spotted crayfish (*Procambarus clarkii*), introduced as a fishery species in the 1960s, thrived.

The prized wildlife, particularly birds, depend on the complex plant-dominated habitat that is characteristic of shallow lakes. Hippopotami (*Hippopotamus amphibius*) and Cape buffalo (*Syncerus caffer*) are still common in the lake and its swamps, but the bird fauna

has been severely depleted. A reduction in the swamp area occurred as areas of former papyrus were reclaimed for farmland during phases of low water level. The introduced crayfish favoured the submerged plants as food, causing a complete loss of these plants and a switch to phytoplankton dominance (Harper *et al.* 1990). This dominance may have been helped by increase in nutrient additions from farming in the catchment, and reduction in algal-grazing zooplankton through overfishing of predatory fish, allowing the smaller planktivorous species to increase. Much of the surrounding former savannah has now been converted to intensive flower farms, producing crops for export and extensively using pesticides, the impact of which on the lake is unknown.

Naivasha raises important questions for small lake management. Its outward appearance is attractive; tourists congregate on its shores and contemplate its placid waters through a screen of tall lakeside trees, although they now see fewer birds and its fisheries are depleted. Recent proposals have been made to introduce yet another fish species to replace those now depleted, and in a lake so altered, there is an argument that whatever its effects such an introduction matters little. On the other hand, would not a return of the lake to its former diverse plant-dominated state through a holistic management plan be more likely to increase its value to the country? There is no simple answer. The plan would involve restrictions on the activities of some of the more powerful present farming users and more investment by an impoverished government faced with many other problems in fisheries management and control.

retained in them for much longer periods than the residence time for rivers, and allows problems to be expressed that would otherwise pass unnoticed to the ocean. Lake sediments deposited under such conditions allow the tracking of the development of problems, yet the relatively moderate retention time also means that problems may sometimes be more easily solved because the water is soon replaced. The tractability of these systems has meant that understanding of how they function has advanced rapidly and produced a variety of remedial measures from the superficial to the fundamental.

The problem always remains, however, of finding a compromise between conservation and exploitation. The ethical rights of other organisms, and the maintenance of

traditional goods, services, amenity and inspiration conflict with the pressures of development, which inevitably come from a rising human population, with undeniable needs for security, happiness and convenience, especially in the developing world (Box 5.1). But developmental pressures also come from the aspirations of those who supply the consumer society of the developed world with services and goods that they are persuaded are necessary. Solutions are not impossible, but are not easy.

ORIGINAL STATUS OF SMALL LAKES

The present state of small lakes can only be understood and their future predicted relative to some reference state.

Inevitably, this has to be the pristine state, with absence of human settlement in the catchment, a state that has not existed over most of the Earth for some thousands of years. However, such a state must be notionally reconstructed, not least because it currently forms the measure by which change in lakes in Europe will be assessed under the European Water Framework Directive (2000).

Lakes and ponds are simply depressions in the Earth's surface, in which water is delayed in its passage from atmosphere to ocean. The origins of natural lake basins are various, from gouging of the landscape by ice to damming of rivers by beavers, landslides or lava flows. They range from gentle tectonic sinking of the Earth's crust, or crater formation in volcanoes, to solution of limestones by percolating rainwater. Perhaps the majority of ponds have been created by human activity, for domestic supply, fisheries and fish culture, or irrigation storage. Hutchinson (1957) lists nearly a hundred different ways in which lake basins have been formed.

In a particular region of climate, the nature of a pristine lake system depends on three main sets of characteristics: firstly, those of its catchment area (watershed, river basin) from which it receives water; secondly, those created by its particular shape (morphometry); and thirdly, those which arise from successive accidents of biogeography (initial colonization, subsequent invasion and local extinction of species). Every lake system is thus unique, though groupings and generalities may be made depending on the more important features.

The first generalization is that the concentrations of total ions in the water of a range of lakes will vary over perhaps two orders of magnitude, from a few tens of milligrams per litre to a few hundred (a further order takes the water into brackish or saline categories). A lower limit will be set by the chemistry of rain water, which has an ionic balance close to that of sea water (because of incorporated marine aerosols ultimately derived from sea spray) and a pH of around 5.5. Rock weathering will add ions and pH will not fall below that of rainwater, except in extensive areas of ombrotrophic mire (peat bog: see Chapter 8) where ion exchange by *Sphagnum* moss may reduce pH naturally to values as low as 3.5.

A second generalization is that the concentrations of certain key substances, likely to limit production, are very low and that their concentrations are largely independent of the total ion concentration. Usually there will only be a few micrograms or tens of micrograms per litre of phosphorus and a few tens or hundreds of micrograms of nitrogen

per litre. There are exceptions where greater concentrations may naturally be found – in pools in areas with great aggregations of grazing mammals, for example, or in lakes with long residence times, where long-term concentration mechanisms exist – but often these are likely ultimately to involve influence of human activities.

Nutrients are scarce because available forms of nitrogen and phosphorus are relatively sparse in relation to the needs of organisms. Natural terrestrial vegetation also requires these nutrients, and has evolved efficient mechanisms for retaining the supplies in soils and minimizing losses by leaching to streams and lakes. Natural lakes are thus relatively infertile because of this nutrient shortage. The amounts of nutrients delivered (the loading) will depend on the local hydrology and its interaction with the natural catchment vegetation and the climate at the time. It will vary with weather – more rain, more nutrient loading in total but lower concentrations; less rain, greater concentrations because the water will be influenced more by percolation through the ground before it enters the lake. Concentrations entering the lake will thus vary from day to day, year to year and century to century because of natural climate variations. The lake may show considerable change over centuries as a result of periods of relative wetness and dryness (Haworth 1972). There will be no single characteristic pristine state, but combinations of values within the ranges of an enormous number of physical, chemical and biological variables. There will also be natural long-term trends in lake development. In regions recently glaciated, such as northern Europe, investigations of the history of lakes by the analysis of sediment cores (Mackereth 1965; Haworth 1969) show progressive reduction in mineral ions in the sediments, reflecting a progressive leaching, over several millennia, of the raw soils created originally by glaciation, a process known as oligotrophication.

The consequence of the naturally low, if variable, nutrient loading will be that only small crops of phytoplankton will be present, and the water will be clear and not obviously turbid. Exceptions may occur in very arid areas of Mediterranean-type climate, where intense rain brings in clay that remains in suspension, though again this may reflect intensification of local land management by people. In clear water, rooted submerged and usually vascular plants, but sometimes also large algae such as charophytes, will grow down to depths where light extinction by the water and its dissolved substances makes net production impossible. This will generally be at a few metres, sometimes more. Natural waters contain thousands of dissolved organic substances, derived from decomposition of vegetation of the catchment or the lake and the surrounding wetlands, and these may pigment the water sufficiently to reduce the depth of macrophyte colonization. In view of the shortage of inorganic nutrients, this externally derived organic matter may also support, in at least some lakes, much of the secondary production of the plankton community through microbial decomposition and grazing of the bacteria produced. Pristine lake waters tend to be supersaturated with carbon dioxide as a result of respiration on incoming organic matter.

Morphometry centrally determines the balance of different sorts of primary producers and the extent of physical structure within the communities. Very shallow lakes may be completely floored by macrophytes and verge on wetlands (e.g. Chapters 8, 9 and 10) and, if the water depth averages only a few metres, macrophyte production will dominate over phytoplankton production. The more complex architecture provided by macrophytes may consequently give a greater biodiversity of associated microorganisms, invertebrates and vertebrates, than in a lake where phytoplankton predominates. The plankton is a very specialist community of comparatively low diversity in fresh waters. Thus the nature of the system is influenced by its catchment and its morphometry. Progressive build-up of peat, by fringing vegetation in particular, and accumulation of sediment, even at rates of only a millimetre or so each year, may cause major morphological changes, and small and relatively shallow lakes develop into mires (wetlands) completely dominated by macrophyte growth in time.

The third determinant of pristine lake state, namely accidents of biogeography, by its very nature contributes much to the uniqueness of a lake system. Because most lakes have now been influenced by human activities, it is very difficult to examine this aspect of natural lakes. However, just as progressive leaching of the catchment may lead to progressive reduction in lake fertility, and climate changes towards dryness may result in natural fertilization (eutrophication) as leached nutrients are dissolved in less water, time lapsed since a major event such as a glaciation will lead to change in the community. Following the melting of the ice, an invasion of species that may or may not be similar to those present before the glaciation (if the lake basin was then present at all) will take place.

Further, lakes and ponds may also be viewed as biogeographic islands isolated in the terrestrial landscape. A number of studies have shown that the well-known relationship between species number and island area also holds for freshwater organisms. Distance between lakes (isolation) and the size of the regional species pool also influence species numbers. The species composition of a specific lake is further determined by differences in habitat requirements and colonization rates between species.

Fish are able to colonize less easily than invertebrates, reptiles and amphibians, or especially microorganisms and birds; and even quite minor events of flooding or drying may lead to the presence or absence of a fish community. A drying phase or an intensely cold period, with deoxygenation under ice, may completely eliminate a fish community. This is especially important because selective predation by fish on zooplankters and larger invertebrates like snails may have consequent effects on phytoplankton, periphyton and macrophyte growth and even water chemistry. The balance of piscivores and zooplanktivores may be particularly important (Box 5.2).

Longer-term evolutionary events may also be crucial. North temperate lakes are depauperate in fish species because the recent glaciation has disrupted their systems; warm temperate and especially tropical lakes of greater longevity are species rich. Food-chain complexities are much greater at low latitudes, where the community may not only be influenced by natural climate changes, leading for instance to extinction through drought, but also in the most long-lived lakes to production of new species and endemism. The larger African lakes and Lake Baikal in central Asia are classic sites for this (Fryer & Iles 1972; Goldschmidt 1996; Chapter 6).

Pristine lake systems thus include both highly specific and broad general features. The latter are crucial in understanding how they have changed and how they might be restored. Particularly important are the low fertility of the water due to retention of nutrients in the catchment, the usual lower pH limit of not less than 5.5, and the existence of natural development (filling-in) of the basin by vegetation and peat deposits. The state of these general features allows an assessment of the extent of overall damage brought about by settled human activity. The detailed uniqueness of the systems, however, often makes it difficult to be precise about the extent of change in a specific case and about the setting of targets for restoration programmes, particularly where specification of changes at almost any taxonomic level is concerned. The problem is further compounded by natural change, particularly as climate changes and influences nutrient loading.

Box 5.2 Strong impact of fish in oligotrophic and hypertrophic small lakes

While 'bottom–up' processes resulting from chemical and physical factors were once thought to drive lake function, 'top–down' effects (acting through the food webs) may be just as important. Zooplanktivorous fish are major players in these processes (Andersson *et al.* 1978; Mazumder *et al.* 1990; Hansson *et al.* 1998; Mehner *et al.* 2002; Jeppesen *et al.* 2003). In very clear ultraoligotrophic small lakes, in mountain and polar regions (Hershey 1985; Paul *et al.* 1995; Gliwicz 2002), predatory fish fare well owing to higher water clarity and a prolonged prey generation time (due to low growth rates) exposing them to predators for a longer period before reproduction. For some zooplankton, higher pigmentation (UV-protection) additionally enhances the predation risk (Sægrov *et al.* 1996). High predation on invertebrates in hypertrophic systems reflects a high biomass of planktibenthivorous fish, mediated in part by loss of piscivores, which are disfavoured by low oxygen concentrations; a lower risk to invertebrates at intermediate nutrient levels reflects higher piscivory and, in shallow lakes, occurrence of plants acting as prey refuges.

Strong evidence for the important role of fish in ultraoligotrophic lakes comes from comparative studies conducted in lakes with and without fish in Greenland and the Sierra Nevada mountains (USA). In species-poor lakes of north-east Greenland, large-bodied crustaceans, such as the tadpole shrimp *Lepidurus arcticus* and *Daphnia pulex*, are abundant in lakes without fish, but largely absent from lakes with fish (the only species found being Arctic charr, *Salvelinus alpinus*), where small-bodied species like cyclopoid copepods and rotifers prevail (Fig. 5.1). When the fish eat the benthic-feeding *Lepidurus*, this releases predation on smaller benthic cladocerans, such as *Alona*, which then become more abundant (Jeppesen *et al.* 2001). Strong impacts of fish on benthic and pelagic invertebrates have also been observed in the more species-rich lakes of the Sierra Nevada (Fig. 5.2a–c). Among the benthic macroinvertebrates (such as caddisflies), abundances were substantially lower in fish-inhabited lakes, while burrowing (for example oligochaetes) and distasteful (for example water mites) taxa were either unaffected by the presence or absence of fish, or even increased in abundance in lakes with fish. Among the pelagic crustaceans, the percentage occurrence (Fig. 5.2d) and abundance of the large predatory species *Hesperodiaptomus* and *Daphnia middendorffiana* were substantially lower in lakes with fish, while those of smaller forms were higher (Knapp *et al.* 2001).

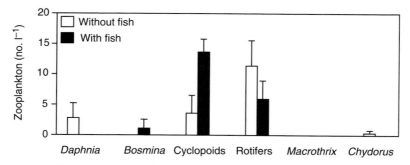

Fig. 5.1. Abundance (mean ±80% confidence limit) of various zooplankton species (rotifers × 0.25) in seven small shallow lakes without fish (white column) and six shallow lakes with landlocked dwarf forms of Arctic charr (*Salvelinus alpinus*, black column) in north-east Greenland. *Daphnia*, *Daphnia pulex*; *Bosmina*, *Bosmina longirostris*; Cyclopoids, cyclopoid copepods; *Macrothrix*, *Macrothrix hirsuticornis*; *Chydorus*, *Chydoris sphaericus*. (From Jeppesen *et al.* 2001.)

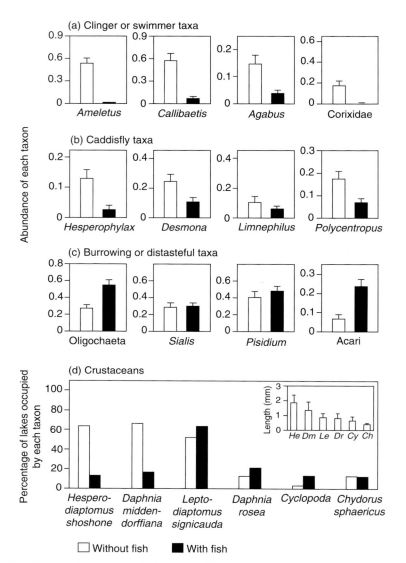

Fig. 5.2. Abundances of benthic invertebrates and zooplankters in small lakes with and without fish. For benthic macroinvertebrates, the abundances (mean ± SE number of individuals per 15 standard net sweeps, expressed as $\log 10(x + 1)$) of (a) clingers/swimmer taxa, (b) case-dwelling taxa and (c) burrowing/distasteful taxa in fishless lakes (open columns) and in lakes with fish (black columns) are shown. (d) The percentage occurrence of crustacean zooplankton taxa in lakes without (open) and with (black) fish. Taxa are arranged from left to right in order of decreasing size and actual sizes are shown in the inset box. (From Knapp *et al.* 2001.)

UNDERMINING THE PRISTINE STATUS

Discussion of change, threats and the present state of small lake systems demands first a definition of the change that

occurs. Conventionally, reductions in naturalness and bio-diversity, and in the ability of the system to produce goods and services that might otherwise be expected, are the criteria of deterioration. Goods and services include domestic water supply, specific fisheries and amenity. Fringing vegetation might, for example, provide a range of products from building materials (reed and timber) to food (wild rice, crayfish and wildfowl). These naturally supplied services and goods are of immense value on a world scale, indeed in total about three times the annual gross domestic product of all the world's artificial economies combined (Costanza et al. 1997).

There is little controversy about the major deleterious factors affecting lakes, which have been amply reviewed by Brönmark and Hansson (2002). Well-established influences include eutrophication, acidification, increased exposure to ultraviolet radiation especially in high mountain lakes, introduced species, toxic pollution (heavy metals, biocides), shoreline development, water abstraction and transfer, and recreational damage. Potentially developing issues include climate change, endocrine disruptors and the escape of genetically modified organisms, particularly cultured fish.

The well-established influences of eutrophication and acidification have received the most attention. Increased nutrient loading began with settled agriculture several millennia ago, but has intensified since the Industrial Revolution with release of phosphorus and nitrogen enriched industrial and domestic effluents into water courses, and since the Second World War with the major intensification of agriculture. The technology to remove nutrients from point sources such as factories and wastewater treatment works is well developed, but diffuse sources such as agricultural land remain problems, as do intensive stock units. The latter potentially can be treated in the same way as domestic effluent, but diffuse sources are presently impossible to treat except by using buffer zones of semi-natural vegetation to intercept nutrients. These work well for nitrogen but less well for phosphorus (Hansson et al. 2005). Indications are that almost as much buffer zone as agricultural land will be needed for effective removal of both nutrients. As the problem is tackled in the developed world, it is worsening in tropical regions, particularly with the burgeoning of major cities and the intensified culture of cash crops for export. It is likely that almost all of the world's lakes and ponds are eutrophicated to some extent because almost everywhere natural vegetation with its inherent nutrient conservation mechanisms has been disturbed, allowing greater erosion of soils and leaching of available nutrients from them.

Box 5.3   The continuing problem of acidification

Burning of sulphur-rich fossil fuel forms sulphuric acid in the atmosphere, leading to reduced pH in the rain (Smith 1872; Odén 1967). During the 1970s and 1980s, effects of acid rain were especially pronounced in parts of Europe and North America. Reduction in pH and increase in aluminium concentration resulted in reproductive failures in fish and invertebrates and considerable shifts in community composition, reduced biodiversity and increased water clarity (Andersson & Olsson 1985; Gunn & Keller 1990; Stenson et al. 1993; Sullivan 2000). Across regions and national boundaries, changes in polices and legislation have resulted in reduced emissions of acidifying substances (a management action directed towards the cause of the problem). Locally, the negative effects of acidification have been circumvented by liming, which gives instantaneous improvements but only treats the symptoms of the problem.

The ultimate cure is to reduce emissions of acidifying substances but this means economic decisions and tedious negotiations, often on an international level. Such negotiations were initiated during the 1970s in Europe and North America and have resulted in considerable reductions in sulphur emissions in Scandinavia (Fig. 5.3) and North America, as well as increases in rainwater pH (Stoddard et al. 1999) (Fig. 5.4), but only slight pH improvements in most acidified systems (e.g. Gunn & Keller 1990; Stoddard et al. 1999). The slow chemical recovery, even though sulphur emissions have been dramatically reduced, is probably owing to continuously high deposition of NO_X. In Sweden, there has been a slight reduction in NO_X since 1990 (Fig. 5.3), whereas in the USA, emissions of NO_X have remained fairly constant (Driscoll et al. 2001). The lower rate of NO_X emission cuts mainly results from increasing emissions from vehicles, industry and agriculture. Sulphur emissions are also rising in most developing countries (Rodhe et al. 1995), especially in Asia (Galloway 1995; Brönmark & Hansson 2002).

Liming has been extensively used as a restoration tool in Scandinavia. In Sweden, of >17 000 lakes affected by acidification, >7000 (90% of the area of acidified lakes) have been treated with lime during the last 10–15 years at a high annual cost (c.20 million per year), and it is expected that liming will have to continue for another 10–20 years. The improvements following lime addition eventually vanish, meaning that repeated treatments are necessary (Fig. 5.5).

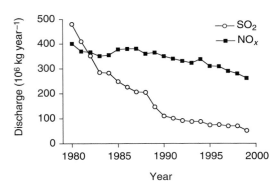

Fig. 5.3. Discharge of SO_2 and NO_x over Sweden 1980–1999, showing a clearly decreasing trend with respect to SO_2, but only a slight decrease in NO_x levels.

Acidification originated as a problem from the burning of fossil fuels in power stations and industrial plants (Box 5.3). This resulted in release of sulphur oxides, which are now retained at many plants by absorption onto limestone. However, even though this has resulted in drastic reductions in sulphur emissions, the chemical and biological recovery of lakes has been very slow. In densely populated regions like the UK, major increases in vehicle usage are now releasing more nitrogen oxides into the atmosphere. These not only provide sources of acidity to lakes in regions of hard and weathering resistant rocks, but also provide nitrate to fertilize the lakes. Again the problem is likely to be worsening in developing countries, with their often ancient, long-leached soils with low buffering capacity and expanding megalopoleis with poorly maintained vehicles and scarce technological modifications to minimize exhaust problems.

Toxic pollutants have been less well studied in lakes than in rivers. Mercury (derived ultimately from volatilization into the atmosphere from industry and fungicides and washed down in rain) remains a problem in lakes because it is accumulated in fish, but heavy metals have in general been discharged to large rivers close to estuaries where most of the world's industrial cities lie. Biocides remain a largely unknown threat; they are certainly present and there is a huge range of compounds. Their effects were likely to have been greater in the mid twentieth century when the hazards, particularly of organochlorine pesticides, were unrecognized and control was meagre. Effects are now likely to be subtle and sublethal, but this does not mean that they should be disregarded. Endocrine disruptors, acknowledged as problematic in rivers, may have effects on lakes too, although dilution into a larger body of water may mitigate these.

In high altitude lakes particularly, effects of increased ultraviolet (UV) radiation are emerging (Schindler *et al.* 1996). The problem may be compounded by climate-induced reduced precipitation, which is leading to reduced leaching into these lakes of pigmented organic substances that are otherwise highly absorptive of UV radiation (Sommaruga-Wögrath *et al.* 1997).

Overfishing is widespread in large lakes but of variable status in smaller ones (Allan *et al.* 2005), and the problems caused by introduced species, often fish, are widespread. Introductions have been for fisheries or for angling (for example, all of the fish now present in Lake Naivasha in Kenya are introduced, either from other African lakes or from North America; Box 5.1). The common carp (*Cyprinus carpio*) has been a frequent introduction because of its ability to grow to large size and its vigorous resistance to being caught. It is however a very damaging fish, which can destroy extensive areas of submerged vegetation. The threat of escape of genetically modified or farm-selected salmonids is serious. Many such fish do escape and affect wild stocks with which they may interbreed (Naylor *et al.* 2005).

Shoreline development and recreational damage are often linked. Lakes provide desirable settings for the building of houses for affluent people. One shore of Lake Windermere in the UK is popularly called 'Millionaires' Row', reflecting a trend which began in the nineteenth century when the rich factory owners of the industrial cities sought refuge in lakeside tranquillity. Elsewhere a 'cottage by the lake' accords status. Shorelines are transformed by such development, replaced by masonry, concrete docks or by gardens and lawns. Some recreational activities, such as power boating, lead to shoreline erosion and noise disturbance of bird communities. These are particular problems of the states of the North American mid-west, such as Minnesota, Michigan and Wisconsin, where lake recreation is highly developed (Box 5.4).

Water abstraction has often led to raising of water levels in natural lakes by dam building and effective conversion to reservoirs. Some lakes are used as storages for river regulation to even out flows between winter and summer and ensure downstream navigation or water supply. Others are used for hydroelectric generation. All these activities tend to lead to abnormal cycles in water levels that may leave a dry and sterile shoreline zone unsuitable for littoral communities and fish spawning. Moreover, during the last century, a large number of ponds and shallow wetland lakes disappeared owing to drainage or artificial infill to increase land available for agriculture or building.

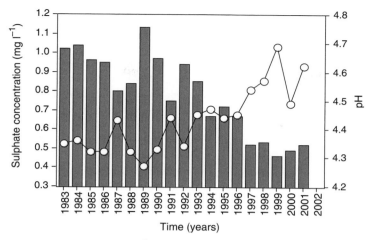

Fig. 5.4. Sulphate concentration (SO4-S, mg l⁻¹) and rainwater pH measured at the meteorological site in Aneboda, southern Sweden, 1983–2001.

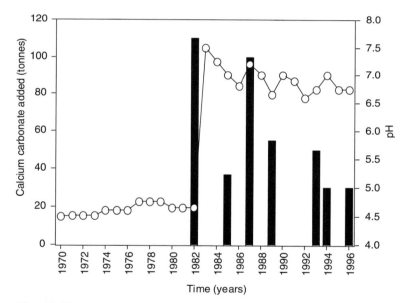

Fig. 5.5. The amount of lime (CaCO₃, tonnes) added to Lake Gårdsjön, Southern Sweden (bars), and subsequent changes in the pH of the lake water, 1970–1996.

Climate change may well prove to be a major influence, affecting temperature and hydrology. There are well-established links between lake phenomena and climate trends, for example in the diminishing periods of ice cover in northern lakes, and between changes in components of the plankton and major global climate phenomena

(Box 5.5). Climate influences the temperature and residence time, mixing characteristics and seasonal dynamics of the plankton of lakes. Effects of established factors like eutrophication and fish introductions are currently much greater than the effects of climate change (McKee *et al.* 2003; Moss *et al.* 2003). However, subtleties can sometimes

Box 5.4   Lake associations and citizen concern

Some of the earliest wetland archaeology concerned the lake villages of Switzerland and France. In England, the shallow wetlands of the Somerset levels supported lake villages, which depended on the fish, fowl and local availability of water, and were connected by a system of intricately constructed wooden walkways. Through water distribution systems, people have become liberated from the necessity of proximity to water supply, but in countries with many lakes, such as Scandinavia, Canada and the USA, living by the lake, either permanently or for vacations, remains a significant aspiration.

The result for North America has been the lining of many small lakes with cottages and build-up of development behind them. Lake-front cottages however are a vulnerable asset, because the activities of cottage dwellers can have major impacts. Boat docks are concreted, fertilized lawns established and domestic waste is discharged to the lake. Aquatic plants are cleared to provide accessible swimming or fishing areas. If the lake becomes significantly eutrophicated, accidental introductions and spread of aquatic weeds, often as strands on the propellers of boats towed from lake to lake, can lead to dense tangles of such plants as Eurasian water milfoil (*Myriophyllum spicatum*) and a perception that all aquatic plants are 'bad'.

Two consequences of modern lake cultures are manifest. One is the rise of lake associations among the residents of particular lakes, dedicated to maintaining the lake values for ethical or financial reasons. The other is the rise in consultancies supplying the market for advice to the lake associations. A glance at issues raised in *Lakeline* (magazine of the North American Lake Management Society) or a visit to its annual meeting will provide rich insights. Some lake dwellers are of the opinion that lakes are for fishing particular species (e.g. heavily stocked bass), plants should be removed, and tidy lawns and concreted docking areas maintained. One group of companies and consultants specializes in weed removal by herbicide, dredging of encroaching emergent plants and calculating the maximum stock of bass that can be introduced. There is even a company that produces mass quantities of dye; blue water is apparently the desirable state.

Other lake dwellers seek maintenance of as natural an edge as possible, minimization of disturbance and nutrient loads, and sensible zonation of conflicting activities. Although relevant consultants apply the available research information with increasing success, the diehard bass fishing community is a force not to be ignored (Hiaasen 1990).

change unexpectedly to major switches in complex ecosystems. The effects of climate change are likely to become increasingly important in future decades (Hulme *et al.* 2002).

RESTORATION OF LAKE SYSTEMS: STRENGTHS AND WEAKNESSES, OPPORTUNITIES AND THREATS

That small lakes suffer from many problems is indisputable. That they will continue to be severely impacted is equally obvious. In contrast, what is controversial is the extent to which mitigation measures can and should be used for restoration. Often, technology is available but immediate costs may be high. There have been many schemes for restoration of lakes that were severely damaged, particularly by acidification and eutrophication, and a variety of remedial measures exists to counter localized effects of shoreline change and recreational damage. The success of these schemes has been variable and often results have fallen well short of aspirations, perhaps because they have been too limited and failed to take into account the links between the lake and its catchment. For guidance, it is useful to assess the s̲trengths, w̲eaknesses, o̲pportunities and t̲hreats (SWOT) of the possibilities for lake restoration to more natural states.

Strengths

Lakes are generally temporary features of the landscape. Even the largest lakes are young in geological terms. The lifespans of most lakes are only a few millennia, often much less. Lake systems have had to respond to frequent disturbance through desiccation, freezing, infilling or drainage, as glaciations and warmer periods, natural succession or river adjustments have taken their toll. Many freshwater species show the characteristics of generalist colonists rather than specialists with narrow niches. Lake organisms must be capable of ready dispersal and of survival in extreme conditions. Extinction and speciation must have been frequent phenomena in their history. Lake microorganisms in

Box 5.5 Case study of potential impacts of climate change on small lakes: the influence of the Gulf Stream on lakes in the English Lake District

Year-to-year changes in the weather have a profound effect on the seasonal dynamics of lakes (George 2002), some quite irregular, whilst others follow a quasi-cyclical pattern related to variations in the atmosphere and ocean (George & Taylor 1995). Monitored for >50 years, the summer mixing characteristics of the larger lakes of the English Lake District are strongly influenced by the trajectory of storm tracks in the Atlantic and the latitude of the Gulf Stream, while interannual variations in mixing influence the seasonal dynamics of zooplankton.

Lake Windermere and Esthwaite Water remain thermally stratified throughout the summer, but there are large year-to-year variations in the depth of their seasonal thermoclines. One key influence on mixing depth is the strength of the wind in early summer, which is influenced by the position of the Gulf Stream. Based on monthly charts of the north wall of the Gulf Stream, an index of position (Gulf Stream Index: GSI) and principal components analysis, northerly movements of the Gulf Stream (positive GSI) are typically associated with calm periods in early summer and a systematic reduction in the abundance of zooplankton in both the lakes (Taylor & Stephens 1980).

Year-to-year variations in the summer abundance of zooplankton are related to the GSI (Fig. 5.6a, b). The Windermere time-series is based on estimates of the summer biomass which were de-trended to account for the progressive increase in productivity. The Esthwaite time-series was based on the average number of *Daphnia* present in the top 5 m during the summer; these data were not de-trended since there was no systematic increase in the abundance of *Daphnia*. Zooplankton has been most abundant when the Gulf Stream has moved towards the south (strongly negative GSI), attributable to interannual variations in the abundance of edible algae. When the Gulf Stream has moved towards the north, mixing has been less intense and the number of edible algae low, and a systematic reduction in zooplankton abundance has occurred. When the Gulf Stream has moved towards the south, mixing has been more intense, and edible algae and zooplankton more abundant. Many factors can influence the growth of these algae, but one of the most important is the frequency and intensity of wind mixing (Reynolds 1993).

Regressions show the extent to which the interannual variations in the depth of mixing have been correlated with the GSI (Fig. 5.7a, b). In both lakes, there has been a strong negative correlation between the depth of the thermocline in early June and the estimated value of the GSI. These 'teleconnections' show the extent to which the dynamics of lakes are often influenced by climate factors that operate on a global scale (Magnusson *et al.* 1990; George *et al.* 2000). There will thus be some significant changes in lake communities as climate changes continue to unfold.

---

particular show great ubiquity, and many genera of aquatic plants are worldwide and have wider distribution than their terrestrial counterparts. Littoral invertebrate communities share many families worldwide. Fish are more regional, reflecting their greater difficulties of dispersal, but some families, namely the salmonids, percids, characins, silurids and cyprinids, are widely distributed. Disturbance owing to human activities is thus endemic when it comes to lake systems and, since they must have frequently recovered in the past from natural disturbance, they must have properties allowing equally effective recovery from human disturbance.

Restoration is also helped by the relatively high turn-over rate of the water mass; lakes are readily flushed through in most cases. They have major aesthetic import-ance in the landscape and hence high public appeal, in contrast to wetlands (Anderson & Moss 1993). This helps build support for the public expense of restoring them, though, as the history of the restoration of Lake Washington shows, there is always a strong resistance among part of the public for spending money other than on immediate personal interests (Edmondson 1991).

Weaknesses

Lakes are also subject to very heavy impacts that mitigate against their successful restoration. The absolute need for fresh water by an expanding human population means that impacts are frequent and increasing. The more attractive and undisturbed the lakeside, the greater is the threat because such sites reach premium status. More fundamen-tally, lakes depend on catchments and cannot be improved or managed if the catchment cannot be controlled. Though

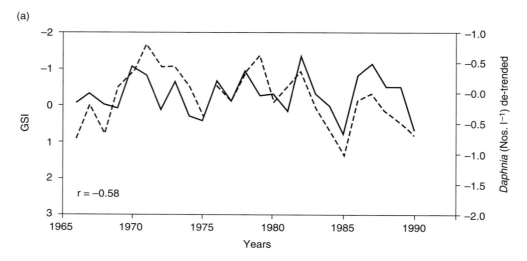

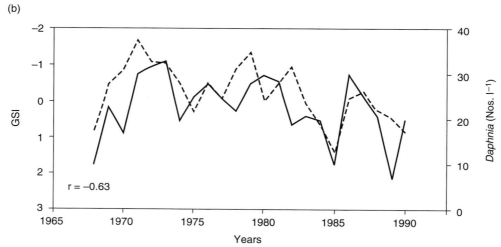

Fig. 5.6. Relationship between the summer abundance of zooplankton in (a) the north basin of Lake Windermere and (b) Esthwaite Water and the position of the Gulf Stream in the Atlantic. Year-to-year variations in the Gulf Stream index (GSI) are indicated by a broken line. Positive GSI values show that the Gulf Stream moved northwards. Correlation coefficients (r) are significant at $p < 0.05$.

this generally affects the larger lakes, sometimes international boundaries cross catchments, creating difficulties because of different national sensitivities and aspirations. There is also a meta-catchment; atmospheric pollution may influence a lake from distances far greater than the hydrological catchment. This is especially a problem for acidification and climate change.

Lakes lie at the bottoms of river basins and all that happens in the catchment ultimately is reflected in the lake under

the inexorable effect of gravity. Because of its partly covalent, partly polar properties, water is the nearest substance known to a universal solvent. The range of catchment activities, and the list of potentially damaging substances that can dissolve in water, is huge and expanding. Moreover, lakes are connected with rivers, and rivers have often been deliberately used as cheap pipes for the disposal of wastes.

Although knowledge of lakes is considerable, it will always be incomplete. Governments and exploitative

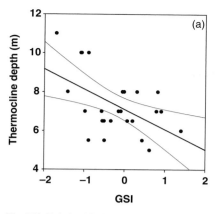
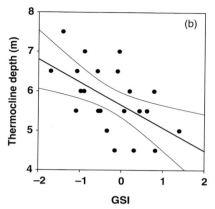

Fig. 5.7. Relationship between the depth of the early summer thermocline in (a) the north basin of Lake Windermere and (b) Esthwaite Water and the position of the Gulf Stream in the Atlantic (Gulf Stream Index: GSI). A positive GSI indicates northerly movements of the Gulf Stream. The lighter lines show the 95% confidence intervals for the fitted regressions.

interests can and do use alleged lack of knowledge as an excuse for doing nothing. More research, it is said, is needed. Research is generally cheaper than action and can be used as a delaying tactic. One weakness in the justification of restoration schemes is that the range of former states is uncertain and the subtler effects of natural and human change cannot be distinguished. This is especially the case with eutrophication, acidification and climate change. There is also a problem with perception of natural succession. Encroaching vegetation, though natural, is perceived as problematic. The lakes mostly affected are the smaller ones, and public appreciation of lakes appears to be proportional to lake size.

There is also the difficulty that where problems have arisen from introduction of species, especially invertebrates and sometimes plants, it may be impossible to reverse the introduction. Conversely, where endemic species have been lost, they are absolutely irreplaceable. Finally, with current world populations, it must be accepted that many areas will have to remain agricultural. Different agricultural systems, resulting in greater or lesser damage, are conceivable, but restoration of catchments even to semi-natural vegetation will usually be impossible. All of these drawbacks weaken the ability to guarantee delivery of high-quality lake restoration.

Opportunities

In the developed world there are many indicators of concern for the state of the environment, among the populace at least. Membership of environmentally concerned non-governmental organizations is high and increasing. Legislation is beginning to recognize the need to look beyond the wetted perimeter of freshwater habitats and to consider the catchment. Two outstanding pieces of legislation are the South African Water Act and the European Water Framework Directive (Box 5.6). The latter insists on river basin management as opposed simply to the river channel management of the past, and has set targets for restoration of aquatic habitats in Europe to 'good' ecological status. This is defined as only slightly different from high status, which is essentially a near-pristine state. In other countries, the term 'integrated catchment management' increasingly enters the lexicon of decision-makers.

There have also been considerable advances in understanding of lake systems, enabling development of a clearer scope for restoration. Since the controversy over the role of phosphates in domestic detergents in the 1960s and suggestions at the time that it was carbon availability that limited algal growth in lakes, much well-directed experimental work has produced a very solid understanding of eutrophication. More recently, whole-lake experimentation has revealed the important effects of fish in determining features of the system through food chain effects (Box 5.7) and this has influenced restoration strategy. Previous understanding was essentially based on agricultural philosophy applied to angling, namely that fish were the products of fertility of the lake, and the more and the bigger fish the better. This attitude has been successfully challenged and indiscriminate stocking programmes are

Box 5.6.   New legislation will help

In Europe, the traditional means of regulating water quality has been control of pollutant discharges, an approach originating in the widespread sewage fouling of rivers as cities expanded to accommodate the mass immigration of workers from the countryside with the Industrial Revolution. The pattern is now being repeated in tropical megalopoleis, where sewage deoxygenates and carries waterborne diseases like typhoid and cholera.

In late-nineteenth-century Wales, the religious revival movement offered prayer meetings and hymn singing to counter the epidemic of deaths. More effective was the creation of treatment works in which the sewage was biologically oxidized, removing pathogens and discharging a largely inorganic effluent, but one that gave rise to eutrophication. Treatment works steadily improved in sophistication to deal with other wastes, which could be regulated by licensing; 'end of the pipe' solutions thus became emphasized. The amount that could be discharged was controlled, but many other problems of natural waters arose from intensified agriculture, air pollution and acidification, heavy recreation, stocking of alien species and overfishing.

The European Water Framework Directive (2000) states that aquatic habitats must be restored to 'good' ecological quality by 2015. The implication is that whatever means are necessary must be developed and used to achieve a target state. The Directive requires that a typology of different river and lake types be established, but this poses problems where habitats vary continuously along the axes of thousands of variables, but with flexibility it is possible to create usable units. In each of these, the ecological quality must then be established with reference to specified components (phytoplankton, aquatic plants, macroinvertebrates, fish, physical structure, hydrology and water chemistry). There must be a high quality reference standard for each type, defined as virtual absence of human influence, a state that probably no longer exists, although it can be reconstructed.

What is practically more important is the definition of 'good' quality to which states must restore their aquatic systems, defined as only 'slightly' different from high quality. Even if most of Europe fail to achieve the intended good quality by 2015, far superior standards than at present prevail are intended.

The issues of arid countries are immediately more crucial. In its National Water Policy (1997), South Africa recognized the past inequalities of water distribution and availability. The *c.*1200 m^3 of water available per person in a population of 40 million may seem more than adequate to support a domestic lifestyle, but it represents severe water stress because of the much greater industrial and agricultural requirements (Chapters 2–4). Extensive damming of water courses for irrigation has in the past ruined river systems and deprived the rural poor of a convenient or adequate water supply.

The National Water Policy has two key features. There are resource-directed measures aimed at protecting the aquatic ecosystem and ensuring adequate flows, and controls on impacts on the available water. There is legislation for support of low-waste and non-waste technologies, and for demand management. Human requirements are placed above commercial requirements. The Working for Water Programme seeks to remove from catchments large numbers of alien trees that are particularly water demanding compared with the native flora, and reduce flows to water courses. Removal is labour-intensive and provides employment for around 40 000 people.

The Water Services Act provides for a minimum of 25 l of water per person per day within 200 m of their homes at a cost commensurate with their income, whilst a water-pricing policy for commercial use will charge the real costs of use, including effluent disposal. The putting of domestic needs above commercial interests will have positive effects on the freshwater environment, for stream flows that have been dried by irrigation projects will be renewed.

becoming less acceptable. There is also an increasing realization of the intrinsic interest of tropical systems and of the important roles they have in protein production in developing countries, where ingenious artisanal fishing methods guarantee a sustainable catch from natural systems without management intervention.

Threats

Despite many local improvements and very positive attitudes on the part of large sections of the populace, net environmental damage to lakes and ponds is almost certainly increasing. Even the most inspired legislation can

Box 5.7 Example of lake restoration after eutrophication: Lake Finjasjön, Sweden

The small (11 km^2) shallow Swedish Lake Finjasjön had a transparency of about 2 m in the 1920s. During the 1940s, the lake quality deteriorated with blooms of the cyanobacterium *Gleotrichia*, owing mainly to untreated sewage from nearby Hässleholm. Despite mechanical treatment from 1949 and biological treatment from 1963, blooming (now *Microcystis*) continued and swimmers suffered from skin rashes. The external loading greatly exceeded the level thought necessary for maintenance of clear water.

To restore the lake, tertiary wastewater treatment reduced the effluent phosphorous content 10-fold; however, internal loading (release from the accumulated sediment) maintained high phosphorus concentrations, and chlorophyll even increased with summer blooms of cyanobacteria. Sediment removal by suction/dredging in 1987–91 led to only minor improvements in lake water quality. To further reduce the external loading to the lake, a 30-ha wetland was constructed below the sewage treatment plant and 5-m buffer zones were established along the most important inlet streams and dykes. At this stage, the in-lake management shifted from dredging to biomanipulation by removing (by means of trawling) 392 kg per ha of planktibenthivorous fish (mainly roach *Rutilus rutilus* and bream *Abramis brama*) between autumn 1992 and spring 1994, estimated to be 80% of the fish stock. These fish feed extensively on large-bodied zooplankton such as *Daphnia*, which otherwise when abundant help maintain a clear state by grazing phytoplankton. Moreover, bream and roach redistribute nutrients from the sediment to the water when feeding on benthic invertebrates.

The short-term effect of the biomanipulation was huge. A dramatic decrease in phosphorus and chlorophyll *a* concentrations occurred and the transparency of the water increased substantially (Fig. 5.8). The average phosphorus concentration of 150–200 µg l^{-1} in the early 1990s decreased to 31 µg l^{-1} in 1996. The reduction in internal loading was attributed to improved redox conditions in the sediment (enhanced binding of phosphate to iron), enhanced benthic algal production and less excretion of benthic-derived phosphorus by fish. Submerged macrophyte coverage increased rapidly from 1% of the lake area in 1993 to 20% in 1996, and the plants reached depths of 2.5 m as compared with 0.3 m before restoration (Fig. 5.8). The predator control of cyprinids increased as the contribution of piscivorous fish rose from 20% to 50% of the fish population, particularly because of an increase in the abundance of large perch (*Perca fluviatilis*). With fewer cyprinids, the zooplankton : phytoplankton biomass ratio increased, suggesting a higher grazer control on phytoplankton by *Daphnia*, which were much larger than before fish removal. Accordingly, facing higher grazer control and fewer nutrients, the cyanobacteria summer blooms disappeared. Biomanipulation had strong cascading effects throughout the food web all the way down to the nutrient level.

After the measures taken to reduce external loading, the inlet concentration was so low (0.04 mg P l^{-1}) that the lake, according to theory, should have been able to maintain the clear state in the long term. However, a partial reversion occurred some years after the fish removal in 1997, leading to an increase in nutrient concentrations, emergence of cyanobacteria and reduced biomass of submerged plants. This coincided with an increase in the numbers of cyprinids and particularly fish fry feeding heavily on zooplankton. Additional control (now requiring much less effort) of the planktivorous fish biomass (90 kg ha^{-1}) took place in 1998/9 with a positive response in water quality in 1999. Transparency then increased to 1.5–2.0 m, chlorophyll levels stabilized at *c.*20 µg l^{-1} (Fig. 5.8), and algal blooms have been absent or very rare. Also the submerged macrophytes have responded with an increase in cover after a few years of decline from 1997 to 1999 (see Hässleholms Vatten 2007). The costs of the restoration amounted to 3 million kronor for the dredging, 0.9 million kronor for the fish removal and 3.5 million kronor for the wetland system.

Sources and further reading about biomanipulation and the restoration of Lake Finjasjön include: Persson (1997), Hansson *et al.* (1998), Annadotter *et al.* (1999), Meijer *et al.* (1999), Nilsson (1999), Strand and Weisner (2001), Jeppesen and Sammalkorpi (2002), and Mehner *et al.* (2002).

---

be undermined by government reticence in practice, especially where it is perceived that booming economies may boom less prolifically. There are worrying signs in some European countries that the spirit of the Water Framework Directive (Box 5.6) will not be adhered to and, as habitats become more degraded, the standards to which younger generations become inured are lower than those of their predecessors. This leads to acceptance, generation by generation, of less-stringent targets for restoration.

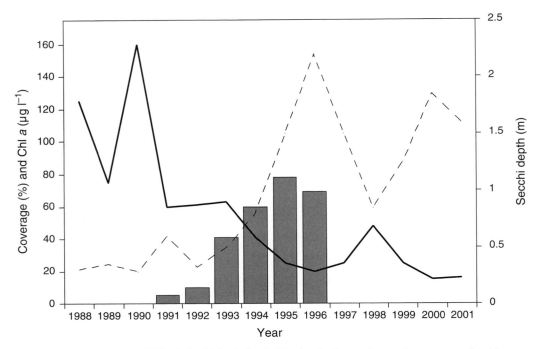

Fig. 5.8. Changes in chlorophyll *a* (solid line*)*, Secchi depth (hatched line) and submerged macrophyte coverage (bars) in Lake Finjasjön (Sweden) before and after fish manipulation conducted from 1992 to 1994. Macrophyte data only available 1991–97. (From various sources: see text.)

The inevitable scientific inability to set an objective description of pristine state (because there is no single such state for any changing system) is interpreted as incompetence, and targets are thus likely to be set to suit economic interests. At the highest government levels in the major powers of the Western world, it is clear that the environment is lower on the agenda than it was at the time of the Rio de Janeiro summit in the early 1990s. As multinational companies acquire greater wealth than the budgets of the majority of small countries, especially in the tropics, this situation is likely to be reinforced. Although natural and semi-natural systems provide goods and services worth three times the global gross national product, at over US$30 trillion, attitudes continue to be towards short-term exploitation rather than indefinite wise use. Sustainability, under the terminology of the land ethic introduced by Leopold (1941), still appears to be unattainable in the foreseeable future. Lakes, as a group, will continue to be degraded, despite notable successes in at least partial restoration. The problems with shallow lakes, as with all natural systems, are indicators of greater global problems (Moss 1995).

6 · Environmental trends and potential future states of large freshwater lakes

ALFRED M. BEETON, ROBERT E. HECKY AND KENTON M. STEWART

INTRODUCTION

Fresh water is the world's most valuable resource. Increased demand for fresh water – as human populations and impacts due to pollution, changes in land use, invasive species, overexploitation of resources, diversion of water and climate change continue to increase – threatens its continued quantity and quality (Chapters 1, 2, 4 and 6). Fresh water exists mostly as ice, groundwater and atmospheric water. Surface fresh water is not uniformly distributed. About 68% of the global liquid surface fresh water occurs in 189 large (>500 km^2) freshwater lakes (Reid & Beeton 1992). These lakes are especially valuable to the countries fortunate to have them and are important to the economies, social structure and viability of the riparian countries (Beeton 2002).

This chapter cannot attempt to deal with all these lakes. Rather it focuses on the African Great Lakes, St Lawrence Great Lakes and Lake Baikal (Fig. 6.1). The African Great

Aquatic Ecosystems, ed. N. V. C. Polunin. Published by Cambridge University Press. © Foundation for Environmental Conservation 2008.

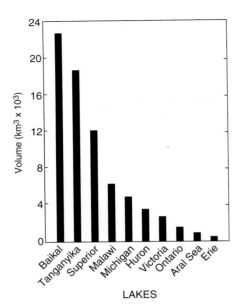

Fig. 6.1. Volumes of the world's great lakes.

Lakes (including Malawi, Tanganyika and Victoria) are broadly representative of tropical tectonic lakes. The St Lawrence Great Lakes (Erie, Huron, Michigan, Ontario and Superior) are representative of glacial scour lakes of North America, northern Europe and Asia. Lake Baikal has similarities to Lake Superior and Great Bear and Great Slave Lakes (Mackenzie River Basin) but is included here because of its many endemic species, great age and status as the largest single volume of liquid surface fresh water. The Aral Sea represents an extreme example of the consequences of high demand for fresh water (see also Chapter 7). The threats and trends observed in these lakes are of global concern for all large lakes.

Large lakes have long been foci of human habitation, and large population centres occur along many of their shores. The St Lawrence Great Lakes provide especially useful transportation into the heartland of North America. These lakes and others have been used for hydroelectric power, navigation, irrigation, waste disposal, commercial and recreational fishing, leisure pursuits and cooling water. They are a major source of domestic and industrial waters.

Lakes Erie, Huron, Michigan, Ontario, Superior, Malawi, Tanganyika, Victoria and Baikal account for nearly 68% of all liquid fresh water on the Earth's surface and rank among the top 18 in volume (Herdendorf 1990). There are 253 saline and freshwater lakes that are >500 km² each in

area. Of these 45% are <1000 km² in area, and fewer than 10% are >10 000 km² (Herdendorf 1982). About 50% of all the large lakes are in North America, and 99 of the 189 large freshwater lakes are in glacial basins in North America and northern Europe, these lakes having very similar physico-chemical characteristics and biota because they originate from large proglacial lakes and a few glacial refugia (Beeton 1984). The so-called glacial relict flora and fauna of these lakes have a circumpolar distribution. For example, Lakes Ladoga and Onega of northern Europe have species associations closely similar to the St Lawrence Great Lakes. North American and Eurasian populations of a glacial relict mysid are considered to be the same species, *Mysis relicta* (Holmquist 1959). Furthermore, species associations of widely separated glacial lakes are strikingly similar, especially the deep-water communities. Great Slave Lake in north-west Canada and Thirty-one Mile Lake in eastern Canada are 2800 km apart, yet they have a community coefficient of similarity of 80% (Dadswell 1974).

The tectonic African Great Lakes, Tanganyika and Malawi, are ancient with large numbers of endemic species. Victoria dried out 15 000 years ago, but the basin has existed as a shallow tectonic depression for over 200 000 years. It also has a large endemic species flock of haplochromine fishes (Johnson *et al.* 1996). Lake Baikal is also tectonic and is the oldest of all lakes. It also has a unique biota of many endemic species. In contrast, the glacial lakes are recent geologically, generally only thousands of years old.

The purpose in this chapter is to give an overview of how these nine important large lakes are changing and consider what states they may be in by the year 2025. Even with this geographical focus, the information base is far from even. The literature is extensive for the St Lawrence Great Lakes, especially for long-term trends in physico-chemical conditions, the biota, land-use changes and action taken to deal with human impacts in the region. Much less is known about the African Great Lakes, although much research is under way. Relatively little information is available on trends in Lake Baikal, largely because little change has occurred given its geographical isolation in a harsh subarctic climate; the literature on its geology and endemic species is most extensive.

TRENDS

Although there are large differences among the nine lakes, the major impacts are similar. Land-use changes, eutrophi-cation and pollution are closely associated. Land-use changes, including that from forestry and agriculture, cause soil erosion and mobilization of nutrients, accelerating eutrophication. These activities also result in some kinds of pollution: suspended solids, sawdust and agricultural pesticides. Major land-use changes associated with urbanization and industrialization result in greatly accelerated pollution. External pressure to divert water from its natural hydrology recurs periodically. Overfishing is also a form of over-exploitation that has plagued the lakes. Invasive species have become an increasing threat and problem, as non-indigenous species are intentionally introduced or gain access largely by greatly increased global transportation.

Land-use change

Changes in land use have long-term impacts on lakes, which are usually irreversible. The St Lawrence Great Lakes region was heavily forested prior to the nineteenth century, but much of the virgin forest was logged by 1910. At the same time, many river mouths became harbours, many streams were dammed to generate power for mills and wetlands were drained to provide agricultural land (Ashworth 1987). As a consequence of these activities, the hydrology and ecology of flowing waters were altered and migrating fishes were negatively impacted, as were various birds and mammals. The disappearance of the Atlantic salmon (*Salmo salar*) from the Lake Ontario basin by 1900 was probably due to the damming of spawning streams (Beeton 2002).

Increased runoff of sediments and nutrients is considered a consequence of accelerated change in land use associated with increasing human population in the Lake Victoria basin. Clearing of riparian vegetation has resulted in erosion and loss of vegetation that formerly retarded runoff, filtered sediments and retained nutrients (Lowe-McConnell 1994). Deforestation and agricultural practices have increased siltation (Jorgensen *et al.* 2003). Human population is growing rapidly in the Lake Tanganyika watershed, but the lake is not severely impacted yet, although some major changes have occurred. About 50% of the central watershed area and 100% of the northern catchment have been cleared of natural vegetation, and siltation is now a problem (Jorgensen *et al.* 2003).

Deforestation and land cultivation are considered to be the greatest long-term threats to Lake Baikal (Kozhova & Silow 1998). Most land-use activities prior to the 1950s were timber harvesting, clearing of forest for pasture and traditional agriculture. In the last 40–50 years, the lake has been altered by the construction of the Baikalsk Pulp and Paper Combine, intense use of mineral fertilizers and modern transportation, growth in coastal populations and increasing consumption of coal for power stations (Jorgensen *et al.* 2003). An increase in the turbidity of the lake's tributaries is a consequence of forest cutting and agriculture (Jorgensen *et al.* 2003). Construction of the Irkutsk hydropower station and the dam on the Angara discharge river in 1956 raised Baikal's water level, affecting the shallow coastal area and creating shore erosion.

Eutrophication

By the 1960s, eutrophication had become a major problem for the St Lawrence Great Lakes as a consequence of excessive nutrients from agricultural runoff, poor waste management and use of phosphate detergents (Beeton 1965). Populations of fish, invertebrates and algae changed, beaches were closed, large blooms of blue–green algae occurred and dissolved oxygen was depleted in some bottom waters (Beeton 1969). The impacts of nutrients and other pollutants occur primarily near shore (Holland & Beeton 1972), although significant inputs can be atmospheric (Eisenreich *et al.* 1977). The early surveys of Lake Erie detected eutrophication near shore, but this was dismissed as purely a 'shore effect' that did not impact the lake (Wright 1955; Fish 1960). It is now realized that the continued and accelerated loading of nutrients, initially only apparent in the nearshore waters, eventually spread into the open lake (Beeton & Edmondson 1972). Because of their large size and diverse habitats, large lakes tend not to show an overall response to nutrient enrichment, as do small lakes (Beeton 2002; Chapter 5). Efforts to control eutrophication in the St Lawrence Great Lakes have been successful. Upgraded sewage treatment and efforts to control non-point-source as well as point-source loadings of nutrients and other chemicals such as chlorides (International Joint Commission 1989), have reduced phosphorus and chloride concentrations in Lake Ontario (Beeton *et al.* 1999). Algal blooms have decreased and the benthic communities have recovered; mayfly (*Hexagenia*) populations have returned to western Lake Erie, Green Bay and Saginaw Bay (Beeton 2002).

Eutrophication is of concern for the African Great Lakes, although Lakes Victoria, Tanganyika and Malawi are experiencing somewhat different forms of environmental impact (Turner 1994; Alin *et al.* 1999). Victoria is the shallowest of the three lakes, and the first to respond to nutrient enrichment, undergoing dramatic change, while Tanganyika is more resilient, probably as a consequence of its greater volume and lower human impact (Bootsma & Hecky 1993). Deforestation, soil erosion, desertification and atmospheric pollution are degrading all three lakes (Hecky & Bugenyi 1992). Eutrophication was under way in Victoria in the 1920s and accelerated in the 1960s with land clearance (Hecky 1993), the evidence being seen in decreases in water clarity, changes in phytoplankton and increases in cyanobacteria. Algal biomass increased five- to eight-fold and productivity doubled (Mugidde 1993). Some of the algal blooms have been associated with fish mortality. Hypolimnetic dissolved oxygen in the deeper waters of Victoria decreased to less than 1 mg per litre over large areas of the bottom of the lake (Hecky *et al.* 1994). Water transparency greatly decreased from 7–8 m in 1960/1 to only 1.3–3 m in 1990/1 (Mugidde 1993) and nitrogen and phosphorus loading have increased (Hecky 1993). The diatom *Melosira* spp. (now *Aulacoseira*) disappeared, and was replaced by *Nitzchia* (Hecky 1993). Cyanobacteria now dominate the phytoplankton. The nitrogen-fixing cyanobacteria are favoured by increased P loading, because tropical lakes have low nitrogen reserves and low N : P ratios as a consequence of low dissolved oxygen and denitrification (Hecky *et al.* 1996). Infestation by water hyacinth (*Eichhornia crassipes*) has been cited as evidence of eutrophication (Jorgensen *et al.* 2003). The great biodiversity of the African Great Lakes, particularly of cichlid fishes, requires maintenance of clear water. Vision and colour are essential to cichlids, and loss of transparency and chromaticity, owing to increased turbidity because of eutrophication, will have led to less sexual selection and more hybridization (Seehausen *et al.* 1997). Benthivorous cichlids probably suffered from reduced benthic productivity as a consequence of reduced light penetration. Consequently, the cichlid populations were already being adversely impacted before introduction of the Nile perch (*Lates niloticus*), although that introduction led to further rapid decline of the haplochromine cichlids (Seehausen *et al.* 1997).

Eutrophication does not appear to be a problem for Lake Baikal, with the exception of some local areas such as the nearshore and harbour region of Kultuk at the southern end of the lake (K. Stewart, personal observation 2003). Concentrations of nutrients and major ions have not changed in 30 years (Granina 1997), and sulphate (SO_4) concentrations have not changed in 53 years (G. Galazy, unpublished data), which is similar to Lake Superior

(Beeton 1969). Some significant shifts have occurred in endemic planktonic diatom assemblages in the sediments of Lake Baikal (Flower 1998), but overall the biota and environment of Lake Baikal have changed very little from the historical record (Kozhov 1963; Grachev 1991; Flower 1994). Nevertheless, Jorgensen *et al.* (2003) consider eutrophication a potential threat to Lake Baikal, owing to input of nutrients and non-toxic organic matter.

Pollution

Pollution has impacted all the large lakes and continues to be a major concern (Beeton 2002). The earliest recorded pollution of the St Lawrence Great Lakes developed from the large amounts of sawdust produced by timber processing in the nineteenth century (Milner 1874). Sawdust clogged tributaries to the lakes for many kilometres, adversely affecting the fishery by eliminating spawning and feeding sites. Waterborne disease, such as typhoid, from domestic sewage in lake waters, was reported in historical public-health records (Mortimer 2004). Pollution was considered a major factor in the decline of fish populations (International Board of Inquiry for the Great Lakes Fisheries 1943). Pollution was evidently considered a manageable problem that when mitigated would result in fish populations returning to former abundance; however the cisco (*Coregonus artedi*) populations never recovered and the blue pike (*Sander glaucum*, formerly known as *Stizostedion vitreum glaucum*) became extinct in Lake Erie (Brown *et al.* 1999). Total dissolved solids, and presumably pollutants, increased long term, in tandem with the growth in the human population in each lake basin; the exception is Lake Superior, where the population remains low (Beeton 1969). The collapse of the cisco population in Lake Erie may show long-term effects of pollution. The commercial catch records for sections of the lake show the fishery collapsed first in the Detroit River and the west end of the lake. The Detroit River and Maumee River were both impacted by pollution and excessive nutrient flow into the western basin of Erie. The fishery declined next in the central Ohio waters of Erie, and then declined progressively eastward in Pennsylvania and New York waters in the deepest eastern basin (Beeton & Edmondson 1972).

Much of the early concern was with visible signs of pollution, namely oil and garbage. After the 1940s, concern turned to the adverse effects of persistent toxic substances affecting all aquatic life, birds and humans (Colborn *et al.* 1990). Toxic substances such as DDT

(dichlorodiphenyltrichloroethane) and PCBs (polychlorinated biphenyls), concentrated especially in large predatory
fish, continue to be of major concern. State agencies dealing
with fishing and public health continue issuing advisory
notices about the dangers, especially to women and children,
of eating larger Great Lakes fishes. Concentrations of
chlorinated organic compounds have declined in some Great
Lakes fishes, but concentrations continue to be high for
all species in Lakes Michigan and Ontario (Rowan &
Rasmussen 1992).

Industry has significantly decreased its release of
hazardous chemicals in the region (Michigan Department of
Environmental Quality 2000). Loading rates and concentrations of chlorine, lead, copper and zinc in the Great
Lakes decreased. The quality of Detroit River water
entering Lake Erie has improved (Michigan Department of
Environmental Quality 2000). PCB concentrations in Great
Lakes fish decreased in the early 1980s (Fig. 6.2), but
Jeremiason et al. (1994) found no reduction occurred after
1986. Large oligotrophic lakes at mid-latitudes may be a
source of PCBs rather than a sink, thereby increasing residence time of PCBs in the atmosphere (Jeremiason et al.
1994). PCBs in Lake Superior water decreased by 26 500 kg
in 1980–92, but only 4900 kg of PCBs had been deposited in
sediments since 1930 (Jeremiason et al. 1994). Volatilization
and not sedimentation may be the major route of loss of
PCBs for Lake Superior. Consequently, PCBs will decrease
very slowly and remain a long-term problem.

The African Great Lakes are especially susceptible to
atmospheric sources of pollution. They have nearly closed
basins and water input is mostly in the form of precipitation

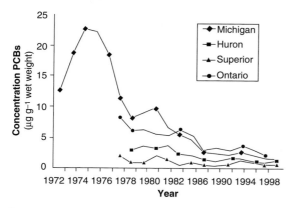

Fig. 6.2. Concentration of PCBs in lake trout in the
Laurentian Great Lakes 1972–98. (Adapted from Michigan
Department of Environmental Quality 2000.)

(Hecky & Bugenyi 1992; Bootsma & Hecky 1993). For
example, atmospheric wet and dry deposition dominates the
phosphorus budget of Lake Victoria (Tamatamah et al.
2004). Outflows are low for these large volume lakes and
water residence times are very long, flushing times being
750 years for Malawi, 7000 years for Tanganyika and
140 years for Victoria (Hecky & Bugenyi 1992). Consequently, pollutants remain in the lakes and accumulate in
the sediments. Lake Victoria is completely turned over each
year, while Malawi and Tanganyika are meromictic, their
deep waters being anoxic and exchanged slowly with surface
waters (Vollmer et al. 2002).

Although apparently not as severe as in the St Lawrence
Great Lakes, pollution is a problem in the African Great
Lakes (Tweddle 1992), and domestic wastewater is not
treated in the latter (Jorgensen et al. 2003). Study of the
most polluted tributary to Lake Tanganyika, the River
Ntahangwa, indicated organic pollution to the lake to be
negligible into the early 1990s (Vandelannoote et al. 1996).
Chlorinated pesticide concentrations in Senga Bay (Lake
Malawi) were similar in air and lower in water and fish
compared to the St Lawrence Great Lakes (Kidd et al.
2001), maybe as a consequence of more rapid chemical
transformation in Lake Malawi (Karlsson et al. 2000). Local
and regional sources appear to provide most of the chlorinated organic compounds to the African Great Lakes.
Mercury pollution is a concern from gold and other mining
activities and biomass burning, but risks from fish consumption remain low in Lake Victoria (Campbell et al.
2003).

Pollution from industrial, domestic and transboundary
sources are a major threat to Lake Baikal (Rose et al. 1998).
Significant pollutants from the pulp mills at Baikalsk and
Selinginsk (via the Selenga River) have entered the lake in
significant quantities. The paper and pulp mill at Selenginsk
has now been converted to a closed system (Jorgensen et al.
2003), but sewage is discharged to the northern lake
area from Severobaikalsk. Atmospheric pollution has been
increasing from industry and significant quantities of heavy
metals are carried to the lake via atmospheric transport from
outside the basin (Kozhova & Silow 1998). Water, fish and
seals all contain measurable amounts of organochlorine
compounds (Kucklick et al. 1994). Concentrations of these
compounds in water and fish in Lake Baikal were similar to
those found in the upper St Lawrence Great Lakes, except
for chlordanes, which were about 50% lower in the latter
(Kucklick et al. 1994).Greater concentrations of DDT were
found in Lake Baikal than in the St Lawrence Great Lakes,

suggesting inputs from continued use of DDT in Asia. It also appears that Lake Baikal is receiving new inputs of PCBs. The death of thousands of Baikal seals in 1988/9 may have occurred partly because their immune systems were weakened by organochlorines, and they therefore could not defeat a viral infection (Iwata *et al.* 1995). Concentrations of oil products in Lake Baikal do not exhibit a consistent trend over time, and fluctuate at less than 1 mg per litre. Nevertheless, oil, phenols and metal concentrations are a source for concern (Kozhova & Silow 1998).

Overfishing

Overexploitation of fishery resources is well documented for the St Lawrence Great Lakes. Commercial fishing is thought to have become important in the 1820s, when whitefish (*Coregonus clupeaformis*) exploitation probably started in Lake Michigan (Brown *et al.* 1999). Commercial catches were first recorded for Canadian waters in 1867 and for the USA in 1899 (Baldwin *et al.* 1979). Canadian data are incomplete for 1879–1914, but much less so after 1914. Nevertheless, these landings probably show population trends for the highly prized species lake trout (*Salvelinus namaycush*), whitefish and cisco (Brown *et al.* 1999). Catch data for carp (*Cyprinus carpio*), burbot (*Lota lota*), sheepshead (*Aplodinotus grunniens*) and other less desirable species probably indicate market demand more than actual population size.

Several deep-water coregonid species were important in the commercial fishery in Lake Michigan (Wells & McLain 1973), but catch records of these were not separated by species. Nevertheless, fishing gear records provide useful insight into the changes in abundance of several species. The largest species are *Coregonus nigripinnis* and *C. johannae*; gill nets with a mesh size up to 11.5 cm were used in the nineteenth century for these species, but mesh size gradually fell to 6.4 cm by 1950, the decrease probably a reflection of declining stocks of the largest coregonids. Fishery surveys show that stocks of these species were depleted by 1930–32, and even perhaps in the first decade of the twentieth century (Smith 1972). Landings of the smallest coregonid species, *C. hoyi*, increased and provide a present day fishery (E. Brown Jr *et al.* 1999). None of the deep-water coregonid populations ever recovered after severe population declines (Smith 1972). The cisco population collapsed in Lake Erie in the 1940s, and in Lake Ontario and the Saginaw Bay of Lake Huron in the 1950s. Cisco was the major part of the Lake Michigan fishery in the nineteenth and twentieth centuries (Wells & McLain 1973). Landings of this species declined in Lake Michigan in the 1940s, increased in 1952, and then collapsed in the 1960s. The collapse of the cisco fishery in Lake Erie was dramatic and alarmed federal and state governments, the industry and many others (Fig. 6.3). The fishery was unstable and fluctuated widely with a catch of more than 14 million kg per year in 1924, abruptly declining to 3 million kg per year in 1925 (Baldwin *et al.* 1979). A brief recovery occurred in 1946–1947, but the decline continued into the 1960s, and production has not exceeded 227 kg per year since then. Overfishing of desirable stocks was undoubtedly a factor in the decline and extirpation of some species in the St Lawrence Great Lakes, but other factors, such as land-use change, eutrophication and invasive species were also important. For example, species now considered open-lake and/or deep-water were formerly abundant at times in shallow nearshore waters (Smith 1972), namely coregonids, lake trout, walleye (*Sander vitreus*, formerly known as *Stizostedion vitreum*) and yellow perch (*Perca flavescens*). Construction of numerous mill dams throughout the region changed the hydrology of tributaries and migration of fishes. The Atlantic salmon, once abundant in Lake Ontario, spawned in streams that were dammed, but collapsed in 1890 (Baldwin *et al.* 1979).

The African Great Lakes contain about 10% of all freshwater fish species worldwide and these are mostly endemic. These fish stocks have been impacted by overexploitation and introduction of non-indigenous species. Lake Victoria has the world's largest freshwater fishery. The Nile perch was introduced to Lake Victoria to provide commercial and sport fisheries and was successful in this, but these and the fisheries for the introduced Nile tilapia (*Oreochromus niloticus*) and an endemic cyprinid *Rastrineobola argentea* are threatened by increase in fishing effort (Lowe-McConnell 1994). In Lake Malawi, increasing exploitation of inshore fishes has resulted in declining catches and loss of biodiversity (Turner 1994). Lake Tanganyika is a source of aquarium fish, and destructive fishing and overfishing have also impacted biodiversity (Jorgensen *et al.* 2003).

All three lakes are subject to overexploitation, landings being more than twice the sustainable yield. The sustainable yield levels have been lowered for the Tanganyika fishery and the catch per unit effort has declined for most valued species in Malawi.

Overfishing could impact Lake Baikal. Overexploitation of comephorid sculpin has resulted in decreased abundance

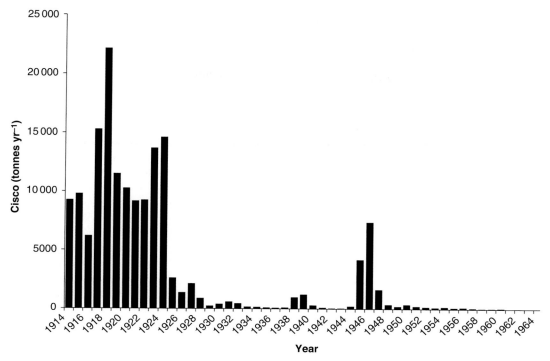

Fig. 6.3. Cisco (*Coregonus artedi*) landings in Lake Erie 1914–64 (data from Baldwin *et al.* 1979).

of its predator the whitefish omul (*Coregonus autumnalis migratorius*), which preyed instead on a copepod (*Epischura baicalensis*) and, as a result, declined in body mass (Grachev 1991).

Biological invasions

Invasive species have caused major long-term changes in the occurrence and abundance of native species, communities and consequently ecosystems of the St Lawrence Great Lakes. An estimated 182 non-indigenous species have now been found in the lakes (Ricciardi 2006), with 40 non-indigenous species being found in the first half of the twentieth century, 76 during the latter half of the century (Ricciardi 2001) and a new invader found about every 8 months since then. Sea lamprey (*Petromyzon marinus*) and dreissenid mussels (*Dreissena polymorpha* and *D. bugensis*) have brought about major changes. Other invasive species, such as carp, smelt (*Osmerus mordax*), alewife (*Alosa pseudoharengus*) and round goby (*Neogobius melanostomus*) have also contributed to significant long-term changes, but not of the magnitude caused by the lamprey and the dreissenids.

The sea lamprey probably entered Lake Ontario in the early nineteenth century and gained access to Lake Erie in the 1920s (Applegate & Moffett 1955) after the Welland Canal was improved. The lamprey was in the upper lakes by the 1930s and, by the 1940s, the lamprey population exploded. Uncertainty remains as to the relative importance of the intense exploitation, habitat degradation and lamprey predation (Hansen 1999), but the consensus is that lamprey predation probably induced complete collapse of lake trout populations in all the St Lawrence Great Lakes except Lake Superior (Brown *et al.* 1999) and decimated populations of other large fish species. In Lake Ontario, the lake trout population may not have recovered because dioxin concentrations in yolk-sac fry would have precluded normal development and survival (Cook *et al.* 2003).

The sea lamprey spawns in at least 433 cold-water tributaries of the St Lawrence Great Lakes. Physical and subsequently electrical barriers were used to prevent spawning and proved useful for estimating abundances of lamprey (Great Lakes Fishery Commission 2001a). About 6000 chemicals were tested in attempting to discover means of chemical control before 3-trifluoromethyl-4-nitrophenol

(TFM) was found in 1958 to kill sea lamprey larvae without significant impact on other fishes (Great Lakes Fishery Commission 2001*b*). Sea lamprey abundances decreased gradually as TFM was used in an increasing number of streams; by the late 1970s, sea lamprey populations had decreased to about 10% of those in the 1950s.

The sea lamprey invasion was largely viewed as a unique problem, but discovery of the zebra mussel (*D. polymorpha*) in Lake St Clair, and its rapid spread to Lake Erie and then Lakes Huron, Michigan and Ontario in the 1980s, focused attention on the wider invasive-species problem (Beeton 2002). The zebra mussel evidently gained access to the lakes via ballast water discharged into Lake St Clair (Leach *et al.* 1999). Shipping accounts for more invasions than any other factor (Ricciardi 2001) and some 77% of invasions since 1970 are likely to be a consequence of transoceanic shipping. The zebra mussel, an efficient filter-feeder, quickly doubled water clarity by reducing the abundance and species composition of the phytoplankton (Holland 1993). Greater water clarity favoured greater abundance and wider distribution of aquatic macrophytes, and rotifers and copepods decreased in abundance (Beeton & Hageman 1994). Concentrations of ammonium, nitrate, soluble reactive phosphorus and silica increased shortly after the zebra mussel became established in Lake Erie (Holland *et al.* 1995), with changes similar to those observed in Lake Erie occurring in Saginaw Bay (Lake Huron) (Nalepa *et al.* 1999). The zebra mussel affected the nutrient dynamics associated with algal blooms (Beeton 2002). Large blooms of blue–green algae were common in western Lake Erie before steps were taken to reduce phosphorus loadings, and these blooms ended in the 1970s and 1980s. However, a bloom of *Microcystis* occurred in western Lake Erie in 1995 and was probably a consequence of changes resulting from zebra mussel activity. The mussel can be selective in its feeding and can reject *Microcystis* (Vanderploeg *et al.* 2001); thus other algae are preferentially consumed leaving adequate nutrients for a *Microcystis* bloom (Beeton 2002). Dreissenids may have re-engineered nearshore transparency and phosphorus dynamics in a number of ways, stimulating benthic algal growth, including that of the nuisance *Cladophora*, which was a sentinel indicator of eutrophic conditions prior to phosphorus control (Hecky *et al.* 2004). The re-emergence of *Cladophora* to nuisance proportions has been reported across the lower Great Lakes, and suggests that dreissenids may be affecting another wave of coastal eutrophication. *Dreissena polymorpha* is abundant in shallow nearshore waters, and the

waters of the Great Lakes below the thermocline appeared to be inhospitable to these mussels (Dermott & Kerec 1997). However, a second dreissenid (*D. bugensis*) invaded the lakes and now thrives at all depths on any substrata suitable for attachment. These mussels may impact the amphipod *Diporeia hoyi* by removing algae essential for the growth of these. *Diporeia hoyi* is a major food of deep-water fishes (Dermott & Kerec 1997) and has declined in Lakes Erie and Ontario and in large areas of southern Lake Michigan (Nalepa *et al.* 2000*b*). In turn, reduction in *D. hoyi* populations probably led to declines in whitefish condition and weight-at-age (Great Lakes Environmental Research Laboratory 2002).

Many of the invaders during the 1990s were euryhaline Ponto-Caspian species resulting from a strong invasion corridor linking the St Lawrence Great Lakes with Eurasia (Ricciardi 2001; Reid & Orlova 2002). Apparently, ballast-water exchange, as now practised, has not been as effective as hoped in preventing invasions. Furthermore, Ponto-Caspian invaders have benefited from habitat alterations caused by the dreissenid mussels. The mussel beds and faecal deposits provide habitat and food for the amphipod *Echinogammarus ischnus*, another invader in Lake Erie. In turn, the juvenile round gobies feed on the gammarid, whereas the adults feed on the mussels. The Ponto-Caspian hydroid *Cordylophora caspia*, which is attached to mussel shells, feeds on mussel larvae (Ricciardi 2001). The invasion history of the St Lawrence Great Lakes can be explained by an invasion meltdown model (Ricciardi 2001), the rate of invasion accelerated by facilitative interactions such as mutualism, commensalism or habitat modification rather than antagonism (Simberloff & Von Holle 1999). The Ponto-Caspian species have become established in a number of habitats in the Great Lakes (Ricciardi & MacIsaac 2000; Vanderploeg *et al.* 2002); *Echinogammarus* became established, as did *Cercopagis pinjoi*, despite abundant competitors, which suggests the invasions were mediated by dispersal opportunity and resource availability. There have been no reported extinctions related to these recent invasions, although several native bivalves have been relegated to refugia.

The aforementioned interactions of zebra mussel, goby, *Echinogammarus* and hydroid suggest that Ponto-Caspian food webs have been transferred with the invasive species to the St Lawrence Great Lakes. In fact, exotic species now dominate the food webs of the lakes. Although negative aspects, especially the ecosystem-change effects of these invasions, have been emphasized, the density and taxonomic richness of macroinvertebrates have increased significantly in shallow waters where dreissenids are present (Ricciardi 2003). The reverse occurs in deep-water communities, probably since less food settles into deep water because the dreissenids have filtered out much of it close to shore (Ricciardi 2003). The lipid content of the deep-water amphipod *D. hoyi* has fallen since establishment of *Dreissena* spp. (Nalepa *et al.* 2000*a*).

No control has been found for zebra mussels and none is anticipated. Emphasis is on prevention of further invasions of non-indigenous species, especially through control of ballast water. A major recent concern is the Asian carp (genera *Mylopharyngodon*, *Ctenopharyngodon* and *Hypophthalmichthys*), which is established in the Chicago Sanitary and Shipping Canal near Lake Michigan. An electrical barrier was installed to halt the spread of this species into the Great Lakes (Stokstad 2003).

Lake Victoria is the only African Great Lake presently impacted by invasive species (Hall & Mills 2000). An early commercial fishery was dominated by two tilapiine cichlid fishes, with a lesser contribution from a variety of endemic species including hundreds of haplochromine cichlids. After gill nets were introduced in 1908, fishing pressure increased, catches declined and increasingly smaller mesh nets were used. Declining fish stocks led to the introduction of the Nile perch and tilapias (Lowe-McConnell 1994). The fishery now is largely for Nile perch, Nile tilapia and *Rastrineobola argentea*. About two-thirds of the endemic haplochromine cichlid fishes disappeared as the Nile perch population exploded (Lowe-McConnell 1994), but fish yields quintupled, and Lake Victoria has the largest freshwater fishery in the world by landings and market value, and a lucrative export market. With the Nile tilapias replacing the endemic tilapias, the food web changed, especially because the detritivorous atyid prawn *Caridina nilotica* replaced the cichlids. Blooms of cyanobacteria accompanied the loss of the phytoplanktivorous haplochromines (Beeton 2002). The invasive water hyacinth, which was introduced into African waterways in the twentieth century, continues to be a problem, although the introduced beetle *Neochetina* has reduced hyacinth standing cover by nearly 90%. The hyacinth has not been a large problem in Lake Malawi, presumably owing to its low nutrient content relative to Lake Victoria (Guilford & Hecky 2000).

Lakes Malawi and Tanganyika appear not to be threatened by invasive species at present, although Tanganyika species have been introduced elsewhere, including the planktivorous sardine *Lamnothrissa* into Lakes Kariba and Kiva (Hall & Mills 2000). This has provided a new pelagic fishery and has been beneficial socioeconomically; however, the fish competes with a characid fish and impacts the zooplankton.

Only few non-indigenous species occur in Lake Baikal (Beeton 2002), including the fishes *Percottas glehni*, *Coregonus peled* and carp, and the plant *Elodea canadensis*, the latter covering the bottom of harbours and shallow bays (Kozhova & Silow 1998) and its possible impact on coastal macrophytes being of concern (Hall & Mills 2000).

Water diversion

Probably the greatest long-term threat to lakes is the likely diversion of water away from them. This threat becomes greater as human populations continue to grow, new areas are developed and fresh water becomes scarce. However, despite the size and volume of large lakes, a surplus of water does not exist which can be diverted from their drainage basins without long-term degradation of their component ecosystems. The Aral Sea is a frightening example of the consequences of diverting water away from a large lake (Jorgensen *et al.* 2003; Chapter 7).

For the North American large lakes, water has been diverted from Lake Michigan via the Chicago Sanitary and Shipping Canal. This outflow has been offset by diverting water from the Hudson Bay drainage into Lake Superior. Lake Baikal and the African Great Lakes appear not to be impacted by diversions out of their basins, but even within-basin withdrawals from the African lakes would reduce their positive, albeit precarious, hydrological balances and convert them to closed basins (Bootsma & Hecky 1993). This would have severe consequences for downstream communities as the lakes are the headwaters of Africa's largest river systems.

POTENTIAL STATES IN 2025

The future states of the world's large lakes must be considered within the framework of growing human populations demanding fresh water. These demands must be dealt with in terms of citizens' environmental concerns, lifestyles and expectations. Consequently, socioeconomic dynamics within lake drainage basins and regions must be addressed, as well as political institutions. The situation for large lakes is usually complicated because two or more countries are within the drainage basins. Citizens of economically undeveloped countries are more likely to consider large lakes as a source of water and fish. Citizens in developed countries are more likely to emphasize recreational and leisure uses of lakes and to be concerned about quality of shoreline property and water, wetland preservation and pollution. Thus lake-front property in cities on the St Lawrence Great Lakes has been and is being converted from industrial, fishing and commercial uses to parks, restaurants and other leisure purposes (Chapter 5).

Large lakes suffer from past uses and abuses, the degradation being a consequence of human population growth within the lake basins. The St Lawrence Great Lakes were being seriously impacted 100 years ago. The African Great Lakes are experiencing extensive land-use change, pollution and overfishing, and as a consequence are undergoing change similar to that which affected the St Lawrence Great Lakes about 60 years ago. Lake Baikal is subject to land-use changes and pollution closely similar to those seen in the Lake Superior drainage prior to 1950.

Many schemes and plans will be developed to divert water from large lakes. This is not new; transboundary diversion is an established practice in the USA (Bulkley *et al.* 1984). When the US Congress was concerned about depletion of the Ogallala aquifer, it authorized a study of large-scale intra-basin diversion to replenish the aquifer. A preliminary engineering study proposed diverting 283 m^3 per second of water from Lake Superior. The cost to do this was estimated to be higher than the cost, at that time, for irrigation water (Bulkley *et al.* 1984). Increased interest in diverting water from the St Lawrence Great Lakes system to areas outside the basin, and consideration by state governors and provincial premiers of a regional approach to water management, led to the Great Lakes Charter of 1985. This specified management principles and procedures for the Great Lakes and, in 2001, Canada amended the Boundary Waters Treaty Act to prohibit bulk water removal from the Great Lakes and other boundary waters (Great Lakes Commission 2003). The 1985 Charter was reaffirmed in 2002, the Annex to it improving the ability to reject environmentally harmful proposals to use the Great Lakes waters.

Uncertainty exists over hydrology, especially groundwater and over-lake precipitation and evaporation, but it is recognized that better data and more accurate assessments of cumulative ecological impacts from multiple stressors are needed in order to deal with possible effects of water withdrawal and/or diversions. By 2025, a decision support system should be in place to aid sound management of the St Lawrence Great Lakes system (Great Lakes Commission 2003).

The total volume of the St Lawrence Great Lakes, Lake Baikal and the African Great Lakes is impressive, but in reality, little water is available if the physical, chemical and biological integrity of these large lakes is to be protected. Precipitation and evaporation data indicate the amount of water available (the outflow), which ranges from 2.7 km^3 per year for Lake Tanganyika to 220.02 km^3 per year for the St Lawrence Great Lakes. These outflows can not be reduced if present uses for navigation and hydroelectric power are to be maintained. The amount of precipitation on the lakes is close to the loss through evaporation, so most outflow comes from the drainage basins and water diversion has implications for the ecosystems of these basins. Withdrawing water in excess of the outflow volume endangers domestic, industrial, fisheries and other uses essential to the viability of riparian countries. A valid concern is whether water is to be viewed as a commodity or a common good by the World Trade Organization. At some time in the future, 50% of humans may suffer from freshwater scarcity; however, data on water supplies and uses for such projections are inadequate, even for the USA (Brown 2002).

The St Lawrence Great Lakes have shown some improvement with respect to eutrophication, and pollution by substances such as PCBs has declined. Overfishing is no longer a major issue, but invasive species remain a major threat. The full impact of *Dreissena* spp. is yet to be realized and species continue to enter the lakes. Restructuring of the shallow and deep waters will probably have occurred by 2025. By 2025, sea lamprey in the St Mary's River may be controlled, enabling the recovery of the lake trout fisheries in Lakes Huron and Michigan. The impacts of invasive species can be expected to decline as effective ballast-water control is implemented. A bloom of *Microcystis* in Lake Erie (Budd *et al.* 2001) and the re-emergence of nuisance blooms of *Cladophora* along many coastlines may indicate a second wave of eutrophication.

The Great Lakes Water Quality Agreements of 1972 and 1978 will greatly influence what the St Lawrence Great Lakes will be in 2025. The Agreements set targets for phosphorus loading (International Joint Commission 1989) with emphasis on control of phosphorus in sewage and detergents, and from non-point sources. The targets were achieved and surpassed by 1988, and some fisheries managers have suggested phosphorus levels are now too low (Gaden 1998).

Major aspects of the 1978 Agreement were persistent toxic substances (Annex 12), contaminated sediments (Annex 14) and airborne substances (Annex 15)

(International Joint Commission 2000). It is significant that the Agreement includes the philosophy of zero discharge and the goal of virtual elimination of toxic substances (Annex 12). Considerable progress has been made in controlling the point-source discharge of toxic substances (International Joint Commission 2000), but airborne pollutants will continue to be a problem well into the future. Stewart and Diggins (2002) have identified contaminated sediments and elevated metals as an ongoing anxiety in some Areas of Concern (AOCs: 43 badly degraded areas in Canadian and USA waters of the Great Lakes) (MacKenzie 1993), although some progress has been made in implementing Remedial Action Plans (RAPs), proposed in 1985 in order to have a logical sequence from problem identification to resolution (Wagner 1994) and based on an ecosystem approach including watershed planning and a broad spectrum of stakeholders. Thus, remediation efforts have reduced inputs of 18 priority toxic chemicals in the Niagara River (Wooster 2000).

The 1978 Agreement defined the Great Lakes Basin ecosystem as 'the interacting components of air, land, water, and living organisms, including man, within the drainage basin of the St Lawrence River at or upstream from the point at which this river becomes the international boundary between Canada and the United States' (International Joint Commission 1989, p. 5). This agreement was 'to restore and maintain the chemical, physical and biological integrity of the waters of the Great Lakes Basin Ecosystem' (International Joint Commission 1989, p. 7). It is significant that this ecosystem approach was adopted, the partial and piecemeal solution to problems being recognized as unsatisfactory (Christie *et al.* 1986). Much of the response to lake problems had previously lacked awareness of wider environmental concerns (Vallentyne & Beeton 1988) and/or addressing problems only when they became severe enough to cause concern; responses were emphasizing management of air, water, land and living resources, not integrating social, economic and environmental concerns (Vallentyne & Beeton 1988). Progress in dealing with problems such as the sea lamprey, eutrophication and reduction in persistent toxic substances was not generally due to an ecosystem approach, but to concerted efforts to deal with specific problems. The area in which the ecosystem approach appears to show promise is in developing and implementing RAPs to deal with AOCs (MacKenzie 1993). RAPs have proved a useful tool to deal with badly degraded sites and heightened interest may result in delisting some of the AOCs by 2025.

Overfishing, biodiversity conservation, eutrophication, land-use change and pollution are likely to continue to trouble the African Great Lakes to the year 2025. Changes in land use will continue to result from increased human population unless more efficient and productive agricultural methods are developed. The population increase is also dependent on small-scale farming, where more land is cleared, but traditional fallowing and limited use of chemical fertilizers and pesticides result in loss of productivity. Because the farmers cannot afford the chemical inputs, increasingly marginal land is brought into cultivation. Erosion will continue to increase and associated nutrients, heavy metals, sediments and other pollutants will contribute to eutrophication and the general degradation of the lakes. Pollution is probably the major long-term threat to Lake Malawi (Tweddle 1992) and will become a major problem for Lake Tanganyika as population and development increase in the basin. Because nutrient loading is primarily from non-point sources in Lake Victoria, achieving reductions will be more challenging than in the St Lawrence Great Lakes. Exploration for oil in Lakes Tanganyika and Malawi (Coulter & Mubamba 1993) is of concern because of potential oil spills and impact on water quality and the biota. Regional and local sources of pesticides dominate in the Lake Malawi area (Karlsson *et al.* 2000). Fortunately, there is the opportunity to control local sources as compared to airborne transboundary sources.

As with the St Lawrence Great Lakes, an ecosystem approach provides a rationale for dealing coherently with the problems affecting the African Great Lakes. Pollution, eutrophication and overfishing can be controlled, but to be effective for the long term there must be commitment and cooperation among citizens, industries and politicians in several countries (Beeton 2002). A major initiative may help regional cooperation (Schneider & Van Dijk 1997) and has already borne fruit in Lake Victoria, where the Lake Victoria Fisheries Organization was created by the riparian states. However, such effective regional mechanisms have yet to be formed for the other lakes. If the requisite commitment and cooperation do not occur soon, accelerated degradation of the lakes will result. Overall effective environmental management may depend on positive economic development in the African Great Lakes region where recent trends have been static or negative. A global issue that is impacting the African lakes is climate warming (Verburg *et al.* 2003); the deep water of the meromictic lakes is warming, becoming less productive and threatening fisheries (O'Reilly *et al.* 2003). This issue is clearly beyond the scope of the riparian countries to address and the global community must bear responsibility and assist in mitigating the consequences.

More productive agriculture, better land management and decline in human population growth by 2025 would help reduce siltation and eutrophication. Overexploitation of the fisheries is likely to continue, but it is hoped that cooperative management in all the African Great Lakes will reduce fishing pressure and slow the decline in biodiversity.

Preservation of the many endemic cichlid fishes has been a major concern locally and internationally because many of these fishes have already been lost. Steps will have to be taken soon if remaining stocks are to be preserved and degradation of their habitats prevented or at least lessened.

Freshwater protected areas (FPAs) have been proposed (Coulter & Mubamba 1993; Pitcher 1995); however, little attention has been given to their development in contrast to the marine environment (Houde *et al.* 2001; Chapter 16). FPAs would not provide instant replenishment of depleted stocks, but they are the simplest solution to managing complex fisheries in areas of overfishing and helping to protect spawning stocks; this may replenish fishing grounds and may thus increase catches in adjacent areas (Roberts & Polunin 1993).

Establishment of FPAs would be beneficial for any large lake, especially those with high endemism and exceptional biodiversity, such as Lakes Baikal, Malawi, Tanganyika and Victoria. Nearshore areas offer the greatest differences in morphometry, vegetation and habitat, and consequently greatest potential for significant fish production, particularly given that they are the areas most used by artisanal fishers.

It is important to have coastal-strip FPAs located at a variety of places around a large lake, to provide habitat diversity and meet the needs of multiple political jurisdictions. On practical and political grounds, it is best to encourage each bordering state to have at least one FPA.

Lake Malawi National Park at the southern end of the lake is currently the only fully protected aquatic area in Africa where fishing is prohibited in all waters within 100 m of the mainland park and included islands. Other FPAs will hopefully be established and protected in the African Great Lakes by 2025, meeting a critical need for covering endemic stocks.

Other than nearshore impacts and nearshore pollution changes, Lake Baikal is likely to remain relatively unchanged to 2025. Lakes Baikal and Superior are similar in their great sizes and water quality. Lake Superior has not experienced long-term changes in water quality (Beeton 1965) or suffered from invasive species to the same extent as the other St Lawrence Great Lakes, and Lake Baikal will probably have a similar future (Beeton 2002). The major threat to the biota and human populations in the region are the organochlorine compounds occurring in water, fish and seals, especially in the southern basin, and the endemic flora and fauna also remain vulnerable to invasive species, especially if resources become overexploited (Grachev 1991).

In Russia, the unique importance of Lake Baikal is recognized, and steps have been taken to protect the lake. The Lake Baikal Coastal Protection Zone was established in 1987 (Jorgensen *et al.* 2003) and tree-felling near to shore ceased in 1986, as well as the practice of log rafting. A Comprehensive Scheme for the Protection of Nature in the Area of Lake Baikal was adopted in 1989, providing for a protective zone around the lake, and control of waste, industry and sub-coastal logging. The Baikal Commission established in 1993 coordinates the protection activities of the federal and regional governments. There are also several local laws and regulations in place to help protect the lake.

Such future scenarios are all subject to uncertainties derived from global and regional climatic change patterns. Efforts have been made to define possible future climate conditions for the St Lawrence Great Lakes region (Quinn & Croley 1999; Lofgren *et al.* 2000). One model, based on Canadian Centre for Climate Modelling and Analysis data, predicts decreases in lake levels of up to 1.38 m for Lake Michigan by 2090 as a consequence of decreased precipitation, and increased air temperature and evaporation. Data from the UK Meteorological Office's Hadley Centre predicts an increase in lake level of 0.35 m for Lakes Huron and Michigan by 2090 as a result of increased precipitation and lower air temperature. Ice cover would be reduced according to the Canadian model (Lofgren *et al.* 2000). Lake warming would affect the thermal structure of all the lakes. Such changes could reduce vertical mixing and possibly result in lower dissolved oxygen concentrations in benthic waters (Quinn & Croley 1999). This could in turn affect sediment chemistry and facilitate the release of toxic substances (Mortimer 2004). Changes have occurred in the freeze and break-up dates of ice in Grand Traverse Bay (Lake Michigan), Toronto Harbour (Lake Ontario) and Lake Baikal over the period 1846–1995 (Magnuson *et al.* 2000).

Climate warming has resulted in a sharpened density gradient, which has slowed vertical mixing and productivity in Lake Tanganyika and probably other deep tropical lakes (Verburg *et al.* 2003). Consequently, rather than enhancing productivity, climate warming in these cases has reduced productivity and altered eutrophication.

CONCLUSIONS

Land-use change, eutrophication, pollution, over-exploitation of resources, invasive species and diversion of water will continue to be major impacts on large lakes. Added to these are the little-known consequences of climate change for water quantity and quality, thermal structure, circulation and biota.

Land-use changes will continue to adversely impact Lake Baikal and the African Great Lakes, by accelerating eutrophication and pollution from erosion and the chemicals used in agriculture. Land-use changes in the St Lawrence Great Lakes basin are likely to have both beneficial and negative impacts, as wetlands are protected and increased attention is given to mitigation of non-point-source pollution. Overfishing is not likely to affect the St Lawrence Great Lakes, but it will continue to be a critical issue for the African Great Lakes, especially overfishing of the Nile perch. A major threat to all the lakes is species invasion. The St Lawrence Great Lakes continue to be invaded by non-indigenous species, mainly through ballast-water transport, and problems with existing invasive species will persist. Steps will probably be taken to conserve endemic fishes through FPAs in the African Great Lakes. Deforestation, cultivation and pollution are likely to favour non-indigenous species already present in nearshore areas of Lake Baikal. Considering the similarities of Lakes Baikal and Superior, by 2025 the only significant changes in Lake Baikal are likely to be in nearshore areas.

The greatest long-term threat to the world's large lakes will derive from their attractiveness as sources of fresh water; however, little surplus water is to be had from the large lakes considered here. For those lakes where precipitation and evaporation are similar, the only surplus water is the outflow, and this will be unavailable if present uses are maintained. Should water be viewed as a commodity or a common good?

Canada and the USA have a long history of cooperation in dealing with the many problems in using and preserving the St Lawrence Great Lakes. The Great Lakes Water Quality agreement included reduction in phosphorus loading, philosophy of zero discharge and virtual elimination of persistent toxic substances. Riparian countries need to cooperate to deal with the many problems involved, to develop and implement institutional arrangements based on sound knowledge of the ecosystem. Research is under way on Lake Baikal and the African Great Lakes, but, unless institutional arrangements are negotiated and implemented for the African Great Lakes, more endemic fish will disappear, overfishing will continue, pollution and eutrophication will accelerate, and unregulated land use will continue, resulting in seriously degraded lakes.

7 · Salt lakes: values, threats and future

ROBERT JELLISON, WILLIAM D. WILLIAMS[†], BRIAN TIMMS, JAVIER ALCOCER
AND NIKOLAY V. ALADIN

INTRODUCTION

Two types of saline water exist on Earth, namely marine waters, including brackish zones of mixing with fresh water, and epicontinental salt lakes. This chapter addresses the values, threats and likely future of salt lakes, which are here defined as permanent or temporary bodies of water with salinities >3 g per litre and lacking any recent connection to the marine environment (i.e. athalassohaline

sensu Bayly 1967). While the use of 3 g per litre to demarcate salt lakes is somewhat arbitrary, it has come into general use and is, coincidentally, the 'calcite branch point'; that is, the salinity at which calcite is precipitated as natural waters concentrate.

The false dichotomy between marine and fresh waters embodied in the titles of institutions, textbooks and conferences obscures the fact that salt lakes are widespread and

Aquatic Ecosystems, ed. N. V. C. Polunin. Published by Cambridge University Press. © Foundation for Environmental Conservation 2008.

Fig. 7.1. Geographical distribution of major areas with saline lakes. Note that saline lakes also occur outside these areas, but at lower density. (From Williams 1998*a*.)

represent significant aquatic resources. They occur on all continents, including Antarctica (Fig. 7.1), under a range of climatic temperature regimes from the coldest to near hottest. Locally, they may be more abundant than fresh waters and, when they are, often dominate the landscape. While they are mostly confined to semi-arid (precipitation 200–500 mm per year) and arid (25–200 mm per year) regions of the world, these regions constitute about one-third of the Earth's total land area (Williams 1998*a*) and inland salt waters are not greatly lower in total global volume (85 000 km^3) than freshwater lakes (105 000 km^3) (Shiklomanov 1990). Salt lakes include the largest lake on Earth, the Caspian Sea (area 374 000 km^2), many other large lakes, lakes at the highest altitudes for any lake (>3000 m above sea level on the Altiplano of South America and in Tibet), as well as the lowest lake on earth, the Dead Sea, at about 400 m below sea level (Williams 1996). Nearly 75% by volume of inland salt waters are in the Caspian Sea. If the 15 largest lakes from both fresh and salt categories are omitted from the comparison (see data from Herdendorf 1990), salt lakes still constitute about a third of inland waters both by area and volume.

Williams (2002*a*) presented a broad overview of the development of salt lakes, threats to salt lakes, their special management requirements and their likely future status in 2025. Here, this analysis is revisited with special emphasis on their values, anthropogenic impacts and the potential for conservation in the twenty-first century. While most of the myriad impacts have long been recognized as important for individual salt lakes, their global extent, and the common trend and rapidity of degradation of a significant and unique component of the biosphere, have not been fully appreciated. Only clear recognition of the extensive damage that salt lakes are now undergoing, and of the likely result that such damage will lead to within the next 20 years, offers any hope that present trends can be changed.

VALUES OF SALT LAKES

Salt lakes encompass a diverse array of aquatic ecosystems with considerable economic and non-economic value. Among these are important scientific, ecological, conservation, cultural, recreational, aesthetic and economic values (Williams 1993*a*, 1998*a*).

Scientific value

The diversity and special characteristics of salt lakes make them of particular interest to many scientific disciplines including ecology, physiology, evolutionary biology, palaeolimnology, hydrology, geochemistry, physical limnology, microbiology and ecosystem modelling. Salt lakes range from 3 to ~300 g per kg total dissolved solids with many regional differences in their chemical composition, in contrast to fresh waters, which are mostly dilute calcium bicarbonate systems. Large regional differences in ionic composition are present due to differing catchment geology and age, and water balance; *in situ* processes determine the evolving chemical composition and salinity within individual lakes. Thus, lakes in adjacent or nearby endorheic basins (a watershed from which there is no water outflow) may vary considerably, providing unique opportunities for comparative studies.

Many early studies focused on describing the salt-tolerant biotic communities and their special adaptations to extreme environments, and this remains an active area of research for both physiologists and evolutionary biologists. The extreme environment and ranges of conditions presented by many salt lakes are of special interest to microbial researchers (e.g. Oren 2002) and studies of biogeochemical cycling. Also, hypersaline soda lakes may provide analogues to early Earth or Martian environments.

The interest of palaeolimnologists and climate researchers stems from the sensitivity of these environments to climatic and tectonic change. The climate record stored in the sediments of widely distributed salt lakes is proving instrumental in assessing past regional climate changes (see Servant-Vildary 2001; Benson *et al.* 2002).

The complex patterns of density stratification resulting from seasonally varying hydrology and thermal stratification provide unique opportunities for physical limnologists and the development of hydrodynamic models (for example MacIntyre *et al.* 1999). Their low species diversity and trophic complexity make these ecosystems attractive candidates for plankton dynamic modelling studies.

While all these different aspects of the scientific study of salt lakes are active, they are as yet not fully realized and salt lakes are still underrepresented in the scientific literature relative to their abundance. Queries of the scientific literature (ISI Web of Knowledge) show that publications with 'salt' or 'saline' as topic words constitute only 2–5% of those with 'lake' or 'lakes' as topic words during the past four decades. However, increasing awareness of the unique characteristics, values, widespread occurrence and threatened status of salt lakes are likely to result in increased scientific study in the coming decades.

Ecological, conservation and cultural values

Where salt lakes occur, they often constitute the dominant aquatic resource and thus are critical components of the natural environment. Many are among the most productive aquatic ecosystems (Melack & Kilham 1974; Jellison & Melack 1993), containing unique communities with endemic species. They are perhaps best recognized for the critical role they play for migrating and breeding waterbirds. While the critical dependence of flamingo populations throughout the world on salt lakes is widely recognized, many other waterbirds are equally dependent on salt lakes, including many species of phalaropes, grebes, gulls, pelicans and plovers. For instance, the Great Salt Lake (USA) has the world's largest staging concentrations of Wilson's phalaropes (*phalaropus tricolor*), the second largest staging population of eared grebes (*Podiceps nigricollis*), the largest breeding populations of snowy plovers (*Charadrius alexandrinus*), white-faced ibis (*Plegadis chihi*) and California gulls (*Larus californicus*), the third largest breeding population of American white pelicans (*Pelecanus erythrorhynchos*), and populations of American avocets (*Recurvirostra americana*) and black-necked stilts (*Himantopus mexicanus*) many times larger than other wetlands on the Pacific flyway. Mono Lake at the western edge of the Great Basin (USA) has the largest staging population of eared grebes, and a large population of migrating phalaropes and breeding gulls. Many other species of waterbirds use both fresh and salt lakes and thus depend on salt lakes in the absence of fresh water. The Ramsar Convention (International Treaty on Wetlands of International Importance) was initially conceived as a means of protecting critical bird habitat. Of approximately 1150 sites listed as wetlands of international importance, 150 contain salt lakes.

Many salt lakes have special significance for traditional and native cultures. For instance, in the western USA, Zuni Lake is a sacred salt-gathering site for several native American tribes, and Pyramid Lake, with its endemic and endangered Lahonton cutthroat trout (*Oncorhynchus clarki henshawi*), is also of special cultural significance. Throughout the world, artisanal salt-making or gathering is associated with hypersaline lakes and traditional fisheries occur in less saline lakes. The Dead Sea is of particular cultural significance to several world religions.

Recreational and aesthetic values

As the dominant aquatic habitat in many arid and semi-arid landscapes, birds and other wildlife congregate in large abundance at many salt lakes, making them popular tourist destinations throughout the world. Thus the lakes of the Eastern Rift Valley in Central Africa (e.g. Bogoria, Nakuru and Elmentia), and in the western USA Mono Lake, the Great Salt Lake and the Salton Sea, are all popular tourist destinations. Many lesser-known salt lakes throughout the world are increasingly becoming sites of ecotourism.

Economic values

Salt lakes have long been mined for a suite of precipitate minerals including sodium and potassium chlorides, borates, nitrates, sulphates of potassium, magnesium, calcium and sodium, calcium and sodium carbonates, and lithium chloride.

Salt-lake biota of commercial value include algal products (for example *Spirulina*), *Artemia* brine shrimps and, in lakes of low salinity, fisheries. Both frozen and dried, *Artemia* and their cysts are important to aquaculture, and large commercial enterprises have developed at major salt lakes. At Great Salt Lake, the mean *Artemia* cyst harvest from 1990 to 1996 was 1.8×10^6 kg dry mass annually with a value of US$35–110 million (Wurtsbaugh & Gliwicz 2001). In less-saline lakes, commercial and recreational fisheries are important. Of particular note is the Caspian Sea sturgeon fishery (Katunin 2000).

THREATS AND IMPACTS

Overview

Environmental impacts on the natural character of salt lakes include almost all human activities that threaten or have already had adverse effects on freshwater lakes (Chapters 5 and 6). An exception is acidification, an impact largely confined to freshwater lakes in the northern hemisphere fed by rain acidified from distant industrial emissions (Likens & Borman 1974; Chapter 5). However, salt lakes are also threatened or already impacted by other human activities less important for freshwater lakes, namely diversions of surface inflows, salinization and mining (Williams 1993*a*), and they are more susceptible to impacts from global climate change than freshwater lakes.

Global climate change

The size and salinity of salt lakes are particularly sensitive to even small changes in any component of their hydrological budgets. Unlike freshwater lakes, where minor changes in inflows, precipitation or evaporation are primarily transmitted into reduced outflows (Chapters 5 and 6), in salt lakes that lack surface outflows, the same changes are quickly reflected in changed size and salinity. While the overall effect of global warming will be a general quickening of the hydrological cycle (evaporation/precipitation), it is the combined effect of increased evaporation due to higher temperatures and the predicted large-scale regional changes in runoff and precipitation (IPCC [Intergovernmental Panel on Climate Change] 2001*d*) that will have profound consequences for salt lakes.

Global climate modelling has not developed to a state where regional trends in precipitation and runoff can be predicted with confidence. Two versions of a Hadley Centre global climate model (GCM), employing slightly different but reasonable assumptions (IPCC 2001*d*, Fig. SPM-4), illustrate this point through the large divergences of predicted runoff that occur between the two models in many regions (for example north-western USA). While site-specific predictions may be uncertain, large regional differences from current conditions are universally predicted (IPCC 2001*d*).

Of particular interest here are predicted changes in the large, endorheic basins of the world that are rich in salt lakes (Fig. 7.1). Both versions of the Hadley GCM predict decreased runoff through much of Australia. In this largely arid country, episodic lakes will fill less frequently and dry more quickly, intermittent (seasonally filled) lakes will be of shorter duration and smaller extent, and permanent salt lakes will shrink and become more saline. In situations where temporary salt lakes remain drier for longer, the biota will increasingly include species with strong abilities to disperse and tolerate long periods of desiccation, as is already the case in episodic salt lakes; in the situation where permanent salt lakes become more saline, the biota will decrease in diversity, in line with the general inverse correlation in salt lakes between salinity and biodiversity (Hammer 1986).

Other broad areas with predicted decreases in runoff include parts of arid central Asia and the Middle East, broad swaths of Africa and the Altiplano region of South America. Any effects of decreased runoff and precipitation in salt lake regions will be compounded by the fact that most climatic models predict increases in temperatures throughout much of these same semi-arid and arid regions (Hammer 1990; IPCC 2001*d*). In the Aral Sea region, for example, temperature rises of up to 6 °C by 2100 are forecast.

Of course, in areas of increased runoff and precipitation salt lakes may grow larger if inflows are not diverted. In contrast to the small (0.03–0.14 m: IPCC 2001*d*) rises in global mean sea level predicted between 1990 and 2025, lake-level rises due to regional climate change may be rapid and exceed the institutional capacity of most countries to respond quickly enough to avoid major economic losses. Within the past few decades, the levels of the Caspian Sea, Lake Van (Turkey), Great Salt Lake and Devil's Lake (USA) have already risen significantly and caused major economic losses due to inundated shore areas (Kadioğlu *et al.* 1997; Ozturk *et al.* 1997; DWR [Division of Water Resources] 2003).

The level of the Caspian Sea rose 2.5 m from 1978 to 1996, the rapid rise damaging industrial and agricultural facilities and inhabited buildings along the shore. The north-west coast of the Caspian (Atirau and Mangistau areas of Kazakhstan) suffered the most; the coastline moved up to 70 km onshore in the Atirau area, flooding *c*.1 million ha of previously agricultural land; the total cost of the damage was *c*. US$150 million (Ozturk *et al.* 1997). Flooding of oil production and transportation facilities resulted further in localized oil spills, which had environmental impacts.

Lake Van in Turkey, the 15th largest lake in the world by volume (600 km^3), rose approximately 2 m between 1985 and 1995, inundating coastal works and low-lying agricultural areas (Kadioğlu *et al.* 1997).

After a century-long decline of 6.4 m (1870–1960), the level of the Great Salt Lake (USA) returned to its historic high in just 27 years and caused substantial damage and economic loss (US$250 million) (DWR 2003). Pumps and evaporation basins were installed at a cost of US$52 million to curtail the rise. An equally rapid decline of 4.6 m over the next decade occurred (1987–1997) and the lake continued to shrink during several years of a drier climatic regime. In North Dakota (USA), the level of Devil's Lake rose by nearly 8 m over the seven years from 1993 to 1999 necessitating costly flood control and mitigation efforts (Todhunter & Rundquist 2004).

These four cases illustrate the potential economic impacts of lake-level changes that will be associated with the predicted widespread changes in regional precipitation and temperature caused by global climate warming.

The level of Lake Issyk-kul in the Tien Shan Mountains (Kyrgyzstan) recently rose 26 cm in four years (Kirby 2002), a remarkable change given that this lake is the 10th largest in the world by volume. The change is attributable to climate warming and increased precipitation, part of the rise also reflecting melting of glaciers. This example suggests regional water resources of the newly independent central Asian republics may change markedly during this century due to global climate change. As many of the major watersheds in this region are transnational, these changes are likely to result in significant political tensions and make basin-wide water management a difficult necessity.

Changes in ultraviolet radiation associated with a decrease in the concentration of ozone in the upper strata of the atmosphere (Williams 1998*b*) are potentially more injurious to salt than freshwater lakes. Ultraviolet radiation (UV-B) is rapidly absorbed in the upper layers of water in lakes but, if lakes are shallow, as many temporary salt lakes are, little absorption can occur before the biota are affected. Lake Cantara South, a shallow salt lake in South Australia, is one example of many that would likely be affected in this way.

In addition to changes in temperature and runoff/ precipitation patterns, salt lakes will be affected by the suite of impacts from global warming identified for freshwater lakes (Magnuson *et al.* 1997; Poff *et al.* 2002), including changes in timing and depth of seasonal mixing, increased water temperatures and productivity, and in some cases bottom-water anoxia.

Surface inflow diversions

Global population increase and economic development in arid and semi-arid regions of the world have led to increasing freshwater scarcity to a degree which may become a crisis early this century (Shiklomanov 2000). While global climate change will necessarily impact salt lakes throughout the world, the most serious impact to permanent salt lakes is likely to be from continuing and increased diversions for irrigated agriculture. Since diversions alter the hydrological budget, and salt lakes respond quickly to such alterations, inflow diversions invariably cause a rapid decrease in lake volume and the physical and chemical features contingent upon volume, especially salinity (Williams 1993*a*). Salt lakes throughout the world are being desiccated due to increases in irrigated agriculture. Apart from several well-known examples (for example the Aral Sea, the Dead Sea, and Mono Lake in

California), the worldwide extent of salt lake desiccation has received scant attention.

CENTRAL ASIA AND NORTHERN CHINA

Inflow diversions have been greatest and had the most profound environmental effects for the Aral Sea (Micklin 1988, 1998; Williams & Aladin 1991). Prior to 1960, the annual volume of inflows from the rivers Syr and Amu Darya was 56 km^3; after diversions for greatly expanded irrigated agriculture, the annual average inflows in the decades that followed were 43 km^3 (1961–70), 17 km^3 (1971–80) and 4 km^3 (1981–90) (Letolle & Mainguet 1993). These massive diversions of water used particularly for cotton and rice cultivation, led to a >15-m drop in the surface level of the Aral Sea after 1960 and an increase in salinity from 10 to 28–30 g per litre when the sea split into two basins in 1989. The water level of the Large (southern) Aral Sea has continued to fall and salinities increased to ~70 g per litre by 2002. In 1960, the fish fauna consisted of over 20 native and introduced fishes while the invertebrate community included >200 species. After parting into the Northern and Large Aral Seas, only seven species of fish, 10 common zooplankton species and 11 common benthos species were present (Plotnikov *et al.* 1991). Increased salinity of the Large Aral Sea has resulted in complete elimination of the fishes and 11 of the invertebrate zooplankton species; only the widely euryhaline rotifer *Brachionus plicatilis* has survived. However, three new halophylic species appeared apparently through aeolian transfer, namely the cladoceran *Moina mongolica*, the brine shrimp *Artemia salina* and the infusorium *Fabrea salina*. Of 10 zoobenthic species, only two euryhaline species of gastropods (*Caspiohydrobia* spp.) and one euryhaline ostracod (*Cyprideis torosa*) remain.

In addition to the total collapse of the Aral Sea fisheries and the towns dependent on them for livelihood, a whole suite of environmental impacts has been associated with desiccation of the Aral Sea including toxic dust-storms originating on the exposed lake bottom, deterioration of the large deltaic ecosystems of the Syr and Amu Darya, drier and more extreme climate, water-table lowering and increased desertification (Micklin 1988). Furthermore, inefficient irrigation practices have led to severe salinization problems throughout the region with 10–15% of irrigated lands in the Kyzyl Orda Oblast being rendered unsuitable for agriculture each year (UN Country Team 2004).

Irrigated agriculture has led to the decline of salt lakes throughout the arid regions of central Asia from Tuz Lake in Turkey to northern China. Kazakhstan's two Ramsar sites (lakes of the Lower Turgay and Irgiz, and Tengiz Lake) are both threatened by water diversions (Wetlands International 2002). Also, Kazakhstan's largest lake, Lake Balkash, has shrunk during the past 20 years, presumably due to upstream agricultural diversions. In northern China, the desiccation of lakes has been especially pronounced (Tao & Wei 1997). In the 1950s, there were 52 lakes over 5 km^2 in area in Xinjiang province, with a total area of 9700 km^2, but the area had decreased to 4700 km^2 by the early 1980s. Lop Nur (3000 km^2) dried in 1964, Lake Manas (550 km^2) in 1960, Lake Taitema (88 km^2) in 1972 and Lake Aydingkol (124 km^2) in the 1980s. Lake Ebinur (1070 km^2) and Lake Ulungur (745 km^2) have respectively been reduced to one-half and one-tenth of their original size since the 1950s. On the Alxa Plateau of Inner Mongolia, Gaxun Nur Lake (262 km^2) dried in the 1970s and the Sogo Nur Lake in the 1980s. In Hubei Province of China, the number of lakes over 0.5 km^2 in area has decreased from 1066 to 309.

AUSTRALIA

Lake Corangamite in south-western Victoria (Australia) has shrunk due to diversion of its major inflowing stream (Williams 1995). By 2002, salinity exceeded 100 g per litre, with marked reduction in its biotic diversity. Although restoration of this Ramsar site is possible and an enquiry has been held, its desiccation has continued. Lake Eyre would have been affected by irrigation plans in its Queensland catchment, but these plans were dropped. Lacking other large terminal salt lakes, the threat of diversions is less in Australia than in other endorheic regions of the world.

However, salinization is a major problem in agricultural areas, and already 30% of such lands are degraded, mainly in south-western Western Australia and in the southern Murray–Darling Basin (Chapter 4). Nationwide, 80 important wetlands are already affected and this number is predicted to rise substantially by 2025. In the wheat belt of Western Australia, 75% of waterbirds have declined in abundance and some 200 aquatic invertebrates are likely to become regionally extinct. The salt waters, both in modified freshwater wetlands and in new sites, are even more depauperate than natural salt waters, and in some areas are being invaded by alien *Artemia* sp., instead of the endemic *Parartemia* spp. However, in the Murray–Darling Basin, some of the evaporation basins, provided to rid the land of salty irrigation wastewater, support many waterbirds.

AFRICA

The demand for fresh water is expected to increase markedly throughout Africa due to increased human population and economic development. In addition to the 14 countries already considered water stressed (<1700 m^3 of renewable water resources per caput), another 11 countries are expected to become water stressed by 2025 (Clark 1999; Johns Hopkins 1998). While some of the most spectacular salt lakes in the East African Rift Valley (for example Bogoria or Nakuru) are not currently threatened by diversions, others have shrunk due to diversions. For example, Lake Abe on the border of Ethiopia and Djbouti has shrunk by 67% since the 1930s, as a result of irrigated agriculture in Ethiopia (UNEP [United Nations Environment Programme] 2000a). In this and many other cases, the transnational character of major river basins throughout Africa makes integrated basin-management difficult. With only 6% of its cropland currently under irrigation, water diversions for irrigated agriculture will certainly increase in all of the endorheic basins. In Malawi, the large shallow saline Lake Chilwa provides one-quarter of the fish caught in Malawi but faces threats of water abstraction within the basin and reclamation of nearby swamps for agriculture or irrigation reservoirs. These swamps are particularly important as fish refugia in times of low lake levels during dry periods when the lake nearly dries.

MIDDLE EAST

Much of the Middle East consists of arid and hyper-arid regions, the dominant surface water resources being the Jordan, Tigris and Euphrates Rivers. The Arabian Peninsula is completely lacking in major river systems and contains very few permanent salt lakes. Salt lakes in the interior consist mostly of episodically flooded *sabkhas* (salt flats), thus water diversions are not economically feasible. However, the level of the Dead Sea at the terminus of the Jordan River has dropped over 13 m since 1981 (Oren 2002) because of upstream diversions in Jordan and Israel for irrigated agriculture. As the Dead Sea has been extremely hypersaline for many centuries, the loss of ecological resources associated with its desiccation is much less than that observed when less saline lakes are desiccated. However, the lake has significant cultural and recreational values; several proposals involving diversions from the Red Sea have thus been made (Beyth 2002) and may be acted on. Iran contains a number of large and productive salt lakes, of which Lake Urmia (483 000 ha) is the most notable with its large populations of pelicans and flamingos and *Artemia* fishery (Scott 1995). While the Mahabad Multipurpose Drainage and Irrigation Project reduced freshwater discharge into the marshes at the south end of the lake in the early 1970s, the overall effect was lessened by return flows. A more serious threat is likely to be pollution from nearby large cities and toxic chemicals used in agriculture. In eastern Iran, Hamun-e-Saberi and Hamun-e-Helmand, two large semi-permanent lakes, spanning from fresh to hyposaline, have been impacted by diversions for irrigated agriculture on streams in neighbouring Afghanistan. Afghanistan has only a couple of notable salt lakes, one of which (Ab-i Istada) is threatened by diversions.

SOUTH AMERICA

In South America, salt lakes lie predominantly in the Bolivian Altiplano, its northern Peruvian extension and the pampas of Argentina. The former contains many ephemeral saline lakes and playas (*salars*) (Hammer 1986) that are not suitable for development of irrigated agriculture. However, Lake Poopo at the terminus of the Desaguadero River, which flows from Lake Titicaca, may be impacted by planned water diversions. In Argentina, the large and variable Mar Chiquita (7000 km^2 in 1987) and the smaller Lagunas y Esteros del Ibera (245 km^2), both Ramsar sites, are threatened by diversions. However, at present most water development in South America is focused on the large exorheic river systems.

NORTH AMERICA

All large salt lakes in the Great Basin of the USA, except the Great Salt Lake, have experienced marked declines because of diversions for irrigated agriculture; these include Mono, Walker, Pyramid, Owens and Winnemucca. The large salt lakes Owens and Winnemucca were completely desiccated in the twentieth century. Before inflow diversions led to its demise, Winnemucca Lake was about 40 km long and 5 km wide and had a salinity of 3.6 g per litre in 1884 (Clark 1920). The lake dried following diversions from the Truckee River and is now merely a flat expanse of dry land next to Pyramid Lake. Owens Lake was about 24 km long, 16 km wide and 10 m deep. Between 1890 and 1914, its recorded salinity ranged from 16 to 214 g per litre (Clarke 1920). Beginning in 1913, diversions of water from the lake to provide domestic supplies to Los Angeles led to its complete desiccation by 1926. The exposed lakebed is one of the single largest sources of particulate aerosols in the USA (Cahill *et al.* 1996), and approximately US$100 million has been spent on mitigation measures aimed at controlling the dust. Mitigation includes shallow ponds, event-oriented sprinklers and establishment of salt grasses.

The level of Walker Lake (Nevada) fell by 40 m from 1882 to 1996 and salinity increased from ~3 to 13 g per litre as it shrank from 280 to 140 km^2 (Beutel *et al.* 2001). Following a brief respite in 1995 as a result of above-average runoff, it continued to decline. In 2004, salinities exceeded 15 g per litre (R. Jellison, unpublished data 2004), and these are considered to exceed the critical threshold for successful reproduction of tui chub (*Gila bicolor*), the primary prey of the endangered Lahontan cutthroat trout. Thus, the recreational fishery may collapse in the near future.

Following the initiation of water diversions in 1941, Mono Lake's level declined by 14 m and its salinity doubled from 48 to about 95 g per litre. At Pyramid Lake, water levels have fallen by about 21 m since 1910 and its salinity increased to 4–5 g per litre. However, both these lakes have management plans in place to prevent further desiccation (see below).

In all of these lakes around the world, increasing salinities have significantly altered the biotic communities. The biological effects of increased salinities depend largely upon the original salinity. They have usually been greatest when the original salinity was low, and least when it was hypersaline. Thus, the effects of the 20 g per litre increase in Lake Corangamite were significant and led to the almost complete disappearance of fish, amphipods, snails and *Ruppia* (ditch grass, Potamogetonaceae), with consequent effects on the associated avifauna that fed on the lake (Williams 1995). Any further increases in the salinity of Walker Lake are expected to lead to collapse of the recreational fisheries. Conversely, the >100 g per litre increase in the Dead Sea had little effect on the lake biota and fundamental processes (Williams 1993*b*). With few exceptions, it is expected that permanent salt lakes with defined surface inflows will be impacted seriously during the developing freshwater crisis in the arid and semi-arid regions of the world.

The effects of falling water levels are not restricted to gross chemical and biological effects; many other physicochemical and environmental changes also follow. They may include changes to the local climate, additional dust blown from exposed lakebeds, falling groundwater levels and the loss of islands, and consequently other effects, as for example in the Aral Sea (Letolle & Mainguet 1993).

Groundwater pumping

Alone or in conjunction with surface diversions, groundwater pumping for agricultural purposes threatens many shallow salt lakes that are essentially surface 'windows' of shallow water tables (Williams 1993*a*). While in the past, this impact has generally been local in nature, increased groundwater pumping in many arid regions during the past several decades has resulted in greatly lowered water tables over large areas (Chapter 3) with the concomitant desiccation of salt lakes. This is particularly true in northwest China, the Middle East and Mexico. Most of the shallow permanent and temporary salt lakes in central Mexico have already disappeared because of over-pumping of groundwater for irrigation, and other deeper lakes have shrunk rapidly (Alcocer & Escobar 1990). In north and central Mexico, four crater lakes in Valle de Santiago (Guanajuato) have undergone rapid desiccation (Alcocer *et al.* 2000). San Nicolás de Paranqueo and Cíntora are already dry (the former in 1979 and the latter sometime between 1980 and 1984), while Rincón de Paranqueo and La Alberca are nearly dry. The original water level in Rincón de Paranqueo and La Alberca shows the lakes to have been around 50 m deep. La Alberca was 35 m deep in 1985, 10 m deep in 1995 and only a few centimetres deep in 2002. Rincón de Paranqueo was 7.5 m deep in 1995 and is now almost dry. Many temporary salt lakes in central Spain are similarly threatened or affected.

In Algeria, *foggaras* (unique systems of subsurface irrigation conduits totalling 1377 km in length), associated oases and salt *sabkhas* are threatened by lowered aquifers because of the large volume of groundwater pumped.

The greatest threat to salt lakes from groundwater pumping may be the increased demand on surface waters as aquifers are depleted. A significant portion of increased world agricultural output over the past 50 years has depended on over-drafting of groundwater aquifers in countries such as China, India, USA, Pakistan, Mexico, Iran, South Korea, Morocco, Saudi Arabia, Yemen, Syria, Tunisia, Israel and Jordan (Chapters 1 and 3). Decreases in agricultural output may occur as major aquifers underlying newly developed agricultural areas in these countries become depleted; signs of this are already present, especially in China (Brown 2003).

Secondary salinization

While one of the effects of surface-inflow diversion from large salt lakes is inevitably increase in the lake salinity, clearance of the natural vegetation and other land-use changes within catchments also increase salinity (Williams 2002*b*). The subsequent salinization involves the mobilization of salts dissolved in groundwater. The salts move towards the surface as the water table rises when the amount of groundwater transpired by deep-rooted plants falls (or after the addition of excess irrigation water to groundwater), and once near the surface, capillary action brings them to the surface. There, evaporation leads to salt deposition. Leaching of deposits, if within the catchment of a salt lake, adds to the natural salt inflows to the lake. This process is referred to as secondary (or anthropogenic) salinization to distinguish it from the process of salinization involved in the natural development of salt lakes. The impacts of secondary salinization are not confined to salt lakes, but are a major threat to all natural water resources in semi-arid and arid regions of the world (Williams 1999, 2001, 2002*b*).

The threat of secondary salinization appears largely to have been underestimated in most dry land countries, Australia being an exception. The extent to which inland waters have already been altered by additional salt inflows is uncertain, but Gleick (1993*b*) estimated that globally, *c*.10 million km^2 of land have already been affected. Secondary salinization has disturbed the natural hydrological and salt cycles in many arid regions, with many salt lakes becoming more saline, and many freshwater lakes turning saline (Williams 2001). In addition, a large number of unnatural salt lakes (so-called evaporation ponds or discharge basins) have been constructed in irrigated areas as basins into which agricultural saline wastewater is discharged (Evans 1989).

The Salton Sea (California, USA) provides an unusual example of salinization of an episodic salt lake now being used as a discharge basin. Deltaic geomorphological processes have caused the Colorado River to alternate outflows between the Salton Basin and the Gulf of California over the past 10 000 years. The basin was dry in 1905 when floods breached a dyke and redirected the entire flow of the Colorado River into the basin, creating the modern Salton Sea. Following the repair of the breach in 1907, the lake shrank due to evaporation, and salinity increased to ~40 g per litre by the 1920s, at which time wastewater from irrigated agriculture largely stabilized the lake level for the remainder of the twentieth century (Schroeder *et al.* 2002). In addition to recreational values, the lake provides critical habitat for large numbers of breeding and migratory bird populations, especially as a result of the large loss of wetlands in western North America during the twentieth

century (Shuford *et al.* 2002). However, the salt load from agricultural inputs is resulting in an annual increase of ~0.3–0.4 g per litre in lake salinity, and threatens the lake's fishes and other biota. Planned mitigation efforts include desalination of inputs and reduction in lake size through construction of a dam across the middle of the lake.

The effects of secondary salinization brought about by human activities on salt lake catchments are chemically similar to those brought about by inflow diversions, that is, increases in salinity and the consequences of this. Direct physical effects are few because, unlike flow diversions, secondary salinization is not associated with large changes in lake volume and water level. It is, nonetheless, equally if not more important since it has a major impact upon temporary salt lakes and is geographically more extensive. Moreover, it has increased the number of salt-water bodies and altered natural hydrological patterns.

Mining

Several human activities physically disturb the beds of dry salt lakes, and of these, mining is the most important, especially for temporary lakes, which are particularly vulnerable when dry. Mining is often for halite, but minerals mined also include trona, calcite, gypsum, borax and, more recently, lithium and uranium salts (Reeves 1978). These minerals are frequently mined from surface deposits, and mining involves the construction of levee banks, causeways and other structures that physically damage the structure of the lake (Williams 1993*a*). Rarely, if ever, is such damage repaired after mining has ceased. Where mining involves subsurface deposits, large quarries referred to as voids, as well as holding reservoirs, may be constructed on the lake bed. In some cases, subsurface mining on salt lakes may be for minerals that are not directly associated with the salt lake as evaporites, clastics or authogenics but are located deep beneath the bed of the lake. Mining is not confined to deposits on or beneath the dry beds of temporary salt lakes. Many minerals are mined from salt-lake brines and a few from beneath the beds of permanent salt lakes, drilling for oil beneath the Caspian Sea providing the most notable example (Kosarev & Yablonskaya 1994).

Apart from the physical disturbance, mining may have impacts on salt lakes in other ways, particularly by adding pollutants. Oil spills from mining rigs in the Caspian, the discharge of mine wastewaters and the location of mine spoil dumps (from which pollutants leach) adjacent to salt lakes provide examples (Dumont 1995).

Mining can also lead to the development of unnatural saltwater bodies in temperate regions. The moderately salt lakes or 'flashes' in Cheshire (UK) are the result of land collapses over salt deposits mined from underground. Quarries containing salt water in Germany developed when the pumping of salt groundwater intrusions stopped (Bohrer *et al.* 1998). In semi-arid regions, the construction of solar salt ponds (from which salt is obtained by the evaporation of seawater or salty groundwater) provides a unique example of unnatural saltwater bodies that have been constructed to 'mine' salt from the sea or underground. Activities other than mining that physically disturb the beds of salt lakes include the construction of canals and other structures designed to drain salt lakes, and in the USA the use of dry lakebeds as racetracks.

The limnological effects of physical disturbance to (dry) salt lakebeds by mining are little known. Levees, causeways and canals will clearly impede the free surface movement of water across the bed of the lake, but the consequences of this are unknown. They may not be significant. In this context, the biota of salt lakes, especially episodically filled bodies, comprises both an aquatic component, present when the lake contains surface water, and a terrestrial component, restricted to the bed of the lake when it is dry. What is clearly significant, however, are impacts on the appearance of the lakes; affected lakes lose much of their aesthetic appeal. Tailing dumps, mining voids, vehicle tracks and other impacts associated with both surface and deep mining at salt lakes likewise detract from and destroy a core part of their aesthetic appeal, namely the visual relief provided by a pristine landscape in a world much altered by humans.

Pollution by mining can have various effects depending upon the pollutants involved. Heavy metals leached from mining dumps act in the same way as toxicants do in all aquatic ecosystems; both biodiversity and biomass are reduced (Moss 1998). Salt wastewater has less profound effects, but will at least alter the natural pattern of salt loading. Hydrocarbons released accidentally by mining for oil in the Caspian Sea are already having an injurious effect on the economically important sturgeon in the lake (Kosarev & Yablonskaya 1994; see below).

Of actions other than mining that disturb salt lakes in a direct physical way, the most important is drainage, which leads to total loss, as occurred in Lake Texcoco, on the bed of which lies Mexico City (Alcocer & Williams 1996). Canalization, dyke and levee construction, road-building, drainage and landfill have all but obliterated the original lake.

Pollution

Inorganic plant nutrients appear not to be major pollutants in salt lakes, though exceptions occur (Williams 1981). For example, Farmington Bay of the Great Salt Lake has become eutrophic as a result of excess nutrients in runoff from urban development around Salt Lake City. Generally more significant is pollution through inputs of agricultural wastewater (often saline), pesticides in runoff and a variety of organic and inorganic wastes from domestic and industrial sources. Because salt lakes are usually regarded as water bodies of little value, they are often also used as sites for dumping solid wastes.

The pollution threats to salt lakes are often assumed to be broadly similar to those pertaining to other lakes (Williams 1993a). All of the sorts of pollutants discharged into fresh waters are also discharged into salt lakes, either directly or indirectly via their inflows. Most of the pollutants now present in the Aral Sea, for example, came from the Syr and Amu Darya (Letolle & Mainguet 1993) and many pollutants in the Caspian Sea come from the Volga River (Kosarev & Yablonskaya 1994).

In nearly all instances where wastes are discharged to salt lakes, it is assumed that the lakes in question will respond in fundamentally the same way as freshwater lakes and rivers (Chapters 2, 5 and 6). Often, the same discharge criteria are used by environmental protection agencies for both salt and freshwater lakes (Williams 1981), however account needs to be taken of the fundamental hydrological differences between them. Salt lakes are more or less closed hydrological systems and thus accumulate and biomagnify many pollutants to a much greater degree than do freshwater lakes and rivers (Williams 1981), and salinity may modify the toxicity of certain pollutants. The effects of pollutants in salt lakes may not be confined to the aquatic biota *sensu stricto*. The accumulation of selenium salts in evaporation ponds constructed to manage salt wastewaters in the western part of the USA provides an example. Selenium was soon transmitted to waterbirds that used the ponds; primary effects were mortality and deformity of adult birds (Schroeder *et al.* 1988).

Overfishing

As with freshwater lakes, salt lake fisheries are often overexploited. Of particular note is the near collapse of the sturgeon fishery in the Caspian Sea. At peak harvests in the 1970s (27 400 tonnes per year), the Caspian Sea provided up to 90% of the world's landings but subsequently declined to *c*.3000 tonnes per year (Ivanov *et al.* 1999). The fishery management problems of salt lakes are similar to those of other lakes and will not be discussed further here, except to note that as salt lakes become more saline, introductions of more salt-tolerant fishes are often attempted with varying degrees of success (for example in the Aral Sea: Aladin & Potts 1992; and the Salton Sea: Riedel *et al.* 2002).

Biological disturbances

The biota of many salt lakes has been unnaturally disturbed by the introduction of exotic species. Fish of recreational interest have been introduced into many moderately salt lakes (e.g. several lakes in Canada and Bolivia, and at least one in Australia (Lake Bullen Merri, salinity about 8 g per litre): Rawson 1946; Hammer 1986). In some cases, fish populations became self-sustaining and of commercial value, as in the Aral Sea, where, beginning in 1927, at least 21 species of fish were, either deliberately or accidentally, introduced, mostly from the Caspian, Baltic and Azov Seas and Chinese lakes (Zenkevitch 1963). All have now become extinct following the rise in salinity of this lake. In the Caspian Sea at least nine of the species introduced have survived (Kosarev & Yablonskaya 1994).

Many invertebrates have been introduced to moderately salt permanent lakes. Thus, of 18 invertebrate species introduced into the Aral Sea either accidentally or deliberately from 1927 onwards, mostly from the River Don and the Caspian and Azov Seas, over 10 established successful populations (Aladin *et al.* 1998). They disappeared when the salinity of the Aral Sea rose beyond their halotolerance. In the Caspian Sea, most introduced invertebrate species appeared after the opening of the Volga–Don Canal in 1954. Over 10 such species are known to have acclimatized to conditions in the lake, with two, the coelenterate *Aurelia aurita* and the ctenophore *Mnemiopsis leidyi*, considered highly likely to have significant impacts on some fish populations (Ivanov *et al.* 2000).

Temporary and/or highly salt lakes are unsuitable habitats for fish and relatively few invertebrate introductions into them have been attempted. However, the widespread and largely *ad hoc* importation of species of *Artemia* brine shrimp into coastal solar salt pans to reduce unwanted algal growths poses a serious threat to the biota of nearby natural salt lakes (Geddes & Williams 1987). Little if any attempt has been made to control these importations, in spite of the hazards involved (Geddes &

Williams 1987). A species of *Artemia* from coastal salt pans in Western Australia has recently invaded inland natural salt lakes (B. Knott, personal communication 2000).

Other forms of biological disturbance apart from exotic introductions exist, for example predation by terrestrial predators on bird species dependent on salt lakes for breeding and food. As the Aral Sea shrank following inflow diversions, many small islands in the south-east of the lake became peninsulas, so allowing predators access to migratory and resident waterfowl populations (Williams & Aladin 1991). Also, the recent expansion of silver gull (*Larus novaehollandiae*) populations in Australia, following the increase and expansion of town dumps which are used as feeding sites by the gull, has increased predation pressure on the banded stilt (*Cladorhynchus leucocephalus*) (Robinson & Minton 1990).

For the most part, the effects of introduced exotic species on the biota of salt lakes are unknown. Introductions are usually made in an *ad hoc* fashion, and any subsequent investigations are more concerned with determining the extent to which introduced species have acclimatized than with any adverse impacts on native species. Nevertheless, there is some evidence that introduced species which become acclimatized may replace native species. In Australian solar salt ponds, for example, introduced *Artemia* species, perhaps because of their ability to produce haemoglobin at high salinities and hence withstand low oxygen concentrations, seem to be able to displace the native *Parartemia* brine shrimp species, at least in the highly salty ponds as in western Australia (Mitchell & Geddes 1977).

Interactions between introduced and native species may not necessarily be confined to those in related taxonomic groups. The most serious effect of introduced *Aurelia aurita* and *Mnemiopsis leidyi* in the Caspian Sea is likely to be their competition for food with planktivorous fish, which may subsequently decrease in abundance (N. V. Aladin, unpublished data 2001).

Other catchment activities

Soil erosion, increased sediment loads and changes in runoff patterns can be the result of other catchment activities, including overgrazing by cattle and sheep and excessive clearance of the natural vegetation (Williams 1993*a*). After rainfall, runoff from overgrazed and/or cleared catchments is usually larger in volume but takes place over a shorter period than it would under natural conditions. Changes to the natural hydrological pattern

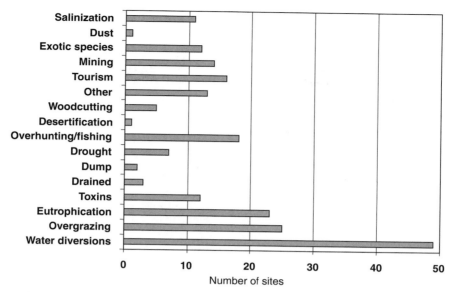

Fig. 7.2. Summary of threats to Ramsar sites including saline lakes (137 total sites), but excluding coastal lagoons. *Source:* Ramsar Site Descriptions (www.ramsar.org.).

have important consequences for seasonal ecological events (Chapters 5 and 6).

In a few cases, urban development on catchments poses a threat to salt lakes. In this context, public opposition has placed on hold the plans for housing developments within the actual crater rims of two volcanic salt lakes, Bullen Merri and Gnotuk, in the Western District of Victoria, Australia. This development threatens the aesthetic and scientific values of both lakes (Timms 1976). Urban development on salt lake catchments has many effects. For salt lakes, the most important impacts may involve localized domestic pollution and loss of aesthetic appeal.

Summary of threats

Overall, the anthropogenic impacts are geographically widespread, mostly irreversible, and degrade values of salt lakes. Certain impacts may be more important on particular sorts of salt lakes, but many are subject to several impacts at the same time. Salt lakes have generally received much less attention and study than freshwater lakes, and it is difficult to derive a representative or summary view of the extent and magnitude of worldwide impacts. The Ramsar Convention on International Wetlands includes all types of inland water bodies within its scope and

salt lakes are fairly well represented by 137 sites within the list of 1308 Wetlands of International Importance (Wetlands International 2002). While many salt lakes are subjected to multiple impacts (Fig. 7.2), including most of those common to freshwater lakes (Chapters 5 and 6), water diversions are the predominant threat, affecting 36% of the Ramsar sites. Because the ecological values of salt lakes are not widely recognized and their waters are not used for human consumption, some threats such as pollution may be underreported and only noted following catastrophic fish or bird kills. Only a few salt lakes, those located in arid regions, still remain relatively unimpacted by anthropogenic activities. Furthermore, overall increases in temperature and regional changes in precipitation and runoff (IPCC 2001d) will dramatically impact salt lakes.

FRESHWATER CRISIS AND CONSERVATION OF SALT LAKES

The ever-increasing human population in semi-arid regions, with the concomitant expansion of activities to support them, notably drainage, irrigation and land-use changes, will make conservation of salt lakes extremely difficult. An imminent freshwater crisis is widely forecast throughout much of the world as water demand rises, primarily for increased agriculture (Cosgrove & Rijsberman 2000). Worldwide, plans for infrastructure improvements (dams, diversions and irrigation) are being made with little attention to the environmental impacts on salt lakes. Furthermore, even infrastructural improvements are not deemed sufficient to avert water shortages.

Objective analyses of the benefits and costs of salt lake degradation are rarely if ever undertaken. The usual situation is one where the relatively easily determined economic benefits derived from lake degradation, which are often of local value, are judged to outweigh the indeterminate costs of conserving and protecting the lake, which is often of wider value. In water-stressed regions the economic value of water may far outweigh economic values of particular salt lakes, and only an appreciation of non-economic values can tip the balance toward conservation.

Given these factors, there is little doubt the worldwide desiccation of salt lakes will continue. Several recent 'vision' statements clearly point in this direction. In central Asia, for example, the 'vision' proposed by UNESCO (2000) for the Aral Sea basin (*sic*) involves almost complete desiccation of the lake itself, and greatly increased 'development' of its catchment to support the growing populations of Kazakhstan, Uzbekistan and Turkmenistan. Also, state planning in China calls for further increases in irrigated agriculture including many large water projects.

The problem is compounded by the failure of international intergovernmental bodies to recognize properly the importance of salt lakes as integral elements of the world's set of inland aquatic ecosystems. The influential 'World Water Vision' advanced at the Second World Water Forum at The Hague in 2000 did not refer to salt lakes (Cosgrove & Rijsberman 2000), and salt lakes were similarly ignored at the Third World Water Forum. Likewise, Groombridge and Jenkins (1998), in a report to the World Conservation Union (IUCN), did not rate salinization as a significant threat to loss of biodiversity.

It is difficult to be optimistic about the conservation of salt lakes. However, recent progress in recognizing the ecological values of salt lakes and gaining nominal protection has been achieved through the Ramsar Convention and the efforts of its partner organizations (Wetlands International, IUCN, Birdlife International, Worldwide Fund for Nature) and many small non-governmental organizations (NGOs) dedicated to conservation of individual salt lakes. The Ramsar Convention defines wetlands broadly as 'areas of marsh, fen, peatland or water, whether natural or artificial,

permanent or temporary, with water that is static or flowing, fresh, brackish or salt, including areas of marine water the depth of which at low tide does not exceed six metres', and thus explicitly includes seasonal, episodic and permanent salt lakes. A number of important salt lakes have been designated as Ramsar sites (Table 7.1) and while the current number is only a small fraction of important salt lakes, more are being listed each year.

The Convention's mission is 'the conservation and wise use of all wetlands through local, regional and national actions and international cooperation, as a contribution towards achieving sustainable development throughout the world'. Among other provisions, contracting parties are obliged to develop national wetland plans. While compliance is largely voluntary, the Ramsar Bureau's unique blend of education, technical assistance and NGO activities has been successful in raising public awareness of wetland values (Bowman 1995, 2003). Because the economic value of freshwater inflows is often considerable, appreciation of the ecological (and non-economic) values of particular salt lakes will often be a necessary prerequisite to establishing legal protection. Thus, the Ramsar Bureau's focus is well suited to the conservation of salt lakes.

Raised awareness is, however, merely the first step. Subsequently, there must be implementation of effective local, national and international management and conservation measures designed to mitigate and minimize the adverse human impacts on salt lakes and, wherever possible, prevent damage to their natural character. Freshwater lakes and wetlands have long been recognized as important natural assets with values over and above their use as a source of water; the need for salt lakes to be similarly recognized is overdue. The inclusion of salt lakes in preparations of a 'World Lake Vision' by several international groups (e.g. the International Lake Environment Committee [ILEC] and Global Water Partnership [GWP]) is essential. Better recognition by the Convention on Biological Diversity and in national wetland strategy documents is also needed. The International Society for Salt Lake Research (ISSLR 2007) should be of value in achieving this recognition.

At the local level, the protection and restoration of Mono Lake provide an outstanding example of what is possible once the total set of values for a salt lake is properly recognized. What factors were important in stopping the diversions? The first and most important was the commitment of a local NGO (Mono Lake Committee 2006) and regional conservation organizations (Audubon

Table 7.1. *Well-known examples of the 137 Ramsar sites that include saline lakes*

Name	Country	Type of saline lakes[a]	Ramsar site size (ha)
Lagunas de Vilama	Argentina	Q, R, Sp, Ss	157 000
Laguna de Llancanelo	Argentina	Q, R	65 000
Banados del Rio Dulce y Laguna de Mar Chiquita	Argentina	Q	996 000
Lake Gore	Australia	R	4 017
Western District Lakes	Australia	Q, R	32 898
Laguna Colorada	Bolivia	Q	51 318
Lagos Poopó y Uru Uru	Bolivia	Q	967 607
Quill Lakes	Canada	Q, R, Sp, Ss	63 500
Partie tchadienne du lac Tchad	Chad	Ss, Sp	1 648 168
Niaodao ('Bird Island') in Qhing Hu	China	Q, Sp, Ss	53 600
Sambhar Lake	India	Q	24 000
Lake Urmia	Iran	Q, Sp	483 000
Lake Bogoria	Kenya	Q	10 700
Lake Nakuru	Kenya	Q, Ss	18 800
Issyk-kul	Kyrgyzstan	Q	629 800
Valley of Lakes	Mongolia	Q, Sp	45 600
Karakul Lake	Tajikstan	Q	36 400
Kourgaldzhin and Tengiz Lakes	Kazakhstan	Q	260 500

[a] Lake types: Q, permanent saline/brackish/alkaline lakes; R, seasonal/intermittent saline/brackish/alkaline lakes; Sp, permanent saline/brackish/alkaline marshes/pools; Ss, seasonal/intermittent saline/brackish/alkaline marshes/pools.
Source: Ramsar Sites Database (www.ramsar.org).

and CalTrout). Their intense efforts were maintained over many years, and involved not only developing new legal concepts in environmental law, but also educating the public as to the importance of scenic, recreational and ecological values which were being impacted by water diversions (Hart 1996). The Ramsar Convention is well suited to the work of local NGOs.

Other successful conservation efforts in the Great Basin (USA) include management plans for Abert Lake (Oregon) that maintain salinities within an appropriate range; the use of the Endangered Species Act to halt the desiccation of Pyramid Lake and restore spawning runs of endemic trout (Truckee–Carson–Pyramid Lake Water Rights Settlement Act); and abandonment of mining plans at Zuni Lake (New Mexico) following national attention by the Sierra Club.

However, such successes will most probably be achieved in well-developed affluent countries with numerous conservation organizations; efforts toward conserving salt lakes have mixed results. The extension of the 'Public Trust Doctrine' (Broussard 1983) at Mono Lake to include non-economic values was a landmark environmental case in California, but it

is not directly applicable to Walker Lake, which is rapidly being desiccated just 50 km away in Nevada; numerous small salt lakes have in fact been impacted by agriculture throughout the Great Basin.

Given the economic value of fresh water and the current shortage and pending crisis in availability, most large salt lakes are likely to undergo significant desiccation. In some cases, efforts may succeed in conserving part of the ecological value of these lakes. At the Salton Sea, efforts are under way to build a dam across the lake to enable maintenance of moderate salinities in the 35–45 g per litre range in a portion of the lake while letting the other half dry. At the Aral Sea, a World Bank project on the Syr Darya in Kazakhstan has been building a dyke across the Berg Strait, which will maintain the Northern Aral Sea at 15–20 g per litre while the Large (southern) Aral Sea dries. In addition to preserving deltaic wetlands, these salinities will allow the fisheries to be maintained in the Northern Aral Sea. Similar dyking projects are likely to be considered at other large salt lakes in an effort to conserve some ecological values while diverting water for irrigated agriculture. In these and other conservation efforts, scientists will play a key role in assessing ecological impacts and proposing management alternatives.

These current and expanding conservation efforts might be capable of conserving a large portion of ecologically important salt lakes throughout the world if the human population were not increasing and if current agricultural practices were sustainable (Brown 2003). Desertification and salinization are annually removing tens of thousands of hectares from agricultural production. Groundwater depletion is likely to reduce agricultural production across large regions to the 2025 time horizon and possibly much sooner. Coupled with increasing population, these factors almost mandate increased diversions of surface water for irrigated agriculture in all the endorheic basins of the world.

Humans use only *c*.10% of the world's annually renewable water resources and 70% of that is used by irrigated agriculture employing, for the most part, inefficient methods (Cosgrove & Rijsberman 2000). Thus the water crisis might be manageable, given sufficient will and international cooperation. However, this ignores the regional and temporal distribution of water resources, the lack of institutional and water management infrastructure, and regional population demography. By 2025, 1 billion people are expected to experience severe and socially disruptive water shortages (Duda & El-Ashry 2000). Even optimistic forecasts, including improved irrigation, full-cost pricing for water and genetically modified crops that require less water, suggest 20–65% more water will need to be diverted for irrigated agriculture. As we enter a period of severe regional shortages of fresh water throughout much of the world, the economic value of salt lakes, which is mostly for fisheries, is likely to pale in comparison to the value of the fresh water required to reduce or reverse the shortages; current progress of conservation efforts is likely to be stalled or even reversed.

LIKELY STATUS OF SALT LAKES IN 2025

For salt lakes as a whole, the future looks certain: by 2025, most salt lakes will have undergone some changes from their natural character, many permanent ones will have decreased in size and increased in salinity, and many unnatural salt lakes will have appeared either as new water bodies or as replacements for freshwater lakes. How far this process will have gone by 2025 depends on many factors and the extent of change will differ among regions and types of lakes involved (Williams 1996*b*). For purposes of discussion, salt lakes are considered below as permanent, seasonally filled or episodically filled water bodies.

Permanent salt lakes

With the exception of those few permanent salt lakes water levels of which are monitored and managed (for example Mono Lake), and the few in areas where secular decreases in aridity have occurred recently (for example Issyk-kul), by 2025 most permanent salt lakes will have become smaller and more saline, with extensive if not complete exposure of their beds to the atmosphere. This regression will certainly be the fate for almost all permanent salt lakes with defined surface inflows. Large-scale water diversions are being planned in many salt lake basins and the argument that the economic value of diverted water exceeds the sum of all other values attributable to a lake is widely, if uncritically, applied. It is being used to justify diversions from Mar Chiquita (Argentina) despite this lake's critical importance to migrant waterfowl in the western hemisphere (Reati *et al.* 1997). In more optimistic scenarios, management actions will be taken to preserve a portion of the ecological values by maintaining some of the lake at lower salinity than the rest.

Not all permanent salt lakes have well-defined surface inflows of economic value. Trends in their limnological features are less well documented, predictions are hence more uncertain. Some intermittent data, however, are available for several salt lakes in Victoria, Australia, and equally indicate decreasing lake sizes. The reasons for their regressions are not clear, but groundwater pumping, land-use changes in the past century and a secular increase in aridity have been proposed. By 2025, all will have become significantly smaller and some of the shallow lakes that now dry only occasionally will become more or less permanently dry. In Mexico, it is clear that groundwater pumping is eliminating permanent salt lakes. This situation is deemed to be similar elsewhere for permanent salt lakes without defined inflows.

The few permanent salt lakes that show no regression at present are likely to remain the same size by the year 2025, providing no marked climatic changes take place over their catchments. This prediction, however, is less firm than predictions advanced for other permanent salt lakes: recall, for example, how quickly the regression of the Caspian Sea in the 1970s was reversed (Kosarev & Yablonskaya 1994) and the recent rapid rise and now fall of the USA's Great Salt Lake. Other sorts of adverse changes are also likely to occur in some of these lakes. The coelenterate and ctenophore introductions into the Caspian Sea are likely to change the nature of this lake's food web in the next two decades, overfishing will continue to be a problem and further adverse changes are likely from oil pollution.

Seasonally filled salt lakes

For seasonally filled salt lakes, in other words most natural temporary salt lakes in semi-arid regions, data on recent trends in hydrological periodicity are few, although many of these lakes are known to have dried more or less permanently following land-use changes, which will continue. Probably, the trends will reflect those shown by permanent salt lakes in the same region; the lakes will be drier for longer periods by 2025, some permanently so.

This simple picture of increasing desiccation is complicated by two events that are already common and are of increasing importance in semi-arid regions, namely salinization and the disturbance of salt and water budgets within drainage basins by diversion of river water. Land-use changes and irrigation, which are expected to increase globally by 50–100% by 2025 (Gleick 1993b), are implicated in both events.

Salinization has already increased the number of salt-water bodies in semi-arid regions and will continue to do so up to and beyond 2025. It has also enlarged natural salt lakes. The effects of water diversions from rivers are likely to be similar, though taking longer to develop; essentially, the diversions redistribute the salt and water load before its discharge to an inland terminus (a salt lake) or the sea. Both endorheic and exorheic drainage basins are involved (W. D. Williams 2001). The catchment itself, therefore, serves as the 'sink' for salts leached from it and so accumulates them. It is noted that salts within the Syr and Amu Darya are now retained within the catchment of the Aral Sea and not discharged into the lake. Similar salt retention within catchments can be assumed wherever significant diversions from rivers in semi-arid regions are made, as in the Murray–Darling River (Australia), the Yellow River (China) and the Colorado River (USA), all of which now have greatly reduced final discharge rates.

Episodically filled salt lakes

Episodically filled salt lakes, most temporary salt lakes in arid regions, are at present the type of salt lake least impacted by human activities, and in absence of global climate change most may retain their relatively natural status to 2025. However, several climate models predict that warming will be particularly great and rapid in certain arid regions (IPCC 2001d), and recent extended droughts in Australia may already reflect this (Karoly et al. 2003). Other arid regions will be less impacted, or will be affected more slowly. Current models predict that Australia will be warmer and drier throughout much of the interior and that large regional changes in runoff and precipitation will occur throughout the Middle East and central Asia. However, considerable differences exist between models concerning regional predictions of precipitation and runoff (Chapter 1). Irrespective of what happens, even small climate changes by the year 2025 could markedly influence the natural status of episodically filled salt lakes. The possible impact of climate change on the periodicity and intensity of El Niño–Southern Oscillation (ENSO) phenomena may be particularly important. ENSO episodes presently have considerable impact on precipitation patterns in many arid regions in North and South America, Africa and Australia (IPCC 2001d).

CONCLUSIONS

Salt lakes are geographically widespread, numerous and a significant part of the world's inland aquatic ecosystems. They are important natural assets with considerable aesthetic, cultural, economic, recreational, scientific, conservation and ecological values. Some features, notably the composition of the biota, uniquely distinguish them from other aquatic ecosystems. Salt lakes develop as the termini of inland drainage basins where hydrological inputs and outputs are balanced. These conditions occur in arid and semi-arid regions (approximately one-third of the total world land area). Many human activities threaten or have already impacted salt lakes, especially surface inflow diversions, salinization and other catchment activities, mining, pollution, biological introduction, and anthropogenically induced climatic and atmospheric changes. By 2025, most natural salt lakes will have undergone some adverse change. Many permanent ones will have decreased in size and increased in salinity, and many unnatural salt-water bodies will have appeared. In certain regions, many seasonally filled salt lakes are likely to be drier for longer periods. The extent to which episodically filled salt lakes will change by 2025 will largely depend upon the nature of climate change in arid regions. Objective cost–benefit analyses of adversely affected salt lakes are rare, and international bodies have not yet recognized salt lakes as important inland aquatic ecosystems. To redress this situation, there is a need to raise awareness of: the values of salt lakes, the nature of human threats and impacts on them, and their special management requirements. More effective management and conservation measures need to be developed and implemented. The conservation of salt lakes will be made much more difficult by the impending freshwater crisis, which will be experienced by many countries in arid and semi-arid regions of the world, and by the anticipated increases in irrigated agriculture. Ultimately, the fate of many permanent salt lakes will depend on how quickly individual countries move towards improved and sustainable agricultural practices and stabilization of their human populations.

Part III
Freshwater wetlands

In this book, the term 'wetlands' identifies ecosystems that, in contrast to deep and open-water systems (lakes, rivers and oceans), possess abundant aquatic herbaceous and/or woody vegetation that thrives in shallow water, on periodically or permanently water-saturated substrate. Most wetlands are sandwiched between land and water and hence are apt to be considered as only transitional habitats or land–water ecotones. Wetlands encompass freshwater systems as well as a number of marine systems that will be addressed later in the book (Chapters 11–13). Inland wetlands cover a wide range of habitats such as seasonal or permanent marshes with herbaceous vegetation, swamps with woody vegetation, and bogs and fens with peat accumulation. Flood plains, dealt with previously (Chapter 4), are also generally recognized as a type of wetland.

Wetlands first attracted attention because of their importance as habitats for wildlife, particularly waterfowl, but their high biodiversity and productivity as well as other ecosystem functions have now been recognized. This recognition has coincided with increased concern about the extent of wetland loss, especially to agriculture by drainage and reclamation, along with degradation due to altered water regimes, habitat alteration, pollution and eutrophication, invasive species, and overharvesting. The Ramsar Convention on Wetlands has helped promote measures for the wise use and conservation of these ecosystems, including their restoration, but they continue to bear the brunt of economic development around the world.

The three chapters in this part deal with freshwater systems grouped roughly by latitudal range into cool temperate bog and mire (peatlands; Chapter 8), temperate (Chapter 9) and tropical (Chapter 10) wetlands. Although flood plains and saltmarshes are addressed as entities in Chapters 4 and 11 respectively, they receive additional attention in this part.

8 · The future of cool temperate peatlands

NILS MALMER, PETER D. MOORE AND MICHAEL C. F. PROCTOR

INTRODUCTION

Peatland is a type of wetland with the water level just below the surface where peat (an organic, autochthonous soil of at least 30 cm depth) has developed. Peatlands globally cover *c*.3.5 million km^2 (Kivinen & Pakarinen 1981; Gorham 1991; O'Neill 2000) and have a mean depth of *c*.2.5 m (Clymo *et al.* 1998). Most of these peatlands lie in the cool temperate zone, between latitudes 45° N and 70° N in Russia, especially east of the Ural Mountains,

Fennoscandia and north-west Europe, Canada and Alaska (Fig. 8.1, Table 8.1). These are all areas that were glaciated or had a very cold and dry climate during the last glaciation (CLIMAP Project Members 1976). Canada is reckoned to contain 25% of the world's peatlands (Dahl & Zoltai 1997), almost 70% of which are forested, and Russia another 25%. In the southern hemisphere, temperate peatlands are located in the lowlands of western Chile, New Zealand, Tasmania and south-eastern Australia, but they comprise

Aquatic Ecosystems, ed. N. V. C. Polunin. Published by Cambridge University Press. © Foundation for Environmental Conservation 2008.

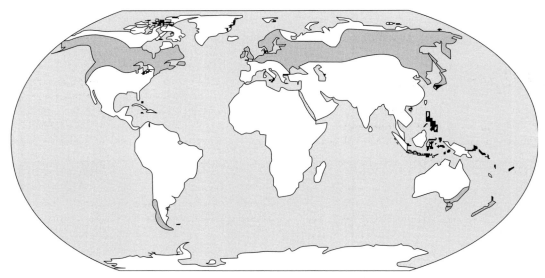

Fig. 8.1. Global distribution of cool temperate peatlands (shaded area). The tundra mires mentioned in the text are found on the North American and Eurasian continents to the north of the cool temperate peatlands. (Simplified from Moore 2002.)

Table 8.1. *Global distribution of peatland areas. From Moore (2002) based on O'Neill (2000).*

Region	Area (Mha)	Global peat area (%)
Eurasia	180.2	44.1
North America	171.1	41.9
South Asia	43.6	10.7
South America	6.2	1.5
Africa	4.9	1.1
Central America	2.6	0.6
Australasia and Pacific	0.3	0.1
Total	408.9	100.0

vegetation forming peatlands is unique in that its gross ecosystem productivity (equivalent to gross primary productivity) exceeds the sum of the total ecosystem respiration and growth in living biomass. The net surplus of carbon compounds is stored within the ecosystem as peat. About 50% of the peat consists of carbon, thus a growing peatland acts as a sink for atmospheric carbon. However, considerable amounts of methane, an important greenhouse gas, are also formed in peatlands and released to the atmosphere. The accumulating organic matter also structurally modifies the ecosystem, particularly elevating the surface in relation to groundwater, so modifying the hydrology (Chapter 3). It also leaves behind a sequential record of its own history in the plant and animal fossils within the stratified layers. Being sensitive to those factors (in part climatic) that influence decomposition rates, the peat stratigraphy can also act as a proxy for past climate reconstruction.

Over the past 10 000 years, the northern peatlands have accumulated a total of 455 Gt of carbon amounting to an accumulation of 0.05–0.07 Gt C per year (Gorham 1991; Clymo *et al.* 1998). The carbon reservoir of the northern peatlands is about two-thirds that of atmospheric carbon and similar to the amount present in all other living things on Earth, which is estimated to be 510 Gt C (Schimel 1995). The plant species accomplishing the first step in the accumulation of carbon and other elements are principally

<0.5% of the total area of cool temperate peatlands. Peat mosses of the genus *Sphagnum* are usually a characteristic component of the vegetation in the most widespread types of peatlands (Masing *et al.* 1990) and contribute most of the organic material forming the peat.

Whereas most terrestrial ecosystems have a carbon output (ecosystem respiration, *sensu* Odum 1969) that is equal to their gross production (unless the ecosystem is in a state of succession leading to living biomass growth), the mire

Sphagnum mosses and dwarf shrubs and sedges, and to a lesser extent trees and herbs. The *Sphagnum* species can be looked upon as ecological engineers (Jones *et al.* 1994) because their specific qualities characterize the ecosystem as a whole. A considerably lower rate of decomposition than most vascular plant litter (Coulson & Butterfield 1978; Johnson & Damman 1993) leads to the *Sphagnum* litter contributing much more to the peat than expected from primary production (Wallén 1992). In fact, these mosses alone contain more carbon in their living and dead tissues than any other genus of plants, including trees (Clymo & Hayward 1982). Moreover, a great share of the carbon in the peat has its source in litter formed by rather few of the *Sphagnum* species, particularly species in the sections *Sphagnum* and *Acutifolia*, which are phylogenetically the two youngest in the genus (Shaw 2000).

The future of the cool temperate peatlands will depend on many varying factors. Increasing human exploitation of the peat resource through burning, drainage, grazing and afforestation, together with impacts from both water- and airborne pollutants, raised global temperatures on productivity and decomposition rates, and changing patterns of precipitation, will all influence these peatlands. Predicting their future state demands information on how this ecosystem has responded to change in the past and how the future global environment will develop. The nature and function of peatlands also need to be grasped so that their management and conservation can be approached in an appropriate way. Building on Moore (2002), the main aim of this chapter is to examine the processes at work in peatlands, and to use this knowledge as a basis for predicting the likely outcomes of global changes for the state of the world's temperate peatlands in the year 2025 and beyond. The feedback effects on atmospheric concentrations of greenhouse gases from likely future changes in peatlands and their vegetation are also considered.

TYPES OF PEATLAND

Most peatlands are oligotrophic and characterized by an acid reaction (water pH usually <5.5). They are divided into different types using morphology, hydrology and vegetation as criteria (Moore & Bellamy 1973; Botch & Masing 1983; Zoltai 1988; Wheeler & Proctor 2000; Charman 2002). For the present purpose two groups are highlighted, namely rheotrophic (flow-fed) and ombrotrophic (rain-fed) peatlands or mires. Since ombrotrophic mires are supplied with water only from precipitation, their

water regime closely follows the pattern of precipitation and evapotranspiration, and for mineral nutrients the vegetation has to rely on atmospheric transport. The flow of water is much greater in rheotrophic mires because they are also supplied with water from the surrounding catchment. The water regime in rheotrophic peatlands mainly follows that in storage in nearby terrestrial soils, and solutes from mineral ground provide an additional source of nutrients for plant growth. Most peatlands consist of complexes formed by various combinations of ombrotrophic and rheotrophic sites.

The most widespread type of peatlands, *aapa* mires (also called patterned fens or string bogs), are rheotrophic gently sloping peatlands with only few ombrotrophic sites formed in an otherwise rather flat landscape (Tolonen 1967) in the interior of continental land masses. They are characterized by raised ridges running along the contours of the mire alternating with elongate pools usually called flarks, both at right angles to the slope. The proportions of hummock ridges and flarks, as well as the degree of tree cover, vary between regions and with climate (Foster *et al.* 1983). Because of the rheotrophic conditions, these mires also vary greatly in their degree of oligotrophy, depending on catchment geology and the water flow.

The most widespread type of ombrotrophic mires (raised bogs) consists of extensive domes of peat, elevated to a height of several metres above groundwater level in the nearby terrestrial soil. In continental interiors, they often have summits covered in conifer forest (Glaser & Janssens 1986), but treeless bogs occur, for instance in the lowlands near Hudson Bay (Glaser *et al.* 2004a, b). Ombrotrophic mires of more oceanic regions, such as eastern North America, the USA Pacific North-west and Alaska, western and Baltic parts of Europe, Chile and New Zealand, are not forested, but are open plateaux and domes, with dwarf shrubs, a few species of sedges and a ground layer of *Sphagnum* mosses. A pattern of hummocks and wet hollows is often seen, usually similar to that on the *aapa* mires, with strings of hummocks oriented perpendicular to slope. In the oceanic areas with an excess of soil moisture (e.g. western Britain and Ireland, western Norway, easternmost Newfoundland and coastal British Columbia), treeless ombrotrophic blanket mires cover plateaux, valleys and all but the steepest slopes in the landscape.

Bordering the Arctic Sea in the northernmost part of the continents is a range of tundra mires. The patterned Arctic polygon mires of the high latitudes are mainly rheotrophic mires, with polygonal pools alternating with flat mire expanses. They arise not only within regions of permafrost north of the Arctic Circle, but also further south in Siberia. A special Arctic mire type is the *palsa* mires, which are most widespread in areas with a continental climate. *Palsa* mires have an irregular surface with elevated mounds of peat (up to 50 m in diameter and several metres high) with permanently frozen cores, interspersed with sedge marshes. These are known to undergo cycles of formation and decay as the ice cores expand and collapse over a timescale of centuries because of changing climate.

For the present purpose these three essentially climatically determined types of peatland complexes (*aapa* mires, raised bogs and tundra mires) can be used to define three different regions of peatland types in the northern hemisphere: the circumpolar tundra zone with complexes of tundra mires in the far north of the continents; the zone with predominating peatland complexes dominated by rheotrophic *aapa* mires in the interior of the Eurasian and North American continents; and the zones with predominating ombrotrophic peatlands in north-western Europe, along the east and west coasts of North America and the east coast of Asia (Fig. 8.1). In addition, a range of topogenous (topography-determined), predominantly rheotrophic mire types have been documented in many locations in the northern hemisphere, but they comprise only a rather small part of the total peatland area. For instance, the domes of raised ombrotrophic mires are usually surrounded by narrow zones of rheotrophic mire (lagg fen). On topographically suitable places, seemingly flat areas of rheotrophic peatforming vegetation occur, which sometimes have trees and may cover extensive areas, and sometimes are smaller and consist of floating carpets (quaking mires). Peat formation may also occur on distinct slopes with an outflow of groundwater, particularly in mountainous areas.

Peatlands are often many millennia old. In western Canada, the majority of peatlands were initiated with the paludification of forests approximately 7000–8000 years ago (Halsey *et al.* 1998; Campbell *et al.* 2000). Many of the Finnish *aapa* mires (Clymo *et al.* 1998) and the raised ombrotrophic mires of oceanic Western Europe have a similar age (for example Svensson 1988; Hughes & Barber 2003). In the case of the North American maritime peatlands, the lack of trees is generally associated with greater age, and the forested condition of the more continental peatlands may be an earlier successional stage that eventually leads to the treeless state, since the forested mires are generally younger than the maritime bogs (Glaser & Janssens 1986). Most of the blanket mires began forming 3100–5100 years ago (Tallis 1998), their initiation and development in Britain and Ireland often being associated with evidence of human activity, particularly burning and grazing (Moore 1993). Development of a peatland beginning with an infilling of a lake (terrestrialization) is also common.

The cool temperate peatlands are thus varied in their nature and their relationships with both the natural environment and the land management activities of human beings. It might be expected, therefore, that their response to environmental change will be equally varied. Then again, they have certain features in common, particularly their peat accumulation processes, and in this respect some general patterns of environmental response can be discerned.

PEATLANDS AND THE ENVIRONMENT

Sensitivity to the physical environment

Since the carbon budget in peatlands is unbalanced (carbon accumulation exceeding carbon output), environmental changes that alter the rate of either productivity or decomposition may influence the overall ecosystem function. It has long been recognized that the upper layer of the organic deposit of a peatland (often around 20–30 cm depth) (Fig. 8.2) is much better aerated than the lower layers (Clymo 1965). This upper layer, the acrotelm, contrasts with the underlying usually much deeper anoxic layer, which is termed the catotelm (Ingram 1978; Ivanov 1981). Living plant roots are generally much more abundant in the acrotelm, and this is also the site of most microbial (especially fungal) activity. The organic material in the acrotelm is a kind of litter similar to that for example in boreal forests. It also passes through the same decomposition processes at rates not much different from those in acid terrestrial soil types (Johnson & Damman 1993; Malmer & Wallén 1993, 1999). Although anaerobic decomposition continues within the catotelm, the overall rate of decomposition is much slower than in the acrotelm. Since peat is defined as very slowly decomposing organic matter in a wet environment, it is only the organic matter included in the catotelm that should be looked upon as peat. Consequently, the accumulation rate of organic matter or carbon in a peatland should refer only to the rate at which it is added to the catotelm.

Since the acrotelm mainly consists of growing mosses and weakly compacted plant litter that is loose in structure, these upper layers in a peatland have a high hydraulic conductivity. In contrast, the catotelm consists largely of compacted decomposed organic material permanently

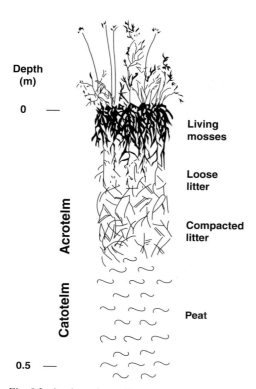

Depth
(m)

0 —

Acrotelm

Catotelm

0.5 —

Living
mosses

Loose
litter

Compacted
litter

Peat

Fig. 8.2. A schematic profile through the surface layer of a *Sphagnum*-dominated peatland with the different soil horizons indicated. (Revised from Malmer 1988.)

saturated with water, and has a hydraulic conductivity around 1000–10 000 times lower than that of the acrotelm (Ivanov 1981). In both ombrotrophic and rheotrophic peatlands, lateral water movement therefore occurs largely through the acrotelm. Further, both the oxygen availability and the redox potential are much lower in the permanently waterlogged catotelm than in the acrotelm with a fluctuating water table. However, many of the vascular plants in peatlands are able to cope with an anoxic environment through aerenchyma tissues that transport oxygen to the roots. Others have their roots concentrated in the acrotelm and might therefore be more sensitive to drought and phytotoxins like sulphide.

Biomass, primary productivity and carbon sequestering

Above-ground biomass in boreal peatlands is variable, often ranging from 400 to 4000 g per m^2 (Moore & Bellamy

1973). For the vascular plants the below-ground biomass has rarely been measured, but 1000–2000 g per m^2 (Wallén 1986, 1992) may be fairly typical; however, values may be four times the above-ground biomass (Malmer *et al.* 2003). High biomass is associated with forested vegetation types, but is also often correlated with the depth of the water table, a larger biomass often being associated with a deeper water table.

Primary productivity in boreal peatlands is generally measured by harvest methods, and root production is normally neglected, although the below-ground productivity of the vascular plants may often be considerably greater than the above-ground productivity (Wallén 1986). Productivity estimates from such methods usually range between 100 and 700 g per m^2 per year of dry matter (see Malmer 1962; Moore & Bellamy 1973; Grigal *et al.* 1985; Thormann & Bayley 1997; Malmer & Wallén 2004). The rate varies with climate and site factors, particularly availability of mineral nutrients, and with vegetation, particularly depending on the presence or absence of forest. For non-forested *Sphagnum* mires, the majority of productivity estimates are in the 100–300 g per m^2 per year range. Since both biomass and productivity also vary with the small-scale variation in microtopography found in many mires, extrapolations of these estimates from square metres to hectares and square kilometres are difficult to perform and the results often uncertain.

Experimental elevations of atmospheric carbon dioxide (usually in conjunction with raised temperature to simulate future climate conditions) have usually resulted initially in enhanced productivity. For example, when sample plots of an *Eriophorum vaginatum* bog in Alaska were supplied with carbon dioxide levels of 680 ppm (roughly double the natural level) together with a 4 °C rise in temperature, carbon fixation by the ecosystem was enhanced in the first year (Oechel *et al.* 1994). However, fixation subsequently declined, the likely explanation being that enhanced vegetation growth swiftly resulted in nutrient limitation. Northern ecosystem productivity is also generally considered to be nitrogen limited (Melillo *et al.* 1993), hence the positive impact of raised atmospheric carbon dioxide levels on productivity may be less than expected, particularly if compared with the effects of an increased supply of nitrogen (Berendse *et al.* 2001).

Sphagnum mosses, and usually also other plants in oligotrophic habitats like peat-forming mires, are well adapted to low nutrient conditions. They often have efficient systems for the mobilization and recycling of elements prior to leaf fall (Jonasson & Chapin 1985) or relocate the elements to the apical meristem as in *Sphagnum* mosses (Rydin & Clymo 1989). Many plants are also able to take up nitrogen in the form of ammonium ions or even as organic nitrogen (Nasholm *et al.* 1998). Deficiency in either or both of nitrogen and phosphorus limits productivity in peatlands. In regions with an enhanced atmospheric nitrogen deposition, a shift from nitrogen to phosphorus as the main growth-limiting element may have taken place during the last decades (Aerts *et al.* 1992).

Decomposition and the carbon balance

The rate of carbon sequestration via photosynthesis relative to carbon loss through respiration and decomposition determines whether a peatland is a source or a sink for atmospheric carbon. The excess of organic material within peatlands is therefore a consequence of slow decomposition rather than unusually rapid primary productivity (Clymo 1984). The two determinants of the decay losses from the litter, before it becomes included as peat in the catotelm, are the decay rate and the residence time for the litter in the acrotelm. The decay rate is directly proportional to the temperature but strongly reduced by a high content of decay-resistant litter like that of many *Sphagnum* spp. The residence time depends on the position of the water level, which is in turn determined by the rate of water supply and the resistance to water flow exerted by the litter in the acrotelm. In addition to the carbon released to the atmosphere as a result of the decomposition processes, a varying proportion is lost in the runoff as dissolved organic matter (Proctor 1997).

In a general although simplistic way, the decomposition in the acrotelm can be described by the usual exponential equation for the decay process:

$$\ln M_t = \ln M_0 - kt \qquad (8.1)$$

where M_t and M_0 designate mass at time t and at the formation of the litter, respectively, and k is the decay constant. The limit between the acrotelm and catotelm appears as a distinct decrease in decay rate (Fig. 8.3). In an example of the present conditions in ombrotrophic bogs with an overall dominance of *Sphagnum* communities, the decay rate was 0.005–0.015 per year (half-lives of 50–140 years) and residence time 60–100 years (Fig. 8.3; Malmer & Wallén 1993, 1999, 2004). The decay loss during the time lag before the organic matter is added to the peat in the catotelm is in the range of 60–80% of the original amount of litter. Both lower and higher decay losses may

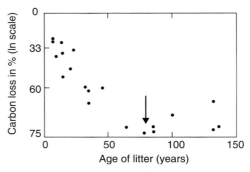

Fig. 8.3. An example of the percentage cumulative decay loss (log scale) of the dead organic matter through the acrotelm and the upper part of the catotelm on an ombrotrophic site with a *Sphagnum*-dominated plant community. Samples (thickness 2.5 cm) combined from four cores (depth *c*.0.3 m, moss layer excluded). The arrow indicates the approximate position of the border between acrotelm and catotelm, although it is better to regard this as a transitional zone than a distinct level (Clymo & Pearce 1995). (Revised from Malmer & Wallén 2004.)

occur, but with decay losses around 90% or more, the carbon input to the catotelm might, in the case of deep peat layers, be less than the decay losses from these layers, in spite of the fact that the decay rate is very low there.

Even a modest rise in the water level will immediately shorten the residence time and rapidly increase the carbon accumulation rate, even without an increase in primary productivity (Fig. 8.3). On the other hand, a lowering of the water level may have only small effects on the total release of carbon, particularly since the organic matter becomes much more recalcitrant with age. The increasing compaction of the litter with increasing decay losses (Johnson *et al.* 1990) strongly reduces the flow of water at the bottom of the acrotelm compared to that near the surface.

Low oxygen availability as a consequence of water logging is an important factor reducing decomposition rates at the transition from acrotelm to catotelm. A deeper water table obtained through drainage therefore increases decomposition rates, as reflected by enhanced soil respiration (Alm *et al.* 1999). Drying and wetting cycles particularly stimulate the carbon output rate from a mire surface (Carroll & Crill 1997; McNeil & Waddington 2003). A deeper or more variable water table in a peatland may thus increase the decay losses. Since decomposition includes enzyme-controlled reactions, the decay rate will generally become slower as a result of any climate cooling and will be enhanced by warming. Low nutrient concentrations also contribute to reduced decomposition rates. Nitrogen limitation, in particular, can result in reduced decomposition (Hu *et al.* 2001), while elevation of nutrient inputs (including phosphorus) is likely to stimulate decomposition.

Among peatlands, much of the variation in the carbon annually accumulated in the catotelm may depend on the local topography. At large scale, however, it seems to be temperature dependent since it decreases with increasing latitude (Clymo *et al.* 1998). For ombrotrophic mires, both the carbon sequestration in the vegetation and litter decay rates decrease inland from the most maritime areas, particularly because of longer periods with frozen peat (Malmer & Wallén 1993). The result is therefore only a slight decrease inland in the carbon accumulation rates. Statistically significant differences in carbon accumulation rate do not seem to exist between ombrotrophic mires and rheotrophic *aapa* mires (Clymo *et al.* 1998), although, in general, decomposition is slower in ombrotrophic than in rheotrophic mires (Johnson & Damman 1993).

Historically, peatlands generally have sequestered carbon as peat in the catotelm at a rate of 190–300 kg C per ha per year (Gorham 1991; Malmer *et al.* 1997; Vitt *et al.* 2000), resulting in an average height increment of approximately 0.5 mm per year. Based on a study of 795 peatlands in Finland, Clymo *et al.* (1998) adopted a figure of 210 kg C per ha per year for the northern peatlands. By extrapolation this gives a value of 0.07 Gt C per year for the past global rate of accumulation of atmospheric carbon in peatlands. The size of this peatland sink for carbon is relatively small when compared with the 6 Gt per year of carbon released into the atmosphere by human activities (Schimel 1995), particularly fossil-fuel burning, but it is nevertheless a significant contributor to carbon sequestration from the atmosphere if seen in a long-term perspective.

Changes over time in the peat accumulation rate in a peatland are usually seen as triggered by variation in climate. For instance, contemporary with a rise in lake water levels 900 years ago, the carbon accumulation rate in an ombrotrophic mire in southern Sweden increased from 150 to 600 kg per ha per year within 100–200 years (Malmer & Wallén 1999, 2004). This peak value was followed by a continuous decrease to about 270 kg per ha per year at the end of the nineteenth century, although the peat-forming vegetation and its productivity remained the same. In the same mire, two similar changes in the peat accumulation rate also occurred during earlier periods (Malmer *et al.* 1997). Such decreases in the carbon accumulation rate are, however, not necessarily climatically determined. It may simply represent a development towards a steady state as a result of establishment of a new equilibrium between litter input and water runoff (Belyea & Clymo 2001; Belyea & Malmer 2004). Therefore, the rate of carbon accumulation in peatlands varies not only because of climate variation, but also because of autogenic processes.

Neither the present nor the future carbon balance in peatlands can be estimated from that during earlier periods. It is necessary to appreciate the potential future global effects that could result from continued reduced carbon accumulation rates in the peatlands due to vegetation changes and increased decay losses. In fact, many peatlands may easily switch to being sources instead of sinks for atmospheric carbon, and thus begin to contribute to an increase in the atmospheric concentrations of both carbon dioxide and methane.

Gaseous emissions

As a result of decomposition processes, a variable but significant proportion of the carbon sequestered as carbon dioxide in a peatland is returned to the atmosphere as methane. In addition, a small portion of the nitrogen in atmospheric deposition is returned to the atmosphere as dinitrogen oxide. In relation to the atmosphere, peatlands can therefore be looked upon as a kind of chemical reactor both accumulating and releasing chemical compounds as a result of the activity of microorganisms. Methane is produced because the anoxic environment and the production rate increases with increasing soil temperature, three-fold or more with a 10 °C warming (Christensen *et al.* 2002). The production of methane is also favoured by the excretion of easily degradable organic carbon compounds of low molecular weight from the vascular plant roots in the lower parts of the acrotelm and in the catotelm (Joabsson *et al.* 1999). The release of methane to the atmosphere is to some extent reduced in undisturbed oxic layers as a result of the activity of methanotrophic bacteria, but the aerenchyma tissue in vascular plants like *Eriophorum* spp. may often form an important passage for the methane directly into the atmosphere (Joabsson & Christensen 2001).

Current estimates of the global methane emissions from northern peatlands vary from 40 Mt per year (Christensen *et al.* 2002) to 72 Mt per year (Masing *et al.* 1990). They may represent as much as 45% of the natural emissions to the atmosphere or, if including the various human sources such as landfill, rice paddies, livestock farming and natural gas leakage, about 15% of the present total global release of methane to the atmosphere. The release of dinitrogen oxide from the northern peatlands is not yet adequately known; however, virgin and partially drained peatlands in Finland indicate that the rate of nitrous oxide emission is not significant, being only 1% of total global output (Martikainen *et al.* 1993).

The production of methane and dinitrogen oxide in peatlands is particularly important because both are greenhouse gases, with more intense effects but a shorter lifetime in the atmosphere than carbon dioxide. Over periods of up to 100 years, methane has, per molecule, 23 times, and dinitrogen oxide 300 times, the radiative forcing effect of carbon dioxide on climate warming. On average, a methane emission of 10–20 g per m^2 per year corresponds to the radiative forcing of 250–450 g CO_2 per m^2 per year. Since the mean annual carbon accumulation as peat in the catotelm is equivalent only to 75 g CO_2, the gaseous exchange of peatlands over time-spans of up to 100 years may be seen as having global warming potential, rather than being a net sink for atmospheric carbon (Christensen & Keller 2004).

Human impacts

In society at large, peatlands have usually been regarded as wasteland. Human activities have also had a generally negative impact on peatlands globally, through drainage for forestry and agriculture, harvesting of peat for various purposes, atmospheric pollution and recreation. However, it is likely that activities of prehistoric people contributed substantially to the development and spread of blanket mires within the appropriate climate regions (Moore 1993).

Mining of peat deposits for various purposes is widespread. Peat represents an important energy source when combusted. It has long been widely used as fuel, particularly in Western Europe. In Britain, around 94% of lowland raised bogs have been severely damaged or destroyed by peat harvesting over the past centuries (Gosselink & Maltby 1990) and, in 1991, as much as 14% of Ireland's energy requirement was supplied by peat. The structure and chemistry of peat has also made it a valuable commodity in horticulture as a soil conditioner.

In southern Scandinavia, the first clear signs of human exploitation of peatlands for agriculture are from the end of the Viking age (Berglund *et al.* 1991), but such exploitation may have started earlier in other parts of Europe. Very large areas of peatlands have since been drained and used

both for agriculture and forestry on both sides of the Atlantic, particularly after the beginning of the nineteenth century. Eight per cent (23 Mha) of the total global area of peatland may be exploited, the greater portion of this being in Europe (Armentano & Menges 1986). Because of burning and increased decay of peat in cultivated soils, the peatlands in these regions may have acted as sources rather than sinks for carbon during the last centuries.

All mire vegetation, and particularly that on ombrotrophic mires, depends on atmospheric inputs for mineral nutrients. The main peat-formers, *Sphagnum* mosses, are generally adapted to low nutrient concentrations and acid conditions, and lose competitiveness with increasing nutrient supply rates. The growth of *Sphagnum* mosses is also reduced if exposed to airborne pollutants like inorganic nitrogen and sulphur compounds (Lee 1998; Gunnarsson & Rydin 2000 and references therein). During the last century, the atmospheric pollution and eutrophication resulting from human activities have inevitably caused subtle changes to peatland vegetation that may prove difficult to monitor and even more difficult to control or reverse.

Today peatlands can represent a kind of wilderness, providing recreational opportunities when situated in the vicinity of densely populated regions. The emphasis of forest managers taking care of peatlands in the densely populated areas in Canada and Europe is now often away from pure forestry and towards multi-purpose use of forests, including hunting, fishing and recreation (Jeglum 1990). One problem in the use of peatlands for recreation is the fragility of the habitat. The delicate structure of the peat surface, so necessary for the hydrological and gaseous exchange processes, is easily damaged by physical pressure like footprints, but outside frequently visited tourist areas the impact of this has, to date, been small.

The influence of humans both in the past and the present on cool temperate peatlands is not entirely negative. In the old agricultural landscape, the exploitation of peat-forming wetlands, particularly in areas with calcareous bedrock, often resulted in grazing land with a high diversity of both plants and animals. The few extant properly managed areas of this kind are much appreciated for their beauty and amenity values. Recent awareness of the rapidity with which other types of peatland have been disappearing from the landscape, particularly in the densely populated regions of western Europe, southern Canada and the northern USA, has led to much effort in rehabilitation (Wheeler *et al.* 1995). The process of peatland rehabilitation requires particularly the restoration and maintenance of a high water table where there has been drainage, and reinstatement of peat-forming vegetation where peat harvesting and losses of *Sphagnum* mosses and other plants have occurred. However, only small areas have so far been treated in this way for the purposes of amenity and nature conservation.

IDENTIFICATION OF LONG-TERM TRENDS

Fossil evidence of peatland development during the Late Holocene

In peatlands, the fossil remains of plants provide a record of successional development (see Godwin 1981), changing hydrological conditions (see Succow & Lange 1984) and nutrient status (see Moore *et al.* 1984) and climate shifts (see Barber 1981; Svensson 1988; Ellis & Tallis 2000). Although care must be exercised in the interpretation of observed changes because a variety of factors can interact and confound the results, the stratigraphy of peatlands provides a solid source of historical information, on the basis of which projections into the future can be made.

Palaeoecological data show that considerable climate changes have been taking place in the last 20 000 years, and will continue to occur. Ombrotrophic mires are clearly most suitable for this type of study because they are expected to respond directly to climate and much less to changes in groundwater movements that can have local causes. The fact that so many raised mires in North America and Europe developed from a rheotrophic to an ombrotrophic state around 7000–8000 years ago suggests a widespread and general cause rather than a local one (Godwin 1981). During that period, conditions in the boreal zone are considered to have been warmer and wetter than at present (Beck 1998), and the situation of 6000 years ago represents the warmest set of global conditions experienced since the last glaciation (Cwynar & Spear 1991), with temperatures up to 4 °C higher than at present. Temperatures subsequently have fallen, and this has had a marked positive effect on cool temperate peatlands. It is during this time that most of these peatlands developed and spread. For instance, blanket mires in north-western Europe have extended southwards during the past 6000 years, which is likely to be in part owing to climate changes, but also as a result of the impact of human activities in forest clearance (Moore 1993). In Minnesota

(USA), ombrotrophic peatlands became established only 2000–2500 years ago (Glaser & Janssens 1986). The emerging pattern of evidence from peatland stratigraphy for climate shifts also ties in well with other evidence for past climate changes. It is thus clear that the cool temperate bogs have changed in extent and in nature as a consequence of the climate changes of the past few thousand years.

Recent changes in peatlands and their vegetation

During the last few decades, several studies in north-west Europe have demonstrated increased tree cover in pristine *Sphagnum*-dominated peatlands, more frequent occurrence of shade-tolerant forest mosses at the expense of peat mosses and an expansion of tall graminoids at the expense of low-growing vascular plants like *Drosera* spp., *Rhynchospora* spp. and *Scheuchzeria palustris* (Åberg 1992; Frankl & Schmeidl 2000; Gunnarsson *et al.* 2002). In this region, deterioration of *Sphagnum* cover, particularly in the hummocks and lawns of ombrotrophic mires, has occurred at many sites (e.g. Twenhöven 1992*a*; Mackay & Tallis 1996; Malmer & Wallén 1996, 1999) combined with improved growth of dwarf shrubs forming heath-like vegetation with liverworts or lichens. Monographic treatments of Swedish ombrotrophic mire vegetation from the beginning of the twentieth century (Osvald 1923; Du Rietz & Nannfeldt 1925) do not mention any of these now-widespread communities. Conversely, in rheotrophic mires with so-called rich fen vegetation (water pH >5.5), *Sphagnum* sect. *Sphagnum* spp. have often replaced the brown mosses (Beltman *et al.* 1995; Gunnarsson *et al.* 2000). Taken together, these changes have usually decreased the species diversity of the plant cover.

These vegetation changes in virgin Swedish mires might have been triggered by (1) drying of the mire surface, (2) increased availability of nitrogen and (3) more acid conditions (Gunnarsson *et al.* 2002). The reason for the drier conditions may either be increased temperature or lowering of the subsoil water level in the surrounding landscape by drainage for forestry and agriculture, because there is no evidence of a decrease in precipitation, and the water level on a mire will always be in a kind of equilibrium with that of its surroundings. The increase in nitrogen availability and acidification both result from airborne pollutants of anthropogenic origin, among which the nitrogen compounds are the most noxious. The critical load of nitrogen for ombrotrophic bogs may be 0.5–1.0 g per m^2 per year (Bobbink *et al.* 2003), which is now exceeded over large areas. Such supply rates will hamper the growth of several, however not all, of the dominating *Sphagnum* spp. (Twenhöven 1992*b*; Risager 1998; Gunnarsson & Rydin 2000). The observed increase in vascular plant productivity may have resulted from reduced competition from the *Sphagnum* mosses rather than increased supply of nitrogen (Malmer *et al.* 2003; Malmer & Wallén 2005). The increase in the grass *Molinia caerulea* in blanket mires to the west of Britain, for example, is a further feature of very recent origin that appears to be correlated with air pollution (Chambers *et al.* 1999). The decline of heather (*Calluna vulgaris*) and of *Sphagnum* species in such regions as the southern Pennine Mountains (UK) is also influenced by nitrogen deposition, but here pollution with sulphur is a further cause of vegetation degradation (Ferguson & Lee 1979).

The establishment of heath-like vegetation on ombrotrophic mires, replacing much of the vegetation with a high abundance of *Sphagnum* mosses, will have strong influence on the carbon balance. The litter from *Sphagnum* mosses contributes much more to the peat formation and rate of carbon accumulation than the much less recalcitrant vascular plant litter. Moreover, the disintegration of the moss cover may open up the peat for erosion and increased carbon losses through the drainage water (Proctor 1997). Calculations of the carbon balance of two ombrotrophic mires, one in boreo-nemoral southern and the other in sub-alpine northern Sweden, have shown that the whole mire surface (or nearly so) must have been covered by a *Sphagnum*-dominated peat-forming vegetation until the end of the nineteenth century to account for both the estimated decay losses and observed peat accumulation rate (Malmer & Wallén 2004). On both mires today, only about 40% of the surface is covered by peat-forming vegetation. The future influx of carbon to the catotelm in these mires may therefore only account for the release of carbon from the catotelm and the total carbon loss from the surface may approximate the carbon in the annually formed litter; thus these mires will not continue to act as sinks for carbon. Similar results have been obtained from Finnish and Swedish *aapa* mires (Klarquist *et al.* 2001; Mäkilä *et al.* 2001) and from tundra mires in Alaska (Oechel *et al.* 1993, 1995).

From 1975 to 1995, the flux of carbon as carbon dioxide increased significantly from one tundra mire in northern Sweden, particularly as a result of a disintegration of the permafrost changing the vegetation and increasing the decay in the deep peat layers (Svensson *et al.* 1999; Christensen *et al.* 2002; Malmer *et al.* 2005). Even more important is the fact that atmospheric methane levels could rise strongly in the event of peatlands both becoming drier (Alm *et al.* 1999) and subject to higher soil temperatures (Christensen *et al.* 2002). Experimental studies with isolated peat cores suggest that raised atmospheric carbon dioxide levels could lead to a doubling of methane emissions from peatland (Hutchin *et al.* 1995).

Potential states of bogs in 2025 and beyond

EFFECTS FROM DESTRUCTIVE EXPLOITATION

By 2025, the global human population may be 25% greater than now (UN [United Nations] 1999; Chapter 1), but in the areas of the world where the northern peatlands are found, the population density may not be very different from the present. The likelihood is therefore that there will be little loss of cool temperate peatland as a direct consequence of human population pressures on space. Urbanization is not very likely, except in centres of oil exploration or forest industry. Increasing exploitation of the peatlands for agriculture and forestry similar to that in Western Europe and eastern North America seems unlikely for the next few decades in less densely populated regions in the interior of continents. However, human demands for resources other than space may place pressures on the temperate bogs. To make use of and transport other natural resources like minerals, oil and gas, it may be necessary to destroy large areas of peatlands, as in western Canada, where peat would have to be removed over large areas to extract the oil from extensive sand deposits.

The demand for peat in horticulture will undoubtedly continue for some years, but peat extraction for these purposes will be limited by transport costs and public pressure from conservation groups. The bogs most at risk are those on the southern fringe of their distribution, particularly in southern Canada, northern USA and Europe (Maltby 1986). Peat supplies for energy may come under pressure when sources of other fossil fuels are either exhausted or so expensive that the peat alternative is economically viable. In North America this is unlikely to occur by the year 2025 because of the continuing availability of conventional fossil fuels and the expense of using peat from boreal Canada. There will, however, be local pressures on peatlands south of the boreal zone in the USA, such as in Minnesota and in North Carolina (Maltby 1986). Projections in northern Europe, e.g. Ireland, suggest that the proportion of energy derived from peat is likely to fall rather than grow in the coming years.

Nevertheless, Russia constitutes a major area of cool temperate bogs in which peat exploitation for power is likely to be a significant threat to peatlands because the transport costs of alternative fossil energy are likely to be relatively great (Moore & Bellamy 1973). The use of biomass fuel based on tree growth in Siberia is the most reasonable alternative. There is no technical reason, however, why peat exploitation and forestry should not proceed in concert, since cut-over bog could provide an appropriate base for the cultivation of fast-growing biomass crops.

In some habitat types, such as forest, piecemeal destruction can result in fragmentation, which in turn can constrain species movements between sites (Collingham & Huntley 2000) and result in genetic isolation and impoverishment. In the case of peatlands, however, fragmentation may already be a part of the system in which both the animal and plant species have evolved (Moore 1990), because the sites develop largely in isolation from one another, and the species that inhabit them must be able to colonize new locations from afar (Moore 1982). The consequence of these geographical features is that fragmentation and subsequent recolonization in isolated locations is unlikely to be a serious problem. However, despite this strength, cool temperate peatlands represent a unique system of wilderness accommodating a highly specialized fauna and flora which is sensitive to all kinds of disturbance.

EFFECTS FROM AIRBORNE POLLUTANTS

Air pollution, including acidification, sulphur enhancement and nitrogen enrichment, will continue to influence cool temperate peatlands, particularly in densely populated industrialized regions. In particular, acidification will lead to rheotrophic mires becoming more oligotrophic; even sites with a pH >5.5 may become acidified (Beltman et al. 1995), despite being buffered by groundwater sources of calcium carbonate. It is possible that there will be more marked contrasts in vegetation with microtopography, because even a few centimetres in elevation could cause a considerable change in local pH and hence in vegetation (Bellamy & Rieley 1967). Ombrotrophic mires are already acidic in reaction, and unless sulphur deposition becomes extremely heavy, are not likely to be significantly acidified.

Acid rain with high concentrations of sulphur will hamper the growth of Sphagnum spp., while even slightly increased concentrations of nitrogen compounds in the atmospheric deposition have already caused great changes in the vegetation, particularly on ombrotrophic peatlands (Lee 1998). These effects will probably be enhanced in the future, although in remote areas Sphagnum spp. have increased productivity with increased nitrogen deposition (Vitt et al. 2003). Because of the sensitivity of many Sphagnum spp. to high nitrogen supply rates, their abundance and productivity will tend to decrease. As a consequence, the part Sphagnum spp. play in the photosynthetic sequestering of carbon will decrease and inevitably result in a more decomposable litter, higher decay rate in the acrotelm and less carbon being accumulated as peat in the catotelm (Malmer & Wallén 2005).

INTERACTIONS WITH THE ATMOSPHERE IN A CHANGED CLIMATE

Although significant climate changes over the next two decades seem likely, precise sets of conditions in the cool temperate zone in 2025 are difficult to predict and will undoubtedly vary geographically within this zone. Projections indicate that regions north of the Arctic Circle are likely to experience the most extreme rise in temperature, of over $2\,^{\circ}\mathrm{C}$ by 2020–2030 (Zwiers 2002; Tao et al. 2003). For the cool temperate zone, the most probable rise in temperature is $c.0.5\,^{\circ}\mathrm{C}$ by the year 2025 (Wigley 1989), but it could be as much as $1\text{–}2\,^{\circ}\mathrm{C}$ (Tao et al. 2003). More important than these annual means may be the degradation of the permafrost, the shortening of the period with frozen peat and the increased length of the vegetation period.

A higher temperature in the atmosphere will increase both its water vapour content and transport of water from the oceans to the continents. Over the past 13 years there has also been a trend towards increasing precipitation over higher latitudes, but only rather small changes may occur in the mean annual precipitation, potential evapotranspiration and surface runoff in the cool temperate zone by 2025, compared to the period 1961–90 (Tao et al. 2003). Models indicate that the continental interior of high-latitude Russia is likely to become drier in summer and rainfall will probably occur in periodic heavy episodes rather than well-dispersed form (Christensen & Christensen 2002). The most critical set of conditions affecting the status of cool temperate bogs might be a combination of higher summer temperature and lower more periodic rainfall, which is likely to lead to increased frequency of summer droughts. Frequent episodes with high rainfall will also increase the erosion of the peat and the amount of carbon lost in the runoff.

The most probable scenario may be that summers will become warmer and drier, springs and autumns longer and wetter, and the winters shorter and milder (see Moore *et al.* 1996). It is salutary to look back in time to seek patterns against which such a scenario could be tested, although the processes are not reversible. The situation 6000 years ago could be helpful in this respect (COHMAP Members 1988). Boreal mires were much scarcer at that time and more confined in their distribution. If the conditions of 6000 years ago (including average temperatures *c*.2 °C higher and lower summer precipitation) had persisted, the northern peatlands would never have reached their present distributions. Any move in the direction of such a climate might thus be expected to initiate a decline in peatland expansion. Feedbacks in the balance of carbon with the atmosphere would enhance such a climate shift and accelerate the decline.

A future rise in temperature alone would be expected to increase both productivity and decomposition rates, because any such rise would increase enzyme activity. Ultimately it could also result in the poleward movement of some species, but this is improbable by the year 2025. Trees might generally be expected to increase in abundance in the peatlands, which would affect the evapotranspiration rate, leading to the drying of surface layers (Chambers 1997). The expected loss from the carbon reservoir in the peat would, to some extent, be compensated for by such an increase in carbon in the overall biomass of the ecosystem. This uptake of atmospheric carbon, however, would occur only while the afforestation of the mire surface was in progress. The development of forest over land that has formerly been devoid of trees also leads to a decrease in albedo, especially where snow-lie is frequent (Betts 2000). On the basis of a 2 °C temperature rise, albedo changes associated with tree invasion of boreal mires could result in heat absorption equivalent to a further 4 °C in spring and 1 °C during other seasons (Foley *et al.* 1994). This feedback effect could enhance the changes associated with temperature rise, particularly in increasing the rate of decomposition.

In general, warmer springs and summers should lead to an increase in growing-season length and in particular to earlier spring starts for plant growth. This in turn should lead to elevated primary productivity, although other factors such as nitrogen or phosphorus availability could set a ceiling on any rise. Vascular plants will presumably react particularly to temperature conditions during the summer, while the water regime and occasional droughts during that period will have less effect on their productivity. In contrast, the growth of *Sphagnum* mosses depends crucially on the supply of water to the moss surface by capillary rise in the sponge-like structure of the densely packed plants. This is

particularly true of hummock-dwelling species, which often have fine leaves and a high degree of capillarity and water retention (for example *S. capillifolium* and *S. fuscum*), while pool species (such as *S. cuspidatum* and *S. recurvum* coll.) possess poorer water-retention properties (Clymo 1973). The productivity of *Sphagnum* mosses will therefore be hampered by summer droughts because of high temperatures, but gain from the longer vegetation period, with good moisture conditions during spring and autumn.

Higher temperatures would increase the decay rates and result in faster decomposition and higher decay losses in the acrotelm, leading to a reduced input of carbon to the catotelm for permanent storage (Christensen *et al.* 2002). This means a changed carbon balance and peatlands would be less efficient sinks than formerly. A deeper mean water level because of a higher temperature and greater evapotranspiration would in principle enhance this effect as the residence time for the litter in the acrotelm would increase. The expected higher decay rate in the acrotelm will increase the rate of mineralization and the cycling of nitrogen and phosphorus. This will contribute to an increase in productivity of the vascular plants but not *Sphagnum* mosses, since the latter do not have access to the mineral nutrients released below ground (Malmer *et al.* 2003). Shading from a denser cover of vascular plants could then reduce the growth of mosses and thus formation of the decay-resistant *Sphagnum* litter. This would in turn further increase the decay rate in the acrotelm and reduce the rate of carbon accumulation in the catotelm. Peat layers devoid of moss cover are also exposed to erosion by wind and water (Malmer & Wallén 1999). Further effects of an increased biomass of vascular plants could be increased production of methane because of excretion of easily degradable carbon compounds from the roots and increased release of methane through plants with aerenchyma tissue, with climatically serious consequences.

Drying of mire surfaces in both the continental and oceanic regions will render them more susceptible to fire (Tallis 1987), with marked influences on species composition and growth of bogs, again reducing their efficiency as carbon sinks (Turetsky *et al.* 2002). Because of the expected drier summers in the interior of the continents, the frequency of fires in the landscape might increase in future. This will also negatively affect peatlands, although it is initially the decomposing surface layer of the acrotelm that will burn, and not the permanently stored peat in the catotelm. Fire in tropical peatlands has been shown to cause considerable injections of carbon into the atmosphere

(Schimel & Baker 2002), so the burning of the northern peatlands would impact the carbon cycle. Lightning is a potential source of fires but there is a generally low likelihood of this throughout the northern peatlands (Clarke 2003). In the case of blanket mires especially, a drying of the mire surface can also contribute to erosion and large-scale peat loss and oxidation (Tallis 1985). Changes in both natural and human-induced disturbance regimes evidently ought to be incorporated in estimates of the carbon balance for peatlands.

CONCLUSIONS

There are many unknown factors in predicting the future state of cool temperate peatlands, some related to the precise nature of future temperatures and precipitation conditions, and others related to the peat-forming process itself. Threats to the cool temperate peatlands over the 2025 time horizon can be regarded as human induced, either directly in terms of habitat destruction and exploitation, or indirectly as a consequence of climate change. The direct impacts are likely to be geographically localized and concentrated in northern Europe, parts of Russia, southern Canada and parts of the northern USA. Climate impacts will be more general, but are likely to be most severe in the continental interiors of Canada and Siberia.

In the regions dominated by peatland complexes consisting of extensive elevated ombrotrophic domes of peat in the coastal areas of the continents, the annual variation in the water regime closely follows precipitation. The changes in the vegetation may therefore depend less on future changes in the temperature than on changes in the amount, regularity and seasonal distribution of precipitation. Oceanic regions are less likely to suffer summer drought; indeed, summer rainfall may increase, in which case the peatlands of these regions should not suffer damage because of climate change. On both sides of the Atlantic, these regions are exposed to an increasing supply of nitrogen and other airborne anthropogenic pollutants. Therefore, the growth of the *Sphagnum* mosses will still be hampered, paving the way for increased growth of trees and other vascular plants, particularly if combined with deeper water levels. The peatlands in these regions have long been heavily exploited, resulting in an overall net release of carbon to the atmosphere during the last few centuries. The expected vegetation changes will all reduce the capacity for carbon accumulation in the remaining peatlands, but it is uncertain whether or not they may become net sources by 2025. However, exploitation may continue, but at a slower rate, and the overall net carbon balance for the peatlands in these oceanic regions might still remain negative.

The most widespread type of peatlands, the gently sloping, non-elevated, predominately rheotrophic peatland complexes found in the interior of the continents, are usually less exposed to atmospheric pollution than peatlands in the coastal areas. For these peatlands, effects of changes in temperature and water regime might therefore predominate. Most probably, trees may be more widespread and show an increased growth in these peatlands in the future, but because of the low fertility of the sites this will hardly be visible by the year 2025. However, seen over a longer period of time, this tree vegetation will result in a drying out of the forested peatlands, because of increased evapotranspiration and interception of the precipitation. During the next few decades, higher temperatures and, particularly, shorter periods with frozen peat may also increase the emissions of carbon dioxide, methane and dinitrogen oxide. If summer droughts hamper the growth of *Sphagnum* mosses more than is compensated for by the longer vegetation period, the productivity of the vascular plants will increase. The overall effect may be a decrease in the net carbon accumulation rate, but is unlikely to be a switch to a net source of atmospheric carbon. However, if greatly increased exploitation of these peatlands occurs, overall carbon dioxide emission will increase and perhaps also give rise to a switch to net carbon flux to the atmosphere.

The third group of peatland complexes, the tundra mires in the Arctic regions, are characterized by permafrost in the catotelm. The expected temperature increase will result in a thawing of much of the permafrost, and thus increased release of gaseous carbon compounds and dinitrogen oxide. At the same time, productivity will probably also increase among the vascular plants. However, the decay rate in both the acrotelm and catotelm might increase so much that the rates of peat formation and carbon accumulation in the catotelm will decrease. These peatlands might then change from sinks to sources of atmospheric carbon as the increases in productivity and carbon sequestering in the plant cover will not compensate for the increased decay loss from the large store of organic matter.

Seen in a global context, the most serious changes expected for cool temperate peatlands up to 2025 and beyond particularly refer to their function as sinks for atmospheric carbon and their release of greenhouse gases. A reduced capacity to accumulate atmospheric carbon is expected, but the magnitude of this reduction cannot be calculated because the present input rate of carbon to the catotelm in these peatlands is unknown. Accumulation rates of >200 years ago can be compared but these refer to climate conditions rather different from those of the present. Neither can we calculate the expected increase in the release of greenhouse gases since no baseline values are available. However, effects on cool temperate peatlands of expected climate changes will have feedback effects on climate, contributing to an increase in the global warming potential of the atmosphere and thus reinforcing the expected changes in the climate.

9 · Temperate freshwater wetlands: response to gradients in moisture regime, human alterations and economic status

MARK M. BRINSON, BARBARA E. BEDFORD, BETH MIDDLETON AND
JOS T. A. VERHOEVEN

INTRODUCTION

Freshwater wetlands in the temperate zones of North and South America, northern Europe, northern Mediterranean, Asia (Russia, Mongolia, northern China, Korea, Japan), southern Australia and New Zealand encompass a wide range of climates and human cultures, not to mention a vast array of wetland types. A previous review of these wetlands encompassing their types, perceived functions and major changes indicated that continued degradation was likely into the future, through loss of area and deterioration in condition owing among others to hydrological alterations, eutrophication and harvesting (Brinson & Malvárez 2002). This chapter extends that analysis by examining sources of variation across moisture regimes and different socioeconomic regions. Climate varies from situations with extreme water deficits to zones with abundant rainfall. This framework is used to identify differences in the relative abundance of wetland types, their

Aquatic Ecosystems, ed. N. V. C. Polunin. Published by Cambridge University Press. © Foundation for Environmental Conservation 2008.

capacity to carry out ecological functions, the extent to which they have been altered by human activities and the way in which restoration may be affected by climate. The developing and developed country duality has shortcomings, but is here used to examine regional socioeconomic status as another source of variation in how wetlands are managed globally. As these issues are not limited to the temperate zone, examples from other climates are included when relevant (see also Chapter 10).

PATTERNS ACROSS MOISTURE GRADIENTS

Wetlands undergo two profound changes along climatic–moisture gradients, namely shifts in the total proportion of the landscape covered by wetlands and in relative abundances of wetland types or classes (Brinson & Malvárez 2002). Because moisture regimes are a function of both precipitation and potential evapotranspiration, 'arid' and 'humid' are here used as terms expressing climatic water balances. The potential evapotranspiration (PET) ratio (potential evapotranspiration divided by precipitation) is >1 for relatively arid climates and <1 for relatively humid ones. The PET ratio normalizes effects of evapotranspiration losses that are controlled largely by temperature (Holdridge *et al.* 1971). Riparian zones adjacent to rivers in arid climates are treated also, whether or not they conform to a particular definition of wetland, because they represent the dry end of the floodplain geomorphic setting where wetlands are common (Chapters 2 and 4). This treatment is somewhat speculative because few studies directly compare wetlands across climatic gradients (Chapter 4). However, there appear to be differences along moisture gradients in terms of functioning, the types of alterations that affect them and issues that arise in restoration of these ecosystems. These comparisons may be useful also in projecting effects of climate change in areas for which changing moisture regimes can be predicted (Burkett & Kusler 2000; Poff *et al.* 2002).

Types

The functioning of wetland ecosystems is driven primarily by their hydrology and geomorphic setting. Hydrologists widely recognize that groundwater flows in and around wetlands change through time, both within and among years (Chapter 3). Seasonal and interannual reversals of flow in both arid and humid regions reflect changes in the relative contribution to wetland water budgets of different components, namely inputs of precipitation, surface water and groundwater, and outputs to groundwater, surface water and evapotranspiration. Seasonal and longer-term shifts in the relative importance of these components are driven primarily by precipitation and temperature patterns, as well as antecedent conditions in a wetland's watershed. Dramatic shifts are more likely in arid and semi-arid regions where precipitation patterns are much more variable than in humid climates.

Wet flats and many depressional wetlands that rely on precipitation as a principal water source are found only where the PET ratio is <1. Organic soil flats (peatlands; Chapter 8) and mineral soil flats are wetlands by virtue of poor drainage owing to low topographic relief, although blanket mires are found on gentle slopes in oceanic climates (Moore 2002). Because flats depend on precipitation as the sole or dominant source of water, they are the first to disappear along the moisture continuum. Depressions that are not maintained largely by groundwater are next to depart, although some persist if their catchments are large enough to capture overland flow to support them. Those that rely on groundwater in semi-arid climates may develop saline waters during dry years (Sloan 1972; Chapters 3 and 7).

Wetlands that receive other sources of water (mainly groundwater discharge) can be sustained where the PET ratio is >1 (slope, riverine and lacustrine fringe). These classes shift in importance depending on local conditions. Lacustrine fringe wetlands are likely to be the least dependent on precipitation. Riverine wetlands on floodplains of perennial streams in arid climates owe their maintenance in part to groundwater discharge and overbank flow driven by upstream sources (Chapters 2, 4). As a consequence, the proportion of flood plain occupied by hydrophytic vegetation diminishes along the climatic gradient from wet to dry (Kroes & Brinson 2004). In addition, the occurrence of rivers decreases as a function of lower drainage density (Leopold *et al.* 1964). Slope wetlands are commonly fed by large groundwater sources, but they cover relatively little surface area. Where wetlands occur in arid climates, they are hotspots of biodiversity, and may even accumulate peat because of almost continuous saturation.

Physiographic setting also contributes to the distribution pattern of wetlands (Winter 2001). For example, rain shadows on the lee sides of mountains in the Andes of Argentina and Sierra Nevada of USA create arid conditions, yet they support riverine wetlands maintained by flows from distant sources at higher altitudes where the

PET ratio is low (<1). The tendency for greater inter-annual variation of precipitation in arid regions (Patrick 1995) may explain a disproportionate reduction in wetland presence beyond that based on average PET ratios.

The species composition of wetland vegetation shifts along biogeographic gradients. Regardless, several core tree genera (*Populus*, *Salix* and *Alnus*) are consistently present and *Fraxinus*, *Betula*, *Celtis* and *Acer* are widely distributed, especially in the riverine flood plains of Europe and the USA (Wiegers 1990; NRC [National Research Council] 2002). Core genera of herbaceous wetlands include *Typha*, *Carex* and *Scirpus*. The diminished area of flood plains in arid climates may have similar geomorphic characteristics to those in more moist climates, but soil properties differ (Friedman & Auble 2000). A comparison of riparian forests across moisture regimes illustrates that disturbance plays a larger role in semi-arid climates and anoxia is a greater selective factor in humid climatic regimes (Fig. 9.1). Both the relative rarity of wetlands in arid climates and their distinctive tree stature (in contrast to surrounding desert and grassland) contribute to regional biodiversity.

Functions

The functions of wetlands differ along moisture gradients in levels of biomass production, decomposition, nutrient cycling, provision of habitat for specialized species and biodiversity. Although enormous variability exists within regions, among hydrogeomorphic settings, and from year to year, some generalizations can be made about climate differences. The most obvious is the extent to which the wetlands accumulate carbon. This function has become increasingly important as the concentration of carbon dioxide (CO_2) in the atmosphere approaches a doubling relative to pre-industrial levels. In general, within the temperate zone, only those wetlands occurring in moist regions have significant carbon stores. Where the PET ratio is low, most wetlands dry infrequently. Under these conditions, production rates exceed those of decomposition,

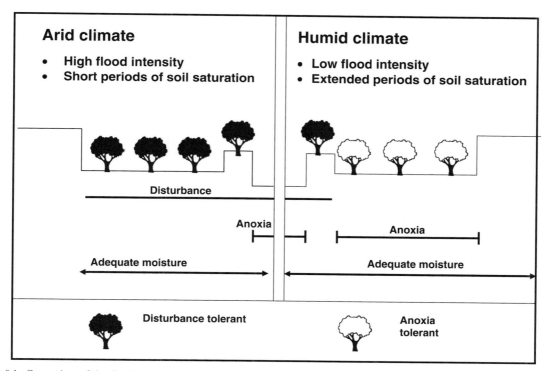

Fig. 9.1. Comparison of the distribution of vegetation in riparian zones from arid climates where disturbance is more influential (left panel) and humid climates where anoxia is more prevalent (right panel). (Modified from Friedman and Auble (2000) with permission from Cambridge University Press.)

which is slowed by the anoxic conditions, and organic matter accumulates. The colder temperatures and shorter growing seasons of higher latitudes and altitudes contribute to slow decomposition and organic matter accumulation. Carbon and nutrients remain tied up in this organic matter. In arid and semi-arid regions, water levels fluctuate widely and water tables frequently drop below the soil surface, thus accelerating system decomposition (Neckles & Neill 1994). This minimizes opportunities for the accumulation of soil organic matter (Schlesinger 1997).

A gradient in rates of nutrient cycling parallels that of organic matter accumulation. Where decomposition rates are slow, nutrient turnover is also slow. In wetlands with large carbon stores, the pools of nitrogen (N) and phosphorus (P) stored in organic form are also large and cycle relatively slowly. Denitrification readily transforms nitrate to nitrous oxides and nitrogen gas (N_2), especially at the interfaces between uplands and wetlands. Extremely small pools of available N and P are cycled rapidly, because microbes and plants use them as soon as they become available. Generally, N and P leave these wetlands only as N_2 or as organic forms that have less potential than nitrate and available P to cause eutrophication in downstream rivers and lakes. When these wetlands are drained, however, they become sources of CO_2, nitrate and available P, both by mobilization through decomposition and export through enhanced flows.

Wetlands in arid and semi-arid regions become P sinks by virtue of their topography or geology rather than biological processes. In depressions that dry frequently, wetlands cycle C, N and P rapidly, but act as physical traps for P adsorbed to soil particles eroded from uplands. If the surrounding terrain is rich in carbonates, P may also be co-precipitated in secondary minerals as groundwater rich in calcium carbonate reaches the wetland surface (Chapter 3). Furthermore, P can be incorporated into amorphous and crystalline structures of iron (Fe) and aluminium (Al), which are affected by pH and the reduction–oxidation status of the soils. In the rich fens of New York (USA), 7–13% of P is calcium-bound and 7–14% is bound to Fe and Al (C. T. Chapin & B. L. Bedford, unpublished data 2003). Typically, P is below detection levels in the shallow groundwater of these fens.

Ecosystem dynamics in both arid and humid regions may be influenced as much by hydrogeomorphic settings and surrounding land uses as by moisture gradients. Hopkinson (1992) found that degree of system closure played a major role in controlling nutrient cycling and retention in wetlands of the humid eastern USA. Wetlands in closed basins retained 90% of nutrient inorganic inputs while those in open riverine settings retained <5%, even though they cycled nutrients more rapidly. In the same region, however, no significant differences were found between floodplain wetlands and those occurring in depressions in terms of accumulation rates of total N, organic C and sediment (Craft & Casey 2000). Total P accumulation was 1.5–3 times higher in floodplain wetlands. Craft and Casey (2000) did not examine inorganic N or P, nor did they construct nutrient budgets that would have allowed them to determine percent retention. They attributed the greater variability in sediment and nutrient accumulation among depressions, rather than between flood plains and depressions, to the degree of human disturbance within contributing watersheds. Although these studies are not entirely comparable, they do caution against making generalizations based solely on moisture gradients.

In arid and semi-arid regions, wetlands are oases for plants and animals that can survive nowhere else (Chapter 4). While the number of plant species tends to be lower than in humid regions, these wetlands support more species of plants, animals and birds than any ecosystem in the surrounding landscape. In humid parts of the temperate zone, other ecosystems fail to match freshwater wetlands in the array of distinctive plant communities and number of plant species. Fully one-third of all plant species of the USA occur in wetlands (US Army Corps of Engineers 2003).

In fact, much of the floristic diversity of herbaceous wetlands worldwide may occur in temperate rather than tropical regions (Crow 1993). This reversal of the typical latitudinal gradient, where diversity reaches its maximum in the tropics, has not been examined extensively, but is true of wetland grasses and butterflies (A. M. Ellison, unpublished data 2002). Species richness of temperate herbaceous wetlands is typically higher than in their tropical equivalent in Costa Rica (Crow 1993). In one comparison, five times more plant species were found in a New Hampshire (USA) bog (108) than in a high-altitude Costa Rican bog (20). The tropical wetlands of southern China have a mean species richness of only 2.6 species per m^2, whereas that of wetlands at higher latitudes or altitudes is 10.4–18.67 species per m^2 (Song & Dong 2002). Herbaceous wetlands in Belize and the Yucatán Peninsula (Mexico) were characterized by only a few species (Rejmánková *et al.* 1996). Some herbaceous wetlands in the temperate zone may support only a few plant species, especially those with high nutrient loading and continuously flooded soils (Keddy 2000).

In contrast, however, some temperate wetlands are exceptionally rich in vascular plants and bryophytes, support a large number of rare and uncommon plant species, or contain an exceptional number of endemic or disjunct species (Walker & Peet 1984; Zedler 1987, 2003; Verhoeven & Bobbink 2001; Bedford & Godwin 2003). Notable among these sites are the rich fens in temperate regions of Canada, Poland, England, the USA and the Netherlands (Wheeler 1984; Nekola 1994; Vitt *et al.* 1995; Almendinger & Leete 1998*a, b*; Verhoeven & Bobbink 2001; Amon *et al.* 2002; Bedford & Godwin 2003), the coastal plain ponds described by Reschke (1990) of eastern North America (Keddy & Wisheu 1989) and California's vernal pools (Zedler 1987, 2003). Rich fens exhibit high diversity within small plots, and high species turnover among plots within and among sites (Vitt & Chee 1990; Olivero 2001; Bedford & Godwin 2003). Fens also contribute to floristic diversity by enriching regional floras. The rich fens of Colorado (USA) contain 10 species of vascular plants and four species of moss that are disjunct from their primary distribution in the Arctic and boreal regions of Canada and the Great Lakes (D. J. Cooper, personal communication 2002; see Hultén 1968 for distribution maps). The vernal pools of California (USA) have long been known to support a large number of rare and endemic plant species. Seven of the genera found in vernal pools are entirely endemic or primarily associated with the California Floristic Province, and another four genera have at least one endemic species (Zedler 2003). These genera are also exceptionally rich taxonomically with between 6 and 15 plant species per genus occurring in vernal pools.

Human-induced alterations

The way that humans alter wetlands can differ between humid and arid climates. Inter-basin transfers (human-induced translocation of water from one watershed to another) occur in both climate extremes, but differ because of the prevalence of gaining streams in humid climates (groundwater discharges to the stream) and losing streams in arid climates (channel water recharges underlying aquifers; Chapter 3). Riparian areas in arid climates will tend to suffer more than those in humid zones because flows continue to decline below the point of water removal. For gaining streams in humid climates at least some

recovery is possible. In the receiving basins of arid climates where water is used principally for crop irrigation, return flows may actually support wetlands by raising ground-water tables and contributing to the flow of otherwise dry streambeds. Either way, inter-basin transfers compete directly for water from the donor basin and can significantly alter riverine wetlands in the receiving basin.

Inter-basin transfer for development has profoundly affected wetlands worldwide (Middleton 1999); 77% of the world's waterways in the northern hemisphere have been altered by inter-basin water transfer, dams or channelization (Dynesius & Nilsson 1994; Chapter 2). Two examples from Illinois (USA) illustrate the phenomenon of too much or too little water for wetlands in a relatively humid climate. In one example, the Cache River has caused the drying of flood plains downstream because flows from its headwaters have been diverted directly to another basin. Conversely, inter-basin water transfers have been related to flooding on a grand scale. Water diverted from Lake Michigan by the city of Chicago is transferred to the Illinois River, where it has been implicated in flooding hundreds of miles away in Missouri (Sparks *et al.* 1998; Smith & Mettler 2002). Furthermore, navigation weirs and levees along the Illinois River create unseasonably high water levels on flood plains, although parts of the river still have seasonal flood pulsing (Sparks *et al.* 1998). Because this portion of the river still retains a flood pulse, the Illinois River has been deemed as one of only three large rivers in North America that could be restored to something of its original character (The Nature Conservancy 2003).

Inter-basin transfers in arid regions for the purpose of crop irrigation are often constructed at the cost of water for wetlands and riparian zones (Ellis *et al.* 2002; Scott & Auble 2002; Stromberg & Chew 2002; e.g. Chapters 2, 3 and 4). This situation is likely to worsen in parts of the world with predicted global warming and increased aridity. Some rivers in the western USA are already subject to so much water removal that almost no water remains downstream of the major abstraction points (Chapters 2 and 4). These dried flood plains cannot maintain native forest species such as *Populus*, and more tolerant exotics such as *Tamarix* are becoming common (Ellis *et al.* 2002; Stromberg & Chew 2002). The extensive marshland of southern Iraq has largely dried out because of diversion of freshwater flows (Chapter 4); partial reflooding promises to restore parts of the former wetlands, but additional threats loom in the future (Richardson *et al.* 2005).

Irrigation return flows to rivers may transport agro-chemicals and other pollutants, and thus have diminished value for wetland restoration and water management (Ellis *et al.* 2002). Along rivers in both humid and arid regions, water quality has been degraded through decades and even centuries of channel modifications, agricultural encroachment and industrial pollution (Chapters 2 and 4). For example, the River Rhine in Europe has lost many important native species, such as salmon, although water quality conditions have improved recently (Schulte-Wülwer-Leidig 1995). Regional water quality problems are also linked to coal and mineral mining, where acid mine drainage lowers pH and increases heavy-metal contamination (Robinson & Robb 1995).

Groundwater that feeds springs and fens has been extracted at all points across the moisture gradient, but in response to different alterations (Chapter 3). In the eastern and especially Midwestern USA, groundwater levels have dropped because of tiling, ditching and channelization for agricultural and urban development. In Story County (Iowa) for example, groundwater is no longer available to supply springs and fens (Hewes & Frandson 1952). Similarly, fens in East Anglia (UK) have been threatened by groundwater abstraction (Fojt 1994). River alteration projects for agriculture and/or urbanization have caused sharp drops in groundwater levels, for example, along the Drava and Rhine Rivers in Europe (Zinke & Gutzweiler 1990; Bonacci *et al.* 1992). Arid regions bear the brunt of declining groundwater tables and associated changes in surface waters (Chapter 3). Aquifers in the western USA are depleted (for example Ogallala Aquifer in Texas: Bolen & Guthery 1982), and springs that are essential for desert nomads are drying up (for example the Azraq Oasis in Jordan: Scott 1993). Both humans and spring-fed flora and fauna suffer when this critical resource disappears (Chapter 3).

Dams along the moisture gradient have supported human cultures for thousands of years for the purposes of irrigation and drinking water, and more recently for hydroelectric power (Schnitter 1994). While dams have provided extremely useful human services, they invariably change the channel and floodplain wetland characteristics above and below the dam (Chapter 4). In arid areas, sometimes little or no water is released from the dam (Petts 1984; Karr *et al.* 1986; Middleton 1999), and the previously maintained riparian vegetation below such dams is diminished or extirpated (Middleton 1999; Ellis *et al.* 2002). Dam removal is becoming more common, particularly because ageing reservoirs become filled with sediment and dams are expensive to maintain (Heinz Center 2002).

Cattle graze in wetlands worldwide (Maltby 1986), but are generally overlooked as major modifiers of wetlands (Middleton 1999). Alteration of streams by cattle can cause perennial streams to become dry or intermittent (Sedell *et al.* 1991), thus drying associated floodplain wetlands. Browsing inevitably causes a reduction in tree recruitment and input of woody debris, a situation that ultimately alters channel morphology and degrades fish habitat (Maser & Sedell 1994). While grazing is generally considered more of a problem in arid settings (Donahue 1999), it can be related to decline in biodiversity of humid regions as well (Vermeer & Joosten 1992; Dumortier *et al.* 1996; Jutila 1997; Middleton 2002*a*, *b*). Furthermore, when cattle grazing ceases, exotic and invasive species often assume dominance, which causes even further management problems, especially for fens (for example expansion of *Cornus sericea* and other woody species: Middleton 2002*a*, *b*), but also floodplain wetlands (McCoy & Rodriguez 1994). However, effects of cattle are reversible upon their removal and are minor in comparison with major hydrological engineering projects.

Restoration

Wetland losses have a long history in the temperate zone and have reached a dramatic stage, particularly in Eurasia (Brinson & Malvárez 2002; Chapter 4). For more than a millennium, river flood plains have been converted to arable land by clearing forests and draining fertile soil. Flood protection, drainage and peat mining also have strongly disturbed the functioning of remaining wetlands by disrupting connections with landscape-scale flows and geomorphological processes. Although in Eurasia major alterations took place prior to 1900, the rapidly increasing human population and more intensive land use and flood protection techniques in the twentieth century have had strong additional impacts. Unaltered depressional, lake-shore and riverine wetlands have become quite rare, while semi-natural wetlands, with some important human influence but still characteristic in terms of functioning and biota, may still cover about 20% of the original wetland area. Wetland losses have been less extreme in temperate North America and occurred mainly after 1850, but they have still affected at least 50% of the original area (Dahl 1990). Only 8% of New Zealand's original wetlands remain (Jones *et al.* 1995).

In recent decades, European conservationists have realized that protection of the remaining wetlands will not be sufficient to successfully safeguard biodiversity in wetlands. There is also increasing interest in enhancing functions such as water-quality improvement and flood-water storage (Maltby *et al.* 1994). Restoration of wetland ecosystems has become a major activity in several regions of Western Europe and North America (Chapters 4 and 5). However, the science of ecological restoration is still in its infancy and most restoration projects have a 'trial-and-error' character (Cairns & Heckman 1996). Restoration is particularly challenging in arid climates, where multiple users compete for what is a limited resource. Even in more humid climates, water management for agriculture or flood protection often alters wetland communities because of changes in water chemistry or internal eutrophication (Lamers *et al.* 1998).

Restoration practices have some universal characteristics regardless of the climate. Prerequisites for success should include clearly established goals and good insight into the ecology of the species and habitats to be restored. For example, if characteristic plant species are not already present in the seed bank, they must be able to reach the site through dispersal or migration. In many cases, wetlands have become isolated patches in a non-wetland (drained) matrix in which the hydrological connections between them have been disrupted, so that the potential for dispersal is low. Plantings may overcome this problem, but at great expense, and may even result in problems if the plant material used is genetically different from native varieties. Most restoration projects therefore restrict themselves to providing seeds or vegetative diaspores, while others adopt nature's 'self-design' principle (see Mitsch 1993).

By definition, any restoration project will attempt to restore a disturbed area back to its unaltered condition. Because of the major effects of climate on wetland structure and functioning, it is important to use nearby reference sites to determine in as much detail as possible the original state and provide guidance on what measures should be taken. If the disturbance has occurred in the distant past, it may be particularly difficult to know what this condition was. Palaeoecological studies of soil cores of the disturbed site may provide clues to its original species composition and structure. In most cases, climates have remained sufficiently similar that palaeoecological information is still relevant for hindcasting reference conditions.

Restoration practices should always consider the larger landscape context of wetlands. With few exceptions, wetlands are connected to larger groundwater flow systems, and in the case of floodplain wetlands to surface flows from the surrounding uplands and from upstream (Winter 1999; Chapter 4). Restoration projects therefore should be oriented towards larger spatial scales (Bedford 1996) and may even target surrounding landscape units for restoration, as discussed for the operational landscape unit (OLU) approach described below.

Another neglected factor in restoration projects is time. No generalizations appear to have been made regarding differences between humid and arid climates; however, the greater interannual variability of precipitation in arid than in humid climates increases the uncertainty of natural plant establishment and survival of plantings. Disturbance regimes also vary; high-intensity floods are more likely to be influential on riparian zones in arid than in humid climates (Friedman & Auble 2000). Where disturbance is frequent, restoration should take into account the temporal variation in plant communities. Succession sequences have often been described through space-for-time substitution by recognizing the change in composition along a chronological sequence. Where successional series are well understood or documented by remote sensing, the whole range of stages could become part of the target rather than the initial stages or the mature stages only.

The time-frame becomes critical in practice because 'project success' is important in building confidence and securing funding for further restoration activities. This is less likely to be a variable across moisture regimes than it is among regional differences in culture and economic status. In any case, good monitoring is essential for recognizing success or failure. Superficially, it would seem that herbaceous wetlands would achieve maturity, and therefore 'success,' earlier than forested wetlands. However, soil development and biogeochemical cycling are seldom evaluated as commonly as vegetation cover. Invasion of unwanted species is particularly troublesome in herbaceous wetlands.

Riverine or riparian wetlands have the advantage of contact with streams or rivers, and thus restoration is enabled by the transport of hundreds of plant species between floodplain areas. In contrast, restoration of the hydrological regime is often virtually impossible, although flow regimes may be modified if compromises with human uses are allowed. This type of flexibility is less feasible in arid regions because of water scarcity. In the highly dynamic riparian zones typical of arid regions, flow regulation and sediment removal by dams hinder the expression of disturbance regimes dependent upon large interannual and seasonal variations in flow. Where artificial levees or dykes prevent overbank flow to flood plains, reconnection with the channel may partially restore wetlands, but removal of thick clay layers may also be necessary (WWF [World Wide Fund for Nature] 1993). Fringe wetlands around reservoirs do not usually have any natural counterpart, especially considering the wide range of water-level fluctuations common in impoundments of arid climates. Environmental conditions of the alternately exposed and submerged shorelines tend to be so extreme that no wetlands or plant community can survive.

Depressional wetlands may have a combination of surface and groundwater sources, but are more likely to depend upon surface-water sources in arid climates. Consequently, surrounding land uses affect both the quality and the quantity of overland flow to them. Extraction of groundwater that lowers water tables in humid climates profoundly affects depressional wetlands, such as the cypress domes in the karst region of Florida (USA) (Spangler 1984). Unless water tables can be returned to their previous levels and ranges, restoration is not possible. However, where depressional wetlands have been drained for agriculture, they can be restored by blocking drainage tiles and ditches. If the groundwater discharge is at least partially recovered and a seed bank is still present (or is distributed at the start of the restoration), reasonable results can be achieved (Seabloom & Van der Valk 2003). Soil salinization may pose an additional problem in arid climates (Chapter 7).

Base-rich and intermediate fens are rare in arid climates. In humid climates, they are known for their high species richness, which is attributable in part to the maintenance of habitat diversity created by varied water depths. Where open-water areas have filled in over time, shallow open-water areas can be created to reinitiate succession and to conserve successional complexity by simulating natural disturbance regimes (Verhoeven & Bobbink 2001). The scale of such successional changes is decades to centuries (Bakker *et al.* 1994). Wet flats may be disregarded because they are absent in arid climates.

PATTERNS RELATED TO ECONOMICS AND CULTURE

While climate and geomorphology may control the distribution of wetland types and their functioning, human activities often overwhelm the influence of natural variables.

Many developed countries have long since converted much of their wetland area to other uses, such as agriculture, silviculture and urbanization (La Peyre *et al.* 2001). At the same time, they potentially have the resources and economic flexibility to preserve their remaining wetlands. Some developing countries have also already lost the majority of their original wetlands (Moser *et al.* 2003), while others, including Brazil, have effective wetland or coastal protection laws (Balgos *et al.* 2005). Simple generalizations do not hold.

Capacity for protection and restoration

The capacity for wetland protection and restoration comes from the legal status of protected areas and also from legislation that regulates private lands. For example, most converted wetlands in the USA are owned by farmers who often are willing to become active participants in wetland restoration projects on their wettest farmland when provided with adequate economic incentives (Lant *et al.* 1995). Developing countries have a lesser capacity to reward private landowner efforts for environmental protection because of competing public needs. Further, developing countries are less likely than developed ones to use alternative energy sources to substitute for the goods and services provided by wetlands, such as water-quality maintenance and flood protection. Consequently, a development-oriented politician in a developing country might consider water-quality improvement and fisheries enhancement as important reasons to pursue wetland restoration and protection.

International development money has commonly been used to convert wetlands to alternative land uses for commodity production. Developing countries that have abundant wetland area and international debt may use conversion of wetlands to service their debt and modernize. In this situation, wetland protection and restoration may interfere with economic development schemes. Further, developing countries are less likely to have the resources and infrastructure to protect and manage natural areas. Cultural priorities appear to be partly responsible when creative solutions have been successful (Rodgers 1988). To compensate local people for losses of traditional grazing or gathering areas in the establishment of national parks, buffer zones have been established that allow plant extraction and cattle grazing (Middleton 2003). In reality, buffer-zones are often encroached upon to unacceptable levels, including the protected area core (Kothari *et al.* 1989; Berkmüller *et al.* 1990; Middleton 2003). Nevertheless, implementing the buffer-zone concept, and other similar approaches, probably increases the capacity of developing countries to protect natural areas.

Developed countries often orient restoration toward recreating natural habitats, with the assumption that these will contribute in a general way toward goods and services. In developing countries, objectives may be oriented toward more explicit uses through the production of food, fibre and water quality. For example, wetlands have been restored below a reservoir on the Benue River in Cameroon (Africa) to enhance local fisheries and provide vegetable gardening (Slootweg & van Schooten 1995; Middleton 1999). In contrast, the restoration of flood plains below the Panshet Dam (India) was mostly for wildlife habitat. The important lesson is that no single approach can be successful everywhere. Much can be learned from the diversity of approaches to wetland management used in a variety of economic and cultural settings. It is much too simplistic to cast the economic/cultural issue as opposite poles represented by developing versus developed countries (Milton 1998), particularly where economic forces external to the countries are partly responsible.

Institutional and cultural barriers to management and protection

Theories abound on national characteristics that promote or hinder environmentally sustainable behaviours. These include national wealth, socioeconomic and educational status, cultural norms, environmental institutions and policies, and economic development pressures (Redclift 1992; Arrow et al. 1995; Hukkinen 1998). If the relative importance of these variables in wetland protection could be identified, barriers to conserving the world's wetlands might be more apparent (La Peyre et al. 2001).

The shift in human perception of wetlands from 'stinking abysses' to waterfowl habitat to protected ecological systems took place largely within the past half century (US Geological Survey 1996; Mitsch & Gosselink 2000). In Europe, many national governments have adopted policies for nature protection, while specific attention to wetlands protection arose with the Ramsar Convention in the 1960s (Everard 1997; Farrier & Tucker 2000; Ramsar 2004). Increasing scientific understanding of wetlands and their ecosystem services led to a growing environmental awareness and concern among the public. Meanwhile, the economic boom that followed the end of

the Second World War caused serious degradation of aquatic habitat. Lakes and rivers fouled with industrial waste and by eutrophication made them incompatible with basic activities such as fishing and swimming. In response, the US Congress passed legislation to restore the physical, chemical and biological integrity of USA waters. Court actions initiated by environmental groups generally upheld and strengthened the authority of wetland protection legislation, affording protection to wetlands.

This history, and similar trends in European countries, would seem to indicate that economic growth, education and institutions have all played a role in promoting wetland protection efforts. La Peyre et al. (2001) examined these and other relationships among wetland management programmes of 90 nations using economic, social, political and environmental characteristics, and land-use pressure. Their model used the United Nations Development Programme's Human Development Index (UNDP [United Nations Development Programme] 1995) to define social capital, an index that takes into account measures of education, quality of life and health of a nation's citizens. They identified social capital as having a dominant and linearly positive relationship with wetland protection, as well as positive effects on a nation's governmental structure and the level of democracy and political stability.

Contrary to expectations, the La Peyre et al. (2001) model also showed that the relationship between wetland protection (calculated as the percentage of a nation's wetlands under Ramsar designation) and economic capital was strongly and linearly negative. This indicates that wetland protection will not occur with economic development, but that the opposite may happen. Given that most countries have goals for greater economic development, independent programmes may be needed to protect wetlands (La Peyre et al. 2001). Thus it seems that efforts to raise the social well-being of a nation's citizens, through improving education, health and quality of life, may help overcome barriers to progress in wetland protection. Environmental education also might promote citizen commitment to protection policies and thereby help bring about stronger governmental policies and programmes. Improvement of economic capital alone, namely economic growth, per caput income, investment growth, and external trade and finance (La Peyre et al. 2001), will not resolve issues of wetland protection if management programmes are not in place. Herein lies one of the greatest uncertainties regarding the future status of wetlands.

NEEDS AND PROSPECTS FOR SCIENCE AND MANAGEMENT

Many countries have converted more than half of their original wetland area to alternative uses (Dahl 1990; Brinson & Malvárez 2002; Chapter 4), and prudent management might consider what happens to the remaining wetlands. Effective management requires that quantity, location and variety of the ecosystem first be established through inventory and mapping. A further strategy is to ameliorate past and avoid future damage. Given an almost universal recognition that wetlands should be protected, and the fact that many are degraded, effective management relies in part on knowledge of how to restore those that have been unintentionally and inadvertently altered. Restoration of degraded conditions is the only way that unavoidable impacts on wetlands can be compensated within the context of the resource itself. All of these activities require training to meet these challenges.

Inventory and mapping

For wetlands, both the scale of inventories and the classification by type need to be well planned to meet the objectives of any assessment programme. Advances in remote sensing and geographic information systems (GIS) make inventories more feasible than ever. However, small-scale detail, such as that mapped in the Mediterranean region (Costa *et al.* 1996) and the USA (Cowardin *et al.* 1979), would be impossible for many developing countries (Scott & Jones 1995). Ramsar has developed criteria for planning a wetland inventory. In addition to a number of administrative suggestions, Ramsar (2006) recommended: (1) stating a purpose and objective, (2) reviewing existing knowledge and inventory methods, (3) determining the desired scale and resolution, (4) establishing a core data set, (5) establishing a habitat classification, and (6) choosing an appropriate inventory method. Several of these components hinge on item 5. Ramsar itself does not offer a hierarchical system appropriate for inventory, but instead uses a list of descriptive types that is useful in compiling a 'directory' (Finlayson 1996) for the purpose of characterizing 'wetlands of international importance', the principal goal of the organization (Scott & Jones 1995). However, more recent efforts in Asia point in the direction of a nested, hierarchical approach at four levels, depending on purpose, in which the first two levels provide the framework for a more detailed

inventory and assessment, the third level offers more information on core attributes of wetland complexes and larger sites, and the fourth gives more information at the site/habitat level (Finlayson *et al.* 2002).

While different countries may have somewhat distinct objectives and desire different scales of inventory, there are advantages to using a common set of building blocks given the inconsistency of inventories among and even within countries (Brinson & Malvárez 2002). A partial solution to attaining a minimum amount of useful consistency would be to adopt several similar types or classes high in the hierarchy that would form the basis for (1) inventory and mapping, with further subclassification depending on purpose and desired resolution, and (2) using the classification as a framework for management, through the recognition of functionally similar types. Examples at this level include Dugan (1993) who suggested seven landscape units (estuaries, open coasts, flood plains, freshwater marshes, lakes, peatlands and swamp forest), Keddy (2000) who described six (swamp, marsh, bog, fen, wet meadow and shallow water), Cowardin *et al.* (1979) with five system levels (marine, estuarine, lacustrine, palustrine and riverine) and the hydrogeomorphic (HGM) approach with five freshwater classes (riverine, depression, slope, lacustrine fringe and flats [organic soil and mineral soil]) for the purpose of assessing wetland condition (Brinson 1993). Each would require additional detail to meet the needs of most inventory/ mapping applications (Fig. 9.2). A functional framework facilitates a more direct link with the goods and services of society supported by wetlands. It also provides a framework for management decisions that require impact assessments and planning for restoration.

Another extension of classification is to recognize aggregations of wetlands at an even higher level than class. These combinations or complexes have been dubbed 'macrosystems' by Neiff (2001) and are consistent with Ramsar wetlands of international importance, many of which are indeed complexes of several wetland classes. This macrosystem level is similar to that identified by Winter (1992) and Bedford (1996) as the hydrogeological setting that controls water flows and chemistry of surface and groundwater sources to wetlands (see below). Recognizing and identifying a macrosystem level has two principal advantages: (1) complexes at this level have unique combinations of wetland types (for example the Everglades of the USA, the Paraná Delta of Argentina, the Guadiana River in the high basin of Spain and the Lake Eyre Basin of Australia) (Brinson & Malvárez 2002), and (2) the

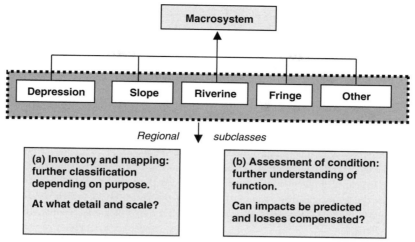

Fig. 9.2. A classification framework based on a core group of classes. Macrosystems are aggregations of classes that form unique combinations at large scales. The core classes may serve as a point of departure for (a) developing subclasses and finer detail for mapping and inventory, and (b) understanding functional characteristics for management.

interconnectedness and interdependency among component parts are based not only on spatial configuration, but also on biological considerations. Macrosystems provide a framework for recognizing cumulative effects at scales larger than a single wetland site or type. Macrosystems can also provide a reference point for restoration where hydrogeological settings can be identified at landscape or catchment scales (Bedford 1996).

Assessment of condition

The impacts described earlier include total displacement of wetlands through drainage, deterioration of condition through alteration and removal of water sources such as in inter-basin transfer, as well as non-hydrological stressors including eutrophication, pollution and cattle grazing. Fragmentation is yet another impact that results in the incremental removal of wetland components that are critical to macrosystem functioning. When policies are developed to compensate for wetland losses, scientists are often asked not only to determine the rate of conversion of wetlands to alternative uses, but also to estimate degradation through losses of condition (NRC 2001). A number of approaches have been developed for this purpose, ranging from the system developed for Europe (Maltby *et al.* 1994) and the Mediterranean (Costa *et al.* 1996), to an array of indices of biotic integrity (Karr 1991) and habitat-based approaches

that use structural components and biotic indicators (Smith *et al.* 1995). Regardless of the approach used and the detail to which it is applied, assessments should have the following attributes.

- The same approach for both impact assessment and restoration. By using a currency and methodology common to both analyses, evaluations of net gains and losses can be more easily compared. Programmes of regulation, restoration and protection will ultimately be required to account for their progress if they are to be politically sustained.

- The need to evaluate not only wetlands, but also the condition of catchment basins that affect them. Principal stressors on wetland condition often originate from outside the boundary of the wetland being assessed. Likewise, individual wetlands are often components of macrosystems that are biologically and hydrologically connected.

Restoration in a landscape setting

Given the extensive loss and degradation of wetlands throughout the temperate zone, their restoration has become a primary goal and challenge in several countries in Europe and North America (Wheeler *et al.* 1995). Numerous examples of both successes and failures exist (Wheeler *et al.* 1995; Beltman *et al.* 1996; Pfadenhauer & Kloetzli 1996;

Detenbeck *et al.* 1999; Kentula 2000; Grootjans *et al.* 2001; Lamers *et al.* 2002; Seabloom & van der Valk 2003). The failures can inevitably be attributed to inadequate attention to the landscape setting that incorporates the regional groundwater flow system determining wetland hydrology and water chemistry (Winter 1988, 1992). Restored sites often fail to show the characteristic environmental features of targeted wetland types in terms of biogeochemical cycling, hydroperiod and erosion/sedimentation. In terms of restoring biodiversity, the entire regional landscape must be considered (Bedford & Preston 1988; Bedford 1996; Zedler 1996). It is the full complement of different wetland types and their distribution in the landscape that support the immense floral and faunal diversity in temperate zone wetlands (Bedford *et al.* 2001).

Several conceptual advances now provide the foundation for restoring wetlands within the context of their landscape settings. These include the concept of HGM setting (Brinson 1993), the hydrogeologists' concept of hydrogeological setting (HGS: Winter 1992; Bedford 1999; Godwin *et al.* 2002), and the earlier mentioned concept of the OLU (J. T. A. Verhoeven, unpublished data 2006). Without moving to the larger geographic scales inherent in the concepts of HGM, HGS and OLUs, more restorations will fail and landscape diversity will be reduced.

The HGM, HGS and OLU concepts are strongly related but differ somewhat in their scale and focus. Brinson's (1993) concept of HGM assigns wetlands to classes that differ in their geomorphology, water sources and hydrodynamics. By drawing explicit attention to these features, rather than to vegetation, the HGM classification of wetlands forced managers to view wetlands within their landscape settings. Application of the HGS concept to wetlands derives ultimately from Winter's (1992) extension of Tóth's (1963) work on groundwater flow systems to lakes and wetlands (see Bedford 1999 and references therein). The HGS concept focuses more explicitly on the landscape than does the HGM concept. Rather than classifying the wetland itself, it classifies those features of the landscape that control wetland hydrology and water chemistry, which in turn determine wetland characteristics. The HGS comprises all those natural features of the landscape that control the flows and chemistry of surface and groundwater to a wetland. This approach forces managers to think about why certain types of wetlands occur where they do and hence what features of the landscape must be restored or recreated before restoration of that type is likely to occur.

Verhoeven's concept of OLUs follows logically from the HGM and HGS concepts, but further considers ways in which humans have altered wetland–landscape linkages (J. T. A. Verhoeven, unpublished data 2006). It recognizes that the natural connections between wetlands and surrounding ecosystems must be an explicit part of any restoration effort. In many parts of the world, however, natural uplands have been altered for agriculture or residential development, the valley slopes have been drained, groundwater recharges limited, and streams straightened and, in some cases, provided with hard fixed banks (Fig. 9.3). In this case, restoration must include not just the wetlands, but also the entire OLU, which consists of the collection of adjacent, hydrologically connected ecosystems in which flows of groundwater or surface water strongly affect their functioning and biota. For example, this would include the upland areas that recharge or drain into the valley slope and stream, and riparian wetlands along the stream, together with the stream and adjacent flood plain (Fig. 9.3). Other OLUs might include small streams with their recharging uplands, wet slopes and banks; mid-size rivers with their marginal wetlands; large rivers with their flood plains, oxbows and river dunes; shallow lakes with their catchments and littoral zone; deep lakes with their littoral zone; and fens with their catchments.

Re-establishing these connections must be a central consideration in all restoration efforts designed and carried out in the coming decades. The challenge is not only to understand better the nature and the dynamics of these connections under undisturbed reference conditions, but also to provide insight into how the connections condition the habitat of biota in the individual ecosystems of the OLU. Studies of the characteristic hydrological, biogeochemical and habitat conditions of reference wetland ecosystems will give clues to the target conditions in a restoration effort. Knowledge of hydraulic engineering will be indispensable in the design of measures to restore characteristic hydrological connections within OLUs.

Training and education

The training of scientists and resource managers builds the capacity for resource protection and restoration. Without investment in training over the next two decades, little improvement can be expected. For generations, would-be natural resource managers have been trained via major study in university curricula, and/or agency-wide training

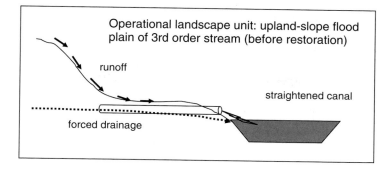

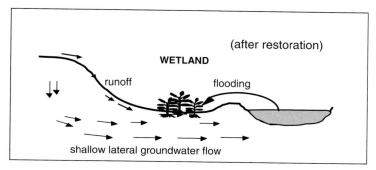

Fig. 9.3. A schematic cross-section of an operational landscape unit (OLU) encompassing a stream valley with a dry–wet gradient, a wet groundwater-fed slope, a flood plain, a stream bank and a stream channel. The upper panel shows how the hydrological connections between the different components have been disrupted in the drained agricultural landscape setting. The lower panel shows the original connections.

programmes specializing in natural-resources management and law enforcement. Some developing countries have only recently added specialized natural-resource courses to the training of their managers.

In India, the Wildlife Institute in Dehra Dun was set up in 1982 to provide training to postgraduates and resource managers, many of whom are posted to national parks and wildlife sanctuaries as park superintendents and management specialists. The Wildlife Institute and USA universities exchange scientists through a training programme (US Fish and Wildlife Service 2003a). The Institute carries out applied research and is building a database for the use of managers and other land-use specialists (Wildlife Institute of India 2003).

European countries offer training, research, policy development and other advisory services worldwide. For example, the Institute of Hydraulic Engineering (IHE, the Netherlands) was established in 1957 to offer international training to practising hydraulic engineers. Over the years, IHE has developed into an autonomous international educational institute that now provides a host of post-graduate courses and tailor-made training programmes in integrated water resource management and research. In 2001, UNESCO's General Assembly declared the existing IHE be established as the UNESCO Institute for Water Education. Within UNESCO's mandate, the new Institute strengthens and mobilizes the global educational and knowledge base for integrated water management and contributes to meeting the water-related capacity building needs of the developing countries and countries in transition (UNESCO–IHE 2003).

The USA/South African Binational Commission focuses on training and capacity building through the efforts of the Nature Conservation and Ecotourism Working Group. South Africans make study tours of environmental and educational facilities in the USA to meet with scientists and managers. This group also develops programmes for rural communities living adjacent to national parks (US Fish and Wildlife Service 2003b).

Technology training and capacity building have become well-worn concepts in natural-resource management and conservation. Because wetland types are so variable within a single biome or political jurisdiction, their management is particularly challenging. The task is made easier and more effective by recognizing the complexity of wetlands and applying management principles that are tailored to differences in functioning and condition. Many of the principles, however, are relatively robust in spite of differences between arid and humid climates, so that technology transfer need not be particularly novel. The real challenge, however, is to match human cultural and socioeconomic conditions with the technical approaches so that the flow of beneficial services to society is achieved. This will require sustained efforts directed toward both public education and the training resources personnel.

Prospects

If current global trends continue, the next several decades will experience an expansion of human populations, intensified international trade and further exploitation of natural resources. This worldwide pattern, however, does not shed light upon individual trends within sovereign countries where management policies are based that might influence wetland status and trends. Regardless, there is no basis for altering the statements (Brinson & Malvárez 2002) that:

> The most industrialized countries are likely to conserve their already impacted, remaining wetlands and move toward selective restoration as opportunities arise and the value of wetland services gains more recognition. Nations undergoing rapid industrialization are accelerating wetland losses now, and may continue to do so for the next several decades, with little attention paid to protection and virtually none to restoration.

In the absence of a clear example of a country in the temperate zone where wetland area and condition have increased substantially over decadal periods, the conclusion is that parallel efforts of increased protection and accelerated restoration are necessary to prevent further losses in area and condition. Both protection and restoration require strong governmental policies and programmes that are related to the social well-being of a nation's citizens, as indicated by improved education, health and quality of life (La Peyre *et al.* 2001). None of this will happen, however, if professionals lack appropriate technical training and the public is not educated to recognize and appreciate the life-support functions of wetland ecosystems.

10 · Present state and future of tropical wetlands

BRIJ GOPAL, WOLFGANG J. JUNK, C. MAX FINLAYSON AND CHARLES M. BREEN

INTRODUCTION

Wetlands are highly diverse and complex ecosystems that include a wide spectrum of aquatic habitats, depending upon the definition adopted for different objectives (Mitsch & Gosselink 2000). They occur in all climatic zones throughout the world (e.g. Chapters 7, 9 and 11). Despite the many functions and values that are now widely recognized, as well as the many national and international initiatives and programmes aimed at their conservation and restoration, wetlands are among the most threatened ecosystems (Williams 1990). During recent years there has been growing concern about their rapid degradation and loss throughout the world, and the factors responsible for their current state are well known (Moser *et al.* 1996; Mitsch & Gosselink 2000; Finlayson *et al.* 2005). The prospects for their future in light of the present state and continuing anthropogenic impacts were reviewed for different kinds of wetlands (Brinson & Malvarez 2002;

Aquatic Ecosystems, ed. N. V. C. Polunin. Published by Cambridge University Press. © Foundation for Environmental Conservation 2008.

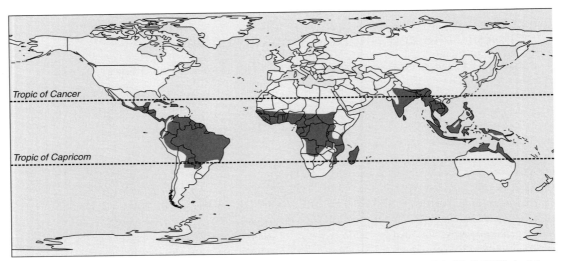

Fig. 10.1. Humid tropical regions (dark shaded areas) of the world. (Adapted from *The Times Atlas of the World* (1999), by John Lowry, Environmental Research Institute of the Supervising Scientist, Darwin, Australia. The political boundaries of the countries are not corrected and not relevant to this chapter.)

Junk 2002). The aim is to complement the overview of Junk (2002) by examining afresh the current state of inland wetlands within the tropical and subtropical regions and their future in relation to various biophysical, social, economic, political and other conditions in different countries. Saline wetlands are only occasionally referred to and are discussed separately (Williams 2002a; Chapters 7 and 11).

Geographically, the areas lying between the Tropics of Cancer and Capricorn are generally considered as tropical (Fig. 10.1), and the climate of these regions is characterized by relatively small seasonal variation in temperature regimes. However, this variation increases with both latitude and altitude, so that temperate climate, characterized by large seasonal variation of temperature and often referred to as montane temperate, is sometimes experienced within the tropical belt, for example in the Himalayas (Fig. 10.1). Importantly, many areas adjacent to these humid zones are arid or semi-arid and experience low or irregular rainfall and large temperature ranges, such as in northern Australia (Finlayson & von Oertzen 1996).

The areas extending from the tropical belt to 35° N and 35° S are commonly referred to as the subtropics, where the winter temperature rarely drops to 0 °C unless under the influence of a strong continental effect. In this respect, the terms 'tropical' and 'subtropical' are not tightly defined geographic concepts, except within the confines of the latitudinal bounds. This is further illustrated by the Ramsar Wetlands Convention that refers to all of the Caribbean islands and the Galapagos, and the continental land mass of America south of Mexico, as the Neotropics, a zone that extends from 20° N to 55° S (Davidson *et al.* 1999). Unless specifically mentioned, the term 'tropical' is used throughout this review to include the areas between the Tropic of Cancer and Capricorn and the subtropical regions extending to 35° N and 35° S.

Within these geographic boundaries, the tropics exhibit a very wide variation in biophysical characteristics such as geomorphology, climate, hydrology, biota and human influences. Of greatest significance to understanding the present state and future of wetlands are the large differences in the history of human impacts and present pressures that occur directly and indirectly as a consequence of human population growth and distribution. In addition, the large social–cultural, economic, political and institutional diversity affects the use and management of wetland resources. The interactions between these human components and the biophysical attributes of wetlands result in a complex dynamic mosaic in which well-defined patterns are not easily discernible. However, it is possible to analyse variation in wetlands along gradients that can be identified separately for geomorphological, climate, hydrological, biological and various human attributes (Chapter 9).

BIOPHYSICAL GRADIENTS
AND WETLANDS

Geomorphologically, tropical regions have a diversity of landscapes across continents that include the world's highest mountain ranges, large rivers and many deep lakes. However, the continents differ greatly: large tectonic lakes, accounting for 95% of total volume of fresh water in the tropical belt, occur mainly in tropical Africa (Serruya & Pollingher 1983). A significant portion of tropical Africa is drained by four large rivers, namely the Congo, Niger, Nile and Zambezi (Chapter 6). In South America, permanent natural lakes are mainly restricted to high altitudes in the Andes in Peru and Bolivia; in Asia and in Australia there are few large permanently inundated lakes (Finlayson & von Oertzen 1996; Finlayson et al. 1998; Chapter 6). The Amazon River drains an extensive area carrying an extremely large volume of water and sediments, but this is not expressed in an extended delta because of a marine current that distributes the sediments northward along the coast (Chapter 13). The Paraguay River floods seasonally across an extensive area in the centre of the South American continent, namely the Gran Pantanal. In Asia, the Ganga and Brahmaputra Rivers join to form the world's largest delta; in Africa, the Niger River has an extensive inland delta, and the Okavango River discharges into a flat dry area forming an inland terminal delta. In Australia, the Burdekin and Fitzroy Rivers provide an example of the extreme variability of flows from an overwhelmingly arid inland.

The types of wetland, and also the abundances of various wetland types, vary along a hydrological gradient that is generally related to precipitation regimes (Gopal et al. 1990; Chapter 9). However, wetlands also occur in areas where the hydrological regime is influenced by water brought by streams and rivers (Chapter 2) from high mountains great distances away, or by the discharge of groundwater (Chapter 3). The tropics include areas that fall along the entire gradient between extremes of precipitation: the world's wettest and driest areas lie within the tropics. The annual precipitation gradient varies from tens of thousands of millimetres to practically no rain at all. Substantial areas of Asia, Africa, Central America and Australia are hot and arid with annual evapotranspiration far exceeding the total annual precipitation. Interestingly, many of the world's large rivers flow through the tropical arid and semi-arid regions. Numerous wetlands have developed naturally in the arid regions along the perennial and seasonal streams and springs, and in aeolian depressions (Gopal 2003).

All tropical and subtropical regions experience large seasonal variations in precipitation that allow for annual or frequent, and in many parts infrequent, flooding of palustrine and lacustrine wetlands, some of which form over extensive areas along the rivers and their numerous tributaries (Chapter 4). Thus, the majority of wetlands in all continents throughout the tropics (and subtropics) are primarily riverine floodplain wetlands and seasonal in character (Welcomme 1979; Ellenbroek 1987; Denny 1993; Finlayson & von Oertzen 1993; Gopal & Krishnamurthy 1993). Lacustrine wetlands, associated with natural lakes, are less common, although they may be locally significant, for example the papyrus swamps of Central African lakes (Denny 1992, 1993), the scattered scrub-like vegetation of many shallow lakes in northern Australia (Finlayson et al. 1998), or palustrine wetlands such as the peat-swamp forests of western Malesia (Anderson 1983).

Wetlands in the tropics are distinguished by the fact that a significantly large number of them are artificial or human-modified and intensively used. In particular, they are a characteristic feature of the landscape in tropical African pastoral systems and in Asia, where humans have responded since historical times to the unpredictable monsoonal precipitation with high interannual variability by creating water storages of all dimensions to ensure water supply for drinking and agriculture (Gopal & Chauhan 2001). Over centuries, most of these water storages have turned into wetlands with significant, often dynamic and abundant biodiversity. A singular example of human-made wetlands is the famous Keoladeo National Park (Bharatpur, India) that was carved out of a periodically inundated floodplain area by the erstwhile ruler (Chauhan & Gopal 2001).

Among biotic factors, grazing by various large herbivores not only significantly affects the type and distribution of wetlands, but also varies considerably between geographical regions (Chapter 9). Whereas there is no grazing in the peat-swamp forests of south-east Asia (Anderson 1983) or in the forested wetlands of tropical south-eastern USA (*Taxodium* swamps) (Ewel & Odum 1984), the buffalo (several introduced breeds of *Bubalus bubalis*: Ohly & Hund 2000) and cattle are major grazers in the Amazonian flood plain. In large parts of Asia and Africa with prolonged dry periods, grazing on flood plains and other seasonal marshes is very extensive and has been practised for many centuries. These wetlands and their functions have been maintained under certain grazing regimes and the regulation of grazing alters the functions and values of wetlands with undesirable consequences for many people (Shukla & Dubey 1996; Middleton 1998). On the other hand, in Australia, the relatively recent introduction of the Asian water buffalo (*B. bubalis*) and domestic cattle has resulted in wide-scale adverse change to the biodiversity and functions of wetlands (Storrs & Finlayson 1997; Finlayson *et al.* 1998).

SOCIOCULTURAL GRADIENTS

Human evolution in the tropics has been influenced by wetlands, which have played an important role in the development and growth of human civilization. The fertile floodplain wetlands of the Nile, Euphrates–Tigris and Indus Rivers gave birth to agricultural systems that shaped the highly evolved cultures of Egyptians, Sumerians and Harappans, 5000–4000 years ago (Hammerton 1972). Seasonal wetlands were used extensively for growing rice (under paddy conditions) and other aquatic plants, for fisheries and other resources. Amerindians colonized wetlands from the time of their arrival in South America about 12 000 years ago (Junk 1995; Roosevelt 1999). They built complex systems of earth mounds, dykes and channels for housing, agriculture, fishery and transport (Parsons & Bowen 1966; Denevan 1966, 1976; Meggers 1987). Throughout the tropics, traditional human societies have depended solely or largely on wetlands and their resources for thousand of years and, in many regions (e.g. northern Australia), continue to depend on them for a variety of resources including food, fodder, fibre, medicine, fuel and fertilizers (Storrs & Finlayson 1997; Finlayson *et al.* 1999; Chapman *et al.* 2001). Rice and fish alone provide subsistence to more than half of the human population (FAO [Food and Agriculture Organization of the United Nations] 2003*a*).

Human population

Today, human population density varies greatly within the tropics. Tropical Asian countries alone account for more than one-third of the world's total human population (UN [United Nations] 2003). The flood plains of the Ganga and Brahmaputra support very high population densities (>1000 people per km^2) whereas most of tropical South America and Africa still have relatively low densities (often <20 people per km^2). Nigeria, Rwanda, Burundi and Malawi (Africa), Guatemala and El Salvador (Central America), most of the Caribbean island countries, and

Indonesia (South-East Asia) have moderately high densities (>100 people per km^2). Very low population densities (<3 people per km^2), such as those in Namibia, Botswana and Australia, are related to the arid climate over most of the country. In fact, in arid and semi-arid countries, most of the human population is concentrated around wetlands. In many tropical countries, especially in Africa, the population growth rate is higher than in the temperate regions. These countries have also the least developed economies.

High population growth rates result in increasing human pressure on available land and water resources. In Brazil, which has a mean population density of 20 people per km^2, the population growth rate is steadily declining. Within the Amazon basin, about 80% of the people live in urban areas, and the population density in and around large wetlands varies between 0.5 and 5 people per km^2. By contrast, in tropical Asian countries, such as India and Bangladesh, population densities are highest in and around the wetlands on which the people depend the most.

Social structures

The different social and cultural characteristics of different communities have a large bearing on their impact on natural resources, and are also related to the biophysical characteristics of wetlands.

In India, wetlands have been an integral part of the sociocultural ethos and find a place of pride in mythology to the extent that they are considered the 'abode of gods'. Numerous wetland plants and animals have been associated with different gods and are held in reverence. Wetlands are sacred sites for several religious rituals, and only wetland plants, such as trapa, lotus and several sedges, are considered suitable for food on religious days. Such practices were in general associated with herbaceous marshes that predominate in the wetlands of India. Although these cultural bonds have weakened with time, and wetlands have been gradually neglected, these practices helped conserve many wetlands over centuries (B. Gopal, unpublished data 1997).

In Africa also, sociocultural life of many local communities has been closely linked with wetlands (such as *dambos*, inland valley swamps), which have been a major source of human sustenance; hence wetlands were until recently protected and their use was managed judiciously (Gordon 1992; Thompson & Polet 2000; Gawler 2002).

In most arid and semi-arid regions, sociocultural practices attach great value to the conservation of water and protection of its quality. In tropical South America, the human population densities are very low and the large annual fluctuations in the water level are probably a major constraint on intensive use of the flood plains (Chapter 4). Further, the usable resources in forested (Amazonia) and savannah-type (Pantanal) wetlands are quite different from those of herbaceous wetlands. Thus, fishing, cattle ranching and agriculture have been the main activities of local communities in the non-herbaceous South American wetlands (Junk *et al.* 2000).

Greater dependence on resources that were available periodically and in greatly varying amounts inculcated in traditional communities a behavioural pattern and sense of responsibility for management that tended to promote sustainable use and lead to development of a conservation ethic based around sustainability and survival (e.g. McGrath *et al.* 1999; Thompson & Polet 2000; Gopal 2002). However, with changing economies, the availability of alternate resources and access to them from other often distant areas, communities were freed from traditional natural bonds and started a phase of neglect followed by changes in land-use practices (Gopal 2005). Once the wetlands were no longer the sole or a dominant source of livelihood, they were disused and abused or, in instances where they were perceived to be resources for income generation, converted to other land uses (e.g. paddy, other agriculture or urban) (Carbonell *et al.* 2001; Tiega 2001; Gawler 2002).

In many parts of Asia and Africa, and in the northern coastal areas of Australia, local communities still manage the wetlands traditionally (Storrs & Finlayson 1997). However, the demands of local populations can lead to increased pressure on wetland resources, for fish, water and timber in particular, and result in major environmental changes. For instance, increasing demand for timber and fuel wood has dramatically depleted and deteriorated the floodplain forests of Amazonian white-water rivers (Parolin 2000).

Large local populations are not necessarily the cause of increased demand on natural resources and their overexploitation. There are many examples of wetland destruction and overexploitation triggered by the demand for resources in developed temperate countries and fuelled by the need for rapid economic growth that requires imports of various kinds from other regions. The expanding shrimp culture industry in South and South-East Asia has been a major cause of wetland degradation in recent years (Landesman 1994; Primavera 1994). Similar impacts occur as a consequence of demands from within a region. For example, Lake Kolleru, an extensive natural interfluvial freshwater wetland in Andhra Pradesh (east coast of India), has been converted to a series of intensively farmed fish tanks solely for the export of fish to other states in India (Gopal 1991; Durgaprasad & Anjaneyulu 2003).

Economic and political gradients

The economic status of a country depends upon many factors, including the availability of natural resources. These factors, however, affect those very resources through (1) greater exploitation to meet the basic needs of the human population, and (2) the lack of financial resources for adequate education, awareness, capacity building and proper management. Although most countries in the tropical belt, except Australia, are classified as developing, they differ greatly in their economies and political systems. Countries such as Venezuela, Mexico and Malaysia are relatively strong economically, while most other countries in South America and Asia and almost all countries in Africa are extremely poor.

Economy and politics are also closely linked. Large economic disparities between communities or regions within a country often create serious political problems, as in several Asian and African countries. In the long term, political stability coupled with democracy is necessary for steady economic development, as well as for the formulation and implementation of sound environmental policies (Folke & Kaaberger 1991). Many countries throughout the tropics (particularly in Africa) suffer from non-democratic political systems as well as unstable governments. This is amply reflected in the level of their economic development, environmental policies and protection of natural resources (see Desai 1998; UNEP [United Nations Environment Programme] 2004). A democratic system generally provides a basis for raising concern and resistance against inappropriate government policies, and for greater involvement of local communities in the management of natural resources. A sustainable management, however, depends greatly upon the level of education and environmental awareness of the population and the scientific and technological capacity within the country to understand, investigate, analyse and communicate the issues concerned. Tropical countries again differ greatly in this respect, with most of the countries in Africa lagging far behind (Conserve Africa Foundation 2005).

Politicians often argue that economic underdevelopment is the cause of poverty, forcing environmental destruction by inadequate and unproductive land use. Strong economic development is deemed to lead to industrialization, efficient use of resources, welfare, decreasing population growth and reduced pressure on the environment (Anon. 1999; UNEP 2002). As part of their contribution to the World Summit on Sustainable Development, several international organizations discussed at length the benefits of environmental management to poverty reduction, and the links between economic development and environment (DFID [Department for International Development]/EC [European Commission]/UNDP [United Nations Development Programme]/World Bank 2002). However, whereas poverty is accompanied by insidious destruction, economic growth can also have negative impacts on the environment through large-scale development projects leading to habitat (for example wetlands) destruction and environmental side effects, which are often insufficiently analysed or ignored. As an example, the planning and implementation of many large development projects, often based on inadequate environmental impact assessments (EIAs), have had serious negative impacts on the environment in Brazil (Laurance *et al.* 2001). Typically, even where EIA may be adequate as in Zambia, developing countries lack the capacity and resources to implement the mitigation requirements (Wehrmeyer & Mulugetta 1999; Wagner *et al.* 2001).

The issues of political and economic drivers of development projects are not simple. Wetlands in the Congo River basin have not been subject to large development projects, presumably because the population density and economic development are very low and political instability limits investment by national and foreign companies (Rothgeb 1996). In developing economies, such as those in Africa, social welfare dominates the political landscape to an extent that mitigates against investment in environmental management, especially where this may increase the vulnerability of citizens. Since need commonly exceeds the availability of welfare, political systems are insecure and subject to abrupt change. Such unstable regimes weaken further political will to engage in environmental issues, and need continues to drive patterns of resource use that are unsustainable.

In the Mekong Basin, the situation is far more complex, with extensive recent destruction of natural wetlands in the delta in favour of rice paddy systems (Walden *et al.* 2004) on one hand and plans through the internationally funded Mekong River Commission for extensive regulation of the river upstream (Hollis 1998) on the other. In 2002, the Mekong River Commission was awarded the Thiess Services'

International River Prize for excellence in river management following the concept of river basin management (Mekong River Commission 2002).

PRESENT STATE OF WETLANDS

An essential prerequisite for an assessment of the state of wetlands is an inventory of the wetlands and their resources. For more than a quarter of a century, considerable effort has been made towards preparing inventories of wetlands both on continental and national scales, but they remain far from complete and do not provide even a reasonably accurate estimate of areas, habitat features, biological diversity levels or states of degradation (Finlayson *et al.* 1999; Chapter 9). A global review of wetland inventory pointed out many inadequacies of past efforts (Finlayson & Spiers 1999). There are also often large differences in the estimates made by different workers using different methods; sometimes these inventories create more confusion than they resolve because of different perceptions of wetland habitats included in the inventory. In most inventories, small or little-known wetlands are excluded. In fact, all continents are covered by a network of small, mostly temporary, wetlands along streams and in depressions that assume important functions in the landscape, mostly for water retention, as filters, sinks and sources of substances, and as habitats for specific organisms (Chapter 9). Such wetlands are of the greatest importance in arid and semi-arid regions where seasonal wetlands may remain without water for several years or decades, yet when water arrives the wetlands support a large diversity of flora and fauna, including many endemics (see Williams 1998*c*, 2000; Chapter 9).

Based on the available information on tropical wetlands, certain patterns in the distribution and abundance of different kinds of wetlands are discernible in different continents. These can be recognized along biophysical gradients. Forested wetlands in the tropics are mostly associated with high rainfall. The tropical South American wetlands are dominated by the forested wetlands of the humid central Amazon basin and by periodically flooded savannahs in the adjacent drier savannah belts (Junk 1993), both of which are rich in biodiversity (Junk 2000; da Silva *et al.* 2001; Junk *et al.* 2002). The subtropical region of the south-eastern USA also has some species-rich swamp forests (Larson 1992; Mitsch *et al.* 1994). The forested wetlands of any major significance in Asia are mostly the peat-swamp forests of South-East Asia, unique in the tropics, and the *Melaleuca* (Myrtaceae) swamps (Anderson

1983; Brünig 1990; Paijmans 1990; Scott 1992; Mitsch *et al.* 1994). The small patches of riverine swamp forests, which once occurred in Kerala, north-east India, and along the foothills of the Himalayas, have mostly been lost (Gopal & Krishnamurthy 1993). The wetlands of the Congo Basin are not yet investigated in depth (Denny 1992). Floating papyrus swamps of Africa represent a specific wetland type, also unique in the tropics (Muthuri *et al.* 1989), although floating mats of aquatic vegetation are found in many countries, at times because of invasion by the water hyacinth (*Eichhornia crassipes*) (Gopal *et al.* 2003; van Duzer 2004). Most flood plains of Africa, Asia and Australia experience only shallow flooding for relatively short periods followed by prolonged dry periods (Chapter 4), and are largely dominated by herbaceous vegetation with some trees (Denny 1993).

Although detailed scientific studies of tropical wetlands are very few, they are known to support very high floral and faunal diversity (Finlayson *et al.* 1998; Gopal & Junk 2001; Junk *et al.* 2006). Tropical wetlands are well known as breeding and overwintering habitats for annual migrations of waterbirds from temperate regions, as well as nomadic movements within continents in response to flooding, providing habitat and abundant food, and contributing greatly to biological diversity (Finlayson *et al.* 2005). Another important and relatively well-documented function of tropical wetlands is the provision of spawning and nursery grounds and food for a large number of riverine fishes (Welcomme 1979, 1985). Tropical wetlands in different continents vary in species richness and in the occurrence of endemic, rare and threatened taxa of fishes and other biota (see general reviews in Finlayson & Moser 1992).

Whereas wetland inventories generally provide information on wetland type, location, size, important biota and some ecological features, this is insufficient to mobilize corrective action where necessary. Inventories rarely provide the information required for understanding problems, management issues and the mitigation measures that may be needed. Wetland management in almost all tropical countries is therefore only rarely based on sound scientific understanding of the system concerned, and often results in unforeseen problems.

FACTORS AFFECTING WETLAND DEGRADATION

The environmental factors that regulate wetland structure, function and biodiversity have been discussed at length (Gopal *et al.* 2000, 2001; Junk 2000, 2002). Hydrology is the single most important driving variable governing all aspects of wetlands (Chapter 9). Most wetlands in the tropics are subject to considerable water-level fluctuations according to dry and rainy seasons. Permanent marshes or swamps are less common and often part of larger seasonal wetlands. The biota in tropical wetlands are well adapted to long periods of drought and usually require large water-level changes for their long-term survival (see Adis & Junk 2002). The frequent and unpredictable occurrence of pluriannual wet and dry periods is particularly characteristic of countries with a monsoonal climate such as India, and the related water-level fluctuations are of fundamental importance for the maintenance of biodiversity in the wetlands (Gopal & Chauhan 2001). Large spawning migrations of fishes occur with the onset of the rainy season in all tropical river–floodplain systems (Lowe-McConnell 1987), and fish catches are positively correlated with the extent of flooding (Welcomme 1979, 2006). Thus, all anthropogenic activities that directly or indirectly alter the natural hydrological regimes cause degradation or loss of wetlands and their biota. These activities include water diversion and abstraction, flow regulation, drainage and reclamation by landfills, mining and land-use changes in the watersheds (Dudgeon 2000*b*; Nogueira & Junk 2000; Dudgeon *et al.* 2006; Chapters 4 and 9). Human activities such as disposal of domestic and industrial wastewaters as well as solid wastes, and intensive agriculture methods that cause changes in water quality (eutrophication and toxic pollution), are also responsible for wetland degradation (Finlayson & Rea 1998). Further important factors include exotic invasive species and overexploitation of wetland resources (sand mining, hunting, fishing, plant harvest for food, forage or fuel: Chapters 4 and 9).

However, some issues related to the degradation and loss of wetlands in the tropics require re-examination. Many analyses have attributed the loss and degradation of various natural resources in developing countries to high human population densities and growth rates, excessive exploitation of resources and inappropriate management owing to lack of scientific, technological and economic resources (Barbier 1997, 2000; Dasgupta 2000). Such analyses fail to recognize the complexity of socioecological systems in developing countries, and have ignored the history of human interaction with the environment in general and wetlands in particular in different continents. Variability within the tropics of climate, hydrological and other environmental factors affecting wetlands, human population distribution, sociocultural and political structures, and levels of economic development, all directly or indirectly impinge upon wetlands and have a bearing on their future. The interactions between these factors give rise to situations of enormous complexity that will require sensitive management measures.

Asia has the longest history of human activity and this can be traced back to irrigation systems as far as wetlands are concerned. Humans settled along river courses, cultivated alluvial flood plains, exploited fisheries, hunted wildlife, diverted water and also stored water (Chapters 2 and 4). Except in the upper reaches of the Nile Valley (see Hassan 2003), human impacts in the African continent remained small, as most communities were nomadic or semi-nomadic (Smith 1992). In Australia, the arid climate tended to limit humans to the fringe of the continent, and it was not until non-Aboriginal settlement occurred that wetlands became largely degraded (Storrs & Finlayson 1997; Finlayson *et al.* 1998). Similarly, in the Neotropics, the human history of extensive manipulation of nature is only about five centuries old, dating from the European discovery of the New World. Harvesting of fish stocks, and irresponsible eradication and destruction of the Amazonian stocks of river turtles and manatees started over a century ago (Verissimo 1895). The formerly abundant populations of caimans, otters and capybaras (*Hydrochoerus hydrochaeris*) are now greatly reduced (Junk & da Silva 1997). The wetlands of the Asian tropics have 'co-evolved' with humans for more than 5000, probably for 8000 years. Wetlands have influenced art and culture in many Asian countries, including India and China (Sculthorpe 1968; B. Gopal, unpublished data 1997). Humans have long modified and extensively used wetland resources, but the phenomena of widespread degradation and loss are recent and largely driven by modern attitudes towards development and consumer affluence. In comparison, in many Western countries, widespread agricultural and industrial development and an attitude that wetlands were wasteland or places of menace have propelled wetland loss rapidly within the past few centuries (Williams 1990; Mitsch & Gosselink 2000; Chapter 9).

A significant proportion of the loss and degradation of wetlands is readily attributable to policies and programmes that cater for the needs of the wealthy and affluent sections of society, both nationally and internationally. The differences must be recognized between those who benefit directly from the use of wetland resources and those who benefit indirectly or from uses that indirectly affect wetlands. Exploitation of wetlands as well as resources,

such as water, that sustain wetlands, nearly always benefits the richer sectors of society that do not hold a direct stake in the wetlands. For example, water storage for irrigation or hydropower that results in the loss of downstream riparian wetlands benefits the people who have the least concern for wetlands. Reclamation of wetlands by drainage and/or landfill also helps only the urban rich. Further, the export of luxury products from the conversion of wetlands can benefit the importing communities by protecting their own environment. Thus, aquaculture in South-East Asia is sustained by the demand for shrimp in Europe and the USA (Primavera 1994; Barraclough & Finger-Stich 1996), and the fish harvested from intensive aquaculture in the former Lake Kolleru in Andhra Pradesh (India) meets the demand for fish in West Bengal and other states. The local fisherfolk, who once had free access to a variety of lake resources, have become labourers employed by a small wealthy group of entrepreneurs (B. Gopal, personal observations 2003). The conversion of peat-swamps in Malaysia and Indonesia to oil-palm plantations (Phillips 1998; Sargeant 2001), and of Sunderban and other mangroves in south Asia to paddy fields (Williams 1990; Blasco *et al.* 2001; Chapter 12) have scarcely benefited local communities, who have been deprived of a variety of wetland resources. The Hidrovia Project in the Pantanal of south America is another example where economic benefits from river transport and soya bean cultivation for export to Europe and the USA would reach only a few, at the cost of losses to local communities (see Hamilton 1999; Brito *et al.* undated). The separation of the subsistence or local economy of wetlands from the larger economy in decision-making marginalizes local people, commonly driving them to increase pressure on marginal resource systems resulting in further degradation.

Another factor must be recognized. Traditionally, natural-resource use in developing countries was widely governed through systems of common property with well-defined rules and the authority to implement them. However, the erosion of traditional authority, mainly through colonial influence and changing economies, led to weakening and in many instances to a collapse of common property, and its replacement with 'open access' where there were no controls over use. The issue of property regimes in the context of the use, management and degradation of wetlands has been discussed (Gopal 2002). Authority and ownership at local level make it difficult for external forces to alienate people from their resources, and therefore promote conservation.

FUTURE STATES

The future of tropical wetlands will depend upon the actions taken by civil society in different countries, the respective policy- and decision-makers and the international community. These sectors can choose to move along the well-trodden path to economic development, at the same or an accelerated pace, without heeding the signals from nature; or they can change course to a sustainable form of development, learning from the mistakes of others in the developed world and their own. It has been shown above that options for directing patterns of resource use differ between the developed and developing economies. In developed economies, there are political options for limiting use of local wetlands through systems of incentives and disincentives, and managing demand for luxury goods obtained from wetlands in other countries. In developing countries, however, wetland goods and services commonly provide food security, and this demands a different approach to policy intervention. There are areas of major concern and these are discussed below.

Water-resources management

The key factor in the sustenance of wetlands is the availability of water; the amount and timing of its availability are prime regulators of wetland ecosystem integrity. As the human population rises and yearnings for economic development on Western models grow, wetlands are likely to be under increasing pressure in most parts of the tropics. Increasing agricultural, industrial, urban and hydropower demand for water threatens the already limited water availability and consequently the wetland ecosystems. Human global water requirements have increased almost 10-fold during the twentieth century (Biswas 1998; Chapters 2 and 3). Only about 31% of the estimated 40 700 km^3 of annual runoff is accessible for controlled human use and only about 16% is currently appropriated for various uses (Postel et al. 1996). However, a substantial part of the annual runoff is stored for irrigation, hydropower generation, flood control and water supply in tens of thousands of large and small reservoirs. Such storages interrupt hydrological connectivity and affect downstream flood plains and estuaries in multiple ways (Chapters 4 and 13), for instance, by changing the flood pulse, modifying sediment transport, changing nutrient cycles and interrupting animal migration routes (Walker & Thoms 1993; Lae 1994; Kingsford et al. 1998;

Kingsford 2000; Pringle 2001; Chapters 2 and 9). Large dams have enormous social and environmental costs, while in many areas, increased siltation and reduced lifespans of many reservoirs have reduced economic benefits as well. Consequently, water quality in rivers has declined and wetlands and their valuable functions have been lost. This has been well documented, but to date has had little impact on water resource managers.

The per caput water consumption in most tropical countries is only 50–100 l per day and may be only 10–40 l per day in some water deficient regions, whereas industrially developed countries of temperate Europe and North America consume 500–800 l per day (Shiklomanov 1999). Industrial and agricultural development in the tropics is bound to increase the demand for water (Postel 2000; Vörösmarty et al. 2001). While the poorer sections of society will suffer from inaccessibility to drinking water which is now being priced as a commodity (Chapter 3), they will also suffer indirectly from the loss of wetlands. As more water is inefficiently used in irrigation to raise cash crops, and large-scale water storage diversion schemes are implemented, water quality will suffer, habitat degradation accelerate, and fisheries and wetland biodiversity decline rapidly. The anticipated benefits of agricultural irrigation projects often do not compensate for environmental degradation, artificial water shortages and increased poverty rates suffered by local communities managing the wetlands by traditional methods (e.g. Mexican wetlands: Contreras-Balderas & Lozano-Vilano 1994).

Tropical countries will not on the whole be able to match the current level of water consumption in temperate developed countries, but a focus on irrigation and urbanization will create serious environmental problems unless water-use efficiency is increased manifold and water recycling is ensured. Central to achieving this increase in efficiency is the appreciation that river and wetland systems are providers of goods and services (Finlayson et al. 2005). This requires a shift from the 'traditional' focus on water as the resource to ecosystems as resources, and ensuring the provision for environmental flows for maintaining their goods and services (Cullen et al. 1996; Finlayson et al. 2005). Such a shift has occurred in South Africa where the National Water Act 36 of 1998 (Government of South Africa 1998) defines rivers as the resource (not only water contained and transported therein) and makes provision for reserving an allocation of water of defined quality (the Environmental Reserve) to sustain the delivery of goods and services from these systems. Further, many aquatic ecosystems in the tropics are shared between countries, and hence efficient use of their resources will require an integrated approach to catchment management.

Numerous wetland habitats are and will be created by water storage throughout the tropics, particularly in arid and semi-arid regions and those with a monsoonal climate. However, constructed wetlands with a focus on the use of their water elsewhere cannot match the functions and biodiversity of natural wetlands (Mitsch *et al.* 1994; Gopal 1999).

Land use and water quality

Further threats to wetlands will increase from a variety of human activities in wetlands and their watersheds. Changes in land use, such as deforestation (or clearing of vegetation), overgrazing, agriculture, urbanization and mining alter the hydrological regimes of wetlands. Exploitation of groundwater also affects the surface water (Chapter 3); in particular, the dry-season flow in rivers is directly affected by groundwater extraction in their flood plains (Chapter 4). Besides altering the flow regimes, changes in land use have resulted in increased transport of sediments, organic matter and nutrients from the watersheds to downstream wetlands. Wetlands are known for their potential for upgrading wastewaters through various nutrient transformations and also for trapping the suspended sediments as the runoff passes through them (Chapter 9), whether in lake littorals or river flood plains, before reaching the open-water systems. However, this potential is limited, and many wetlands are adversely affected by nutrient enrichment and siltation. Agrochemicals used in intensive agriculture are the major source of pollution by nutrients and pesticides in wetlands. Heavy deforestation and subsequent habitat modification are the main reasons for freshwater fish species decline in Madagascar (Stiassny 1996). Export-oriented agro-industries are major threats to South American wetlands, as shown for the *cerrado* belt (Wantzen 1998). Aquaculture is similarly causing serious degradation of natural wetlands (e.g. Chapter 12), owing to the extensive use of organic feeds and a variety of chemicals, resulting in rapid loss of their biodiversity and multiple functions. Lake Kolleru, an interfluvial freshwater wetland on the east coast of India, is an important example of such degradation within the past 20 years (Durgaprasad & Anjaneyulu 2003).

Further degradation of water quality will occur because of the discharge of domestic and industrial wastewaters directly into wetlands. Today, wastewater treatment systems in most tropical countries (except Australia) are in a precarious state, despite the water shortage and the danger for public health. The facilities for the treatment of municipal wastewater in these countries are either nonexistent or cater for only a very small portion of the total population, even in large urban areas. More than 90% of wastewater is discharged directly into rivers and streams without treatment (Johnson *et al.* 2001). Where sewage treatment systems have been provided, they are generally inefficient and the treated effluents rarely meet the prescribed standards; the same is true for industrial effluents which are at best only partially treated. The problems caused by water pollution will become aggravated by the paucity of water and prolonged dry periods, such as those in arid and semi-arid regions, or as caused by the diversion and withdrawal of water. In many tropical countries, increasing human population densities will counteract efforts to control water pollution.

Wetlands will also be strongly impacted by the disposal of solid wastes as these habitats are considered suitable sites for reclamation by landfill. Overexploitation of wetland resources, including sand mining, hunting, fishing, timber exploitation and plant harvest, is also likely to increase in many countries by the year 2025.

Invasion by exotic species

Invasive exotic species, both plants and animals, are a serious threat for many wetlands (Chapter 9). Water-level fluctuations such as periodic floods can disturb wetlands and facilitate dispersal of many ruderal plants, as shown for the upper Amazon River flood plain (Seidenschwarz 1986). Among the exotic invasive plants, water hyacinth is probably the most noxious species that has spread throughout the tropics (Gopal 1987). It now occurs in practically all river basins; its spread has occurred from downstream introductions to upstream areas. Despite efforts made for over a century to control the spread of *Eichhornia* through all possible means, it has continued to expand its range into previously unaffected areas and has impacted native biota (Gopal 1987).

The introduction of the Nile perch (*Lates niloticus*) caused severe reduction and partial extinction of the endemic cichlids in Lake Victoria and associated wetlands (Kaufman 1992; Schofield & Chapman 1999; Chapter 6), and exotic fishes have threatened the native fish fauna of Madagascar (Stiassny 1996). In tropical South America, the mussel *Limnoperna fortunei*, probably introduced via ship ballast water from Asia to the Rio de la Plata in 1993, in two years reached Santo Tome at the middle Paraná River (Darrigan & Ezcurra de Drago 2000) and, by 2001, the Pantanal of Mato Grosso in central South America. Growing rapidly on hard substrates, the mussel chokes water pipes in hydropower and drinking-water systems, although its impact on wetland biota has not yet been investigated.

Despite many precautionary measures (including legal provisions in many countries) and an increasing international effort (see Mooney & Hobbs 2001), humans have failed to check further introductions or control the growth and spread of already-introduced species. Growing intercontinental trade and tourism are expected to add to the gravity of the problem; thus invasive exotic species will continue to adversely impact wetlands throughout the tropics to the year 2025 and beyond.

Global climate change

By the end of the twenty-first century, the mean global surface air temperature may rise by 1.4–5.8 °C, with the temperature increase larger in polar regions than in the tropics (IPCC [Intergovernmental Panel on Climate Change] 2001a). The greatest warming in higher latitudes may be in autumn and winter; precipitation could increase by 7–11% locally but will generally decrease, and the central parts of continents may be warmer and drier (IPCC 2001a). Understanding of impacts of global climate change on aquatic ecosystems (Firth & Fisher 1992) has yet to encompass tropical wetland biota (van Dam *et al.* 1999, 2002; Gitay *et al.* 2001, 2002).

There is likely to be increased spatial and temporal variability in precipitation even if there is no clear unidirectional change (IPCC 2001a). It is expected that drier areas will become drier and wetter areas may receive more precipitation. Similarly, dry and wet years may occur with altered frequency, causing heavy droughts and floods. The amplitude and frequency of multiannual extreme climatic events like El Niño–Southern Oscillation (ENSO) episodes have already increased since the mid 1970s (IPCC 2001a). The combined impact of changes in temperature and precipitation has important consequences for the flow regimes in rivers and streams (Chapter 2). Extremes in flow conditions and the temporal distribution of extreme flows will have important bearings on floodplain wetlands (see Poff 1992; Chapter 4); and streams of arid lands are likely to be more sensitive to changes brought about by global warming because they represent extremes in hydrology, temperature and productivity (Grimm & Fisher 1992). However, owing to their resilience to disturbances such as flash floods and drying, desert streams in particular and stream riparian systems in general may be the last to respond to global climate change (Minshall 1992).

Palaeoclimatic studies show that tropical wetlands are very sensitive to changes in hydrology. A rise in temperature will increase evaporation rates and reduce water availability for wetlands, mainly in semi-arid regions. Fire stress will also increase during low-water periods, thereby affecting the structure of many biotic communities. However, while the responses and adaptation to such climatic changes by the wetland biota are expected to be slow as has probably occurred through Earth's history, the consequences will be more severe because of human response to these changes. For example, a vicious cycle of landscape degradation, desertification and rural displacement set in motion in the Senegal River valley by the Sahel-wide drought beginning in the 1970s was exacerbated by rural development policy that favoured large-scale state-imposed irrigated rice production (Venema *et al.* 1997).

The projected rise in global mean sea level by 9–88 cm between 1990 and 2100 (IPCC 2001a) could lead to extensive loss of freshwater wetlands in many tropical regions (Bayliss *et al.* 1997; Eliot *et al.* 1999; Nicholls *et al.* 1999). However, alterations in water-resource management practices will greatly influence the wetlands, and changes in economic development will affect wetlands to a much larger degree than will climate changes (Vörösmarty *et al.* 2000). There is a danger of wetland biota being lost more rapidly as a consequence of human reaction to climate change.

Sustainable use and management

In the wake of the many pressures on limited land and water resources, and impending global change, the future of tropical wetlands hinges on the ability of national governments to develop and sustain management systems that can promote adaptive management in the use of wetland goods and services. The term 'sustainable development' (World Commission on Environment and Development 1987) has recently been defined by the Consultative Group on International Agricultural Research (CGIAR) as the 'successful management of resources ... to satisfy changing human needs while maintaining or enhancing the quality of the environment and conserving natural resources' ([Technical Advisory Committee] TAC/CGIAR 1989). This definition can also be used for wetlands without pointing specifically to agriculture because wetland resources are also managed for many other purposes. The requirement that 'sustainable development should be ecologically sound, economically viable, socially just, culturally appropriate and based on a holistic scientific approach' reflects the concern of environmentalists not to separate society from environment and economics from ethics (Becker 1997). These aspects are of specific importance for tropical wetlands that are still managed by local communities with traditional methods. Considering the fact that wetlands are closely linked to their catchment areas, the requirement for a holistic scientific approach should lead to the elaboration of catchment area management plans, as agreed under the Ramsar Convention on Wetlands (Ramsar 2004).

However, the major challenge to the future of wetlands in the tropics comes from the approach to their management. Recognizing the long history of traditional use and management of wetlands in tropical countries by local communities, the Ramsar Convention on Wetlands has placed much emphasis on 'wise use' as a strategy for management and conservation. However, keeping in mind the complex dynamic characteristics of wetlands, including their interactions with human societies, there is a need for a fundamental change in the approach to achieving wise use. For this to occur, governance and management need to be responsive to both social and ecological issues at spatial and temporal scales relevant to both.

It is people's needs, perceptions and desires that manifest in actions that determine prospects for wetlands. Some of these actions have direct consequences for wetlands, as for example does harvesting resources such as water, peat and food from a wetland by members of a household or local community. Others arise indirectly as actions taken in response to political and policy decisions made remotely from any particular wetland, for example becoming a signatory to conventions such as the Ramsar Convention, Convention on International Trade in Endangered Species (CITES) and the Convention on Biological Diversity (CBD) (Bartley 2000), or implementing national policies in support of economic and structural reform. In this way, the actors and issues that affect wetlands are connected across a range of scales from global to household and through direct and indirect interactions and feedbacks.

PRESSURES AND FEEDBACKS

Global and national scale

At the largest spatial scale, pressure for political and economic reform in developing countries has come from global institutions, such as the International Monetary Fund and the World Bank, through the structural readjustment and development projects they have funded. Although economic reform in developing countries, such as those in Africa, was justified by the premise that the projects would significantly improve the welfare of the rural poor, there is considerable disquiet over the effectiveness of such projects (Reed 2001, 2002). In particular, there are extensive social, economic and environmental costs. These costs have taken many forms, with examples from across Africa: whilst some individuals have made gains, the majority of the rural population seems to have lost the most; shifts of power and influence favoured some individuals and/or households, but disadvantaged others; institutional reforms have weakened the abilities of rural communities to defend their interests against more powerful companies and political elites.

Examples of downward pressure originating from government actions being propagated through the system, and often with unpredicted outcomes, include the Pongolo and Zambezi River flood plains (Heeg & Breen 1982; Nyambe & Breen 2002). On the Pongolo River flood plain, flow regulation led to emergence of new power groups that manipulated flow releases from an upstream impoundment in ways that altered patterns of resource use, favouring some and disadvantaging others (Merron *et al.* 1993). Among the Barotse people of Zambia, central government attempts to exert control over the fishery and weaken traditional control was contested by the traditional leadership, leading to an unexpected strengthening of traditional governance and a weakening of central government influence (Nyambe & Breen 2002).

People of developing countries in the tropics are increasingly forced to adjust their livelihoods in response to pressures, many of which arise remotely from their locations, and which at the same time weaken their abilities to adjust. As livelihoods lose their self-sustaining properties, so the feedbacks are felt in national governments and even global civil society to seek more stringent protection, weakening local people further.

Local and household scale

More than anyone else, the economies of poor rural people, especially those in developing countries, depend on the life-support functions provided by wetlands, including water, food, fibre for crafts and construction, and lands for cultivation (Mbewe 1992). Only rarely are economies, whether national or household, strong enough to replace the goods and services provided by wetlands once these are lost. The consequences of wetland loss are therefore most severe for the poorer people of developing countries. At best, an increased proportion of the household budget has to be spent on subsistence, while in many cases it means a lower-quality diet, a decline in total food intake, and even rising mortality and emigration. It is not surprising that wetland conservation appears on the agendas of the World Bank and bilateral aid agencies (Mbewe 1992).

Agriculture is one of the most important activities linking individual households with wetlands. Many developing countries have a strong agricultural base, and wetland cultivation is particularly important for food security, especially during periods of drought (Wiseman et al. 2003; McCartney et al. 2005). During the 1969–70 drought in Zimbabwe, 84% of the farmers with agricultural lands in *dambo* lands were able to support their families, while only 21% of farmers without such lands were able to do so (Lambert et al. 1990). Wetland soils are better supplied with water and are often inherently more fertile than associated non-wetland soils. Provided that the constraints of waterlogging and the difficulty of working in what are often very heavy soils are overcome, the returns to rural household economies of wetland cultivation may be very significant. The FAO has long recognized the potential that cultivation of wetlands has for improving food security in Africa (FAO 1998; Frenken & Mharapara 2002). Agriculture, however, can also be integrated with aquaculture in various ways for greater benefits and sustainability (see several papers in Sugunan & Gopal 2006).

The benefits of wetland cultivation must, however, be seen in the light of potential negative consequences for both catchment water quality and supply (Willrich & Smith 1970; Hemond & Benoit 1988), and biodiversity (Kotze & Breen 1994, 2000; Mitsch & Gosselink 2000). Traditional agriculture requiring little external input is potentially far more compatible with wetland functioning and integrity than large-scale capital-intensive agricultural development that reflects the influence of global institutions. To assist local people to make informed choices amongst practices and technologies, both traditional and modern, in support of sustainable livelihoods, the pressures and feedbacks operating at these local and household scales need to be better understood and integrated with those of the larger scales. This is likely to require incorporation of local skills, resources, forms of cooperation, participation and flexibility in planning and management of projects that affect tropical wetlands (Giesen et al. 1991; Pretty & Guijt 1992; Pimbert & Gujja 1997; Gichuki 2000; Adams 2001). It is further emphasized that project development should entail initial and ongoing constructive dialogue among stakeholders which can identify the most beneficial solutions to the inherent and emergent challenges facing the local community and provide a mechanism for incorporating external expertise (Finlayson & Eliot 2001).

Using ecological economics

Wetlands provide many non-marketable services that are difficult to value, for instance, periodic water storage and release, water purification, sediment trapping, stabilization of local and regional climate, maintenance of species diversity and aesthetic values such as landscape beauty that affect the quality of life of local populations.

These non-economic goods suffer because they are used for free by the community but nobody takes sufficient care of them (Hardin 1968). Often the value of these goods and services is ignored in the economic evaluation of development plans although several methods are available for their measurement (e.g. Whitehead & Blomquist 1991). Sustainable management has to equally consider economic and non-economic goods that together form the 'natural capital' of a wetland (Daly 1991; Goodland 1991; Daily & Ehrlich 1996). The bias between the requirements of a detailed environmental impact analysis and the technocratic arguments of planners and politicians is shown in the discussion about the rectification and deepening of the upper Paraguay River in Brazil, to facilitate ship traffic for the Hidrovia Project. This project may yet seriously affect the integrity of the Pantanal, one of the world's largest and still rather pristine wetlands (Ponce 1995; Hamilton 1999; Gottgens et al. 2001).

Furthermore, cost–benefit analyses cannot consider unpredictable changes in political, economic and socioeconomic conditions. For instance, until the middle of the twentieth century, agriculture and animal ranching were important economic activities in the flood plain of the

River Rhine and led to large-scale, costly flood-control measures. Since then, floodplain services such as buffering of floods, water purification, maintenance of biodiversity and recreation have become much more important and now require the costly reconnection of the remaining floodplain areas to the main channel (see URL http://www.ikse.org). In regions with a pronounced dry season, water availability will become the limiting factor for human welfare, development and wetland protection. The cultivation of water-demanding crops, such as paddy rice which is actually stimulated by some governmental projects in the Sahel region (Barbier & Thompson 1998), will soon become economically unwarrantable, when considering the real costs of water for irrigation (Hoekstra & Hung 2003). In the case of the Hidrovia in the Brazilian Pantanal, alternatives for shipping of agricultural products are already being developed and will modify the basis of the current cost–benefit calculations (Huszar *et al.* 1999). These examples show that major modifications of structures and functions of wetlands should be avoided to minimize restrictions for alternative management concepts of future generations.

The 'natural capital', including ecosystem goods, services, biodiversity and cultural considerations of rivers and wetlands worldwide, is estimated to be worth US\$8498 per ha per year (Costanza *et al.* 1997; although see de Groot *et al.* 2002), and values of wetlands in North America and Western Europe are shown by the large amounts of money expended for wetland restoration. Key projects such as rehabilitation of the Everglades and parts of the Mississippi River flood plain have cost billions of US dollars and demonstrate that (1) economic benefits of wetland destruction are often overrun by the costs of negative side effects, (2) the economic framework changes quickly and modifies cost–benefit analyses of development projects, often in favour of the values of intact wetlands, (3) only parts of the former wetland area can be recovered to near-natural conditions under very high costs and (4) maintenance of wetlands is always much cheaper and more effective than rehabilitation after degradation (e.g. Chapters 4, 7 and 9). The last aspect is of specific importance for tropical countries; it is not too late to invest in wetland maintenance and protection. The alternative is to face a situation where typically countries have neither the money for expensive rehabilitation projects, nor the political and economic conditions to resettle the large numbers of people that increasingly occupy and degrade wetlands. The challenge is to recognize the value of tropical wetlands before they are further degraded or lost; economic instruments are one way of demonstrating this value. Fortunately, appropriate national policies are now being developed in several tropical countries, for example in Colombia (Ministerio del Medio Ambiente 2001).

Part IV
Coastal wetlands

The coastal wetlands constituted by saltmarshes, mangroves and estuaries rank among the most productive, dynamic and ecologically important of coastal environments. Each of them provides nursery, feeding and refuge functions for many animal species of human importance. Coastal wetlands also influence environmental quality along the land–river–sea continuum, such as biogeochemical cycling, transformation of nutrients and filtering of contaminants. The three ecosystem types dealt with in this part overlap to some extent; for example, many estuaries are bordered by luxuriant saltmarsh (particularly temperate waters) or mangrove (tropical and subtropical waters) vegetation.

Saltmarshes consist of grasses, herbs and low woody vascular plants on intertidal shores worldwide, in estuarine, barrier-island and open-coastal environments (Chapter 11). The longevity of this type of ecosystem is related to changes in relative sea level, tidal range and its capacity to accrete sediment. Mangroves, the only woody halophytes found at the confluence of land and sea, cover $c.181\,000\ \mathrm{km}^2$ of coastline worldwide, and occur in somewhat distinct estuarine, deltaic and marine formations (Chapter 12). Commercially important, especially as sources of timber and fuel, mangrove forests have been widely degraded and in many areas replaced by alternative land uses such as aquaculture. Based on geomorphology, estuaries vary in form between drowned river-valley, bar-built, fjord-type and tectonic types (Chapter 13). As major conduits of runoff and waste from the land, and access points to river systems for transport and other purposes, estuaries are as vulnerable as the other wetlands to economic development and human impacts.

11 · Saltmarsh

PAUL ADAM, MARK D. BERTNESS, ANTHONY J. DAVY AND JOY B. ZEDLER

INTRODUCTION

Coastal saltmarshes are ecosystems with characteristic vegetation, geomorphology and habitat conditions. At least three different settings for saltmarsh systems can be distinguished, namely estuarine (including coastal lagoons: Chapter 13), open low-wave-energy coast (foreland salt-marshes: Dijkema 1987) and barrier islands. Whilst all are recognizable as saltmarsh, they differ in the degree to which they are influenced by marine or fresh water, and hence in their vulnerability to different stressors.

Spatial variation in the composition of saltmarsh floras and vegetation occurs at a number of scales. Globally, there is a latitudinal gradient, with vascular plant species richness being lowest in the tropics. Within estuaries, there is a longitudinal gradient from the head to the mouth. In individual marshes, the physical gradient from sea to land

Aquatic Ecosystems, ed. N. V. C. Polunin. Published by Cambridge University Press. © Foundation for Environmental Conservation 2008.

is reflected in a visible zonation. Dominant plant species in saltmarshes often spread vegetatively, and their clones have sharp boundaries which are visually striking and strengthen the impression of discrete zones. Within zones, there may be a mosaic of variation associated with microtopographic variation, particularly related to tidal creeks (Beeftink 1966; Zedler *et al.* 1999).

Saltmarshes are sedimentary environments and their existence reflects the interplay between sediment input, the biota and erosion. The various factors involved in this interplay are subject to change, especially in response to human activity. In some circumstances, saltmarshes can develop from bare mudflat to apparently mature stages within a few years; erosion can be even faster (Adam 2000). In considering possible future states of saltmarshes, the effects of human modifications have to be superimposed over the natural dynamism. Although rapid change can occur, the number of cases where change has been followed by direct observation or experimentation is small (Packham & Liddle 1970; Snow & Vince 1984; Zedler *et al.* 1992; Figueroa *et al.* 2003). Mostly, change is inferred from the current distribution patterns of species and communities, but substitution of spatial differences for temporal change must be treated with caution if biotic composition or environmental conditions have changed during the development of a marsh (Adam 1990). While early succession may be rapid, the later stages of marsh development may be stable for long periods. Extensive upper marshes on the North Norfolk coast (UK) developed after the period of rapid sea-level rise following the last glaciation, and have persisted for more than 6000 radiocarbon years, accommodating the generally rising relative sea level (Funnell & Pearson 1989; Funnell & Boomer 1998). Such ages are not unusual for upper marsh plains on the Atlantic and southern North Sea coasts of Europe (Allen 2000); however, future environmental alterations may trigger change even in these old marshes.

FUNCTIONS AND VALUES

Saltmarshes provide a number of important ecological goods and services, including: biogeochemical cycling and transformation of nutrients; nursery, feeding and refuge areas for juvenile stages of fish and crustaceans, many of commercial importance (Beck *et al.* 2003); provision of habitat for birds, many of them migratory and subject to international conservation concern; shoreline protection through their ability to dissipate the energy of incoming waves; and passive recreation and aesthetic, cultural and spiritual values.

Saltmarshes in many countries have been given protected area status but, while important, this does not address the maintenance of the ecological services. If saltmarsh conservation is to be achieved within the framework of policies which aim to maintain the Earth's life support systems, then saving isolated fragments of saltmarshes will not be sufficient (Percy 2000).

While saltmarshes are readily studied within their vegetated boundaries, they are linked to, and affected by, much wider landscapes and seascapes (Valiela *et al.* 2000). Linkages between saltmarshes and their catchments have been little studied, including those mediated through groundwater flows (Chapter 3). Groundwater abstraction may not only affect water and nutrient flows under and into saltmarshes, it can also cause subsidence, raising the relative sea level and increasing the depth and frequency of tidal flooding, as observed around Chesapeake Bay (USA) (Stevenson *et al.* 2000).

Managers need to be aware of the open nature of the saltmarsh ecosystem, linkages provided by migratory species and some pollutants sometimes being transnational or even transcontinental. Nevertheless, decision-making cannot await completion of in-depth studies of every possible linkage. Research will undoubtedly be needed to implement management, but, given tight budgets and timelines, it will need to be focused on particular problems rather than indulge an open-ended agenda.

The flora and vegetation of coastal saltmarshes bear similarities to those around inland salt lakes and other inland saline habitats (Chapman 1974; Chapter 7). Wading birds and waterfowl may use both coastal and inland saline ecosytems, providing ongoing opportunities for the spread of plant propagules.

With increasing salinization of terrestrial habitats at landscape (Ghassemi *et al.* 1995; Cullen 2003) and local scales (Environment Canada 2002), there are opportunities for the colonization of new habitats by coastal halophytes (Galatowitsch *et al.* 1999; Preston *et al.* 2002). Saltmarsh species have been used in attempts to reclaim and rehabilitate salinized land, and there is growing interest in the domestication potential of halophytes (Glenn *et al.* 1991, 1998), although the spread of species outside their natural range could result in weed problems.

SALTMARSHES AND CHANGE

Saltmarshes have endured a long history of human impacts, which has included embankment to create agricultural land, infilling for urban and infrastructure development, grazing, haymaking, and harvesting of wildfowl and other species. Although frequently imagined to be amongst the few examples of 'wild nature' surviving in heavily industrialized regions, saltmarshes are as much a cultural construction as the rest of the landscape.

In regions with a long history of land-claim, simply adding the area of identified reclaimed marshland to the current area of extant marsh does not provide a useful measure of the original extent of saltmarsh. Particularly in the case of land-claim for agriculture, continuing or accelerated sedimentation outside the embankment may permit new marsh colonization (Kestner 1962; Percy 1999). Where land-claim is for infrastructural development, such as ports, further expansion of saltmarsh is normally limited.

Over time, reclaimed saltmarshes may develop into habitats of considerable conservation value (Gray 1977). The conservation value of freshwater grazing marshes developed on the sites of reclaimed saltmarshes is now formally recognized by the European Union. However, intensification of agriculture is now causing the loss of much of the diversity of reclaimed land.

Some forms of impact have historically been more geographically restricted. In the eastern and southern USA, ditching and draining of saltmarshes for insect control have occurred for 150 years. Insects breeding in saltmarshes, primarily mosquitoes, were perceived as a nuisance and threat to human health. During the Great Depression, ditching was carried out in public-works projects (Dreyer & Niering 1995), modifications to saltmarshes being a factor that potentially facilitated the establishment of the European genotype of the reed *Phragmites australis* (Burdick & Konisky 2003). From the 1940s to the 1960s, spraying with dichlorodiphenyltrichloroethane (DDT) and other insecticides was extensive (Walters 1992) and had a substantial impact on marshes, but did not eradicate the problem. Many techniques continue to be used (Carlson *et al.* 1994) and, in coastal Australia, the threat to human health posed by a number of arboviruses spread by saltmarsh-using mosquitoes has resulted in greater use of measures, including habitat modification and chemical use (Dale 1994). The runnelling technique developed in Australia may facilitate the spread of mangrove propagules into saltmarsh (Breitfuss *et al.* 2003), which, while reducing habitat for the mosquitoes, may exacerbate loss of saltmarsh to mangrove (Saintilan & Williams 1999; Chapter 12).

Several wetland-associated diseases of humans such as those caused by the West Nile virus in the USA (Griffing 2003), and others in Australia (Van Buynder *et al.* 1995), are becoming more prevalent. In addition, with increased coastal human population density and the effects of global warming on insects and pathogens, the control of insect populations in saltmarshes is likely to become a major management issue over the next few decades.

The trend to increasingly larger ships has resulted in more dredging of shallow water to maintain access to ports. In the USA, large quantities of dredge spoil have been used to fill saltmarshes, although spoil has also been used to create new marshes. In Louisiana, where subsidence is a serious problem, the fine-textured dredge spoil can be used in moderation effectively to compensate for saltmarsh loss. Excess addition of spoil, however, rapidly shifts saltmarsh to upland, favouring less-valued species. High precision is required in the implementation phase if dredge spoil is to be used successfully to prevent wetland loss.

In southern California, dredge spoils are coarser in texture, and sediment accumulation, rather than subsidence, is a cause of saltmarsh loss. Here, dredge spoils have not proven effective in restoring biologically diverse saltmarshes; few dredge-spoil areas support dense and tall cordgrass (*Spartina foliosa*: Langis *et al.* 1991; Phinn *et al.* 1996). One of the motivations for saltmarsh creation in this region is to provide habitat for the endangered light-footed clapper rail (*Rallus longirostris levipes*), but this species does not nest in the short *S. foliosa* on dredge spoil (Zedler 1993; Zedler & Callaway 1999) (Fig. 11.1). Compared with natural marsh reference sites, dredge spoil is low in nitrogen (N) and organic matter (Langis *et al.* 1991), and the composition of the epibenthic invertebrate community differs (Scatolini & Zedler 1996).

Experimental addition of urea to compensate for the low N levels made *Spartina* grow taller than 90 cm, and did not cause scale-insect outbreaks (Boyer & Zedler 1996, 1998); but the effects were short-lived, because belowground tissues did not accumulate enough reserves to sustain tall plants once urea addition ceased (Boyer *et al.* 2000). Even four years of urea addition could not reverse this degradation threshold (Lindig-Cisneros *et al.* 2003) while at the larger scale, adding urea led to the C_3 succulent *Salicornia bigelovii* outcompeting the C_4 grass *S. foliosa* (Boyer & Zedler 1999).

FUTURE PROSPECTS

Two types of saltmarsh change may be usefully distinguished, namely those involving immediate and absolute

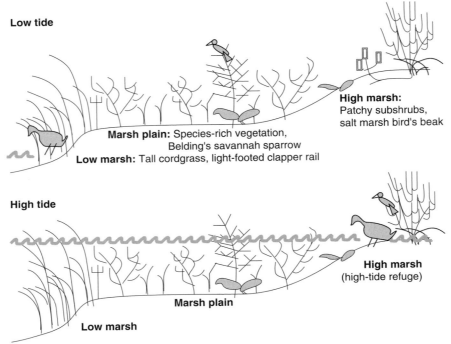

Fig. 11.1. Depiction of southern California saltmarshes. (a) The low marsh, marsh plain and high marsh linked by two endangered birds, which nest in the marsh and move upslope during high tide; (b) the high marsh linked to the upland by the ground-nesting bee pollinators of saltmarsh bird's beak (*Cordylanthus maritimus* ssp. *maritimus*, an endangered plant) and Belding's savannah sparrow (*Passerculus sandwichensis beldingi*), which feeds beyond its nesting territory, especially during the non-nesting season. (Cartoons by J. Zedler.)

loss of habitat through conversion, and those altering the habitat where some form of intertidal saltmarsh ecosystem remains. Where these changes result in reduction in diversity or ecosystem functions, the outcome is considered to be mostly undesirable. Not all changes are so easily characterized, and whether or not management intervention should be contemplated in these cases will require debate. For example, from an ecosystem perspective, the widespread invasion of saltmarshes by mangroves in south-eastern Australia (Saintilan & Williams 1999) threatens long-term marsh survival at particular sites. But is it likely to adversely affect site productivity or alter exchanges with the adjacent estuary? Against the background of continuing decline of mangroves (Wilkie & Fortuna 2003; Chapter 12), mangrove spread, even at the expense of saltmarsh, might be viewed from many perspectives as 'good'. Similarly, perspectives on the threat posed by the spread of *Phragmites australis* in the USA depend on value judgements (Ludwig *et al.* 2003).

Given the history of loss and degradation of all coastal habitats (e.g. Chapters 13, 16 and 18), coastal-zone managers will be faced with difficult decisions as society mandates action to repair past damage. Restoring one type of habitat may involve loss or damage to another. There are likely to be many interacting factors driving change. The complexity of the interactions constrains the capacity for precise predictions of future states in saltmarshes. There will be change, and general directions of change may be indicated by past or current trends, but detailed predictions for individual sites will remain elusive. Two of the most widespread and interrelated forcing factors are rising relative sea level and global warming.

Sea-level rise

The effects of sea-level rise on saltmarshes will depend on relative sea-level rise, which will vary considerably among regions (Adam 2002). Rising relative sea level has already had impacts such as in New England (USA), where over the past half century low-marsh vegetation has replaced high-marsh vegetation (Warren & Niering 1993; Donnelly & Bertness 2001; see also Morris *et al*. 2002). Sea-level rise has accelerated since the end of the nineteenth century, coinciding with the Industrial Revolution, and reached a rate that is unprecedented in the past 2500 years (Donnelly & Bertness 2001).

While halophytes are by definition salt tolerant, species vary in their tolerances and responses to changes in soil salinity. Salinity in saltmarsh soils is determined by interactions between climate, elevation and tidal access (controlling tidal inundation) and vegetation (determining evapotranspiration rate). Salinity and other factors affect *Spartina alterniflora* (Bradley & Morris 1991, 1992; Hwang & Morris 1994; Morris *et al*. 2002). The productivity of *S. alterniflora* stands is a function of relative elevation of the marsh surface. At its upper limit, *S. alternifolia* is limited by salinity stress (Morris 1995), and presumably by hypoxia at its lower limit. As the relative elevation of the marsh surface decreases with increase in sea level, hence decreasing salinity, increasing productivity will result in the trapping of more mineral sediment, so maintaining the relative elevation of the marsh surface. If the elevation cannot be maintained, increased sea level will result in decreased productivity (and hence lower sedimentation) resulting eventually in complete loss of vegetation.

Increased carbon dioxide and temperature (global warming)

The direct effects of climate change on saltmarshes could be considerable. Although detailed ecophysiological studies to explore causation are few, the latitudinal limits of salt-marsh species and communities are correlated with temperature. Species' distributions and abundances can be expected to change owing to warming alone, although the rates and nature of responses will vary and warming effects will interact with other environmental changes.

Temperate and tropical saltmarsh communities contain mixtures of C_3 and C_4 species. Northern-hemisphere salt-marshes are noteworthy for the dominance of C_4 *Spartina* species at relatively high latitudes. Increased temperature in itself might be expected to favour C_4 species, but increase in carbon dioxide (CO_2) is likely to stimulate photosynthesis in C_3 species but not C_4 species. Experimental CO_2 enrichment in chambers within brackish saltmarsh communities (Drake 1992; Drake *et al*. 1996*a*, *b*) has allowed the assessment of plant

and ecosystem responses to possible future atmospheric CO_2 concentrations, but not of interactions with warming.

Exposure of a brackish marsh dominated by the C_3 *Schoenoplectus americanus* (formerly *Scirpus olneyi*, chair-maker's bulrush) to 700 ppm CO_2 (twice the current ambient concentration) resulted in increased photosynthesis and reduced respiration (Arp & Drake 1991; Drake 1992; Drake *et al*. 1996*a*). Shoot number and root growth increased, but root and shoot N concentrations declined (Curtis *et al*. 1989*a*, *b*, 1990; Drake *et al*. 1996*b*). Increase in carbon assimilation was not shown in an adjacent marsh dominated by a C_4 species treated similarly. Elevated CO_2 in both C_3- and C_4-dominated marshes reduced transpiration and hence increased water-use efficiency (Drake 1992), the former possibly limiting summer increase in salinity within the rhizosphere. The effects of these changes in plant growth and N content are likely to be complex. Under elevated CO_2, the C_3 plants had less insect and fungal damage, while the C_4 plants had more fungal damage. The lesser damage to C_3 plants probably reflects the plants' lower N concentration and attractiveness to pathogens and herbivores, while the fungal infection in C_4 plants is related to improved water status (Thompson & Drake 1994).

Elevated atmospheric CO_2 levels may delay senescence of C_3 plants but had no effect on decomposition in either C_3 or C_4 species, which recycle N prior to senescence (Curtis *et al*. 1989*a*; Norby & Cotrufo 1998). However, rhizosphere bacteria associated with C_3 roots fixed more N into plant-available compounds under elevated CO_2 levels than under the ambient conditions (Dacey *et al*. 1994), the greater bacterial activity possibly being a response to increased root biomass or exudation. There was no such effect in C_4 plants. An increase in C_3 relative to C_4 plants under high CO_2 in Maryland (USA) suggests this (Arp *et al*. 1993).

Biodiversity as well as ecosystem functioning may be affected by warming and CO_2 increase, but experiments are unlikely ever to manipulate all relevant factors simultaneously. Monitoring of natural sites will provide data on change but without providing information on causes. The most useful approach to identifying the factors responsible for change will be modelling. Simas *et al*. (2001) modelled the saltmarshes of the Tagus estuary (Portugal), which experience a sub-Mediterranean climate and mesotidal regime, and support a mixture of C_3 and C_4 species. The model fitted the present marsh state well and suggested that the marshes are only likely to suffer damage from sea-level rise under the worse-case scenarios of Houghton *et al.* (1990), IPCC (Intergovernmental Panel on Climate Change) (1995) and Houghton (1999). Saltmarshes in areas with high tidal ranges may be less vulnerable to sea-level rise owing to greater sediment transport and accretion, although this prediction might need tempering on a local basis to recognize changes in storm incidence or rainfall distribution (Simas *et al.* 2001).

Where hinterland topography and land use permit, the response of saltmarsh to rising relative sea level would be expressed as a landward retreat. However, the current spatial distributions and relative abundances of species are unlikely to be preserved intact. The realized distributional limits of species reflect not only physiological constraints, but also the outcome of competitive interactions between species. Each species will have an individual response to the changed environmental conditions and this may alter competitive hierarchies. At a broad geographical scale, the latitudinal distribution of species and communities (Adam 1990) reflects species' responses to climatic conditions, particularly temperature. As climatic warming occurs, it would be anticipated that species' distributions would extend into higher latitudes, but each species' response will be different. There will still be a latitudinal pattern, but the relative distribution of species is likely to be different from that of today, with today's communities proving to be ephemeral over longer time periods. As the environment changes there will be a sorting of the biota rather than a migration within and between marshes of existing assemblages. It is the individual nature of species' responses to environmental conditions, and the consequent kaleidoscopic arrangement of species' interactions, that limit the ability to predict the future states of marshes. Nevertheless, established long-lived clonal plants may exhibit resilience in the face of environmental change until critical thresholds in particular factors are crossed, so that in the short term (which may extend beyond 2025), processes of inevitable change may not be detectable.

Changes to other climatic factors

In some regions, there may be changes both in rainfall amount and seasonality. Increased rainfall may counter effects of increased temperature on soil salinity, whereas decreased summer rainfall and increased temperature may result in higher soil salinity in infrequently flooded upper marsh zones, with consequent vegetation dieback and creation of bare pannes. At higher latitudes in the northern hemisphere, ice rafting is a major disturbance (Dionne 1989). Warmer winter temperatures may reduce the extent and severity of damage, as well as the availability of habitat for early recolonists of open patches, but they could also increase damage by increasing the number of freeze–thaw events, increasing frost heaving and damage to roots or rhizomes.

Changes in the severity and seasonality of storms may affect the balance between erosion and accretion, and also affect the amount and deposition of dead plant material across marshes. In temperate saltmarshes, large accumulations of wrack (deposited plant detritus) can be rafted by tides and storms into the high marsh and stranded for many months until they decay or are swept away. Stranded wrack can be 10–25 cm deep and can kill underlying vegetation if in place for an extended period, the new bare spaces being available for recolonization (Bertness & Ellison 1987). Wrack deposition will also be affected by changes in source, so that factors affecting the extent and productivity of seagrass and algal beds will have repercussions for nearby marshes.

In southern California, there is a current trend towards dominance by pickleweed (*Sarcocornia pacifica*), driven by sediment accumulation and increasing soil salinity (Zedler *et al.* 2001). In this Mediterranean climate zone, changes to rainfall affecting soil salinities could enhance or reverse this trend; reduced soil salinities could enhance growth of species other than *S. pacifica*. Storm activity could add more sediment to already choked channels, raising the marsh plain out of the tidal zone, lowering soil salinities sufficiently to allow invasive species to dominate and causing mortality of stenohaline fauna.

Intermittently, open coastal lagoons, which are characteristic of many coasts including southern Australia, South Africa and southern California, may be substantially altered if frequency and strength of storms change. Dune washover events may extend the period of estuarine closure (Chapter 13), leading to lagoons becoming hypersaline. Even if channels are dredged and flushing restored, ecosystem recovery may never be complete (Zedler *et al.* 1992).

Invasive species

Invasive species are one of the most serious threats to biodiversity globally (Mooney & Hobbs 2000; e.g. Chapters 12–14). Of particular concern are 'transformer' species, 'which change the character, condition, form or nature of ecosystems over substantial areas' (Richardson *et al.* 2000).

Saltmarshes provide examples of both plants and animals acting as transformers.

The classic plant transformers in the saltmarsh environment are members of the genus *Spartina*. The majority of *Spartina* species are indigenous to the Americas, but *S. maritima* is native to Europe and Africa. The chance hybridization in southern England between *S. maritima* and the introduced North American *S. alterniflora* resulted in the production of the sterile *S. townsendii*, which later gave rise to the fertile amphidiploid *S. anglica* (Marchant 1968). *Spartina anglica* has been very widely planted, with the intention of creating new marshes and stabilizing channels; with plantings acting as nuclei for subsequent spread and invasion of other sites, initial plantings in England and mainland Europe were followed by those in Australasia and China (Chung 1985, 1994; Chung *et al.* 2004). Through colonizing mudflats below the previous seaward limit of saltmarsh spread, *S. anglica* has caused loss of habitat for wading birds (Millard & Evans 1984; Goss-Custard & Moser 1988; Hacker *et al.* 2001).

Spartina anglica, *S. alterniflora*, *S. patens* and *S. densiflora* have all been introduced to the west coast of North America. Like *S. anglica*, *S. alterniflora* has displaced both the infauna and birds that forage on the mudflats (Sayce 1988). In California, *S. alterniflora* has hybridized with the native *S. foliosa*, creating a vigorous hybrid which invades low marsh and mudflats, while continued introgression between the hybrid and *S. foliosa* is leading to local extinction of the native species (Callaway & Josselyn 1992; Antila *et al.* 1998, 2000; Ayres *et al.* 1999, 2003).

Hybridization, creating new taxa capable of exploiting different niches from those of either parent and threatening the genetic integrity of parent species, may be a particular problem in saltmarshes. Many wetland genera are very widespread, and opportunities clearly exist for the introduction of one species into the range of a congener or for breakdown of barriers between sympatric species. Such 'genetic invasions' are by no means confined to the several instances in *Spartina*. In south-western Spain, there is pollen flow from high-marsh *Sarcocornia fruticosa* to the stigmas of *S. perennis*, the dominant species on successively developing clumps lower on the marsh. The resulting hybrids themselves become dominant and promote continued sediment accretion. Thus, the hybridizations can be seen as facilitating rapid successional change on these marshes (Castellanos *et al.* 1994; Castillo *et al.* 2000; Figueroa *et al.* 2003). Other cases of hybridization amongst halophytes have occurred in *Typha* and *Tamarix* (Pearce & Smith 2003).

A particularly insidious form of invasion to which wetlands may be prone is from exotic genotypes. Widespread wetland species are often made up of very many ecotypes (Adam 1990). *Phragmites australis* is indigenous but was rarely dominant in North American saltmarshes; however, a non-native genotype introduced in the 1950s (Saltonstall 2002, 2003) has become dominant over large areas, displacing other native species (Chambers *et al.* 1999). Given the wide distribution of many saltmarsh species, such cryptic invasion may be more common than recognized. Caution should be applied to proposals to use out-of-range genotypes in restoration and rehabilitation.

Many invertebrates have been introduced to estuaries around the world (Carlton 2000), and a number of these may have impacts on saltmarshes. The Australasian burrowing isopod *Sphaeroma quoyanum*, introduced by ships to California in the mid nineteenth century, has substantially modified saltmarsh habitat on the west coast of the USA by forming dense anastomosing burrow networks in firm *Salicornia* sediment rather than in softer low-marsh *Spartina* sediment (Talley *et al.* 2001). The burrows alter sediment shear strength, thus increasing erosion rates.

The number of introduced plant species categorized as transformers is relatively small, and most introduced species in saltmarshes tend to be classified as 'benign invaders' (Richardson *et al.* 2000). However, two matters need consideration. Firstly, in other ecosystems, many species have had a long latency before their 'weed' potential was realized. Secondly, the interactions between introduced and native species may be more complex than the visible dominance assumed for transformers. The effects of introduced plant species on the long-term survival of the endangered *Cordylanthus maritimus* ssp. *maritimus* (salt marsh bird's beak) in California (Noe & Zedler 2001) should raise concern about any introduced species (Fellows & Zedler 2005). The exotic annual grass *Parapholis incurva* is a host for the hemiparisitic annual *Cordylanthus* but, being an annual, dies early in summer before *Cordylanthus* reaches reproductive maturity, and thus acts as a habitat sink.

Eutrophication

Eutrophication of coastal water is widespread (McComb 1995), the two major inorganic nutrient problems in estuaries being caused by increased N (Howarth *et al.* 2000*a*; NRC [National Research Council] 2000*a*; Boesch *et al.* 2001; Castro *et al.* 2003) and, more locally, phosphorus (P) (McComb & Lukatelich 1995; Chapter 13). In terms of biological response, the N : P ratio may be more important than absolute concentrations. Increased nutrient levels can have substantial impacts in adjacent coastal waters, as in the Gulf of Mexico where a hypoxic zone is caused by nutrients from agricultural runoff (Goolsby *et al.* 1999; NRC 2000*a*). Impacts of increased nutrients are particularly severe in lagoons that are only intermittently flushed, as in Australia and on the west coast of North America (McComb & Lukatelich 1995). In the USA, the dominant source of N to estuaries varies between catchments; for most, the source is agricultural, for some it is urban, and for a few it is atmospheric, indicating that catchment-specific programmes will be required to reduce N loadings (Castro *et al.* 2003).

Increased N is of particular significance to saltmarshes, which historically have been N limited (Jefferies & Perkins 1977; Cargill & Jefferies 1984; Kiehl *et al.* 1997). Anthropogenic increases in the availability of N to saltmarshes may thus be creating environmental conditions without historical precedent. The growing demand for global food production (Vitousek *et al.* 1997*b*) has meant fertilizer production has continued to grow; but much of the N applied is in excess of that immediately required and enters both surface and groundwater from runoff and airborne deposition (Chapter 3). Eutrophication will increase the growth of competitive dominant plant species, leading to declines in community species richness; and through altering tissue, N content may affect palatability with consequences throughout the food chain.

The recent expansion of *Phragmites* in North America was contributed to by anthropogenic eutrophication that favoured the expansion (Chambers *et al.* 1999; Meyerson *et al.* 2000; Bart & Hartman 2003; Burdick & Konisky 2003). *Phragmites* was long thought to be limited in invading saltmarshes by high soil salinities; but spreading into marshes flooded by normal seawater is possible through clonal integration (Amsberry *et al.* 2000; Minchinton & Bertness 2003), and the potential for further continuing spread is considerable (Burdick & Konisky 2003). Development adjacent to saltmarshes, leading to nutrient enrichment and salinity reduction, may promote the spread of *Phragmites* as in New England (USA) (Bertness *et al.* 2002). The lower marsh in this region was dominated by *S. alterniflora*, but there was a striking zonation (Miller & Egler 1950; Redfield 1972), driven by competition for N. With shoreline development and eutrophication of the saltmarshes there is a dramatic simplification of the system as the diverse mid and upper marsh is squeezed out by the expansions upwards of *S. alterniflora* and downwards of *Phragmites* (Bertness *et al.* 2002) (Fig. 11.2). In this region, there is an almost three-fold decrease in plant species diversity and concomitant reductions in animal diversity.

Other species with growth form and responses to eutrophication and lower salinity similar to *Phragmites*, such as *Typha* spp. (Zedler *et al.* 1990), *Bulboschoenus* spp. and *Schoenoplectus* spp., also form dense monospecific stands in saltmarshes when conditions in, or adjacent to them, are modified by human activity. Catchment runoff includes many chemicals other than nutrients, although there is little evidence of direct impacts on saltmarshes. However, triazine herbicides and their degradation products have been detected at significant concentrations in saltmarshes in Essex (UK) (Meakins *et al.* 1985; Leggett *et al.* 1995). These herbicides are very widely used in agriculture. Sublethal concentrations decreased growth rate and photosynthetic activity of diatoms and vascular halophytes (Mason *et al.* 2003), the former being a major component of the micro-algal biofilm which is important in stabilizing the sediment surface of saltmarshes (Coles 1979). When the saltmarsh sediment surface was exposed to the chemical Simazine, diatoms migrated deeper, reducing the surface sediment stability (Mason *et al.* 2003). In Essex, sublethal herbicide concentrations could be leading to loss of sediment stability and reduced vegetation cover, contributing to long-standing saltmarsh erosion (Mason *et al.* 2003). Given the global use of triazine herbicides, this chain of consequences could be of widespread occurrence.

Consumer pressure

The extensive saltmarshes of eastern North America were long considered an ecosystem controlled primarily by bottom–up forces, such as nutrient levels, soil salinity and oxygen levels (see Mendelssohn & Morris 2000), except where migratory waterfowl caused extensive damage (Smith & Odum 1981). Herbivory by smaller grazers was considered minor and hence the major energy outflows were exported to adjacent waters or to decomposers (Nixon 1980; Pfeiffer & Wiegert 1981). In Europe, the importance of grazing by livestock in determining the composition of saltmarsh vegetation has long been recognized (Adam 1978; Bakker 1989).

However, on Sapelo Island (Georgia), the gastropod *Littoraria irrorata* is capable at high densities of turning

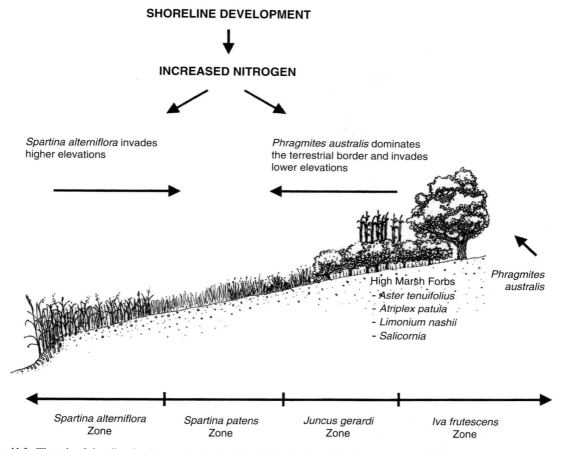

SHORELINE DEVELOPMENT

↓

INCREASED NITROGEN

Spartina alterniflora invades higher elevations

Phragmites australis dominates the terrestrial border and invades lower elevations

High Marsh Forbs
- *Aster tenuifolius*
- *Atriplex patula*
- *Limonium nashii*
- *Salicornia*

Phragmites australis

| *Spartina alterniflora* Zone | *Spartina patens* Zone | *Juncus gerardi* Zone | *Iva frutescens* Zone |

Fig. 11.2. The role of shoreline development in altering New England saltmarsh landscapes. (Drawing by M. D. Bertness.)

dense *S. alterniflora* stands into mudflat in less than a year (Silliman & Bertness 2002). Although *Littoraria* can be found at densities of 500–1000 animals per m² in the high marsh where predators are unable to penetrate the tall *S. alterniflora*, the snails are virtually excluded from the seaward border of the marsh by crab, turtle and fish predators, which play an important role in preventing the gastropod from entirely reducing the *Spartina* to an unvegetated mudflat (Silliman & Bertness 2002). Blue crabs are one of the major predators of *Littoraria* but have been greatly depleted by fishing (Lipcius & Stockhausen 2002). Overfishing of blue crabs may lead to consumer control of *Spartina* (Silliman & Bertness 2002). Given the global extent of overfishing, there is potential for similar effects elsewhere.

In North America, the snow goose (*Chen caerulescens caerulescens*) was in decline for much of the twentieth century from hunting and destruction of temperate saltmarsh winter feeding grounds. In the 1980s, there was a shift from feeding on saltmarshes to crop lands and golf courses (Jefferies *et al.* 2002), in response to the greater availability of resources following heavy fertilizer use and changed agricultural practices. The resources enhanced winter survival, leading to a population explosion and greater pressure on high-latitude marshes used during the summer breeding season (Jefferies 1997; Handa *et al.* 2002; Jefferies *et al.* 2002). The consequences for the saltmarshes are devastating. Geese feed on marsh plants by grubbing out roots and, since the root mat in high-latitude saltmarshes is superficial, a barren mudflat can result (Fig. 11.3). Without

(a)

(b)

(c)

Fig. 11.3. The impact of foraging lesser snow geese (*Chen caerulescens caerulescens*) on Hudson Bay saltmarshes. (a) Relatively intact vegetation, (b) denudation of marsh by foraging geese and (c) exclusion plot in otherwise denuded area. (Photographs courtesy of R. L. Jefferies.)

vegetation, soil salinities increase, soil oxygen levels decrease, recolonization by seed under these conditions is impossible, and reinvasion by vegetative growth is slow (Handa *et al.* 2002). In 1997, more than 2500 ha of Hudson Bay saltmarshes were denuded by snow geese, with consequences potentially felt upwards through the food chain.

In north temperate Europe, a comparable explosion of the overwintering populations of several geese species

(Esselink *et al.* 1997; Esselink 2000) has resulted in very heavy grazing pressure on saltmarshes, and in pressure on summer grounds in Arctic Russia likely to be similar to that experienced in Canada.

FISH USE OF SALTMARSHES

Saltmarsh creeks and pans provide nursery habitat for many fish (Adam 1990), and even rarely inundated marsh plains may be used by a diversity of fish at high tide (Connolly *et al.* 1997; Connolly 1999).

In southern California, killifish (*Fundulus parvipinnis*) gain access to the marsh plain via small creeks (Desmond *et al.* 2000) and obtain several times more food once there (West & Zedler 2000), resources that are critical to growth (Madon *et al.* 2001). Killifish growth may be positive during most of the year until early spring, when tides of low amplitude fail to flood the marsh; killifish cannot feed on the marsh plain, lose weight and ultimately die. Reduced access to the marsh during springtime explains this species' annual life cycle (Madon *et al.* 2001). Small creeks have implications for saltmarsh restoration, as small creeks are difficult to construct with the large machinery often employed (Williams & Zedler 1999).

For natural marshes, rising relative sea level may eliminate areas of saltmarsh with established creek systems, while dense vegetation could prevent formation of new creeks. The remaining marsh plain would be inundated by sheet flow rather than overflow from creeks, and fish use might be impaired. Fish populations in saltmarshes may also be at risk from invading species, evidenced by increasing populations in southern California saltmarsh channels of the introduced yellow-fin goby *Acanthogobius flavimanus*, capable of altering food webs (Williams & Zedler 1999).

Site-specific actions

The factors discussed above, although manifested at individual sites, act or may act at large geographical scales. Decisions involving the destruction or conversion of sites, while having wider implications, will be made on a case-by-case basis. The case for the conservation of saltmarshes has been strongly made in the Western world for more than four decades. The perceived relationship between saltmarshes and the sustainability of such activities as fishing and wildfowling has ensured that the value of saltmarshes is popularly recognized, even if some of the arguments in favour were not well founded (Nixon 1980).

Saltmarshes are in theory protected by a range of legal measures (Shine & de Klemm 1999). However, little of the protection is absolutely mandatory; it inevitably has conditions that will continue to allow loss of wetlands 'in the national interest'. The task of weighing long-term environmental values against economic and social interests is political, although the extent to which the process is transparent and open to public participation clearly varies between jurisdictions.

Even in countries with well-developed wetland protection policies and strong conservation movements (such as the UK), governments may continue to advance proposals involving wetland destruction. In 2003, the North Kent marshes at Cliffe (UK) were one of a number of proposed sites for a new international airport for London, in spite of including several conservation areas (RSPB [Royal Society for the Protection of Birds] 2003). These saltmarshes and mudflats of the Thames estuary are one of the most important overwintering sites in the UK for migratory waders and wildfowl. If the airport were built, active measures would be required to discourage flocks of birds within a 13 km radius. Even with such measures, the bird-strike risk would have remained high and the proposal has now been dropped; but the important point is it was not immediately rejected when first suggested.

One of the difficulties of guaranteeing protection of saltmarshes is that terms like 'the national interest' are left open to interpretation on a case-by-case basis. In the Gulf of Maine (USA) for example, where only 50–75% of pre-colonial coastal wetlands remain, saltmarshes remain vulnerable to coastal development (Morgan & Short 2002). From a government planning-authority perspective, necessary coastal development may be restricted to major items of national infrastructure which require water access (port facilities and associated industry). How far should it extend to recreational facilities such as marinas, for which there may be high demand, or even to coastal real-estate development, which may yield high returns to both government and developers? In countries such as the UK, many easily developed potential marina sites were taken years ago. New development proposals are likely to require substantial construction costs, so association with onshore property development is likely (Sidaway 1991).

Development threats to saltmarshes are even more likely in developing countries, because of human population growth and infrastructure development leading to demand for coastal land. In many cases, the conservation organizations have little influence and environmental laws are not as strong as in developed countries. In the tropics, saltmarshes are poorly known, their values little appreciated, and they may be at a particular risk of loss.

Where saltmarshes have been degraded, limited opportunities for rehabilitation may exist, depending on the original cause of damage. Where tidal flushing has been impaired, re-establishing original hydrology by removing levees and floodgates may be relatively easily achieved, the main impediment often being public opposition to reversing what may be perceived as appropriate management. The effects of eutrophication will be more difficult to address. Reduction of inputs will require management of catchments and alteration of agricultural and industrial practices. Expensive engineering solutions, such as that employed in the Peel–Harvey estuary (Western Australia) to alter the flushing characteristic of the system, are unlikely to be feasible in many localities. Removal of nutrients through harvesting biomass (which must be disposed of elsewhere) may be possible at some sites (see Hodgkin & Hamilton 1998), but the original pre-eutrophication environmental conditions are unlikely to be restored. Addressing the symptoms of degradation, but not the cause (e.g. manual removal of invading dominant species) might be justified at sites with particular conservation values, but will require long-term commitment if it is to have any significance.

Control of invasive plants in saltmarshes is likely to be difficult. Manual digging out of *Spartina* is labour intensive and unlikely to be effective except in the earliest stages of invasion. Herbicides have been used to control *S. anglica* in New Zealand (Bascand 1970) and England (Truscott 1984), and have been tried on *S. alterniflora* (Edwards & Davis 1974). Aerial spraying is costly, involves the risk of damage to non-target species, and public concern exists over the use of chemicals in the environment (Grevstad *et al.* 2003). Biological control agents would need extensive research before they could be released in the field (although see Wu *et al.* 1999 and Grevstad *et al.* 2003). Biological control measures have recently been reviewed by Lacambra *et al.* (2004).

Despite the experiences of other countries, further planting of *Spartina* is still being actively promoted in China (Chung 1994; Chung *et al.* 2004). Although early experimental plantings were unsuccessful, *S. alterniflora* is now well established, with continuing spread from *in situ* seed production and germination. It is anticipated that continuing spread will promote the eventual reclamation of extensive areas of offshore sands (Chung *et al.* 2004).

Considerable effort is expended on attempting to control the spread of *Phragmites australis* in the eastern USA, herbicides, mowing and burning being the main techniques used (Turner & Warren 2003). As invasion is by introduced genotypes within the native range of *Phragmites*, biological control is unlikely at present. However, of the 26 herbivorous arthropods known to feed on *Phragmites* in North America only five are thought to be native (Casagrande *et al.* 2003); the remainder are accidental introductions. Some of these are spreading and may be locally abundant (Blossey 2003). The impact of one of the introduced herbivores appears slight (Casagrande *et al.* 2003), however introduction of further specialized herbivores from Europe might be more beneficial than the control measures currently applied with limited success (Blossey 2003).

Given the increased awareness of the environmental values of saltmarshes, governments are likely to respond to concerns about losses by requiring mitigation, either through rehabilitation of degraded sites or creation of new marshes. Legislation requiring mitigation has been in place within the USA for some decades, and the concept is being looked at favourably by governments elsewhere (Department of Land and Water Conservation 2002). The concept is attractive because it allows approval to be given to projects while maintaining that there is no net loss of wetland area or functions. There is likely to be an increasing number of mitigation or offset proposals.

Adoption by governments of such policies presents ecological science with severe tests of its predictive ability. While the challenge will lead to greater research effort and hopefully better outcomes, there is currently undue optimism both about the success of past projects and the prospects for the future. One of the lessons to be learned from the past is the need to monitor projects (e.g Ambrose 2000), and to apply adaptive management to both the approval process and on-site works. At the simplest level, this means ensuring that what was approved actually happens. More importantly, it requires setting explicit measurable goals and monitoring performance against these goals with sampling designs that have rigorous spatial and temporal controls. The enthusiasm for marsh restoration may be tempered in the future by concerns about the potential for increased mosquito populations and accompanying increased risk of disease transmission to humans living close to the restoration sites (Willott 2004).

A major potential threat to the survival of saltmarshes is rising relative sea level. Unless this is exceeded by the sedimentation rate, or there is opportunity for saltmarsh to extend inland, habitat will be lost. One management option is to remove barriers to landward movement and to promote saltmarsh establishment. On coasts with a long history of saltmarsh land-claim, there may be extensive areas of low-lying land, separated from saltmarsh or an estuary by embankments. There are many historical instances where embankments have been breached (Crooks *et al.* 2002), in storms or as a result of maintenance failure, and saltmarsh has become re-established naturally. In northwest Europe, there is currently considerable interest in a measure called managed realignment, whereby embankments are deliberately breached and saltmarsh re-established (Pethick 2002; Blowers & Smith 2003; Boorman 2003) (Fig. 11.4). This will permit the maintenance of saltmarsh area and ecosystem functioning in the face of rising sea level, and provide a living buffer to more landward embankments, offering economical and sustainable asset protection against both rising sea level and increased storm damage. There are also strong economic drivers. The cost of maintaining existing sea defences as sea level rises, and of draining land which may be lower than sea level, is disproportionate to the value of the agricultural land protected. However, where the asset values are higher, justifying managed realignment may be difficult. The proportion of the coastline subject to realignment is likely to be relatively small and in many cases there may be strong local opposition, notwithstanding the environmental and economic arguments in favour (Pethick 2002).

The complexity of the issues involved and number of agencies and interest groups likely to be consulted during the preparation of shoreline management plans is high (Winn *et al.* 2003). While any plan needs to have a long-term perspective, there also need to be mechanisms for regular review in the light of experience, changing community expectations and more refined predictions of future environmental change (i.e. adaptive management).

The number of managed realignment sites in the UK is still small and the total area involved was less than 200 ha in 2000 (Pethick 2000) and around 600 ha in 2006 (H. L. Mossman, personal communication 2006). Many of the areas potentially available for saltmarsh restoration are, as a result of sediment consolidation and oxidation of organic matter, considerably lower than the marshes outside the enclosing embankments (Crooks *et al.* 2002; Pethick 2002). Sediment recharge to increase surface elevation before restoration is likely to be crucial to eventual successful restoration (Pethick 2002). In all the UK sites, sedimentation and halophyte colonization have occurred (Crooks *et al.* 2002),

Fig. 11.4. Managed coastal realignment in Tollesbury, Essex, in the UK. An area of about 21 ha was re-exposed to tidal influence when a breach was made in the old sea wall in August 1995. (a) The breach in the old sea wall, photographed in 1997. (b) Trees and hedges killed by seawater, with the original agricultural drainage system, in summer 1998. (c) Limited saltmarsh development on low-lying sediments inside the breach, summer 1998. (Photographs: A. J. Davy.)

although local erosion has taken place around the breach sites. One of the major questions to be resolved is whether managed realignment should involve costly complete embankment removal or localized breaches (Pethick 2002).

A particular complication for progressing managed realignment is conflict between different environmental policies. The European Union has issued a number of Habitat Directives intended to secure the conservation of species and their habitats. In many cases, saltmarsh on the seaward side of embankments, and fresh and brackish marshes on the landward side, are both subject to Habitat Directives. It is therefore illegal to allow development which might threaten either habitat, a fact that has halted saltmarsh restoration at many proposed sites in the UK (Pethick 2002).

In the USA, governments, responding to public concern, have given priority to saltmarsh restoration and creation for several decades. One of the largest ongoing restoration projects is in San Francisco Bay, where salt-marshes which had been lost to agriculture and salt ponds are being restored through the breaching of sea walls in a process very similar in technique to the UK planned realignment approach (Williams & Orr 2002). Even if managed realignment and mitigation schemes go some way to limiting future loss of saltmarsh area, and to maximizing retention of ecosystem functioning, the fate of particular communities and species will still require consideration. Distinctive local assemblages of species occur on some saltmarshes, their presence reflecting particular habitat conditions and the quirks of biogeographical history. In response to changing climate conditions, the ecological and geographical distribution of each species will be affected differently. A distinctive habitat is likely always to have a distinctive assemblage of species, but the composition of that assemblage will change over time, and it may not be possible to preserve it unaltered.

While most of the saltmarsh flora has a wide geographic distribution, and would be predicted to have a similarly wide, if different, distribution in the years to come, a number of species currently have much more local distributions. Many are restricted to the high marsh, a zone vulnerable to disturbance and loss. The autecology of few of these species has been studied, but it is unlikely that most would naturally recolonize mitigation or planned retreat sites, or even that their specialized habitat requirements would easily be replicated. If these species are to survive, intensive interventionist management may be required. There is a case for deliberately seeking to expand populations of rare species and even to introduce selected aliens in the upper marsh (Ranwell 1981). In the face of evidence from around the world of the unexpected consequences of introductions and the even greater uncertainty of predicting the fate of transplants under environmental conditions different from those of today, there will be understandable reluctance to adopt such an approach. While transplantation of rare halophytes in 'wild' habitats may be discouraged, emergency measures may be required to save the genetic resource these species represent. Some species may be candidates for introduction into horticulture, in other cases they might be used in landscaping of such environments as median strips in highways.

One species for which deliberate introduction is proposed is *Atriplex* (*Halimione*) *pedunculata*, an upper-marsh annual plant that has always been rare in Britain and was thought to be extinct (Leach 1988). Since 1993, an active programme of removal of competing perennials and manual spreading of seeds has been undertaken (Wyatt 2003). Additionally, plant introductions to other sites occurred, but none has become self-sustaining, requiring continuing habitat manipulation to retain the species.

ACTIONS REQUIRED

Within the 2025 time horizon, it is likely that many of the changes in saltmarshes will result from site-specific management and planning decisions. However, the consequences of changes to global climate and atmospheric CO_2 concentration, which are already under way, will not magically stop by 2025, even if remedial action were taken immediately. The slow rate of oceanic mixing means that the sea-level rise owing to thermal expansion will lag considerably behind the onset of atmospheric warming. It is likely that by the year 2025, effects of rising sea level on saltmarshes, although detectable, will be generally minor compared with the impact of habitat loss through infill and land-claim. This in no way lessens the urgency of taking action now to limit the release of greenhouse gases in order to reduce the magnitude of future anthropogenic climate change.

Other action is also required to address the more immediate threats to the survival of saltmarshes. The Ramsar Convention on Wetlands (Ramsar 2004) obliges the signatories to nominate at least one site for inclusion on the Ramsar list of wetlands of international significance, to manage all wetlands according to principles of wise use, and to implement a national wetlands strategy. Only a minority of countries have so far adopted national strategies, and whether wise use is the guiding principle in most countries is debatable. Nevertheless, increasing the number of signatories and national strategies could greatly improve the prospects for saltmarsh survival.

Acknowledgement of the role of saltmarshes in providing ecosystem services within the context of both catchments and offshore waters will require management strategies and legislation which recognize the importance of all saltmarshes. Nevertheless, it is important that particularly significant sites are given status as formal protected areas, preferably including adjacent waters as part of broader marine and estuarine protection schemes. In most Western countries, coastal saltmarshes are well represented in the protected area system; however, there is an urgent need to raise the public profile of saltmarshes and increase formal conservation in tropical areas.

In spite of much legislation applicable to saltmarshes (Shine & de Klemm 1999), it is often fragmented between different agencies, applied inconsistently and provides limited opportunity for public involvement in the legal process. One of the potential advantages of developing national wetland strategies as required by the Ramsar Convention would be to provide a uniform focus and set of objectives for legislation to underpin policy. Desirable features of international strategy, policy and supporting legislation should include:

- recognition of the ecosystem functions provided by saltmarshes
- recognition of the vulnerability of saltmarshes to loss and degradation
- establishment of a regime to conserve biodiversity and ecosystem functions of saltmarshes, including transnational functions such as provision of habitat for migratory species
- definition of circumstances when further destruction of saltmarsh may be permitted
- establishment of a regime to promote rehabilitation of degraded sites and to make provision for managed realignment
- a clear statement of the circumstances under which mitigation (compensation) will be permitted, and establishment of a mitigation regime that provides for the clear setting of objectives and monitoring of outcomes
- requirements for public consultation in the determination of policy, the strategic planning process, the assessment and approval of development proposals, and ongoing management and monitoring.

CONCLUSIONS

It is likely that many saltmarshes will be lost to development by the year 2025. Which sites will be affected cannot be predicted, but once approval has been given for embanking or filling, the fate of a particular site will be sealed. Even where there is high awareness of the values of saltmarshes, there will be losses to infrastructure and recreational development. Increasingly, there will be requirements to balance such losses with rehabilitation of previously damaged sites or creation of 'new' marshes. Mitigation projects will need to be monitored and managed, and appropriate regulatory frameworks will be needed.

Many of the changes to saltmarshes will result from the interaction of a number of factors, which makes it difficult to predict the future of individual sites. The various examples discussed here have been chosen to illustrate the complexity of interactions and the likely site-specificity of outcomes; the broad scenarios offered are uncertain in detail. Even when general trends might be anticipated, actual outcomes at individual sites may be determined by stochastic events such as major storms or rainfall events.

As with virtually all ecosystems, invasions of saltmarshes by exotic species can be expected to increase. Given the ease with which organisms may be transported relatively rapidly around the globe, it is impossible to predict what particular species will provide new threats in the future, or where they will be experienced. However, we confidently predict that in many cases, control will be difficult or impossible. Invasives, interacting with indigenous species, are likely to create new assemblages on saltmarshes, different from any historical precedents.

Whether or not particular species are successful invaders may depend on the extent to which sites become eutrophic, and the growth and productivity (and to some extent the resilience) of natural vegetation will also be affected by nutrient inputs. Management of nutrient and pollutant inputs will rarely be possible on a site-specific basis, but will require integrated catchment approaches.

Rising relative sea level could cause erosion and loss of saltmarsh, although this is likely to be relatively localized towards 2025. Implementation of potential ameliorative measures such as managed realignment will require major attitudinal change. Current and projected schemes are largely tokenistic, and policy and social changes sufficient to allow projects of adequate scale to address the rate of loss seem unlikely in the near future.

Global warming in itself is likely to cause substantial changes in the distribution and relative abundance of organisms. However, there will be substantial interaction between the effects of temperature, atmospheric CO_2 concentration, nutrient concentration, grazing pressure and competition from introduced species. It is currently difficult to predict outcomes, other than to suggest that by 2025, the composition and distribution of saltmarsh communities will be different from that of today. Whether these changes are regarded as undesirable, to be countered by management intervention, or accepted as the new status quo, will depend on society making decisions on issues that have not yet been defined, let alone seriously considered.

Existing approaches to coastal management have, in general, not served saltmarshes well. A new dialogue between scientists, the public and politicians is required if society's expectations of environmental sustainability are to be met. Even with recognition of the factors leading to change, maintenance of present regulatory, economic and planning frameworks offers at best an uncertain future for saltmarshes globally.

12 · Future of mangrove ecosystems to 2025

RICHARD S. DODD AND JIN E. ONG

INTRODUCTION

Saltmarsh vegetation is ubiquitous along sheltered coastlines, where freshwater discharges result in sediment accumulation (Chapter 11). In temperate regions of the world, coastal saltmarsh is mostly low-stature perennial grasses and herbs, whereas the tropical equivalent is woody vegetation that reaches shrub or tree height. The woody plants occupying the tropical intertidal mudflats are referred to as mangroves, and the plant community is commonly called a mangrove community, or more simply mangrove. Mangroves are the keystone species of a broader ecosystem that includes a number of additional plant taxa and animal species that are dependent on, but not necessarily exclusive to, the intertidal mangrove ecosystem.

In addition to requiring adequate sedimentation for colonization, mangroves are limited by temperature and humidity, so that globally, they are found within the tropics and subtropics, with some incursions into lower

Aquatic Ecosystems, ed. N. V. C. Polunin. Published by Cambridge University Press. © Foundation for Environmental Conservation 2008.

latitudes of the temperate zones. Under optimal conditions of high and uniform precipitation and temperature, such as those found in equatorial regions, true forests develop with trees attaining heights of 50 m and above-ground biomass that is comparable with terrestrial lowland tropical forest. A general attenuation in plant size towards the latitudinal extremes of mangrove distributions is associated with increased seasonality of climate and reduced wet-season precipitation (Blasco 1984; Duke *et al.* 1998). At the extreme, these climate conditions may result in very low freshwater inputs, as is the case for the very arid-zone mangroves of the Arabian Gulf (Dodd *et al.* 1999). Locally, under adverse conditions of high salinity, aridity and high calcium concentrations in sediments, mangroves form a dwarf community up to 1 m in height.

The extreme stresses associated with high soil interstitial salt concentrations, anaerobic and unstable soils, and high soil temperatures have resulted in a number of unique adaptations that are more or less characteristic of mangrove species. Different forms of aerial roots, such as the stilt-roots of *Rhizophora*, pneumatophores of *Avicennia* and *Sonneratia*, peg-roots of *Xylocarpus moluccensis*, sinuous buttresses of *X. granatum* and knee-roots of *Bruguiera* serve in aerating the underground roots growing under anaerobic conditions. Physiological adaptations to deal with high salt concentrations include exclusion by root cells, elimination of salt by secretion and tolerance of high intracellular salt concentrations (Popp 1995). Reproductive dispersal units are commonly propagules, rather than seeds. This unusual mode of dispersal, known as vivipary and cryptovivipary, is surprisingly frequent among mangrove taxa, compared with plants in general, and has evolved several times. Its significance to mangrove success is not well understood, but it may serve in promoting rapid establishment in the soft sediments and minimize the risks of being washed away by tides.

The mangrove ecosystem is floristically poor compared with most terrestrial tropical forests. Approximately 84 species of mangrove belonging to 26 families are recognized worldwide (Saenger 2002), suggesting that relatively few species have adapted to these harsh environmental conditions. However, phylogenetically divergent groups are represented, including ferns (*Acrostichum* spp.), palms (*Nypa fruticans*) and a wide range of magnoliid angiosperm families. Most of these families include both mangrove and non-mangrove species, suggesting that adaptations to tolerate the coastal marsh environment have evolved several times through evolutionary history in independent lineages (Dodd & Afzal-Rafii 2002).

Mangroves are mostly restricted to a narrow coastal fringe throughout tropical and subtropical regions of the world, consequently their cover as a proportion of total land area is small, about 0.12% (Ong *et al.* 2001). They also occur in a zone that is the most human-impacted in the world. The important values of mangroves to coastal zone protection, ecosystem functioning and carbon sequestration were being recognized in the second half of the twentieth century (Thom 1967; Odum & Heald 1972; Lugo & Snedaker 1974; Ong 1993). However, this coincided with increasing human activities such as coastal development for urban and recreational use and aquaculture, and with burgeoning human populations along the coastal zone. As a result, destruction of mangroves reached alarming rates during the twentieth century and continues today in some parts of the world.

The ecology and biology of mangroves have been the subject of several recent reviews (Li & Lee 1997; Hogarth 1999; Ellison & Farnsworth 2001; Kathiresan & Bingham 2001; Alongi 2002; Saenger 2002). In this chapter, the aim is to briefly review characteristics of the mangrove ecosystem, then examine trends around the world and attempt to develop a prognosis for the system by the year 2025. Predicting future trends is predicated on historical cause and effect and requires that the factors that are most likely to threaten mangrove success in the future be identified. Mangrove vegetation is of different types, and the level of threat posed by environmental constraints will vary with geographical location and type of mangrove formation. The long-term viability of ecosystems thus also needs to be considered in a historical perspective. How have mangroves fared under historical climate changes? Recent trends in mangrove cover are examined here and put in the context of palaeohistorical fluctuations. The major mangrove formations are then classified and threats likely to have the most important impact on each identified. Some of the most threatening pressures over the next quarter century are outlined and possible ways of mitigating mangrove degradation and loss are discussed.

MANGROVE AREA AND DISTRIBUTION

Current mangrove area and worldwide distribution

Estimates of the total surface area of mangrove made in the early 1990s range from 100 000 km^2 (Bunt 1992) to 240 000 km^2 (Twilley *et al.* 1992). Much of this cover is located in developing countries, where accurate statistics are difficult to obtain. Improved methods for forest resource assessment (Blasco *et al.* 1998; Dahdouh-Guebas 2002) have provided what are likely to be more accurate estimates. Spalding *et al.* (1997) reported mangrove cover to be 181 000 km² and the Food and Agricultural Organization of the United Nations (FAO) estimated that total mangrove cover in the year 2000 was 146 500 km² (FAO 2003*b*).

The FAO estimates indicate that the Indo–West-Pacific region (eastern Africa to Oceania) accounts for 57% of the world's mangrove cover, with 40% in Asia alone (Fig. 12.1, Table 12.1). Approximately 75% occurs in developing countries, where national programmes for the protection of natural resources are the least developed (Saenger *et al.* 1983; Hamilton & Snedaker 1984).

Mangrove area trends

Increasing human pressure on coastal resources has resulted in loss of 50% of the world's mangrove forests (Kelleher *et al.* 1995) with losses of over 70% during the last 50 years in some countries such as Thailand (Burke *et al.* 2001). Reliable data at national and regional levels are few, but time series of mangrove area, as for the Merbok mangrove in Peninsular Malaysia, based on old ordnance survey maps (which go back 50 years) and satellite images (M. D. E. Haywood, F. J. Manson, N. R. Loneragan, W. K. Gong & J. E. Ong, 2001) show a linear decline (Fig. 12.2). If this trend ($y = -1.2425x + 2508.9$, where y = area (km²) and x = year) continued, there would be hardly any mangroves left by 2025, when Malaysia anticipates reaching developed nation status. In the Merbok mangrove, the major threat has changed from conversion to agriculture (rice growing) in the 1950s, to conversion to aquaculture (tiger prawn ponds) in the 1980s, to the present threat of conversion to suburban housing development (M. D. E. Haywood, F. J. Manson, N. R. Loneragan, W. K. Gong & J. E. Ong, personal communication 2001). At a local level, prediction of the

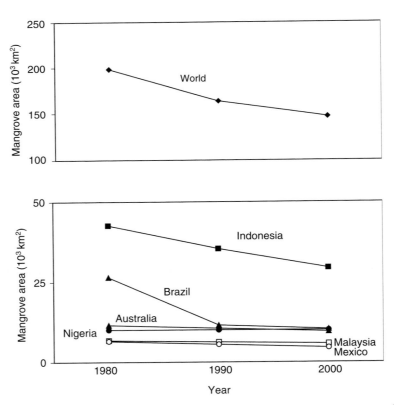

Fig. 12.1. World mangrove area in 1980, 1990 and 2000 and in the six nations with the largest mangrove areas. (Data from FAO 2003*b*.)

Table 12.1. *Mangrove area trends by nation and region from 1980 to 2000*

Nation/region	Mangrove area (10^3 km^2)			Annual change (%)		Percent of world mangroves	
	1980	1990	2000	1980–1990	1990–2000	1980	2000
Indonesia	42.54	35.31	29.30	−1.8	−1.8	21.5	20.0
Brazil	26.40	11.50	10.10	−8.0	−1.3	13.3	6.9
Australia	11.50	10.50	9.55	−0.9	−0.9	5.8	6.5
Nigeria	9.99	9.98	9.97	ns	ns	5.0	6.8
Malaysia	6.69	6.21	5.72	−0.7	−0.8	3.4	3.9
Mexico	6.40	5.43	4.40	−1.6	−2.1	3.2	3.0
Bangladesh	5.96	6.10	6.23	0.2	0.2	3.0	4.2
Myanamar	5.31	4.80	4.32	−1.0	−1.0	2.7	3.0
Cuba	5.31	5.30	5.29	ns	ns	2.7	3.6
Papua New Guinea	5.25	4.92	4.25	−0.6	−1.5	2.7	2.9
India	5.06	4.93	4.79	−0.3	−0.3	2.6	3.3
Colombia	4.40	3.97	3.55	−1.0	−1.1	2.2	2.4
Africa	36.59	34.70	33.51	−0.5	−0.3	18.5	22.9
Asia	78.57	66.89	58.33	−1.6	−1.4	39.7	39.8
Oceania	18.50	17.04	15.27	−0.8	−1.1	9.3	10.4
N. and C. America	26.41	22.96	19.68	−1.4	−1.5	13.3	13.4
South America	38.02	22.02	19.74	−5.3	−1.1	19.2	13.5
World	198.1	163.6	146.5	−1.9	−1.1	–	–

Source: Data from FAO (2003*b*).

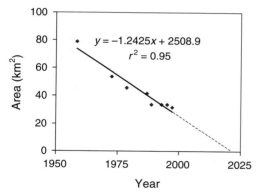

$$y = -1.2425x + 2508.9$$
$$r^2 = 0.95$$

Fig. 12.2. Plot of area of the Merbok mangrove from 1959 to 1998. (Data from M. D. E. Haywood, F. J. Manson, N. R. Loneragan, W. K. Gong & J. E. Ong, personal communication 2001.)

rate of loss as well as the threat can be made with reasonable confidence. A similar dataset is available for the Penang mangrove in Malaysia (Penang State Economic Planning Unit & DANCED [Danish Cooperation for Environment and Development] 1999). Obtaining reliable

data on mangrove area over time is difficult because of the varying methods used (such as ground surveys, remote sensing and expert estimates) to estimate mangrove cover in different countries. FAO (2003*b*) data for 1980, 1990 and 2000 indicate that global mangrove area declined by 26% from 198 000 km^2 to 146 500 km^2 during the period 1980–2000 (Table 12.1, Fig. 12.1). During this period, the annual overall rate of deforestation fell from 1.9% for the decade 1980–90 to 1.1% for the decade 1990–2000 (Fig. 12.1); however, to see this as an apparent global trend of declining rate of deforestation would be overly optimistic. In fact, reductions in the rate of mangrove loss in South America account entirely for the worldwide reductions, thus for 86.5% of the world's mangrove forests the deforestation rate remained constant over the last two decades of the twentieth century. Assuming the lower annual deforestation rate of 1.1%, world mangrove cover can be expected to reach 111 108 km^2 by the year 2025, representing 76% of that in 2000 and 56% of the world cover in 1980 (Fig. 12.1).

The FAO (2003*b*) data are the first that provide time trends in mangrove area change at global, regional and national levels. The 12 nations with the largest mangrove cover (ranging from 2.2% to 21.5% of world mangrove area in 1980, with a combined total of 68%) have changed little in their relative rankings over the two decades 1980–2000 (Table 12.1). Only in Bangladesh has mangrove area increased, while Nigeria and Cuba recorded no change. Mangrove reforestation in the Sunderbans is reflected in a 0.2% annual rate of increase of mangrove area in Bangladesh in both decades. For all other nations, with the exception of Brazil, mangrove deforestation continued with no sign of a reduction in the rate of loss between 1980–90 and 1990–2000. The annual rate of mangrove loss in Brazil declined from an exceptionally high rate of 8% per year during 1980–90 to 1.3% per year during 1990–2000. Indonesia, the nation with the largest area of mangrove worldwide, saw an annual rate of deforestation of 1.8% per year for both decades, while in Mexico, the nation with the sixth largest area of mangrove worldwide, the rate of loss increased from 1.6% to 2.1% per year.

A palaeohistorical perspective

Vegetation cover must be seen as part of a dynamic process which under natural conditions spans scales of time ranging from generations to millennia. The distribution of mangroves today is a single snapshot over evolutionary time, the result not only of contemporary forces such as human exploitation, temperature and aridity, but also of historical processes such as tectonic events and major climate changes. Tectonic events that resulted in relative shifts among continents and the opening and closing of marine passages have left their imprint on the distribution of mangrove taxa (Duke 1995; Saenger 1998; Ellison et al. 1999; Plaziat et al. 2001). For example, disjunct species distributions between the Indo-West-Pacific and the Atlantic-Caribbean-East-Pacific regions for genera such as *Avicennia* and *Rhizophora* can only be explained by isolation and differentiation following closure of marine passages for dispersal (Saenger 1998).

A series of major climate cycles from glacial to warm interglacial periods during the Pleistocene had major effects on mangrove distributions through temperature, aridity and sea-level changes. During the Holocene, mangroves moved landward when sea levels were rising (Woodroffe 1987, 1990; Fujimoto et al. 1996). However, this is likely to have varied regionally according to coastal geomorphology and the balance between rates of sedimentation and rates of sea-level rise. In the Congo basin, a series of episodes of

mangrove expansion coincided with lower sea levels during glacial maxima (Dupont et al. 2000). However, along the Colombian coast, where the continental shelf is narrow, mangrove colonization appears to be recent, following sea-level rise about 2600 years before present (BP) (Jaramillo & Bayona 2000). Extensive mangrove communities developed in river valleys of northern Australia about 6800–5300 years BP (Woodroffe et al. 1985) and by 1300 years BP, these had been transformed to freshwater wetlands (Clark & Guppy 1988). These historical changes are reflected in the genetic structure within and among populations today (Duke et al. 1998; Maguire et al. 2000; Dodd & Afzal-Rafii 2002; Dodd et al. 2002).

Historically, mangroves have fluctuated in extent both latitudinally, as a result of cooling and warming cycles, and regionally and locally, as a result of tectonic processes and sea-level changes. Even after massive anthropogenic losses, mangrove area may be greater today than it was at some times in the past, but the present rate of change and fragmentation are unprecedented and are of serious concern for ecosystem viability and success.

MANGROVE FORMATIONS

Mangroves exhibit a great diversity of forms and species composition according to physical and chemical environmental conditions in different geographical regions. Several attempts have been made to draw order out of this complexity. Thom (1982) identified geomorphological settings that characterize coastal mangrove systems, whereas Lugo and Snedaker (1974) described physiographic settings that determined the physiognomy of mangrove systems. Riverine discharge and bidirectional tidal flow are the two dominant processes that determine the functional types of mangrove forests (Woodroffe 1992). In a simplified classification (Ewel et al. 1998), these different functional forms lie along gradients among three extremes, namely tide-dominated, river-dominated, and interior or basin mangroves. This more simplified view of mangrove system diversity and identification of the dominant physical processes provide a clearer understanding of what are likely to be the most important threats to mangrove health (Fig. 12.3).

Tide-dominated mangroves

Tide-dominated mangrove formations, also known as fringe mangroves, are the most exposed to tidal forces. Strong

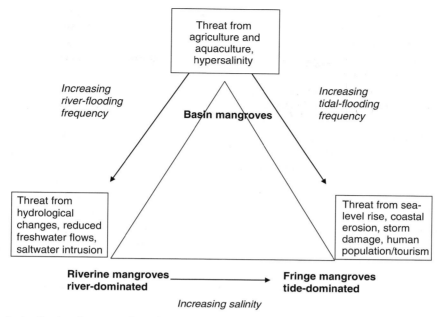

Fig. 12.3. A simple classification of mangrove formations showing relationships to dominant physical processes and sources of major environmental degradation. (Modified from Woodroffe1992 and Ewel *et al.* 1998.)

erosional forces remove sediments, so the system is highly dynamic with bidirectional fluxes. Tide-dominated mangroves may fluctuate in extent or presence according to the balance between forces of sedimentation and erosion. In general, mangroves grow only in sheltered coastal environments, where very heavy tides are uncommon. However, fringe mangroves are exposed to hurricanes and may offer protection from coastal erosion. Fosberg (1971) claimed that conversion of mangroves to rice paddies in the Ganga delta aggravated the losses of lives and property in the floods of 1970 following a hurricane and tidal surge. In the Saloum River estuary in Senegal, breaching of a protective sand bar resulted in sediments being buried by wave-borne sand. *Rhizophora mangle* behind the sand bar was destroyed, and further back in the delta, trees were stressed as a result of evaporation of ponded water leading to hypersaline conditions (Diop 1997).

Fringe mangroves occupy the coastal front, where human population growth and development are highest, and sea-level changes and high-energy storms may have greatest impact. Rising sea levels and increasing frequency of storms (IPCC [Intergovernmental Panel on Climate Change] 2001*b*) are expected to have serious impacts on fringe mangrove systems.

River-dominated mangroves

Riverine systems are characterized by high freshwater flows, particularly during the wet season, and heavy loads of terrestrial sediment that have not been recirculated by ocean currents. Typically, salinity levels are relatively low and decrease with increasing distance upstream. However, exceptions occur in arid regions where reverse salinity gradients may produce higher salinities upstream than near the river mouth, as in northern Australia (Ridd *et al.* 1988; Wolanski 1989; Wolanski *et al.* 1992) and Senegal (Diop 1990). The vast flood plains associated with the deltas of major rivers (Chapters 2 and 4), such as the Ganga, Brahmaputra and Mengha that flow into the Bay of Bengal, the Indus delta of Pakistan and the Merbok and Matang deltas of Malaysia, the Orinoco River delta in Venezuela, the Amazon delta of South America and the Fly and Purari river deltas in Papua New Guinea are classified towards the riverine extreme (Fig. 12.3). Tidal range tends to be relatively low in these delta systems and decreases upstream, so that depositional processes are commonly more important than erosional processes, leading to a system that can support extensive and tall mangrove forests. Nevertheless, major channels in a large delta system may shift as a result of sedimentation and wave action, and some channels may cease to function in a deltaic system, but still provide habitat for mangroves.

Because riverine systems are dominated by freshwater discharge, they are strongly influenced by activities occurring upstream that modify river hydrology and sediment load but may occur a considerable distance from the mangrove forest. A classical example is the damming of river systems such as the Indus River (Pakistan), where annual water flow has decreased five-fold since construction of the last dam in 1955 and the amount of sediment delivered downstream has declined four-fold (Hogarth 1999). Mangrove area is declining in the Indus delta by as much as 2% per year (Hogarth 1999) and a large proportion of the mangroves consists of stunted vegetation because of environmental stress. Barrages constructed on the St Louis River (Senegal) and freshwater diversion on the Gambia River decreased freshwater flows, resulting in changed hydrological regimes and mangrove loss (Blasco 1983).

Basin mangroves

Basin or interior mangroves occur on the landward margin of fringe and riverine systems. These may be infrequently inundated by tides and by river wash, so that sedimentation and erosion are typically low. Interstitial soil salinities may vary considerably according to local conditions. In arid areas, these basin mangroves occupy highly saline conditions produced by evaporation and water uptake and transpiration by plants (Chapter 11). In extreme conditions, the landward margin of basin mangroves may be denuded of plants because of the hypersaline conditions, producing salt pans typical of the Sahelian region of West Africa. In other areas of moderately high rainfall, soil salinities may be lower and conditions can support productive mangrove forests.

Because of the relative protection of these interior mangroves from wave action and heavy freshwater discharge, they have been the preferred sites for agriculture (between the mangrove trees?) and aquaculture.

THREATS TO MANGROVES TO THE YEAR 2025

Human population demography

Although the potential threats to mangrove vegetation and habitat include natural processes such as hurricanes, tsunamis, long-term cyclical changes in sea level and climate, human populations have had and will have the most profound influence on ecosystem health. These human influences will be exerted directly on the ecosystem from exploitation or conversion to other uses, and indirectly through remote environmental changes that have a cascading effect on the coastal zone. Most of the changes in mangrove cover that have occurred over the last century can be traced to direct or indirect effects of human interference. Despite growing awareness of the importance of protecting natural resources, the geographical distribution of mangrove ecosystems is such that they are particularly at risk from human pressure for two reasons. Firstly, approximately three-quarters of the world's mangroves lie in developing nations where resources for their conservation are least favourable. Secondly, population pressure is likely to increase disproportionately in the coastal zone.

In 1990, approximately 2 billion people lived within 100 km of the coast; in 1995 this became 2.2 billion. To put this in a geographical perspective, in 1990, if the coastlines of Antarctica and the Arctic are excluded, only 19% of the world's land area supported 38% of the world human population, rising to 39% by 1995. The trend of increasing coastal occupancy is likely to continue. If the world human population reaches 8.7 billion by 2025 (US Census Bureau 2003) and the population increase in coastal zones is at a rate of 1% every 5 years, an estimated 3.9 billion people, representing 45% of the world population, will occupy coastal areas by 2025. This is likely to be a conservative estimate, bearing in mind that 14 of the world's 17 largest cities and two-fifths of cities with populations of 1–10 million people are located near coastlines (Tibbetts 2002).

Nations with mangroves had a human population of 4.6 billion in 2000, and this is projected to rise to 6.1 billion in 2025 (US Census Bureau 2003). Assuming 40% of the population of these nations occupied the coast in 2000 (Burke et al. 2001) and 45% will occupy the coast in 2025, the coastal populations of mangrove nations will rise by about 50% from 1.8 billion in 2000 to 2.7 billion in 2025.

Urban development, landfill and coastal fringe squeeze

With much of the global human population inhabiting the coastal zone, one of the most serious threats to mangroves is their conversion to provide increased urban areas for industrial development, port facilities and housing. In some parts of the world (e.g. Mexico, Caribbean), this has been exacerbated by tourism, where mangroves have been reclaimed for hotel resorts, golf courses, airports and road construction. Statistics concerning the loss of mangroves to these types of reclamation activities are not easy to obtain and losses are included in the overall figures for deforestation. However the city of Cartagena (Colombia) was carved out of mangrove swamp, mangroves around Lake Maracaibo (Venezuela) have been decimated for construction of oil refineries and tourism, and mangrove in the Parque Nacional Laguna Tacarigua (Venezuela) has been destroyed by deforestation, urbanization and dredging (Ellison & Farnsworth 1996). Mazatlán in north-west Mexico has seen a burgeoning tourist industry and a doubling of the urban area between 1973 and 1997 at the expense of natural vegetation, agricultural lands and reduction in the area of two lagoons north of the city (Ruiz-Luna & Berlanga-Robles 2003). Although a marina has now been developed in one of the lagoons, which permits tidal flow, pollution is reported to be a problem. This illustrates a common sequence accompanying urban development; natural vegetation is first converted to agriculture, then further urban development occurs at the expense of agricultural lands (Ji *et al.* 2001).

Ecotourism is a growing industry that could offer both the interest and the financial incentive to protect mangroves from exploitation. However, it remains to be seen if the ecological benefits outweigh the ecological costs of these activities. Bringing large numbers of tourists to mangrove swamps will require an infrastructure and other amenities that may have unforeseen impacts on the ecosystem. Bank erosion caused by the wakes of tourist boats for example could prove to be a significant problem.

Shrimp farming

Mangroves have been used for shrimp farming for a long time. This tradition involved the enclosure of mangrove areas and the water (including the shrimps therein) during a spring tide. The water was held until the next few spring tides, allowing the trapped shrimps to grow naturally without addition of feeds, before the shrimps were harvested during a low spring tide. This is known as extensive shrimp farming, where both the effort and the density of stocking are low compared to semi-intensive and intensive shrimp farming. In semi-intensive and intensive farming, mangrove trees are cleared and ponds up to *c.*2 m depth are dug in the cleared areas. Predators are removed with piscicides and the ponds stocked with shrimp larvae from the wild or from culture. In Asia, the tiger prawn

Penaeus monodon has been the most commonly farmed species because the larvae are easy to obtain for culture and the non-cannibalistic nature of these prawns allows stocking rates of >20 000 larvae per hectare in intensive shrimp farming. Initially, shrimps were fed using minced fish, but most farms now use specially formulated feeds that contain nutrients and disease prophylactics like antibiotics.

Fish ponds in mangroves essentially preceded shrimp ponds. Milk fish (*Chanos chanos*) has been reared for centuries in Indonesia and was grown on a relatively large scale in Philippines mangroves from about the mid 1900s. As nutrients in the ponds were depleted, they were abandoned and new ponds dug, resulting in a shifting pattern of aquaculture that was to the detriment of the mangroves. Much of the mangrove in the Philippines was either destroyed or became degraded by the late 1970s when the shift to intensive shrimp culture started; even the FAO recommended the use of mangroves for shrimp aquaculture (for example Merbok Mangrove in Malaysia: Ong *et al.* 1980).

It soon became evident, however, that mangroves were not the most suitable sites for pond aquaculture because almost all the soils are potentially prone to acid sulphate conditions (oxidation of ferrous sulphides that leads to the formation of sulphuric acid, reducing the pH to a level unsuitable for growing shrimp, fish or rice). The remedy was to lime the ponds, which was economically viable in spite of the cost, resulting in a worldwide boom in mangrove-pond shrimp farming. The industry only slowed down in the mid 1990s when it was hit by a virus pandemic, the result of poor hygiene and overcrowded conditions. Most of the mangroves in Thailand, for example, have been converted to shrimp ponds and because of the high density have been hardest hit by disease and contamination.

The seriousness of this matter prompted the UN Joint Group of Experts on the Scientific Aspects of Marine Environmental Protection (GESAMP) to produce a manual on sustainable aquaculture (GESAMP 2001), offering guidelines for future coastal aquaculture development. However, if the disease problem is overcome and mangrove land continues to be cheaply available, this major threat to mangroves will be revived.

Apart from the loss of mangroves as a result of conversion to shrimp ponds, the pond-building process results in the oxidation of sequestered soil organic matter, which releases very significant amounts of carbon into the atmosphere (Ong 1993).

Timber extraction for rayon

One of the lesser-known but major uses of mangrove timber (mainly *Rhizophora*) is for the manufacture of rayon (Ong 1982, 1995), a fibre used for making high-quality fabric, mainly in Japan. Sites, usually with tens of thousands of hectares of *Rhizophora* mangroves, are identified, and a harvesting and wood-chipping plant is then established. There is usually enough timber for the chipping plant to operate for about 15 years, which is equivalent to about the lifespan of the chipping plant. Companies apparently treat this as a 'mining' operation; the chipped timber is sent to Japan by ship for processing and exploitation shifts to other sites when the available timber has been depleted. From experience in the Malaysia states of Sabah and Sarawak, there appears to be no attempt by the industry to ensure any long-term operation, or any regeneration of the mangrove being left to the local forestry authorities.

The royalties and other fees paid by the industry are very small and are tied to the price of the raw material, over which the industry has a monopoly. There is very little benefit to the local economy except for a few favoured individuals.

It is most disheartening that the rayon industry has not ensured that the harvesting of *Rhizophora* timber is on a sustainable basis, because *Rhizophora* forests can be sustainably managed for timber production. The Matang Mangrove management system (Box 12.1) is an excellent example, having gone through its third harvest rotation (Ong 1982, 1995; Gan 1995). This is not a case of the absence of technological restraints, but of unwillingness of an industry to act responsibly.

Hydrological change

Changes in hydrology, either as a result of upstream activities, or as a result of downstream modification of tidal flow, can have serious consequences for the health of mangrove ecosystems. The upstream activities include diversion of water for urban or agricultural use and construction of dams. In either case, changes in freshwater flows may change sedimentation rates and lead to seawater intrusion further into deltas and rivers. The Indus delta in Pakistan is an example of severe mangrove degradation as a result of reduced freshwater inputs. Since the latest dam was

---

Box 12.1   The forestry management system of the Matang Mangrove (Malaysia)

The Matang Mangrove is situated in north-west Peninsular Malaysia. This 40 000-ha mangrove has been managed for timber production since the early 1900s, a rare example of sustainable use of a tropical forest (Ong 1982).

Colonial British foresters devised the original forest management plan, based essentially on rule of thumb. Over the years, there have been a number of revisions for the latest working plan (see Gan 1995). The present plan involves a 30-year rotation of some 30 000 ha of productive forest reserve. This means that the annual coupe (i.e. the area of forest harvested each year) is 1000 ha. The trees are harvested in patches of a few hectares so that propagules from surrounding areas will help in natural regeneration. These harvested trees are about 30 m tall and have girths (at breast height) of about 65 cm. The stumps and slash (leaves, branches and propagules) are left to decompose naturally, a process that takes about 3 years. An inspection is carried out after a year or so to determine whether natural regeneration is adequate. Areas with inadequate natural regeneration (about half of the area harvested) are manually planted with propagules collected from the surrounding mature forest. The plants are then left to grow for 15 years before they are thinned. During this process, the tree density is reduced to about half. The thinning process produces poles that are now mainly used as foundation piles (many of the buildings in Singapore constructed in the early twentieth century sit on foundations of such mangrove piles). This is more an economic consideration than a silvicultural one, as natural thinning would have occurred a few years earlier (Gong *et al.* 1984). The trees are allowed to grow for another 5 years when a second thinning is carried out (at 20 years). Again, about half the trees are removed and sold as poles. The trees are left for another 10 years and then harvested, completing the 30-year rotation. The final harvest is used for the production of charcoal.

The Matang Mangrove has now gone through three rotations, and there does not appear to have been any significant drop in timber yield between the second and the third rotation (the first harvests were of very old mature trees so cannot be used in comparison of yields). The major drawback of the system is that the Matang Mangrove has become a monoculture plantation since *Rhizophora apiculata* is the preferred species for charcoal production.

constructed on the Indus River in 1955, annual water flow has declined in some years to one-fifth of past flow rates (Hogarth 1999) and this has been accompanied by up to 75% reductions in amounts of sediment being transported. As a result, the Indus delta has become increasingly saline and mangrove area has declined by 2% per year. With shoreline regression due to reduced sediment supply, mangroves will probably not be able to keep pace with anticipated sea-level rises resulting from climate change.

Reduced water flows may result in longer residence time of water in an estuary and reduced flushing of organic pollutants. According to a model of large-scale mangrove conversion to shrimp ponds in the major Guayas River estuary (Ecuador), with 90% conversion of mangroves into ponds, total nitrogen concentrations increased five-fold. However, by reducing freshwater discharge from the river to 10%, the same conversion to shrimp ponds resulted in a 60-fold increase in nitrogen concentrations (Twilley et al. 1998a).

There are many examples of mangrove decline following construction, such as of highways and levees, which impedes seawater flushing. Hypersaline conditions develop that lead to mortality. In the Ciénaga Grande de Santa Marta lagoon–delta system fed by the Magdelena River (Colombia), construction of a coastal highway altered the natural flow of marine and fresh water, resulting in c.60% of mangrove forests being destroyed (González 1991). Connecting the lagoon to the Magdelena River and to the ocean through culverts under the highway was expected to reduce the hypersaline conditions, but natural regeneration of mangroves has been slow and reforestation is being attempted (Twilley et al. 1998b; Elster et al. 1999).

Pollution

At the interface between marine, freshwater and terrestrial environments, mangrove vegetation is exposed to a wide range of pollutants, including those from point sources (e.g. local oil spills, urban sewage and thermal discharge) and those from non-point sources (e.g. pesticides, fertilizers in river discharge, industrial runoff, and oil discharge from general shipping operations). Some pollutants, such as nitrogenous and phosphorous nutrients, may promote growth and productivity of mangrove trees, whereas others, such as heavy metals, may be benign to the trees, but detrimental to other ecosystem components and, yet others, such as oil, may be toxic to the entire mangrove community. The risks of pollution are closely linked to

human population density and national affluence and developmental status. Developed nations are imposing increasingly more stringent rules on development and operations that may lead to environmental hazards. However, an estimated three-quarters of the world's mangroves are in developing nations, which generally possess few resources for such management. The most urgent concerns to the 2025 time horizon are from oil pollution, nutrients and heavy metal contamination.

OIL POLLUTION

Although major oil spills from tankers and pipelines are best known because of media coverage, the insidious contamination of fresh and marine waters from non-point sources is likely to pose the greatest risk to mangroves in the future. With increasing international attention and regulation, the number of major oil spills and the total volume of oil spilled have declined since the late 1970s (International Tanker Owners Pollution Federation Limited 2000). However, with increasing size of tankers, single catastrophic spills have disproportionate effects. Oil spills are point sources and, although they may have considerable impacts locally, their global effects are less significant.

Coating of mangroves with oil usually results in rapid defoliation, followed by mortality (Getter et al. 1985; Fagbami et al. 1988; Duke et al. 1997). Subsequent microbial breakdown of the hydrocarbons in sediments is slow (Oudot & Dutrieux 1989; Scherrer & Mille 1989), so that mangrove recovery is a long process. In the US Virgin Islands, no recovery of mangroves was observed 7 years after an oil spill in 1971 (Getter et al. 1985), and at Galeta (Panama) at least 20 years may be needed before mangroves recover from the toxic impact of catastrophic oil spills (Burns et al. 1993). Although heavy coating with oil commonly results in mortality, sublethal doses of oil may cause loss of ecosystem function. Unusually open canopies in the mangroves of Bahía las Minas near Galeta following an oil spill in 1986 were attributed to elevated oil concentrations in surrounding sediments (Duke et al. 1997). The total area of damaged mangroves was five to six times greater than that deforested.

The severity of damage to an ecosystem depends on a number of factors, including the volume of oil spilled, the nature of the sediments and species composition. No large-scale damage occurred in mangroves following a small localized spill in north-eastern Queensland, and sediments had lost a large proportion of the lighter, more toxic fraction within a month of the spill (Burns & Codi 1998). However, with larger spills, such as the Galeta spill, the lighter aromatic fraction remained in sediments for 5 years (Burns & Yelle-Simmons 1994) and Corredor *et al.* (1990) found oil in sediments decades after a Puerto Rico spill. Differences in species' susceptibility to oil toxicity may result in changes in species composition following oil damage. Lamparelli *et al.* (1997) reported 64.5% defoliation of *Avicennia schaueriana*, 43.4% defoliation of *Laguncularia racemosa* and 25.9% defoliation of *Rhizophora mangle* following an oil spill along the São Paulo coast in Brazil. However, *Rhizophora* does not recover following defoliation, whereas *Avicennia* and *Laguncularia* are likely to be more successful (Duke *et al.* 1997).

NUTRIENTS AND HEAVY METALS

Pollution from nutrients, mainly of nitrogen (N) and phosphorus (P), and heavy metals is likely to increase in mangrove ecosystems to the 2025 time horizon as a result of increased use of fertilizers and pesticides, increased urban and industrial waste, and deforestation. World fertilizer use is projected to increase from 150 million tonnes per year in 1990 to 251 million tonnes per year by 2020 (Bumb & Bannante 1996). The greatest increase is expected in fertilizer use in developing countries, which will increase their share of the world total to 59% in 2020 (40% in 1990). Agricultural activities in developing nations are replacing tropical forests without adequately conserving riparian zones that would protect streams and rivers from fertilizer and pesticide runoff.

In the tropics, a series of N pulses is expected with (1) deforestation, (2) slash and burn, (3) fertilizer application during intensive agriculture and (4) human waste and effluent discharge during urban and industrial development (Downing *et al.* 1999). Through these different stages of tropical development, N : P ratios are expected to decrease in response to a combination of denitrification, volatilization of N during burning, the solubility of P in warm waters and the higher N : P ratios typical of urban and industrial waste (Downing *et al.* 1999). These trends are thought to have more profound influences on the coastal zones of tropical than temperate latitudes. Disturbance of pristine tropical forests is likely to result in a switch from P to N limitation in coastal waters, potentially leading to large changes in the composition of mangrove communities.

There has been great interest in the use of mangroves as a natural system for the treatment of sewage effluent (Nedwell 1975; Clough *et al.* 1983). They may perform a useful role in the natural treatment of wastewater through direct uptake of nutrients by roots and by chemical and bacterial transformation in the soil and rhizosphere (Wong *et al.* 1995, 1997). Several trials have suggested that mangroves are tolerant of organic pollution. No accumulation of nutrients occurred in mangrove tissues despite elevated N and P concentrations in sediments in the Futian National Nature Reserve in the People's Republic of China (Wong *et al.* 1997). Although mangroves do not appear to suffer from sewage effluent and therefore may be effective in wastewater treatment, great care is needed to ensure that the treatment areas are adequately secure from leakage into freshwater supplies (Ewel *et al.* 1998). Furthermore, care is needed to ensure that industrial wastewater containing heavy metals is not allowed to enter the treatment areas, as the heavy metals may be accumulated in mangrove tissues forming a potentially dangerous sink (Tam & Wong 1997; Clark 1998).

Climate change

Projected rises in atmospheric carbon dioxide (CO_2) by 40 ppm, global temperatures by 0.4–1.1 °C and sea levels by 0.03–0.12 m to the 2025 time horizon (IPCC 2001*b*) may significantly affect mangroves in some areas more than others (Table 12.2). For example, increased aridity predicted for the west coast of Africa is likely to have serious consequences for the mangroves in this region. These predicted environmental changes could influence plants in general and mangroves in particular in three ways. Firstly, direct physiological responses to elevated CO_2 may result in modified plant growth. Secondly, direct responses to increased temperature may also lead to higher growth rates and, in addition, new environments for colonization may become available. Thirdly, sea level rise may result in landward progression of mangrove ecosystems.

Effects of elevated CO_2 on metabolic processes have long been an area of interest to plant physiologists. In woody plants, net CO_2 assimilation increases with increasing CO_2 partial pressures under light saturation, and this is associated with an increase in total above-ground and below-ground biomass (McGuire *et al.* 1995; Curtis & Wang 1998). Whether this response continues through the life cycle of the plant is uncertain. Photosynthetic acclimation or downregulation may lead to changes in physiological processes that reduce photosynthetic capacity (Ceulemans & Mousseau 1994; McGuire *et al.* 1995) under

Table 12.2. *Historical and predicted changes in global climate (from IPCC 2002) and effects on coastal ecosystems*

Observed changes	Predicted changes	Effects on coastal ecosystems
CO_2 increased by 31% since 1750; CH_4 increased by 151% since 1750	CO_2 likely to increase by 75–350% above 1750 levels by 2100	Elevated CO_2 could promote net photosynthesis and primary productivity
Global mean surface temperature increased by 0.6 °C over the last 100 years. Sea-surface temperature increased by about 50% of land temperatures since 1950	Global mean surface temperature expected to rise by 0.4–1.1 °C by 2025. Effects are likely to be greatest in northern land masses	Latitudinal shifts in species range. Changes in temperature-dependent physiological processes. Death of coral reefs
Precipitation increased 5–10% at mid and high latitudes during twentieth century. Rainfall decreased by 3% over subtropical land areas	Global average precipitation will increase during twenty-first century, with changes of from 5–20% regionally	Precipitation increase: increased productivity, increased flooding of deltas, increased sedimentation
	Precipitation will decline in most land masses bordering tropical and subtropical latitudes, except northern Indian Ocean, South-East Asia and Pacific South America	Precipitation decrease: decreased productivity, species loss owing to aridity, salt intrusion upstream as a result of decreased freshwater flows, lower rates of sedimentation
Global average sea levels have risen by 1–2 mm during the twentieth century	Global sea levels will rise by 0.03–0.14 m by 2025. Regional variation is likely to be substantial	About 20% of coastal wetlands could be lost by 2080. Marine flooding and erosion
Warm episodes of the El Niño–Southern Oscillation (ENSO) events have been more frequent, persistent and intense since the mid 1970s	Increases in hot weather expected, extreme rainfall events are likely to increase. Peak wind intensity of tropical cyclones will increase	Increase in numbers of severe flooding events. Increases in severe storm damage. Coastal erosion

elevated CO_2 partial pressures. However, there is little support for photosynthetic acclimation to elevated CO_2 levels, except in potted plants (Curtis & Wang 1998). In mangroves, elevated CO_2 has also resulted in positive effects on photosynthetic rates and growth rates. *Rhizophora mangle* may grow faster when the ambient CO_2 concentration is doubled, and this is associated with enhanced photosynthesis rates, reduced stomatal conductance and greater root : shoot ratios (Farnsworth *et al.* 1996). Stomatal conductance fell and water-use efficiency increased under elevated CO_2 in four species of mangrove from Florida (USA), however, no changes in net primary productivity occurred in *Rhizophora mangle*, *Avicennia germinans* and *Conocarpus erectus*, and net primary productivity decreased in *Laguncularia racemosa* (Snedaker & Araújo 1998). Responses to CO_2 concentrations may vary in combination with other environmental factors. In *Rhizophora stylosa* and *R. apiculata*, responses to higher CO_2

concentrations were positive under limiting conditions of humidity, but only at low saline concentrations, with the latter effect being more important for the less salt-tolerant *R. apiculata* (Ball *et al.* 1997).

Overall, the direct responses to rising CO_2 concentrations are not likely to have major effects on mangrove productivity, but may be important under conditions of environmental stress, such as increasing aridity (but not increased salinity) and low nutrient status. Elevated CO_2 may also lead to changes in species composition as a result of variations in competitiveness among species.

While higher CO_2 concentrations will affect plant growth and interspecific competition, climate change is likely to result in redistributions of populations (Rehfeldt *et al.* 1999, 2002) and in some cases of entire species. Historical changes in plant distributions during cyclical climate fluctuations of the Pleistocene suggest that organisms are relatively successful over the long term in the face of environmental change. However, the maximum temperature change predicted for the next 100 years is comparable with the net change since the last glacial maximum in the northern hemisphere (18 000 years BP). Mangroves should be capable of exploiting environments at higher latitudes than they do today. However, how successful will migrations be in the face of such rapid rates of climate change? In addition, deforestation for alternative land use has created a fragmented landscape with potential barriers to natural migration. Do plantings of mangroves in suitable environments at higher latitudes need to be planned to avoid the risks associated with natural migrations?

Sea levels have been rising at an estimated global rate of about 2 mm per year over the last 100 years (Peltier & Tushingham 1989; Trupin & Wahr 1990; Douglas 1991), but there is considerable uncertainty about future patterns (Chapter 1). A 6-cm rise per decade (Field 1995) would result in a 18-cm global rise in sea level by 2025, but the predictions of the IPCC (2001b) are more conservative (0.03–0.14 m by 2025), albeit still as much as 2.5 times greater than the recent historical rate of rise. The natural trend following rising sea levels will be a landward progression of mangroves and a gradual loss of the seaward edge, assuming that rates of sediment deposition will lead to vertical accretion sufficient to offset rising sea levels. This may well be the case in major deltas, where freshwater discharges remain at current levels or increase as a result of increased precipitation. However, in areas where increased aridity is anticipated, sediment loads in rivers may decrease, with the result that the mangrove band will become narrower as the seaward edge becomes submerged. Clearly, progressional changes can only be assessed at a local level, taking into account differences in landform, land use, sediment deposition and tectonic changes, and anthropomorphic modifications of hydrology. Over the last 50 years, conversion of land behind the coastal fringe, particularly for agriculture and aquaculture, has not only resulted in mangrove deforestation, but has also potentially removed this land from future colonization, leading to a coastal squeeze of mangrove area (Ong 1995).

Reconstruction of historical mangrove distributions based on pollen analyses indicates that major changes can take place, and that these may be positive or negative in terms of favouring mangrove forests. Major flooding of the coastal plain of northern Australia about 6800–5300 years BP, following sea-level rise, resulted in development of extensive mangrove forests. Stratigraphic studies of the Caribbean suggest that during the Holocene mangrove vegetation kept pace with a sea level rise of 8–9 cm per 100 years, was under stress with a rise from 9–12 cm per 100 years and collapsed completely at higher rates of sea-level rise (Ellison & Stoddart 1991). However, responses of mangrove systems to sea-level change are poorly understood, but it appears that mangrove ecosystems have survived in Florida following sea-level rise at twice the critical rate of 23–27 cm per 100 years (Ellison & Stoddart 1991; Snedaker et al. 1994).

Economic market failure

THE VALUE OF MANGROVES

On rarity or scarcity alone (only 0.12% of the world's total land area) mangroves occupy very valuable land. On top of that, various mangroves provide many goods (such as timber as fuel wood or source of charcoal and rayon, shellfish, finfish and other sustenance products) and services (for example carbon sequestration, nutrient absorption, fish feeding, breeding grounds and maintenance of estuarine channel depth for convenient navigational use) (Ewel et al. 1998). Only some of the goods (and none of the services) have market value, so that in reality world mangroves are grossly undervalued. There have been brave attempts to attach values to services (see Costanza et al. 1997). These values are at best contentious and are completely ignored by mainstream economists, who, whilst agreeing that there are non-market values, would not attach a figure to these, but rather admit that there has been an economic market failure for mangroves. When economic market failure occurs it is usually the role of governments to ensure equity.

EQUITY

Comparing the extent of mangroves, human population density, population growth and the per caput gross domestic product between selected developed (Australia, Japan and the USA), newly developed (Singapore and Hong Kong), nearly developed (Malaysia) and developing countries (Indonesia, Brazil, Nigeria and Bangladesh), the developed countries have afforded almost complete protection to their mangroves, while the developing countries have not, being perhaps too poor or otherwise occupied to provide any such equity (Ong et al. 2001). Conversely, the newly developed countries, which lost much of their mangroves in their quest for development, have been willing to devote considerable funds and effort to the restoration and establishment of new mangroves. This

Table 12.3. *Mangrove and land areas, population density, population growth and per caput gross domestic product (GDP) as purchase-power parity (PPP; based on World Bank ratios) of 'selected' countries*

Country	Mangrove area 10^3 km^2	Mangrove area % world	Land area 10^3 km^2	Land area % land	Population density (n km^{-2})	Growth (% yr^{-1})	Per caput GDP (PPP US$)
World	181	100	148 000	0.122	40	–	–
Indonesia	42.6	23.5	1 905	2.24	117	1.6	2.685
Bangladesh	5.8	3.2	144	4.03	897	2.2	1 410
Nigeria	10.5	5.8	924	1.14	134	2.9	930
Brazil	13.4	7.4	8 512	0.16	20	1.9	6 418
Subtotal	72.7	40.4	11 485	0.63	–	–	–
Australia	11.5	6.4	7 713	0.15	3	1.2	23 145
Japan	0.005	0.003	378	0.001	335	0.3	23 780
USA	2.0	1.1	9 373	0.021	29	1.0	33 872
Subtotal	13.5	7.5	18 084	0.08	–	–	–
Singapore	0.006	0.003	0.62	0.97	6407	0.7	25 353
Hong Kong	0.003	0.002	1.04	0.29	6250	2.3	20 485
Malaysia	6.4	3.5	330	1.94	67	2.4	2 685

Source: Data mainly from Spalding *et al.* (1997) and Asiaweek (2000).

suggests that if the economies of developing countries possessing large areas of the world's mangroves (for example Indonesia, Nigeria and Brazil) improve in the next two decades, the rate of loss of the world's mangroves may be abated (Table 12.3).

The 1997 financial crisis pushed 22 million people in the Asian region into poverty, and Indonesia, with a quarter of the world's mangroves, was particularly badly affected (Anon. 2004). Avoiding the 1997 crisis would have benefited emerging-market countries by US$107 billion per year, and a mechanism for avoiding the crisis would have cost only US$545 million annually (Anon. 2004). This is a small sum for the international community, the World Bank or World Trade Organization compared to the amount it would save in addition to the considerable global environmental benefits.

MITIGATION

Rehabilitation

The term 'rehabilitation' is used here in preference to 'restoration' because purists will claim that restoration is a practical impossibility and falsely implies a claim for a viable alternative to conservation. It is preferable and certainly cheaper to conserve natural stands of mangroves than to mitigate their destruction with restoration plans or promises. Rehabilitation has its role but cannot be an alternative to conservation.

As a result of the huge losses in mangrove areas, there have recently been increased efforts at mangrove rehabilitation. Mangrove rehabilitation is however not a recent phenomenon, as seen from the enrichment planting of managed forest to ensure sustainable supply of timber (see Box 12.1). Various methods for rehabilitation have been reviewed (Field 1996, 1998, 1999).

Many of the recent mangrove rehabilitation attempts are not confined to areas where mangrove previously existed, mudflats in front of mangroves being the main target areas (Walters 2000). Proponents often do not realize that this leads to the destruction of an adjacent natural ecosystem that is functionally as vital as mangroves to the health of the general environment. A project has even included the use of fertilizers that may have an adverse impact on coral reefs, although fresh water is more likely to be the major limiting factor (Pearce 2003). Many such projects are funded by international agencies.

Most mangrove rehabilitation projects involve the planting of one single species. Many of the projects on mudflats in Asia use *Rhizophora mucronata* because of its long propagules, making this the only species that stands a chance of surviving extreme tidal inundation. Mangrove forests managed for timber production are also planted with the preferred species (see Box 12.1). Thus rehabilitation tends to produce extensive, almost monospecific stands of mangroves, with reduction of mangrove species diversity.

Whilst the replacement of trees may sustain a timber harvest, the loss of plant species diversity may not fully restore other critical ecological services. For example, crabs that are considered as keystone species in mangroves (Smith *et al.* 1991) may be associated with a particular plant species or set of environmental conditions that the plant species provide (for example Lee & Kwok 2002; Ashton *et al.* 2003).

Rehabilitation projects are certainly good public-relations exercises for their advocates. However, funding agencies should understand that planting in areas that previously did not support mangroves is not only a waste of funds, but is also counterproductive as it will more likely than not cause adverse ecological and social impacts.

Conservation areas

Ong (1995) discussed two broad categories of conservation, namely (1) conservation for posterity with no intervention and (2) sustainable use for long-term prosperity. Recognizing an economic market failure for mangroves and to ensure equity mangroves in some countries have been given complete or partial protection, but this is in developed countries for the most part. Many of the developing countries may also recognise market failures, but most are too poor to initiate equity measures. The conservation approach taken by developed nations is widely unrealistic for the developing countries, but since it is not technically difficult to use mangrove on a sustainable basis, developing countries should at least conserve their mangroves based on concepts of sustainable use. In addition, developing countries should be strongly encouraged by the international community to identify unique or special mangrove sites to be set aside as protected areas. Internationally recognized sites (such as those recognized by UNESCO or the Ramsar Convention) would have greater value because they attract international tourists.

However, the draw of tourists alone is often insufficient as a financial incentive for conservation. To encourage developing countries to conserve some of their mangroves as heritage sites, many developing countries would require other financial incentives. One new form of incentive may be in the form of carbon trading, given that mangroves are long-term sinks for atmospheric carbon (J. E. Ong & W. K. Gong, personal communication 1991). However, under the present rules of the Kyoto Protocol (UN 1998), carbon sequestered by natural or pristine forests, or even by sustainable-yield plantations, does not qualify for carbon trading. The substantial carbon sequestered in the soils both of sustainable plantation and old-growth mangroves should be made eligible for carbon trading. Disturbance of mangrove soils, as when mangroves are converted to aquaculture ponds, results in oxidation and release of previously sequestered mangrove carbon back into the atmosphere. If developing countries are prepared to keep their mangrove-sequestered carbon intact and continually add to this pool through the conservation of their mangroves, the Conference of Parties to the Kyoto Protocol should allow this carbon to be traded as a financial incentive for establishment of protected areas of various types or sustainable exploitation.

Education and participation of local communities

Mangroves were long considered to be unproductive unhealthy swamps, yet this view remains surprisingly entrenched in the minds of the public and government officials. Improving the awareness among the wider community of the ecological role that mangroves play in protecting the physical environment and coastal food webs, is critical for future mangrove conservation. Attempts at raising public awareness are fortunately being made, and range from education in the classroom to provision of information at visitor centres around the world.

An extension of the educational role in ecosystem management is the participation of local communities in management. Non-governmental organizations have played a crucial role in participation that has broadened the commonly sectoral view of management, from a single product to the multiple products and services that mangroves provide. Involvement of local communities in a strategic planning process helps the identification of the sources of problems and of the means to overcome them (Ramsar 2004).

In 1979, Costa Rica placed its mangrove forests under the Directorate of Forestry, which required a management plan for any forest use. Following the decline of the banana industry in the Sierpe area on the Pacific coast in 1984, the number of people illegally exploiting the Terraba–Sierpe Forest Reserve increased dramatically. The only group legally authorized to exploit mangroves was a cooperative of local people who made their living from charcoal and tannin extraction from bark. In 1989, an attempt was made to work with local people towards a management plan for the mangrove forest. A socioeconomic analysis identified five causes of increased pressure on the resource. Firstly, low efficiency of conversion processes resulted in a larger area of forest being exploited. Secondly, lack of transport capabilities meant that producers sold to intermediaries and income to the producers was relatively low, again leading to increased exploitation. Thirdly, poor characterization of the resource resulted in loss of potential of other higher-value uses. Fourthly, illegal exploitation caused damage and wastefulness such that products were sold at a lower price, again leading to increased exploitation. Fifthly, lack of resources available to the Directorate of Forestry meant inadequate supervision and management of the forest Reserve. To address these issues, the local cooperative was encouraged to accept illegal harvesters as members and was identified as an organization capable of managing the mangroves. Barges and a truck were bought to allow more efficient transport, and technological improvements were made to the processing for charcoal production. Biological surveys are aimed at quantifying the products extracted and the growth rates of mangroves as a basis for determining sustainable use levels. The Directorate of Forestry has an educational programme targeting children in the local school. Results from this type of approach can only be evaluated after a number of years; however, it serves as an example of an attempt at forward-looking mangrove management.

International guidance

The international community can play an important role in providing education, expertise and financial aid to nations to protect their natural resources. The major international agencies, such as IUCN (World Conservation Union), UNDP (United Nations Development Programme), UNESCO, WWF (World Wide Fund for Nature), Wetlands International and the European Union have been key players in persuading governments to attach importance to mangrove conservation. The Ramsar Convention on Wetlands (1971) provides the framework for national action and international cooperation for the conservation and wise use of mangroves and all wetlands and their resources. There is growing recognition that sustainable use may be the most effective way to protect mangroves, providing that use implies all values of mangroves, including non-monetary benefits such as biodiversity and coastal protection. Mangroves have been recognized as underrepresented among Ramsar wetland sites. Of the 138 nations that are signatories to the Ramsar Convention, 55 have sites in which mangroves are significant and 29 have 50 sites in which mangroves are dominant (Ramsar 2004).

At the World Summit on Sustainable Development held in Johannesburg in 2002, the UN Secretary General identified five key areas for the future of the planet (Johannesburg Summit 2002). One of these key areas was to protect biodiversity and improve ecosystem management. At the Summit, commitments were made to reduce biodiversity loss by 2010, to restore fisheries to their maximum sustainable yields by 2015, to establish a representative network of marine protected areas by 2012, and to improve developing countries' access to environmentally sound alternatives to ozone depleting chemicals by 2010. These commitments were supported by 32 partnership initiatives submitted to the UN, with US$100 million in additional resources, and a USA announcement of US$53 million for forest management in 2002–2005. This was an indication of international interest, but it remains to be seen whether the commitment will be sufficient to turn around the degradation of mangroves by the year 2025.

13 · Environmental future of estuaries

MICHAEL J. KENNISH, ROBERT J. LIVINGSTON, DAVE RAFFAELLI AND
KARSTEN REISE

INTRODUCTION

Estuaries are biologically productive coastal ecosystems where fresh water from land drainage mixes with seawater (Kennish 1986, 2001*a*). They are coastal indentations with restricted connections to the ocean and remain open at least intermittently (Kjerfve 1989). The system can be subdivided into three regions, namely (1) a tidal river zone characterized by lack of ocean salinity but subject to tidal rise and fall of sea level, (2) a mixing zone, the estuary proper, characterized by water mass mixing and strong gradients of physical, chemical and biotic components, reaching from the tidal river zone to the seaward location of a river mouth or ebb-tidal delta, and (3) a nearshore turbid zone in the open ocean between the mixing zone and the seaward edge of the tidal plume at full ebb tide.

The global coastal population may approach 8 billion people by 2025 (Chapter 1) and anthropogenic activities affect nearly all estuaries (Weber 1994; Hameedi 1997). Developing

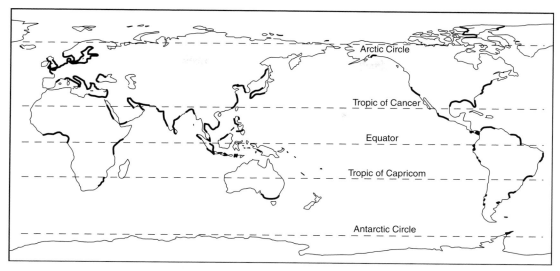

Fig. 13.1. World map showing existing coastal areas (in bold) significantly impacted by human activities. Note coastlines bordering heavily populated, industrialized nations of the northern hemisphere exhibit more extensive impacts than those of the southern hemisphere. (From Alongi 1998.)

countries widely lack government environmental regulations or have less stringent controls than most developed countries, and their technology to mitigate environmental impacts in estuarine embayments is also poor. However, currently the most significant human impacts are concentrated along heavily populated industrialized coasts (Fig. 13.1).

Anthropogenic impacts on estuaries include point and non-point source pollution, habitat loss and alteration, shoreline development, altered hydrologic regimes, sea-level rise, land subsidence, energy/resource operations, boating and shipping, overfishing, introduced/invasive species and sediment input/turbidity (Table 13.1).

ESTUARIES AS ECOSYSTEMS

Estuarine systems are among the most temporally variable of all systems on Earth and are characterized by an array of habitats with common physical, chemical and biological attributes marked by abrupt changes in space and time. Each estuarine system exhibits a unique combination of these habitats. Primary productivity also differs widely both within and among systems. The biological attributes and productivity of each estuary are influenced by the adaptive responses of indigenous species. Seasonal and interannual cycles of controlling factors, together with intermittent storm events, provide a highly variable background.

Within individual drainage basins, the complex combinations of habitats form highly productive systems that are interconnected physically, chemically and biologically. High nutrient levels, multiple sources of primary and secondary production, shallow depths, organically rich sediments, energy subsidies from wind and tidal currents, and freshwater inflows together elevate the natural productivity. Although similar patterns occur in many estuaries, unique arrays of habitats in different systems lead to broad variations of trophodynamic processes (Livingston 2000, 2002). Autochthonous and allochthonous sources of nutrients result in high primary productivity and substantial secondary production that supports fisheries. Anadromous (marine organisms that migrate to breed in fresh water) and catadromous (freshwater organisms that breed in the sea) species use major parts of these areas. Most estuarine species spawn offshore, the young-of-the-year populations migrating to inshore nurseries. Different combinations of controlling factors, along with specific patterns of inshore–offshore migration of marine forms and offshore movements of euryhaline species, contribute to the area-specific characteristics of any estuarine system.

The basic characteristics of a given estuary are partly determined by physiographic conditions (depth, surface area, connections to the open ocean) and freshwater inputs of dissolved and particulate matter (Livingston 2000, 2002).

Table 13.1. *Physical and biological impacts of human activities on estuarine and coastal marine environments*

Sources	Impact categories
Habitat loss and alteration	Wetland reclamation, dyking, ditching, canal and dam construction, silviculture, dredging and filling, bulkheading, domestic and industrial development, marina and mariculture operations
Shoreline development	Shorefront construction, shore-protection structures (bulkheads, sea walls, sloped revetments, jetties), industrial installations, harbour development
Energy/resource operations	Oil, gas and mineral extraction, electric power generation
Altered hydrological regimes	Freshwater diversions, reservoirs, dams, channelization projects
Sea-level rise	Habitat modification, water-quality changes, altered biotic communities
Land subsidence	Shoreline retreat, altered habitat
Boating and shipping	Propeller cutting, propeller scarring, vessel groundings, oil spills, chemical contaminant releases
Recreational and commercial fishing	Hydraulic dredging, tonging, raking
Overfishing	Depleted fish and shellfish stocks
Introduced species	Non-indigenous invasive species that disrupt endemic biotic communities
Debris/litter	Environmental hazards, degraded habitat

Seasonal and interannual cycles of temperature, rainfall and river flows drive recurrent changes of primary and secondary productivity. Location-specific ecological responses are tempered by myriad adaptive responses of estuarine species. Inputs from freshwater and saltwater wetlands (allochthonous and autochthonous), *in situ* phytoplankton productivity and submerged aquatic vegetation all contribute to the rich coastal food webs. With the exception of a few species that spend their entire life cycles in estuaries, most species migrate inshore as larvae and juveniles derived from offshore spawning. Many commercially important species such as oysters, penaeid shrimp, blue crabs and various finfishes are adapted to rapid habitat changes. These species use the abundant food resources of coastal systems while remaining relatively free from predation by the stenohaline offshore marine species. Thus, estuaries are often physically stressed but exist as a highly productive sanctuary for developing stages of offshore forms, many of which are used directly and indirectly by humans. Gradients of physical variables, together with food availability and use, provide the basic components of the highly complex food webs of inshore marine systems. Predator–prey interactions thus combine with a dynamic habitat to produce productive coastal food webs.

Area-specific combinations of the rapidly changing physical conditions and intermittent (seasonal, interannual) cycling of organic production, together with biological components that vary continuously (e.g. predator–prey interactions and competition), largely control the final structure of the food web of each estuary. Biological interactions, in response to gradients of salinity and primary producers, determine the specific characteristics of any given coastal system. With increasing salinity, population distribution becomes more even, with a corresponding decrease in relative dominance and an increase in species richness and diversity (Livingston 2002). Seasonal changes of predation pressure as well as interannual recruitment characteristics modify this general condition at any given time. Hence, coexisting estuarine assemblages reach different levels of equilibrium that depend largely on the exact temporal sequence of key habitat and productivity factors. The high adaptability of coastal species often prevents direct (linear) relationships with specific physical/chemical driving functions. In this way, non-linear processes contribute to what may appear as chaotic conditions, even though there is often an underlying organization that is grounded largely in the trophic responses of existing species to a continuous series of physical/chemical interventions.

ENVIRONMENTAL FORCING FACTORS

Human activities and conflicting uses in coastal watersheds and adjoining estuaries have been responsible for impairment of water quality, loss and alteration of habitat, diminished productivity and reduction of resources in many regions around the world (Clark 1992; Goldberg 1994, 1995; McIntyre 1995; Viles & Spencer 1995; Kennish 1997, 2000, 2001a, b; Alongi 1998; e.g. Chapters 2–5). Point- and non-point-source pollution significantly impair water quality via nutrient and organic carbon loading, pathogen input and chemical contamination of estuarine waters (Clark 1992; Kennish 1997, 2001a). Sewage influx remains a serious problem in many countries, particularly those that are economically less developed. Large-scale modifications of coastal watersheds and estuarine basins cause physical alteration of habitats that are often detrimental to estuarine organisms (Table 13.1; e.g. Chapters 11, 12 and 14). Unregulated commercial and recreational fishing often culminate in overharvesting of finfish and shellfish populations, which can severely deplete resources beyond the limits of sustainability and result in shifts in biotic community structure.

Habitat impacts

HABITAT LOSS AND ALTERATION

Physical disturbances associated with human activities have caused substantial loss and alteration of habitat with detrimental consequences for estuaries around the world. Habitat destruction in estuarine embayments and adjacent watersheds has decimated biotic communities and helped deplete fishery resources (Rose 2000). The anthropogenic activities that impact the physical integrity of estuaries are diverse and pervasive. They may be divisible into three principal categories, namely those on watersheds, shorelines and embayments.

WATERSHED CHANGES

The most conspicuous watershed activities with serious physical impacts on estuaries include residential and commercial development, dyking and ditching, canal construction and channelization, water diversions and impoundments, wetland reclamation, coastal subsidence (coupled with groundwater, oil and gas withdrawal), mariculture operations and dredge-and-fill activities (Boesch et al. 1994; Hopkins et al. 1995; Simenstad & Fresh 1995; Kennish 1997, 2000, 2001b; Bryant & Chabreck 1998; Chesney et al. 2000; Adam 2002; e.g. Chapters 2–4). The ecological functioning of estuaries is adversely affected not only by the habitat transformation, but also by modification of hydrological, nutrient and sediment regimes. For example, the increase in areal

coverage of impervious surfaces in coastal watersheds hastens surface runoff, promoting pollutant and nutrient transfer into estuaries (e.g. Chapters 2, 4 and 9). While human activities in some catchments accelerate sediment loading to some neighbouring estuaries (Edgar & Barrett 2002), river diversions and dam construction reduce sediment delivery to others (Aubrey 1993; Chapter 2).

Activities in adjoining wetlands also bring serious physical changes in estuarine habitat (Kennish 2001b; Adam 2002). Reclamation and impoundments have destroyed c.50% of the original tidal saltmarsh habitat in the USA (Pinet 2000; Kennish 2001b). Wetland habitat destruction is responsible for up to 50% of depleted fishing stocks in the USA (Chesney et al. 2000). Reclamation, dyking and conversion to polders have eliminated nearly all the surrounding wetland and mudflat habitat along the mainland coast of the Dutch Wadden Sea. At some locations in south-east England, saltmarsh habitat decreased by 10–44% between 1973 and 1985/1988 (Burd 1992). Saltmarsh in northern Europe has been altered by an array of human influences, including tidal power, tidal barriers, water storage, evaporative salt production, aquaculture, pollution, insect control, introduced species and land reclamation (Adam 2002; Chapter 11). More than 50% of the mangrove forests in South-East Asia and c.75% of the mangrove forests in Puerto Rico have been lost because of human activities, most notably mariculture, silviculture and habitat reclamation (Eisma 1998; Chapter 12). Nearly 50% of the mangroves in the Philippines have been converted to mariculture ponds (Hopkins et al. 1995).

From the mid 1950s to the mid 1970s, destruction of coastal wetlands in the USA occurred at a rate of c. 225 000 ha per year (Ciupek 1986; Moy & Levin 1991). Domestic, industrial, agricultural and recreational activities accounted for much of this habitat loss. Physical decimation of wetland habitat can significantly reduce biodiversity (Dobson et al. 1997). Dredge-and-fill operations and construction of levees, dykes, ditches and impoundments physically destroy wetland habitat and modify hydrologic conditions that can alter water quality and biotic communities in estuarine basins (Turner & Lewis 1997). Dredge-and-fill activities convert wetland areas to open water or upland habitats, directly impacting the resident biota (see Chapter 11). Dyking, ditching, leveeing and the formation of impoundments influence tidal flooding and drainage to estuaries which can elicit acute changes in the structure of biotic communities. Impoundments are generally used to manage hydrologic regimes, such as in the coastal marsh habitat of Louisiana (USA) (Boesch *et al.* 1994; Bryant & Chabreck 1998).

SHORELINE DEVELOPMENT

Particularly in urban areas, estuarine shorelines are commonly dotted with docks, piers, boat ramps and marinas, which provide users with easy access, but also mark sites of physically disturbed natural landscapes. Buildings, roadways and utilities change the landform configuration as well as local hydraulic regimes and drainage patterns, typically increasing runoff and non-point-source pollution inputs. Shore protection structures (such as bulkheads, revetments and retaining walls) in these areas also often interfere with bay-water circulation, sediment flows and shore bathymetry (Nordstrom & Roman 1996; Nordstrom 2000). Bulkheads may stretch uninterrupted for many kilometres on the open shore and within protected lagoons. Construction of buildings, shoreline protection structures and recreational features places fixed structures in a dynamic estuarine environment. A considerable amount of estuarine habitat in close proximity to the shorefront constructions has been lost and altered. Thus many breakwaters, bridges, retaining walls, jetties and pontoons added to the waterways around Sydney (Australia) have led to long-term habitat disturbances (Glasby & Connell 1999). Wooden shoreline structures treated with chromated copper arsenate or other protective compounds are sources of heavy metals to the water that ultimately become concentrated in sediments (Weis *et al.* 1993). Many shoreline developments and human structures also frequently and inadvertently create unnatural habitats for the proliferation of epibenthic communities. It is uncertain what long-term consequences will arise from the many structural modifications of estuarine shorelines. However, dredging of estuarine shorelines to maintain navigable waterways and create lagoons and various permanent structures profoundly modifies or destroys benthic habitat.

Port Phillip Bay in south-eastern Australia provides an example of an embayment heavily impacted by shoreline development (Fig. 13.2). Bordered by the cities of Melbourne and Geelong, Port Phillip Bay is a system intensively used for a wide array of recreational activities such as swimming, fishing, camping, boating and sailing (Bird & Cullen 1996). Much of the 256-km shoreline has been altered to protect beaches, moor boats and enable public access to bay waters. Artificial shoreline structures (e.g. sea walls, jetties and breakwaters), piers and docks are common features. Numerous marinas have been con-

structed, and beach restoration projects have been undertaken in many areas to combat erosion and preserve beach habitat (Bird 1990). Watershed development and excessive recreational use of the bay have contributed to extensive habitat loss and pollution problems. Effective remediation can only be realized through responsible, long-term coastal management strategies (J. R. Clark 1995).

ESTUARINE EMBAYMENT MODIFICATION

In open estuarine environments, dredging and dredged material disposal impact benthic habitats and communities in several ways. Dredging removes bottom sediments and benthic organisms. Mortality of the benthic organisms increases dramatically during dredging and disposal processes owing to mechanical injury by the dredge and sediment smothering. Recovery of the dredged site for habitation by the benthic community often requires a year or more. Dredging roils bottom sediments, mobilizing nutrients and chemical contaminants that may impair water quality (Kennish 1997, 2001*a*). For eutrophic systems, the release of nutrients can exacerbate impairment of water quality by stimulating plant growth, raising the biochemical oxygen demand and potentially contributing to dissolved-oxygen depletion.

Boat activity can alter water and sediment quality and benthic habitat structure over extensive areas (Kennish 2002*a*). Chemical contaminant emissions from motorboat and watercraft engines including uncombusted fuel and oil, polycyclic aromatic hydrocarbons (PAHs) and non-aromatic structures accumulate in sediments and elsewhere (Albers 2002). Antifouling paints and primer bases on vessel hulls are sources of trace metals to estuarine waters. Turbulence from pressure waves and propeller wash remobilizes, resuspends and facilitates dispersal of contaminants from estuarine floors (Kennish 2002*b*). Such contaminants potentially pose lethal and sublethal threats to estuarine organisms, most prominently in the benthos.

The most serious benthic-habitat impacts of boat use in estuaries are attributed to propeller scarring and vessel groundings (Kenworthy *et al.* 2002; Whitfield *et al.* 2002). Propeller cutting of the substrate produces narrow (0.5–1 m wide) linear single or twin scars that may extend for tens to hundreds of metres along the estuarine floor. Vessel groundings create excavations known as 'blowholes', which commonly disturb hundreds to thousands of square metres of estuarine bottom habitat. Propeller scarring and vessel groundings most evidently affect seagrass meadows, where propellers can redistribute bottom sediments and tear and

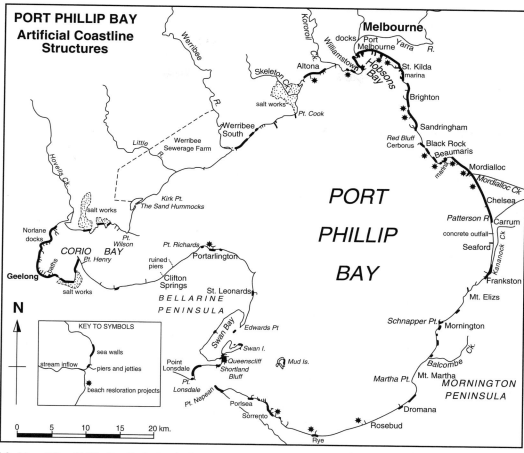

Fig. 13.2. Map of Port Phillip Bay displaying the location of artificial coastline structures and beach restoration projects. (From J. R. Clark 1995.)

excavate leaves, roots and rhizomes (Dunton & Schonberg 2002; Kenworthy *et al.* 2002; Orth *et al.* 2002); recovery may take 5–10 years (Kennish 2002*a*). Boating damage of estuarine habitats is escalating, especially in developed nations where more affluent societies have greater access to recreational pursuits.

Fishing-gear impacts are also detrimental to benthic habitats in estuaries, as in the case of hydraulic clam dredging (Godcharles 1971), use of bull rakes and pea–digger rakes for hard clam harvesting (Peterson *et al.* 1983, 1987), cockle dredging (De Jonge & De Jonge 1992), scallop dredging (Fonseca *et al.* 1984) and trawling (Guillén *et al.* 1994). Recovery of scarred beds may take 3 or more years.

Aquaculture is also potentially damaging to estuarine embayments. For example, shrimp-farm effluents may cause hypernutrification of estuarine waters, and the chemicals may adversely affect endemic flora and fauna. The introduction of aquaculture stocks can disrupt indigenous shrimp stocks via escapement and the spread of diseases. Excessive water use and entrainment of estuarine organisms at shrimp farms are also detrimental (Hopkins *et al.* 1995). Suspended culture of mussels (e.g. *Mytilus edulis* and *M. trossulus*) and net-pen culture of salmonids enrich the surrounding benthic substrate with organic carbon through deposition of faeces and pseudofaeces (Grant *et al.* 1995), resulting in benthic microbial and macrofaunal community shifts (Findlay *et al.* 1995). Addition of gravel, shell and other substrate to mudflats or sandflats to enhance clam production physically alters the bottom habitat and can also disrupt benthic communities (Simenstad & Fresh 1995). The adverse effects of aquaculture activities on estuaries are expected to increase substantially in future with the growth in aquaculture.

Pollution impacts

Because estuaries are heavily used for recreational and commercial activities, they are subject to pollution (GESAMP [Joint Group of Experts on the Scientific Aspects of Marine Environmental Protection] 1982). Most marine pollutants derive from the 20% of the world's population living in industrialized countries where societal affluence is greatest (Taylor 1993). The commonly reported pollutants enter the environment by six principal pathways: (1) non-point-source runoff from land; (2) direct pipeline discharges; (3) riverine inflow; (4) atmospheric deposition; (5) maritime transportation; and (6) waste dumping at sea (McIntyre 1992, 1995; Goldberg 1994, 1995; Kennish 1997). Land-based sources of pollution predominate, with runoff responsible for 44% of the inputs, atmospheric deposition for 33%, maritime transportation for 12%, waste dumping for 10%, and offshore production for 1% (GESAMP 1990).

Nutrient enrichment, organic carbon loading (for example sewage), oil spills and leakages, and toxic chemicals (for example PAHs, halogenated hydrocarbons and heavy metals) account for many of the most acute and insidious pollution problems in estuaries (Table 13.2). Together with pathogens, these pollutants can significantly compromise water quality, degrade habitats, have lethal and sublethal impacts on biota and alter the functioning of estuarine systems. Other pollutants such as litter are more aesthetically displeasing, but they are also potentially hazardous to marine mammals, birds and other organisms that ingest or become entangled in them (Shaw & Day 1994).

NUTRIENT ENRICHMENT AND ORGANIC-CARBON LOADING

Estuarine eutrophication is an emerging global problem associated with nutrient enrichment and organic-carbon loading (Dederen 1992; Rabalais 1992; McComb 1995; Nixon 1995; Valiela 1995; National Estuary Programme 1997*a*, *b*; Valiela *et al.* 1997; V. H. Smith *et al.* 1999; Livingston 2002). Non-point-source inputs of nutrients to estuaries, such as agricultural and urban runoff, contribute significantly to observed eutrophication problems (Carpenter *et al.* 1998). As human coastal populations have increased, the severity of nutrient enrichment and organic-carbon loading

from watersheds has worsened, with impacts in estuaries commonly persisting for years (Balls *et al.* 1995). Acute imbalances in the trophic structure of estuarine systems can arise from hypereutrophication (Ingrid *et al.* 1996; Livingston 1996, 2000, 2002). Nitrogen and phosphorus enter estuaries via surface water inflows (streams, rivers and direct land runoff), ocean water, groundwater discharges and atmospheric deposition (both wet and dry). Anthropogenic sources include wastewaters, sewer overflows, malfunctioning septic systems and fertilized lawns and farmlands. Untreated or partially treated sewage also substantially raises concentrations of organic carbon, heavy metals and other contaminants in estuaries (Kennish 1992, 1997). In addition, it degrades water quality by increasing estuarine biochemical oxygen demand and densities of pathogenic microorganisms (Chapman *et al.* 1996; Kennish 1997; Eganhouse & Sherblom 2001) as, for example, in San Francisco Bay (Livingston 2002).

Treated municipal wastewater and urban stormwater runoff may contain more than 100 enteric pathogens (NRC [National Research Council] 1993) and pose a serious threat to human health, being responsible for cholera and hepatitis, as well as gastroenteric diseases.

Various acute and insidious impacts are coupled with nutrient enrichment and organic carbon loading. For example, nutrient-enriched growth and biomass of microphytes and macrophytes are followed by accelerated organic carbon decomposition and benthic respiration which promote widespread and recurring hypoxia (<2 mg per litre dissolved oxygen) and anoxia (0 mg per litre dissolved oxygen). These changes often adversely affect the structure and function of biotic communities (Dauer *et al.* 1992; Diaz & Rosenberg 1995; Ritter & Montagna 1999).

Dissolved oxygen is declining in many estuarine systems worldwide (Holmer 1999) and hypoxia is a widespread problem (Kennish 1997). Other biotic impacts linked to nutrient and organic-carbon enrichment include algal blooms, shading effects, mortality of benthic and pelagic species, reduced biodiversity and habitat use, diminished secondary production and diminution of recreational and commercial fisheries, as well as altered species compositions and distributions (Weston 1990; Costello & Read 1994; Alongi 1998; Kennish 1998*a*, 2000; Holmer 1999; Chesney *et al.* 2000; Howarth *et al.* 2000*b*). Harmful algal blooms have been linked to shellfish contamination, fish kills and human health impacts (Burkholder 1998). Phytoplankton blooms in response to nutrient enrichment create shade conditions unfavourable

Table 13.2. *Common pollutants recorded in estuarine environments*

---

(1) Excessive nutrients causing progressive enrichment and periodic eutrophication problems

(2) Sewage and other oxygen-demanding wastes (principally carbonaceous organic matter) which promote anoxia and hypoxia of coastal waters

(3) Pathogens (e.g. certain bacteria, viruses and parasites) and other infectious agents often associated with sewage wastes

(4) Petroleum hydrocarbons originating from oil-tanker accidents and other major spillages, routine operations during oil transportation, effluent from non-petroleum industries, municipal wastes and non-point runoff from land

(5) Polycyclic aromatic hydrocarbons entering estuarine and marine ecosystems from sewage and industrial effluents, urban stormwater runoff, oil spills, creosote oil, combustion of fossil fuels and forest fires

(6) Halogenated hydrocarbon compounds (e.g. organochlorine pesticides) principally originating from agricultural and industrial sources

(7) Heavy metals accumulating from smelting, sewage-sludge dumping, ash and dredged-material disposal, antifouling paints, seed dressings and slimicides, power-station corrosion products, oil-refinery effluents and other industrial processes

(8) Radioactive substances generated by uranium mining and milling, nuclear power plants, and industrial, medical and scientific uses of radioactive materials

(9) Thermal loading of natural waters, owing primarily to the discharge of condenser cooling waters from electric generating stations

(10) Debris/litter and munitions introduced by various land-based and marine activities

(11) Fly ash, colliery wastes, flue-gas desulphurization sludges, boiler bottom ash and mine tailings

(12) Acid mining wastes

(13) Drilling muds and cuttings

(14) Pharmaceuticals and alkali chemicals

(15) Pulp and paper-mill effluents

(16) Suspended solids, turbidity and siltation/sedimentation

---

for the proliferation of submerged aquatic vegetation, and affected locations may gradually become phytoplankton dominated with the ongoing loss of critically important benthic habitat.

Nutrient overenrichment of estuarine systems of the north-eastern Gulf of Mexico (Rabalais 1992) has over several decades triggered phytoplankton blooms, destabilization of plankton populations, dissolved-oxygen depletion, unbalanced food webs, general reduction of useful productivity and profound alteration of system function (Livingston 2000, 2002). Estuarine eutrophication has been reported for the Dutch Wadden Sea (Netherlands) (De Jonge 1990; Kennish 1998a), the Harvey–Peel Estuary (Australia) (McComb 1995), and for portions of the Maryland coastal bay system (Boynton et al. 1996), Chesapeake Bay (Malone et al. 1996) and the Barnegat Bay–Little Egg Harbor Estuary in the USA (Kennish 2001c). Shallow poorly flushed

estuaries with long residence times are generally the most severely impacted by eutrophication.

CHEMICAL CONTAMINANTS

Halogenated hydrocarbons, PAHs and heavy metals are three classes of persistent particle-reactive contaminants. Being rapidly attached to fine-grained sediments and other particles (Thomas & Bendell-Young 1999; Turner 2000; Kennish 2001b), they tend to accumulate in the system (Kennish 2002b, c). As a result, benthic organisms may be exposed to elevated concentrations of the contaminants at heavily impacted sites and in the sea surface microlayer (see Kucklick & Bidleman 1994).

Because estuarine sediments are a major repository of chemical contaminants, they have been the target of various monitoring and assessment programmes. In the USA, for example, programmes of the National Oceanic and

Atmospheric Administration (NOAA) and Environmental Protection Agency (EPA) have investigated chemical contaminants in sediment and biotic media to detect changes in environmental quality of estuarine and coastal marine waters (US EPA 2001). Sediment grabs and cores have been used to derive information on current and historical chronological records of past contaminant levels in estuaries (Bricker 1993; Valette-Silver 1993; Daskalakis & O'Connor 1995; Moser & Bopp 2001), and estuarine sediment toxicity has also been extensively investigated (Long *et al.* 1996; Kennish 1997; Thompson *et al.* 1999; O'Connor & Paul 2000; Winger 2002).

Heavy metals and persistent organic contaminants elicit an array of adverse responses from estuarine and marine organisms. Heavy metals have been coupled with feeding, digestive and respiratory dysfunctions; aberrant physiological, neurological and reproductive activities; tissue inflammation and degeneration; and neoplasm formation and genetic derangement (Kennish 1992, 1997). In particular, high-molecular-weight PAHs and halogenated hydrocarbons have substantial biological impacts and possess significant carcinogenic, mutagenic and teratogenic potential (Eisler 1987, 2000; Kennish 1992; Fernandes *et al.* 1997).

It is not merely the concentration of chemical contaminants that must be considered when assessing biotic impacts, but also their bioavailability. This is affected by the physical and chemical characteristics of environmental media (Hamelink *et al.* 1994). Sediment pore-water chemistry, redox potential, pH, organic matter content, acid volatile sulphides and sediment particle size all strongly influence the bioavailability of contaminants in estuarine sediments and pore waters (Winger 2002). Laboratory toxicity testing of estuarine organisms can effectively augment data derived from measurements in field environmental matrices.

Some contaminants, notably halogenated hydrocarbons such as dichlorodiphenyltrichloroethane (DDT) and polychlorinated biphenyls (PCBs), are also biomagnified in food chains in addition to being very widespread and persistent.

Among heavy metals, mercury is a ubiquitous contaminant, elevated mercury concentrations occurring in many estuaries in the USA (O'Connor & Beliaeff 1995; Locarnini & Presley 1996; Kannen *et al.* 1998; Livingston 2002). Organometal methylmercury is the most toxic form, posing significant danger to estuarine and marine organisms, even at parts per billion (ppb) levels of exposure.

Methylmercury also bioaccumulates in aquatic food chains (Wren *et al.* 1995), and humans may be at risk from this process (Kennish 1997). Aside from natural sources of the element (e.g. volcanic, hydrothermal and geothermal emissions), various anthropogenic activities, such as combustion of fossil fuels and certain industrial operations, contribute to the contaminant pool.

Without also dealing with the myriad land-based sources of estuarine PAHs and heavy metals, remedial actions to address their impacts are unlikely to be effective (Eisler 1987, 2000; Bryan & Langston 1992; Kennish 1992, 1997, 1998*b*; DeGroot 1995; Dickhut & Gustafson 1995; Wild & Jones 1995; Bothner *et al.* 1998; Wania *et al.* 1998; Cearreta *et al.* 2000). Chemical contamination problems have been most acute in urban industrialized estuaries such as in the USA and Europe, for example the Tees and Tyne Rivers (UK) (Tapp *et al.* 1993; Mathiessen *et al.* 1998), the Bay of Cádiz, and the Barbate, Odiel and Bilbao Rivers (Spain) (Drake *et al.* 1999; Cearreta *et al.* 2000).

Fisheries overexploitation

Estuaries have suffered severe depletion of fisheries resources (Sissenwine & Rosenberg 1996; Botsford *et al.* 1997; Rose 2000; Kennish 2002*c*). Thus, in the USA, various recreationally and commercially important finfish and shellfish species have dramatically declined in abundance owing to intensive fishing. For example, several species of fish in the Albermarle–Pamlico Sound system of North Carolina (Dean 1996), oysters (*Crassostrea virginica*) in Chesapeake and Delaware Bays (J. Kraeuter, Rutgers University, personal communication 2001), hard clams (*Mercenaria mercenaria*) in the Barnegat Bay–Little Egg Harbor Estuary (New Jersey) (Kennish 2002*a*), red drum (*Sciaenops ocellatus*) and sea trout (*Cynoscion arenarius* and *C. nebulosus*) in Tampa Bay (Florida) (Kennish 2000), blue crabs (*Callinectes sapidus*) in Corpus Christi Bay (National Estuary Programme 1997*c*) and major fish species in San Francisco Bay (Monroe & Kelly 1992; San Francisco Estuary Project 1998; Kennish 2000) have all been fished down.

Overfishing also generates imbalances in biotic community structure and ecosystem function (Menge 1995; Jennings & Kaiser 1998). It can induce indirect trophic (food web) interactions that may have far-reaching consequences for biotic systems in estuaries beyond mere removal of target species (Jennings & Lock 1996; Botsford *et al.* 1997; Hall 1999; Pinnegar *et al.* 2000). In northeastern Gulf of Mexico estuaries, fishing has caused shifts in trophic interactions as a consequence of both top–down and bottom–up food-web effects (Livingston 2002). It is difficult to understand the cumulative impact of overfishing on multiple species in a region when evaluating biotic community responses in specific estuarine systems, but fishing may be the anthropogenic factor of overriding importance in ecological extinctions (Jackson *et al.* 2001; Chapter 1).

Freshwater diversions

Development of coastal watersheds typically alters the amount of fresh water entering estuarine basins. The increase of impervious cover associated with building and roadway constructions accelerates freshwater runoff. Stormwater management systems, notably catchment and retention structures, decrease freshwater discharge rates. Reservoirs and dams in upland drainage basins likewise decrease downstream flows. The implementation of flood control measures such as channelization also modifies water-flow regimes (Chapters 2 and 4). Impounded marsh and destruction of wetland habitat can contribute to changes of natural water-storage capacity in areas bordering estuarine shorelines. Wastewater discharges, water withdrawal for domestic uses and water diversions for agricultural, municipal and industrial purposes also impact estuaries. Excessive groundwater withdrawal in coastal watersheds reduces base flow to tributary streams and rivers (Chapters 2–4), significantly decreasing freshwater inputs to estuaries (Bryant & Chabreck 1998; Carpenter *et al.* 1998; Lane *et al.* 1999; Chesney *et al.* 2000; Kennish 2000, 2001*b*, *c*).

The water flow from coastal watersheds influences the physical, chemical and biological conditions in adjoining estuaries. Diversion of freshwater flows reduces the residence time of water in estuaries, which leads to reduction of the area of habitat suitable for biota (Valiela 1995). Salinity in estuarine embayments fluctuates in response to variable freshwater input, and with it, estuarine circulation patterns and biotic communities change. The loading of nutrients, chemical contaminants, sediments and other particulate matter also varies with the volume of freshwater inflow, and the flux of these substances can dramatically affect the species composition and abundance of estuarine organisms, as well as the trophodynamics of the system.

These effects are most conspicuous in estuaries, where large volumes of fresh water are diverted to meet the domestic, agricultural and industrial demands of coastal and inland communities. Human use diverts more than 1.7×10^{10} m^3 of fresh water from the supply of San Francisco Bay each year primarily for irrigation of farmlands during the May–October dry period. In dry years, freshwater diversion reduces inflow to the bay by 60–70% (Nichols *et al.* 1986). Biotic effects of the diminished inflow include decreased phytoplankton biomass and zooplankton abundance in the upper estuary, diminished reproductive success and variable abundance of some fish species, and a depressed pelagic food web (Kennish 2000).

In watershed areas surrounding Charlotte Harbor (Florida), and elsewhere in the USA, diversion of fresh water for agricultural and domestic needs has significantly reduced discharges to estuaries, particularly during the dry season, resulting in substantially higher salinities and near-marine conditions. Hydrologic modification (i.e. channelization) also contributes greatly to the flux of freshwater flow to Narragansett Bay (Rhode Island). In Tampa Bay (Florida), altered freshwater inflow is mainly attributed to the damming of rivers for flood control (Chapters 2 and 4) and hydrologic modification of tidal creeks (Kennish 2000). Freshwater flows into Galveston Bay (Texas) are largely ascribed to surface-water withdrawals and stormwater runoff, both of which are substantial in nearby watersheds (National Estuary Programme 1997*b*).

Altered freshwater flows in coastal watersheds are linked to biotic impacts in adjoining estuaries, as in Apalachicola River and Bay in the Gulf of Mexico coastal system (Livingston 1997, 2000, 2001, 2002). In the Apalachicola system, freshwater inflow is correlated with nutrient and organic-carbon loading, salinity levels, phytoplankton production, and the distribution and abundance of benthic invertebrates and finfish (Livingston 2001, 2002). Trophic relationships in the estuary are also indirectly affected by freshwater inputs, changes in salinity, nutrient concentrations, turbidity and other physical–chemical factors (Livingston *et al.* 1997). Altered freshwater flows in coastal watersheds appear to be an emerging global problem (e.g. Chapters 2–4) with serious implications for the ecological well-being of estuaries worldwide.

Major alteration of freshwater and seawater inflow to estuarine systems in the south-west Netherlands has been implemented to avert flooding, such as that which killed more than 1800 people in 1953. As a result of the 'Delta Plan', three estuaries in the region have

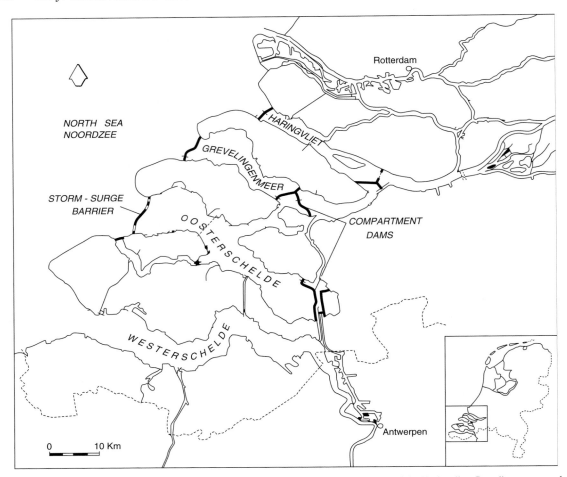

Fig. 13.3. Map of the delta region in the south-west Netherlands illustrating dyke-altered areas of the Haringvliet, Grevelingenmeer and Oosterschelde systems. (From J. R. Clark 1995.)

been closed off by dykes, notably the Haringvliet, Grevelingenmeer and Oosterschelde (Fig. 13.3). Subsequent to dyking, the Haringvliet basin became a highly polluted freshwater body, the Grevelingenmeer became a saltwater system and the Oosterschelde remained a productive estuary bordered seaward by a storm-surge barrier. Acute hydrologic modification in the Oosterschelde associated with construction of a storm-surge barrier across the estuarine mouth and compartment dams up estuary clearly demonstrates how anthropogenic controls can severely alter freshwater and seawater inflow to an estuary and disrupt natural conditions. For example, saltmarshes and tidal flat habitats have declined appreciably in the Oosterschelde owing in large part to the

aforementioned human-induced changes in the system (J. R. Clark 1995).

Introduced/invasive species

Biotic invasions of estuarine and coastal marine environments grew markedly during the past century with expansion of marine transportation, commerce, mariculture and other human activities (e.g. Chapters 10–12). Numerous organisms have entered new environments accidentally or on the hulls and in the ballast water of ships; others have been intentionally introduced via mariculture. Few estuaries are not affected in some way by introduced/invasive species (Carlton 1989; Carlton & Geller 1993).

The introduction of exotic species can have profound effects on estuarine ecosystems, causing: (1) decreases in abundances and survival of native species; (2) local extinction of susceptible native flora and fauna through competition, grazing, predation and habitat modification; (3) spread of parasitic diseases; and (4) alteration of nutrient cycling, hydrology and energy budgets of native ecosystems (Mack *et al.* 2000). The trophic structure of heavily invaded systems has often changed, with dominant species commonly being outcompeted or replaced by non-indigenous forms. In some cases, invaders have radically transformed habitats. For example, the European periwinkle (*Littorina littorea*) has been responsible for transforming mudflat and saltmarsh habitats to rocky shores along coastal inlets in the north-east USA (Bertness 1984; Mack *et al.* 2000). Invasion of non-endemic marsh grasses has modified wetland processes, including sediment transport and deposition, organic-carbon flux and associated faunal communities. Human-induced biotic invasions have potentially significant global impacts as well because they may collectively pose a threat to biodiversity over extensive regions or provinces.

Introduced/invasive species have frequently increased dramatically in abundance in their adapted estuarine environments because of the lack of natural biotic controls. Some of these species have been intentionally introduced to estuaries to establish new recreational or commercial fisheries (Nybakken 1988; Day *et al.* 1989; Kennish 2000). An example is the import of the Pacific oyster (*Crassostrea gigas*) from Japan to the state of Washington (USA). However, the remarkable success of introduced/invasive species can decimate historically important fisheries. A viable strategy to control the spread of non-indigenous species is to prevent their entry into a new range (Mack *et al.* 2000). Post-entry control of invasive species remains an expensive and generally impractical approach to minimize estuarine impacts.

Although most estuaries harbour non-indigenous species, those that are sites of shipping ports or centres of aquaculture and commercial fishing generally exhibit the greatest number of biotic invasions (National Estuarine Research Reserve System 2001). San Francisco Bay, with nearly 250 introduced species, ranks among the most heavily invaded estuaries on Earth (Cohen & Carlton 1998; San Francisco Estuary Project 1998). Non-indigenous species now comprise 40–100% of the benthic and fouling communities in different areas of the Bay, with most macroinvertebrate species along the inner shallows

consisting of introduced forms (Kennish 2000, 2001*a*). Among the most detrimental invaders in this heavily used estuary is the Asian clam (*Potamocorbula amurensis*), which has greatly impacted the phytoplankton community and may have significantly altered bottom–up controls in the system. It has also outcompeted more favourable bivalve species such as *Macoma balthica* and *Mya arenaria*, attaining densities as high as 30 000 individuals per m^2, and disrupting planktonic and benthic communities and finfish assemblages (San Francisco Estuary Project 1998). Impacts of the Asian clam cost *c*. US$ 1 billion each year (Orsi & Mecum 1986; Cohen & Carlton 1998).

Other prominent invasive species in San Francisco Bay are the striped bass (*Morone saxatilis*) and the smooth cordgrass (*Spartina alterniflora*). Invasive plants include *S. alterniflora* elsewhere in the USA and in England (Thompson 1991) and *Phragmites australis* in the USA (Weinstein & Kreeger 2000; see Chapter 11), and animals include the bivalves *Perna perna* and *Dreissena polymorpha* (Carlton & Geller 1993; Kennish 2000). The parasitic protist *Haplosporidium nelsoni* has nearly destroyed the American oyster (*Crassostrea virginica*) fishery in a number of mid-Atlantic estuaries of the USA. Subtropical shipworm species (*Teredo bartschi* and *T. furcifera*) have destroyed many untreated wooden structures in New Jersey (Kennish & Lutz 1984). The European green crab (*Carcinus maenas*) and Chinese mitten crab (*Eriocheir sinensis*) have become established in various estuaries of the USA, disrupting native communities.

Sea-level rise

More than 100 million people now reside within 1 m of mean sea-level, and thus global sea-level rise poses a great threat to coastal communities during the twenty-first century (Douglas & Peltier 2002). Many coastal watersheds bordering estuaries are under attack by rising sea level associated with anthropogenically induced global warming, as well as altered hydrological and geological processes in the coastal zone induced by an array of more localized human activities (Nuttle *et al.* 1997; Kennish 2001*b*). By the year 2050, global warming may result in an overall 10-cm rise in relative sea level (Titus & Narayanan 1995), but the extent of movement will depend on vertical displacements of the coastal zone (Nuttle *et al.* 1997). In the USA, relative sea-level rise has been 2–4 mm per year along the Atlantic coast (Stevenson *et al.* 1986) and up to *c*.10 mm per year in coastal Louisiana (Boesch *et al.* 1994). Superimposed on changes in relative sea-level rise are responses of shorter duration, such as sea-level variations generated by El Niño, hurricanes and other major storms (Nuttle *et al.* 1997).

Despite the eustatic rise in sea level, many estuarine wetland systems in the USA and elsewhere are maintaining their position or expanding because of high accretion rates fostered by sediment supply, relatively low wetland elevation, frequent tidal inundation, nutrient supply, and high *in situ* organic-matter production, generally accounting for more than 90% of the sediment volume (Bricker-Urso *et al.* 1989; Nyman *et al.* 1993).

The eustatic sea-level rise was 10–25 cm over the past century (Douglas & Peltier 2002) and *c.*1.8 ± 0.3 mm per year during the 1980s and 1990s (Douglas 1991; Baltuck *et al.* 1996; IPCC [Intergovernmental Panel on Climate Change] 1996). The increase in the global mean surface temperature (0.6 ± 0.2 °C) has been largely responsible for the rise (IPCC 2001*b*). With global mean temperature expected to rise by 0.5–1.7 °C by 2040 (Kennish 2002*c*), eustatic sea level is projected to rise by 5.6–30.0 cm over the same period (IPCC 2001*b*). This is likely to threaten saltmarshes (Chapter 11) and some mangroves (Chapter 12), lacking sufficient accretion.

The landward progression of estuarine shorelines owing to rising sea level could substantially reduce the areal coverage of intertidal habitat and significantly modify the configuration of the basins involved. Some wetlands will be converted to open-water habitat as estuaries widen and deepen. These physiographical changes will increase the tidal prism, tidal range and salinity in the system. The altered wetland and intertidal habitats, along with modified salinity regimes, could have devastating consequences for biotic communities, trophic interactions and fishery resources. For example, birds that depend on benthic intertidal fauna for forage would probably decline in abundance. In addition, the loss of wetland nursery habitat could lead to declining stocks of important finfish and shellfish species.

Estuaries located between the equator and 20° N and 20° S latitude will be more impacted by future global warming effects than others (see Ledley *et al.* 1999), specifically experiencing more extreme weather conditions including severe droughts and periods of excessive precipitation (Kickert *et al.* 1999). Freshwater inflow to estuaries between latitudes 20° N and 20° S will vary markedly as a consequence of climatic shifts expected to the 2025 time horizon, thereby affecting nutrient, sediment and contaminant inputs. Fluctuating salinities will create unstable conditions for estuarine organisms. Greater frequency of severe storms and storm surges will accelerate shoreline erosion, reduction of estuarine beaches, loss of wetland habitat and coastal flooding. Advancing seas will increase the probability of saltwater intrusion in coastal groundwater supplies. They will also affect the composition of plant and animal communities, promoting the establishment and proliferation of salt-tolerant species.

Sea-level effects will be a function of the local environmental conditions and also of the physical characteristics of estuaries, such as their size and shape, orientation to fetch and local currents, occurrence of barrier islands, extensiveness of coastal wetlands and state of development of adjoining watersheds (Jones 1994). Human intervention may delay or obscure effects of sea-level rise, as with the construction of sea walls and other engineering structures to minimize wave and current impacts.

Subsidence

Subsidence problems are emerging in many coastal cities of developing countries, such as Manila, Rangoon and Jakarta, but also of developed countries. Houston (Texas, USA), the Po Delta (Italy) and Tokyo (Japan) have respectively subsided by 2.7 m, 3.2 m and 4.6 m (Baeteman 1994). The frequency of coastal subsidence impacts on estuaries is on the rise worldwide, mainly because of poor watershed development and other human activities.

Subsidence effects are local or regional in nature, but impacted estuaries often exhibit major modifications such as acute shoreline retreat, loss of fringing wetland habitat, and increase in basin volume and open-water habitat. Habitat alteration is a major concern.

The subsidence results from sediment compaction, crustal (tectonic) movements, groundwater withdrawal, and oil and gas extraction. Areas surrounding coastal metropolitan centres, where excessive groundwater withdrawal has occurred to meet domestic and industrial water demands, are common sites of subsidence (Stevenson *et al.* 2000). Gas and oil withdrawal has exacerbated the subsidence over an area *c.*130 km in diameter in Galveston Bay (Texas), where shoreline erosion and retreat have taken place, mudflats and open bay waters have expanded, fringing saltmarsh habitat has been lost, and sediment distribution and bay circulation have changed (Shipley & Kiesling 1994; Kennish 2000). Relative sea-level rise in the Dutch Wadden Sea has increased owing to extraction

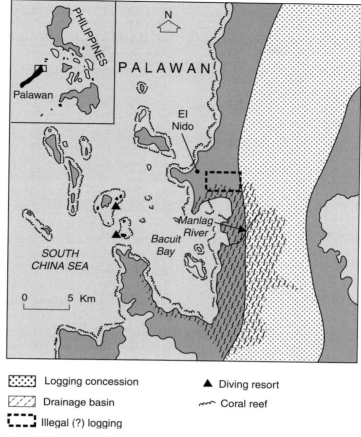

Fig. 13.4. Map of Palawan Island, Philippines, delineating timber harvesting and logging areas in proximity to Bacuit Bay. (From J. R. Clark 1995.)

of natural gas, and the south-east coast of England has likewise subsided for several reasons (Adam 2002).

Sediment compaction is the primary agent of subsidence in coastal Louisiana (DeLaune & Pezeshki 1994), where relative sea-level rise has been as high as 2 cm per year over the past century (Boesch *et al.* 1994), the average rate of relative sea-level rise exceeding the global rate over five-fold (Penland & Ramsey 1990).

Sediment input/turbidity

Most often associated with silviculture operations and construction/development in coastal watersheds, high inputs of sediment can smother benthic communities and create high turbidity which limits light penetration in the water column. The net effect is frequently diminished primary production.

Estuarine shellfish beds (e.g. oysters and clams) can also be significantly impacted by high rates of sedimentation, commonly shown by yielding reduced harvests. Elevated rates of deposition at tidal inlets and river mouths may generate tidal deltas and sand bars that often affect estuarine current patterns and alter physical–chemical conditions.

Timbering and logging operations on Palawan Island (Philippines) have led to serious soil erosion in watershed areas and sedimentation problems in Bacuit Bay (Fig. 13.4). The high turbidity and sedimentation in the bay have had devastating consequences for the biotic communities. For example, extensive mortality of corals has occurred, and the food chain has been seriously disrupted as evidenced by declining reef fish populations. Offshore pelagic species also appear to have been impacted by the deforestation and logging activities (J. R. Clark 1995).

FUTURE ANTHROPOGENIC THREATS

Habitat loss and alteration

Human impacts on estuaries are expected to increase considerably to the year 2025 owing to the burgeoning coastal human population (Weber 1994; Hameedi 1997), more than 75% of which may live within 60 km of the coast by the year 2020 (Roberts & Hawkins 1999), many of these people in urban areas (Postel 2000). The population growth and urbanization is projected to be more rapid in developing nations of the tropics and in the southern hemisphere. Although a disproportionately large number of estuarine anthropogenic impacts now occur in urbanized systems of developed nations in the northern hemisphere, the geographical shift in population growth indicates that estuaries in the southern hemisphere will be subject to far greater impacts in the years ahead (Alongi 1998).

Based on published studies (e.g. Day *et al.* 1989; McIntyre 1992, 1995; Windom 1992; Yap 1992; Jones 1994; Goldberg 1995, 1998; Kennish 1997, 1998*a*, 2000, 2001*a*, *b*, 2002*a*, *b*), the 10 most important future anthropogenic stresses on estuarine environments include the serious 'Tier I' stressors and less threatening 'Tier II' stressors (Table 13.3). Many of the impacts are directly coupled to human activities in coastal watersheds, but others are the result of human uses of the estuarine water bodies themselves. An expanding coastal population will invariably hasten habitat degradation. The loss and alteration of estuarine habitat are projected to be the most serious threat to the future health and viability of estuaries worldwide because they cause environmental fragmentation and functional deterioration as well as the decimation of biotic communities. They also promote water-quality impairment by facilitating nutrient and contaminant transfer. In addition, the destruction of fringing habitat renders coastal communities more vulnerable to stochastic natural events, such as hurricanes, storm surges and coastal flooding (Kennish 2001*a*; Chapters 11, 12 and 16). Habitat destruction differs from most other stressors in terms of intensity and scale of ecological consequences, being capable of restructuring the function and controls of estuarine systems at all temporal and spatial scales (Valiela 1995).

Eutrophication

Eutrophication is similar to habitat destruction in that it can thoroughly restructure the function and controls of estuarine systems and, on an aggregate basis, may elicit global scale changes. With about half of the total nutrient influx to the sea deriving from anthropogenic sources including municipal and industrial wastewaters (Windom 1992), nutrient enrichment is an emerging problem in estuaries worldwide (McIntyre 1995). The most serious sewage pollution occurs in developing countries that lack sewage-treatment capacity (Yap 1992). In developed countries such as the USA and UK, where government controls restrict the release of sewage waste to aquatic systems, sewage and associated organic and nutrient loading are less and should decline further to the year 2025 and beyond (Office of Technology Assessment 1986; National Academy Press 1993; National Estuary Program 1997*b*, *c*; Kennish 2000, 2001*a*).

Excessive nutrient loading will promote algal blooms, benthic primary production, and hypoxic and anoxic events in estuaries that increase mortality of benthic and pelagic organisms (Goldberg 1995; Nixon 1995; Livingston 2000). Estuarine eutrophication has already become more prevalent for example in Korea, Japan, China, India, the Mediterranean and the Caribbean (Galloway *et al.* 1994; Valiela 1995; Alongi 1998). Moderate to high eutrophic conditions exist in more than 80 estuaries in the conterminous USA (Bricker *et al.* 1999). Eutrophic conditions are likely to worsen in 86 estuaries in the USA by 2020 (Bricker *et al.* 1999). With greater intensity of agriculture, mariculture, livestock rearing and industrial activity in many developing countries, eutrophication will increase further during the next 20 years in these already nutrient-enriched estuaries. The most serious eutrophic conditions are anticipated in estuaries of developing nations in South America, Africa and Asia, where sewage waste disposal from rapidly growing coastal human populations will probably escalate substantially.

Fisheries overexploitation

Overfishing will become an even more serious problem to the 2025 time horizon as more countries adapt the latest technology to their commercial fishing. Many fisheries in developed countries today are heavily overcapitalized, with numerous and well-equipped vessels capable of depleting fishery resources. Important elements in the long-term protection of these resources are effective fishery management and regulation programmes. In addition, measures must be undertaken to stem the degradation of estuarine habitat vital to the survival of numerous fish species.

The fisheries of estuarine systems appear to be producing at or near their maximum sustainable yield.

Table 13.3. *Ranking of principal anthropogenic stressors to estuarine environments based on assessment of published literature (see McIntyre 1992, 1995; Windom 1992; Yap 1992; Jones 1994; Kennish 1997, 1998a, 2000, 2001a, b; Goldberg 1995, 1998).*

Stressor	Principal impacts
Tier I stressors	
(1) Habitat loss and alteration	Elimination of usable habitat for estuarine biota
(2) Eutrophication	Exotic and toxic algal blooms, hypoxia and anoxia of estuarine waters, increased benthic invertebrate mortality, fish kills, altered community structure, shading, reduced seagrass biomass, degraded water quality
(3) Sewage	Elevated human pathogens, organic loading, increased eutrophication, degraded water and sediment quality, deoxygenated estuarine waters, reduced biodiversity
(4) Fisheries overexploitation	Depletion or collapse of fish and shellfish stocks, altered food webs, changes in the structure, function, and controls of estuarine ecosystems
(5) Sea level rise	Shoreline retreat, loss of wetlands habitat, widening of estuary mouth, altered tidal prism and salinity regime, changes in biotic community structure
Tier II stressors	
(6) Chemical contaminants Higher priority synthetic organic compounds	Adverse effects on estuarine organisms including tissue inflammation and degeneration, neoplasm formation, genetic derangement, aberrant growth and reproduction, neurological and respiratory dysfunction, digestive disorders and behavioural abnormalities; reduced population abundance; sediment toxicity
Lower priority oil (PAHs)/metals/radionuclides	
(7) Freshwater diversions	Altered hydrological, salinity, and temperature regimes; changes in abundance, distribution, and species composition of estuarine organisms
(8) Introduced/invasive species	Changes in species composition and distribution, shifts in trophic structure, reduced biodiversity, introduction of detrimental pathogens
(9) Subsidence	Modification of shoreline habitat, degraded wetlands, accelerated fringe erosion, expansion of open-water habitat
(10) Sediment input/turbidity	Habitat alteration; reduced primary production; shading impacts on benthic organisms

Hence, the harvest of fishery species here cannot be sustained, and continued fishing pressure will likely lead to collapse, especially of high-value species. Depletion of piscivorous fish is likely to promote population growth of prey and competitor species (Day *et al.* 1989; Valiela 1995), and may dramatically alter estuarine food webs. Overfishing is another anthropogenic factor capable of mediating global-scale change to the biotic structure of estuaries.

The rapid human population growth in coastal watersheds does not bode well for the future of estuarine fisheries. By 2025, the world demand for fish will exceed 110–120 Mt per year (FAO [Food and Agriculture Organization of the United Nations] 1996). It is highly unlikely that fishery landings can meet this demand, unless they can be augmented substantially by aquaculture production. Because of weak or non-existent fisheries management programmes during the next 20 years, stock depletion and collapsed fisheries in developing countries may occur at an even greater rate than in developed countries, with imbalances likely in biotic community structure and ecosystem function of estuaries.

Sea-level rise and climate change

Eustatic sea-level rise is expected to continue unabated through the twenty-first century, increasing by an estimated 9–88 cm (IPCC 2001*b*). On a local or regional scale, tectonic and isostatic adjustments (i.e. land subsidence, emergence and ocean-basin deformation) also cause relative sea-level changes. The anticipated eustatic sea-level rise (IPCC 2001*b*) will affect essentially all estuaries, although their responses will vary depending on factors such as the configuration, size and depth of the estuarine basin, the occurrence of barrier islands, and the morphology of adjoining watersheds.

Physical effects of a global sea-level rise of 2.6–15.3 cm by the year 2020 (IPCC 2001*b*) will include increases in estuarine basin areas, shoreline retreat, reductions in estuarine beach habitat, losses of fringing wetland habitat and altered tidal prisms. These changes will also affect biotic communities, notably those inhabiting intertidal and fringing wetland habitats.

Chemical contaminants

Estuarine and coastal marine environments will continue to receive chemical contaminants from agricultural and industrial activities and urbanization, especially in developing nations. While a wide array of contaminants will enter estuaries via land runoff, river discharges and atmospheric deposition, the most important in terms of potential biotic and habitat impacts will include halogenated hydrocarbon compounds, PAHs and heavy metals, with adverse effects on estuarine organisms (Kennish 1992; McDowell 1993). The most severe effects will continue to be manifest in biotic communities locally subjected to high levels of contaminants near urban centres (O'Connor & Beliaeff 1995; Long *et al.* 1996; Kennish 1997, 1998*a*, 2001*a*), the effects being however less prevalent than those of overfishing, eutrophication and habitat degradation.

As agricultural production and industrial applications increase in developing countries, many pristine and low-impacted estuaries will be exposed to greater inputs of these contaminants. Tighter controls on chemical-contaminant releases in developed countries will reduce contaminant accumulation in many estuaries, particularly in North America and Europe. Developing countries in tropical regions and elsewhere now use many contaminants banned in developed countries, and new chemical compounds will be introduced at least to the year 2025. Based on organochlorine-pesticide use increasing at a rate of 7–8% per year (Tolba & El-Kholy 1992), amounts of some of the most toxic and persistent compounds will quadruple in the environment of developing countries by 2025. Estuarine bottom sediments will provide a sink for many of these contaminants.

Other contaminants such as heavy metals and radionuclides are less threatening to estuaries than halogenated hydrocarbons (GESAMP 1990), but will remain a more localized problem, being found in highest concentrations near source inputs (e.g. smelters, mines, nuclear power stations and fuel reprocessing plants) (McIntyre 1992). Fossil-fuel combustion and subsequent atmospheric transport and deposition will be responsible for the more widespread regional distribution of certain contaminants such as cadmium, lead and mercury.

Freshwater diversions and inputs

Although coastal freshwater diversions are on the rise, their impacts appear to be local and regional rather than global. Many developing countries do not have the technology or the financial base necessary to divert freshwater flows on a large scale (e.g. Chapters 2–4). The most significant freshwater diversions now occur in developed countries, which will remain the principal sites of impact during the next several decades. However, the volume of diverted freshwater flows will also increase in some developing countries, especially those where development pressures are excessive and freshwater supplies are limited.

As development continues in these watersheds during the twenty-first century, human modification of freshwater flows to estuaries will take on an increasingly important role that may have devastating consequences for the biotic organization and function of estuaries.

Introduced/invasive species

Changes in abundance and distribution of invasive species can be expected to continue to result in dramatic shifts in the trophic organization of estuaries with potential ongoing repercussions (Carlton 1989; Cohen & Carlton 1998; Kennish 2000). It is anticipated that the number of estuaries impacted by introduced species will increase considerably worldwide to the 2025 time horizon and beyond, concomitant with escalating shipping activity, mariculture ventures and recreational pursuits. Although introduced species have the potential to reduce biodiversity (via ecological extinction) and cause other biotic problems in estuaries, their impacts are unlikely to be of greater global significance than those associated with habitat degradation, eutrophication and fisheries overexploitation.

Coastal subsidence

Compaction of subsurface sediments, tectonic (crustal) movements and withdrawal of groundwater, oil and gas have had greater effect on relative sea-level changes than eustatic sea-level rise in some coastal regions. Subsidence will play a greater role in relative sea-level rise in some regions up to the year 2025 as demands of burgeoning human populations for groundwater and oil and gas supplies increase. This will result in an expansion of open-water habitat and alteration of shoreline and wetland habitats in affected estuaries due to submergence. Biotic communities will be impacted most notably in fringing habitats.

Sediment input/turbidity

It is anticipated that sediment inputs/turbidity will increase most significantly in developing countries where activities such as deforestation and urbanization will lead to accelerated erosion. However, the problems will also continue in developed countries where the construction of coastal industrial facilities and homes will continue. The overall effect will be greater alteration of estuarine habitats and related impacts on biotic communities to the 2025 time horizon and beyond.

MULTIPLE STRESSORS AND THE ESTUARINE SYSTEM

Habitat loss and alteration, eutrophication, sewage, fisheries overexploitation and (now) sea-level rise represent Tier I stressors, because they can have far-reaching ecological consequences, potentially modifying the structure, function and controls of estuarine systems, and contributing to a decrease in biodiversity (Table 13.3). Tier II stressors include chemical contaminants (especially synthetic organic compounds), which will continue to be most problematic in urban–industrialized regions, and altered freshwater flows coupled to freshwater diversions, which will cause changes in the structure and function of biotic communities over broader geographical areas. Stressors that will be less acute, but still damaging, are introduced/invasive species, coastal subsidence and sediment input/

turbidity, but all of these factors together will result in shifts in the composition of estuarine biotic communities and/or degradation of valuable estuarine habitat at global scales.

Interactions between different stressor types are likely to affect the biological and ecosystem responses to anthropogenic change to the year 2025 and beyond, changing the perceived relative importance of individual stressors. Impacts regarded as less acute, or even benign, may become significant when they come to operate in combination.

Multiple stressors now impact estuarine and other aquatic systems (Breitburg et al. 1998, 1999; Folt et al. 1999; Lenihan et al. 1999; Ruiz et al. 1999; Schindler 2001), and the theory underpinning stressor interactions is simple in concept. Interactions between stressors can be synergistic, where the biota is affected more than could be predicted from additive or multiplicative models of the effects of the individual stressors, antagonistic where the combined effects of several stressors is less than expected from the models, or neutral where no interactions can be detected.

Examples of antagonistic effects include the enhancement of primary production of estuarine phytoplankton through increased nutrient input (eutrophication), and its reduction through toxic effects on phytoplankton by elevated levels of copper, arsenic, cadmium, nickel and zinc (Breitburg et al. 1999). Synergistic effects include: (1) reduced flow rates and effects of an invasive protozoan parasite Perkinsus on the growth rate of the eastern oyster Crassostrea virginica in the Neuse estuary (North Carolina) (Lenihan et al. 1999); (2) combined effects of thermal stress and invasive invertebrate taxa facilitated by thermal change in North America (Ruiz et al. 1999); (3) biological competition between invertebrates and their consequent vulnerability to metal contamination in Victoria (Australia) (Johnston & Keough 2003); and (4) influence of organic enrichment of sediments on the susceptibility of benthic infauna to macroalgal blooms in the Mondego estuary (Portugal) (Cardoso et al. 2004). There seem to be no reports of neutral effects, and they are probably under-represented.

Multiple-stressor effects may also occur through different stressors acting at different phases of life cycles, often in different sections of a river–sea continuum. Migratory fish species will be the most susceptible to these interactive effects, including the many species that spawn in estuaries or use them as nursery grounds but spend most of their adult life offshore. The issue is well illustrated by the fish totoaba (*Totoaba macdonaldi*), which spawns in the Colorado estuary (USA) where adults are overexploited by fisheries, and the damming of the Colorado River has reduced water flows to the extent that juvenile survival is low (Cisneros-Mata *et al.* 1997; Roberts & Hawkins 1999). Such diadromous species serve to reinforce the argument that estuaries should be viewed as part of a much larger system for management to be truly holistic.

There seem to be fewer documented examples of multiple-stressor effects in estuaries compared to other aquatic systems upstream (e.g. Chapters 2, 5 and 10). This could be due, in part, to a lack of research in this area or because the transitional and fluctuating physical nature of estuaries favours a robust biota in which such interactions are less easily manifested. Nevertheless, the examples given highlight the existence of multiple stressor effects in estuaries worldwide and the need to address them in management policies, in spite of the outcomes of such interactions not being easy to predict. Whether stressors act synergistically, antagonistically or are neutral seems to be context dependent, contingent on the stressor, biological taxon and background environment (Breitburg *et al.* 1999). In the face of such complexity and uncertainty, it would be prudent to take a highly precautionary approach to the management of multiple stressors in estuaries.

The concept of multiple stressors also shows the direction that environmental management must take for estuaries. For example, where sea-level rise causes habitat squeeze at defended coasts, hard marine structures would further eliminate natural shoreline habitats. However, a sediment supply derived from an offshore source could buffer against further erosion, recreate lost habitats and restore estuarine filter functions (Reise 2002).

The ecosystem paradigm remains a viable approach to resource management issues in coastal areas. Because ecosystem research is often used to answer questions that are not necessarily asked at the beginning of the programme, research relevant to management must remain open-ended. The hypothesis development that underlies reductionist approaches to estuarine research may need some revision. Ecosystem processes cannot be defined entirely by *post hoc* accumulation of information that is often taken in a disparate manner with no real application to the system as a whole. Accordingly, the usual methodology for ecosystem research, both in terms of data collection and analysis, remains somewhat incomplete when attempts are made to apply research results to practical ecosystem problems. In many ways, the application of even comprehensive scientific data to the management of coastal systems is not direct, and requires a multidisciplinary approach that often goes beyond the mere assembly of facts. This application can be successful (Livingston 2002), but it requires a broad array of approaches that incorporate educational, economic and political factors.

CONCLUSIONS

Estuaries are highly productive environments subject to a wide array of human impacts from a burgeoning population in coastal watersheds. Coastal development is accelerating at an alarming rate globally, with the coastal human population expected to be 8 billion people by 2025. This large population pool will exert considerable pressure on estuarine resources. Because of their exceptional recreational, commercial and aesthetic value, estuaries will be a target of ongoing human activities that in some cases will further threaten their long-term health and viability.

Human activities must be assessed in both estuarine basins and adjoining coastal watersheds. Impacts originating from each source area can have devastating consequences for biotic communities and habitats in estuaries. Ten anthropogenic stressors deemed to be the principal threats to estuarine environments up to the year 2025 are in decreasing level of importance as follows: habitat loss and alteration, eutrophication, sewage, fisheries exploitation, sea-level rise, chemical contaminants, freshwater diversions, introduced/invasive species, subsidence and sediment input/turbidity.

Habitat loss and alteration, eutrophication, sewage and fisheries overexploitation are Tier I stressors that will continue to modify the structure, function and controls of estuarine systems and reduce biodiversity at a global scale. As Tier II stressors, chemical contaminants, freshwater diversions, introduced species, sea-level rise, subsidence and sediment input/turbidity will have more local or regional impacts, arising in urbanized estuaries in relatively close proximity to metropolitan centres. To protect estuarine environments effectively in the future, it is necessary to address the Tier I stressors today by proposing and implementing management strategies that will mitigate their impacts. A concerted effort on the part of government agencies, academic institutions, business, industry and the general public to collaborate on the resolution of these issues is required to ensure that estuaries are protected and maintained for future generations.

Part V
Rocky shores

The rocky-shore ecosystems addressed in this part are hugely important in terms of biodiversity and ecosystem services yet they cover less than 0.5% of the global marine surface area. They all have hard foundations that confer a certain physical resilience; however, the biogenic structures provided by kelps and corals are proving particularly vulnerable to environmental change, especially so in the case of coral reefs.

Rocky intertidal systems occur at the interface of the land and sea in every climate zone, and are typically open ecosystems with steep environmental gradients that support a wide range of organisms (Chapter 14). Their accessibility to people renders them susceptible to use and abuse, but also amenable to management, while a relative lack of biogenic structures and general openness of character confer resilient qualities on them. The kelp forests that dominate shallow temperate coasts are comprised primarily of structure-producing brown algae, the global distribution of which is physiologically constrained by light at high latitudes and by nutrients, warm water and macrophytes at low latitudes (Chapter 15). Mammals, fishes, crabs, sea urchins, lobsters, molluscs and algae reflect the high productivity of the kelp forests, for which pertinent archaeological records offer a window on past exploitation patterns and effects over hundreds of years at some locations. Tropical coral reefs are built by anthozoan polyps with symbiotic zooxanthellae and other organisms secreting calcium carbonate that thrive in conditions of warm water ($>18\,°C$), high light intensity and seawater aragonite saturation, stable full salinity (35‰) and low dissolved nutrient concentrations (Chapter 16). Recent average sea-surface temperatures $c.1\,°C$ higher, CO_2 concentrations $c.50\%$ higher, and sea levels 40–80 m higher than in the last few million years mean that present-day corals are living ever closer to their upper tolerance limits. Chapters 14–16 all conclude with recommendations and actions to sustainably manage these rocky shores and promote conservation, while considering both local and global scales.

14 · Rocky intertidal shores: prognosis for the future

GEORGE M. BRANCH, RICHARD C. THOMPSON, TASMAN P. CROWE,
JUAN CARLOS CASTILLA, OLIVIA LANGMEAD AND STEPHEN J. HAWKINS

INTRODUCTION

What are the current human impacts on rocky shores and what will the ecosystem be like by the year 2025? This assessment builds on the review of Thompson *et al.* (2002), but provides a more global perspective and outlines actions that are needed to counter threats to rocky shores. The chapter begins by briefly describing the characteristics of rocky shores, the natural driving forces controlling rocky-shore communities and the human and ecosystem services provided by rocky shores. It then provides an overview of past and present human impacts, and projections are made of their intensities to the year 2025. The chapter concludes with recommendations and actions to sustainably manage rocky shores and promote conservation. This assessment includes both anticipated improvements such as more efficient pollution control, and likely future negative impacts such as the irreversible effects of invasive alien species, and considers both local and global scales. An attempt is also

Aquatic Ecosystems, ed. N. V. C. Polunin. Published by Cambridge University Press. © Foundation for Environmental Conservation 2008.

209

made here to identify those societal pressures and consequent impacts that are of an overarching nature (such as global warming) because they apply to all marine and coastal ecosystems.

CHARACTERISTICS OF ROCKY SHORES AND THEIR CONSEQUENCES

Because intertidal rocky shores form a narrow fringe around the coastlines of the world, they are accessible and influenced by events that occur both at sea and on land and extremely vulnerable to human activities compared to many other marine ecosystems (e.g. Chapters 20–22). This vulnerability is tempered, however, by their defining feature, namely the substratum of hard rock, which is not altered by most human actions; this is in contrast to coral reefs (Chapter 16). The relatively two-dimensional nature of rocky shores compared to many other marine habitats constrains physical escape from prevailing conditions, whereas sediment-dwelling organisms can burrow away from the surface. Conversely, rocky shores do not harbour pollutants in the same way as do soft sediments.

Rocky shores are also among the most exposed to extreme sets of physical conditions because of the twice-daily exposure to air and submersion by the rising tide. Coupled with their dependence on a rocky substratum, this means that rocky shore organisms are adapted to variable and unpredictable conditions. Broad-scale dispersal of propagules and larvae means populations have the capacity to recover after impacts such as overharvesting or acute pollution over timescales of less than 10 years. Except when physical conditions are permanently altered, as when sea defences change the hydrodynamic regime, there is normally no need for active intervention to enable restoration.

The capacity for recovery of intertidal rocky shores is exemplified by the flooding of the Orange River, which caused mass mortalities along 45 km of the west coast of South Africa (Branch *et al.* 1990). In the 6 years that followed, three patterns emerged. Firstly, all open-coast sites recovered to resemble (within 10% similarity) the original community composition. Secondly, the rate of recovery depended on distance from a source of recruits. Sites within 5 km of undisturbed communities that could supply propagules took 2 years to recover; those 45 km away took 4–6 years. Thirdly, physical alteration by the development of a small harbour at one site transformed it from an open-coast wave-beaten site into one protected from wave action. At this site, the community never returned to its original

condition; instead it came to resemble that of natural sheltered coves (Fig. 14.1).

Compared with lakes, estuaries and lagoons (e.g. Chapters 5, 7 and 13), rocky shores tend to be relatively open systems (Gaines & Roughgarden 1985; Underwood & Fairweather 1989; Menge 1991; Small & Gosling 2001), although this is not the case universally (Todd 1998; Todd *et al.* 1998; Cowen *et al.* 2000). The larvae and adults of most rocky-shore organisms disperse and migrate over distances spanning centimetres to thousands of kilometres, and they are influenced by inputs of materials such as nutrients and particulate matter from afar. Both factors mean that rocky shores can vary considerably in time and space (Lewis 1976; Bowman & Lewis 1977; Underwood *et al.* 1983; Hartnoll & Hawkins 1985; Johnson *et al.* 1998; Underwood 1999; Jenkins *et al.* 2001), often making it difficult to distinguish human-induced from natural changes.

These features have major consequences for management, notably vulnerability to human activities, tolerance

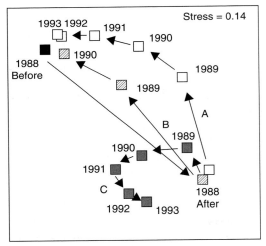

A ☐ 45 km from source, on open coast
B ▨ 5 km from source, on open coast
C ■ 45 km from source, in harbour

Fig. 14.1. Recovery of rocky shores on the west coast of South Africa, following flood-induced mass mortalities in 1988. The conditions on shores that were 5 km versus 45 km from any source of recruits was assessed over a 5-year period by multi-dimensional scaling analyses of entire communities between low and high spring tide levels, based on log-transformed data of percentage cover. (G. M. Branch, unpublished data.)

Table 14.1. *Natural characteristics of rocky shores and their consequences for management*

Natural characteristics	Consequences for management
Accessible Small in extent Impacted by events at sea and on land	Vulnerable to human activities
Stable physical rock matrix Physical gradient steep, variable	Tolerant and resilient to change
System open owing to larval dispersal and migration plus inputs of materials Variable in time and space	Impacts and outcomes of management difficult to predict

to varying physical conditions, resilience in the face of change and considerable spatial and temporal variability (Table 14.1).

ECOSYSTEM SERVICES

Rocky shores provide human society with many goods and services. Shellfish, finfish and seaweeds are collected and caught on rocky coastlines and have provided subsistence foods since prehistory. Living and dead marine organisms plus fossils have generated curios and ornaments, constituting currency in some societies. Pressure has increased because of global tourism. Bait for line fishing and traps is also collected from rocky shores. These goods are crucial to sustaining the lives of many people in less-wealthy countries of Latin America, Africa and Asia. Even in richer countries, foods previously used for subsistence are increasingly becoming luxury items (for example limpets in the Azores: Hawkins *et al.* 2000), putting further pressure on diminishing resources as supplies fail to match demand. Commercialization, the development of global markets and rapid international transportation have all intensified demands for food.

Rocky coastlines are natural sea defences and headlands can provide sheltered anchorages. Soft-rock coastlines, especially of chalk or sandstone, are often reinforced artificially to retard erosive processes in heavily populated areas where land is expensive. Except when at the base of cliffs, rocky shores are generally accessible to people from land; they are used extensively for recreation, although probably not to the same extent as sandy shores (Chapter 17). Activities include bird-watching, fishing and food collection for personal use, and

appreciation of nature, although rising labour costs mean that even the most well-heeled amateur naturalists no longer have access to the 'stout-backed quarrymen' who helped find cryptic fauna and were eulogized by Kingsley (1890). Early recreational use of shores often had a strong element of self-improvement by appreciation of nature, and their high biological diversity makes rocky shores an invaluable educational resource that is accessible to people of all ages and levels of education. Shores are also superb areas for engaging public interest in both the natural world and science as a whole.

Research on rocky shores has included ground-breaking field experimental ecology (for example Hatton 1938; Jones 1946; Connell 1961; Paine 1969), much of it richly used in teaching. There is concern, however, that heavy use of sites for research and education can be damaging (Hawkins 1999).

The most important ecosystem service of rocky shores is the redistribution of algal detrital material from both the intertidal and subtidal zones (Chapter 15) which fuels many inshore and coastal systems. As highly productive and structurally complex systems, rocky shores are important feeding areas for birds and fish, and nursery grounds for fish and mobile invertebrates. Materials and energy are thus imported and exported by waves and currents, and by both tidal and seasonal movements of animals.

Enclosed bays, rias and fjordic coastlines can be bounded by extensive rocky substrate supporting large numbers of filter-feeders that can influence water quality (Hily 1991). Similar effects can occur on the artificial rocky shores of former dock basins redeveloped for amenity and recreational usage (Conlan *et al.* 1992; Hawkins *et al.* 1999, 2002*a*).

FACTORS THAT NATURALLY INFLUENCE ROCKY-SHORE COMMUNITIES

Rocky shores are bounded to the landward by terrestrial systems, and to their seaward side they are part of a continuum with subtidal reefs or sedimentary shores (see reviews by Lewis 1964; Stephenson & Stephenson 1972; Little & Kitching 1996; Raffaelli & Hawkins 1996). Thus they have steep environmental gradients between land and sea and between exposed headlands and sheltered bays and inlets (for review see Raffaelli & Hawkins 1996). The vertical gradient is essentially unidirectional with increasing stress owing to greater exposure to air at higher shore levels. There are also horizontal gradients associated with exposure to different intensities of wave action. This stress gradient is not unidirectional; some organisms such as suspension feeders function better in wave-swept conditions, whilst others such as some of the large algae thrive in shelter. Salinity is another major environmental gradient that is evident in rock pools. Climatic influences operate over much larger biogeographical scales, associated with latitude and modified by ocean currents and upwelling regimes. Therefore, biota found at any given location are determined ultimately by their ability to colonize sites and tolerate conditions, and proximately by smaller-scale physical influences, propagule supply and interactions with other organisms (see reviews by Lewis 1964; Connell 1972; Stephenson & Stephenson 1972; Little & Kitching 1996; Raffaelli & Hawkins 1996).

Offshore hydrographic conditions, including upwelling, strongly influence most rocky shores (Menge *et al.* 1997, 2003; Menge 2000). Waves, tides and currents transport material and hence energy onto, away from and within rocky shores. Mobile marine animals, including fish and crustaceans, as well as birds, reptiles and mammals, extensively exploit rocky shores as feeding, resting, spawning and nursery areas (Rangeley & Kramer 1995; Bradshaw *et al.* 1999; Burrows *et al.* 1999; Coleman *et al.* 1999; Thompson *et al.* 2000*a*). There are also major functional links between rocky shores and other inshore habitats and the land itself.

Thus the interaction of many physical and biological factors influences the structure and functioning of rocky-shore communities, operating at scales ranging from global to local effects (Table 14.2). Superimposed on these natural factors is a set of human impacts, including overarching sociopolitical issues that impinge on all ecosystems.

Interactions among the natural factors complicate predictions about their combined effects (see Chapter 13). For instance, around the 2600-km coastline of southern Africa, differences in biomass are primarily attributable to local productivity interacting with wave action. In the case of filter-feeders, wave action plays an overwhelming role, leading to biomass values that differ more over short distances of tens of metres between sites with differing levels of wave action than over a gradient of productivity spanning hundreds to thousands of kilometres (Bustamante & Branch 1996*a*, *b*). Conversely, grazers respond more to differences in productivity and the input of subsidies from other systems than they do to wave action (Bustamante *et al.* 1995*a*, *b*). Thus, simple predictions about the role of biotic interactions are often not easy to make because of complex interactions between physical and biological processes.

HUMAN IMPACTS: THE PRESENT AND FORECAST TO 2025

Pollution

Pollution can have impacts at all levels of biological organization from molecules to ecosystems: impairment of molecular immune responses (Harvell *et al.* 1999); endocrine disrupters exerting multi-level influences on organisms (from cells up to populations); oil spills and toxic algal blooms directly and indirectly affecting populations and communities; eutrophication directly influencing communities and ecosystems; and mining causing severe and lasting community disruption (reviewed in Thompson *et al.* 2002).

Marine organisms can develop tolerance to pollution via phenotypic responses such as metallothionein production (Bebianno & Langston 1991, 1992, 1995; Bebianno *et al.* 1993) and sequestration in inert structures (Brough & White 1990). In species with limited dispersal, tolerant strains can develop locally, for example in algae (Fielding & Russell 1976) and in polychaetes (Grant *et al.* 1989; Hateley *et al.* 1992). In the case of the periwinkle *Littorina saxatilis* (which has direct development), specimens from sites historically contaminated with heavy metals from mining at the Isle of Man (UK) (Daka *et al.* 2003) have proven tolerant when subjected to toxicity testing (Daka & Hawkins 2004), and this tolerance is suspected to be genetically determined.

Pollution has been combated and continues to be abated in mature industrialized economies. Many traditionally dirty

Table 14.2. *Summary of natural factors, human impacts and societal context influencing rocky shore ecosystems*

Scale of operation	Natural factors	Human impacts	Overarching socioeconomic effects
Physical			
Global	Climate	Global change	Population growth
	Latitudinal gradients		Macroeconomics
	Biogeographical effects		Globalization
Regional	Productivity	Pollution	War and famine
	Currents	Modification of coastal processes	
	Tidal regime		
	Riverine input		
	Substrate type		
Local	Waves	Power generation	
	Sand inundation	Mining	
Biological			
Regional/local	Recruitment and dispersal	Genetic modification	Societal and demographic shifts
Local	Adult–recruit interactions	Pollution	Wealth and poverty
	Competition	Harvesting	Export and transport
	Predation	Introduction of alien species	
	Disease and parasites		

industries are in decline in richer countries but are being relocated or otherwise developed in poorer countries that have cheaper labour costs and often less stringent environmental legislation and enforcement. As standards of living rise, awareness of pollution and the financial ability to impose controls can follow, but with some lag. There can therefore be optimism about pollution effects to the 2025 time horizon, providing there is no major economic collapse or widespread warfare.

ENDOCRINE DISRUPTERS

The ability of anthropogenic substances to disrupt normal endocrine function in animals has raised concern. One of the best-publicized examples is from the worldwide effects of tributyl tin (TBT) pollution, which originates from the leachates of antifouling paints. Whelks have proven particularly susceptible to TBT because they respond to much lower concentrations than other marine organisms (Ellis & Pattisina 1990; Matthiessen & Gibbs 1998), developing imposex (superimposition of male sex characteristics on females), leading in badly affected cases to female sterility. This has devastated populations and led to local extinctions in sheltered bays and estuaries (Bryan *et al.* 1986; Gibbs & Bryan 1987; Spence *et al.* 1990; Hawkins *et al.* 1994) and sublethal effects on open coasts.

Dogwhelks are long-lived, and because imposex is irreversible it persists in populations for extended periods. Recovery in the UK, following legislation in 1987 to ban TBT paints on small boats, has taken at least 10–15 years (Thompson *et al.* 2002). Other substances have also been recognized as having endocrine disruptive effects for a variety of marine organisms including other molluscs, crustaceans, echinoderms and polychaetes (Depledge & Billinghurst 1998), but here also enactment of legislation is likely to lag behind detection of impact.

OIL SPILLS

The effects of oil spills are some of the best-recorded community-level impacts of anthropogenic stress (Clark *et al.* 1997; Hawkins *et al.* 2002*b*). Some species such as barnacles (Southward & Southward 1978) and mussels (Southward & Southward 1978; Newey & Seed 1995) can be remarkably tolerant of oiling. Many are affected more during the clean-up than by the oil itself (see reviews of Southward & Southward 1978; Foster *et al.* 1990). Grazing molluscs seem to be particularly susceptible to both oil and chemical dispersants (Smith 1968; Hawkins & Southward 1992; Newey & Seed 1995). Considerable disturbance can also be caused by physical cleaning, especially if high water pressures or temperatures are used (for example after the *Exxon Valdez* spill: Shaw 1992; Peterson *et al.* 2000*a*).

In the well-studied *Torrey Canyon* oil spill in Cornwall (UK), the major damage was not caused by the estimated 14 000 tonnes of oil that came ashore, but by excessive treatment with over 10 000 tonnes of dispersants. These killed the principal gastropod grazers, *Patella* spp. (mainly *P. vulgata*), *Osilinus* (*Monodonta*) *lineata* and *Littorina* spp., leading to dense growths of algae on many shores (Smith 1968; Southward & Southward 1978; Hawkins & Southward 1992; Hawkins *et al.* 1994). This macroalgal cover subsequently provided a favourable environment for early survival of *P. vulgata*. Treated shores recovered over the next 10–15 years through a series of damped oscillations, while an untreated shore recovered within 2–3 years (Southward & Southward 1978; Hawkins & Southward 1992). The *Exxon Valdez* spill had similar patterns of impact and recovery. Here the clean-up did not involve dispersants, though methods including manual wipe-up, removal of oiled rocks and seaweed, bioremediation and pressurized hot-water washing were used. The last method kills animals as effectively as dispersants, and recovery of rocky shore assemblages treated this way was still slow. The brown alga *Fucus gardneri* had still not recovered to its original levels some 7 years after the spill (see reviews of Paine *et al.* 1996; Peterson *et al.* 2000*a*).

Oil spills are one of the most visible and newsworthy forms of pollution and will no doubt continue to occur on shores adjacent to major shipping routes and oil refineries. The frequency of spills is likely to decline with increases in the size of vessels and improvements in design and legislation, but the potential for large-scale devastation from incidents with supertankers will probably increase to the 2025 time horizon.

EUTROPHICATION AND TOXIC ALGAL BLOOMS

Eutrophication is a significant problem in enclosed seas in many parts of the world (for example the Baltic: Bonsdorff *et al.* 1997; the Irish Sea: Allen *et al.* 1998; Canada: Meeuwig *et al.* 1998; and the USA: Cloern 2001), driven largely by intensive agricultural runoff (Iversen *et al.* 1998) contaminated by high fertilizer loads and, to a lesser extent, sewage discharges (Nixon 1995). Nitrogen deposition from the atmosphere is also important (Paerl & Whitall 1999). In the Baltic, the main effect of eutrophication has been a decline in the perennial macroalgae (*Fucus vesiculosus*), including reduction in depth range (Kautsky *et al.* 1986) and increases in the abundance of ephemeral algae (Schramm 1996; Worm *et al.* 1999). Eutrophication can also directly affect microbial communities on rocky substrata (Meyer-Reil & Koster 2000), with potential indirect effects on grazers.

Management of sewage and fertilizers may reduce eutrophication in some developed regions of the world by 2025, and some dramatic improvements have already been observed (Cloern 2001). Developing countries, however, are expanding their use of inorganic fertilizers and increasing sewage discharges into the sea. Levels of eutrophication are therefore expected to follow a trajectory similar to that observed in developed countries (Nixon 1995). However, oligotrophic tropical systems are expected to exhibit stronger responses to eutrophication than those in the temperate zone (Corredor *et al.* 1999). Atmospheric inputs of nitrogen are increasing (Paerl & Whitall 1999) and forecasted increases in rainfall may also exacerbate eutrophication by leaching terrestrial organic nitrogen (Hessen *et al.* 1997).

In addition to direct effects, eutrophication has been linked to increases in the incidence of harmful algal blooms (Smayda 1997; Paerl & Whitall 1999; Wu 1999; Cognetti 2001). Warm water associated with El Niño–Southern Oscillation (ENSO) events was also implicated in initiating blooms on the coasts of China in 1997–8 (Yin *et al.* 1999), and the cysts of harmful algae can be spread in ballast water (Hallegraeff 1998). When toxic blooms occur, substantial mortality of rocky-shore filter-feeders, grazers (Southgate *et al.* 1984) and predators (Robertson 1991) has been observed. For example, when a bloom of *Chrysochromulina polylepis* was washed ashore over large areas of the Scandinavian coast in 1988, it decimated populations of dogwhelks and other marine invertebrates in some areas (Bokn *et al.* 1990; Robertson 1991; Wu 1999). The impacts on the whole community can resemble that of a badly treated oil spill, with a proliferation of algae produced by decreases in the abundance of grazers (Southgate *et al.* 1984).

MINING

Both the direct effects of mining and the side effects of dumping unwanted mine tailings can radically alter the nature of rocky-shore communities. Much of the west coast of South Africa and Namibia is mined for diamonds, affecting rocky shores by two different processes. Firstly, shallow-water divers work from the shore operating suction pipes that suck up diamond-bearing gravel which is screened on the shore and then fed back into the sea.

In the process, the shore is abraded, grazers and filter-feeders are reduced in numbers and algal growth enhanced, converting shores into seaweed-dominated ecosystems (Pulfrich *et al.* 2003*a*). Secondly, land-based mining produces large quantities of waste (tailings), and in certain areas this is deposited into the sea, building up beaches and smothering or scouring adjacent rocky shores. This increases the abundance of filter-feeders but depletes grazers and again leads to algal domination (Pulfrich *et al.* 2003*b*). The effects are acute, but mercifully local.

In northern Chile, several copper mines are located in the Andes, but the waste tailings from these are diverted to the coast. Between 1976 and 1989, one mine alone deposited 130 million tonnes of solid waste on the coast, with a copper content of about 6000–7000 µg per litre. This transformed a previously diverse intertidal community to one dominated by the 'sentinel' green algal species *Enteromorpha compressa* (Castilla 1996; Correa *et al.* 1999; Farina & Castilla 2001). From 1990, waste was purified by settling and only the 'clear water' was then released at the coast. Despite this, the depauperate community has persisted (Fig. 14.2).

These mining activities tend to be local in their effects, but their impacts are severe. Moreover, although effects of mining on rocky shores are the focus here, mining takes place offshore at far larger scales (Chapter 22). The present forecast is that mining will increase in intensity, particularly offshore where lucrative deposits await the development of technology that will allow economically viable extraction, and where the effects will be less visible and less easy to monitor. By its very nature, mining disturbs entire communities and therefore should not be permitted in marine protected areas. On rocky shores, the alteration of biotic communities is usually obvious, and it is predicted here that increasingly stringent environmental legislation will mitigate against the worst effects, particularly in cases where human health is threatened.

Human harvesting

In some areas rocky shores are used intensively for food. Although it is difficult to separate fisheries statistics into species collected from the intertidal versus subtidal and pelagic, both gastropods and bivalves, which constitute 2% and 29% respectively of world marine mollusc catches, are harvested from the intertidal zone. The world gastropod fishery shows clear variations among regions with collection in South America, Asia and Oceania far exceeding that in Europe and Africa (Leiva & Castilla 2001). The principal species harvested and the techniques used vary considerably between regions. In the Americas, the principal species are *Concholepas concholepas* in Chile, strombid species of conch in the Mexican Caribbean, and *Haliotis* spp. in Baja California (Mexico). Japan, the Republic of Korea and Australia accounted for 95% of the total catches in Oceania and Asia between 1979 and 1996, the main species here being *Haliotis* spp. and *Turbo truncatus* the horned turban shell. France, the UK and Ireland account for over 90% of the gastropod fishery in Europe, the main species being the common periwinkle *Littorina littorea* and the subtidal whelk *Buccinum undatum*. The harvest in

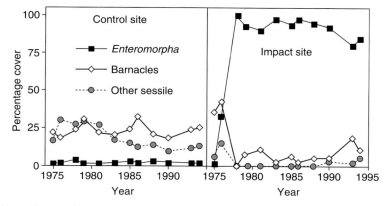

Fig. 14.2. Condition of intertidal assemblages on the coast of Chile (1975–95) at control sites and at sites where copper tailings were disposed of. (After Castilla 1996.)

Africa appears relatively low, although only three African countries are registered in world gastropod fishery records. In addition, species collected for their ornamental value as marine curios (Newton *et al.* 1993) have not been included in these statistics. Methods of harvesting vary considerably from artisanal subsistence gathering to large-scale commercial collection (Kingsford *et al.* 1991; Kyle *et al.* 1997; Crowe *et al.* 2000). However, even small-scale fisheries can exert considerable pressure on stocks where there are large numbers of collectors, as for example in Chile, where there are some 45 000 registered artisanal fishers engaged in small-scale fisheries.

Harvesting has direct effects on target species and can also have distinct indirect effects on other intertidal organisms. In some regions, the effects are so extensive that whole shoreline landscapes may be modified (Paine 1994). Direct effects include reduced abundance and average size of individuals (Hockey & Bosman 1986; Underwood 1993). When intensive collection occurs at a large spatial scale, recruitment overfishing may result. There are several examples of this on isolated islands, such as overexploitation of intertidal limpets in Hawaii (*Cellana* spp.), the Azores, Madeira and Canaries (*Patella* spp.) (for review see Hawkins *et al.* 2000).

In Chile, exclusion of humans from a Marine Protected Area at Las Cruces led to an increase in the body size and abundance of the predatory muricid gastropod *Concholepas concholepas* (Fig. 14.3a), which had previously been exploited (Moreno *et al.* 1984; Duran & Castilla 1989; Castilla 1999, 2000, 2001). This was followed by a reduction in the abundance of mussels, which are its main prey. Keyhole limpets also increased in abundance. Loss of the mussels created areas of open space, which became colonized by barnacles (Castilla & Duran 1985). Subsequently there was a decline in the abundance of *Concholepas* and *Fissurella*, which was attributed to food shortages. Hence, reducing collection by humans led to predictable direct and less predictable indirect changes in community structure (Hawkins *et al.* 2002a).

In South Africa, intense harvesting of the brown mussel *Perna perna* by subsistence and recreational fishers (Tomalin & Kyle 1998) has depleted the natural stocks and affected the community structure as a whole. Where mussels are depleted, foliose algae proliferate. In particular, beds of articulated coralline algae develop, and have the ability to exclude mussels for prolonged periods of time (Lambert & Steinke 1986). As erect algae outcompete encrusting corallines, the latter also decline in abundance

at localities where mussels are depleted. The net effect of intense subsistence harvesting is that communities become less diverse, converging on a common homogenized state (Fig. 14.3b). From the point of view of the mussel stocks, there are two additional adverse effects. Firstly, removal of adult mussels inevitably removes a by-catch of unwanted juvenile mussels. Secondly, mussel beds are the preferred settlement site for recruits. Thus, depletion of mussels not only diminishes the adult stocks, but also diminishes future recruitment into the fishable stock and transforms the community structure into one that retards recovery of mussels (Harris *et al.* 2003).

Harvesting of rocky-shore organisms is likely to intensify to the 2025 time horizon as a consequence of the increasing demand for food resources. In the Third World, this is likely to be dominated by subsistence and artisanal fishing, linked to poverty and the need to obtain food. However, use of non-traditional harvesting methods and improved transportation and storage will all increase the potential for export markets to be developed for further commercialization of harvesting in these regions. In economically more affluent countries, harvesting on rocky shores may well decline in intensity overall. However, exploitation of more lucrative species is likely to increase, and has already led to substantial illegal activities and depletions of stocks. In both cases, the immobility and accessibility of the resources makes them particularly vulnerable and also makes it difficult to enforce control measures.

Alien species

There are many examples of non-native ('alien') species that have invaded biogeographical provinces from which they were previously absent. The barnacle *Elminius modestus* has colonized much of the European coastline from Australasia. The alga *Sargassum* has spread from the Pacific coast of Asia to both the north American Pacific coast and to Europe. Many species have been exported from Europe, including the common shore crab (*Carcinus maenas*), which has found its way to both coasts of the Americas, South Africa and Australia. This is a multi-way traffic, the frequency of which seems to be increasing. Invasions occur frequently and are more likely in some areas: Europe, North America and south-western Australia seem to be particular hotspots. Proximity to major harbours, volume of trade and intensity of aquaculture operations increase the likelihood of such events (e.g. Chapters 10, 11 and 13).

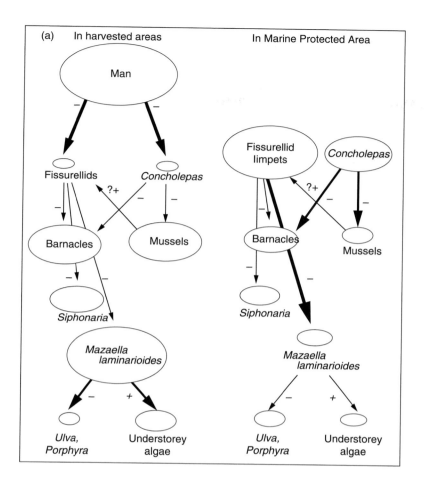

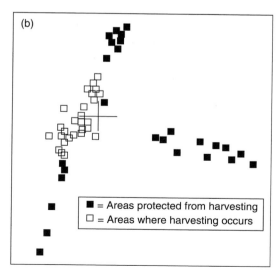

Fig. 14.3. Effects of human harvesting on rocky shore communities. (a) Comparisons between community structure and negative (−) and positive (+) biological interactions, in areas where harvesting takes place versus those inside a Marine Protected Area on the Central Chilean coast. The sizes of the ellipses indicate the relative biomass of different organisms, and the thickness of the arrows shows the strength of the interactions (modified from Branch & Moreno 1994). (b) Correspondence analysis of lower balanoid communities at protected and harvested areas in Transkei (South Africa) (modified from Hockey & Bosman 1986). Harvested communities converge to a common state, whereas areas protected from harvesting are represented as 'satellite' communities.

Major canals such as Panama and Suez can provide short-cuts enabling transfer directly among tropical regions without extensive sea journeys through cold temperate areas. There is also some evidence that low-diversity or disturbed ecosystems are more susceptible to invasion than those less disturbed and more diverse. Many of the species that invade are eurythermic and euryhaline, and occupy a wide range of habitats such as both soft and hard substrata, and are thus ideally adapted for low-diversity disturbed ecosystems (Carlton 1999; Grosholz *et al.* 2000). *Littorina littorea* and *Carcinus maenas* are examples. In cases such as *L. littorea* in the USA, there is dispute about whether invasion or range extension has occurred, although molecular genetic tools can now be used to unravel past species' distributions (Wares *et al.* 2002). The consequences of range expansion and invasions are similar, though impacts are likely to be greater if the immigrant comes from a different biogeographical province with a different suite of competitors and consumers (as is the case for *C. maenas* in Australia: Grosholz 2002; Thresher *et al.* 2003).

The Mediterranean mussel *Mytilus galloprovincialis* was first recorded on the west coast of South Africa during the 1970s, but has now spread to occupy almost 2000 km of coastline. In the process, it has transformed communities on exposed shores, displacing the endemic mussel *Aulacomya ater* (Hockey & van Erkom Schurink 1992). In addition, it competes with the limpets *Scutellastra granularis* and *S. argenvillei* for primary space. Being a relatively small species, *S. granularis* can survive on the alternative substratum provided by the mussels themselves. Indeed, its densities and reproductive output at population level have increased in areas where mussels have invaded, as settlement on mussels is preferable to bare rock. However, large *S. granularis* cannot survive on mussels, so maximum size and reproductive output per individual are reduced (Griffiths *et al.* 1992). The situation for the much larger *S. argenvillei* is more perilous. *Mytilus galloprovincialis* displaces all individuals of this limpet that are large enough to reproduce. Taking such interactions a step further, physical factors may moderate biotic effects. For example the competitive dominance of *M. galloprovincialis* over *S. argenvillei* is moderated by wave action (Fig. 14.4). At low intensities of wave action both species are absent, while on extremely exposed shores both are present but are under stress, and neither achieves domination. Between these extremes, however, the interaction is powerfully influenced by wave action. On semi-exposed shores, *M. galloprovincialis* is recruited at low levels

and grows slowly, so it has almost no impact on the abundance, size structure and attainment of sexual maturity by *S. argenvillei*. On more exposed shores, however, *M. galloprovincialis* becomes dominant because of its faster growth and higher settlement, and occupies 80–95% of available space (Steffani & Branch 2003*a*). Although the limpet can inhibit mussel settlement on bare rock, it is incapable of preventing lateral encroachment, which leads to the exclusion of most of the limpets; those that do co-occur with the mussel are small and virtually none attain sexual maturity (Branch & Steffani 2004).

In the future, there is likely to be an increase in the frequency of introductions, particularly in areas that receive heavy shipping traffic and on coasts that have sheltered waters, because organisms that survive transportation in ballast are likely to be adapted to calm waters. Harbours provide such a haven, but further spread of aliens is most probable where there are natural calm-water bays that provide stepping stones. The amount of shipping, the size of vessels and the speed of ships are all increasing. The faster the travel, the greater the likelihood that organisms attached to hulls or carried in ballast water will survive the journey. Stricter international laws will be required to prevent substantial increases in the rate of introduction of alien species.

Species introductions also occur as a consequence of mariculture. For example, attempts to introduce the South African abalone *Haliotis midae* into mariculture ventures in California resulted in the disastrous accidental introduction of a sabellid epizoite that burrows into the shell and stunts growth. Its effects on its normal host *H. midae* are usually minimal, but it causes severe stunting of Californian species of *Haliotis*. The sabellids also escaped from mariculture and infected other molluscs in the wild, necessitating the closure of most abalone mariculture sites in California and strenuous efforts to locate and eliminate molluscs that had become infested in the wild (Leighton 1998; Ruck & Cook 1998). Thus the impacts of introductions via mariculture are not confined to the target species; accidental introduction of associated species, parasites and diseases can wreak havoc on native species.

Mariculture is rapidly expanding, and genetic modifications associated with mariculture are also likely to multiply. Undesirable side effects of mariculture are likely to increase, and some may prove extremely difficult to contain, partly because they are often unpredictable.

One of the greatest potential risks to rocky shores is from the introduction of alien species that can lead to permanent

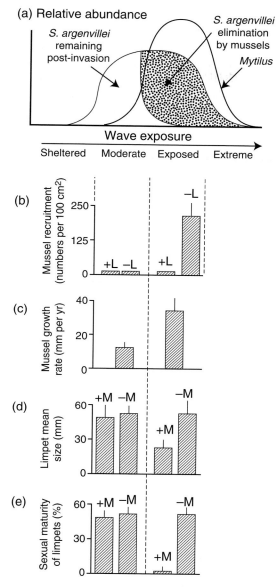

Fig. 14.4. Synopsis of interactions between an alien mussel (*Mytilus galloprovincialis*) and an endemic limpet (*Scutellastra argenvillei*) on sheltered, moderate, exposed and extremely exposed shores on the west coast of South Africa. (a) Relative abundances of the two species; (b) intensities of mussel settlement in the presence (+L) or absence (−L) of the limpet; (c) growth rates of mussels; (d) mean sizes of limpets; and (e) percentage of the limpet population reaching sexual maturity in the presence (+M) or absence (−M) of mussels. Data are derived from Steffani & Branch (2003a, b, c) and Branch & Steffani (2004).

changes in community structure. This is especially so for invasive pathogens, which may find hosts lacking natural resistance (see Harvell *et al.* 1999), and high-ranking competitors or predators such as *M. galloprovincialis* in South Africa (Griffiths *et al.* 1992) and *C. maenas* in California (Grosholz *et al.* 2000).

Some non-native species, however, can also have positive effects, especially if the species is an ecosystem engineer providing new or additional habitat. For example, *Pyura praeputialis*, a non-native species on the Chilean coast, is an ecosystem engineer and provides additional habitat in the interstices amongst individual animals, thus increasing biodiversity (Cerda & Castilla 2001). Similarly, higher diversity of invertebrates has been found amongst the alien *Sargassum muticum* than amongst native algae (Norton & Benson 1983). In some cases, the alien species provides additional food resources for local people. Examples include *M. galloprovincialis* in South Africa and *P. praeputialis* in Chile.

Alteration of coastal geomorphological processes

SEA DEFENCES
Many sedimentary coastlines worldwide adjoin low-lying land or are threatened by flooding or coastal erosion. This has led to construction of sea defences, for example in the Netherlands, on the Adriatic Sea and in Japan (Koike 1993). Virtually all of what were originally sedimentary coastlines in these areas have now been 'hardened' by sea defences. Most artificial rocky shores support communities similar to, but less diverse than, natural shores (Southward & Orton 1954; Hawkins *et al.* 1983; Chapman & Bulleri 2003). These artificial shores can provide stepping stones between populations presently isolated by distance. This has consequences for population genetics (see Kimura & Weiss 1964; Keenan 1994) and may lead to range extensions of species with restricted dispersal. For example, littorinids have extended their range along the Belgian coastline, which lacks natural rocky shores, via breakwaters (Johannesson & Warmoes 1990).

The severity of flooding and coastal erosion is likely to increase during the next few decades because of sea-level rise and the increasing frequency of storms (Rodwell *et al.* 1999; Grevemeyer *et al.* 2000). This will lead to increasingly 'hardened' coastlines over the 2025 time horizon as more sea defences such as offshore breakwaters and sea walls are constructed. The connectivity between rocky shores will increase as a result, as will the extent of habitat available for rocky-shore organisms. However, changes in hydrodynamics will have consequences for sediment-dwelling organisms and will lead to increased fragmentation of sedimentary habitats (M. Frost, P. S. Moschella, R. C. Thompson & S. J. Hawkins, unpublished data).

SEDIMENTATION

The amount of suspended sediment in the water column is expected to increase by 2025 as a consequence of increased coastal erosion, riverine inputs, artificial replenishment of coastlines and dumping of sediment overburden from mining. When deposited, this material may smother rocky habitats, particularly in sheltered locations. This will lead to an increase in the abundance of sand-dwelling organisms such as polychaetes and sand-tolerant species such as the anemone *Anthopleura elegantissima* and the red alga *Chondrus crispus* (Daly & Mathieson 1977; Taylor & Littler 1982). Organisms with opportunistic life histories, such as ephemeral and some turf-forming algae, will also be favoured on rocks that are subject to intermittent inundation (Airoldi 1998). Hence diversity at the 'shore' scale of resolution may increase as a consequence of an increase in the variety and fragmentation of habitats (Littler *et al.* 1983; McQuaid & Dower 1990).

GENERATION OF POWER

The La Rance tidal barrage scheme in France, where tidal amplitude has been reduced, hydrographic exchange altered and the distribution of sheltered rocky shore organisms changed (see Little & Mettam 1994; Retiére 1994), emphasizes the view that if physical conditions are changed, rocky-shore communities will be irreparably altered. There has been a recent resurgence of interest in tidal barrages. Other renewable energy sources such as offshore wind turbines and wave-energy machines are likely to expand rapidly. These devices will inevitably lead to more artificial hard structures and, in the case of wave-energy machines, will certainly reduce wave action to the landward side. There will also be the risk of occasional wreckage of structures on the shore causing localized disturbance.

RECREATION, RESEARCH AND EDUCATION

Recreational gatherers collect a wide variety of organisms, mainly invertebrates, for use as fishing bait, for their ornamental value or for food. Although the daily take may be small, the cumulative effects can be substantial (Underwood 1993). Additionally, habitats can also be damaged by human trampling and disturbance associated with collection (Newton *et al.* 1993), and by other forms of recreation often associated with tourism, such as access to the sea for scuba diving (Hawkins & Roberts 1992), surfing and swimming as well as educational visits (Fletcher & Frid 1996; Bellan & Bellan-Santini 2001).

Disposable income, leisure time and availability of personal transport have all increased dramatically over the last 50 years in the developed world. This has been associated with a decline of subsistence gathering of food and collection of algae, but an increase in recreationally related impacts (Fletcher & Frid 1996), including those in areas of the Third World that are being developed for international tourism (Hawkins & Roberts 1992). Paradoxically, these impacts can be particularly heavy in conservation areas where public access is encouraged to promote awareness of marine wildlife (Fletcher & Frid 1996). Increases in the amount of time available for leisure are likely to lead to an increase in this form of disturbance by 2025. Despite some localized negative impacts of research and educational activities (Hawkins 1999), these activities will be beneficial in influencing perceptions and attitudes to coastal environments and in providing information for management.

Global change and large-scale phenomena

Global changes in many climatic variables (temperature, insolation, ultraviolet radiation (UV), sea level and wave action) anticipated by 2025 (IPCC [Intergovernmental Panel on Climate Change] 2001a; Hulme *et al.* 2002) will no doubt lead to shifts in the geographical distribution of some intertidal organisms. Most intertidal organisms have morphological or physiological adaptations to cope with changes in environmental stresses greater than those anticipated by 2025. Consequently, although there may be changes in species abundance at the fringes of their distribution (see for example Herbert *et al.* 2003), these are only likely to affect organisms at local to regional scales and will depend on the gradient concerned. For instance, shifts might be expected to occur along the vertical emersion gradient at a scale of centimetres to metres, along horizontal gradients of wave exposure at a scale of tens to hundreds of metres, and along geographical climatic gradients at a scale of tens to hundreds of kilometres.

Changes in the distributions of species are likely to have significant effects where they are associated with shifts in the relative proportions of functional groups or in the abundance of organisms that have a key role in structuring communities. Similarly, considerable shifts in species distributions would occur if small changes in climate led to large-scale changes in ocean circulation.

WARMING

Climate can directly influence the distribution of intertidal species. This operates through shifts in competitive abilities at the northern and southern distributional boundaries, and at the upper vertical distributional boundary (see Denny & Paine 1998), where temperatures are either too low or too high for growth and reproduction. For example, in response to elevated temperature, species will become increasingly stressed at the equatorial margins of their distribution, leading to local extinctions. At higher latitudes, if low temperatures are also a major factor limiting their distribution, species with broad dispersal will be able to colonize polewards quite rapidly (see review of Clarke 1996), while those with limited dispersal will face a bottleneck, their distribution being squeezed by increasing temperature from the equator and their restricted ability to disperse towards the poles.

Organisms may respond to an environmental challenge by (1) moving somewhere else, (2) staying and adapting to the changes or (3) going extinct (Clarke 1996). Although historical examples of all of these responses can be demonstrated for climate change (for example range expansion and morphological evolution: Hellberg *et al.* 2001), there is no real understanding of the balance between them. Given the gradual nature of changes in the biogeographical axis, the steepness of vertical gradients and the ability of shore organisms to tolerate environmental fluctuations, the likelihood that rocky-shore animals and algae will survive and eventually move seems high.

Different species will, however, move in distinct ways depending on their dispersal characteristics (Hiscock *et al.* 2004). The influence of climate on the distribution and abundance of invertebrates is also mediated through reproductive output (Southward & Crisp 1954; Southward 1967; Kendall *et al.* 1985; Lewis 1996). For example, southern species of barnacles reaching their northern geographical limits in the British Isles have fewer broods than further south in Europe, and the success of early broods released into the plankton is very low (Burrows *et al.* 1992).

Because of the connectivity of rocky-shore habitats, rocky-shore organisms will probably respond to climate change by shifts in distribution and abundance along environmental gradients over biogeographical scales. Hence local extinctions will generally be matched by colonization of new areas. For example, abundances of the boreal cold–temperate barnacle *Semibalanus balanoides* and the subtropical *Chthamalus* spp. fluctuate markedly over time, with numbers being strongly positively correlated with sea-surface temperature with a time lag of 2 years (Southward 1991; Thompson *et al.* 2002, Fig. 6). Similar fluctuations have been seen in northern (*Patella vulgata*) and southern (*P. depressa*) limpet species (see reviews of Southward *et al.* 1995, 2005). Consequently, rocky-shore plants and animals may provide valuable indicators of more extensive change offshore (Southward 1980, 1991; Southward *et al.* 1995).

The effects of changes in temperature may be amplified or reduced by interactions with other physical factors. For example, based on thermal tolerances alone, the ranges of some tropical and warm–temperate macroalgal species are expected to extend to higher latitudes as a consequence of increases in global temperature. Apart from temperature, however, the distribution of many algae is also regulated by day length. Hence expansions in algal distribution from lower latitudes may be limited by their inability to adapt to prevailing photoperiods rather than by temperature (Beardall *et al.* 1998).

Interactions with other organisms will also modify the direct effects of temperature, further restricting prediction of the consequences of climatic changes. For example, small changes of about 2 °C in air temperature across the Cape Cod peninsula appear to fundamentally alter the influence of *Ascophyllum nodosum* canopy on survival of *S. balanoides*. In the south, macroalgal canopy has been shown to provide a refuge from thermal stress and to enhance barnacle survival (Leonard 2000; Thompson *et al.* 2002). Further north, where shading is less important, the algae provide refuge for a predatory whelk that feed on the barnacles and reduce barnacle survival relative to areas higher on the shore above the *Ascophyllum* zone (Leonard 2000).

Climate change will enhance the success of some non-native species invasions and they will be able to spread further. For example, *Crepidula fornicata*, an immigrant to the UK, was initially found only on the south and west coasts of England and Wales, but it has now been found as far north as Scotland (J. Davenport, personal communication 2004). It is likely that climate change will lead to greater variability of temperature as well as rising averages. This may favour species transfer from climate regions strongly influenced by the continent (American East Coast, Eastern Pacific) to more equitable regions ameliorated by a strong oceanic influence including warming currents such as the Gulf Stream (USA Pacific and European coasts). Invasion seems to be less likely in high-diversity communities and so human activities that reduce diversity (eutrophication, chronic pollution or harvesting of living resources) will probably increase the risk of further invasions.

Based on predictions of a 1.4–5.8 °C rise in temperature by 2100 (IPCC 2001*a*), typical horizontal shifts in species distributions in the region of tens to hundreds of kilometres and small changes in vertical distribution (depending on the tidal range at a given location) are predicted by 2025. Different species will respond in different ways depending on life-history traits and habitat requirements (Helmuth *et al.* 2006), with recent range extensions of southern species being recorded (Herbert *et al.* 2003; Mieszkowska *et al.* 2005; Lima *et al.* 2006). Assuming the current pattern of gradients in sea and air temperature remains, the speed of horizontal shifts in distribution will tend to be more rapid in regions where isotherms are widely separated. However, there will be exceptions, and there is potential for local extinctions and major shifts in community structure and ecosystem processes in some regions (see Harley *et al.* 2006, Helmuth *et al.* 2006 for reviews).

ULTRAVIOLET RADIATION

Ultraviolet radiation, principally UV-B, can have inhibitory effects on photosynthetic performance, growth and nutrient uptake, and can also damage DNA in algae (Beardall *et al.* 1998). It can also alter behaviour (for example in the frequency with which sea urchins cover themselves with debris in response to elevated light levels: Adams 2001), shift sex ratios and reduce survival in invertebrates (Chalker-Scott 1995). Although UV-B radiation is maximal in the tropics, the greatest ecological effects are likely to occur at higher latitudes, where organisms may lack adaptive mechanisms such as screening compounds, methods of repair or behavioural strategies that help reduce the deleterious effects of UV-B (Beardall *et al.* 1998). Gaps in the ozone layer also occur at higher latitudes. To date, most work has focused on individual responses; more research is required on potential effects at the level of communities and ecosystems (see review of El-Sayed *et al.* 1996).

SEA-LEVEL RISE, STORMS AND EXTREME WEATHER EVENTS

Only small changes in sea level are anticipated by 2025 (IPCC 2001*a*), the 'best estimates' for 2020 being 6–7 cm. Direct consequences for rocky-shore organisms are likely to be minimal compared to the effects of changes in climate or storminess, and relative to natural fluctuations in tidal rhythms associated with short-term (18.6-year) changes in the astronomical cycle (Denny & Paine 1998). An exception to this might occur in regions where a small change in tidal level leads to a substantial shift in the extent of horizontal versus vertical rocky shoreline with associated consequences for the area and aspect of intertidal substrata (Graham *et al.* 2003).

Also predicted are greater frequency of extreme events such as storms and increased rainfall in winter or hot spells in summer (IPCC 2001*a*). An increase in wave action associated with storms will lead to an increase in the relative abundance of filter-feeders, which do well in these conditions, and a reduction in the abundance of grazers. These changes are likely to be associated with an increase in biomass and a reduction in species diversity (see Bustamante & Branch 1996*b*; Ricciardi & Bourget 1999). Extreme thermal events are likely to result in kills at the upper limits of distribution of many intertidal species during both hot (Schonbeck & Norton 1978; Hawkins & Hartnoll 1985) and cold weather (Todd & Lewis 1984). The abundance of intertidal epilithic microalgae is also strongly influenced by weather conditions, declining during summer. Hence climatic extremes will also affect primary productivity leading to alternating conditions of feast or famine for the molluscs that graze these microbial films (reviewed by R. C. Thompson *et al.* 2000). There will be consequent changes in the balance between bottom–up regulation versus top–down control of intertidal communities (Thompson *et al.* 2004). These features would all lead to increased temporal variability in the structure of rocky-shore communities, with extreme conditions leading to periodic mortality of some species and an associated increase in the frequency with which space is made available for recolonization. The frequency and magnitude of disturbance are likely to increase.

EL NIÑO–SOUTHERN OSCILLATIONS

As natural shifts in atmospheric and oceanic circulation across entire ocean basins, the ENSO and North Atlantic Oscillation (NAO) (Jaksic 1998) may be linked to climate change driven by increased anthropogenic carbon dioxide in the atmosphere (see Urban *et al.* 2000). On rocky shores, the most significant impact of ENSOs is to reduce the frequency and extent of upwelling of cold nutrient-rich water along the west coast of the Americas and other upwelling coastlines. ENSO events lead to warmer nutrient-poor water and changed currents and storm regimes (Allison *et al.* 1998), and have been associated with changes in species distributions and abundances, prevalences of marine diseases and marine primary production levels (see Fields *et al.* 1993; Harvell *et al.* 1999). Recorded changes have been particularly striking at localities such as the Galapagos (Ecuador) that lie astride complex and contrasting currents (Vinueza *et al.* 2006).

Impacts of the ENSO similar to those on pelagic fisheries and subtidal kelp forests (Dayton *et al.* 1992, 1998) are apparent in intertidal communities, with both positive and negative effects depending on the species concerned. For example, in Chile, the 1982/3 ENSO was thought to have caused massive die-offs of brown algae (Soto 1985) and of littoral invertebrates (Tomicic 1985). Settlement of a keystone predator (*Concholepas concholepas*) was low in years associated with either El Niño or La Niña (negatively related to the ENSO Index: Moreno *et al.* 1984), with potential consequences for both intertidal communities and food gathering (Castilla & Camus 1992). In California, recruitment of intertidal barnacles (Connolly & Roughgarden 1998) and tide-pool fish (Davis 2000) was affected during the 1997 ENSO.

ENSO events have been a persistent feature of late Quaternary climate variation and have been occurring since at least the Pleistocene (Keefer *et al.* 1998; Bull *et al.* 2000). There has been an apparent trend towards more severe El Niños at the end of the twentieth century (Jaksic 1998; Stone *et al.* 1999). If this forms part of a longer-term trend, impacts on affected rocky shores are likely to become more frequent and intense over the next few decades.

PROJECTING INTO THE FUTURE

Impacts on rocky shores will range from sublethal effects on individuals through to populations and community-level responses. Many of these impacts will be localized and stem from point pollution sources and local human usage. Collection of living resources can affect entire coastlines, and both commercial and recreational usage may have widespread effects. Creation of new hard substrata as part of sea-defence schemes will also occur on a

large scale in many parts of the world. These changes will have direct and indirect effects on individuals, populations, communities and ecosystems, leading to changes in primary and secondary productivity, with consequences for both commercially exploited and unexploited species, together with ecosystem goods and services.

In trying to anticipate the future it is important to view the past. Many problems already existed in the 1960s (e.g. Thompson *et al.* 2002; Table 14.1), but perhaps were not appreciated at the time. Some, such as oil spills, sprang into public consciousness after major events such as the *Torrey Canyon* wreck off the coast of Cornwall (UK) in the 1960s. Scientific detective work in the 1980s unravelled the effects of TBT on gastropods and oysters in the face of scepticism by industry (Ludgate 1987). Declining yields and concerns about the need for and scope of marine protected areas prompted work on intertidal resources in the 1970s and 1980s, which showed the extent and scale of human predation on rocky shores in southern Africa (Branch & Odendaal 2003), Chile (Castilla & Duran 1985) and New Zealand (Towns & Ballantine 1993).

Looking forward, what if any will be the surprises? Maybe a new wonder pesticide (as TBT was hailed in the 1970s) will be seen to have a host of unforeseen side effects. An area of current concern is the side effects of various preparations used to treat ectoparasites in aquaculture (e.g. Collier & Pinn 1998). Given the pace of biotechnology, genetically modified organisms could be developed for algal and animal mariculture. Fast-growing strains of seaweeds, mussels, oysters, crabs and shrimps could all be developed and escape from culture. These would influence biodiversity at the within-species level in terms of genetic variation but could also influence population and community processes (e.g. Kapuscinski & Hallerman 1991; Hallerman & Kapuscinski 1995).

The traditional historical human afflictions of war, pestilence and famine will no doubt feature over the next few decades. The consequences of even quite small regional wars have been illustrated in the Arabian Gulf where oil pollution destroyed many marine communities (Jones *et al.* 1994). Even small-scale insurrections lead to breakdown of law and order, and an early casualty is enforcement of conservation and anti-pollution legislation.

The ability to predict the consequences of changes in a single impact vary from reasonable certainty, in the case of some pollutants on single species, to considerable uncertainty, for example in terms of ecosystem responses to changes in global climate or the introduction of non-native

Table 14.3. *Actions available to help preserve rocky intertidal habitats together with a qualitative assessment of their relative importance and the regional scale at which action would be required in order to be effective*

		Major threats			
Scale	Actions[a]	Species invasions	Harvesting	Changes to coastal geomorphology	Global change
Local	Regulations		**		
	Co-management		***		
	Marine Protected Areas		***	*	
	Education/awareness	**	***	*	**
National	Equitable access		***		
	Policy and legislation		***	***	***
	Eliminate subsidies		***		
International	Research and monitoring	***	***	***	***
	Regulation required:				
	Aquaculture and aquarium trade	***			
	Shipping operations	***			
	Greenhouse gases				***

[a] *, useful; **, important; ***, essential.

species (Thompson *et al.* 2002; Table 14.1). Unfortunately, the ability to forecast the combined interactive effects of several environmental factors is at best fairly modest. Hence, unpleasant surprises can be expected to 2025, but their nature remains largely unpredictable. The greatest ecological surprises are likely to occur where (1) environmental change induces shifts between alternate states (Barkai & McQuaid 1988; Paine *et al.* 1998; but see Dayton *et al.* 1998, for an example of stability in algal communities despite loss of considerable biomass); (2) an organism is particularly susceptible to a pollutant (for example dogwhelks and TBT); or (3) an exotic species has a much more prominent role in an invaded community than at home (as is the case for *Sargassum muticum*).

WHAT CAN BE DONE?

Gaps in scientific knowledge about rocky shores are outlined in Thompson *et al.* (2002). Here, the focus is on policy and management priorities. Rocky shores must not be studied in isolation. Table 14.3 summarizes the four major effects of human activities on rocky shores, and the actions necessary to mitigate them. However, part of

the danger lies in developing such lists without also ensuring that appropriate actions are then taken to resolve them.

It is worth focusing on one of these activities here, namely harvesting of rocky shores by subsistence communities, a ubiquitous activity in most developing countries. There are many examples where this has led to unsustainable depletion of stocks. Solutions are not easily found in conventional fisheries management, but there have been some success stories for co-management, namely the involvement of users and central authorities in joint responsibility and decision-making for the management of renewable resources. One illustrative case has been mussel harvesting on the eastern coast of South Africa (Harris *et al.* 2003). Under previous regulation, subsistence fishers were given no recognition and were entitled to harvest only amounts allowable to recreational fishers, which were never enough to provide a regular source of food, and this forced subsistence fishers into illegal harvesting. Confrontation with authorities became violent, and to reduce detection the fishers turned to nocturnal and quick but destructive methods of stripping the mussels. The impasse was only broken when co-management was initiated. This allowed

Scientists   **Managers**

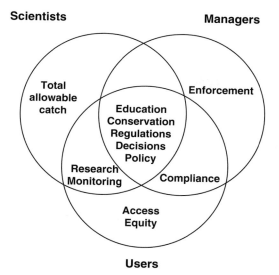

Fig. 14.5. Summary of the roles, responsibilities and benefits that can be jointly attained among scientists, managers and users in a system of co-management.

the fishing community equal participation in making decisions about regulations, involved them in research and improved enforcement. Benefits were mutual, although considerable effort and time were necessary to achieve them. Determination of a sustainable total allowable catch was based on scientific experiments that actively involved the community and led to joint agreement on desired levels of harvesting. Enforcement remained the responsibility of the governmental authority, but was based on cooperation by the community and consequent improvements in compliance. The community gained knowledge, training, legal and exclusive access to the local resources and an equitable way of sharing them. There were gains for conservation, decision-making and the development of policies and regulations.

Co-management (Fig. 14.5) should not be regarded as a panacea, but it clearly has huge advantages over previous top–down regulations imposed by edict (Berkes *et al.* 2001), and the principles are particularly relevant to rocky shores for five reasons. Firstly, harvesting is one of the major factors affecting rocky shores. Secondly, they are extremely accessible, and can be harvested by low-tech means by almost anyone. Thirdly, rocky shores are difficult to police and manage. Fourthly, it is relatively easy to assign 'ownership' to local communities because the physical habitat lies close to user communities, and most of the resources are sessile or slow-moving so that local users can benefit from the control measures to which they agree. Finally, most of the resources that are harvested have relatively low commercial value, and are better used as a source of food than for commercial sale, thus reducing the incentive to break rules and cheat for short-term gains at the expense of long-term sustainability. Perhaps more so on rocky shores than anywhere else, biological, social and economic factors conspire to make co-management a good bet. Winning local communities around to agreeing on and setting controls is essential if management is ever to achieve sustainability.

15 · Current status and future trends in kelp forest ecosystems

ROBERT S. STENECK, RODRIGO H. BUSTAMANTE, PAUL K. DAYTON,
GEOFFREY P. JONES AND ALISTAIR J. HOBDAY

INTRODUCTION

Kelp forests dominate shallow rocky coasts of the world's cold-water marine habitats. They are comprised primarily of brown algae in the order Laminariales and produce one of the largest biogenic structures found in benthic marine systems. Kelp forest ecosystems include structure-producing kelps and their associated biota such as marine mammals, fishes, crabs, sea urchins, lobsters, molluscs, other algae and epibiota that collectively make this one of the most diverse and productive ecosystems of the world (Mann 1973, 2000; Dayton 1985a, b). Economically, kelp forest ecosystems have been significant to maritime peoples for thousands of years (Simenstad et al. 1978; Erlandson 2001). The aim of this chapter is to review how kelp forest ecosystems have changed at very large spatial scales over very long periods of time (decades to millennia). This perspective then provides a strong basis for assessing how different modern kelp ecosystems are from those in the

Aquatic Ecosystems, ed. N. V. C. Polunin. Published by Cambridge University Press. © Foundation for Environmental Conservation 2008.

past, and thus in turn allows an appreciation of possible future states.

LARGE-SCALE PATTERNS: GLOBAL INTERPLAY OF DIVERSITY, PRODUCTIVITY AND DISTURBANCE

Kelp forest ecosystems are distributed on rocky shores at mid-latitudes (Fig. 15.1). Taxonomically, the kelps are not diverse. The southern California coast of North America has the highest diversity of kelp; however, this comprises only 20 species of kelp distributed among 16 genera, most genera being monotypic. While total species diversity among kelp forest ecosystems can vary among taxonomic groups (for example, the algal diversity is greatest in south Australia: Bolton 1996), the species diversity of functionally important groups such as herbivores and large carnivores is greatest in southern California. In fact, North America provides excellent case studies for comparison because it has both the world's lowest- and highest-diversity kelp forests along its western North Atlantic and southern California coasts, respectively (Fig. 15.2).

The physical structure, algal biomass and organisms associated with kelp forests profoundly alter local and adjacent environments and ecologies. Kelp canopies dampen waves, and this influences water flow and associated processes of coastal erosion, sedimentation, benthic primary and secondary productivity, and recruitment of associated animals (Duggins *et al.* 1990; Mork 1996; Jackson 1998; Lvas & Trum 2001). The canopies reduce light, creating understorey conditions favourable for a suite of species adapted to low light intensity (Santelices & Ojeda 1984*a*) and thus potentially influencing interspecific competition among algae (Dayton 1985*a*). Kelps are substratum for numerous sessile animals and algae (Duggins 1980; Reed & Foster 1984; Dunton & Schell 1987) and habitat for mobile organisms specialized in living and feeding directly on the kelp or its associated assemblages. For example, trophically specialized limpets depend upon kelp for their existence (Steneck & Watling 1982; Estes & Steinberg 1988; Bustamante *et al.* 1995). Kelp forest architecture provides habitat, nursery ground and food for myriad mobile pelagic and benthic organisms (Bernstein & Jung 1980; Bologna & Steneck 1993; Levin 1994). Few fish species

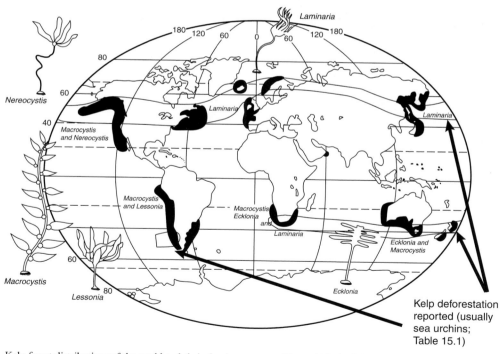

Fig. 15.1. Kelp forest distributions of the world and their dominant genera. Note mid-latitude bands in both hemispheres where kelp deforestation by sea urchins is common. (From Raffaelli & Hawkins 1996, modified by Steneck *et al.* 2002.)

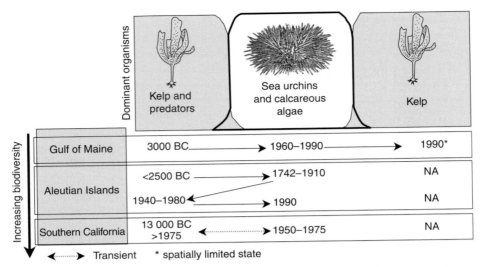

Fig. 15.2. Timing of phase changes in community state of kelp forests of North America. Kelp with vertebrate predators, sea urchins without kelp and kelp without predators have been identified for some or all of the case-study locations. Kelp forests are listed from the greatest number of trophic levels on the left to fewest trophic levels on the right. Case studies are listed from lowest species diversity in Maine to highest diversity in southern California. (From Steneck *et al.* 2002.)

feed directly in kelp (but see Andrew & Jones 1990; Jones & Andrew 1990); however, algal canopies serve as a nursery habitat for a range of rocky reef fishes feeding on microinvertebrates (Jones 1988; Carr 1989; Holbrook *et al.* 1990). Since predatory fishes use canopies as habitat, canopy loss can translate to increased survivorship of resident prey organisms and their larvae (Bray 1981; Gaines & Roughgarden 1987). Thus, if kelp canopies change, the ecosystem processes associated with them will be altered.

Kelps concentrate biomass and are a significant source of nutrition for coastal marine ecosystems via food webs based on macroalgal drift and detritus (Dunton & Schell 1987; Duggins *et al.* 1989). They commonly detach owing to the effects of storms and herbivores that damage kelp stipes, causing dislodgement. Usually, herbivores consume less than 10% of the living biomass (Mann 2000) thus leaving considerable detritus as food for benthic detritivores and microbes (Linley *et al.* 1981), which in turn make the carbon available to coastal suspension feeders. In effect, kelp forests concentrate and magnify secondary production, thereby supporting complex food webs in adjacent coastal zones (Duggins *et al.* 1989; Bustamante & Branch 1996; Mann 2000).

THE ECOLOGY OF KELP FORESTS

Kelp forests persist in a balance between ecological processes driving succession and stability (Dayton *et al.* 1984). They wax as a result of greater recruitment and net productivity, and wane with biomass loss from disturbances both physical and biological. Several strong ecological interactions control forest development and deforestation. The successional processes include spore dispersal, germination, gametogenesis, growth and intra- and interspecific competition. The deforestation also results from many ecological processes that can release populations of herbivores. All these processes operate at varying spatial and temporal scales and they will be the focus of this section, which will consider where kelp can live and develop forests, under what conditions kelp deforestation takes place, and what the out-of-system consequences of deforestation are.

Kelp forest distribution and development

Globally, kelp forests grow on shallow rocky shores in a mid-latitude band where light and oceanographic conditions allow the development and persistence of this growth form (Fig. 15.1). While kelps can grow in Arctic and sub-Antarctic regions (see Dunton & Dayton 1995), their abundance and diversity are low there (probably owing to light limitations: Dunton 1990; Henley & Dunton 1997) and thus they rarely develop forests above about 60° latitude. Similarly, warm temperatures and low nutrient concentrations generally prevent kelp forests from developing in subtropical or tropical regions (Bolton & Anderson 1987; Gerard 1997). The lowest-latitude kelp beds usually correspond with ocean-current anomalies in latitudinal gradients of warm temperatures and/or low nutrient conditions. For example, kelp forests are found near the tropics of Cancer and Capricorn only along the western coasts of southern California to Mexico, northern Chile to Peru and western South Africa (Fig. 15.1). In each of these cases, ocean currents, directed by Coriolis forces, flow toward the equator advecting cool and often nutrient-rich water. Low-latitude kelps (usually <40° latitude) are often diminutive and share or lose community dominance to fucoids such as *Sargassum* and other large brown algae that become more diverse and abundant toward the tropics (Hatcher *et al.* 1987).

In general, the three interacting processes that control the development of kelp forests are recruitment, growth and competition (North 1994). Locally, kelp forests are established and maintained by successful settlement of zygotes, which grow and are thinned by mortality from intraspecific competition during their benthic life (Dayton *et al.* 1984; Reed & Foster 1984; Chapman 1986). Recruitment is often seasonal and influenced by environmental conditions at the time of settlement. In complex kelp forests tiered with multiple levels (for example canopy, stipitate and prostrate forms: Steneck *et al.* 2002), such as the California kelp forest, kelp recruitment and growth is regulated by light available through breaks in the kelp canopy (Dayton *et al.* 1984; Reed & Foster 1984; Graham *et al.* 1997) as well as by available nutrients (Dayton *et al.* 1999). Following intense storms that deforest or thin kelp canopies, recruitment is usually strong, but the kelp species that grow to dominance will depend upon nutrient conditions at the time (North 1994; Tegner *et al.* 1997).

Kelp growth depends on interactions among nutrient availability, temperature and light. Kelps dominate coldwater coastal zones (Fig. 15.1) but can become physiologically stressed at high sea temperatures, particularly when nutrient availability is low (Tegner *et al.* 1996a; Gerard 1997). In some regions without upwelling, periods of low nutrient concentrations correspond with warm summer temperatures when the water is stratified (Chapter 14). The combined effects of low nutrients and high rates of respiration result in kelp plants that erode more rapidly than they grow (Gagne *et al.* 1982; R. S. Steneck, unpublished data 1984). In kelp forests driven by the upwelling of new nitrogen, such as those of southern California, warm surface water is a surrogate for low nutrient availability (Tegner *et al.* 1996a). In this system, when El Niño events disrupt coastal upwelling, kelp plants become nutrient-starved and die back (Tegner & Dayton 1991). As a result, the distribution, abundance and size of kelp plants decline as seasurface temperatures increase (Dayton *et al.* 1999).

Kelp deforestation

Widespread kelp deforestation can result from disease, herbivory and physiological stress, or interactions among those processes. In lower-latitude (usually <40°) kelp forests, periodic deforestation results from oceanographic anomalies in temperature, salinity or nutrients that either kill kelp directly or trigger diseases that become lethal to physiologically stressed plants. At mid-latitudes (c.40–60°), herbivory from sea urchins is the most common and important agent of kelp deforestation in both hemispheres. Latitudinal differences in patterns and processes shaping kelp forests have resulted in different researchers working in the same kelp forest system but reaching different conclusions (Foster 1990). The geography of kelp deforestation patterns and processes will be addressed here.

Kelp-free patches have probably always occurred at some scale, but those created by physical factors tend to be relatively small and short-lived. The oldest term for algal deforestation is the Japanese word *isoyake*, which literally means 'rock burning' (D. Fujita, personal communication 2002). The word was coined by Yendo (1902, 1903) to describe algal deforestation in coastal zones of central Japan where the algal decline was thought to have resulted from salinity anomalies (Yendo 1903, 1914) rather than grazing because herbivorous sea urchins were rare. The *isoyake* killed all foliacious algae first and then all encrusting coralline algae before recovering several years later (Yendo 1914). Other mass mortalities of *Ecklonia*- and *Eisenia*-dominated kelp forests resulted from incursions of the Kuroshio Current along the central Japan coast (D. Fujita, personal communication 2002). On Honshu Island, near the southern limit of Japanese kelp, anomalous incursions of the warm Tsushima Current periodically create *isoyake* conditions. Such oceanographically induced kelp deforestations are usually short-lived and reversible, as was the first described *isoyake* case in Japan (Yendo 1914).

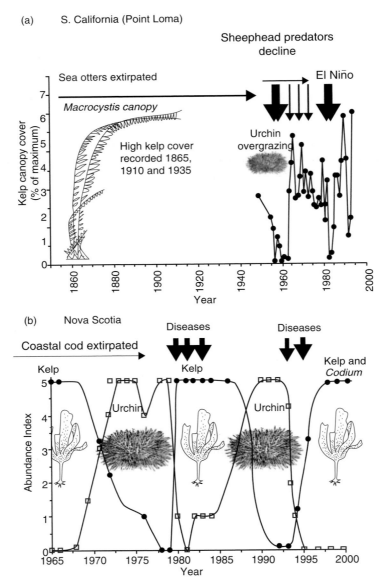

Fig. 15.3. Temporal trends in kelp forested and urchin-grazed deforested periods in (a) Southern California and (b) Nova Scotia. Arrows indicate the timing and magnitude of change in major community-altering forcing functions. Width of arrowheads indicates magnitude of the forcing function's impact. The strong fluctuations in Nova Scotia's sea urchin population result from periodic disease-induced mass mortalities. Kelp abundance fluctuates inversely with urchin abundance (Scheibling *et al.* 1999). (From Steneck *et al.* 2002.)

Kelp deforestations also result from El Niño events. Strong El Niños halt coastal upwelling of nutrient-rich water and cause surface waters to warm (Dayton *et al.* 1992; Chapter 14). These anomalies in California (Fig. 15.3a) have caused patchy deforestation followed by rapid recovery (Tegner & Dayton 1987; Dayton *et al.* 1992). Such physiological stresses are likely to be more common towards the low-latitude limits of kelp ranges. For example, the northern limit of three species of brown algae in northern Chile shifted south towards higher latitudes following the El Niño event of 1982/3 (Peters & Breeman 1993). Such stresses may make kelps more susceptible to disease. Low-latitude kelp in northern New Zealand succumbed to a disease, which may have resulted from a physiological stress (Cole & Babcock 1996; Cole & Syms 1999).

Within mid-latitudes, where kelp development is less likely to be limited by physical processes such as temperature, nutrients and light, deforestation most often results from sea-urchin herbivory (Leighton *et al.* 1966; Lawrence 1975; Himmelman 1980; Dayton 1985*a*, *b*; Duggins 1980; Estes & Duggins 1995; Mann 2000; Steneck *et al.* 2002). This is most evident in the northern hemisphere where the most widespread and long-lasting herbivore-induced kelp deforestations have resulted from sea-urchin grazing (Table 15.1). These primarily *Laminaria*-dominated kelp forests (Fig. 15.1) have been reduced in historical times to coralline-dominated 'urchin barrens' (urchin-induced kelp deforestation) in the Aleutian Islands of Alaska, the Gulf of Maine, Canadian Maritimes (to Newfoundland: Himmelman 1980), northern Japan (Hokkaido Island), Iceland and northern Norway (Table 15.1). South of those regions, forests either remain intact such as in southern California (Fig. 15.3a; Dayton *et al.* 1984) or are deforested patchily in relatively few regions such as southern Norway (Sivertsen 1997), Ireland (Kitching & Thain 1983), Britain (Kain 1975) and southern Japan (Honshu Island: Fujita 1998).

Kelp deforestation from sea urchins is less common in the southern hemisphere (Table 15.1). In Chile, there has been extensive sea-urchin deforestation in the latitude range 46°–54° S (Dayton 1985*b*), while only patchy deforestation has been reported in north central Chile (i.e. 32° S: Ojeda & Santelices 1984) (Table 15.1). Dense *Macrocystis* forests with few urchins have been described in the southernmost regions of South America (*c*.55° S), in Chile (Castilla & Moreno 1982; Santelices & Ojeda 1984*b*), and in Argentina (Dayton 1985*b*). While southernmost Chile has four sea urchin species (*Loxechinus albus, Pseudechinus magellanicus, Arbacia dufresnei* and *Austrocidaris canaliculata*), their densities are low and they subsist on drift kelp and rarely graze attached *Macrocystis* plants (Dayton 1985*b*). Along the east coast of southernmost South America to the northern limit of kelp in Argentina (42° S: Barrales & Lobban 1975), the sea urchin *Arbacia dufresnei* is the dominant herbivore but its grazing impact on kelp forests is minimal (Barrales & Lobban 1975).

Elsewhere in the southern hemisphere, sea-urchin deforestation is patchy or restricted to particular depth zones. In New Zealand, a band of sea urchin 'barrens' exists at intermediate depths at around 10 m (Choat & Schiel 1982). Urchins there may be prevented by wave turbulence from grazing shallower zones. In South Australia, kelp deforestation is restricted to regions of high spatial heterogeneity that provide shelter for nocturnally grazing sea urchins (Andrew 1993). Recent *Macrocystis pyrifera* deforestation in Tasmania, largely affected by the past El Niño events, has resulted in an area reduction of at least 50% over the last 10 years (Edyvane 2003). At the same time, high-density populations of the 'range-expanding' sea urchin *Centrostephanus rodgersii* have become established, increasing grazing pressure and forming extensive 'barrens', while the alien Japanese kelp *Undaria pinnatifida* now occupies regions of the former kelp forest habitat (Edyvane 2003; Valentine & Johnson 2003). In South Africa, sea urchins alone do not overgraze kelp forests (Velimirov *et al.* 1977), but grazer-induced phase shifts can occur from a diversity of herbivores (G. M. Branch, personal communication 2001).

Whatever regulates sea-urchin abundances or grazing behaviour also often controls the distribution and abundance of kelp forests (Steneck *et al.* 2002) (Table 15.1). Predators are commonly strong interactors (*sensu* Paine 1980) and, as such, are the single most important agent controlling sea-urchin populations (Dayton 1975; Cowen 1983; Duggins 1983; Tegner & Levin 1983; Estes & Duggins 1995; Sala *et al.* 1998; Steneck 1998; Shears & Babcock 2002). When sea-urchin predators are dramatically reduced or eliminated through intense and unsustainable fishing, subsequent hyperabundance of the sea urchins is common, and kelp deforestation results (Lawrence 1975; Estes & Duggins 1995; Steneck 1998). Exceptions to this pattern are found at lower latitudes where diverse guilds of kelp forest herbivores and predators compensate for the loss of single predator species. Sea-urchin abundance can also be influenced by other factors such as disease (Fujita 1998; Scheibling *et al.* 1999), turbulence (Choat & Schiel 1982) and storms (Dayton 1985*a*; Ebling *et al.* 1985) that can locally or periodically reduce sea-urchin abundance and thus control kelp forest development indirectly (Foster & Schiel 1985).

Widespread, long-lasting kelp deforestation from sea urchins may be a relatively recent phenomenon. In the Aleutians of Alaska, the transition may have occurred

Table 15.1. *Comparison of subtidal kelp forest ecosystems of the world*

Site and latitude	Kelps and their controlling agents[a]				Spatial and temporal scale of deforestation				References
	Dominant kelps	Deforesting herbivores	Predators or diseases of kelp herbivores	Drift kelp, disease, oceanography	Regional distribution	Depth range	Local distribution	Duration deforested	
Western North Atlantic									
Nova Scotia 43–45° N	*Laminaria* (1), *Agarum* (1)	Echinoid (1)	Fishes (2), urchin disease	Drift, disease	Widespread	Broad	Homogeneous	Decades	Steneck *et al.* (2002)
Maine 43–44° N	*Laminaria* (1), *Agarum* (1)	Echinoid (1)	Fishes (2), (crabs)		Widespread	Broad	Homogeneous	<Decades	Steneck *et al.* (2002)
Eastern North Atlantic									
North Iceland 65° N	*Laminaria* (1)	Echinoid (1)	?		Widespread	Broad	Homogeneous	?	Hjorleifsson *et al.* (1995)
North Norway 65–71° N	*Laminaria* (1)	Echinoid (2)	Seabirds		Widespread	Broad	Homogeneous	Decades	Hagen (1983), Sivertsen (1997), Bustnes *et al.* (1995)
South Norway 55–64° N	*Laminaria* (1)	Echinoid (1)	?		Restricted	Broad	Patchy	Decades	Sivertsen (1997)
Britain and Ireland 52–55° N	*Laminaria* (3)	Echinoids (2)	Crabs		Restricted	Broad	Patchy		Kain (1975), Kitching and Ebling (1961), Ebling *et al.* (1966)
Eastern North Pacific									
Alaska (Aleutians) 50–55° N	*Alaria* (1), *Laminaria* (3), *Thalassiophyllum* (1), *Agarum* (1)	Echinoid (1)	Sea otter		Widespread	Broad	Homogeneous	>Decades	Steneck *et al.* (2002)
Southern California 30–35° N	*Macrocystis* (1), *Laminaria* (1), *Pterygophora* (1)	Echinoids (3), gastropods (8), fishes (2)	Sea otter (1), fish (1), lobster (1)	Drift, oceanographic (ENSO) events	Restricted	Broad	Patchy	<Decade	Steneck *et al.* (2002)

Region	Dominant algae	Grazers	Predators	Drift / event	Spatial extent	Depth	Pattern	Time scale	References
Western North Pacific									
North Japan (SW Hokkaido) 39–46° N	*Laminaria* (2)	Echinoids (1–3)	Crabs, urchin disease	Oceanographic event	Widespread	Broad	Homogeneous	Decades	Fujita (1998), D. Fujita, personal communication (2001)
South Japan (W. Honshu) 36–38° N	*Undaria* (1), *Eisenia* (1), *Ecklonia* (1)	Echinoids (3), fish (1)	?	Oceanographic event	Restricted	Broad	Patchy	?	Fujita (1998), D. Fujita, personal communication (2001)
Eastern South Pacific									
North Chile 18–42° S	*Lessonia* (1), *Macrocystis* (1)	Echinoids (2), fishes (1), gastropods (2)	Asteroids (3), Fishes (3)	Drift	Widespread	Shallow	Patchy	Decades	Vasquez (1993), Vasquez and Buschmann (1997), Ojeda and Santilices (1984)
South Chile 46–54° S	*Macrocystis* (1), *Lessonia* (2)	Echinoids (1), gastropods (1)	Asteroids (1)	Drift	Restricted	Shallow	Patchy	Decades	Dayton (1985b)
Southern-most Chile 55° S	*Macrocystis* (1), *Lessonia* (2)	Echinoids (4)	Asteroids (1)	Drift	None				Castilla and Moreno (1982), Santelices and Ojeda (1984b), Vasquez et al. (1984)
Argentina 42–55° S	*Macrocystis* (1), *Lessonia* (1)	Echinoid (1)	?	?	None				Barrales and Lobban (1975)
Western South Pacific									
Australia (New South Wales) 32–35° S	*Ecklonia* (1)	Echinoids (1), fishes (1)	Fishes (2)		Widespread	Moderately deep	Patchy	?	Andrew (1993), Andrew and Underwood (1993), Andrew (1994), Andrew and O'Neill (2000)
Australia (Tasmania) 43° S	*Macrocystis* (1), *Ecklonia* (1)	Echinoid (1)	Fish (1), lobster (1)		Restricted	Broad	Homogeneous	Years	Edyvane (2003)

Table 15.1. (cont.)

Site and latitude	Kelps and their controlling agents[a]				Spatial and temporal scale of deforestation				References
	Dominant kelps	Deforesting herbivores	Predators or diseases of kelp herbivores	Drift kelp, disease, oceanography	Regional distribution	Depth range	Local distribution	Duration deforested	
New Zealand (North Island) 34–37° S	*Ecklonia* (1), *Lessonia* (1)	Echinoids (2), gastropods (2)	Fishes (?), lobster (1)	Kelp disease	Widespread	Mid-depth	Homogeneous	Decades	Choat and Schiel (1982), Andrew and Choat (1982), Choat and Ayling (1987), Schiel (1990), Cole and Babcock (1996), Babcock et al. (1999), Cole and Syms (1999)
New Zealand (South Island) 41–47° S	*Ecklonia* (1), *Lessonia* (1), *Macrocystis* (1)	Echinoids (1)	?		Restricted	Broad	Patchy	?	Schiel (1990), Schiel et al. (1995)
Eastern South Atlantic South Africa 30–35° S	*Ecklonia* (1), *Laminaria* (1), *Macrocystis* (1)	Echinoids (1), gastropods (1)	Lobster (1), fish (?)		Widespread	Deep only	Patchy	?	Anderson et al. (1997), G. Branch, personal communication (2001)
Eastern Indian Ocean W. Australia 28° S	*Ecklonia* (1)			Oceanography (high temperature and low nutrients)					Hatcher et al. (1987)

[a] Numbers in parentheses denote number of ecologically important species in subtidal kelp forests for specified taxa.
Source: After Steneck et al. (2002).

early in the twentieth century (Estes & Duggins 1995). In Japan, fishers observed deforestation and patches of corallines first in the early 1930s (Fujita 1987, 1998), although sea urchins were not mentioned at that time. Later, enlarging coralline patches grazed by sea urchins were reported during the 1950s and 1960s (Ohmi 1951; Fujita 1998). In California during the 1960s, the term 'barrens' was coined to describe urchin-induced kelp deforestation (Leighton *et al.* 1966). In the North Atlantic, the first gaps in kelp forests were reported in the 1960s for the Gulf of Maine (Lamb & Zimmerman 1964), Nova Scotia (Edelstein *et al.* 1969; Breen & Mann 1976), Ireland (Ebling *et al.* 1966) and the British Isles (Jones & Kain 1967). Gaps in kelp forests in the western North Atlantic coalesced and expanded during the 1970s and 1980s in Nova Scotia (Mann 1977) and the Gulf of Maine (Steneck 1997). Expansive coralline 'barrens' existed in Newfoundland in the late 1960s (Himmelman 1980) and were possibly present there earlier (Hooper 1980). In the eastern North Atlantic, widespread urchin-induced deforestation was first observed in northern Norway in the early 1980s (Hagen 1983; Sivertsen 1997) and in the early 1990s in Iceland (Hjorleifsson *et al.* 1995). By the mid 1970s, sea urchins were viewed as the major cause of kelp deforestation (Lawrence 1975) such that by the mid 1980s conferences were held to discuss (among other things) how sea urchins could be eradicated (Pringle *et al.* 1980; Bernstein & Welsford 1982). Urchin-induced kelp deforestation was widely reported in mid-latitudes of the northern hemisphere from 40° to 60° N (higher in the eastern North Atlantic owing to the Gulf Stream) during the 1960s and 1980s. At the time, some researchers openly wondered if kelp deforestation was an 'irreversible degradation' (Mann 1977).

Biodiversity, trophic cascades and rates and consequences of kelp deforestation

In the North American case studies, the extirpation of predators led to increased herbivory by sea urchins resulting in kelp deforestation at local to widespread spatial scales (Fig. 15.3b). In the western North Atlantic and Alaska, where predator diversity is low (Steneck *et al.* 2002), the transition between kelp forests and coralline communities was rapid, frequent (Fig. 15.3b), widespread and in some cases long-lasting (Tables 15.1 and 15.2). These patterns differ from southern California, where the diversity of predators, herbivores and kelps are high

(Steneck *et al.* 2002), deforestation events have been rare or patchy in space and short in duration (Harrold & Reed 1985), and no single dominant sea-urchin predator exists (Tegner & Dayton 1997) (Fig. 15.3a). The biodiversity within functional nodes, such as trophic levels, is critical to the structure and functioning of kelp forest ecosystems. Nevertheless, even the most diverse systems can lose, and are losing, their functional diversity as overfishing reduces the ecologically effective population densities of important species, rendering them ecologically extinct (Estes *et al.* 1989). All of this suggests that the fragility and rate of change in kelp forest ecosystems may depend on local biodiversity. It remains an open question whether diverse kelp forests will persist or if serial disassembly and instability will inevitably result.

The direct consequences of kelp deforestation among kelp-dependent communities will be profound. In southern Africa, where extensive forests of the kelps *Ecklonia maxima* and *Laminaria pallida* exist (Velimirov *et al.* 1977; Field *et al.* 1977, 1980) (Fig. 15.1), kelp is the principal primary producer contributing to the energetics of the shallow rocky subtidal food webs (see Field *et al.* 1977; Newell *et al.* 1980; Wickens & Field 1986). In addition, kelp decay plays a significant role for microheterotrophs through the regeneration of nutrients available to a range of subtidal producers and consumer species (for example Steele 1974; Fenchel & Blackburn 1979; Newell *et al.* 1988; Painting *et al.* 1992). Kelp deforestation can also affect surrounding marine and terrestrial communities (Graham *et al.* 2003, box 2). Drift from giant kelp (*Macrocystis pyrifera*) contributes 60–99% of beach-cast autotrophic detritus in the southern California bight (Zobell 1971) and similar amounts in eastern Nova Scotia (Mann 2000). Offshore contributions are facilitated by gas-filled floats and stipes which, when adult sporophytes are detached from the bottom due to grazing or physical disturbance, provide for long-distance dispersal by rafting (Harrold & Lisin 1989; Hobday 2000). When floating kelp rafts are deposited on the shore, the floats break, and they wash into shallow nearshore habitats and ultimately into offshore basins (Graham *et al.* 2003). Secondary productivity of both shallow (Vetter 1995) and deep-sea (Harrold *et al.* 1998) soft-sediment systems is consequently driven in a large part by allochthonous food subsidies from regional kelp resources (e.g. Chapters 17, 19 and 22). Kelp detritus can also make its way into nearby intertidal food webs through the capture either of fine kelp particles by filter-feeders such as mussels (Duggins *et al.* 1989) and clams

Table 15.2. *Spatial and temporal scale of change in kelp forest ecosystems in three regions of North America*

Event	North-west Atlantic	West Aleutian Islands	Southern California
Pristine state (prior to human contact)	Kelp forest	Kelp forest	Kelp forest
Scale (patch size)	100 > 500 km (Steneck *et al.* 2002)	200–1000 km (Estes *et al.* 1989)	<10 km (Harrold and Reed 1985 and Tegner *et al.* 1996a)
Human contact/occupation	10 000 ybp (Bourque 1995)	8000 ybp (Simenstad *et al.* 1978)	12 000–13 000 ybp (Erlandson *et al.* 1996)
Marine organisms in diet	5000 ybp (Bourque 1995)	4500 ybp (Simenstad *et al.* 1978)	11 600 ybp (Erlandson *et al.* 1996)
First known phase change	?–40 ybp (Adey 1964)	2500 ybp (Simenstad *et al.* 1978)	4000–6000 ybp (Erlandson *et al.* 1996, Erlandson *et al.* 2004)
First European contact/exploitation	460/400 ybp (Steneck 1997)	260/260 ybp (Simenstad *et al.* 1978)	460/200 ybp (Simenstad *et al.* 1978)
Kelp bed re-establishement rate	0.5–4 yrs (Steneck *et al.* 2002, Jones and Kain 1967)	2 yrs (Estes *et al.* 1989)	0.5 yrs (Harrold and Reed 1985 and Tegner *et al.* 1996a)
Alternate (kelp-free) state	Coralline/urchin	Coralline/urchin	Coralline/urchin
Storms remove dominant kelp	Small scale (Steneck *et al.* 2002)		Large scale (Dayton *et al.* 1992)
Recent apex predators	Crabs (Leland 2002)	Killer whales (Estes *et al.* 1998)	
Introduced competitors	Bryozoan and *Codium* (Lambert *et al.* 1992)		

ybp, years before present
Source: After Steneck *et al.* (2002).

(Soares *et al.* 1997) or of large pieces of drift kelp by limpets (Bustamante *et al.* 1995) and sea urchins (Day & Branch 2002; Rodriguez 2003; Chapter 14). All these kelp consumers are in turn strong interactors in their respective food webs and communities, and any perturbation to their populations will alter their community structures. Kelp detritus enhances the inherently low productivity of terrestrial ecosystems on arid islands (Polis & Hurd 1996). During dry years, carbon and nitrogen from marine bird and mammal faeces and beach-cast marine detritus fuel terrestrial productivity, with the greatest impact on islands with large shoreline-to-area ratios (Graham *et al.* 2003). The importance of marine subsidies declines during rainy years when high precipitation increases terrestrial production. Excellent examples are found in southern California where numerous low-productivity islands (such as the Channel Islands) are imbedded within a highly productive marine system. In addition to localized areas of high guano and pinniped excrement accumulation, the shoreline is loaded with large quantities of kelp detritus (Graham *et al.* 2003). All of this suggests that changes in the volume of organic material exported from kelp forests in temperate arid islands and coasts (for example western coasts of America and Africa) will probably affect the community structure and composition of intertidal rocky shores (Chapter 14), sandy beach infauna (Chapter 17), coastal food webs (e.g., Chapters 11–13) and benthic assemblages in deep-water canyons (Chapter 22).

Few have examined the consequences of declining kelp for the biodiversity of mobile organisms using these habitats. Responses of kelp-associated fishes indicate the potential ecological price that the 'habitat-user' pays for habitat degradation. While some fishes play key functional roles as ecosystem 'drivers', by far the greatest biodiversity

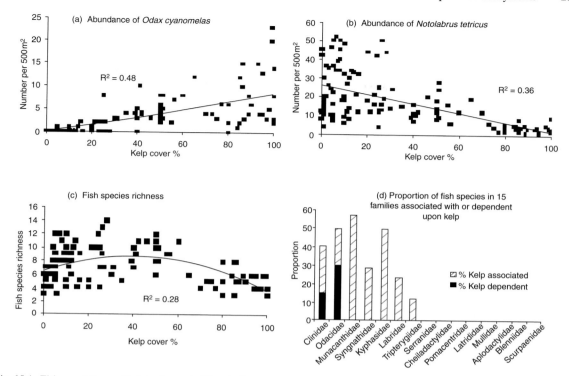

Fig. 15.4. Fish associations with kelp forests in Victoria (Australia): (a) kelp-feeding fish *Odax cyanomelas*, (b) turf-feeding wrasse *Notolabrus tetricus*, (c) fish species richness as a function of kelp cover and (d) proportion of fish species associated with or dependent upon kelp. (From G. P. Jones, unpublished data 2001.)

of fish species are ecosystem 'passengers'. Rocky-reef fish assemblages typically comprise species that are a mix of those associated with kelp stands and those associated with open reef areas, which are either urchin-grazed or dominated by algal turfs. The biogenic structure of habitats has a substantial influence on associated fish assemblages (Jones 1988; Carr 1989; Holbrook *et al.* 1990, 1993; Jones & Andrew 1993; Syms & Jones 1999), and the abundances of individual passenger species are often explained by the availability of their preferred habitats (see Kingett & Choat 1981; Bodkin 1988; Anderson 1994; Levin & Hay 1996; Hartney & Grorud 2002). For example, on the southern coast of Victoria (Australia) the abundance of the kelp-feeding fish *Odax cyanomelas* is positively correlated with kelp cover, while that of the wrasse *Notolabrus tetricus*, which is associated with algae turfs, is negatively correlated with kelp (Fig. 15.4 a, b). Processes that maintain spatial and temporal heterogeneity in kelp-dominated habitats will be vital to maintain fish biodiversity. Fish species richness

is often correlated with habitat complexity (Angel & Ojeda 2001; García-Charton & Pérez-Ruzafa 2001). In the case of southern Australia, greatest fish diversity occurs at sites of intermediate kelp cover where patch complexity is the greatest (Fig. 15.4c). Any process resulting in a homogeneous habitat structure, whether because of total deforestation or invasive species (Levin *et al.* 2002b), will be likely to have a negative impact on reef fish biodiversity (Harmelin-Vivien *et al.* 1999). Of 237 fish species across 15 families in southern Australia, only seven species (3%) may be totally dependent on kelp for food or living space, including one herbivore and a few species closely associated with kelp canopies (Fig. 15.4d). However, a further 41 species (17%) are strongly kelp-associated, using kelp as their primary recruitment habitat. Thus, while total deforestation will probably result in population declines and some local extinctions, the major effect of a phase shift from kelp-dominated to deforested habitat will be a phase shift in the structure of fish communities.

POTENTIAL STATES IN 2025

Extrapolation of known trends

It is likely that climate change, human population growth, coastal development, oil spills, fisheries-induced impacts, disease and invasions of non-native species will continue and possibly accelerate over the 2025 time horizon. All of these may well contribute to a continuing disassembly of kelp forest ecosystems. It is difficult to extrapolate known trends into the future, because non-linear thresholds and complex interactions can cause ecosystems to behave unpredictably (e.g. Scheffer *et al.* 2001). Some activities may change the nature of functional relationships. For example, as discussed, overfishing of sea-urchin predators can cause kelp forests to decline, but overfishing of sea urchins themselves can have the reverse effect.

It is tempting to speculate that if sea-urchin populations usually increase as their predators are extirpated, then perhaps the sea-urchin fishery exploitation will expand globally and reverse these trends. However, fishing pressure on sea urchins hinges on the marketability of the urchin roe. Since the primary market for sea-urchin roe is Japan along with a growing list of primarily Asian countries, often the only way harvesters can make a profit collecting these animals, which usually contain no more than 15% usable product by volume, is to receive a high price per unit mass. In the USA, the domestic price per unit mass would be too great to make the fishery viable. Thus foreign currency exchange rates functionally regulate fishing pressure on this driver of many kelp forest ecosystems. However, because it is impossible to predict global currency exchange rates, it will be equally difficult to predict the fishing pressure on sea urchins over the next several decades.

While global temperature is expected to increase to the year 2025, related patterns of droughts, fires, heatwaves, storms and precipitation are expected to increase in some areas but decline in others (Houghton *et al.* 1996). This underscores the limitations of attempting to generalize about kelp forest ecosystems globally. However, the three cases described offer insights into risks that may in future befall these ecosystems with widely varying diversities and scales in space and time. This review will now consider trends from the largest spatial and temporal scales to the smallest.

Ocean-climate change: global warming, regime shifts and ENSO

Ocean temperature regulates the physiology and biogeography of marine algae (Adey & Steneck 2001). Global warming, regime shifts, and El Niño–Southern Oscillation (ENSO) events are climate-driven thermal effects that can variably impact kelp forest ecosystems at a wide range of temporal and spatial scales. Global warming operates at the largest temporal and spatial scales, but the projected changes to the 2025 time horizon are modest compared to regime shifts at the scale of whole ocean basins. Regime shifts cause temperature fluctuations nearly an order of magnitude greater and persist at the temporal scale of decades (Steneck *et al.* 2002). Superimposed on both of those changes are ENSO events, which can cause the greatest temperature anomalies but impact coastal zones at smaller scales and over periods of only a year or two. The interaction among these three ocean climate effects is complicated because each varies differently in space and time. However, when two or more thermal anomalies coincide, the compounded perturbations to kelp forests can be very substantial (Paine *et al.* 1998).

The rate and magnitude of global temperature rise during the past century is unprecedented over at least the past 600 years (Houghton *et al.* 1996; Mann *et al.* 1998). As a group, kelps are limited to cold-water coastal zones (Fig. 15.1). Thus, kelp living close to their upper thermal limits will be likely to recede to higher latitudes during protracted warming periods.

Temperature effects are complicated by a variety of ocean–atmosphere interactions. While the long-term frequency of strong storms in the eastern North Pacific Ocean has changed relatively little since the year 1625 (Enfield 1988), recently the frequency and intensity of extreme cyclones has increased markedly (Graham & Diaz 2001). Two storms in the 1980s were described as 'storms of the century' (Seymour *et al.* 1989). The recent steady (rather than stepwise) increase in storm activity may be related to rising sea-surface temperatures in the western tropical Pacific (Graham & Diaz 2001). Thus, abiotic storm-induced disturbances of kelp forests could increase commensurately with global warming and would be likely to affect negatively the upper and nearshore local limits of kelps (Graham 1997). However, kelp forests usually recover rapidly from such disturbances (Tegner *et al.* 1997).

In 1982/3, a 4–5 °C ENSO warming halted coastal upwelling and created one of the most severe disturbances of a giant kelp forest ever documented (Paine *et al.* 1998) (Fig. 15.3a). Nitrogen concentrations in seawater influence the welfare of kelp and vary with ocean temperature (Gerard 1997). When surface waters become stratified owing to ENSO or as might be expected with global warming, nitrogen concentrations decline and kelps become nitrogen limited and may die (Dayton *et al.* 1999). *Macrocystis* beds are particularly vulnerable because they possess limited nitrogen storage capacity (Gerard 1982). Although groundwater and atmospheric inputs of nitrogen to coastal oceans have increased, the impact is modest or undetectable relative to natural fluxes in all but enclosed basins with limited flushing (Jickells 1998). Thus, trends towards increasing temperature and decreasing nitrogen availability in kelp forests are likely to continue.

The combination of warming coastal oceans and increased stratification will be likely to shrink the geographical range of kelp beds living closest to the tropics. Indirect effects of temperature on species that influence kelp forests will be discussed below. Sea-level rise of 20 cm by the year 2050 as a result of global warming (Houghton *et al.* 1996) could cause increased coastal erosion and turbidity which then could affect kelp depth distribution (Vadas & Steneck 1988) and possibly other processes such as photosynthesis and recruitment (see Graham 1996). Increased sedimentation from erosion could also reduce the area of substrate available for kelp settlement, as is currently postulated in Tasmanian kelp declines (Edyvane 2003).

Regime shifts and ENSO events can cause thermal anomalies of shorter duration but potentially greater impact. In general, ENSO events create warmer sea temperatures along the southern California coast, but the events are somewhat more frequent during regime shifts when warmer conditions prevail on that coast (Steneck *et al.* 2002). El Niño warming is usually followed by La Niña cooling (McGowan *et al.* 1998) and associated violent storms (Tegner & Dayton 1987, 1991; Seymoure *et al.* 1989; Tegner *et al.* 1997).

One of the best-studied climatic regime shifts occurred in the North Pacific in the mid 1970s, and the regime apparently ceased at the end of the 1990s (Steele 1998). If the duration of the post-1999 regime is similar to the last two, then cooler than average temperatures and perhaps lower than average El Niño frequency (and higher frequencies of La Niña and storm events) may be expected over much of the next quarter century (McGowan *et al.*

1998; Fei Chai, personal communication 2001). However, the duration of storm impacts in California is relatively brief and the ecosystem recovery rapid, so storms should have little lasting impact (Tegner *et al.* 1997) (Fig. 15.3a).

Changing coastal biodiversity: new apex predators and competitors

Changes in kelp forest biodiversity that affect functional components of kelp beds can disrupt the system in both predictable and unpredictable ways. The most conspicuous changes to the kelp forest result from overfishing key drivers such as apex predators and sea urchins (Estes & Duggins 1995; Jackson *et al.* 2001). Significant reductions in either can lead to trophic-level dysfunction, and alternate stable states or large-scale instabilities (Steneck *et al.* 2004). For example, extirpation of sea-urchin predators led to hyperabundances of sea urchins that had been stable for decades in Alaska (Steneck *et al.* 2002) and remarkably unstable in Nova Scotia (Fig. 15.3b) owing to epizootic disease cycles (Scheibling *et al.* 1999). Arguably, predator loss led to hyperabundances of species, setting the stage for disease-related mass mortalities (McNeill 1976).

Apex predators in pristine kelp forests were probably relatively large vertebrates in the northern hemisphere, while large invertebrates dominated many southern hemisphere kelp forests. The case studies show a consistent trend of fishing down food webs such that large vertebrates were often targeted and extirpated relatively rapidly. Today, fish are the most commonly identified predator, but sea otters and sea ducks are also important predators in some northern regions (Table 15.1). In all cases, these vertebrate predators are smaller in body size and/or fewer in number than they were 5000–10 000 years ago at first human contact (Steneck *et al.* 2002, 2004). Crustaceans such as spiny lobsters and crabs are among the most important invertebrate predators (Table 15.1). Extrapolating from the known trend in the Gulf of Maine, where extirpation of large predatory finfish led to the dominance of crabs as apex predators, it is possible that crab predation elsewhere is the result of a disrupted trophic cascade. Crabs are dominant predators of sea urchins in Japan and in the UK (Table 15.1), and both regions have a long history of fishing that targeted coastal groundfish. In Europe, coastal groundfish stocks were fought over in the thirteenth and fourteenth centuries, and their depletion is thought to have contributed to the development of distant-water fisheries in Iceland and eventually North America (Kurlansky 1997). It is possible that the very early extirpation of apex predators in Europe's coastal zones led to the rise of crabs as predators just as has more recently happened in the Gulf of Maine (Leland 2002).

Overfishing in kelp forests leads to ecological (Estes *et al.* 1989) and possibly absolute (Tegner *et al.* 1996b) extinctions. This loss in biodiversity may make these systems more susceptible to invasion from non-native species (Stachowicz *et al.* 1999). Recently, kelp forests in the western North Atlantic have been invaded by the green alga *Codium fragile*, an introduced competitor that could replace the kelp species in this region that have a long history of resilience and dominance (Chapman & Johnson 1990; Levin *et al.* 2002b) (Fig. 15.3b). The possible replacement of *Laminaria* by *Codium* requires sufficient breaks in the canopy for the latter to take hold. This has been facilitated by the introduction in the 1970s of a non-native encrusting bryozoan that coats, embrittles and opens the kelp canopy every summer (Lambert *et al.* 1992; Levin *et al.* 2002b). However, these two invaders join a long list of invading species that have become important players in kelp forest ecosystems of the western North Atlantic (Steneck & Carlton 2001). Species such as the common periwinkle (*Littorina littorea*), and the green and shore crabs (*Carcinus maenas, Hemigrapsus sanguineus*) have not only invaded but in many cases have come to dominate the ecosystem (Steneck & Carlton 2001). Whereas other marine systems have a history of invasion, few have seen the large-scale changes in dominance evident in the western North Atlantic. The successful series of invasions there stands in stark contrast with patterns observed in the species-rich southern California kelp forests, where introduced species remain subordinant to native dominants.

Declining water quality

Development on land often reduces the permeability of soil in the watershed, resulting in greater runoff and increased turbidity from plankton and particulates (e.g. Chapters 2, 13 and 16). Where this has occurred, kelp beds have declined because of the shallowing of the photic zone (Edyvane 2003; Valentine & Johnson 2003). Although nitrogen compounds could increase owing to increased supply from land runoff, sewage disposal and atmospheric sources globally, anthropogenic inputs of nutrients are evident only in areas of restricted water exchange. Thus most coastal waters 'appear still to be dominated by inputs from the open ocean' (Jickells 1998).

In heavily urbanized areas of Japan, terrestrial deforestation and damming of rivers is thought to have starved coastal waters of iron and humic substances necessary for kelp development (Suzuki *et al.* 1995; Matsunaga *et al.* 1999). This is hypothesized to have created a phase shift from kelp dominance to coralline dominance without any changes in herbivory, sea temperature or macronutrients (Matsunaga *et al.* 1999). There seem to be no other examples of chemically induced phase shifts from kelps to coralline algae.

Point-source pollution is often very conspicuous but rarely has it resulted in widespread and long-lasting deforestation of kelp ecosystems. In southern California, raw sewage in the 1950s resulted in the loss of kelps and the creation of sea-urchin 'barrens', but these and other impacts were short lived (North 1994) (Fig. 15.3a). The massive 1989 *Exxon Valdez* oil spill in the vicinity of kelp forests of south-west Alaska, however, minimally impacted kelps and, for most components of the ecosystem, full recovery took 2 years or less (Dean & Jewett 2001). Oil coated sea otters in the spill area and, while the impacts are debated, the greatest impact was a decrease from around five otters per km of shoreline to two or three otters per km (Paine *et al.* 1996), and otters may have suffered a long-term and lingering impact from oil in at least parts of the spill area in Prince William Sound (Monson *et al.* 2000). Although thousands of sea otters and seabirds died following the *Exxon Valdez* oil spill, fewer than 10 dead fish were reported (Paine *et al.* 1996). Thus, fish may be inherently less susceptible to oil spills, but are by no means immune.

Given the likelihood that both non-point-source and point-source pollution will rise with increased human population growth, water quality is expected to continue to decline. It is not yet known whether thresholds of accelerated mortality exist with declining water quality as they apparently do for increasing fishing pressure (Myers *et al.* 1997). However, it is currently top–down impacts on ecosystem drivers (such as sea urchins and their predators) that most consistently denude kelp forests.

CONCLUSIONS

Kelps are the largest bottom-dwelling organisms to occupy the euphotic zone. Their size and photosynthesis-to-biomass ratio constrain their distribution globally and locally. Kelp forests fail to develop at high latitudes as a result of light limitations and at low latitudes owing to nutrient limitations, high sea temperatures and competition from other macrophytes. On shallow mid-latitude rocky marine shores worldwide, phyletically diverse, structurally complex and highly productive kelp forests develop. These forests are uniquely capable of altering local oceanography and ecology by dampening wave surge, shading the sea floor with their canopies, providing a physical habitat for organisms above the benthic boundary layer and distributing trophic resources to surrounding habitats. In this context, the three kelp forest case studies from North America represent ecosystems along a continuum of natural biodiversity and human interactions. The archaeological literature may be used to estimate an ecological baseline of the structure and function of kelp forests prior to contact with modern humans (Steneck *et al.* 2002) (Fig. 15.2, Table 15.1).

Kelp forests of the eastern Pacific may have facilitated an early coastal migration of humans into the Americas (Erlandson 2002). The concentration and high productivity of vertebrates and invertebrates along this coast would have provided early human settlers with a stable source of food between 15 000 and 10 000 years ago. Archaeological data indicate that coastal settlements exploited organisms associated with kelp forests for thousands of years, occasionally resulting in the localized loss of apex predators, outbreaks of sea-urchin populations and deforestation (Erlandson *et al.* 2004). However, these human impacts on kelp-bed systems were probably localized and relatively ephemeral.

Consumer animals structure kelp forest interactions via two primary drivers, namely herbivory by sea urchins and carnivory from predators of sea urchins, but other forcing functions can be important. For example, kelps are prone to destruction and thinning by storms and competitors. Further, their growth and survival are sensitive to temperature, light and nutrient availability. However, the spatial scale and magnitude of these impacts on kelp forests are small relative to those of consumers. Kelp deforestation worldwide results from sea-urchin grazing, which is controlled by predation in kelp forests where human harvesting impacts have been minimal.

Over the past two centuries, the commercial export of kelp forest consumers led to the extirpation of sea-urchin predators such as the sea otter in the North Pacific and groundfish such as cod in the North Atlantic. In those systems, sea-urchin abundances increased and kelp forests were denuded over vast stretches of coast. In southern California, the high diversities of predators, herbivores and kelp appear to have buffered this system from systemic deforestation.

Biodiversity of kelp forests may also help the ecosystem resist invasion from non-native species. In the species-depauperate western North Atlantic, introduced algal competitors carpet the benthos and threaten the dominance of kelp. Other introduced herbivores and predators have taken hold and have come to dominate components of the system.

Global and regional climate changes have measurable impacts on kelp forest ecosystems. Increasing El Niño frequencies, oceanographic regime shifts and violent storms cause deforestation. This, in combination with the serial loss in biodiversity from overfishing, appears to be the greatest threat to structure and functioning of these systems over the 2025 time horizon.

Kelp forests cannot be considered in isolation. They have vital ecological linkages with deep-sea, subtidal, sandy and rocky intertidal and coastal ecosystems (e.g. Chapters 14, 17 and 22). The high interconnectedness and dependence on kelp organic exports highlight the significant trophic functions these plants play for adjacent ecosystems. This regional trophic connectivity magnifies the importance of kelps to communities well beyond their distribution. While kelp forest ecosystems should persist, they may do so with entirely altered food webs; for example, they may be devoid of key fish and invertebrate species. Some of these alterations will reduce stability not only of kelp forests, but also of neighbouring coastal and marine ecosystems.

Management of kelp forest ecosystems should focus on restoring biodiversity and especially on minimizing fishing on predators. In particular, northern hemisphere species such as sea otters, and the fish sheephead and cod should be restored to fulfil their functional role in the Alaska, California and western North Atlantic systems, respectively. While sea otters are already protected, other commercially valuable species such as Atlantic cod are unlikely to be preserved for the ecosystem role they perform. Southern hemisphere invertebrate predators such as lobsters and wrasses should also be restored to levels where they can function in their ecological roles. Ultimately, human values and political resolve will determine the conservation agenda. Significant investment in education of stakeholders, the general public and policy-makers will be necessary for this conservation goal to succeed.

16 · Projecting the current trajectory for coral reefs

TIM. R. MCCLANAHAN, ROBERT W. BUDDEMEIER, OVE HOEGH-GULDBERG
AND PAUL SAMMARCO

INTRODUCTION

Coral reefs are shallow-water marine ecosystems formed at the edge of the land and sea that provide numerous resources for millions of people. Reefs have persisted as an ecosystem although their species composition has changed subtly over time. Most of the present reef species originated about 1–10 million years before the present (Budd 2000). This time has been sufficient for multiple cycles of global climate change with an estimated 22 ice ages over the past 1.8 million years (Muller & MacDonald 1997). Past glacial cycles have been driven by changes in the Earth's eccentricity and obliquity, or Milankovitch frequencies, with atmospheric carbon dioxide concentrations lagging around 600 years behind the rise in temperature (Zachos *et al.* 2001). An unprecedented experiment involving raising atmospheric carbon dioxide independently of the Earth's natural orbital cycles is now being undertaken and we are seeing the initial ecological responses to this change.

Aquatic Ecosystems, ed. N. V. C. Polunin. Published by Cambridge University Press. © Foundation for Environmental Conservation 2008.

Coral reefs have persisted through changes in seawater temperature and sea level that have accompanied glacial and other cycles (Pandolfi 1996). Species extinction, however, has occurred over this period and has been associated with synergistic losses in habitat and climate change (Pandolfi 1999; Budd 2000). The Scleractinia (modern stony corals), for example, have undergone a major radiation since the Cretaceous 65 million years before present (BP), with species numbers increasing until recently (Veron 1995). Much of the recent reduction is considered to have been caused by changing climate conditions in areas such as the Caribbean.

The present atmosphere is one of the most carbon-dioxide-rich in recent geological history, with the current level being greater than at any time in the last 420 000 years. The level projected for the year 2100 is greater than that seen in the last 20 million years (Sandalow & Bowles 2001). The slow cooling that has been occurring over the past 50 million years (Lear et al. 2000) is being reversed by greenhouse gases such as carbon dioxide (Levitus et al. 2001), which has increased from about 280 to nearly 370 ppm by volume (ppmv) (IPCC [Intergovernmental Panel on Climate Change] 2001a). This has had a concurrent effect on the carbonate chemistry of surface ocean water (Kleypas et al. 1999a). Greenhouse-gas concentrations will continue to increase over the next 50–100 years; they are predicted to be 470–580 ppmv by 2050 (IPCC 2001a). The average temperature of the Earth has risen 0.4–0.8 °C since the late nineteenth century, and the warmest temperatures of the last 150 000 years were only about 1 °C above today's. Monsoon intensity is currently the strongest in the record of the past 1000 years (Anderson et al. 2002a) (Fig. 16.1a). By the year 2050, global average temperature is predicted to increase 1.0–2.2 °C, and global average sea level is expected to rise 0.15–0.18 m (IPCC 2001a). In addition, the rate of environmental change is likely to accelerate considerably in the near future. Multiple synergistic environmental disturbances are likely to interact to create stressful conditions unique to the current epoch; these include human population growth and resource use, warming seawater and the rising concentration of carbon dioxide in surface seawater.

Coral reefs are formed under conditions of warm water (>18 °C), high light intensity, high aragonite seawater saturation, stable full salinity (35‰), and low dissolved-nutrient concentrations (Table 16.1). The average natural conditions for reefs in the last few million years include a tropical surface temperature c.1 °C lower than the recent past, an atmosphere with about two-thirds of the present-day carbon dioxide concentration and a sea level 40–80 m below present (Fitt et al. 2001). This explains assertions that present-day corals live near the upper limit of their temperature tolerance (Fitt et al. 2001; Coles & Brown 2003). Understanding the past and present environmental influences is important because in future, coral-reef organisms and reef distributions will be affected by changes that may depend on how these environmental factors vary over the current age of global climate change. The accumulation of greenhouse gases is most likely to influence seawater temperatures, monsoon and El Niño–Southern Oscillation (ENSO) climate systems, and calcium carbonate saturation states in seawater. These longer-term changes in the climate and oceans are also influenced by human population growth and resource use. We will briefly discuss these environmental and human forcing factors, their current trends and likely influence on coral reef ecology by the year 2025 (McClanahan 2002a), and recommend areas for future research and management.

Environmental forcing factors

PHYSICOCHEMICAL FACTORS

The abundance and distribution of coral-reef organisms are influenced by physicochemical factors through their tolerance of variation in those factors. A summary of environmental factors collected from c.1000 reefs indicates that there is a significant range for each variable, and variations around the means are relatively minor (Kleypas et al. 1999b) (Table 16.1). For example, temperatures in both space and time range from 16 to 34 °C, but standard deviations around the mean values of maximum, minimum and average temperatures are <2 °C. This suggests that some variations are tolerated or adapted to, but that coral reefs occur within relatively well-defined ranges of ambient temperature, salinity, phosphorus level and aragonite saturation. With increasing variability in physicochemical factors such as temperature, coral communities have persisted but often with reduced species richness (Veron & Minchin 1992; McClanahan & Maina 2003) and changed ecological functions such as reef calcification and growth (Cortes 1993). Seawater nitrate concentration and light penetration are more variable, and high variation is more likely to be tolerated in these compared with factors having low variation. Depending on the factor that is changing, environmental variation will have variable consequences for reef ecology.

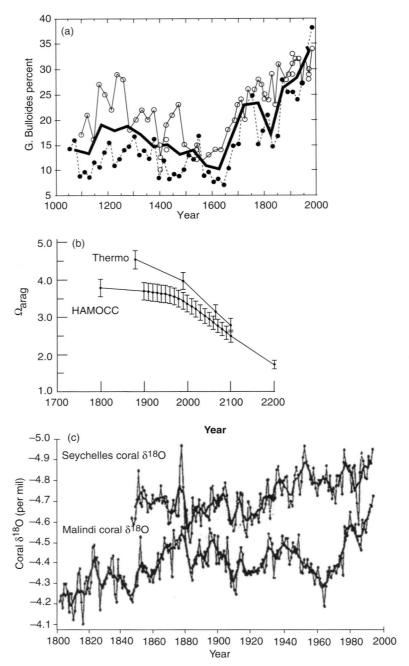

Fig. 16.1. Time-series graphs of (a) monsoon intensity (measured as G. Bulloides per cent) over the past 1000 years in the Indian Ocean (reproduced with permission from Anderson *et al.* 2002*b*); (b) estimated global aragonite saturation concentration (mean ± SD) of seawater (HAMOCC, Hamburg Ocean Carbon Cycle Model) (reproduced with permission from Kleypas *et al.* 1999*b*); (c) mean and running averages of seawater temperature trends in the Seychelles and Malindi (Kenya) indicated by $\delta^{18}O$ (oxygen isotope) data (reproduced with permission from Cole *et al.* 2000); and (d) the number of reef provinces reporting severe incidences of coral bleaching.

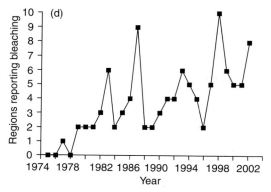

Fig. 16.1. (cont.)

Table 16.1. *Environmental variables, the minimum (Min), maximum (Max), mean and standard deviation (SD) of sites for a global database of c.1000 coral reef sites*

Variable	Min	Max	Mean	SD
Temperature (°C)				
Mean	21.0	29.5	27.6	1.1
Minimum	16.0	28.2	24.8	1.8
Maximum	24.7	34.4	30.2	0.6
Salinity (PSU)				
Minimum	23.3	40	34.3	1.2
Maximum	31.2	41.8	35.3	0.9
Nutrients ($\mu mol\ l^{-1}$)				
NO_3	0	3.34	0.25	0.28
PO_4	0	0.54	0.13	0.08
Aragonite saturation (Ω_{arag})				
Mean	3.28	4.06	3.83	0.09
Maximum depth of light penetration (m)				
Mean	−9	−81	−53	13.5
Minimum	−7	−72	−40	13.5
Maximum	−10	−91	−65	13.4

PSU, parts per thousand.
Source: Based on data from ReefBase (taken with permission from Kleypas *et al.* 1999*b*).

Temperature is often a good indicator of the latitudinal limits of reefs, but is poorer at predicting reef distributions within those limits. Differences in light and aragonite saturation state (Ω) may be more important within the latitudinal range of corals (Kleypas *et al.* 1999*a*, *b*). Aragonite saturation is a significant limitation on reef growth, since both corals and calcifying algae require aragonite-supersaturated seawater and, at least from short-term experiments, calcification rates appear to be proportional to the level of saturation (Gattuso *et al.* 1998; Langdon *et al.* 2000). Many of the large-scale regional differences in the distribution of reef communities may be influenced by the spatial variability of temperature and aragonite saturation. Reefs in low aragonite saturation or upwelling conditions grow slowly and are poorly cemented (Cortes 1993; Kleypas *et al.* 1999*b*), and less-dense calcification also occurs in areas of nutrient enrichment (Risk & Sammarco 1991). The eastern Pacific and high-latitude reefs have low aragonite saturation levels that reduce calcium carbonate production and reef formation.

One hundred years ago, mean aragonite saturation state in the tropics was 4.6 ± 0.2 (± 1 SD; $\Omega = 1$ indicates chemical equilibrium and values >1 indicate supersaturation); it is presently 4.0 ± 0.2, and is predicted to be 3.1±0.2 by 2065 (Kleypas *et al.* 1999*a*). These values indicate that aragonite and calcite precipitation rates on reefs should have already decreased by *c*.8% from pre-industrial levels, potentially resulting in a 25% decrease in calcification rates by the end of this century (Fig. 16.1b). There is some evidence to support this hypothesis from the Florida Keys (USA) (K. P. Helmle, G. M. Wellington, R. E. Dodge & P. K. Swart, personal communication 2000), but most multidecadal studies of coral cores indicate that calcification is actually unchanged or increasing (Lough & Barnes 1997; Carricart-Ganivet & Beltran-Torres 2000). Given that there is only a predicted 8% change in precipitation to the present, this may simply be due to high variance in the rates with relatively small changes in the mean. However, while most reef organisms deposit aragonite, others such as red coralline algae secrete a skeleton of high-magnesium calcite, which is more soluble than aragonite. These algae are more affected than corals by increased seawater acidity (Langdon 2002). Coralline algae are important for cementing reefs and corals, and their loss or low calcification can lead to brittle and poorly cemented reef structures (Cortes 1993).

Shipboard measurements and oxygen isotope chemistry of coral cores indicate that surface seawater temperature has increased by about 1.5 °C over the last 100 years, slightly higher than the Intergovernmental Panel on Climate Change (IPCC) global summary of 0.7–1.4 °C (Parker *et al.* 1995; Cole *et al.* 2000; Barnett *et al.* 2001; IPCC 2001*a*). Most of this warming has occurred since the mid 1970s (Fig. 16.1c). This indicates an accelerating temperature trend, but further field data are required to test this, as most models do not predict an exponential rise (IPCC 2001a). Seawater temperature is relatively cyclical on different timescales (Tudhope *et al.* 2001; Zachos *et al.* 2001). For example, during the last century, the ENSO in the Indian Ocean appeared to oscillate between 3.5-year and 5.3-year periods, with the shorter cycle dominant before the 1920s and after the 1960s, and the longer cycle dominant during the interim. Both cycles are contained within a decadal cycle of 11.8–12.3 years (Charles *et al.* 1997), associated with strong Indian Ocean dipole events (Saji *et al.* 1999). The most extreme warming may occur when decadal and ENSO oscillations are temporarily in phase. A lower-frequency oscillation that can influence global temperatures is the North Atlantic Oscillation (NAO) (Hurrell 1995). These oscillations will influence the temperatures on both a regional and global basis and complicate the ability to distinguish global and long-term from local and short-term oscillations.

The apparent predictability of ENSO-associated patterns is contradicted by evidence that both ENSO events and coral bleaching are spatially patchy and sometimes show surprising relationships. A number of 150–200-year temperature records from coral cores indicate strong connections between Indian Ocean and Pacific oscillations. However, there have been diversions from this association (Charles *et al.* 1997; Cole *et al.* 2000) (Fig. 16.1c) and differences among locations (Kuhnert *et al.* 1999). Despite the spatial and temporal variability, century-scale temperature records from the Indian Ocean are showing the predicted rise in water temperatures. Measured rises include 0.8 °C in a Seychelles core and approximately 1.2–1.4 °C in both western Australia and East African coral cores (Charles *et al.* 1997; Kuhnert *et al.* 1999; Cole *et al.* 2000). Similar trends exist in the tropical waters of the Great Barrier Reef and Pacific waters generally (Lough & Barnes 1997; Hoegh-Guldberg 1999). There are, however, differences to be expected between regions, latitudinal positions and hemispheres, as discussed below.

Sea level has been a critical influence on the development of reefs. Sea-level rise is associated with the global warming phenomenon through both the thermal expansion of surface seawater and the melting of the Antarctic ice cap. Sea level has risen *c*.0.15 m over the past 100 years and is predicted to rise a further 0.09–0.88 m by the year 2100

(IPCC 2001a). The maximum sustained rate of reef growth is *c*.10 mm per year (Buddemeier & Smith 1988); this should be sufficient for reef maintenance, especially if reef growth is stimulated by sea-level rise, but is likely to vary considerably with coral and algal species composition and other factors. The considerable variation around the estimated mean values, with possible reduction of calcification and increase in erosion, may mean that this net growth rate is not sustained everywhere. Historically, sea-level rise has had a positive influence on coral development, since it expands the area that corals can colonize (Pandolfi 1999), although there should be a loss of habitat in the deeper areas of a species' range as the shallows are populated. Nonetheless, reefs at high latitudes, such as Hawaii, have drowned during periods of rapid sea-level rise (Grigg 1997). Sea-level rise is an opportunity for the expansion of corals and calcifying organisms, but the current stresses and declining aragonite saturation state make this expansion unlikely. Increases in temperature as a result of climate change in concert with other anthropogenic factors are likely also to see a reduction in the health of corals (Hoegh-Guldberg 1999).

Human development of tropical coasts, combined with changed land and water use, associated river discharge and sediments and changed seawater salinity, can induce ecological changes in coral reefs. Sudden large drops in salinity (from 35 to 15–20‰ in 24 h) associated with heavy rains have caused coral bleaching and death (Jokiel *et al.* 1993), as has intense deposition of sediments (Rogers 1990). More modest and gradual changes in these factors are more common, and corals appear to have a greater ability to tolerate and recover from such changes (Moberg *et al.* 1997). For example, modest changes in taxonomic composition of reef biota in a marine protected area were attributed to the pulsed or seasonal nature of the river discharge and lack of other interactive effects, such as fishing (McClanahan & Obura 1997). Warming from the mid 1970s has increased evaporative flux in the tropics by about 5% (Graham 1995), and global warming is expected to further increase the tropical hydrologic cycle (Chen *et al.* 2002).

Associated with runoff from land are substances, some of which are important for growth, such as phosphorus, iron and nitrogen (Dubinsky & Stambler 1996). Such nutrients improve conditions for fast-growing algae with high organic but little or no calcium carbonate productivity (Littler *et al.* 1991; Lapointe 1999; Ferrier-Pagès *et al.* 2000). Some algae can competitively exclude the slower-growing corals, which produce more inorganic carbon (Hughes & Tanner 2000), and this may lead to reefs that no longer appreciably accrete calcium carbonate. A third concern is that nutrients favour infaunal organisms that erode reefs (Highsmith 1980; Rose & Risk 1985; Risk *et al.* 1995; Holmes *et al.* 2000). Phosphorus and nitrogen may not alone increase the abundance of microboring organisms (Koop *et al.* 2001) but may interact with herbivory (Chazottes *et al.* 1995; Carreiro-Silva *et al.* 2005). It is likely that nutrients and bioeroder abundance are influenced by high levels of organic matter associated with planktonic productivity (Highsmith 1980), and water-column carbon often associated with inorganic nutrients. Reefs in waters rich in plankton and organic matter are therefore likely to have high erosion rates and lower net reef accretion.

BIOLOGICAL FACTORS

Biological interactions on reefs also determine species abundances and the production rates of both organic and inorganic carbon. For example, the grazing of algae by herbivorous fishes has been suggested to control net algal production on coral reefs and is seen by some investigators as a greater influence on algal growth than nutrients or light in the Caribbean (Hatcher 1983; Carpenter 1988; Miller *et al.* 1999), certainly since the die-off of the long-spined sea urchin *Diadema antillarum* in the early 1980s (Lessios 1988; Hughes 1994). Prior to this, algal control was primarily exerted by grazing echinoids (Sammarco 1982). Grazing also encourages the more herbivore-resistant calcifying algae (Lewis 1986) and mediates competition between corals and erect fleshy algae (Tanner 1995); therefore herbivores indirectly influence reef accretion. The balance between calcification and erosion is critical to the net growth of reefs, and both of these are influenced by biological interactions.

Reefs are rich in predator–prey relationships. In particular, fish often influence the abundance of invertebrates (McClanahan *et al.* 1999) and plant species (Sammarco 1982; Hay *et al.* 1983). Therefore, fishing can influence these interactions, the ecological processes of which they are a part and reef production. By decreasing herbivorous and invertebrate-eating fishes, fishing can also indirectly influence reef processes of calcification and erosion (Risk & Sammarco 1982; Sammarco *et al.* 1987). More obvious forms of resource degradation are achieved through destructive fishing methods such as drag nets, explosives, poisons and 'ball-and-chain' techniques, which have become increasingly common (Muthiga *et al.* 2000).

Determining and ranking the importance of each of the above factors (Table 16.2) in coral reefs is difficult owing to lack of research, lack of consensus among reef scientists, regional variation and a continual unfolding of observations on reef dynamics during this period of rapid environmental change. Until recently, most reef scientists agreed that heavy fishing and organic pollution were the two dominant environmental problems facing coral reefs (Ginsburg 1994). These factors affect reef health; diseases and warming sea-surface temperatures (SSTs) are also devastating reefs regionally and globally (Goreau *et al.* 2000; Aronson & Precht 2001). Diseases appear to be increasing, and their rapid spread can cause large changes in reef communities (Porter *et al.* 2001). Diseases that have drastically changed Caribbean reef ecology include that which killed most of *D. antillarum* in the early 1980s (Lessios 1988), and that which killed most of the *Acropora* spp. in the mid 1980s (Aronson & Precht 2001). Patterns of human natural-resource use and waste disposal have constituted major problems and will continue to do so, but this trend may become less pronounced as resources decline in abundance and greater efforts are made to protect the environment. Nonetheless, stresses associated with global warming and marine diseases are likely to increase in frequency and intensity with time, and be more difficult to control.

IDENTIFIED TRENDS

Few regions can presently be claimed to possess 'pristine' coral reefs (Jackson 1997; Hughes *et al.* 2003; Pandolfi *et al.* 2003). The development of scuba, scientific diving and a heightened interest in the marine environment have provided a view of current trends in coral-reef ecology and degradation; however, information from before the 1960s is patchy and incomplete making a historical outline difficult to construct (Jackson 1997; Gardner *et al.* 2003). Here, some trends in each of the major regions will be briefly reviewed and an overview given of global environmental trends as a basis for projections to the year 2025.

Caribbean and Atlantic

The Caribbean and Atlantic contain only about 9% of the Earth's coral reefs (Bryant *et al.* 1998), but have some of the best-studied reefs owing to a number of permanent marine stations, a relatively benign working environment and accessibility to many visiting scientists. The reefs are under moderate direct anthropogenic threat and have

Table 16.2. *Important environmental factors likely to influence coral reefs and their projected direction and relative rate of change to the year 2025*

Environmental factor	Direction of change	Relative rate of change	Explanation
Human population growth and movement	+	−	Expected to increase, but at an increasingly slower rate for the next 50 years
Fishing (overfishing)	+	−	Should follow current population trends
Habitat destructive fishing	+	+	Competition for dwindling fish resources will increase destructive fishing
Coral mining and collection	−	−	Replacement of coral with cement and other synthetic materials
Tourism (anchoring and coral contact)	+	−	Expected to rise, but level with the expendable income of developed nations
Vessel grounding	+	−	Increase with ship traffic, but should decrease with improved navigation technology and lawsuits associated with grounding
Waste emissions	+	−	Expected to follow human population growth, but may be reduced faster with limiting resources and improved recycling
Coastal development, runoff, sediments and changing salinity	+	−	Continued deforestation and high-intensity agriculture and pastoralism; in some cases increased damming may reduce these factors
Herbicides and pesticides	+	−	Expected to follow increased intensification of land use
Industrial and urban pollution	+	−	Expected to follow human population growth
Oil pollution	−	−	Improved tanker design and public pressure
Greenhouse gases	+	−	Expected to follow energy consumption
Xenobiotics and diseases	+	+	Increased movement of people, boats and atmospheric circulation with increased soil erosion
Environmental education	+	+	Increased awareness of environmental problems and recognition of the importance of education and community support for conservation
Laws, regulations and enforcement	+	+	Increased recognition of the importance of restrictions on uncontrolled resource use
Pest populations (coral-eating invertebrates and sea urchins)	+	−	Overfishing will change ecological state towards dominance by unused species with strong competitive ability
Algal overgrowth	+	−	Loss of higher trophic levels will improve conditions for fast-growing lower trophic levels
Erosion of reefs	+	+	Combination of increased organic pollution and reef-eroding pests
Warm water	+	−	Expected to follow greenhouse gas emissions
Seawater acidity	+	−	Expected to follow greenhouse gas emissions
Hurricane and storm damage	+	+	Expected to follow greenhouse gas emissions
Sea level	+	+	Expected to follow greenhouse gas emissions
Coral bleaching	+	+	Expected to follow greenhouse gas emissions
El Niño frequency	0	0	Presently at maximum frequency
El Niño intensity	+	+	Expected to follow global temperatures

Note: +, positive, increased rate of change; −, negative, decreased rate of change; 0, unchanged.

moderate coastal human population densities of *c*.64 persons per km^2 (Bryant *et al.* 1998), but coral cover has declined by 80% in the past 30 years (Gardner *et al.* 2003), whilst turf and erect fleshy algae and octocorals have become dominant since the mid 1980s (Sammarco 1996; Ostrander *et al.* 2000; Gardner *et al.* 2003; Shinn *et al.* 2003) (Fig. 16.2). Approximately five hypotheses have been proposed to explain this shift (Shinn *et al.* 2000; Lessios *et al.* 2001; Porter *et al.* 2001).

(1) The shift was caused by a disease-induced pan-Caribbean loss of an important algae grazer, the sea urchin *D. antillarum*, in the early 1980s. The pathogen that caused the mass mortality has never been identified. *Diadema antillarum* may have been naturally abundant (Lessios *et al.* 2001), but some investigators believe it was unnaturally abundant (>5 per m^2) owing to heavy fishing that reduced its predators (Hay 1984; Hughes 1994). This is consistent with the process of 'fishing down' food webs (Pauly *et al.* 1998). Fishing is hypothesized to have reduced important herbivorous fishes such as parrotfish and, when *D. antillarum* died, there were few large herbivores to compensate for their loss. The result was massive overgrowth of reefs by erect fleshy algae. Few reefs have shown signs of coral community recovery since this change (Connell 1997; Gardner *et al.* 2003), unless *D. antillarum* has also recovered (Edmunds & Carpenter 2001).

(2) High levels of nutrients were largely or partially responsible for the shift (Lapointe 1999). Many of the large losses of live coral cover in coastal waters have been associated with nutrient enrichment from runoff.

(3) A loss of the common coral *Acropora cervicornis* induced by white-band disease affected many Caribbean reefs. This occurred shortly after the mass mortality of *D. antillarum*, opening up additional space for colonization by algae (Aronson & Precht 2001).

(4) Coral bleaching and associated coral mortality opened up space that was colonized by algae (Ostrander *et al.* 2000) and these escaped intensive grazing (Williams & Polunin 2001).

(5) Diseases and coral mortality may have been caused by dust carrying pathogens from West Africa via high-altitude transoceanic equatorial winds. Introduction of unusual pathogens could have resulted from extended droughts in northern Africa and desertification of the Sahara Sahel (Shinn *et al.* 2000).

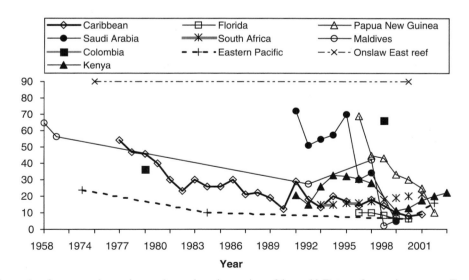

Fig. 16.2. Time series of representative coral cover changes in various regions of the world. Data are from various sources: Eastern Pacific (Wellington & Glynn 2007; sites with small sample size removed), Caribbean (Gardner *et al.* 2003), South Africa (Schleyer & Celliers 2002), Maldives (McClanahan 2000*b*; Zahir 2002), Kenya (T. R. McClanahan unpublished data), Florida (Causey *et al.* 2002), Saudi Arabia (Wilson *et al.* 2002) and Papua New Guinea (Sweatman *et al.* 2002).

Any of the above may have contributed to the observed shift. However, algal dominance has also been identified in remote areas where *A. cervicornis* was rare or nutrients were unlikely to have increased over time (Ostrander *et al.* 2000). Small-scale experimental studies indicate that herbivory still promotes the growth of erect frondose algae (McClanahan *et al.* 2002*a*). Loss of coral can lead to algal dominance; when additional space is created and herbivory is stable algal biomass can accumulate until herbivory increases to compensate for this additional space and resource (Williams *et al.* 2001). Coral bleaching and mortality are often associated with warm calm periods (Goreau *et al.* 2000), and the loss of coral makes space available for rapidly colonizing organisms (Ostrander *et al.* 2000). The combination of continuing coral mortality through bleaching and disease (Richardson 1998), low herbivory and nutrient enrichment has probably interacted to promote erect-algal dominance. These changes are not restricted to a few reefs heavily influenced by humans; they have occurred on reefs with various levels of human influence throughout the region (Hughes 1994; Ostrander *et al.* 2000; Williams & Polunin 2001).

Eastern Pacific

Eastern Pacific reefs occupy a small percentage of the Earth's total reef area but represent an extreme environment for coral reefs. Reefs in this region are typically of low diversity with small continental shelf area, experience seasonal coastal upwelling and are isolated from centres of diversity (Glynn & Ault 2000). Modern reefs originated <6000 years ago, occur patchily and are considered to possess a relict fauna of Tethyan species (Glynn & Ault 2000). Forty-one species of zooxanthellate scleractinian corals and hydrocorals inhabit this region, and many coastlines have yet to be explored (Glynn & Ault 2000). Hermatypic corals are generally restricted to one or a few species, and skeletal growth rates can be similar to those of reefs in other regions (Cortes *et al.* 1994). The structural integrity of these reefs is however low, and submarine cementation is rare. Crustose coralline algae produce poorly developed pavements and bioerosion is high (Cortes 1993). The low aragonite saturation of the seawater ($\Omega = 3.0$–4.0) and high abundance of bioeroding organisms associated with cool plankton-rich waters are probably at least partially responsible for this.

All of the above factors make these reefs vulnerable to environmental change, and the vulnerability is exacerbated by their isolation from the larger and more diverse western Pacific fauna, a narrow continental shelf, the recent origin of the reefs and low species richness (Glynn 2000). The 1982/3 El Niño seriously disturbed many of these reefs (Fig. 16.2). Reefs in areas such as the Galapagos have largely disappeared (Glynn 1994) because of extreme coral mortality and rapid erosion of the carbonate framework by sea urchins (Eakin 1996; Reaka-Kudla *et al.* 1996; Glynn 2000).

The 1982/3 El Niño was considered a rare event at the time, but it was followed by another event in 1998 that produced less coral mortality (Glynn *et al.* 2001). The precise conditions of the 1982/3 and 1998 events are only partially known. Some species that nearly became locally extinct after the 1982/3 El Niño were indeed driven extinct by the 1998 event. One site-restricted species, *Acropora valida*, and three species of *Millepora* (*M. platyphylla*, *M. boschmai* and *M. intricata*) are now either locally extinct or represented by only a few remaining colonies (Glynn & Ault 2000), while another six coral species have declined dramatically in abundance (Glynn 2000). These species were not the dominant reef-building corals, but the crown-of-thorns starfish (*Acanthaster planci*) threatens the remaining reef-builders (Fong & Glynn 1998, 2000).

The Pacific, Asia and Australia

The Australian and Asian Pacific regions contain *c.*70% of the Earth's reefs (Bryant *et al.* 1998) and possess the highest coral species diversity. They also possess the greatest number and diversity of reef management institutions, ranging from traditional and localized tenure, common in many Pacific islands including Papua New Guinea, to complex national zoning programmes, currently used on Australia's Great Barrier Reef. Human population densities vary widely in this region; some regions have low populations (central Pacific and Australia) and others have some of the highest coastal densities (Indonesia, South-East Asia). Local threats vary from extremely high (South-East Asia) to low (Central Pacific and Australia) (Bryant *et al.* 1998).

In the mid 1990s, South-East Asia, with $68\,100$ km^2 (a quarter of the Earth's) reef area, was identified as the region with the greatest frequency of threats; 56% of the reefs are in the 'high threat' category (Bryant *et al.* 1998). Destructive fishing is one of the most serious local-scale threats. Techniques include dynamite fishing, cyanide fishing, ball-and-chain fishing and other methods that break coral in order to chase out fish for the live-fish food trade (McManus *et al.* 2000). Pollution and dredging of the sea floor are intense around major cities like Hong Kong, Jakarta and Manila (Morton 1996; Tomascik *et al.* 1997). Extensive deforestation has led to high levels of sedimentation in many near-shore areas (Hodgson & Dixon 1992). There has also been large-scale removal of coastal mangroves for wood pulp, aquaculture ponds, tidal agriculture and plantations. For example, since the 1980s, mangrove cover has been reduced by 60% in Indonesia and by 90% along the coast of Java (Tomascik *et al.* 1997). The net result is high levels of sediment and organic matter in the water column that reduce light penetration, coral diversity and the coral depth distribution (Edinger *et al.* 1998). Based on threats to biodiversity, the Philippines, Sunda Islands (Borneo, Sumatra, Java, Celebes and Timor) and southern Japan are three areas where conservation activities are urgently required (Roberts *et al.* 2002).

Seventy per cent of the Australian subregion reef (which contains 48 000 km^2 or 20% of the Earth's reef area) are in the 'low threat' category (Bryant *et al.* 1998), largely because of low coastal population density (12 people per km^2 compared to 128 people per km^2 in South-East Asia) and many marine protected areas (including associated habitats in addition to coral reefs: 374 967 km^2 as compared to 36 263 km^2 in South-East Asia). Nonetheless, outbreaks of the coral-eating starfish *A. planci* occur periodically on the Great Barrier Reef and other reefs in this region (Moran 1986), and greatly reduce coral cover (Done 1992). The outbreaks on the Great Barrier Reef appear to occur at a frequency of 10–15 years, which approximates the rate required for recovery of certain affected corals (Done 1987).

Outbreaks of *A. planci* in the western Pacific are attributed to natural predator–prey cycles (Bradbury *et al.* 1985), disturbance generated by the loss of predators (Ormond *et al.* 1988) and increased larval survivorship owing to increased inorganic nutrient levels in the water column (Birkeland 1982). Predator–prey cycles may also have been dampened, leading to the loss of both coral and starfish abundance (Bradbury & Seymour 1997). The dampening cycle may be accelerated when other factors such as coral bleaching, pollution or dynamite fishing increase coral mortality or slow recovery.

Poor water quality in part of the Australasian region is considered to represent a major threat, especially to inshore reef systems (D. M. Williams 2001). Changing patterns of land use within river catchments (McCulloch *et al.* 2003) have led to major changes in sediments, nutrients, herbicides and pesticides originating from the land. As a result, coral communities have disappeared from inshore regions or have undergone major changes in distribution and abundance (D. M. Williams 2001).

Coral bleaching has occurred through most of the region and much of the coral mortality reported may be because of these warm-water events (Hoegh-Guldberg 1999; Goreau *et al.* 2000). The interaction of multiple sources of coral mortality and predator–prey impacts is likely to produce complex responses to global climate change (Hughes *et al.* 2003), but the island geography, associated oceanography and high species diversity in this region may provide greater resilience to further ecological degradation. The changes to the zoning of the Great Barrier Reef Marine Park (with an increase in no-take areas from 4.6% to 32.5%) will serve to increase ecological resilience to threats like climate change that will drive major changes, even in relatively unpopulated regions such as Australia.

Western Indian Ocean

Western Indian Ocean and Red Sea reefs contain *c.*14% of the Earth's reefs, most of which are fringing reefs off Africa and Arabia and on oceanic islands, notably the Maldives, Chagos, Seychelles and Mascarenes (Bryant *et al.* 1998; McClanahan *et al.* 2000). Fishing pressure is intense in many areas of Africa, notably southern Kenya to northern Tanzania, western Madagascar and some of the more populated islands of the Seychelles and Mascarenes (McClanahan *et al.* 2000). Destructive fishing techniques, such as drag nets with small mesh and dynamite fishing, are common in these areas (McClanahan & Mangi 2001). Fish catches in these areas rarely exceed 4 kg per fisher per day (McClanahan & Mangi 2001). In contrast, fisherfolk in a number of areas, such as the Maldives, Chagos and some areas of Arabia, use highly selective fishing techniques (Medio *et al.* 2000; Sheppard 2000). Catches in the Chagos are typically around 50 kg per person per day (Mees *et al.* 1999). Nonetheless, even the most remote reefs of the Chagos are showing effects of shark overfishing (Sheppard 2000). Pollution is relatively slight and occurs generally around moderately sized coastal cities. In the Arabian and Red Sea region, oil spills such as those around oil depots (Loya & Rinkevich 1980) are a source of damage. The Gulf of Oman has among the highest reported levels of beach tar concentrations (Coles & Al-Riyami 1996).

The largest source of damage to Western Indian Ocean reefs in recent time, however, has been coral bleaching associated with periods of warm and still water or warm El Niño conditions (Goreau *et al.* 2000) (Fig. 16.2). Bleaching occurred on these reefs in 1987/8 and 1994/5, and the largest and most intense bleaching in 1997/8 affected most of the region except the northern Red Sea and South Africa, producing coral mortality of 40–99% (Goreau *et al.* 2000). This mortality was most pronounced on some of the most pristine and undisturbed reefs, characterized by abundant, bleaching-sensitive branching coral genera such as *Acropora*, *Stylophora*, *Pocillopora*, *Seriatopora* and branching *Millepora* (McClanahan *et al.* 2001). In addition, soft corals were greatly damaged and benthic algae increased on many of these reefs (McClanahan *et al.* 2001). The initial effects on the fish were less detrimental than this; some herbivorous fish increased, but losses of butterflyfish, wrasses and damselfish occurred in Kenya over a 3-year period (McClanahan *et al.* 2002*b*). This extensive bleaching damage is probably greater than all the other current sources of coral mortality.

TRAJECTORY OF CORAL REEFS TO 2025

An essential part of the scientific learning process is to construct hypotheses that can be disproved (Popper 1972). Below is a series of hypotheses or predictions, based on current coral-reef science, followed by a more detailed rationale, description and caveats in the text. The predictions are ecologically conservative in that they are made to increase awareness and future monitoring of potentially detrimental ecological changes and to provoke the application of the precautionary principle (O'Riordan & Cameron 1994). We present a number of predictions focusing on external forcing factors (Table 16.2) and potential outcomes for important components of the coral-reef ecosystem (Table 16.3). The effects of each factor will greatly depend on the position of individual reefs in terms of latitude, proximity to shore and human populations, and position within an ocean basin.

Physicochemical factors

CLIMATIC OSCILLATIONS AND DISTURBANCES
Recognition of the importance of moderate-scale oscillations in climate and oceanographic conditions including

the ENSO, Pacific and Indian Ocean Decadal Oscillations and NAO, and their influence on reefs is a reasonable basis for predictions up to the 2025 time horizon. The smallest-scale oscillations, such as the Indian Ocean biannual oscillations and El Niño, are likely to be repeated frequently during the next 20 years. Changes in their frequency and intensity will result in measurable changes in reef ecology. Therefore, an increasing intensity in the intra-annual monsoon oscillation (Chen *et al.* 2002) is likely to have one of the greatest and most predictable influences on reefs in equatorial areas up to 2025.

Isotope studies of recent and ancient corals indicate that ENSO has existed for at least the past 130 000 years, and ENSO events of the past 120 years and of a period in the seventeenth century are among the most frequent and intense on record (Tudhope *et al.* 2001; Cobb *et al.* 2003). ENSO intensity and variability are correlated with each other, and the weakest and least variable events are found during glacial periods. During the past 100 years, the ENSO has exhibited an oscillation frequency of 3.5–5.3 years. Prior to 7000 BP, ENSO periodicity was approximately 15 years (Rodbell *et al.* 1999) and, in the nineteenth century, the oscillation was approximately 10 years (Urban *et al.* 2000). Consequently, the current frequency may be at its known maximum, and may stay at this high frequency in the near future. Nevertheless, the intensity and variation of oscillations may increase as judged by the intensity of the 1983 and 1998 oscillations relative to the historical record (Tudhope *et al.* 2001). The intensity of El Niño need not change for bleaching events to be associated with El Niño events, if the underlying sea temperature continues to increase over time. The interaction of increasing El Niño frequency and rising temperatures will have an additive or multiplicative effect on the frequency of bleaching events (Hoegh-Guldberg 1999).

Oscillations with a frequency equal to or greater than a decade will only repeat themselves once or twice in the coming quarter century and their effects may be less easy to predict. The Indian and Pacific Oceans' near-decadal cycles are likely to produce two more warm periods in the coming quarter century. These may coincide with ENSO events and produce some of the warmest conditions experienced in the recent geological record. They are also likely to be as intense as the 1982/3 and 1997/8 events, but reefs may have already experienced changes in species composition and dominance by this time, and this may dampen the ecological response to these events.

Table 16.3. *Ecologically conservative predictions for the coral reef environment, function and components in the year 2025*

Ecological variable	Direction of change	Explanation and caveats
Physicochemical environmental factors		
Phosphorus	−/+	Reduced in offshore areas of the subtropics due to reduced winds and water-column mixing but increased in equatorial areas due to increased monsoons. Nearshore areas will experience increased water cycle, human and domestic animal waste production, soil erosion and water column mixing. Temporal variation will be high
Nitrogen	−/+	Offshore water-column mixing in subtropics will be reduced resulting in increased nitrogen fixation and may also increase in nearshore waters due to increased erosion and water cycle. Decrease in equatorial areas where monsoon conditions increase
Trace elements	−/+	As for phosphorus
Seawater acidity	+	Increased atmospheric CO_2
Currents and mixing	−/+	Increase in equatorial areas and decrease in subtropics
Light penetration	?	Difficult to predict due to interactions between cloud cover, aerosols and phytoplankton concentrations, but light at sea surface will increase in subtropics and decrease in equatorial areas as above
Ultraviolet light	+	A small increase
Dissolved and particulate organic matter	+	Increased plankton and fleshy algae
Ecological functions		
Gross organic production	+	Increased temperature and nutrient concentrations, lower hard coral cover, but high erect algal production may reduce net compared to gross production
Inorganic production	−	Decreased calcifying coral and algae abundance due to increased acidity and loss of coral cover through bleaching
Organic/inorganic ratio	+	Increased non-calcifying algae and decreased calcifying algae and coral
Reef components		
Primary producers		
Turf algae	−/0/+	Increased space and nutrients but increased competition with fleshy algae
Erect fleshy algae	+	Reduced herbivory and more colonizable space
Coralline red algae	−/+	Increased seawater acidity but higher nutrients
Erect calcareous green algae	−/+	Increased nutrients, sediments and decreased herbivory can increase but increased seawater acidity can decrease abundance
Seagrass	+	Increased sand and nutrients
Symbiodinium in corals	−/0	Changes in taxonomic composition but will decrease with loss of coral
Consumers		
Soft corals	−	Increased bleaching mortality will reduce abundance unless increased organic matter can compensate by increasing growth

Table 16.3. (*cont.*)

Ecological variable	Direction of change	Explanation and caveats
Sponge	−/+	Increased bleaching, diseases will cause losses, but increased organic matter, space and reduced predation will increase encrusting and endolithic species
Hard corals, animal	−	Bleaching mortality and competition with erect algae
Hard corals, algae/animal	+	Higher nutrients will increase the density of *Symbiodinium* in the animal host
Sessile invertebrates		
Commercial invertebrates	−/0	Increased use by humans but decreased use by predatory fish
Non-commercial invertebrates	+	Ecological release from predatory fishes
Fishes		
Herbivorous/detritivorous fishes	−	Increased fishing
Invertebrate-eating fishes	−	Increased fishing
Piscivorous fishes	−	Increased fishing
Planktivorous fishes	−	Increased fishing
Detritivores		
Epifaunal invertebrates	−/+	Commercial species decrease while non-commercial species increase
Infaunal bioeroders	+	Increased inorganic nutrients and organic matter

Note: +, positive, increased rate of change; −, negative, decreased rate of change; 0 = unchanged.

DISTURBANCE FREQUENCY AND REEF RECOVERY

The frequency and intensity of disturbances, whether caused by weather, predators or pollution, produce a regeneration niche which will be occupied by species that have regeneration times that fit within the period of disturbance. Because of the high species diversity in coral reefs, there is a large number of associated regeneration times, ranging from days to decades (Nyström *et al.* 2000). An important example from coral reefs is the benthic production components of algae and corals that have regeneration times ranging from tens of days to decades (Steneck & Dethier 1994; Ninio *et al.* 2000). Turf algae have the fastest regeneration times, of the order of 20 days, followed by fleshy and coralline algae, and lastly corals. Corals have regeneration times ranging from a few years to decades, depending on the species, depth, type of reef and growth form (Ninio *et al.* 2000). The fast-replacing organisms increase ratios of organic to inorganic carbon production relative to the slow-replacing species. We discuss the likely ecological impacts of predicted changes in the forcing functions below (Table 16.2).

Oceanographic and environmental change

Increased heating in the tropics is projected to increase the intensity of tropical atmospheric circulation, annual monsoons and seawater mixing, and produce more extreme hurricanes and ENSO events (Meehl 1992). Reefs are most affected by monsoons largely between 0° and 10° N and S latitude and tropical storms between 10° and 20° N and S. Equatorial convective areas are becoming cloudier and wetter, and tropical and subtropical areas are experiencing subsidence circulation, becoming less cloudy and dryer (Chen *et al.* 2002). Doubled carbon dioxide levels in the atmosphere are predicted to increase the intensity of hurricanes by 5–12% (Knutson *et al.* 1998). In general, currently dry areas will continue to become drier and wet areas wetter (Meehl 1992).

Very different environmental changes and associated ecological responses are expected between tropical and subtropical regions. A problem with the general model predictions is that associated with greenhouse gases are increased aerosol concentrations above land, particularly over Asia, which can decrease rather than increase the land–sea gradient and result in less intense monsoons and rainfall (Ramanathan *et al.* 2001). Rainfall over Asia in the last century has been negatively influenced by warm ENSO periods but, since the mid 1970s, rainfall has become independent of ENSO events, probably owing to the higher winter and spring temperatures that override the ENSO influence (Kumar *et al.* 1999). This suggests further spatial heterogeneity in the effects associated with global climate change in future, with some areas experiencing more and others less intense monsoons, ENSO events and rainfall. The variance will depend upon the balance between greenhouse-gas warming, aerosol cooling and seasonal variability, which complicates simple predictions for any individual locality and the Earth as a whole.

The increased hydrologic cycle, human population density and associated land use and waste release in tropical countries are likely to increase the concentration of nutrients and trace elements in near-shore waters (Wilkinson 1996; Tilman *et al.* 2001; Chapters 11–13). Subtropical areas, however, may be expected to experience more intense tropical storms that will result in greater interannual pulsed climate disturbances. Nutrient concentrations are likely to increase in terrigenous and deep-water sources in the tropics, with an increase in monsoon intensity and runoff from land (McClanahan 1988). Many of these nutrient elements are currently at high levels in near-shore waters and may increase further with the projected increase in the hydrologic cycle and coastal erosion. Nitrogen fixation, however, represents the largest source of offshore seawater nitrogen and this is greatly influenced by the amount of water-column mixing in tropical waters; the less mixing, the more fixation there is (Smith 1984; also see Wilkinson & Sammarco 1983). The predictions for the levels of nitrogen and phosphorus compounds can be quite different depending on a reef's distance from shore or whether changes in climate will increase or decrease mixing (Smith 1984). In subtropical Hawaii during warm ENSO phases, there are increases in stratification of the water column, concentration of nitrogen-fixing *Trichodesmium* and the ratio of nitrogen to phosphorus (Karl 1999).

Coral reefs will be influenced by changes in adjacent planktonic ecosystems that absorb light (Yentsch *et al.* 2002). When nitrogen and phosphorus are made available through coastal erosion, plankton blooms can be expected in near-shore waters. Phytoplankton in the water column are more limited by nitrogen and iron than by phosphorus,

and this is attributed to the high water movement in open seas relative to freshwater lakes (Chapters 5 and 6) and more isolated estuaries (Smith 1984; Coale *et al.* 1996; Chapter 13). In contrast, benthic and coral-reef production rates are more limited by phosphorus, but this can vary with season and nitrogen inputs (Delgado & Lapointe 1994). High water temperatures will increase not only gross production, but also metabolism, particularly of microbes that produce dissolved organic matter (DOM). This could result in lower net and secondary production of species consuming flagellates, dinoflagellates and eukaryotic plankton (Karl 1999). In areas with reduced monsoon intensity, there may be more stability in the water column, a greater frequency of N_2-fixing plankton blooms, more phosphorus limitation, and greater metabolism and production of DOM. Depending on the plankton requirements of larvae, this could enhance or decrease recruitment rates to coral reef populations.

Phosphorus is made available to shallow waters by mixing and runoff and is therefore likely to increase with stronger winds, greater runoff, increased intensity of the hydrologic cycle and more waste emissions. Inorganic elements from land are derived from land conversion and use of fertilizers (McCulloch *et al.* 2003). Many tropical regions underwent land conversion in the middle part of the twentieth century. Fertilizer use is expected to increase (FAO [Food and Agriculture Organization of the United Nations] 1990), and export of nitrogen to the sea is closely associated with human population density and fertilizer use (Vitousek *et al.* 1997*a*). Wet and dry cycles are probably critical to soil erosion and the variability in these cycles is projected to increase (McCulloch *et al.* 2003). This suggests that, while the future input of terrestrial elements and phosphorus into near-shore waters will be high relative to pre-agricultural conditions, the inputs will vary substantially with changes in fertilizer use and wet and dry cycles, but will continue to rise at a declining rate above the present. In oceanic areas experiencing intensified monsoons, phosphorus levels are expected to rise owing to increased seawater mixing. In non-upwelling and extra-tropical areas that depend on fixed nitrogen and water-column stability, phosphorus will decrease and nitrogen will increase. Because coral-reef calcification is more sensitive to phosphorus than nitrogen (Ferrier-Pagès *et al.* 2000), those areas that have increased phosphorus are likely to experience reduced reef growth.

Water flow plays an important role in delivering nutrients and flushing waste products, particularly during warm periods (Hearn *et al*. 2001; Nakamura & van Woesik 2001). Where monsoons remain intense, the conditions for nutrient supply, flushing and coral-reef growth will improve. However, one of the most notable effects of the 1997/8 warm ENSO was a weakened northeast monsoon which caused massive coral bleaching and mortality (Goreau *et al*. 2000). The lack of stirring caused stratification of seawater temperatures and kept temperatures above 30 °C for many months. Consequently, even the slow-growing massive coral species were affected (Mumby *et al*. 2001) because of the synergistic effect of warm water and reduced water flow (Nakamura & van Woesik 2001). There is evidence that water motion can influence the effects of ultraviolet (UV) radiation. Diminished currents can reduce the transfer of gases and waste products that can in turn increase the stress associated with warm water and high light intensity (Kuffner 2001). The frequency of weakened monsoons associated with warm ENSO events and rising water temperatures are likely to be among the forces of global change most detrimental to coral reefs.

Water-column mixing can also influence light and UV penetration depending on phytoplankton concentrations (Yentsch *et al*. 2002). If water-column phytoplankton concentrations increase in near-shore waters as predicted, then UV penetration to the bottom is likely to decrease in coming decades. The opposite may occur in offshore areas influenced by intensified monsoons. Ultraviolet light at sea level may be expected to increase slightly with reduced ozone levels in the coming decades, but expected reductions will probably be dampened in the tropics and most temperate regions. For a worst-case scenario of tropical ozone depletion, every 10 years there will be a 2.3–3.2% increase in UV-B (308 nm) intensity at 1 m water depth (Shick *et al*. 1996). Some Indo-Pacific corals produce mycosporine-like amino acids, which act as sunscreens that can shield them from the effects of UV (Shick *et al*. 1999; Kuffner 2001). The pocilloporin pigments that produce the bright colours of many corals (Dove *et al*. 2001) further act to reduce excess light and bleaching while increasing photosynthetic efficiency (Salih *et al*. 2000), but pigmented corals may be more susceptible to heat stress (Dove 2004). There are interspecific differences in the susceptibility of corals to UV-B (Siebeck 1988); this suggests that in the future there could be selection for species more tolerant of UV, although the complex nature of the relationship between thermal and photic stress dictates that the selection outcomes are likely to be complex.

Production

Organic production on reefs is primarily influenced by herbivory and the species or functional composition of benthic algae (Steneck & Dethier 1994). Calcium carbonate deposition results from the abundance and species composition of calcifying algae and corals, and these are influenced by rates of disturbance such as herbivory and physicochemical conditions (Littler *et al*. 1991). Predictions therefore require a holistic view rather than one based on singular changes in physicochemical rates or species interactions. For many benthic primary producers, there is an inverse relationship between organic and inorganic carbon production (Littler & Littler 1980). With increased bleaching and coral mortality, and reduced saturation of calcium carbonate in seawater, organic production may be expected to increase and inorganic carbon production to decrease.

Increased inorganic nutrients and organic matter in seawater and reduction of calcifying algae, coral and predatory fishes will produce reefs with high levels of bioeroding organisms (Risk & Sammarco 1991; Risk *et al*. 1995). These will include both infaunal species such as various microbes, worms, sponges and bivalves, and epifaunal organisms such as sea urchins (Holmes *et al*. 2000), and the increase is expected to result in the loss of reef framework and an increase in sand production (Eakin 1996). Increased production of sand and reduced herbivory will improve conditions for colonization by seagrasses, and many areas previously dominated by coral-reef organisms are likely to be converted to seagrass ecosystems (McClanahan & Kurtis 1991).

Currently, the most obvious and damaging effect on coral reefs is coral bleaching through loss of the coral-symbiont dinoflagellate *Symbiodinium*. From the early 1980s, the frequency and number of regions reporting bleaching and associated mortality have been increasing (Hoegh-Guldberg 1999; Goreau *et al*. 2000) (Fig. 16.1d), and bleaching events may be expected to occur once every 3–5 years in coming decades, resulting in a significant ecological reorganization of reef communities. Hoegh-Guldberg (1999) predicted that coral bleaching will become a nearly annual phenomenon by 2020, but this assumes that bleaching will continue at the current threshold temperatures and that the same species of coral and *Symbiodinium* will be dominant on reefs. Past bleaching events have resulted in low and others in high coral mortality (Goreau *et al*. 2000), largely a function of the early stage of climate-driven changes. Baker *et al.* (2004) argued that this may be partially due to a shift in dominance of *Symbiodinium* towards those taxa that are more tolerant of higher water temperatures. An alternative view is that there is high specificity between coral hosts and their symbionts and weak evidence for the advent of truly novel symbioses within this study (Hoegh-Guldberg *et al.* 2002; Goulet 2006).

Consumption

FISHES

Primary production and its transfer into the consumer food web will depend on rates of herbivory, which are partially controlled by fishing; through fishing, herbivore abundance may be reduced or possibly enhanced by depletion of carnivores. Coral reefs exhibit complex food webs composed of a variety of commercial and non-commercial species. The presence of non-commercial herbivores, notably sea urchins, can to some extent compensate for the loss of commercial herbivorous fishes as grazers (McClanahan 1995). In East Africa, fishing and competition with sea urchins result in a reduction of both herbivorous fish and species of erect algae. A similar pattern has been suggested for Caribbean reefs (Hughes 1994; Williams & Polunin 2001), with the exception of the pathogen that largely eliminated the dominant sea urchin *D. antillarum* in most of this region in 1983, resulting in a proliferation of erect algae (Lessios 1988). Fishing, coral diseases (Aronson & Precht 2001), sea-urchin diseases and other sources of mortality of herbivores and algal competitors are likely to continue and result in the ongoing dominance of many reefs by late-successional fleshy algae (Sammarco 1982; Szmant 2001). Among the consequences of fleshy algal dominance are reductions in net primary production (Carpenter 1988), fish abundance (McClanahan *et al.* 2002*c*) and hard coral cover (Hughes & Tanner 2000).

Without effective fisheries restrictions, fishes will continue to be influenced primarily by heavy fishing in near-shore waters (Munro 1996). In many regions where coastal populations exceed 50 people per km^2, notably the Caribbean, South-East Asia and East Africa, fishing is already intense, and competition for fishes and the desire for profit have promoted the use of destructive fishing techniques (Muthiga *et al.* 2000). Fishing for high-value species, such as for the live fish trade and shark fins, will continue to expand into more remote reef areas (Sheppard 2000). The lack of refuge from fishing and habitat destruction will

jeopardize the sustainability of these fisheries. Some species with restricted ranges may be threatened with extinction (Roberts *et al.* 2002). The consequences of the loss of these consumers will most likely include trophic changes in reef ecosystems as described above for algae, sea urchins, snails and starfish. In some cases, the ecological role of the commercial species can be replaced by non-commercial or low-preference target species (Shpigel & Fishelson 1991), but in many cases there are limits, for example to the grazing efficiency of these species (McClanahan 2000*a*). Recovery of coral-reef fishes and biomass indicate that diversity takes about a decade, though many groups can take two or three decades to recover (McClanahan 2000*a*; Russ 2004; McClanahan *et al.* 2007).

INVERTEBRATES

The response of sessile reef invertebrates to resource use is likely also to be complex because of the interactions between human exploitation, losses of predators of target species through fishing (McClanahan 2002*b*) and changes in the physicochemical conditions of seawater and their effects on recruitment and growth. There may be fundamentally different responses, depending on the commercial value of the species. Even this categorization, however, will not provide a strong indication of the direction of change. Many of the highly valuable commercial invertebrate species with slow growth, such as sea cucumbers and large-bodied molluscs, are likely to experience continued overfishing, resulting in large population declines (Uthicke & Benzie 2001; McClanahan 2002*b*). Many giant clam, large-bodied gastropod and sea cucumber populations are presently at dangerously low levels in many regions, requiring the development of marine protected areas (MPAs) and aquaculture to protect them from local extinction (Munro 1989). Some smaller fast-growing invertebrates may, however, benefit from the current overfishing of finfish, which appears to reduce mortality on these species through reduced predation (McClanahan 2002*b*). In general, populations of large-bodied commercial species are most likely to be influenced by direct collection, while small-bodied species may be most influenced indirectly by fishing.

A number of non-commercial sessile invertebrates, such as some reef-eroding sea urchins and coral-eating starfish and snails, are likely to continue to have a controlling effect on reef community structure. Loss of their predators can result in increased populations of such species (Bradbury & Seymour 1997). However, diseases that reduce these species or their prey, such as coral in the case of the crown-of-thorns starfish predation (Bradbury & Seymour 1997), may ultimately limit their populations. In some cases these species are tolerant of the loss of their prey and persist by their low metabolic demands or ability to rely on alternate prey (McClanahan 1992); their presence on the reef can then significantly suppress the recovery of their prey. Coral-eating snails, such as *Drupella cornus* in the Indo-Pacific and *Coralliophila abbreviata* in the Caribbean, may benefit from a loss of predatory fishes and thus suppress the recovery of corals damaged through bleaching, pollution or overfishing (Knowlton *et al.* 1988).

DISEASES

Coral-reef ecologists have documented a rise in the frequency and variety of diseases affecting corals (Porter *et al.* 2001). Declining water quality, increasing incidents of exotic species introductions and climate change have all been implicated in the increasing incidence of coral disease (Harvell *et al.* 1999).

Devastating diseases in coral reefs have also been commonly reported for coralline algae, sponges, soft corals, sea urchins and fishes (Peters 1997; Richardson 1998). Diseases require a virulent pathogen, susceptible host and an environment stressful to the host in order to thrive. Future conditions of warm water, high nutrients, poor land use, increased dust and sediments, toxic chemicals, slow flushing or water movement, and increased connectivity of marine environments are likely to lengthen the present list of coral-reef diseases and expand their distributions and virulence.

The effects of these diseases can be devastating to reef communities. Geological evidence suggests that the white-band disease impact on once-common *Acropora* species in the Caribbean over the last 10 years (Aronson & Precht 2001) is unique in the past 4000 years. In many cases, the loss of these corals has been associated with increased algal cover; in others, there has been replacement by the corals *Agaricia* or branching forms of *Porites* (Greenstein *et al.* 1998). The disease-induced loss of coral and the sea urchin *D. antillarum* (Lessios 1988) has shifted dominance of many Caribbean reefs from corals to erect algae (Szmant 2001). Recent cases suggest that diseases will continue to infect reef organisms and cause surprising ecological changes to coral reefs.

CONCLUSIONS

Coral reefs are undergoing change on a global scale, and this change is associated with a number of synergistic disturbances. Many of these changes began before scientists were

formally quantifying and documenting the state of reef ecosystems (Jackson *et al.* 2001; Pandolfi *et al.* 2003). Current trajectories are from localized small-scale disturbances such as fishing, river discharge and pollution effects, towards regional and global-level disturbances associated with warm water, diseases and changes in seawater chemistry. Regional-level ecological change was first noticeable in the Atlantic and Eastern Pacific in the early 1980s, and the Indo-Pacific and Indian Oceans followed these. Changes in the future will depend on the location of reefs, but a general shift in benthic dominance from late-successional coral to late-successional algal taxa can be expected, with higher organic and lower inorganic production, and a consequent reduction in reef growth and substratum complexity. Losses in coral species at the local level and possible local extinction in regions with small shelf size are likely to continue. The consumer community will shift from large-bodied, edible, slow-growing and ornamental species to various unused invertebrates and some fast-growing and colonizing generalist fishes. In many places, reefs will be replaced or colonized by seagrass, rubble and sand ecosystems (McClanahan & Kurtis 1991).

Recommendations for further research

Identifying reefs that will or will not be extirpated in the current age of global change will help prioritize conservation within the financial limits of support for coral-reef research and conservation (Salm & Coles 2001). It is important to identify the environmental, ecological and biological properties of coral-reef communities and organisms that will allow them to persist into the future. Perhaps the most obvious distinguishing environmental factor is to identify diverse coral reefs in stable cool water and low light conditions that are projected to persist under those conditions in the future. In contrast, warm retention areas with high temperature variation can also act as refugia as these conditions promote acclimation and resistance to extreme temperature events (McClanahan *et al.* 2005). Global climate change might be mitigated by factors that (1) reduce temperature stress, such as depth, upwelling and mixing, (2) enhance water movement, such as tides, winds and waves, (3) decrease light intensity, such as reef slope, depth and turbidity, (4) promote acclimation to extreme conditions and (5) correlate with reduced bleaching, including temperature variability and salinity stability (Vermeij 1986; West & Salm 2003). Some of these factors might result in lower tolerance to extreme variation or in poorer recovery potential from disturbances. Therefore, research on the short-term and medium-term responses of reef taxa to combinations of these factors is needed. Factors hypothesized to influence the recovery potential include (1) connectivity and currents among reefs, (2) spatial heterogeneity and refugia from disturbance, (3) recruit availability and success, (4) diversity, keystone species' and associated species' ecological redundancy, (5) persistence of herbivores and higher trophic levels through isolation from resource use or through successful management, (6) physicochemical environmental conditions that promote coral growth and (7) low population densities of population-controlling species such as disease-causing microbes, corallivores or bioeroders (McClanahan *et al.* 2002*d*). Interactions among the above factors may be more important than the single factors alone.

Recommendations for management

Global climate change and globalization of trade have made management of coral reefs a global concern and responsibility. Management is no longer a simple issue of managing local pollution and maintaining fisheries; it must also address the global context. Management efforts are needed from local to global levels. At the global level, efforts to reduce greenhouse gases now will have little effect on coral reefs to the 2025 time horizon, but there is a need to support the ongoing efforts (Sandalow & Bowles 2001) in order to reduce climate change effects beyond 2025. Restrictions on the global trade of reef resources such as sharks, ornamental species and the live-fish trade are one area where immediate action at the global level can reduce further degradation. Management of watersheds and water quality are important issues at the regional level. A greater effort to reduce the influx of inorganic and organic nutrients and sediments generally is needed to promote reef resistance and recovery after disturbances.

Regionally and globally, efforts to identify and manage species that are prone to extinction are needed (Glynn 2000; Roberts *et al.* 2002). Consequently, one priority for the 2025 time horizon is to identify specific reef areas and species that will require special protection in order to persist (Salm & Coles 2001). Species that are likely to be extinction-prone generally include those that (1) have the narrow environmental limits common to many coral reef organisms, (2) are endemic or have restricted ranges, (3) have small population sizes often caused by restricted habitat requirements or being high in the food chain, (4) have asexual reproduction and low dispersal capabilities,

(5) grow slowly and (6) are endemic to biogeographic regions with small continental shelves (Roberts & Hawkins 1999; Glynn 2000; Hughes *et al.* 2002). On coral reefs, however, all species hosting *Symbiodinium* (Cnidaria, Foraminifera, Porifera, Mollusca), calcifying organisms (i.e. Corallinaceae, Halimedaceae) and their obligate symbionts and hosts are of special concern. They are all potentially sensitive to warm water and changes in the seawater carbonate system. As part of the cataloguing of coral-reef species and their vulnerability, data on factors that influence susceptibility to climate change need to be collected to assist the identification of these extinction-prone species and to preserve them and their habitats.

Regionally and locally, there is a need to renew efforts for sustainable fisheries management through restrictions on areas, species, sizes, gears and effort levels. Area restrictions will include more MPAs and other fisheries reserves and financing the existing network (McClanahan 1999). Restrictions on fishing should also include elimination of programmes that seek to extend fisheries into previously unfished areas. Many tropical nations have near-shore areas that are overfished, but they are fished sustainably because there are sources of recruits that are too distant or dangerous to be fished by artisanal fishers. Where this is not the case, such as on many Caribbean islands, losses of breeding stock and catches have occurred (Berkes 1987). Unfished areas can act as refuges for breeding populations, becoming sources of vagile adults and larvae for nearby fisheries (McClanahan & Mangi 2000). Increasing the capacity of fishers to exploit these populations could lead to large-scale collapse of both near- and offshore fisheries. Programmes to increase fishing capacity may, however, be justified if closed areas of substantial size near to shore are successfully established.

Restrictions on destructive fishing will challenge conservation as human numbers increase in tropical countries and competition for dwindling resources accelerates. In addition to legislation and enforcement to eliminate destructive fishing methods, there is a growing need to alleviate poverty, so that fishing is not one of the few major economic options for the peoples of tropical nations. This requires economic planning and implementation, and improved attempts to realistically balance populations, resources and economic growth. Both tropical and temperate nations need to take responsibility for balancing ecological and economic production, and the control of carbon, toxic and nutrient wastes to provide an improved future for coral reefs and their economic services.

It is critical that the ultimate cause of global change is treated and not just the symptoms; that cause is human overpopulation. Most countries that have the highest rates of human population growth are in the tropics, and many of those countries possess coral reefs (Sammarco 1996). Until the problem of growing human population is addressed, those factors causing coral-reef ecosystems to become degraded or shift phase cannot be expected to be satisfactorily controlled. It is the ever-increasing demand for food, natural resources and economic growth that is the basis of anthropogenic global change. The solution to this problem will require increased international cooperation, and achieving this goal is vital.

Part VI
Soft shores

The soft shore ecosystems span all of the Earth's climatic zones, and have for many centuries been a source of food and livelihood services for humans. Each has unique ecological features, whilst the stresses acting on them are often similar.

Sandy shores are dynamic harsh environments, the action of waves and tides largely determining species diversity, biomass and community structure (Chapter 17). There is an interchange of sand, biological matter and other materials between dunes, intertidal beaches and surf zones. This transport is critical to the maintenance of the sand budget and consequently to the landforms and habitat on which sandy-shore organisms depend. Storms and associated erosion present the most substantial universal hazard to the fauna. Human activities interfere with these sediment exchanges and directly alter, restrict or replace landforms and biota. The attention of Chapter 17 is focused on subaerial beaches on wave-dominated open coasts, coastal dunes, low-energy beaches and human-restored environments. Covering <0.2% of the global ocean, and occurring in a range of salinities around all continental masses except Antarctica, seagrasses comprise about 60 angiosperm species in 12 genera that complete their life cycle in the marine environment (Chapter 18). Seagrasses are highly productive plants that assimilate and cycle nutrients and other chemicals; their extensive biomass traps sediments, reducing coastal turbidity and erosion, and provides habitat for a rich faunal assemblage. The soft-bottom continental shelf that underlies approximately 7% of the global marine surface supports very substantial fisheries production, exerts important controls on marine productivity and contains rich and varied marine communities (Chapter 19). These benthic ecosystems are important in the economies of many coastal states through the provision of food, non-living resources (e.g. aggregates) and environmental services, benefits that are subject to changes in climate, oceanography, hydrology (e.g. river discharge), land use, waste disposal, fishing, aquaculture and extraction of non-living resources.

17 · Sandy shores of the near future

ALEXANDER C. BROWN, KARL F. NORDSTROM, ANTON MCLACHLAN,
NANCY L. JACKSON AND DOUGLAS J. SHERMAN

INTRODUCTION

Most sandy shores consist of coupled surf zone, intertidal beach and dune systems (Short & Hesp 1982), which together constitute a littoral active zone of sand transport. Many of these systems were formed during Holocene deglaciation which increased sediment discharge to the shore. This also drove a rapid rise in global sea level to approximately its present elevation by c.5000 years before present (Bird 1993). On open coasts subject to oceanic swell,

waterborne sediment transport may extend well beyond the surf zone to a depth of up to 20 m, while aeolian transport of sand extends landward to fully vegetated dunes. In low-energy and estuarine sandy-shore systems both the magnitude and cross-shore extent of transport are smaller. Transport both along and across the shore is critical to the maintenance of the sand budget and consequently to the landforms and habitat on which sandy-shore organisms depend (Clark 1983). Human activities interfere with these

Aquatic Ecosystems, ed. N. V. C. Polunin. Published by Cambridge University Press. © Foundation for Environmental Conservation 2008.

sediment exchanges and directly alter, restrict or replace landforms and biota. Global climate change adds further stresses by accelerating rates of change in natural processes and generating human actions to resist these changes.

Brown and McLachlan (2002) provided the basis and background for assessment of the future of sandy shores, concentrating attention on subaerial beaches on wave-dominated open coasts. This chapter synthesizes this information with evaluations of coastal dunes, low-energy beaches and human-restored environments, documenting trends and identifying implications for the future.

NATURAL CHARACTERISTICS OF BEACHES AND DUNES

Beaches

Sandy shores are dynamic environments with unstable substrata and thus present hostile conditions to biota. The total absence of attached macrophytes in the intertidal and immediate sublittoral zones is a conspicuous result of these hostile conditions. Plants are unable to overcome the instability of constantly shifting sand to which they would otherwise lend stability. Over 20 species of invertebrate macrofauna may be resident intertidally, often in very large numbers. Other macrofauna invade the intertidal beach from the surf zone as the tide rises, and terrestrial mammals, snakes and arthropods such as spiders may invade it at night in search of food (Brown & McLachlan 1990). The vast majority of resident intertidal sandy-beach animals are tiny and live between the sand grains, often at a considerable depth. This interstitial meiofaunal component may comprise over 600 species on some beaches (Brown 2001). Virtually all animal phyla are represented. Bacteria and Protista also live between and on the grains, frequently in large numbers. Diatoms may be present as well, especially on relatively sheltered shores.

Open beaches with relatively gentle slopes and swash, termed dissipative beaches (Wright & Short 1984), present less hostile conditions to the fauna than ocean beaches with steep slopes and more turbulent swash (reflective beaches). Reflective beaches are chiefly inhabited by a semi-terrestrial macrofauna dependent on stranded wrack or kelp. Increasingly dissipative conditions lead to increasing species diversity and food-chain complexity. The surf zone plays an ever more important role in the bionomics of the systems as conditions become more dissipative, especially where circulating cells of water support surf diatoms, such

as *Anaulus*, which then drive much of the food web. A semi-closed ecosystem results (McLachlan 1980). Reflective beaches are, in general, net importers of material from the sea, while dissipative systems are exporters (Brown *et al.* 2000). Beaches in estuaries (see also Chapter 13) are often characterized by a steep foreshore, with little microtopographic variation, and a broad flat low-tide terrace (Nordstrom 1992). These beaches possess many features that are analogous to reflective beaches on exposed coasts, but they are of smaller scale and have lower wave energies. Waves breaking directly on the steep foreshore result in greater sediment-activation depths of the beach matrix (Jackson & Nordstrom 1993; Sherman *et al.* 1994), and there is far greater variety of species on the subtidal flats fronting estuarine beaches than on the wave-dominated foreshores.

Dunes

Coastal dunes form where there is sand on the beach and a wind that is strong enough to move it. They are far more prevalent than reported in the literature, in part because only the larger dune fields, or the dunes that have human importance, have been studied in any detail. Increases in sand supply on the beach and increases in the frequency and speed of onshore winds result in larger dunes. Dunes may take the form of an isolated foredune ridge where sand supplies or onshore wind speeds are limited (i.e. on estuarine shorelines), or develop into dune fields kilometres wide where sand is plentiful and strong onshore winds are frequent. Foredunes are part of the same sediment-exchange system as beaches and, under natural conditions, the form and dimensions of dunes are determined by beach morphodynamics (Short & Hesp 1982).

Disturbance influences the composition and richness of vegetation communities in dunes (Keddy 1981; Moreno-Casasola 1986; Ehrenfeld 1990). Fewer species can tolerate the stresses associated with sand mobility and salt spray near the beach, and richness is diminished in this portion of the cross-shore gradient (Moreno-Casasola 1986). Pioneer plants (e.g. *Cakile edentula*) that are tolerant of salt spray and sand blasting form embryo dunes in seaward-most portions of the backbeach; grasses (e.g. *Ammophila* and *Spinifex*) form foredune ridges. Portions of the dune that are landward of the foredune and protected from salt spray and sand inundation are colonized by woody shrubs (e.g. *Myrica*) in the seaward portions, and by trees and upland species in the landward portions. Wet or moist dune slacks may occur in low portions of the dune (swales) that are close to the elevation of the water table. The transition from pioneer beach plants to fully mature forests can extend over environmental gradients of hundreds to thousands of metres, depending on the frequency and magnitude of onshore winds that drive the physical stresses.

Beaches and dunes are two discrete ecological systems, and few, if any, resident species of plants or animals are shared (McLachlan 1980), but they are components of a single sediment-transfer system, and fauna may make use of both components for feeding or escape from predators. Accordingly, human actions that separate, truncate or compress the gradient across which sediment moves represent threats to both ecological systems.

Natural threats

Short-term changes in beach morphology in response to fluctuating wave regimes or weather conditions are conspicuous and fairly well studied (Brown & McLachlan 1990; Short 1996). Superimposed on these are slow long-term trends in accretion or erosion that may be only apparent after decades or centuries. Long-term retreat of the coastline with diminishing beach volumes is a common worldwide trend (Bird 1985), but the significance of these changes is difficult to assess on a site-specific basis. Diminishing beach volumes on many natural coasts may represent only temporary stages in erosion–accretion cycles that accompany displacement of the shore, with eventual re-establishment of beach characteristics at new locations further inland. In these cases, erosion is not a threat to the quantity of sandy beach habitat. There is no evidence that the wholesale elimination of beaches and dunes on natural coasts composed of unconsolidated sediments will occur unless the landward migration is prevented by human structures. These features may migrate inland at a greater rate, and their dimensions, morphology and associated biota will change (particularly in the dune zone), but they are likely to exist. Shoreline retreat may be accompanied by greater losses where beaches are backed by uplands composed of resistant formations that cannot replace sediment removed from the beach. The quality of sandy beach habitat may be altered where sandy barriers transgressing marsh surfaces alter the sedimentary characteristics of the intertidal foreshore, particularly in estuarine environments (Botton et al. 1988; Chapters 11 and 13).

Storms are an integral part of a natural cycle of beach and dune change, and they represent the most universal, recurrent natural hazard faced by sandy-shore animals. Storms often result in great mortality of organisms, but the ability of certain organisms to survive by behavioural means is a key feature of many sandy-shore animals (Brown 1996), and interspecific competition is minimized because few macrofaunal species can tolerate the harsh conditions. Thus storms create and define the ecosystem; they do not threaten it. Driving factors that change the magnitude and frequency of the storms or the ability of the beach to absorb wave impacts are controls on the physical characteristics of beach ecosystems. These controls are caused by direct and indirect human impacts exacerbated by global climate change.

Global warming is changing sea and air temperatures, contributing to sea-level rise, and changing patterns and intensities of storms and rainfall. El Niño events may increase in frequency and/or severity, adding thermal stress to the other consequences of physical changes. There may also be changes in the intensity of upwelling. Reduction of the ozone layer and resulting increases in ultraviolet radiation have presented a threat to organisms (including shallow water species) through a variety of effects. Increased ultraviolet (UV) radiation impacts surf-zone biota, including phytoplankton, bacteria, crustaceans and their larvae and fish (Browman et al. 2000; Gustavson et al. 2000; Wubben 2000). An important result of these impacts is reduced bioproductivity (Hader 1997; Browman et al. 2000). The fauna of the intertidal beach is much less likely to suffer ill-effects because most species are cryptic, living within the sand, and most species that do emerge do so only at night (Brown 1983).

HUMAN ACTION CHANGING BEACHES AND DUNES

In 1994, approximately 37% of the human population lived within 100 km of the coast (Cohen et al. 1997) and, by 2020, over 60% of global population may reside within 60 km of the coast (UNCED [United Nations Conference on Environment and Development] 1992). Based on activities in Western Europe and North America, a generalized representation is given of past trends in the intensity of development, along with the principal human actions (Fig. 17.1). The changes have evolved from incidental or accidental actions to direct modification in response to changes in population pressures, perception of resources,

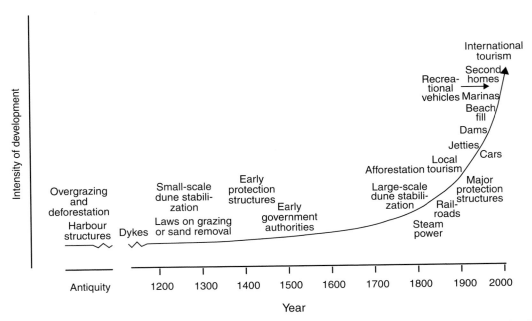

Fig. 17.1. Trends in the intensity of human development (arbitrary units) affecting coastal landforms and habitats. (Reproduced from Nordstrom 2000, by permission of Cambridge University Press.)

income, leisure time and technological advances. Other locations may have gone through similar phases of landscape conversion, but the phases may have been at different dates or have had different durations. The trend line is smooth (Fig. 17.1) because it represents a broad regional portrayal. A site-specific curve would show considerable short-term fluctuations related to periods of declining human population owing to piracy, war and disease, and changes in economies. There is a pronounced change in the slope of the curve beginning about two centuries ago, corresponding to the availability of steam power, which enabled large modifications to the landscape. The slope of the curve has increased with increases in the size and availability of machinery, development of the internal combustion engine and growth of tourism as a major industry (Nordstrom 2000).

Beaches and dunes change because of both planned and unplanned human actions (Table 17.1). The most destructive alterations are those that regionally eliminate, or reduce substantially, beaches and dunes, usually via decreased sediment supply in a littoral cell. At a local scale, the most destructive alterations are those that completely replace landforms and habitat with buildings, transportation facilities or engineering structures such as groynes,

revetments and breakwaters. Some changes may retain the shape of the landform and provide habitat, although the surface characteristics, growth conditions and biotic interactions may vary.

Altering through use

Coastal and ocean-based forms of recreation are rapidly growing areas of the tourism industry (Hall 2001). Recreational activities (Table 17.1) can reduce the physical stability of beach and dune surfaces. Pedestrian footpaths and vehicle tracks increased by 600% and 265% respectively over a 10-year period in the Aberffraw dune system (Wales, UK) (Liddle & Greig-Smith 1975). Pedestrian uses on the beach may have only slight impact on invertebrates relative to the physical effects of changes in wave climate, although these activities may frighten fauna and change use patterns (Burger & Gochfeld 1991; Burger *et al.* 1995; Brown & McLachlan 2002). Vegetation in dunes is more sensitive to trampling, which can: compact and change the bulk density, organic matter and moisture content of soil; reduce the cover and height of vegetation; decrease production of biomass; reduce the number of flowering species; cause disappearance of vulnerable species; introduce weeds and

Table 17.1. *Ways in which beach landforms and habitats are altered by human actions*

Modification category	Reason for modification
Eliminating beach for alternative uses	Constructing buildings, transportation routes, promenades
	Constructing alternatives surfaces (golf courses, landfills)
	Mining for minerals, construction aggregate (removal and dumping)
Altering through use	Recreational activities
	Swimming
	Trampling (walking, game playing)
	Off-road vehicle use
	Horseback riding
	Fishing and harvesting (food, bait, wood, seaweed, flowers, fruits)
	Grazing
	Extraction of oil and gas
	Extraction and recharge of water
	Military (active uses and buildings)
Reshaping	Grading beaches and dunes to increase levels of protection
	Removing sand that inundates facilities
	Breaching barriers to control flooding
	Dredging channels to create or maintain inlets
	Enhancing recreational use
	Widening beaches to accommodate more visitors
	Creating flat platforms for cabanas, pavilions on beach
	Eliminating topographical obstacles to access
	Grading dunes down to provide views of the sea
	Enhancing environmental values
	Creating more naturalistic landscapes
	Altering environments for wildlife
Altering faunal viability or use patterns	Introducing pets or feral animals
	Nature-based tourism
Altering landform mobility	Constructing shore protection and navigation structures
	Shore-perpendicular structures (groynes, jetties)
	Shore-parallel structures
	Onshore structures (sea walls, bulkheads, revetments, dykes)
	Offshore structures (breakwaters, headlands, sills, reefs)
	Constructing marinas
	Constructing buildings
	Introducing more or less resistant sediments into beach or dune
	Stabilizing landforms using sand fences, vegetation plantings or resistant materials
	Remobilizing or changing habitat characteristics by altering existing vegetation
	Controlling density (mowing, grazing, fires)
	Planting species to increase diversity
	Planting forests
	Introducing exotics (intentionally or unintentionally)
	Clearing the beach of litter

Table 17.1. (*cont.*)

Modification category	Reason for modification
Creating new or larger landforms and habitat (nourishment and restoration)	Providing protection to human structures
	Creating larger recreation platforms
	Creating new habitat
	Burying unwanted or unused structures
	Managing sediment budgets (bypass, backpass)
Altering external conditions	Damming or mining streams
	Pollution
	Oil spillage from vessels, terminals
	Burial of waste followed by exhumation or underground seepage
	Chemical and thermal factory effluent
	Radioactive pollution from power plants and reprocessing plants
	Organic enrichment from poorly treated sewage
	Fertilizers and herbicides from farmlands
	Diverting or channelizing runoff

Source: Modified from Nordstrom (2000); Brown and McLachlan (2002).

exotic plants; interfere in natural vegetation succession; cause loss of biodiversity; and disrupt fauna (Chapman 1989; Andersen 1995*a*). Fishing and harvesting aquatic organisms for food and bait can drastically reduce populations and change predation patterns but seldom eliminate them because the effort of collecting fails to justify the reward (Brown & McLachlan 2002). The physical disturbances associated with fishing and harvesting may be more deleterious than the actual collection of animals (Wynberg & Branch 1997).

Grazing of dunes by domesticated species can increase available light at the ground, reduce the stock of organic matter and nutrients due to decreased input of vegetation litter and accelerated rates of decomposition (Kooijman & de Haan 1995), and increase rates of aeolian transport, leading to greater dune mobility. Grazing, as a way of controlling ground cover, can be considered detrimental or beneficial depending on type and intensity of use and specific location in the dune environment (Boorman 1989).

Extraction of oil and gas or groundwater (Chapter 3) can lead to subsidence, local increases in sea-level rise and increased rates of beach loss (Inman *et al.* 1991; Bondesan *et al.* 1995; Nicholls & Leatherman 1996) or accretion. Extraction of water from dunes for domestic or agricultural purposes can lower the water table, leading to serious effects on the dune ecosystem that in turn may affect the intertidal beach (Brown & McLachlan 1990). Artificial recharge with

water from other sources (as occurs in the Netherlands) may add nutrients and disturb the direction and volume of groundwater flow leading to unnatural water-table fluctuations (Geelen *et al.* 1995). Recharge can also occur from watering lawns and gardens on private properties. Watering and wastewater disposal above coastal bluffs can add weight to the cliff materials, weaken them (by solution in some cases) and lubricate surfaces, along which slides develop (Kuhn & Shepard 1980; Shuisky & Schwartz 1988).

Military activities may be destructive to landforms and biota at specific sites in the short term (for example impact on vegetation at bombing ranges), but they may provide the opportunity for long-term evolution by natural processes in intervening areas. Military use of dunes for manoeuvre or firing ranges can prevent urbanization or agricultural uses in them and allow for natural development of dunes after this use ceases (Doody 1989). The impacts of war on beach, dune and nearshore systems are only partially understood, but there are immediate and potential long-term threats to human uses and ecosystem health (Lanier-Graham 1993; Grunawalt *et al.* 1996).

Reshaping

Reshaping landforms by grading (bulldozing and scraping), using earth-moving equipment, is common, but its effects are poorly studied. Many of the ways in which grading is practised are still controversial in terms of their benefits or adverse impacts on geomorphic systems. Beaches and dunes are reshaped primarily to: (1) create a barrier to prevent wave overwash and flooding; (2) uncover human facilities inundated by wave overwash or wind drift; (3) create or maintain channels to facilitate navigation, relieve flooding or alter circulation patterns; (4) enhance beach recreation by creating wider, higher or flatter platforms; (5) eliminate dunes that restrict access or obscure views of the sea; and (6) create new landforms to enhance environmental quality (Nordstrom 2000). Grading is a common practice of local governments because many municipalities have their own equipment and because other options are too costly, are prohibited or take too long to occur under natural conditions. Grading beaches and dunes for shore-protection purposes may create environments that are superficially similar to natural environments, but the locations, shapes, internal structure and evolutionary trajectories may be quite different from those of natural environments. Grading to enhance recreation is not likely to result in landforms and habitats with much natural value; it has been conducted to counteract adverse impacts of humans or create new natural environments, but these projects are usually small scale (Nordstrom 2000).

Altering landform mobility

Coastal landforms can be made less or more mobile by direct human actions or by changing rates or patterns of sediment transport caused by waves, winds and currents. The most extensive alterations in landform mobility are associated with shore-protection and navigation projects. Many shore-protection structures (groynes, bulkheads and sea walls) have pronounced effects, but surprisingly few studies have been devoted to investigating their impacts on the biota of adjacent beaches and dunes. Some studies have been conducted in estuarine environments where shore-parallel structures have altered available sandy beach habitat (Thom et al. 1994; Spalding & Jackson 2001; Jackson et al. 2002).

In New Jersey (USA), a highly developed shore, protection methods in both ocean and estuarine environments have changed in type and frequency of implementation, revealing an early preference for groynes, followed by a period of construction of shore-parallel structures, and the currently favoured period of beach nourishment (Fig. 17.2). Groynes are shore-perpendicular structures designed to trap sand moving alongshore. These structures change patterns of wave breaking and surf-zone circulation, redirect sediment offshore, and create differences in beach widths and sediment characteristics on the updrift and downdrift sides. Use of groynes has diminished recently relative to shore-parallel structures and beach nourishment (Fig. 17.2), in part because of accelerated erosion on their downdrift sides; but new groynes are still being constructed, often to designs that allow for some bypass of sediment (Kraus & Rankin 2004). Many scientists are averse to groynes, but there is no evidence that groynes cause more harm to the environment than alternative structures built to protect eroding shores. These structures change patterns of erosion and deposition within beach and dune environments, but they do not replace those environments, and they can favour dune development by increasing sediment source areas and serving as aeolian traps on their updrift sides (Nordstrom 2000).

Shore-parallel structures built offshore of the beach are designed to break up waves or hold beach sand in place. They are more common on shores where wave energies and tidal ranges are low and their size and construction costs can be kept low. Shore-parallel structures built landward of the beach provide a barrier between the dynamic beach and locations of human development. They truncate the landward portion of the beach that would be reworked by storms and restrict or prevent exchanges of sediment and biota between the beach and dune or upland. These structures prevent the coastal environments from migrating landward in response to sea-level rise or reductions in the sediment budget, and they change the boundary between nature and human development from a zone to a line. They also allow human structures to survive closer to the water than other forms of shore protection. Unlike groynes, which rearrange sediment transported by natural processes, they are designed to resist natural processes. As such they are less environmentally friendly than groynes.

Protection and navigation structures can function as artificial reefs and can be colonized by a great diversity of reef-dwelling organisms (Hurem 1979); they can also benefit certain categories of fauna by providing sheltered areas or traps (Botton et al. 1994), although such uses have seldom been planned. There is no long-term evidence that the changes are better than what would have occurred in the absence of the structures.

Many marinas have been built in recent years. Marinas and port facilities act as artificial headlands that break up

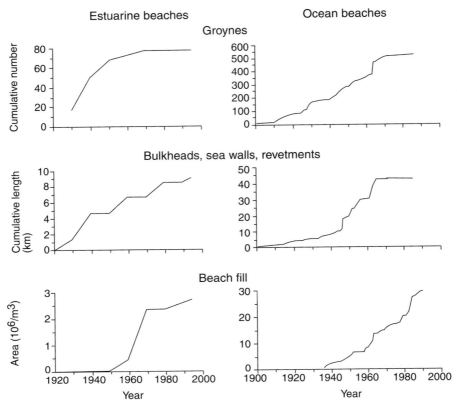

Fig. 17.2. Cumulative trends in implementation of shore protection projects through time in New Jersey (USA). (Modified from Jackson 1996 and Nordstrom 2000.)

the orientation of the shoreline, change refraction and diffraction patterns, trap sediment, deflect sediment offshore and starve adjacent downdrift beaches. New beaches may be intentionally created adjacent to these same marinas, but they are likely to be managed for beach-related activities, and provisions have to be made to protect or augment natural values to ensure that maximization for recreational use does not preclude development of natural features (Nordstrom 2000).

Buildings provide barriers to sediment transport, but they can also constrict and accelerate wind flow between them, increasing the likelihood of local scour. Buildings can be made more compatible with natural processes, and the passive effects of buildings need not be a threat to coastal resources. Direct actions taken by shore-front owners to make use of their properties (by constructing sand barriers, removing aeolian and overwash deposits and

planting exotic species) result in greater changes to landforms than the passive effects of the buildings.

Sand fences, including manufactured forms and natural brush, are used to trap sand (moved by wind) to create or repair dune ridges or prevent inundation of cultural features landward. Sand fences are one of the most important human adjustments affecting landforms on developed coasts because they are one of the few structures permitted seaward of the dune crest in many jurisdictions; they are inexpensive and easy to construct; and their deployment usually occurs at the highly dynamic boundary between the beach and dune. Many coastal dunes that are perceived to be naturally functioning owe their physical and biotic characteristics to use of sand fences.

Dune surfaces have been stabilized using a variety of methods, including straw or other biodegradable and non-biodegradable matting or spray. These efforts at stabilization are often directed at threats that are more imagined than real, resulting in unnecessary environmental losses (Angus & Elliott 1992). Afforestation (planting trees) has been conducted to protect crops from sea winds, establish a forest industry and stabilize dunes. The practice was carried out on a relatively large scale in the past, but ecological problems, including loss of flora and fauna, changes to soil characteristics, lowering of the water table and seeding into adjacent unforested areas, have led to attempts in some areas to remove woodland and restore dune habitat that has greater conservation value (Sturgess 1992).

Beaches can be made more stable by introducing coarser sediments, such as gravel, to a primarily sandy beach, but there are differences in kinds of habitat and values of habitats associated with the different types of gravel used in nourishment operations. Adding coarser sediments to sandy beaches has not been common in the past, but it is likely to occur more frequently as suitable sources of sandy beach-fill are exhausted and technologies for transporting gravel improve. It does, however, change beach morphodynamics and result in loss of some intertidal species (McLachlan 1996).

Dunes may be remobilized (destabilized) by removing surface vegetation to change sediment budgets, stimulate growth conditions for desired species, or reinstate natural processes or appearance (van Boxel *et al.* 1997). Introducing more mobility into coastal landforms has been a recent phenomenon, and it has normally been conducted at small scale and in locations where facilities are not immediately threatened, but the existing projects are important for the precedent they set and the experience they provide for future implementation at larger scales.

One of the principal means of changing the morphology and mobility of landforms by changing vegetation is through accidental or intentional introduction of exotic vegetation. Plant invasions can result in decreased biodiversity, loss of authenticity of vegetation, interference with successional processes, undesirable appearance of landscape and reduction of value of land for conservation and recreation (Andersen 1995b). Important exotics that owe their introduction or spread to human actions include the Japanese sedge (*Carex cobomugi*), introduced to coastal dunes in the USA; pine trees (for example *Pinus nigra* ssp. *laricio* and *P. contorta*), used to stabilize dune systems throughout the world; the bitou bush (*Chrysanthemoides monilifera*), introduced to Australia from Africa; *Acacia cyclops*, introduced to Africa from Australia; the Australian *Casuarina equisetifolia*, used to stabilize mobile dunes in

Mexico and elsewhere; and European beach grass (*Ammophila arenaria*), introduced to stabilize dunes on the Pacific coast of the USA, South Africa and Australia. Stabilizing dunes with exotic vegetation is a major threat to ecosystem health that is often disguised as a beneficial green alternative.

Beach raking to eliminate organic litter has become a common practice in resort communities to create a clean environment. Raking eliminates plant growth and the wave-deposited wrack lines that serve as sand traps or sources of nutrients. It thus eliminates topographic and vegetation diversity and the incipient dunes that could grow into naturally evolving foredunes. Raking is allowed in many municipalities because it is perceived as important for human health and aesthetic appeal, but it turns beaches into artefacts with little natural value. Raked beaches are considered attractive by many tourists, and many residents and visitors now view natural beach litter lines, incipient vegetation, beach fauna and even dunes as undesirable features. Raking may be the greatest threat to subaerial biota on beaches in economically developed areas.

Creating landforms and habitat

BEACH NOURISHMENT

Beach nourishment is a soft engineering alternative that has value for: (1) protecting buildings and infrastructure from wave attack; (2) improving beaches for recreation; (3) creating new natural environments; (4) eliminating detrimental effects of shore protection structures by burying them; and (5) retaining sediment volumes during sea-level rise. Bypassing and backpassing operations that remove sediment from accreting beaches and deposit it on nearby eroding beaches are local-scale sand-management operations that can also be considered forms of beach nourishment. Beach nourishment is practised extensively in developed countries and is the long-term strategy to combat erosion in some countries (Hamm *et al.* 2002). Hard engineering is still important in countries like the UK where flooding and sea-level rise threaten public infrastructure and private settlements (Turner *et al.* 1998). In cases where removal of structures and retreat landwards is not possible, beach nourishment appears to offer the best available alternative to increase the resilience of coastal systems without loss of ecological functions or capacity to provide recreational and other services.

Fill sediments can be placed in deep water offshore, in the nearshore, on the subaerial beach, on top of dunes and on the seaward or landward sides of dunes. Nourishment of the upper beach has been a common practice in the past. The result of nourishment on the upper beach is a widened and often overly steep beach with a morphology and sediment composition that may be out of equilibrium with natural processes. The nourished beach may be physically different from the original beach in terms of sand compaction, shear resistance, moisture content, and size, sorting and shape of sediments (NRC [National Research Council] 1995). These changes can negatively impact the macrobenthos (Rakocinski et al. 1996; Peterson et al. 2000b); nevertheless, beach nourishment has proved effective in combating erosion (Hamm et al. 2002), and, if done properly, may also enhance the habitat of selected species of biota.

Beach nourishment has had a positive impact on threatened species, including seabeach amaranth (Amaranthus pumilus), piping plover (Charadrius melodus) and several species of sea turtles, by providing habitat that would otherwise be unavailable (Crain et al. 1995; NRC 1995). Nourishment also provides a source of sand for dune building and space that allows new dunes to survive wave attack, allowing well-vegetated dunes to be rapidly established (Freestone & Nordstrom 2001). Beach nourishment operations will not categorically improve conditions for re-establishment of nature if active recreational uses take precedence, and considerable environment costs may be incurred owing to: (1) removal of habitat and death of biota in the borrow area; (2) increased turbidity and sedimentation in both the borrow area and nourished area; (3) disruption of mobile species that use the beach or borrow area for foraging nesting, nursing and breeding; (4) enhancement of undesirable species; (5) change in wave action and beach morphology (for example from dissipative to reflective); (6) change in grain size characteristics; (7) higher salinity levels in sediments emplaced or in aerosols associated with placement by spraying; and (8) change in community structure and evolutionary trajectories resulting from new conditions in the borrow and nourished areas (Roelse 1990; Looney & Gibson 1993; Nelson 1993; Löffler & Coosen 1995; NRC 1995; McLachlan 1996; Rakocinski et al. 1996; Rumbold et al. 2001; US Army Corps of Engineers 2001).

Detrimental environmental effects of nourishment are often considered temporary (US Army Corps of Engineers 2001), but many subtle or complex impacts are unknown (Gibson & Looney 1994; Rakocinski et al. 1996; Gibson et al. 1997), and prediction of the long-term cumulative implications of large-scale projects is premature (Lindeman & Snyder 1999; Greene 2002). Most biological studies quantify the elimination and recovery of fauna, but thresholds for species and environments should also be specified. The longest recovery times for biota appear to occur when there is a poor match between the grain size characteristics of the fill materials and original substrate (Reilly & Bellis 1983; Rakocinski et al. 1996; Peterson et al. 2000b; US Army Corps of Engineers 2001). The great cost and long time-frame required to conduct high-quality studies of ecological impacts, and the difficulty of applying results from one location to another, will result in many unanticipated and unwanted environmental impacts. However, these kinds of impacts are not deterring nourishment projects from being undertaken in most locations where protection and recreation projects are economically justified.

DUNE NOURISHMENT AND ENHANCEMENT

Nourishment of the upper beach can alter aeolian transport, dune growth and vegetation change by: (1) increasing beach width, which increases the likelihood of aeolian transport and decreases the likelihood of marine erosion of the foredune; (2) changing grain size characteristics and mineralogical composition of sediments, which affects the likelihood of aeolian entrainment and moisture conditions; and (3) changing the shape of the beach or dune profile (van der Wal 1998, 2000). Dunes may be nourished directly. The sediments often come from the same sources as those used to nourish beaches, causing the substrate to resemble the backbeach rather than the better-sorted, finer-grained substrate that would form by sediment blowing off the backbeach. Direct emplacement of fill creates a substrate not found in either natural dunes or dunes created by accretion at sand fences or vegetation plantings. A dune dyke constructed from poorly sorted sediments, obtained from a tidal delta and upland source of glaciofluvial sediments, had immobile sediments on the surface, with dramatically reduced rates of sediment transport (Baye 1990). Edaphic conditions were established that favoured plant species from stable dunes and inland sites while discouraging growth of native foredune dominants, such as Ammophila (Baye 1990). Fill materials borrowed from backbay environments may contain seeds and rhizomes of marsh plants, leading to establishment of salt-tolerant plants such as Phragmites and Spartina in dunes where these species were previously uncommon.

Altering external conditions

DAMS AND STREAM MINING

Natural reduction in sediment supply, driven by global climate change through the Holocene, established a millennial-scale backdrop of widespread beach erosion. This was partly offset in the past by widespread deforestation for cultivation, grazing and construction of settlements, which resulted in delivery of vast quantities of sediment to coasts by streams. The damming of rivers, which was widespread in the twentieth century, dramatically reversed this trend, resulting in increased rates of erosion. In 1950, there were 5270 large dams in the world; there are currently more than 36 500 (WRI [World Resources Institute] 1998). Construction of dams has been the principal cause of coastal erosion in many locations throughout the world (Awosika et al. 1993; Innocenti & Pranzini 1993). It is difficult, however, to determine the magnitude of this impact (Sherman et al. 2002) or to distinguish between the effects of the dams and effects of concurrent quarrying, land reclamation, urbanization, afforestation and agricultural use (Nordstrom 2000).

POLLUTANTS

The most spectacular pollution events on beaches result from oil-tanker accidents, although accidents on other vessels are far more common (Brown 1985), and oil slicks from accidental spillages and from cleaning out tanks at sea also occur (Brown & McLachlan 2002). Spillage from oil terminals and rigs, and runoff and natural seepages of petroleum hydrocarbons are also significant. Crude oil has a toxic component; it can clog delicate filter-feeding mechanisms and appendages, and act as a barrier reducing oxygen tensions in the sand in water flow through beaches (McLachlan & Harty 1981). Meiofauna and bacterial populations may recover fairly soon, but most macrofaunal species take longer to become re-established, and opportunistic polychaete worms may increase and temporarily dominate beaches (Southward 1982).

Organic enrichment may result from the discharge of raw or partially treated sewage, leading to a lowering of oxygen tensions within the sand and encroachment of anoxic layers. Excessive additions of phosphorus and nitrogen can occur from sewage, from industrial waste and use of fertilizers on farmlands (WRI 1998). For most of the developed world, nutrient pollution of nearshore waters is a substantial threat to ecosystems (Howarth et al. 2002). Consequent eutrophication, distorting nutrient cycles and leading to algal blooms and oxygen deprivation, is a major threat to the biota of beaches, especially in sheltered bays and lagoons (Gowen et al. 2000).

Factory effluents vary greatly in their impact on sandy shores, but they tend to have greater impact in confined areas such as shallow sheltered bays, which are preferred locations for factory construction. Small industries may discharge effluents into sewers or storm water drains that can open onto the beach (Brown et al. 1991). Factories and power stations frequently discharge effluent at a higher temperature. Effects may be subtle, but some animal-population densities may decline over the long term, and size and rate of growth may change, and abnormal reproduction may occur (Naylor 1965; Barnett & Hardy 1969; Nauman & Cory 1969; Barnett 1971; Hill 1977; Siegal & Wenner 1984). Radioactive pollution affects relatively few beaches, but is a source of concern near nuclear power stations and reprocessing plants (McKay et al. 1986; Brown 1994).

Pollution of intertidal beaches and surf zones most commonly arises from seaborne materials, but pollution of the dunes is more likely from land-based sources. Fertilizer residues from agricultural land behind the dunes have been found to effect changes in dune plant communities (Ranwell 1972), and pesticides may be a problem. Airborne pollution can be important, as well as runoff. More attention needs to be paid to movements of groundwater with regards to both pollution and nutrient transport (Uchiyama et al. 2000).

Accumulation of human litter brought to the shore by waves and currents from other locations is a growing problem for aquatic organisms. Non-biodegradable litter, such as plastics, is especially problematic owing to entanglement and ingestion (Ryan 1988; Brown & McLachlan 2002; Derraik 2002). Dunes are often convenient places to dispose of refuse (Mather & Ritchie 1977), and they are traps for paper and plastic blown inland from beaches; however, litter in dunes may be more of an aesthetic issue than a hazard.

Altering faunal viability or use patterns

Fauna can be affected by humans without altering the ground surface or vegetation. Pets and feral animals are threats to fauna, particularly birds. Nature-based tourism may also have severe negative impacts on coastal bird populations. The presence of people affects behaviour, reproductive success and population levels of both breeding and migratory birds (Burger 1991; Burger *et al.* 1995), while frequent intrusion by human observers leads to avian habituation and learning. Access to remote areas for nature-based tourism often requires use of vehicles (Priskin 2003), greatly increasing the impacts on fauna beyond those of a passive observer.

CURRENT TRENDS

Pressure of human activities

Human birth rates have been reduced in the most developed countries (Potts 2000), but the world population continues to grow rapidly (Chapter 1). Future pressures on sandy shores cannot be estimated solely by multiplying present pressures by predicted increase in human coastal populations because tourism is economy-driven rather than population-driven. Pressure on the sandy-shore ecosystems is intensified by the tendency of people to move to within a few kilometres of the coast (Roberts & Hawkins 1999) either for tourism or to avoid conflicts or population pressures inland (Brown & McLachlan 2002). Human development of sandy shores appears to be inevitable, and there is no reason to expect that the pace and scale of coastal development will decrease. Development is determined by growth in demand, fed by rising income and employment in inland areas. Coastal tourism is a rapid growth industry in developed and developing countries, and governments often become catalysts in resort developments by passing laws that encourage tourism and providing access routes and infrastructure (Nordstrom 2000).

Improved coastal zone management

Developmental pressures have been partially mitigated by increasing awareness of environmental issues and by practical steps to conserve ecosystems. Dramatic changes in coastal land-use regulations have occurred since the 1960s in some countries (Nordstrom & Mauriello 2001; Brown & McLachlan 2002), and marine and seashore protected areas have been established. Commercial activities, such as mining, often have to undergo rigorous environmental impact assessment, and firms are usually committed to restoring the beaches or dune systems that they adversely affect. Regulations provide greater control over activities on the beach and in dunes, resulting in more areas of dunes being protected, with less alteration of the surface of dunes, more setbacks from primary dune crests, and in some cases, reduced density of development on the landward sides of dune crests. Government programmes for protecting endangered species have revealed great potential for restoring naturally functioning sandy-shore environments. These programmes require local jurisdictions to ensure that nesting birds are not adversely affected by human activities. Elimination of human disturbances may then lead to accumulation of litter in wrack lines, colonization by plants and growth of incipient dunes (Freestone & Nordstrom 2001). Thus, the natural dynamism and topographic diversity are restored as by-products of single-species protection. Education and outreach programmes help maintain awareness of the significance of natural environments, the role participants can play and the rationale for regulations that at first seem unappealing or unnecessary. Non-governmental organizations have supplemented and complemented governmental activities and have proved successful where governmental regulations do not exist or are poorly enforced. More and more countries are moving towards adoption of international regulations, such as the London Dumping Convention and the European Union Directives concerning marine pollution (Figueras *et al.* 1997). It is possible that in the near future the discharge of raw or semi-treated sewage to sea will be banned in all countries aspiring to sustainable development, although at present Namibia is the only coastal country in Africa to have done this.

Evolution of beach and dune environments

Beaches and dunes managed by humans represent a spectrum of landforms and habitats, ranging from large natural environments to virtually featureless sand platforms. Dune growth is prevented in many locations by beach cleaning or mechanical grading to maintain beaches for recreational use. Other locations may have some semblance of dunes, although the dunes may not be built by aeolian processes, colonized by indigenous coastal species or free to migrate inland. Coastal erosion and attempts to retain a fixed shoreline result in loss or truncation of beaches and dunes. Beach nourishment operations can replace lost sediment, but nourishment is usually intended to protect buildings and provide recreation space, not to restore natural systems. If dunes are rebuilt, they are often low, narrow and linear because they are designed to form a dyke against wave attack and flooding, and not to take up recreational beach space or interfere with views of the sea. Dunes of a developed coast are usually restricted to a much narrower space than natural dunes, and they are often located where only the dynamic seaward section of a natural dune would occur. Accordingly, the species found on the backdune environment of natural dunes can only exist in many developed areas if growing conditions are enhanced by providing a relatively stable environment that is protected from inundation by sand, water and salt spray. This is done by restoring the fronting beach or maintaining sacrificial foredunes.

Coastal evolution

The kinds of coastal environments identified above represent stages that can evolve or be retained in a given condition through human efforts. A time-line of changes in relative dimensions of shoreline environments based on an idealized composite of changes observed on the east coast of the USA (Fig. 17.3) is typical of many low-lying shores; this includes a vision of what is possible in the future. Most of the changes depicted in Fig. 17.3 have occurred in several locations (such as portions of the coast of the USA and the Netherlands), but many sites are still in earlier

phases. A view along the coast of many countries reveals a series of segments that are now at different states in the temporal continuum. Only a few of the developed countries have shore segments in Phase 4, but all countries with coasts have segments in Phase 2.

Phase 1 (Fig. 17.3) depicts natural conditions on a transgressing shore, where dimensions of the natural beach and dune are initially unhindered by humans, but surfaces become increasingly altered by trampling, harvesting or grazing. Many locations that are currently migrating under near-natural conditions are being modified by non-intensive tourist activities, such as nature-based tourism in remote areas and controlled tourism in protected areas near developed areas. Many locations subject to nature-based tourism are evolving, in that increasing numbers of visitors, supported by increasing investment in support infrastructure, will eventually lead to elimination of much of the natural-resource base and conversion to Phase 2 (Fig. 17.3). Seashore protected areas may not undergo intensive development (Phase 2), but many such areas are threatened by erosion because of rising sea levels and

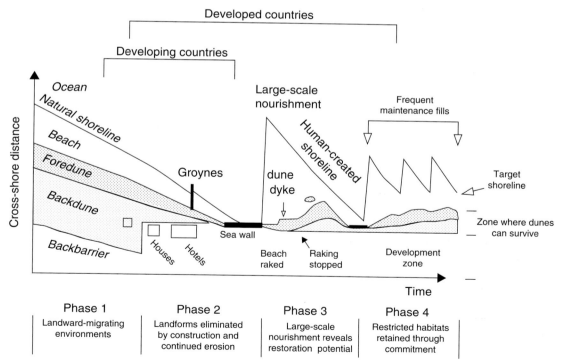

Fig. 17.3. Common scenarios of development on the east coast of the USA.

restricted sediment inputs owing to human development in adjacent areas, and they may require artificial nourishment, just as the developed areas do in Phase 4 (Fig. 17.3).

Phase 2 represents the conversion of landward portions of the dune to cultural environments, with increasing intensities of development through time. Increasing levels of investment make larger-scale shore-protection projects more feasible. The end of Phase 2 depicts the loss of beach and dune habitat that is one consequence of using static protection structures without beach-fill (Hall & Pilkey 1991). Small-scale beach nourishment may be conducted within this phase, but such operations provide only short-term relief of erosion problems. Many segments of shore in developing countries are within Phase 2 (Fig. 17.3) and now face decisions on whether large-scale shore-protection projects are feasible and whether sea walls or beach nourishment are the appropriate solution.

Phase 3 represents the initial commitment to large-scale nourishment projects (Fig. 17.3). The construction of a wide beach creates the potential for full environmental restoration, but the effect differs according to subsequent management actions. The new sand may be treated solely as a recreation platform, graded and raked, preventing growth of natural dunes. Sand fences may be placed on the backbeach to provide flood protection or prevent sand drift, creating the linear sand dyke often associated with a developed coast. Suspension of beach raking leads to a wider dune with greater topographical and biological diversity (Freestone & Nordstrom 2001), but loss of created environments can occur if administrative delays prevent nourishment from occurring in a timely fashion (end of Phase 3).

Fencing and vegetation plantings can be used to build dunes in a sand-deficient environment but, inevitably, beach nourishment may be required to maintain a healthy well-vegetated dune (Mendelssohn *et al.* 1991). Phase 4 depicts the need for nourishment to retain dune integrity, given erosion and competition for space for recreation and construction of user facilities (Fig. 17.3). Beach and dune environments are prevented from migrating landward, so they must be maintained on the seaward side of private properties and public promenades and thoroughfares. A significant aspect of the management approach in Phase 4 is that the restored dune landscape (not just the recreation beach or the human development) is considered a resource and is protected by nourishment. Preservation of existing landforms and habitats argues for small frequent nourishment operations (Dette & Raudkivi 1994; van Noordwiijk &

Peerbolte 2000). Design studies for projects in New Jersey (USA) include nourishment at the relatively high frequency of once in every 3 years (US Army Corps of Engineers 1989), which should be adequate to retain the natural environments that form to seaward of cultural features. Restored dunes in developed areas are not as wide as landforms in the undeveloped enclaves that remain in Phase 1, but judicious use of sand fences and vegetation plantings could recreate the types of habitats lost, if not the area and spatial relationships (Freestone & Nordstrom 2001; Nordstrom *et al.* 2002).

SANDY SHORES TO THE YEAR 2025

Changes in the status of sandy-shore ecosystems are likely to occur to the year 2025 because of the effects and implications of global warming and increasing human pressures, all of which are already being felt. Increased UV radiation could have some effect, but the effect would be less on intertidal sandy beaches than on most other ecosystems owing to the cryptic nature of the biota. The biota of the surf zone and the flora of the dunes would be at greater risk, but because of reduction in the emission of aerosols, further depletion of the ozone layer is unlikely (Friedrich & Reis 2000).

Global warming

There have been significant advances in the development of mitigating scenarios for greenhouse-gas emissions, including costs of mitigation (IPCC [Intergovernmental Panel on Climate Change] 2001*c*). However, even if emissions become stabilized or actually reduced, which is unlikely, global warming and its consequences will continue to increase during the present century because of the lag in climate response (Wigley 1995).

Changes in sea temperature can have severe effects on marine biota, as witnessed during events such as El Niño. The impacts of El Niño on some South American beach ecosystems give an indication of changes that might be expected from rapid global warming. On Peruvian beaches, the abundance of many species plummeted during El Niño events, but this was followed by rapid recovery when conditions returned to normal (Arntz *et al.* 1988; Tarazona & Parendes 1992). Subtidal areas that had been anoxic saw increases in species abundances and diversity, and extension of vertical distribution in many species during El Niño events (Tarazona *et al.* 1985, 1988). These changes are largely related to changes in productivity, indicating that effects of elevated temperature may be indirect.

In comparison with El Niño events, sea-temperature changes by the year 2025 due to global warming are predicted to be very gradual, allowing marine populations time to adjust. Predicted temperature increases are unlikely to totally disrupt sandy-shore ecosystems. An average atmospheric temperature rise of 1–5 °C is predicted by the year 2100 (IPCC 2001a). Temperature rise for the oceans as a whole is likely to be about one-half of these values, given a lag of some 20 years between air and sea temperatures (Shannon et al. 1990), although some lagoons and shallow bays may mirror the atmospheric temperature rise more closely. The worst scenario for the year 2025 would appear to be a water temperature increase of 1 °C, and significantly less for open ocean beaches. These are small changes compared with those experienced by some beaches close to discharges from power stations or those subject to El Niño. Aquatic sandy-beach animals are often subject to relatively rapid changes in temperature, particularly in areas of upwelling (Brown & McLachlan 1990).

Sandy-shore animals seldom experience temperatures close to their upper tolerance levels, an exception being some ocypodid crabs (Fishelson 1983). All sandy-beach animals are capable of burrowing into the substrata (Brown 1983) and thus of escaping hostile surface temperatures. However, if the temperature rise due to global warming were added to natural warm-water events such as El Niño, the combination could have severe negative impacts.

Increased water temperatures may affect proteins, muscle function, cardiovascular performance, reproduction, development and growth, metabolism and increased sensitivity to pollution in fish (Wood & McDonald 1997), and may do so also in invertebrates, although many experimental temperature increases greatly exceeded those anticipated by the year 2025. Effects of temperature rise due to global warming up to 2025 may be subtle rather than dramatic. Some redistribution of species may occur, with biota from the tropics and subtropics tending to invade higher latitudes. For example, the distribution of Donax mussel species may change, as the genus appears to be limited by the 5 °C sea-surface isotherm (Brown & McLachlan 1990). Some distributional contraction may occur; regions such as Western Europe may become cooler because of disruption of currents such as the Gulf Stream (IPCC 2001a). Changes in temperature regimes may affect the growth rates and breeding seasons of some sandy-shore species. Among the more subtle effects of temperature

change are alterations in the speed of burrowing into the sand by invertebrates (McLachlan & Young 1982). However, given the timescale and the relatively small changes in ambient temperature for most beaches, compensatory metabolic adaptations to temperature change can be predicted for most species.

Rises in sea level are more serious. Average sea-level rise is likely to be below 30 cm by 2025 and, as with temperature changes, this rise is slow (Brown & McLachlan 2002). The mobile, highly adaptable, sandy-shore biota will not be at direct risk from it. The most significant threat to these biota is loss of habitat, especially if sea-level rise is accompanied by increased storminess. The rate of beach erosion will generally be increased, and some beaches displaying long-term accretion may suffer a reversal of this trend. Some narrow beaches may be eliminated, while others lacking dune systems will become severely restricted. Systems that are backed by extensive dune systems should suffer less, the habitat remaining essentially unchanged, albeit moving landwards. Erosion may be partly mitigated by artificial beach nourishment, but eroding natural environments are not likely to be primary targets of nourishment operations. It is possible that new beaches may form in some areas.

Among the less publicized features of enhanced atmospheric carbon dioxide is the possible increase in primary production through enhanced photosynthesis (Melillo et al. 1993). Surf-zone diatoms might increase in density and result in enhanced organic input to intertidal beaches. Any changes in nearshore oceanographic conditions will have consequences for sandy-shore ecosystems, and perhaps particularly for the surf zone. Not all these consequences are necessarily negative but they may be complex (Shannon et al. 1990). For example, increased upwelling would enhance the supply of nutrients and actually lower the average annual water temperature of the surf zone, but increased wind velocities or durations might have adverse effects, including offshore advection and turbulence. There might be a negative correlation between upwelling events and larval recruitment; during upwelling, larvae may be transported away from the coast into deeper water and never reach the coast (Roughgarden et al. 1988).

Direct human pressures

Direct human fishing pressure along sandy coastlines is likely to increase (Brown & McLachlan 2002), and this may lead to ecosystem degradation. Much of the pressure can be controlled, but small-scale artisanal fisheries are frequently not amenable to such management and present a growing regional-scale threat to sandy beach macrofauna. This is especially so in the context of predictions of growing human populations and food demand. The predicted increase in global human fish consumption by 25% by the year 2030 (FAO [Food and Agriculture Organization of the United Nations] 2002) is likely to be partially met through aquaculture in developed nations. In many of the poorest states it will be met through increased take by artisanal fishers. In the mid 1990s, about 200 million people depended directly on artisanal fisheries, and their catch accounted for about 50% of all human fish consumption (Pauly 1997).

In developed areas, a conflict exists between the tendency for a healthy natural beach system to be dynamic and the human desire for a system that is stabilized to make it safe, maintain property rights or simplify management (Nordstrom 2003). Given the human preference for stability, any acceleration in rate of shoreline erosion is likely to generate a rapid response from coastal managers. In developed industrialized countries, where mitigation efforts are cost effective, it is likely that accelerated changes owing to natural causes may have less impact than human actions to modify sediment supply or protect property and maintain a stable or predictable resource base (Titus 1990; Jones 1993; Midun & Lee 1995; Nordstrom 2000).

Managed retreat from the coast (i.e. removal of structures and withdrawal landwards) would resolve problems of erosion and provide the space over which new landforms and biota could become re-established, but a retreat option appears unlikely by the year 2025. It is likely that only sparsely developed shores would be abandoned. Abandonment may not occur on coastal barriers where creation of new inlets poses a threat to houses bordering estuaries or to existing inventories of commercial shellfish; in such cases, sparsely developed coastal barriers may be maintained as dykes (Bokuniewicz 1990). Most local governments and property owners would probably advocate management options that approach the status quo under accelerated sea-level rise (Titus 1990). The great value of land and real estate is the driving force; too much may be invested on highly developed shores to consider anything short of holding the existing shoreline (Nordstrom & Mauriello 2001).

Many measures are in place to protect and enhance natural environments in developed countries, and these measures can be employed to counteract many negative impacts of climate change, tourism and increasing human populations, expected up to the 2025 time horizon. In many economically less-developed countries, few measures have been taken to protect coastal or other environments. More than half the countries in Africa have been involved in civil war in the last few decades and political instability is the norm rather than the exception. Depressed or volatile economies in other areas severely restrict options. In such circumstances, environmental conservation is hardly ever mentioned and what legislation there might be is not enforced. Sandy shores are generally low on any list of conservation priorities. Some countries may attain a measure of stability over the next two decades and begin to devote time and energy to conservation matters, but in most cases this hope appears to be remote. Exploding human populations, poverty, AIDS, unemployment, economic ills and crises of governance are likely to continue and dominate agendas until at least 2025, even if armed conflicts cease.

Protection of existing human-use structures appears appropriate for intensively developed areas, but it is unclear whether hard engineering structures such as sea walls, or environmentally compatible alternatives such as beach nourishment, will be considered the more cost-effective solution (Neves & Muehe 1995). Sea walls may be considered appropriate where sandy beaches now occur, although nourishment may be preferred in resort areas (Dennis *et al.* 1995; French *et al.* 1995; Volonte & Nicholls 1995). The past preference for engineering solutions to rising sea levels may result in more and larger sea walls (Bird 1993; Brunsden & Moore 1999), particularly because these structures do not require major institutional land-use changes (Titus 1990).

Maintenance of ecosystems will require proactive human efforts to maintain sediment budgets, preserve space for biota to grow and thrive, and generate appreciation of the intrinsic, recreational and educational aspects of nature. It will be prudent to consider alternatives that enhance the capacity of beach systems to respond to perturbations in addition to simply resisting change (Klein *et al.* 1998; Nicholls & Branson 1998). Solving existing problems using solutions that maintain biodiversity will increase flexibility in the face of climate change (Nicholls & Hoozemans 1996). Many problems of erosion and habitat loss may be mitigated by beach nourishment, and it is likely that this practice will become more widespread in the future (Brown & McLachlan 2002). The cost of nourishment will be considerable, but could be trivial relative to the economic losses that would accompany potential erosion (El-Raey *et al.* 1995), or to the value of beaches for tourism (Houston 2002). Restoration of natural environments in locations enhanced by beach nourishment indicates that habitats can be maintained in the face of sea-level rise, but ecological values may not be a by-product of nourishment operations unless beaches are managed to achieve these values (Nordstrom 2000).

CONCLUSIONS

Sandy shores are dynamic, with exchanges of material occurring across dunes, beach faces and surf zones. The fauna of these shores is of necessity highly mobile and adaptable to changing conditions. Humans may initially be attracted to the shore for its dynamic qualities, but their attempts to maintain fixed facilities and a predictable resource base often restrict natural mobility and contribute to loss or truncation of natural environments.

The present conservation status of sandy-shores is spatially variable. Some remote areas are virtually pristine, with low human population levels and little alteration by tourism. Others, in industrial or urban areas, lack macro-fauna and are seriously degraded due to chronic pollution or the hardening of surfaces, destruction of dunes and construction of sea walls and other structures. Between these two extremes is a range of sandy-shores suffering impacts from a variety of causes.

Few generalizations can be made about the future of sandy-shore ecosystems as a whole, but the chief long-term threat facing them, virtually worldwide, is loss of habitat resulting from increasing erosion attendant on sea-level rise, changes in storm patterns associated with global warming, and construction and use of human facilities. On a regional basis, sandy-shore ecosystems may experience increasing pressures from artisanal fisheries in developing nations, especially as other food sources become scarce.

Temperature rise by the year 2025 is likely to have relatively subtle effects on the biota, but projected changes in sea level and current patterns may significantly alter sand transport, enhancing erosion and loss of habitat in many areas but possibly favouring accretion in others. Human pressures on the coast are expected to increase, in some cases dramatically, but increased pressure on sandy shores may be mitigated by improved legislation and management resulting from better understanding of sandy-shore processes. Underdeveloped countries, notably in Africa, are expected to lag behind in this regard, with human pressures increasing on coastal ecosystems for the foreseeable future.

Many management actions can be employed to preserve natural environments and expedite return of environments that have been lost. These include: maintaining stricter control over access routes, particularly through dunes; limiting the number of people on beaches or excluding them where rare, endangered or threatened species occur; banning off-road vehicles or granting permits for their use only under special circumstances and only on the beach foreshore; eliminating or severely restricting mechanical cleaning of the beach; controlling collection of beach animals for food and recognizing that activities associated with collection may be more damaging than removal of the animals; avoiding practices that negatively impact surf zones which are important nursery areas for fish species; devoting special attention to threatened species of shore-birds and turtles that come ashore to breed; creating more nature reserves, especially for threatened species; and keeping beach managers, would-be developers, ecotourists and the general public well informed of legislative and non-legislative environmental initiatives and the rationale for them.

The immediate global reduction of greenhouse-gas emissions is a critical goal, but proper management at a more local level may help significantly to at least retard the degradation of sandy-shore ecosystems. Improved legislation to protect the environment and its biota is still required in many countries. This legislation needs to be enforced and supplemented by education programmes. In many instances, it is assumed that the only reason for maintaining sandy shores in a near-pristine state is to attract tourists; this attitude must be revised. Better communication between scientists, legislators and beach managers is still required (Brown *et al.* 2000), but the messages delivered by scientists may be different from those delivered in the past, in that they may involve compromise solutions to achieve both natural and human goals in restricted space.

Priority must be given to the avoidance of structures or activities that may reduce natural sand transport either alongshore or across the shore, including exchanges between beaches and dunes. Artificial beach nourishment is likely to be increasingly employed in developed countries, and should significantly influence sandy-shore ecology in the future. Artificial beach nourishment is infinitely preferable to the construction of sea walls or other sand-retaining structures, but beach and dune profiles and sediments should be similar to those that occurred naturally. The full potential of nourishment operations will not be realized without a multi-objective management approach that addresses habitat improvement in addition to the traditional goals of protection from erosion and flooding, and provision of recreation space.

18 · Seagrass ecosystems: their global status and prospects

CARLOS M. DUARTE, JENS BORUM, FREDERICK T. SHORT AND DIANA I. WALKER

INTRODUCTION

Seagrasses form important underwater marine and estuarine ecosystems on all continents except Antarctica. Seagrass ecosystems are highly productive, forming extensive habitats which support highly diverse communities. The seagrasses themselves assimilate and cycle nutrients and other chemicals. Their extensive above- and below-ground biomass traps sediments, reducing coastal turbidity and erosion, as well as providing habitat for other organisms both attached and free-living (Hemminga & Duarte 2000).

Globally, seagrasses are in decline, almost entirely because of human impact. Causes range from changes in light attenuation due to sedimentation and/or nutrient pollution, to direct damage and climate change. Over the last two decades, the loss of seagrass from direct and indirect human impacts amounts to 18% of the documented seagrass area (Green & Short 2003). Seagrass

Aquatic Ecosystems, ed. N. V. C. Polunin. Published by Cambridge University Press. © Foundation for Environmental Conservation 2008.

ecosystems are in need of active management to ensure their persistence and long-term survival, requiring education, increased awareness, management and conservation on a global scale.

Seagrass meadows are coastal benthic ecosystems found submerged in nearshore waters; the group of plants collectively known as seagrasses consists of about 60 similarly functioning angiosperm species (den Hartog 1970; Hemminga & Duarte 2000). These are all rhizomatous monocotyledons, within two families (Potamogetonaceae and Hydrocharitaceae) encompassing 12 genera, which are restricted to the marine environment to complete their life cycle. Seagrasses are flowering plants that pollinate on or under the water's surface and produce fruits and seeds which are important to maintaining the habitat, although their primary mode of expansion is clonal growth. Seagrasses are rooted in shallow sandy to muddy coastal sediments receiving at least 10–20 % of irradiance incident at the water surface (Hemminga & Duarte 2000; Green & Short 2003). A few species (such as genera *Phyllospadix*, *Posidonia* and *Thalassodendron*) can grow on rocks or very thin sediments (den Hartog 1970; Hemminga & Duarte 2000). Seagrasses thrive in a range of salinities, from 5 PSU (practical salinity units), or even lower, to 60 PSU (McMillan & Moseley 1967; Walker 1989).

Where they occur, seagrasses represent the dominant component of the shallow marine landscape. Most seagrass meadows are monospecific, although tropical and subtropical meadows may contain up to a dozen species, particularly in the Indo-Pacific region, which contains the most diverse seagrass flora (Fig. 18.1). Seagrass meadows form lush ecosystems and, on an area basis, are some of the most productive communities on the planet with an average total seagrass biomass (dry) of 460 g per m², half of which develops below the sediment as rhizome and root material, and an average net primary production (dry) of 5 g per m² per day (Duarte & Chiscano 1999). The primary production of seagrass meadows is augmented by the contribution of micro- and macroepiphytic and benthic autotrophs, which contribute about as much to ecosystem production as the seagrass itself (Hemminga & Duarte 2000). Seagrass habitat interacts with other critical coastal habitats, namely saltmarshes (Chapter 11) and bivalve reefs in temperate regions, and mangrove forests (Chapter 12) and coral reefs (Chapter 16) in the tropics.

Mapping the extent of the global seagrass ecosystem is far from complete, the most recent estimate of the global seagrass area being 177 000 km² (Green & Short 2003), but this is without doubt much too low. In most countries, no generalized spatial mapping of seagrass distribution has been conducted and seagrass locations are only known from localized observations. In many areas of the globe, seagrass observations have not been made. Over the past decade, increased investigation has led to the discovery of seagrass distribution in areas where it was previously unknown (for example the deep-water seagrass beds in the Gulf of Mexico: Hammerstrom *et al.* 2006), and also to documentation of absence of seagrass where it might be expected. For instance, seagrasses are often absent from coastal waters in the vicinity of cities and towns both in

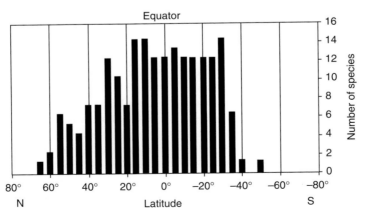

Fig. 18.1. Global species richness of seagrasses by latitude. (Based on Green & Short 2003.)

developed and developing countries. More seagrass mapping and monitoring are needed to document the status and trends of seagrass populations. Current knowledge suggests that seagrasses are being lost at a greater rate than is being documented.

Seagrasses are important as habitats supporting high biodiversity, yet they occur at the land–sea interface and are particularly vulnerable to anthropogenic impact. Their lush canopies provide habitat to a rich faunal assemblage, which uses seagrass meadows as nursery and recruitment grounds (Thayer *et al.* 1975; Green & Short 2003). The seagrass fauna encompasses various trophic guilds, including large herbivores in both tropical (for example dugongs, manatees and sea turtles) and temperate (for example swans and geese) waters, and invertebrates. Seagrasses are also food for sea urchins and fish (Pollard 1984; Heck & Valentine 1995). Worldwide, seagrasses provide crucial links in the food web for animals and people, including subsistence gleaning for protein on tropical reef flats by villagers, nursery resources for commercially important finfish species and habitat for commercial and recreational bivalve fisheries.

Seagrasses develop dense canopies that impede water movement, damping current and wave energy (Grizzle *et al.* 1996) and turbulence, reducing sediment resuspension (Short & Short 1984). Seagrass meadows thus increase sedimentation rates (Gacia & Duarte 2001; Agawin & Duarte 2002), acting as a filter of coastal waters. The high total organic inputs from both the water and the plants fuel sediment microbial activity, which is typically enhanced in seagrasses relative to adjacent sediments that are bare of vegetation (Hemminga & Duarte 2000; Holmer *et al.* 2001). Calcifying organisms such as coralline algae, molluscs and Foraminifera are important components of seagrass meadows, contributing substantially to the formation of carbonate sediments (Walker & Woelkerling 1988).

Seagrass meadows provide important services such as nutrient and gas cycling, through the intense biogeochemical processes within the meadows; habitat provision for a diverse assemblage of plants and animals, often of commercial value; and erosion control, through the dissipation of wave energy, the stabilization of underlying sediments and the delivery of important amounts of organic and inorganic material to other parts of the marine environment (see Hemminga & Duarte 2000). Seagrass meadows rank amongst the ecosystems giving greatest added value in terms of the services they provide to marine ecosystem processes and ultimately to society (Costanza

et al. 1997), which should inspire conservation efforts to preserve them.

Seagrass meadows are and have been highly dynamic ecosystems. Because of the narrow fringe that seagrass meadows occupy in the coastal zone, these ecosystems have experienced major, undocumented regression and progression with sea-level change through geological time. Although recent rates of sea-level change have been in the order of, at most, a few millimetres per year, these changes may translate to significant rates of horizontal inundation or desiccation with increasing or decreasing sea level, respectively. However, seagrasses cannot keep up with human-imposed change as it is now happening globally.

While seagrasses are recognized as priority subjects for conservation efforts in international (such as the Convention on Biological Diversity and the EU Habitats Directive) and national frameworks, there is evidence of significant widespread decline (Short & Wyllie-Echeverría 1996; Hemminga & Duarte 2000; Green & Short 2003; Spalding *et al.* 2003). These declines have multiple and sometimes synergistic sources, many of which are related to direct and indirect anthropogenic pressures including climate change (Short & Neckles 1999; Duarte 2002). As many of these pressures are expected to continue and even increase over the coming decades, there is a need to examine the present status and likely prospects for seagrass meadows globally, thereby helping to inspire effective conservation efforts. In this chapter, the current status of seagrass worldwide is examined, including what is known to date of seagrass ecosystem distribution. Threats to seagrass are discussed that are largely human-induced but sometimes natural, as well as the measures that can be adopted to control, or ideally reverse, seagrass decline. Finally, the prospects for seagrasses to the 2025 time horizon are considered.

SEAGRASS STATUS

The extent of the seagrass habitat is documented for only a fraction of the world's coastline (Green & Short 2003). Major losses of seagrass have been documented in Europe (for example De Jonge & De Jong 1992; Marbà *et al.* 1996; Hily *et al.* 2003), in the developed parts of the USA (for example Orth & Moore 1983; Koch & Orth 2003; Short & Short 2003), as well as in Australia (Walker & McComb 1992) and in localized areas of the developing world (Spalding *et al.* 2003). The loss of seagrass has in some locations led to loss of fisheries, changes in bottom substrata and breakdown of ecosystem structure. World-wide, vast healthy seagrass meadows still exist in areas not impacted by human activity. The magnitude of seagrass loss has been greatest in developed countries, but the current rate of loss is highest in the developing world.

Regional seagrass status and trends

AFRICA

Very little is known about seagrasses in West Africa, with most of the reports pertaining to *Cymodocea nodosa* and *Zostera noltii* in the Canary Islands and on the Mauritanian coast (Van Lent *et al.* 1991; Vermaat *et al.* 1993; Reyes *et al.* 1995). In East Africa, there is high species diversity in mixed species stands along the central coast (Bandiera & Gell 2003; Ochieng & Erftemeijer 2003), which are heavily used by humans for gleaning and artisanal fisheries, both of which result in significant loss of seagrasses. In many areas, seagrasses grow adjacent to mangrove forests (Chapter 12). The main human impacts on seagrass result from sedimentation following watershed clearing and development as well as direct human activities in the coastal zone, but losses have not been measured.

MIDDLE EAST

In the Red Sea, there are 11 seagrass species, often growing in mixed beds (Lipkin *et al.* 2003). One species (*Halophila stipulacea*) can penetrate to great depths (70 m). Nutrient pollution and coastal development are impacting seagrass beds in the Red Sea. In the Arabian Gulf, there are extensive seagrass beds supporting large dugong populations, but also more direct industrial impacts, because of oil extraction, than in other parts of the region (Phillips 2003).

INDIA

Many seagrass species are found extensively around the subcontinent of India and its islands, although little information exists on their spatial distribution (Jagtap *et al.* 2003). Upland deforestation, mangrove destruction, shoreline construction and storms have all had important impacts on seagrasses. Dugong populations were recently eliminated in India after a 50-year period of decline. Large human population pressures along the coast continue to threaten seagrasses through vessel traffic and sewage disposal.

AUSTRALIA

Australia hosts the highest number of seagrass species of any land mass in the world (about 30 species: Walker 2003;

Coles *et al.* 2003*a*), with large multispecies meadows across vast shallow areas of the coastal fringe. The major threats to seagrass are industrial development, nutrient loading near population centres, port development, coastal agriculture and fisheries. However, seagrasses are better protected by legislation and enforcement in Australia than anywhere else, and because of this and small human populations, seagrass losses are lower than on other continents.

SOUTH-EAST ASIA

South-East Asia is the global centre of biodiversity for seagrasses, as it is for coral reefs and other biota (Bujang & Zakeria 2003; Kuriandewa *et al.* 2003; Supanwanid & Lewmanomont 2003). There are extensive seagrass beds throughout the area, but very few have been mapped. Mining of metals and sand, as well as watershed clearing for agriculture, aquaculture and industry, all negatively impact seagrasses (see Fortes 1988; Terrados *et al.* 1998).

EAST ASIA

Seagrass beds in the countries of temperate Asia (Japan, Korea, China and northern Vietnam) are dominated by *Zostera* spp. which occupy much of the coastal soft substrate (Aioi & Nakaoka 2003; Lee & Lee 2003). No actual distribution maps are available. Seagrasses have suffered severe losses in industrialized parts of the coast, in many cases from hardening of the shoreline and land reclamation through filling of tidal areas for industrial development. Direct fisheries impacts also contribute to seagrass loss.

PACIFIC ISLANDS

The many islands of Micronesia, Melanesia and Polynesia are typically surrounded by coral reef flats with extensive seagrass habitat and high seagrass biodiversity (Coles *et al.* 2003*b*). However, the extent of seagrasses across the region is poorly documented. The greatest impacts come from coastal development, artisanal fishing and gleaning on the reef flats, and tourism infrastructure development.

NORTH AMERICA

Seagrass is found extensively along the coast except in high-population areas where the habitat has been lost owing to development and pollution discharge (Wyllie-Echeverria & Ackerman 2003; Short & Short 2003; Koch & Orth 2003). The dominant North American seagrasses are eelgrass (*Zostera marina*) in temperate waters and turtle-grass (*Thalassia testudinum*), accompanied by additional species, in subtropical areas (Onuf *et al.* 2003). Eelgrass experienced a severe die-off during the 1930s on the east coast of North America and in Europe; although it recovered in many places, human impacts have prevented it from re-establishment in parts of its range. In the USA, seagrass distribution is well documented; major losses have been identified on both coasts and in the Gulf of Mexico. Canada and Central America have experienced fewer declines but are faced with threats owing to increased coastal development.

CENTRAL AND SOUTH AMERICA

Seagrasses in the Caribbean are well documented but only sporadically mapped and are facing threats particularly from agricultural runoff, tourism and urban development (Creed *et al.* 2003). Very little is known about seagrasses in the rest of South America (Creed 2003). No one has yet investigated whether seagrass occurs for long stretches of the South American coast. Brazil has several tropical species along its coastline, and seagrasses there experience threats from urban development and watershed deforestation.

EUROPE

The Mediterranean Sea is dominated by *Posidonia oceanica*, a reef-forming seagrass that persists over centuries (Procaccini *et al.* 2003). Heavy urbanization has impacted seagrasses, such as through eutrophication, organic inputs from aquaculture and coastal engineering, and 30–40% of *P. oceanica* in the western Mediterranean has been lost in the last few decades. In northern Europe (Hily *et al.* 2003; Bostrom *et al.* 2003), where *Zostera* species dominate, losses of seagrass have also been extensive and include indirect impacts from nutrient pollution and increased turbidity as well as direct impacts from coastal modification, fishing, boating and aquaculture. For example, in the Baltic Sea, shifts in eelgrass distribution from deeper to shallower water reflect loss of water clarity over time (Fig. 18.2).

Global impacts and losses

Globally, the estimated loss of seagrass from direct and indirect human impacts amounts to 33 000 km^2, or 18% of the documented seagrass area, over the last two decades

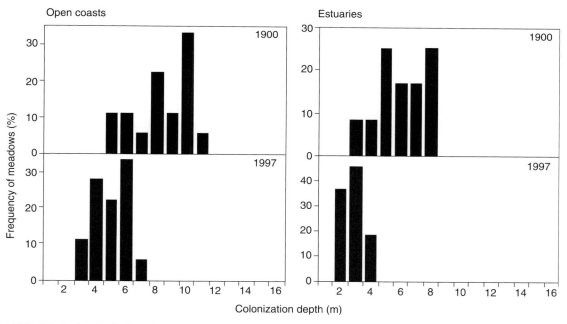

Fig. 18.2. Colonization depth of Danish eelgrass meadows along open coasts and in estuaries, 1900–97. (Modified from Boström *et al.* 2003.)

(Green & Short 2003), based on an extrapolation of known losses (Short & Wyllie-Echeverria 2000). Reported losses probably represent a small fraction of those that have occurred and many losses may remain unreported; indeed actual losses may never be known because most seagrasses leave no long-term record of their existence. Seagrasses exist at the land–sea margin and are highly vulnerable to pressures from human populations, which live disproportionately along the coasts (Nicholls & Small 2002). Human population growth, with concomitant increased pollution, hardening and alteration of coastlines and watershed clearing, threatens seagrass ecosystems and has resulted in substantial and accelerating seagrass loss over the last 20 years (Hemminga & Duarte 2000; Short & Wyllie-Echeverria 2000; Spalding *et al.* 2003).

The likely primary cause of seagrass loss globally is reduction in coastal water clarity, both from increased nutrient loading and increased turbidity. The primary cause of nutrient enrichment in coastal waters is anthropogenic loading from coastal watersheds (Valiela *et al.* 1992; D'Avanzo & Kremer 1994; Duarte 1995; Short & Burdick 1996; Short & Wyllie-Echeverria 1996; Tomasko *et al.* 1996; Borum 1996; McMahon & Walker 1998). In general, pristine coastal seas are nitrogen limited, and nitrogen inputs from point and non-point sources cause eutrophication (Ryther & Dunstan 1971; Nixon & Pilson 1983) and alter ecosystem structure and function (Valiela *et al.* 1992; Borum 1996). Increased nutrient inputs are also occurring adjacent to industrialized regions of the world through direct atmospheric deposition of nitrogen (Paerl 1985). Equally, losses of water clarity come from increasing inputs of nitrogen and phosphorus from waste discharge, atmospheric deposition and land runoff along temperate more-industrialized coasts.

In contrast, in tropical areas, the major impact on water clarity is the discharge of vast quantities of sediment into coastal waters as a result of poor land-use practices, watershed deforestation and coastal clearing (leading to erosion), and lack of erosion controls (Fortes 1988; Duarte *et al.* 1997; Bach *et al.* 1998; Terrados *et al.* 1998). The relatively high light requirements of seagrasses make them vulnerable to reduced light penetration of turbid coastal waters and thus erosion and sediment transport lead to seagrass elimination. This is one of the major threats to seagrass ecosystems in South-East Asia (Fortes 1988; Bach *et al.* 1998; Terrados *et al.* 1998), where deforestation leads to sediment yields to the coastal ocean ten times higher than in any other region of the world (Miliman & Meade 1993). The deposition of 681 g dry per m^2 per day experienced in a *Thalassia hemprichii* bed in Bay Tien (Vietnam) (Gacia *et al.* 2003) caused a level of burial that cannot be sustained by most seagrass species (Duarte *et al.* 1997).

Direct human impacts on seagrasses threaten seagrass habitat, particularly in areas with dense human populations. The direct impacts include: (1) fishing and aquaculture, (2) introduced exotic species, (3) boating and anchoring and (4) habitat alteration (dredging, reclamation and coastal construction). Fishing methods such as dredging and trawling (Riemann & Hoffmann 1991) may significantly affect seagrasses by direct removal and by modifying the benthos (Fig. 18.3). Damage to *Z. marina* by scallop dredging reduces shoot density and plant biomass (Fonseca *et al.* 1984), and digging for clams can also exert extensive damage (Orth *et al.* 2002). Many of these impacts remain unquantified as yet, and their long-term effects are unknown. In the Mediterranean, the use of certain types of fishing gear like bottom trawls has detrimental effects on seagrass beds; in some areas, trawling marks cover 18% of the meadow surface (Pasqualini *et al.* 2000) (Fig. 18.3a). Mussel harvest in the Dutch Wadden Sea is believed to be a major factor in the loss of *Z. marina* and *Z. noltii* there (De Jonge & De Jong 1992).

Worldwide, coastal areas are being targeted for aquaculture developments. Aquaculture of fish and algal biomass has been shown to produce major environmental impacts, particularly because of shading, eutrophication and sediment deterioration through excess organic inputs (Seymour & Bergheim 1991; Ackefors & Enell 1994; Shireman & Cichra 1994; Dosdat *et al.* 1995; Holmer *et al.* 2001). The effects of fish farms (Fig. 18.4) and other aquaculture developments are of concern as areas of productive seagrass habitats are often targeted for such developments, as in the Philippines (Holmer *et al.* 2001), the Gulf of Thailand (T. Ruangchoy, personal communication 1996) and along the Mediterranean coast (Delgado *et al.* 1999; Pergent *et al.* 1999). Fish pens can cause seagrass loss (Delgado *et al.* 1999; Pergent *et al.* 1999), and mussel culture adversely affects *Z. marina* and *Z. noltii* beds in France (De Casabianca *et al.* 1997). Extensive and intensive aquaculture developments are expanding worldwide, increasing the risk of more seagrass loss.

The introduction of exotic marine organisms, from accidental release, vessel ballast water, hull fouling and aquaculture, remains an area of concern, particularly where the introduced species are competitors for soft-bottom substratum such as the alga *Caulerpa taxifolia* (Meinesz *et al.* 1993) and the fan worm *Sabella spallanzanii* (Lemmens *et al.*

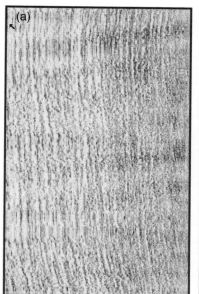 

Fig. 18.3. Side-scan sonar echograms of two *Posidonia oceanica* beds at *c*.7 m depth in the Spanish Mediterranean littoral showing (a) impacts of trawling as parallel scars running vertically along the image, and (b) scars left by anchor displacement along the meadow. Each image corresponds to an area *c*.35 m × 15 m. (Images courtesy of Mediterráneo Servicios Marinos.)

1996). Large-scale engineering projects have also resulted in species invasion, such as that by *Caulerpa racemosa*, introduced to the Mediterranean through the Suez Canal (Lipkin 1975; Piazzi *et al*. 1994), which also allowed the introduction of the seagrass *Halophila stipulacea* (Lipkin 1975).

Boat-propeller damage to seagrass communities is prevalent in shallow areas with heavy boat traffic, particularly in the Florida Keys (Zieman 1976). Boat anchoring leaves scars in *P. oceanica* landscapes (Francour *et al*. 1999) (Fig. 18.3), as do boat moorings (Walker *et al*. 1989). Return of large temperate meadow-forming seagrasses to mooring scars may take decades, and docks and piers shade shoreline seagrass, an effect that may fragment the habitat (Burdick & Short 1999). Boating may also be associated with organic inputs in areas where boats do not have holding tanks (Marbà *et al*. 2002).

Development of the coastline, particularly related to increased human population pressure, leads to alteration and fragmentation of coastal seagrass habitats. Coastal development (construction of ports, marinas and groynes) is usually localized around human population centres. Housing developments impact coastal water quality, the number of houses in a watershed being directly correlated with the rate of seagrass loss (Short & Burdick 1996; Short *et al*. 1996). Construction of the causeway at the southern end of Cockburn Sound (Western Australia), in combination with industrial pollution, destroyed existing seagrass (Cambridge *et al*. 1986). Construction of roads through shallow waters which modifies water flow may affect circulation and lead to seagrass loss, such as observed in Cuba where coastal waters were rendered hypersaline by the effects of road construction over shallow lagoon areas.

Dredging and reclamation of marine environments, either for extraction of sediments or as part of coastal engineering or construction, can remove seagrasses. Land reclamation directly eliminates seagrass habitat and results in hardening of the shoreline, further eliminating seagrass habitat, as seen throughout Tokyo Bay (Japan). Groynes alter sediment transport in the nearshore zone. Dredging removes seagrass habitat as well as the underlying sediment, leaving bare sand at greater depth, resulting in changes to the biological, chemical and physical habitat values that seagrasses support (Walker *et al*. 2001). Beach nourishment (see Chapter 17) may impact adjacent seagrasses by

Fig. 18.4. Heavily epiphytized seagrass (*Posidonia oceanica*) under fish cages in Sicily. (Photograph by S. A. Sørensen.)

delivering sediment that may shade or bury the seagrasses. Beach nourishment can also impact seagrasses growing in areas where sediments collect, often at depths <30 m (N. Marbà & C. M. Duarte, unpublished data 2001).

CLIMATE CHANGE

By 2025, climate change will not yet have had a major impact on seagrasses globally. In some areas, there will be warming with accompanying species shifts and sea-level rise. For the near future, the impacts of climate change are dwarfed by direct and indirect human impacts (above), all of which will be negative. Climate change looms as a possible major impact to the 2100 time horizon and some impacts of climate change on seagrasses may be positive.

Global climate changes derived, at least partly, from anthropogenic combustion of fossil fuels, emission of greenhouse gases and changes in land use with increasing concentrations of carbon dioxide will most likely have substantial long-term impacts on seagrass ecosystems (Brouns 1994; Short & Neckles 1999; Duarte 2002).

Climate-related changes of potential importance for seagrass growth and distribution include global warming, rising sea-level, increase in atmospheric and oceanic carbon dioxide, and the increasing frequency and strength of storms (IPCC [Intergovernmental Panel on Climate Change] 2001). While the increase in carbon dioxide can be predicted with relatively high precision, global warming and especially its meteorological implications are more difficult to foresee, climate changes interacting with other human-caused changes in the marine environment making such prediction even more complex.

The expected increase in global temperature may in due course have numerous effects on seagrass performance (Short & Neckles 1999). Temperature affects almost every aspect of seagrass metabolism, growth and reproduction (see Biebl & McRoy 1971; Evans *et al.* 1986; Durako & Moffler 1987), and also has important implications for geographical patterns of seagrass species abundance and distribution (Walker 1991). Progressively increasing temperature may be a major threat to local populations of seagrasses, especially where they live close to low-latitude borders of distribution (Spalding *et al.* 2003). Seagrass distribution shifts could be even greater if oceanic circulation were to change in response to global warming, leading to abrupt changes in water temperature beyond those directly resulting from warming, as water masses shift at the edge of present biogeographical boundaries between seagrass floras (Hemminga & Duarte 2000; Green & Short 2003).

While rising temperature may have major negative effects on local seagrass beds, there seems to be less reason for concern for seagrasses on the global scale. In the literature on effects of temperature on seagrasses, there seems to be a bias towards the detrimental effects of high temperature and less focus on the negative impact of low temperatures. Seagrasses probably evolved in warm waters, suggested by the high species diversity of seagrass beds in the tropical Indo-Pacific region (Duarte 2001), and although a few genera such as *Zostera* have had great success colonizing cold temperate waters, most species grow in warm waters (Fig. 18.1). There may be no apparent physiological limitations latitudinally constraining seagrass distribution (Duarte *et al.* 2002). Nevertheless, it is reasonable to expect that increasing global temperature will enhance species diversity in subtropical regions and allow cold-water species to expand their geographical distribution further towards higher latitudes, thereby increasing the importance of seagrass ecosystems at the global scale (Fig. 18.1).

Warming to the year 2025 is projected also to raise sea level by 10–15 cm through thermal expansion of the ocean and, to a lesser extent, melting of glaciers and ice sheets (IPCC 2001). The rise in sea level has implications for circulation, tidal amplitude, current and salinity regimes, coastal erosion and water turbidity, each of which could have major negative impacts on local seagrass performance (Short & Neckles 1999).

The present atmospheric level of carbon dioxide is relatively low considered over a geological timescale, and the photosynthetic systems of most types of autotrophic organisms evolved during times of much higher levels of carbon dioxide than at present (Raven *et al.* 1995). With climate change, increasing levels of dissolved carbon dioxide in seawater may increase the competitive advantage of seagrasses over algae because seagrasses are currently more carbon-dioxide-limited than algae; increases in carbon-dioxide will stimulate seagrass productivity (Beer & Koch 1996; Zimmerman *et al.* 1997; Invers *et al.* 2001), as it will that of terrestrial higher plants (see Bowes 1993;

Amthor 1995), rooted aquatic macrophytes (Madsen & Sand-Jensen 1991) and oceanic phytoplankton (Hein & Sand-Jensen 1997).

Increase in the frequency and strength of storm events will result in increased turbidity of coastal water and poorer light conditions for benthic plant communities (Duarte 2001). Many species of seagrasses can survive periods of low light and partial burial (Duarte *et al.* 1997), but storm events often reduce growth and survival and require new colonization by seeds to re-establish seagrass beds (e.g. Cabello-Pasini *et al.* 2002). Conversely, physical disturbance represents an energetic subsidy and may be of advantage to species diversity and improve growth conditions for climax plant species (Odum *et al.* 1979). Uprooting of slow-growing seagrass species forming dense canopies opens space for colonization of more opportunistic species such as *Halophila* spp. (Duarte 2002). Overall, the net effect of increasing frequency and strength of storm events on seagrasses is unclear.

Losses as a result of natural causes

A marine slime mould caused the large-scale wasting disease seen on both sides of the Atlantic in the 1930s (Renn 1936; Rasmussen 1977) which destroyed 90% of eelgrass and affected scallop, waterfowl and fish populations. A smaller-scale recurrence of the wasting disease in the 1980s was caused by the slime mould *Labyrinthula zosterae* and not evidently by human agency (Short *et al.* 1986, 1988; Muehlstein *et al.* 1988, 1991). Wasting disease is endemic in eelgrass; large-scale die-offs have some association with increased salinity and water temperature (Burdick *et al.* 1993), and the 1930s die-off occurred at the time of a warming period (Stevens 1936).

Seagrass is removed by large storms, cyclones and hurricanes (see for example Patriquin 1975; Poiner *et al.* 1989; Preen *et al.* 1995), but impacts are variable and there is no relationship between the strength of cyclones and their impact on seagrass beds (Poiner *et al.* 1989). Increasing frequency of storm events will probably increase the damage inflicted on seagrass beds. Other extremes of climate, such as desiccation (Walker *et al.* 1988), erosion (Marbà & Duarte 1995) and ice scour (Robertson & Mann 1984), can also remove seagrass.

Grazing by swans and other water birds can remove large amounts of seagrass (Supanawanid *et al.* 2001).

Large-scale herbivory impacts may follow other causes of decline. Outbreaks of sea-urchin grazing have been reported from Cockburn Sound (Western Australia) (Cambridge *et al.* 1986) and Botany Bay (New South Wales) (Larkum & West 1990), where seagrasses were already in decline from eutrophication effects and were unable to recover from grazing. Dugongs and turtles can also graze seagrass beds (Bjorndal 1980; Lanyon *et al.* 1989), reducing leaves down to lengths of a few centimetres, repeated cropping often leading to seagrass loss. These animals are themselves under threat from human activities.

CUMULATIVE IMPACTS

There have been few coherent, broad-based studies in either time or space of the cumulative impacts of multiple stressors on the structure of marine communities, including pollution, siltation, habitat fragmentation and introductions of invasive species. Further effort is needed to investigate the influence of these human activities on seagrass communities, although such multidisciplinary studies will require a long-term funding commitment.

The most potent danger to the likely future of seagrass ecosystems is that synergistic effects of human-caused disturbances may arise. For instance, the increase in precipitation due to climate change and concomitant increases in runoff of nutrients and silt from land, partly caused by changes in land use, are potentially severe combined threats to the distribution and performance of seagrass beds. The loss of eelgrass from Waquoit Bay (Massachusetts, USA) resulted primarily from nutrient enrichment of this enclosed coastal pond (Short & Burdick 1996), but the rapid decline of eelgrass that occurred in shallow beds along the shoreline was exacerbated by increasing numbers of residential boat docks which fragmented the already stressed beds (Burdick & Short 1999). In the Dutch Wadden Sea, extensive losses of eelgrass occurred in the 1970s from the cumulative effects of turbidity in conjunction with increased shellfisheries, construction activity and nutrient loading (Hily *et al.* 2003). Local human impacts on *Posidonia oceanica* beds combined with large-scale erosion, derived from urbanization and sea-level rise, have caused widespread decline of north-west Mediterranean seagrass beds (Marbà *et al.* 1996; Marbà & Duarte 1997). Anthropogenic pressures also render seagrass ecosystems more vulnerable to negative impacts from climate change. For instance, modification of the coastline by the construction of harbours and other structures on the shore leads, in the presence of sea-level rise, to considerable beach erosion (see Chapter 17), which propagates down-slope to cause losses to seagrass ecosystems (Marbà & Duarte 1997).

While examples of the effects of cumulative impacts exist, a framework within which to forecast the responses of seagrasses to multiple stresses is still lacking, so that the preceding discussion of likely impacts of direct and indirect human pressures and climate change on seagrass ecosystems must be considered a conservative scenario of the possible losses.

PROSPECTS

The current rate of seagrass loss illustrates the imperilled status of the ecosystem and the need for increased public awareness, expanded protective policies and active management. In order to achieve such goals it is important to focus resources to monitor seagrass habitat trends, conserve existing seagrass resources, act to attenuate the causes of seagrass loss, and develop knowledge and technologies to reverse ongoing seagrass decline.

The widespread loss of seagrasses is largely a consequence of the rapid growth in human activities and transformation of the coastal zone. Global human population growth is concentrated in the coastal zone (Cohen 1995; Nicholls & Small 2002), which also harbours a disproportionate fraction of the world's wealth. Some rapidly growing industries are linked to the marine environment, such as tourism, maritime transport and aquaculture. Consequently, human activity in the coastal zone is likely to continue to increase, with a potential for even greater impacts on seagrasses (Duarte 2002). Future development-derived impacts on seagrass ecosystems are likely to be greatest in developing countries, which contain much of the extant global seagrass area (Green & Short 2003). There, human population growth is forecast to be fastest, and the potential for adopting sustainable approaches to development is constrained by resources available for management. Moreover, should resources become available, these are likely to be diverted to ensure the conservation of ecosystems for which there is greater public awareness both locally and globally, such as coral reefs (Chapter 16). Indeed, only recently have seagrass ecosystems been targeted in designing marine protected areas (MPAs). The prospect for seagrass ecosystems is therefore not positive and this prognosis can only be reversed through the development of a new attitude, involving proactive policies and efforts to manage, protect and restore seagrass ecosystems.

MANAGING, PROTECTING AND MONITORING SEAGRASS ECOSYSTEMS

Increasing awareness of the importance of seagrass has led many countries to enact legislation to prevent destruction of these habitats. Their value is recognized in the Convention on Biological Diversity (Coles & Fortes 2001). Environment Australia (EA) has federal powers to manage dredging across Australia to minimize seagrass damage. The Environmental Protection and Biodiversity Conservation Act legislates against seagrass destruction. State agencies, such as those in Western Australia and Queensland, have policies to prevent removal or destruction of seagrasses and a mandate, where necessary, to replace them. Strong measures to protect seagrass have also been implemented through the Habitats Directive of the European Commission, which specifically recognizes seagrass meadows as preferential habitats for conservation policies. Many countries in Europe (such as France and Spain) have implemented legislation to protect seagrass meadows. However, some of these laws restrict coastal zone use to such an extent that they are impractical to apply and are therefore ineffective for management.

Even under these 'best-case' scenarios of management and protection, losses still occur, and the global status of seagrass management and protection is poor. Most seagrass habitat is unprotected, and where existing policies are protective they are rarely enforced. Active management of seagrass habitat occurs in only a very small proportion of the world's seagrasses. In many areas where management does occur, conflicting uses of the coastal zone nonetheless often result in seagrass decline or loss. As these are underwater habitats that are not easily seen from shore, they are often overlooked or damaged before their existence is noted. Seagrass protection policies, protective measures and enforcement are very variable globally (Table 18.1).

Increasing awareness and understanding of the importance of seagrasses is critical to their survival. Educating the public to appreciate the importance of aquatic resources and effective ways to manage them, in combination with establishment of effective MPAs, is one way of ensuring seagrass survival. In addition to legislation that protects seagrass beds across countries or continents as in Europe and Australia, efforts to protect seagrass beds through the designation of seagrass meadows as MPAs are growing worldwide. The implementation of the Natura 2000 network of sites, aimed at preserving a fraction of the territory of European Union member states, has resulted in a major increase in the area of Mediterranean seagrass beds under protection. Some seagrass beds are now central elements of World Heritage sites, such as the *Posidonia oceanica* meadow in Ibiza-Formentera (Spain). The Shark Bay World Heritage Site (Western Australia) contains some 5000 km^2 of seagrass beds of high diversity, as well as populations of turtles, dolphins and more than 10 000 dugong (Walker 1989). The Great Barrier Reef World Heritage Site (Queensland, Australia), contains lagoonal and deep-water seagrasses (Poiner *et al.* 1989). In the USA, the Florida Keys National Marine Sanctuary has been declared to help protect and conserve seagrass communities, as well as the coral reefs, hard-bottom habitats, mangroves and the marine life (NOAA [National Oceanic and Atmospheric Administration] 2006). The development of MPAs must form part of the framework for sufficiently comprehensive and representative marine conservation.

Clearly, management of seagrasses is inadequate on a global scale. Even in areas with a high degree of management, development and commercial use of coastal waters and upland watersheds create indirect impacts on seagrass ecosystems. Moreover, evaluation of the effectiveness of protection of seagrass in MPAs is still insufficient, so that the effectiveness of current conservation practices remains poorly evaluated. In the Cabrera Archipelago National Park (Spain), while *P. oceanica* beds are recovering, a slow process that will require centuries to be completed, their continued recovery is jeopardized in places by organic inputs from visitors to the Park (Marbà *et al.* 2002). Identifying such responses requires efficient monitoring programmes.

An unresolved issue in most monitoring programmes is their power to detect change. Most monitoring programmes use estimates of seagrass cover or density, which are generally poor at detecting change (Heidelbaugh & Nelson 1996), and substantial seagrass loss may remain undetected. Improved monitoring methods are required.

Several seagrass monitoring programmes have been initiated across the globe at various levels. Monitoring programmes at local scales are many, including the French programme started in 1984, which surveys the upper and lower limits of *P. oceanica* meadows at 33 sites in the Mediterranean, and which has shown the beds to be increasing in size as wastewater treatment plants have reduced nutrient inputs to coastal waters (Boudouresque *et al.* 2000). Citizen monitoring using community volunteers

Table 18.1. *Status of seagrass protection by region*

Area	Information on seagrass protection	Legislation	Marine protected areas that include seagrasses	Enforcement of policies	References[a]
North Pacific	Mostly poor except USA	Some in USA, no net loss policies and fish habitat protection	Country dependent	None in some areas to moderate (USA)	Wyllie-Echeverria *et al.* (1994), UNEP (1999*a*)
Chile, south-west Atlantic	Poor	Not reported	Not reported	Not reported	Not cited
North Atlantic	Good in USA and EU	Yes, varies between states and countries	Few, but high protection not guaranteed	Ranges from poor to high	Wolff (1997), NOAA (1998)
Caribbean	Good in some countries	Threatened animal species and fisheries	Yes, but high protection not guaranteed	None in some areas to moderate (USA)	Clark (1998), NOAA (1998)
Mediterranean	Patchy	Inter-country agreements in EU	Yes, but high protection not guaranteed	None in some areas to moderate (EU)	UNEP (2000*b*)
South-east Atlantic	Poor	Not reported	Not reported	Not reported	Not cited
South African	Poor	Not reported	Not reported	Not reported	Not cited
Indo-West Pacific	Poor except Australia and the Philippines	Yes, in Australia, and the Philippines, and in some Pacific island nations	Yes, in SE Asia and the Pacific including Australia	Poor to non-existent, except high in Australia	Coles (1996), Leadbitter *et al.* (1999)

[a] UNEP, United Nations Environment Programme; NOAA, National Oceanic and Atmospheric Administration.
Source: Adapted from Coles and Fortes (2001).

can be a valuable and effective mechanism for increasing public awareness of seagrass habitat and tracking large-scale changes. In Queensland (Australia) and the Western Pacific, Seagrass-Watch is an outstanding example, with many participants, a regular newsletter and ongoing tracking of habitat extent and change (McKenzie *et al.* 2000; Coles *et al.* 2003*b*). Similar volunteer programmes have been established in Spain.

Thorough monitoring programmes at the national level exist in a few countries, such as that run by Danish county authorities, which since 1989 has provided excellent data on changes in seagrass communities at the national level (Ærtebjerg *et al.* 2003). In 1992, in the USA, the NOAA's Coastal Change Analysis Program (C-CAP) was initiated to monitor seagrass distribution and establish ongoing assessment using aerial-photography distribution maps

(Dobson *et al.* 1995). C-CAP's goal is to map seagrass distribution at 5-year intervals for the coast of the USA, using standardized protocols, and has demonstrated significant seagrass loss (Short & Short 2003).

Regional-level efforts are also in place, such as the Caribbean Coastal Marine Production Network (CARICOMP) started in 1985 to monitor seagrass, coral reefs and mangroves. It consists of marine laboratories and conservation groups using standardized techniques, and aims to determine the dominant influences on coastal productivity, monitor ecosystem change and detect human impacts (Creed *et al.* 2003).

Individual scientists around the world are monitoring seagrass ecosystems in their areas. These efforts provide a valuable assessment of the seagrass resource, but there is no central data repository and methods are often not comparable, making it difficult to assess seagrass status at the global scale. A global monitoring programme to assess the status and trends of the seagrass ecosystem (SeagrassNet) has since 2001 collected quarterly data from fixed transects on seagrass species composition, biomass, distribution and depth and environmental variables in 14 countries of the Western Pacific, Australia, Brazil, Africa and the USA, using a standard protocol (Short *et al.* 2002c). The aim is a 'global report card' on the status and prospects of seagrasses worldwide.

Restoring seagrass ecosystems

Reduced anthropogenic nutrient loading and elimination of sediment discharge into coastal waters will result in improved water clarity and allow seagrasses to begin to re-establish their former distribution, achieve historical depth limits and reverse the downward spiral of habitat decline (Duarte 1995). The current trends in human population growth and distribution make achievement of these goals a very great challenge with major demands on financial investment, political change and environmental awareness.

Direct improvement of seagrass distribution may be achieved through restoration, which has progressed from early transplant efforts to scientific investigations of site selection and improved transplant methodologies (Short *et al.* 2002a). The scientific tools for seagrass restoration are: (1) a quantitative site selection model (Short *et al.* 2002a), (2) a protocol to identify a sustainable source of planting stock, (3) a reliable planting method (Davis & Short 1997; Harwell & Orth 1999; Granger *et al.* 2000; Short *et al.* 2002b), and (4) monitoring using scientific

criteria to identify the outcome of the restoration (Short *et al.* 2000). The most important tool is site selection, ensuring adequate water clarity, low bioturbation and appropriate sediment and physical conditions (Short *et al.* 2002a). Determining success of seagrass restoration requires rigour and demonstrated replacement of habitat function. Research on all these aspects of seagrass restoration is under way in many parts of the world.

Sound science is important as a basis for restoration implementation and practice, ranging from providing rigorous methodology for fulfilling the statutory requirements for compensatory mitigation to offset impacts on seagrass (Davis & Short 1997; Fonseca *et al.* 1998; Paling *et al.* 2001) to simplifying the techniques. In addition, costs have to be reduced and procedures developed that community-based volunteer groups can use to restore coastal environments (Short *et al.* 2002b). Both kinds of restoration, namely mitigation and community-based, are needed to slow the loss and reverse the degradation of seagrass habitat.

Revegetation efforts are expensive, and have proved effective only at small spatial scales (areas <1 km^2), and mainly in sheltered environments. It has only been attempted with *Z. marina*, *Z. noltii*, *Halodule wrightii*, *T. testudinum*, *Syringodium filiforme*, *Posidonia australis*, *P. coriacea* and *Amphibolis griffithii* (Fonseca *et al.* 1998; Lord *et al.* 1999). Revegetation cannot fully restore lost seagrass areas, but can stimulate natural revegetation processes. Seagrass recolonization is slow, requiring from a few years for the fast-growing species to centuries for the slow-growing species (Duarte 1995; Marbà & Duarte 1998; Kendrick *et al.* 1999; Marbà *et al.* 2002). The seagrass area that can be recovered through revegetation in the most optimistic scenarios is but a small fraction of the area lost annually. Hence, while a useful management option, revegetation efforts must be combined with effective conservation and improved water clarity (monitoring has no direct effect on seagrass loss) if seagrass loss is to be reversed.

OUTLOOK FOR SEAGRASSES IN 2025

Worldwide, seagrasses are being lost faster than they can regrow or recolonize. By 2025, seagrass losses on all continents and in the oceanic islands will be greater than they are now and more noticeable, the extent of this loss depending on the implementation of immediate steps to eliminate the global threats to seagrass habitat. Despite some increased awareness and protection, there is no sign that global losses of seagrasses will be slowed in the next 20 years. Most impacts on seagrass ecosystems to the year 2025 will be from increasing human population densities, rather than predicted climate change or sea-level rise. Human impacts, particularly through eutrophication and increased sedimentation, will continue to cause seagrass losses worldwide.

Climate change is likely to have long-term, albeit not large, impacts before 2025. Increased sea level may *per se* have neutral effects on seagrasses, but lead to losses in already eroding coastal areas, particularly those with the prospect of increased storm frequency. In addition to promoting submarine erosion, hardening of shorelines will limit the progression of seagrasses facing sea-level rise, restricting the ability of seagrasses to migrate shorewards and thus reducing available habitat. Increased concentrations of carbon dioxide may favour seagrass growth.

Only increasing awareness of the need for development to be sustainable, and the need to protect and conserve ecosystems, will help mitigate expected seagrass losses. A precondition is that the values and services of seagrass ecosystems become more widely acknowledged in order to foster efforts to promote legislation and enforce protection measures for seagrasses worldwide.

19 · Continental–shelf benthic ecosystems: prospects for an improved environmental future

STEPHEN J. HALL, STUART I. ROGERS AND SIMON F. THRUSH

INTRODUCTION

The value of continental-shelf systems has been estimated to be US$1640 per ha per year (total worldwide value: US$11 546 billion per year) (Costanza *et al.* 1997). By comparison, the average farm income for the USA is US$200 per ha per year. Therefore, whether the concern is for human welfare or the conservation of marine natural heritage, the future of these ecosystems deserves careful consideration.

Although less charismatic than coral reefs (Chapter 16), and less familiar than wetlands or mangroves (e.g. Chapters 11–13), continental-shelf benthic systems support a major component of global fisheries production, exert important controls on marine productivity and contain rich and varied marine communities. The purpose here is to look forward and consider the future of these ecosystems to the 2025 time horizon.

Aquatic Ecosystems, ed. N. V. C. Polunin. Published by Cambridge University Press. © Foundation for Environmental Conservation 2008.

A wide array of factors affect the biodiversity of continental-shelf ecosystems, a number of which are the consequence of human activity. Generally, the importance of different stressors varies from place to place, and the resident biological communities show variation in sensitivity to these impacts as a result of environmental or biogeographical constraint and disturbance history (Hall 2002). Although impacts are often immediate and local in scale, the potential for broader-scale chronic and degradative change is now recognized in some, if not many, areas. The nature of these broad-scale changes and the agents that cause them are now beginning to be understood.

When an impact is severe enough to dominate all other factors, consistent global patterns emerge. The overall reductions in biomass of harvested species as a result of fishing is, perhaps, the exemplar of such a global pattern (see Pauly *et al.* 1998). For the most part, however, factors affecting continental-shelf ecosystems vary around the planet, even though they may be locally and regionally severe. This situation makes globally averaged values of stress and disturbance meaningless; it would make no sense, for example, to consider the stresses placed on enclosed shelves with high population densities in the surrounding catchments to be directly relevant to other shelf ecosystems. Nor would it be appropriate to assume that large-scale impacts are restricted to shelf areas adjacent to regions of high population density; for example poor forestry and agricultural practices also pose a significant threat in regions where the population is low, as does the globalization of fisheries. On the Antarctic continental shelf, for example, there has been a long history of exploitation of top-level predators and more recently increasing exploitation of demersal fish stocks (Chapter 21). In fact, extensive impacts in regions of low population density may have the most profound effects owing to the often shorter history of disturbance and exploitation. Globally, these regions may well be among the best places to focus conservation efforts.

One solution to predicting ecological response to future trends lies in developing an understanding of the scales of disturbance and the scales of ecological recovery or succession that follow these disturbance events. Over much of the continental shelf, natural disturbance events covering a large area (tens to hundreds of square kilometres) are rare. For example, although there are exceptions, storms rarely stir the seabed below about 60 m depth, and the effects are sometimes surprisingly limited. For example, despite wave heights exceeding 3 m for more than 40 hours, and a maximum recorded wave height of 11.82 m in a North Sea force 9 gale, at most only the top 1 cm of sediment at 25 m depth was eroded (Green *et al.* 1995). Other broad-scale natural disturbances of continental shelves include earthquakes and volcanic activity, iceberg scour (which is limited to the polar continental shelves) and hyperpycnal sediment flows (which are restricted to areas of extreme terrigenous sediment loading) (Foster & Carter 1997; Wheatcroft 2000). Toxic algal blooms and anoxic/hypoxic events are other broad-scale disturbance phenomena that are also restricted to specific locations, although in many locations the extent and frequency of disturbance appears to be increasing (Verity *et al.* 2002).

This chapter seeks to (1) examine the key human-induced drivers of change for soft-sediment continental shelves, (2) examine society's aspirations for these ecosystems and (3) discuss the impediments and opportunities for meeting these aspirations within a 2025 time horizon. The purpose is not to provide a comprehensive review of all the controls on shelf–sea ecosystems, which has been done elsewhere (e.g. Hall 2002). Rather, it is to explore key issues influencing the likely future of continental shelf ecosystems at a global scale.

Although many factors must be taken into account when considering the future of any ecosystem (Hall 2002), this chapter focuses on two major issues, namely water quality and fishing. These two issues were specifically chosen because of their significance to continental-shelf ecosystems in many parts of the world. They also reflect different types of anthropogenic impacts, one linking land and sea and the other a predominantly marine threat, and this contrast has important implications for drivers of change in these ecosystems. These two factors are widely considered dominant human drivers of change in shelf systems, with fishing considered to be the factor that has the largest direct impact (Dayton *et al.* 1996; Pauly *et al.* 1998; Hall 1999; Verity *et al.* 2002). While at a local scale, other impacts may well be more important, from a global perspective these two are without doubt the principal concerns (Peterson & Estes 2001; Norse & Crowder 2005). This focus choice is also supported by a detailed evaluation of 32 human activities or 'pressures' in an assessment of the North Sea, where the first five activities in the top-ranking class of pressures related to commercial fisheries (removal of target species, seabed disturbance, discarding and non-target species effects) and inputs from land (organic micropollutants and nutrients) (OSPAR 2000*a*).

Table 19.1. *Change in objectives and the policies required to achieve them for future fisheries management*

Present		Future
Objectives		
Sustaining stocks	→	Sustaining assemblages and ecosystems
Maximizing annual catches	→	Maximizing long-term welfare
Maximizing employment	→	Providing sustainable employment
Ensuring full resource use	→	Ensuring efficient resource use (minimizing waste)
Tending to short-term interests	→	Addressing both short- and long-term interests
Addressing local considerations	→	Addressing both local and global considerations
Policies		
Open and free access	→	Limited entry, user rights and user fees
Sectoral fishery policy	→	Coastal zone inter-sectoral policy
Command and control instruments	→	Command and control and macroeconomic instruments
Top–down and risk-prone approaches	→	Participative and precautionary approaches

Source: From Garcia and Grainger (1996).

FISHING

Commercial fishing is now recognized for the profound changes it can induce in marine ecosystems. The continental shelves of all countries are exploited and many stocks are at, or beyond, their capacity to sustain exploitation. Commercial fishing has changed the density and size structure of many exploited populations, with consequent effects on the fluxes of materials through the food webs of continental-shelf ecosystems (Pauly *et al.* 1998; Hall 1999).

There is growing evidence of significant ecological change to the sea floor owing to habitat disturbance by bottom-fishing gear (Dayton *et al.* 1995; Jennings & Kaiser 1998; Watling & Norse 1998; Auster & Langton 1999; Hall 1999; Kaiser & de Groot 2000). Despite the often-significant challenges in conducting these impact studies, some consistent responses to habitat disturbance have emerged across broad spatial scales (Thrush *et al.* 1998; Kaiser *et al.* 2000; McConnaughey *et al.* 2000; Cryer *et al.* 2002). Common changes in sea floor communities that have occurred across a variety of habitat types include reduced habitat structure, lower diversity and loss of large and long-lived sedentary species.

In the light of these effects on fish communities and on the benthic system that supports them, many countries are now trying to increase their efforts to manage fisheries for ecosystem sustainability and find a balance between exploitation and the conservation of marine biodiversity (Table 19.1) However, as fishery resources have been overexploited close to the main human population centres, fishing activity has shifted to exploit resources in developing countries that often have limited means to effectively manage their fisheries. The fishing industry has also moved to exploit deep-water fish stocks, including those of previously inaccessible areas such as rocky seamounts and reefal environments (Koslow *et al.* 2000, 2001).

Fishing-induced changes in shelf systems

At the global scale, fishing has dramatically reduced target-species biomass in many continental-shelf systems, and there has tended to be a pattern of sequential exploitation of species, beginning with the top predators and gradually 'fishing down the food web' (Pauly *et al.* 1998; Chapter 1). Exactly how these changes have altered the competitive and predatory relationships within these systems, and what the ecological consequences of these changes are, is open to considerable debate (see Chapter 20). There is little doubt, however, that fishing has dramatically altered the ecology of shelf fish communities. Examples of the changes that have occurred are provided by analyses of how demersal fish communities in the North Sea, Georges Bank (USA)

Table 19.2 *Candidate criteria for a definition of ecosystem overfishing and the status of three demersal fishery systems*

Criterion	Study		
	Gulf of Thailand	North-east USA	North Sea
Biomass of one or more important species falls below minimum biologically acceptable limits	✓ Important demersal fish species abundance one-tenth of their levels in 1960s	✓ Exploitation rates on principal groundfishes reduced recently, but harvest rates on other components increased to non-sustainable levels	✓ No signs of persistent recruitment overfishing, biomasses of important resource stocks are below minimum acceptable levels
Biological diversity declines significantly	✓ Decline in diversity, owing to loss of important components, but continuing high total yields	✓ Dominance of species groups changed as a direct result of excessive fishing and sequential depletion	✗ Diversity of the system has fluctuated without trend
Harvesting leads to increased year-to-year variation in populations/ catches	✗ No increase in interannual variation in aggregate landings	✓ Greater interannual fluctuations in landings owing to increased dominance of species with more variable recruitment characteristics	✗ No apparent increase
Significant decrease in resilience or resistance of the ecosystem to perturbations	✗ No apparent trend	? No data available	? No data available
Lower cumulative net economic or social benefits than might be obtained with less intense fishing	✓ Net benefits from the fish community would be higher with less fishing effort, but increased shrimp landings provide substantial alternative benefits	✓ Rebuilding of depleted resources and their efficient management would result in very large additional benefits	? No data available
Fishing impairs long-term viability of ecologically important non-resource species	? Unclear; almost all species are used	✓ Small pelagic prey species remain abundant and underexploited. By-catch of turtles and marine mammals are of significant concern	✓ Concern about viability of some elasmobranch species

✓, criterion supported by available data; ✗, criterion not supported by available data; ?, unclear pattern.
Source: Adapted from Murawski (2000).

and Gulf of Thailand have changed during a period of intense fishing (Hall 2002) (Table 19.2), and by a comprehensive summary of the fishing-induced changes observed in the North Atlantic (Pauly & Maclean 2002).

Changes occur in the structure of fish communities and benthic communities impacted by fishing activity (reviews of Dayton *et al.* 1996; Jennings & Kaiser 1998; Hall 1999; Collie *et al.* 2000); however, such changes in diversity and abundance of species are not the only concern when attempting to forecast the future of continental-shelf eco-systems. In the case of the benthos, for example, the fauna also plays an important role in the functional attributes of shelf seas by exerting control over the fluxes and recycling of energy and matter (Thrush & Dayton 2002). Although studying these functional aspects of benthic communities is extremely difficult and complex, their importance or the impact that fishing might have should not be underestimated.

There is an expectation, for example, that because trawling reduces the abundance of the bioturbating mac-rofauna that plays a key role in remineralization processes, and because the physical mixing by trawling (unlike the mixing by macrofauna) does not contribute directly to community metabolism, fishing can have a profound effect on biogeochemical cycles. Duplisea et al. (2001) con-structed a simulation model to examine effects of trawling disturbance on carbon mineralization and chemical con-centrations in a soft-sediment system. Contrasting a nat-ural scenario, where bioturbation increases as a function of macrobenthos biomass, with a trawling scenario, where physical disturbance results from trawling rather than the action of bioturbating macrofauna (which are killed by the action of the trawl gear), the model suggested that the effects of low levels of trawling disturbance will be similar to those of natural bioturbators (Duplisea et al. 2001). However, at high levels of trawling the system became unstable owing to large carbon fluxes between oxic and anoxic environments within the sea floor sediments. The presence of macrobenthos in the natural disturbance scenario stabilized sediment chemical storage and fluxes, because the macrobenthos are important participants in the total community metabolism. This simple model suggests that, where physical disturbance due to waves and tides is low, intensive trawling may destabilize ben-thic-system chemical fluxes, with the potential to propa-gate more widely through marine ecosystems (Duplisea et al. 2001).

One possible way to predict the effects of fishing on functional attributes is to consider how the overall trophic structure of communities changes, rather than just species composition. One of the few studies that has attempted this for benthos is that of Jennings et al. (2001), who found that, although chronic trawling disturbance led to dramatic reductions in the biomass of infauna and epifauna, these reductions were not reflected in changes to the mean trophic level of the community, or the relationships between the trophic levels of different size classes of epifauna. Indeed, despite order-of-magnitude decreases in biomass of infauna, and a shift from a community dom-inated by bivalves and spatangoids to one dominated by polychaetes, the mean trophic level of these communities differed by less than one trophic level between sites, and differences were not linked to levels of fishing disturb-ance (Jennings et al. 2001). This stability in the trophic structure of the benthos could imply that species less vulnerable to disturbance are taking the trophic roles of larger more vulnerable species. This is difficult to imagine in the pelagos, where there is a strong relationship between trophic position and body size, but perfectly possible in the benthos, where this relationship breaks down.

It should be stressed that the above results are only applicable to the free-living fauna of mobile substrates and cannot be extrapolated to areas with lower natural dis-turbance, or where habitat-forming species are found. For example, it is known that larger filter-feeding bivalves can consume a substantial proportion of primary production; Hermesen et al. (2003) estimated a scallop production for Georges Bank (up to 280 kcal per m^2 per year) which would result in the consumption of a substantial propor-tion of the water-column productivity (Sissenwine 1984; Steele 2005).

WATER QUALITY

Fluxes of nitrogen and phosphorus have been altered over large sections of the globe (e.g., Chapters 2, 3 and 13), and the mobility and availability of these nutrients have increased in the sea (e.g. Vitousek et al. 1997a). Here, two important water-quality issues that affect the nearshore shelf environment are considered, namely the input of contaminants leading to eutrophication and elevated sus-pended sediment concentrations. The main source of both these problems is poor land use and, in the case of eutrophication, inadequate treatment of human, industrial and agricultural waste.

While environments can be subject to naturally high rates of sediment loading or be predisposed to the adverse effects of eutrophication, both problems are exacerbated by human land use, particularly agriculture, forestry and urbanization. Thus, with the predicted rise in human population density, particularly in coastal regions, the environmental footprints of eutrophication and sediment impacts are likely to extend onto the continental shelf.

Eutrophication

On a global scale, point sources are generally considered to be less important than non-point nutrient sources; however, where large coastal metropolises occur, point sources can have a very significant impact. Input of wastewater from New York City, for example, contributes 67% of the nitrogen to Long Island Sound each year (Rabalais 2005). In developing countries, an estimated 90% of wastewater is discharged into rivers and streams without treatment, and in China, 80% of the 50 000 km of major rivers are too polluted to sustain fisheries (Tockner & Stanford 2002). In this context, it is significant that by 2015 more than half the world's population is expected to be urban dwelling and the number of people living in mega-cities (>10 million people) will more than double to 400 million.

With respect to non-point nutrient sources, fertilizer application remains the principal culprit. Although input rates have stabilized in some watersheds, on a worldwide basis fertilizer use continues to increase (Seitzinger *et al.* 2002). Transport by rivers is a principal route by which nutrients enter coastal seas, which in some areas (such as the Baltic Sea and the Gulf of Mexico around the Mississippi River) far exceeds the inputs from any other source. One of the most rapidly increasing sources of input, however, is the atmosphere, with up to 40% of the industrial, agricultural and urban nutrient fluxes entering coastal waters via this route (Paerl 1995).

Atmospheric emissions are linked in particular to industrial activity and energy consumption, and China and India are expected to see the most dramatic rises in these in future. China's economic development will be mainly in the coastal provinces and it is there that the most dramatic increases can be anticipated. However, agricultural inputs via the atmosphere are also important. For example, 40% of inorganic nitrogen fertilizers are volatilized as ammonia, either directly following application or from the wastes of animals fed on the crops grown.

Coastal areas with reduced water exchange are particularly prone to alterations in the relative availability of nutrients, and hence semi-enclosed seas (e.g. Baltic, Black and Adriatic Seas), fjords, lagoons and bays exhibit the most visible effects of eutrophication. However, when loading is high and/or oceanographic conditions lead to poor water exchange and stratification, eutrophication can impact the open shelf (Rabalais *et al.* 2002; Rabalais 2005). In the early stages, nutrient additions to the coastal zone typically result in increased production. Further eutrophication, however, results in changed species composition and indirect feedback effects (e.g. shading effects by phytoplankton, induction of hypoxia and physiological and behavioural responses), followed by extreme effects as a result of smothering of the sediment surface by macroalgal mats, or the large-scale induction of anoxia and consequent mass mortality of benthos and fish (see Baden & Pihl 1990). Initially, these effects do not impact the sea floor uniformly; anoxic patches determined by topography, history, hydrodynamics and depositional events are the precursors to broad-scale effects. Drifting macroalgae can also play an important role in connecting benthic habitats in estuaries and nearshore regions with effects further offshore.

The largest coastal hypoxic zones recorded in the world are in the Baltic Sea (84 000 km^2), the northern Gulf of Mexico (22 000 km^2) and the north-western shelf of the Black Sea (40 000 km^2) (Rabalais 2005). However, following the economic collapse of the Soviet Union and the consequent declines in fertilizer application, the 1990s witnessed greatly reduced nutrient inputs to the Black Sea. In 1996, the anoxic region failed to develop for the first time in several decades and, by 1999, the area was less than 1000 km^2 in size (Mee 2001).

Sediment loading

Geological and oceanographic evidence shows that sediment loading to estuaries and coasts has increased concomitantly with human population growth and the development of coastal margins. For example, sediment loads in rivers draining the Atlantic seaboard of the USA would have been four to five times lower than current values if the landscape had remained undisturbed by humans (Meade 1969).

The impacts of elevated suspended sediment concentrations and catastrophic depositional events have not received the same level of attention as eutrophication, partly at least because in many areas these impacts are associated with the development of coastal land, and ecologists beleaguered with the problems of eutrophication and pollution have tended to ignore sediment as a contaminant.

A number of factors predispose rivers and their catchments to delivery of high sediment loads, notably easily erodible sediment, steep terrain and moderate to low annual freshwater discharge (Mulder & Syvitski 1995). The Eel River in northern California (USA) exhibits these characteristics, and clay-dominated terrestrial sediment deposits resulting from a single depositional event have been documented over a large (30 km × 8 km) area of the adjacent continental shelf (Wheatcroft *et al.* 1996). A similar phenomenon has been noted off the continental shelf of Poverty Bay (New Zealand) (Foster & Carter 1997). The characteristics of rivers and catchments that produce large episodic depositions of terrestrial sediment on the sea floor are common in many Pacific Rim countries (Milliman & Meade 1983; Alongi 1998). Naturally high rates of sediment delivery through estuaries can be further exacerbated by land uses such as deforestation, farming and urbanization. Sediment runoff from land and rapid sedimentation events within coastal regions may become more common owing to climatic variability intensified by changes in resource use and exploitation by humans (Fowler & Hennessey 1995; Wheatcroft 2000).

Deposition of terrestrial sediment has been shown to significantly influence benthic communities (Peterson 1985; Norkko *et al.* 2002; Thrush *et al.* 2003), with the potential for indirect effects. Changes may arise, for example, when increased turbidity, as a result of an altered sediment input and resuspension regime, clogs the feeding structures of suspension feeders, thereby influencing their role in benthic–pelagic coupling (see Ellis *et al.* 2002; Norkko *et al.* 2002). There is also the potential for subtler but nonetheless degradatory changes in sediment porosity and stability.

SOCIETY'S GOALS FOR CONTINENTAL-SHELF ECOSYSTEMS

It is reasonable to assume that the future of continental-shelf ecosystems to the 2025 time horizon will be strongly influenced by society's goals for the whole environment. In broad terms, these goals relate both to the pressing need to improve human welfare, and to maintain or restore marine biodiversity. Broad principles embodying these concepts have now been widely adopted in various forms under the general banner of ecologically sustainable development (ESD). Clearly, however, at local scales and in the shorter term, human welfare and environmental protection of shelf ecosystems may not always be mutually compatible.

To what extent do the societal aspirations embodied in the concept of ESD provide a genuine framework for real progress with integrated management and the protection of biodiversity? In this section, the most important and influential global and sectoral agreements relating to the continental shelf are summarized, and the likelihood of these commitments being honoured in different global regions is assessed.

In accordance with the Charter of the United Nations and the principles of international law, all states have the sovereign right to exploit their continental-shelf resources in accordance with their own environmental policies, and the responsibility to ensure that activities within their control do not cause damage to the environment beyond the limits of national jurisdiction. Despite this overarching framework, there has been sufficient concern over the past two decades with the current state of the marine ecosystem that a plethora of additional international agreements and programmes have been established. These have addressed environmental concerns that include direct and indirect effects of fishing activity in coastal waters and on continental shelves, human population growth, climate variability and climate change, the effects of runoff, land-based and atmospheric discharges and the direct consequences of disposal and dredging at sea. A key point to emphasize is that, while each of these impacts will occur to a greater or lesser extent in all global regions, local priorities differ. In Africa, for example, the most pressing issues include food security, governance, poverty and health, while in Asia and the Pacific, deforestation and land degradation, freshwater resources, climate change, sea-level rise and regional air pollution are also important. It is very difficult, therefore, to generalize global priorities, and this may in turn limit agreements on high-level international aspirations from reaching fruition.

In 1992, more than 100 heads of state met in Rio de Janeiro (Brazil) for the United Nations Conference on Environment and Development (UNCED). This 'Earth Summit' was convened to address urgent problems of environmental protection and socioeconomic development. The assembled leaders signed the Convention on Biological Diversity (CBD), endorsed the Rio Declaration and adopted Agenda 21, a 300-page plan for achieving sustainable development in the twenty-first century. Agenda 21 reflected a global consensus and political commitment at the highest level to development and environmental cooperation, and made it clear that progress could only be implemented through national policies achieved through international cooperation. Thus, although agreements under this convention have wide-ranging implications for ESD on a global scale, they also provide the basis for all current commitments and obligations by coastal states to the conservation and sustainable use of their continental shelf ecosystems. Programme areas relevant to such shelf environments include the integrated management and sustainable development of coastal areas, including exclusive economic zones, marine environmental protection, and the sustainable use and conservation of marine living resources.

A 5-year review of 'Earth Summit' progress was undertaken in 1997 by the United Nations General Assembly, followed in 2002 by a 10-year review by the World Summit on Sustainable Development (WSSD). Despite the framework established during the previous decade, it was generally felt that progress in implementing sustainable development had been disappointing and that clear commitments and actions were necessary at the beginning of the new millennium.

As a result of the review, a number of key commitments were identified at the WSSD to highlight those issues relevant to continental shelf ecosystems on which action was most urgently needed. These included commitments to:

- Encourage the application by 2010 of the ecosystem approach for the sustainable development of the oceans.
- Develop and facilitate the use of diverse approaches and tools (including the ecosystem approach), the elimination of destructive fishing practices and the establishment of marine protected areas consistent with international law and based on scientific information, including representative networks, by 2012.
- Achieve, by 2010, a significant reduction in the current rate of loss of biological diversity.
- On an urgent basis and where possible by 2015, maintain or restore depleted fish stocks to levels that can produce the maximum sustainable yield.
- Put into effect the FAO (Food and Agriculture Organization of the United Nations) international plans of action for the management of fishing capacity, and prevent, deter and eliminate illegal, unreported and unregulated fishing.
- Eliminate subsidies that contribute to illegal, unreported and unregulated fishing and to overcapacity.

Achievement of these goals would go a long way towards mitigating the threats posed to continental-shelf benthic systems to the 2025 time horizon. However, on the global stage, these goals can only be achieved through national progress and legislation, a factor that is fundamental to the success or failure of these sustainable development initiatives. Realizing the benefits of regional coherence and

support, the United Nations Environment Programme (UNEP) established a Regional Seas Programme to encourage groups of countries sharing common seas to find regional solutions to their particular problems. Two of the major priority issues that are currently addressed by Regional Seas Agreements include biodiversity conservation and integrated coastal area management. There are now more than 140 coastal states and territories participating in such arrangements, and with the signing of the North-East Pacific Regional Seas Agreement in March 2002, the programme covers nearly all of the planet's continental-shelf ecosystems (UNEP 2000*c*). While the practical benefits of such a programme have yet to be realized and a full assessment remains to be done, it is reasonable to assume that these Regional Seas Agreements will form the basis for all future cooperation in those areas of the continental shelf system that fall outside national jurisdiction.

Many of the aspirations described in the WSSD environmental goals have been adopted and modified by other regional governments and administrations. As noted earlier, this is an essential step that translates these generic international agreements into national and regional policy and legislation. For example, the EU (European Union) has developed a Marine Strategy as a contribution to its Strategy for Sustainable Development, which encompasses many of the principles outlined at the Earth Summit (EC [European Commission] 2002). The Strategy recognizes that setting specific sectoral or issue-based objectives with time-lines for their achievement is ambitious, and will require substantial levels of integration and commitment. In terms of the loss of biodiversity and destruction of habitats, the EU is already committed, following the European summit in Gothenburg in 2001, to halt biodiversity decline by 2010. In order to progress this target, the Natura 2000 network of natural protected areas has been established throughout the EU to ensure the long-term survival of Europe's threatened species and habitats. Similarly, the EU is committed to a reform of the fisheries management process to reverse the decline in stocks and ensure sustainable fisheries and a healthy ecosystem, both in the EU and globally. The recent draft Marine Strategy Directive (EC 2005) will also require the achievement of 'good environmental status' throughout continental-shelf seas. These are ambitious and challenging objectives that will guide the state of European continental-shelf ecosystems to the year 2025 and beyond. Other objectives

Table 19.3. *Eleven objectives of the EU Marine Strategy relating to continental shelf ecosystems*

Objective
(1) Halting biodiversity decline by 2010 and ensuring sustainable use of biodiversity through protection and conservation of natural habitats by applying an ecosystem-based approach.
(2) Changing fisheries management to reverse declining stocks and ensure sustainable fisheries both in the EU and globally.
(3) Eliminating pollution by dangerous substances.
(4) Preventing pollution by radioactive substances by 2020.
(5) Eliminating eutrophication problems caused by man by 2010.
(6) Eliminating pollution by litter by 2010.
(7) Phasing out illegal discharges of oil by 2010 and all discharges of oil by 2020.
(8) Reducing the environmental impact by shipping through the development of the 'clean ship' concept.
(9) Raising the quality of seafood without risk to human health.
(10) Implementing the commitments made in the Kyoto Protocol regarding the reduction of greenhouse gases.
(11) Improving the knowledge base on which marine protection policy is founded.

relate to the need to deal with hazardous substances, eutrophication, chronic oil pollution, radionuclides, micro-biological threats and climate change (Table 19.3). The Oslo and Paris (OSPAR), Helsinki (HELCOM) and Barcelona Conventions are major instruments through which many of these environmental goals are implemented more widely in the Mediterranean, north-east Atlantic and the Baltic Sea areas.

In a global context, many of the same aspirations have been incorporated into the FAO (2001*b*) Code of Conduct for Responsible Fisheries, which sets out non-mandatory principles and international standards of behaviour for responsible fishing practices for the effective conservation, management and development of living aquatic resources, taking account of the ecosystem and biodiversity. Paramount amongst these is the expectation that states should apply the precautionary approach widely to conservation, management and exploitation of living aquatic resources to protect them and preserve the aquatic environment. The Code also warns that the absence of adequate scientific information should not be used as a reason for postponing or failing to take conservation and management measures (FAO 2001*b*). Although both these examples provide an international context for national legislation that already exists, or can subsequently be developed, they do not of themselves have authority to implement real change.

Goals for human welfare: implications for continental-shelf ecosystems

In parallel with the preparation of marine environmental objectives has been the development of an equal number and range of aspirations for human welfare, to address the ongoing problem of global poverty. While it is outside the scope of this chapter to provide a comprehensive overview of these goals, several of these objectives may slow the rate of progress towards improved quality of the continental-shelf environment. In particular, work undertaken by the UN (United Nations) and World Water Council (WWC 2003*b*) highlights progress to deal with the pressing need for water and sanitation.

At the Millennium Summit in September 2000, the UN reaffirmed their commitment to sustainable development and elimination of poverty, and prepared 'millennium development goals' from the agreements and resolutions of world conferences organized by the UN in the past decade. The goals focus the efforts of the world community on achieving significant, measurable improvements in people's lives. Importantly, the goals recognize both that the environment provides goods and services that sustain human development, and that better natural-resource management increases the income and nutrition of poor people. Improved resource management also reduces the risk of disaster from floods, improved water and sanitation reduce child mortality and better drainage reduces malaria.

The first seven 'millennium development goals' are mutually reinforcing and are directed at reducing poverty in all its forms. Two objectives relevant to continental-shelf ecosystems under the broad headings 'Eradicate extreme poverty and hunger' and 'Ensure environmental sustainability' are to (1) halve, between 1990 and 2015, the proportion of people who suffer from hunger, and (2) halve, by 2015, the proportion of people without sustainable access to safe drinking water.

In 2000, the WWC identified a pressing need for improved water supplies to large parts of the developing world, and realized that the continued impoundment of reservoirs was inevitable and necessary. It identified the clear need by 2025 to provide water, sanitation and hygiene for all.

ACHIEVING THE OBJECTIVES: CHALLENGES AND APPROACHES TO IMPLEMENTATION

Understanding the social dimensions

There are many impediments to mitigating environmental threats to the world's continental shelves and these differ both geographically and for different threats. These challenges include information needs in general and, specifically, developing an understanding of the complexity and interconnectedness of natural systems, and how this can affect ecosystem response to change; these issues are the natural focus for science and scientists. Perhaps more than anything else, however, it is the social, economic and political dimensions of environmental management that present the greatest challenge. While everything may not be known about the processes that control marine systems, or the nature and magnitude of anthropogenic change (and threat of change) may not be documented fully, more often than not enough is known to determine what actions would mitigate impacts; the question of how to change societal behaviour to take those actions is much more difficult. Fishing is perhaps the classic example; more fisheries biology is not needed to highlight that many stocks are depleted beyond levels that are wise. Nor is more research needed to prove that levels of fishing mortality need to be reduced in most of these fisheries. What does need to be known, however, is how to reduce the levels of mortality in socially and politically acceptable ways. It is understanding the social dimensions of the problem and integrating this

understanding with scientifically defensible and acceptable solutions that is perhaps the biggest challenge.

With the exception of fishing, the human footprint on continental shelves results largely from land-based activities. It is here, then, that solutions to most of the threats to continental shelves will need to be sought. Arriving at such solutions will require not only scientific understanding, but also an understanding of the social and economic setting in which solutions must be found. Although global geopolitical and economic considerations are undoubtedly important, particularly with respect to the divide between developed and developing countries, the well-worn adage 'think global act local' probably applies. Success will come in small steps rather than giant leaps.

An illustration of how research into the social dimensions of a problem has the potential to help identify solutions can be found in a recent analysis of fishing activity in Australian waters. Traditional Indonesian fishers have access to Australian waters through a Memorandum of Understanding (MoU) that was signed between the two countries in 1974. Surveys in the late 1990s revealed significant depletion in the benthic invertebrate stocks (trochus [gastropod] and trepang [sea cucumbers]) to which the fishers had access (Birkeland *et al.* 1982). As a result of this fishing, significant concerns were expressed about the threats to the biodiversity of the region. As the invertebrate stocks declined, fishers turned to the populations of sharks, which are caught for their fins. Shark stocks declined in their turn and fishers are now seeking sharks outside the MoU box, creating a problem with illegal fishing in marine reserves.

In a search for a solution, sociological research revealed that the vast majority (93%) of all vessels fishing in one part of the MoU box came from a single small island in the Indonesian archipelago, and that 69% of all voyages originated from a single village on that island (Fox 1998). These findings and other research pointed to the critical role this village played in the sustainability of the fisheries in the region and clearly showed where effort to find a solution needed to be placed (Fox 1998). The focus of this effort will need to be on the generation of alternative livelihoods, perhaps through the development of low-technology sustainable aquaculture.

The work of the FAO (e.g. FAO 2004, 2006) illustrates how minor changes in resource allocation and engagement of key participants in a fishery can make a substantial difference to outcomes. In the Bongolon region of Guinea (West Africa), resources for fisheries patrols are limited.

With the ability to undertake only six or seven patrols per month, the National Centre for Fisheries Surveillance and Protection was unable to control the high number of incursions along the 300-mile coastline by illegal trawlers. As a result, fisheries resources were becoming severely depleted. Recently, however, a new system has been adopted in which small-scale fishers in motorized canoes report by radio to the enforcement officers when an incursion is detected. This partnership has reduced infringements by 59% and has also led to a reduction in fatalities through accidents between illegal industrial trawlers and small-scale fishers. The improved catches by the traditional fishing community has greatly improved the economic security of the sector and is being viewed as a model for replication elsewhere.

These examples illustrate the kind of sociological research that will be needed if the networks and motivations underlying behaviour in a region are to be understood. It is the integration of this understanding with the underpinning scientific knowledge of the ecology and biology of the system that will most likely lead to identification of lasting solutions.

Financial drivers

Another fundamental requirement to ensure that goals for both human welfare and the environment are met is the international political will to develop an open and non-discriminatory trading and financial system. Without this in place, the financial resources of the developing world to deal with even the basic issues of poverty and health will be insufficient. The international community will need to make a commitment to help develop good governance and reduce poverty, and to ensure that the debt problems of developing countries are solved sustainably in the long term. The UN has taken the initiative by identifying a high-level goal which recognizes that many of the poorest countries will need additional assistance and must look to the rich countries to provide it.

It is also argued that, if developing countries can take advantage of new technologies and the reduced trade barriers that come with an increasingly global economy, this could lead to both environmental and societal benefits. Important recent initiatives in the World Trade Organization (WTO) and the Organization for Economic Cooperation and Development have provided the legal basis for an expanded global trading system that takes into account important social and environmental concerns, in addition to the achievement of core economic goals (WTO 2003).

A key element of any effective trade system, however, will be the recognition and internalization of environmental costs. Such internalization will ensure that industries make investment decisions based on the full cost of production, rather than a marginal cost in which society carries the additional environmental overhead. Carbon trading and emissions taxes are approaches designed to bring such environmental dimensions into the market system. At present, however, there remain many areas where such costs are not internalized and where perverse financial incentives operate. This is perhaps most obvious in the fishing sector, where support can take a variety of forms, all of which fall under the general heading of fishing subsidies. These subsidies include grants for the construction of new vessels, grants for vessel modification, preferential credit and tax treatments, payments to foreign countries for access to fisheries, and price subsidies or tax breaks for things like fuel and ice. Public expenditure on the infrastructure and services used by industry also count as subsidies.

Based on 1989 data, the FAO estimated that there was an approximately US$54 billion annual deficit between fishing revenues and costs, most of which was presumed to be as a result of subsidies. It is generally agreed that reductions in fishing subsidies would have a positive impact on the sustainability of fish stocks and on the environmental performance of the sector (Mattice 2003). Such reductions have been argued for strongly by organizations such as the World Wide Fund for Nature. Institutions such as the World Bank are now endeavouring to avoid lending money for projects that would increase capacity or effort in marine capture fisheries.

An interesting perspective on these matters is that, contrary to the generally held impression that it is human irresponsibility that lies at the core of biodiversity loss, it is in fact a consequence of human investment decisions and therefore an issue of human responsibility (Swanson 1996). Using marine fisheries as an illustration, Swanson (1996) pointed out that the well-recognized problems associated with open-access fisheries were only resolved when nation–states made the investment decision to restrict access and bear the costs of ensuring that restriction is enforced. It was only when depletion of resources on the continental shelf became apparent that institutional investment in the establishment of 200-mile limits came to be viewed as worthwhile. The basic idea here is that all decisions at state

Table 19.4. *Marine Stewardship Council Principles for Certification*

Principle 1	A fishery must be conducted in a manner that does not lead to over-fishing or depletion of the exploited populations and, for those populations that are depleted; the fishery must be conducted in a manner that demonstrably leads to their recovery.
Principle 2	Fishing operations should allow for the maintenance of the structure, productivity, function and diversity of the ecosystem (including habitat and associated dependent and ecologically related species) on which the fishery depends.
Principle 3	The fishery is subject to an effective management system that respects local, national and international laws and standards and incorporates institutional and operational frameworks that require use of the resource to be responsible and sustainable.

Source: Marine Stewardship Council (www.msc.org).

level concerning the regulation of natural resources are investment decisions. In other words, the state decides implicitly whether a particular resource or region is worthy to be allocated scarce societal investment funds to ensure that the access regime is complied with.

Recognizing that different access regimes carry with them very different societal costs for enforcement, this perspective sheds light on a situation that has arisen in a number of areas, concerning the sale of fishing rights by developing countries to the fleets of industrialized nations. When a developing nation government considers where it should invest the much-needed revenues gained from providing access, it will often be in areas other than in enforcement of access provisions. As a consequence of the decision not to invest in enforcement, the opportunity for overexploitation and illegal fishing is considerable and resource depletion more likely. In essence, it is the choice of human society not to invest in the species that results in its depletion. It is to be hoped, of course, that society knows about the resource depletion and is fully informed about the trade-offs with other priorities that it is making; in many instances, however, this is unlikely to be the case.

Creating genuine value propositions

Perhaps the most important paradigm shift necessary for effecting genuine change in patterns of human resource use and waste disposal is for sustainable practices to be seen as a financially profitable opportunity, rather than a business cost. There is now a growing body of argument that companies should focus on the financial value that is created or destroyed from issues surrounding ESD (Gilding *et al.* 2002). The basis for this argument is that there are now powerful market forces that put value on sustainability

for most industries and companies, and that action or inaction on sustainability will lead to the creation or destruction of company value. Proponents argue that focusing sustainability efforts to maximize the financial value of an enterprise, rather than on doing it because they 'should', will be the best way to integrate ESD concepts and effect real change.

A tangible example of this line of thinking can be seen in the area of environmental certification. In fisheries, for example, the growing influence of the Marine Stewardship Council (MSC) is notable (e.g. MSC 2004). The MSC aims to promote environmentally sustainable fisheries by creating a worldwide certification system for fish products from fisheries that are managed in an environmentally responsible manner. 'Eco-labelling' allows consumers to make conscious decisions about the products they purchase, based on environmental considerations. The fishery products that the MSC endorses are independently assessed against rigorous standards, and the Council ensures that they meet the criteria for environmentally sustainable fishing practices. Fisheries apply for MSC certification on a voluntary basis and must conform to a set of clear principles and criteria to be eligible for certification (Table 19.4). While suspicion remains in some quarters regarding such eco-labelling, transparent processes and clear principles and criteria should ensure that this approach gains credibility and delivers conservation and financial benefits.

CONCLUSIONS

Scientific understanding of the nature and value of continental-shelf ecosystems remains woefully inadequate for many parts of the world, yet sufficient is known to be concerned about the changes wrought by human activity. In particular, the first-order effects on shelf systems of degraded water quality from poor land-use practices or the impacts of fishing are now reasonably well understood and documented. It is some comfort to report that the need to manage these issues is now also recognized in the high-level policy documents of national and international bodies, all of which articulate a desire to reduce the threats and balance short-term economic benefit with long-term sustainability.

An illustration of progress is the issue of ecosystem effects of fishing. This is a subject that was on the fringes of fisheries management thinking 10–15 years ago, both in terms of policy and practice. Now it is deeply embedded in policy and fisheries are increasingly adopting practices to mitigate effects. Fishing-gear restrictions and improvements, spatial resource management and reserve areas are now core tools in the fisheries management armoury. Economic incentives to prosecute fisheries in a more sustainable manner are also becoming more important, as the discrimination and value systems of consumers in developed countries change. While it cannot yet be claimed that the environmental performance of fisheries is satisfactory, there are some encouraging signs. Similar signs are evident with respect to improvements in water quality. At the very least, recognition of these problems reflects a desire to do something about them at the highest levels.

A primary consideration with setting any desired goal, however, is how progress towards it can be measured. The management mantra that 'if you can't measure it, you can't manage it' is no less true for environmental issues than any other, and effective measures of environmental performance are essential. In recognition of the importance of this issue, the literature is becoming increasingly well populated with proposals for and analyses of environmental performance indicators (Baan & van Buuren 2003; OSPAR 2003). This literature shows that there are clearly many possible approaches, but that none of them is perfect. In developing an appropriate set of performance measures for any given situation, therefore, compromise and continuous improvement will be important principles. In general, measures are used either to assess attainment of political objectives (e.g. extent to which legislation is implemented or area of continental shelf protected), or to monitor the state of the marine environment (e.g. measures of biological diversity or concentration of contaminants in sediment). Both have a place in a portfolio of monitoring and assessment methods, but they must be part of an overarching framework to ensure that they complement each other and together form a coherent set of measures. Achieving this objective is still a long way off.

However, the imposition of new programmes of marine protection will require much better developed and harmonized monitoring and assessment methods at the level of regional seas, and will need to be much more closely linked to marine research needs and activities. Without this underpinning science, the understanding needed to develop indicators, set thresholds, monitor progress against objectives and report success or failure in a meaningful way will not keep pace with the aspirations of society.

While there are several examples of acceptance of high-level aspirations in the developed world (Western Australia, Government of, 2002; OSPAR 2002), globally the implementation of WSSD goals has been patchy. One fundamental reason for this relates to the practicalities of implementing change, and can be illustrated by the example of the EC. As already described, the ambitions and geographical scope of the EU Marine Strategy are broad, with the main focus on the European regional seas (the Baltic, the north-east Atlantic, the Mediterranean and the Black Sea). Even within the relatively sophisticated management process of the Commission, however, there is still doubt about whether existing legislative measures are sufficient to provide the desired level of protection and conservation of the marine environment. Over 20 years of EC legislation on the environment, a complex mix of policy and legislation has been created, and this aggregation still has gaps that do not specifically address the marine environment. For example, the long series of disconnected measures on water protection was replaced by a new catchment-based Water Framework Directive, while the Common Fisheries and Common Agricultural Policies continue to lack a proper environmental perspective. On the positive side, at least this legislation can be enforced, which is a distinct advantage over the regional marine conventions such as HELCOM and OSPAR, which are more directly focused on the marine environment and lack the 'teeth' of EC legislation.

In addition, the European Commission cannot develop or implement a marine strategy on its own, but must work in cooperation and consultation with all stakeholders, particularly with the well-respected, but relatively powerless, existing regional marine bodies. For this to work effectively, these regional organizations must have consistent approaches to marine environmental management. As there is no mechanism to ensure that this standardization occurs, there are real practical difficulties for those contracting parties that belong to more than one regional organization (i.e. OSPAR and HELCOM in the north-east Atlantic and Barcelona Convention in the Mediterranean). The recent development of a similar strategy for the Arctic region by the Programme for the Assessment of the Marine Environment (PAME 2003) reinforces the need to have consistency of approach.

In the final analysis, the threats to continental-shelf ecosystems can be substantially mitigated and the sustainable provision of goods and services from them ensured. In some cases, however, it will take many decades for the benefits to be realized. With respect to water quality, for example, the buffering capacity of the soil ecosystem will mean that changes in land-use practice in some areas may not result in benefits for many years. There is little doubt, however, that the social and political challenges to achieving this situation are formidable and will loom largest for developing countries in the coming decades.

To underpin the decision-making process and ensure that all stakeholders have a clear understanding of the trade-offs that must necessarily be made, a continued investment in science will be essential. Although only a tiny fraction of the world's shelf environment has been sampled, knowledge of the functioning of these ecosystems has increased markedly in recent decades. The view that shelf sea floor ecosystems are homogeneous expanses of sandflats and mudflats dominated by widely distributed species has been superseded. Of 35 sites on the Norwegian continental shelf, for example, 39% of 508 recorded species were restricted to one or two sites, and only three species spanned the entire 2700-km^2 study area (Ellingsen 2002). It is only with further investment to understand the biodiversity and ecology of continental-shelf systems that the trade-offs being made will be understood, and it will be possible to judge whether management actions are having the desired effect.

Part VII
Vast marine systems

Very extensive marine ecosystems such as the open ocean, polar seas and deep sea together cover >90% of the ocean surface and >60% of the Earth's. They dominate the biomass and maybe biodiversity of aquatic ecosystems and play major roles in the Earth's climate. These vast ecosystems have long been a primary source of human food and they harbour many of the largest and most charismatic animals, including great whales, polar bears, emperor penguins and the giant tube worms of hydrothermal vents. Yet their productivity is typically controlled by organisms largely invisible to the human eye (i.e. phytoplankton and bacteria). These marine systems include some of the most exotic and remote habitats on the planet and undoubtedly cradle extraordinary undiscovered species and evolutionary novelty. Because of their very great spatial extent and occurrence in polar regions, far out to sea or beneath thousands of metres of water, these ecosystems seem remote from most human activities and thus perhaps unlikely to have suffered substantially from deleterious anthropogenic impacts. However, the basis for describing any such changes and predicting future anthropogenic impacts remains poor. For the most part, these ecosystems remain frontiers in terms of exploitation, science and environmental management.

Across the subtropical euphotic zone (Chapter 20), polar and ice-edge seas (Chapter 21) and abyssal seafloor (Chapter 22), many, possibly most, of the vast areas involved may pass through natural cycles of change over periods of years to decades. These changes may be driven by large-scale ocean–atmospheric forcing such as El Niño–Southern Oscillation (ENSO) events, the Pacific Decadal Oscillation, the North Atlantic Oscillation and the Antarctic Circumpolar Wave. Separating natural ecosystem variations from anthropogenic change is thus a major challenge, especially where long-term data sets are lacking. Non-linear responses and regime shifts in response to environmental changes, both natural and anthropogenic, may be common and thus further highlight human ignorance.

20 · The marine pelagic ecosystem: perspectives on humanity's role in the future

PETER G. VERITY, JOHN H. STEELE, T. FREDE THINGSTAD AND
FEREIDOUN RASSOULZADEGAN

INTRODUCTION

Hensen (1887) proposed that food supply controlled variations in adult fish stocks, and therefore quantitative studies of plant and animal production in the sea might permit predictions of annual fish yields. If fish could be 'harvested' by man, Hensen (1887) argued that relationships similar to agriculture existed between primary production and fish yield. The original conceptual basis for quantitative marine ecology has a terrestrial origin (Verity *et al.* 2002). In retrospect, however, it is now appreciated that natural vegetation on land, such as that of trees, has long time-scales compared to marine primary producers (Steele 1991). Unlike terrestrial systems, marine pelagic communities appear to be much more adaptable to natural drivers and anthropogenic pressures, at least at timescales of years to decades.

The differences between marine and terrestrial systems are more important than the similarities. These

Aquatic Ecosystems, ed. N. V. C. Polunin. Published by Cambridge University Press. © Foundation for Environmental Conservation 2008.

differences include the way that marine systems respond to change, differences in the ways that humans attempt management and differences in how the terrestrial concept of sustainability relates to similar practices in the marine pelagic ecosystem. For example, the long timescales of the natural vegetation on land are reflected in the regeneration of forests, which can take decades to centuries. One potential concern with global warming is that plant communities could not adapt to the relatively rapid latitudinal shift in temperature and so could not maintain optimal conditions for growth. This is not really a problem in the sea, where the base of the food web, the phytoplankton, has lifetimes of days not decades. Fish, residing at the top of the marine web, mainly live for a few years, although some in the deep sea (Chapter 22) and off coral reefs (Chapter 16) live for decades. These factors have important implications for the effective application of scientific understanding to management of all marine resources and furthermore to human policies that affect the pelagic ecosystem in other ways, such as pollution, bioinvasions, diseases and eutrophication.

Humanity evolved in the terrestrial world, which is therefore intuitively better understood compared to aquatic environments. But human activities, planned and unplanned, are now causing dramatic changes in the latter, even systems as physically huge as the marine pelagic realm. This chapter will provide an overview of the status of anthropogenic impacts on the marine pelagic ecosystem by comparison with natural phenomena and, using marine fisheries as an example, suggest strategies for sustainability in the context of balancing social needs with environmental conservation.

STATE OF THE ECOSYSTEM

The status and trends of the environment and organisms in the marine pelagic ecosystem, focusing on the upper 200 m, were recently evaluated in detail (Verity *et al.* 2002). Salient conclusions from that analysis are briefly summarized below as a preface to recommendations for the future.

The harvesting of at least 90–100 million tonnes of fish per year globally, and perhaps as high as 130–190 million tonnes per year (including by-catch), is the largest anthropogenic impact on aquatic biota. Fisheries are switching to younger fish, increasing their by-catch, and targeting smaller species lower on the food web. Coupled

with long-term declines in those very stocks of fish of low trophic level, more and more regions are likely to experience fisheries collapses in the future (Pauly & MacLean 2003; Chapters 1 and 19).

At the ocean margins, an epidemic of nuisance phytoplankton blooms is occurring, accompanied by marine mammal, fish and invertebrate die-offs, human deaths and illness, and significant financial loss to aquacultural, fisheries and tourism industries, causing food-web dysfunction. Once considered to be rogue blooms, they are now commonplace. Regions previously free from harmful algal blooms now suffer such events, species previously benign have become toxic or nuisances and, in many regions, the frequency and intensity of red-tide outbreaks have been increasing (Smayda 1997; Steidinger *et al.* 2004).

Bioinvasions of alien species (plankton, macroalgae, benthic invertebrates and vertebrates) are now commonplace (Ruiz & Carleton 2003), and often the result of rapid and repetitive transoceanic or regional dispersal via transit in ballast water or by shellfish transplantation. Ecosystem invasions by non-indigenous jellyfish, medusae and ctenophores are increasingly commonplace. These takeovers by exotic gelatinous organisms are often associated with eutrophication and/or uncontrolled fishing (Purcell *et al.* 1999; Mills 2001).

It is becoming evident that the number of severely depleted species of marine mammals, turtles and fish is unprecedented in human history. Many marine species, including some sharks and tunas, are considered endangered (see Hilton-Taylor 2000), and many are likely to be rendered extinct (Roberts & Hawkins 1999; Ellis 2003; Chapter 1).

Chronic illnesses, disease epidemics, morbidity and mass mortality events are being observed across an array of taxonomic groups. A variety of pathogens, invasions of alien and often toxic species and human illnesses appear to be increasing in frequency and spatial extent (Harvell *et al.* 1999). These are provocative indicators at the ocean margins of a decline in marine pelagic ecosystem health (see Verity *et al.* 2002 for a thorough treatment).

These vectors often interact in ways that are not easily predicted and difficult to eliminate once established. They signal change, degradation and ongoing cryptic reorganization of marine communities, leading towards a new equilibrium driven by the need to adapt to multiple marine ecological disturbances.

SUSTAINABILITY OF HUMAN INFLUENCE: AN EXAMPLE

It is particularly inappropriate to put fisheries in the general category of natural-resource management along with forestry, not only because of the timescale problem already discussed, but also because the sea is considered as 'the common heritage of mankind' with a lack of any ownership for the sea, the seabed and the fish (Steele & Hoagland 2003). Unlike those on land, the resources and physical characteristics of the ocean and seabed are typically considered to be either *res communis*, belonging to the public, or *res nullius*, belonging to no one. In either situation, the absence of private rights to a scarce resource, such as fish, often leads to the 'tragedy of the commons' (Hardin 1968), a concept that arose not from forestry but from farming. During the seventeenth and eighteenth centuries, the establishment of long-term property rights in land, and their allocation by owners to tenant farmers, became the sustainable solution to an overexploitation of the commons.

The comparison of fisheries management with farming suggests two options that can be considered generically as 'American' and 'European' (Steele & Hoagland 2003). In the former, market forces have led to very large farms often owned by even larger corporations who necessarily take a long-term view. Such farms are, typically, very dull aesthetically but financially efficient. The 'European' alternative is to maintain a geographically and culturally traditional landscape. This is aesthetically pleasing, but requires central support by subsidy and import controls. Each approach is a solution for a sector of the economy which, although significant politically, is not critical for highly industrialized countries.

The developed world now depends more and more on developing countries for fish supply so that the mature fisheries at high latitudes are even less critical economically but are still culturally visible. For example, Europe and North America try both to require economic efficiency and expect cultural continuity (Kurlansky 1997). It is not surprising this leads to failure. Attempts to implement programmes of marketable property rights in fish, through such measures as individual tradable quotas (ITQs), have been subject to moratoria in countries like the USA (NRC [National Research Council] 1999). Alternatives involving effort limitation are recognized to aggregate or integrate fishing effort into larger management blocks that necessarily take a longer-term view but may deny independence to individual fishers; they are also likely to decrease the number of boats and people employed. At present the fishers, rather than the fish, are regarded as endangered. As the costs of enforcement of present policies increase and the fish stocks decline, it will be interesting to see whether free-market forces begin to trump the proponents of cultural preservation. Certainly in countries such as Iceland or New Zealand, where the fishing economy is critical to national economic production, institutional changes in the direction of property rights and market allocations have been seen (NRC 1999).

The European agricultural solution often results in part-time farmers. This is not feasible in open-sea fishing given the nature of the work and the substantial capital depreciation. It may be successful in nearshore fisheries for high-priced shellfish. The logical outcome for all these problems seems to be fish-farming. In northern Europe, the rearing of salmonids has converted an expensive delicacy into a supermarket commodity comparable to chicken. It may seem that, when the open sea has become bereft of directly edible fish, it will then supply an enhanced source of protein for the fish farms, from small rapidly growing fish further down the food web (Chapter 1). This form of 'enclosure' is an obvious consequence of the inability to evolve a management protocol for the open sea. Ironically, such a combination, namely harvesting at lower trophic levels to feed higher-level farmed stocks, could still lead to greater natural flexibility in marine ecosystems, such as having an abundance of 'jellies' rather than small fish at the top of the food web (Chapter 1).

The present focus on sustainability, especially of individual fish stocks, ignores the natural variability of marine ecosystems at decadal scales (Steele & Hoagland 2003). Attempting to damp out these interannual trends by year-to-year management practices may even exacerbate the oscillations, particularly by compounding them with economic processes at the same timescales. The natural physical and ecological causes of these 'cycles' need to be understood and sufficiently long-term management then revised to ameliorate rather than amplify the economic consequences.

NATURAL OSCILLATIONS AND ANTHROPOGENIC IMPACTS TO THE YEAR 2025

For many purposes, the ocean could once be regarded as a vast reservoir relative to human activities; anthropogenic impacts could usually be assumed to be restricted to local areas. With the combination of human population increase, changes in consumption patterns and technological changes in activities such as fishing, obvious impacts have spread into large offshore regions. For example, the collapse of Atlantic cod fisheries (Hutchings 2000) and anoxia events in North Sea bottom waters (Karlson *et al.* 2002) are warnings of future trends in the absence of appropriate actions.

Human activities have a direct impact at both ends of the ocean's pelagic food chain. At the upper end of the food chain, the impact of large-scale fisheries and hunting on marine vertebrates (mammals, reptiles and birds) potentially cascades down through the food web, creating indirect influences at lower trophic levels (Verity & Smetacek 1996). Likewise, changes in nutrient load and nutrient stoichiometry on the lowest trophic levels potentially cascade upwards through the food chain to create changes in rates, biomasses and biodiversity at the high end. This pelagic food chain contains life forms from viruses to whales, covering eight orders of magnitude in size, forming a dynamic system where important processes cover timescales from fractions of a second to decades. Even if good conceptual models existed to cover these ranges, the net effect of forces working from opposite ends of the food web would probably be hard to predict.

Despite such difficulties, there is a relatively well-accepted paradigm for the qualitative changes expected in the structure of the lower part of the food web along natural gradients from oligotrophy to eutrophy. In oligotrophic waters, primary production is dominated by picoplanktonic (size <2 µm) organisms, while larger-celled algae dominate in more eutrophic conditions. Combining the phytoplankton with their microbial predators (protozoans) and with the degraders of organic matter (heterotrophic bacteria), this can be summarized in a kind of ladder-like food-web structure. Such a picture is an elaborated view of the apparent counter-intuitive observation that nutrient-rich environments are dominated by short food chains, such as the 'classical' diatom–mesozooplankton–fish pathway (Fig. 20.1), while nutrient-poor environments are dominated by long food chains (e.g. autotrophic picoplankton–heterotrophic-flagellates–ciliates–copepods–fish) (Ryther 1969).

Even from highly simplified views of the pelagic food web (Fig. 20.1), it is obvious that human activities have an impact at the bottom of the food web. Not only will the amount of dissolved nutrients released to the environment around the ocean margins have an impact through stimulation of microbial growth, but the qualitative composition of such releases to the ocean is also important. Organic pollutants, even when non-toxic, change the environmental forcing by fuelling substrate into the heterotrophic bacteria (Fig. 20.1), potentially shifting the flow of matter, energy and nutrients toward 'long' food chains from bacteria via heterotrophic flagellates and ciliates to copepods, before

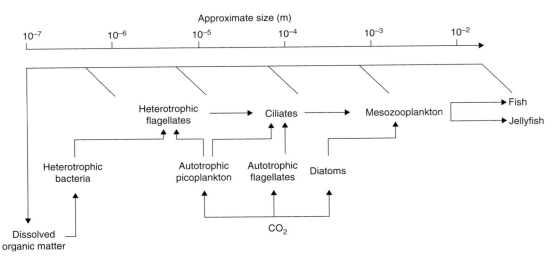

Fig. 20.1. Idealized representation of carbon flow through the pelagic food web, indicating how organic carbon may follow a long pathway, starting from small autotrophic picoplankton, or a short 'classical' pathway, starting with diatoms before passing through mesozooplankton to enter the higher parts of the food web ending in either fish or jellyfish.

reaching trophic levels of potential commercial interest. Dam-building in rivers has the effect of reducing silicate transport to the ocean (Conley *et al.* 2000). Silicate limitation will restrict the potential for establishment of the 'short' food chain from silicate-dependent diatoms via copepods to fish. Dam-building in the Danube River appears to have had such an effect on the Black Sea ecosystem (Humborg *et al.* 1997).

Modern agricultural practices in industrialized countries tend to release large amounts of nitrate relative to phosphate in runoff waters, whereas releases of untreated domestic sewage are relatively enriched in phosphate. In places like northern Europe, the consequence is a linking of coastal ocean chemistry to meteorological conditions: winters with high precipitation and large water flow in the rivers lead to excess transport of nitrate to the North Sea (Hickel *et al.* 1993), with phosphate and silicate-limited growth of the primary producers as the expected consequence. Harmful algal species often exhibit diverse nutritional strategies including direct consumption of organic nutrients as well as predation on competing phytoplankton. Hence, coastal zone management strategies may not only change ecosystem functionality, but also the stoichiometric relationship between elements and the way the elements are tied up in inorganic and organic forms, with potentially severe ecosystem consequences (Anderson *et al.* 2002*a*).

Aeolian inputs to the ocean may contain substantial inputs of nitrogen, phosphate and iron. In the Mediterranean Sea, northerly winds may carry nitrogen-enriched aeolian inputs characteristic of the industrialized areas of Europe, while winds from the south carry Saharan dust enriched in phosphate and iron (Bethoux *et al.* 1998; Herut *et al.* 1999). Long-term changes in dominating wind patterns, for example by extended changes in the pressure difference between the Azores and Iceland, may thus influence the biogeochemistry of the Mediterranean basin. Adding to the complexity of climate change effects on the pelagic ecosystem is the interaction between dust-transported iron availability and the high iron requirement of the nitrogen-fixing enzymes. The iron in Saharan dust transported into the Atlantic has thus been suggested as a mechanism stimulating nitrogen fixation in the Sargasso Sea ecosystem to a degree that may shift this system to phosphate limitation (Wu *et al.* 2000).

One of the predictions of climate models, in addition to a temperature increase, is increased instability of weather patterns. Upwelling events where nutrients are brought up into the photic zone are frequently wind-driven. Typically, such events are believed to stimulate the 'classical' food chain of diatoms–mesozooplankton–fish (Fig. 20.1), potentially enhancing fish production, but also being a short-lived phenomenon lasting only until most of the imported nutrients are re-exported via mechanisms such as sinking diatoms and copepod faecal pellet production (Wassmann 1998; Boyd & Newton 1999). In such a scenario, low-pressure passages in resonance with the food-web transfer of energy and matter through the 'classical' part of the food chain could be very beneficial for fish production. Mixing events that were insufficiently frequent would push the system towards long pathways (Fig. 20.1), while too-frequent events would not allow the upwelled nutrients to be translated into biological production in the photic zone (Boje & Tomczak 1978; Payne *et al.* 1992).

Socioeconomic decisions may thus influence the oceanic environment, not only through land use and wastewater treatment, but also via the longer loop, where greenhouse gas emissions potentially change wind and/or precipitation patterns and thus nutrient load and stoichiometry in coastal seas. In the North Pacific Central Gyre, a decrease has occurred in silicate and phosphate over the last three decades (Karl *et al.* 2001). This has been suggested to be caused by long-term, climate-driven changes that have induced a 'domain shift' in the microbial food web towards dominance of small prokaryotes (dominance of the left side of Fig. 20.1).

Also spectacular is the regime shift higher up in the food web in the Black Sea, where a large part of the production has been shifted into the commercial dead-end of jellyfish biomass (Berdnikov *et al.* 1999). As opposed to fish larvae that seek out their prey by visual means, jellyfish are non-visual predators. One hypothesis for such shifts has thus been that decreased water clarity (e.g. caused by humic substances or increased content of inorganic or biogenic particles) may induce such a regime shift towards jellyfish dominance (Eiane *et al.* 1999). However, understanding of the driving forces, and knowledge of when to expect such shifts, is poor.

The collapse of populations at the high end of the food chain, such as the North Sea herring, the Grand Banks (USA) cod stock (Hutchings 2000) or the large whales (Roman & Palumbi 2003) are well known (Chapters 1 and 19). The potential cascading effects down to the lower levels in the food web remain, however, mostly speculative. Part of the reason for this is perhaps the dichotomy suggested by the simplified food web (Fig. 20.1). In the world of unicellular organisms at trophic levels lower than those of most copepods, characteristic timescales for most processes are those of days or less. In many cases this is short relative to the hydrographical mixing processes of the water column, as demonstrated by the fact that stable vertical structures such as deep chlorophyll maxima become established (Steele 1978, 1995). At the level of copepods, the typical timescales for reproduction increase to weeks or months, while the timescale of vertical motion may be of hours to days. The result is that the coupled processes of copepod feeding and growth in the microbial food web may be separated in time and space (Hansen *et al.* 1997). These differences in characteristic timescales between trophic levels increase again to the next predator levels, with the result that coupled processes become much more difficult to demonstrate and model above the microbial level.

WHERE NEXT: GLOBAL SCIENCE OR LOCAL ISSUES?

Science appears to have come to the end of the first generation of global ocean programmes (World Ocean Circulation Experiment [NODC 2003], Joint Global Ocean Flux Study [JGOFS 2008] and Global Ocean Ecosystem Dynamics [GLOBEC 2004]). Although the original impetus, two decades ago, for a coherent suite of programmes came from the USA's National Science Foundation, these programmes became international in content and participation. Their development increased the scale of the science, required greater integration of management and became dominant themes in marine pelagic research (NRC 2000*b*). But there is now increasing emphasis on the application of the science to societal issues, including pollution, climatic change and fisheries (Field *et al.* 2002). These require better integration across the sectors. As past work is built on, but these growing concerns are also responded to, are global programmes with their sectoral foci the best way forward? Or, as has been suggested (NRC 2000*b*), should the emphasis shift to a loose coupling of regional or local problem-based programmes?

Current studies of the functioning of pelagic marine systems are based largely on modelling the responses of particular populations to physical forcing. These studies have been very successful in demonstrating the first-order role of physical processes in determining the distribution and abundance of copepods, krill and fish larvae (Steele & Collie 2005). However, the biological controls within marine ecosystems that act through density-dependent feedbacks across trophic levels need to be understood.

These processes are essential for the longer-term persistence of systems and especially their ability to switch among different, but ecologically coherent, regimes (Verity & Smetacek 1996; Steele & Collie 2005). The comparison of ecosystems in contrasting locations should not only demonstrate the different physical controls, but can also help elucidate possible underlying ecological similarities or constraints. For this, data and concepts from all trophic levels need to be incorporated.

One outstanding feature of past and current 'global' projects has been the lack of real interaction or overlap between studies of the upper and lower parts of marine food webs. This may have been because: (1) the research interests vary greatly among groups (e.g. global carbon cycle versus fisheries studies); (2) the regions studied do not usually overlap (e.g. the open ocean versus continental shelf); (3) the methodologies are distinct (e.g. biogeochemistry versus taxonomy and population dynamics); and (4) the research groups are often separated by location and funding sources.

However, research as well as social concerns have brought these groups closer together. For fisheries, addressing the vexing questions of causality of fish population declines, namely overfishing (top–down effects) versus environmental change (bottom–up and top–down effects) (Chapter 1), requires much more attention to lower trophic levels than has been traditional, as well as more integrative studies combining ecological and food-resource approaches. Similarly, questions about the ecological consequences of longer-term climate variability require more study of the effects of changes in carbon and nutrient export on recycling through mesopelagic and benthic food webs. These issues make it imperative that future projects consider marine food webs in their entirety, from phytoplankton and other microscopic organisms all the way to top carnivores.

A key aspect is the export of organic matter into the ocean's interior. The fraction of total primary productivity (PP) that is exported to deeper water or to higher trophic levels (including fish) depends on, but is not directly related to, the rate at which 'new' nutrients are introduced to the euphotic zone (Iverson 1990; Steele & Collie 2005). This 'new production' (or the corresponding f-ratio = new production/total production) is difficult to determine experimentally. Many estimates of transfer to higher trophic levels or 'export' to deeper water are extrapolations from a few measurements or rely on estimates from models (e.g. Pauly & Christiansen 1995). Also, the definition of new production depends on the relevant timescale as well as on the physical structure of the water column. Nearly all the recent *f*-ratio determinations (e.g. JGOFS) have been for open-ocean waters and should not be extrapolated to coastal regions (e.g. Chapter 19), for which there are few data. This is particularly true for regions subject to heavy fisheries exploitation, where the main concern is the availability of new production to higher trophic levels (Steele & Collie 2005). If management is to become eco-system-based, good estimates will be needed of the portion of primary production ultimately available to higher trophic levels, principally fish, but also marine mammals, sea turtles and seabirds.

Great cycles occur in the natural abundances and geographic locations of sardines and anchovy, with periods of a few decades (Worm & Myers 2003). Dramatic switches in the abundance of major ground fish stocks have been witnessed on Georges Bank off New England (USA), where cod were replaced by dogfish and mackerel, while off Newfoundland crustaceans became abundant after the cod went (Chavez *et al.* 2003). These switches can have great economic consequences, especially on fishing communities. Yet, these large rearrangements in species composition appear to be ecologically acceptable responses, using alternative pathways within the higher trophic levels of the ecosystem. There is little evidence of any significant effects on primary productivity.

Climate change may increase ocean temperature and stratification, thus reducing new production by nutrient input from depth. Together, these physical trends could create significant non-linear alterations in export to both the deep ocean (Chapter 22) and upper trophic levels. Thus, changes in production do not translate simply into corresponding changes in yields of organisms at higher trophic levels, for example pelagic and demersal fish. Conversely, there are issues concerning the consequences of the dramatic changes in abundance of fish stocks to other food-web components. Key questions include: do these changes have impacts on the microbial components in food webs through release from top–down controls? Or do 'transports' from the large zooplankton (copepods) and by detritus act to partially decouple feedback between upper and lower food-web levels? Or is this distinction a result of the dichotomies in the way the research is structured (Verity *et al.* 2002)?

Physical forcing on marine systems is related to winds, circulation, density fronts, gyres and tides. All these forcing factors are associated with very different time and space scales, and their biological consequences can occur at a considerable distance in time and in space from the site where the energy input occurs. Further, responses of marine food-web components to these factors are not necessarily linear. This has led to the concept of 'regime shifts' (Beamish 1993), in which gradual long-term changes in ocean physics can lead to relatively rapid switches from one community structure to another. This has major implications not only for the potential to forecast fish stock changes, but also for attempts to infer climatic changes from palaeo-oceanographic records. Rigorous testing of the underlying hypothesis of multiple equilibrium states (Steele & Henderson 1984; Collie & DeLong 1999) requires much more information on food-web dynamics, especially at intermediate trophic levels.

The consequences of physical forcing in terms of secondary production may differ greatly between coastal and open-ocean areas. In coastal waters, the mixed layer can extend to the bottom, affecting benthic processes (see Chapter 19). In the open ocean, the mixed layer is shallow with respect to the sea floor, so processes that occur in the mixed layer will influence the benthos less immediately than they do inshore.

Heavy fishing tends to remove the larger, more commercially valuable fish, leaving primarily the smaller, less-valuable fish (Pauly *et al.* 1998; Myers & Worm 2003). There is observational and theoretical evidence that such large changes at the top of marine food webs can induce switches in equilibrium states at lower trophic levels (Spencer & Collie 1997). Such effects can lead to top–down regime shifts in which the system sustains a different community of fish and other species. Such an alternate state can be stable and may need a strong perturbation to shift it back to its previous, or other, state (Collie & De Long 1999). Historically, these switches appear to occur on decadal to centennial timescales but, under strong forcing, could become more frequent (Steele & Henderson 1984).

CONCLUSIONS

Scientific perspective

All of these issues come together in the context of human impacts on the diversity of marine life (Steele & Collie 2005). How can human use of marine resources and the maintenance of adequate species richness to sustain marine ecosystem structure and function be balanced? In fisheries, for example, three apparently simple questions need to be answered. (1) How diverse are the resources available to humans? (2) How much can be removed by humans? (3) How sustainable is the yield? Answers to such questions require an understanding of the statics (food-web structure), the kinematics (energy and nutrient fluxes) and the dynamics (system stability) of ecosystems. Yet, effects of differences in marine biodiversity on marine food-web kinematics remain unknown (Grassle 2000). The main changes observed in marine species are at the larger end of the size spectrum in the macrozooplankton and fish. Yet most of the marine biodiversity lies at the other end of the size spectrum: in the viruses, bacteria, algae and microzooplankton of the microbial loop. Very little is still understood about how external factors influence biodiversity.

Interest in longer timescales has consequences for the spatial scales of study. In the ocean, both physical and ecological scales of time and space are inextricably linked (Steele 1995). The processes of mixing and transport of water masses change with increasing time and space scales; studying biological processes on longer temporal scales implies the need to study processes at larger spatial scales. Thus, scales of interest now have to extend from the micro- and mesoscale to basin and global scales (Schwartzlose *et al.* 1999). This increasing spatial range imposes greater demands on theory and field programmes. In particular, the longest temporal scales require retrospective data, such as those obtained from fish scales (Baumgartner *et al.* 1992).

The focus on physical–biological coupling at different trophic levels in JGOFS and GLOBEC has brought scientific benefits. At the same time, the development of linearized food-web models (Vezina & Platt 1988; Christensen & Pauly 1993) can provide comprehensive pictures of trophic structure by ignoring the non-linearities imposed by the physics and community dynamics. Yet, there is increasing evidence that these non-linearities may be expressed by abrupt or discontinuous changes in community structure, namely regime shifts (Steele 2004). How pervasive such hysteresis loops are in natural systems remains to be seen (Scheffer *et al.* 2001). But they can have severe consequences for management, as witnessed by the lack of revival of cod stocks in the north-west Atlantic. Thus a 'precautionary approach' would assume that such alternative stable states would occur in marine ecosystems under stress; once changed, ecosystems may stay that way.

Philosophical perspective

The secular impacts of humans on the oceans represent a staggering diversity of vectors and accumulated injustices (Verity *et al.* 2002). There can be little doubt that, with burgeoning human population, continued evolution of human society on a global scale is predicated on mutually beneficial moral ethics and altruistic behaviour. There exists a continuum of perspectives on how humanity should balance social needs with those of environmental conservation (Gorke 2003).

At one end of the spectrum, the fundamental conservationist view is that the pelagic ecosystem is crying out for help from the activities and by-products of a species that, for the first time in the history of the planet, is capable of altering evolution. Conservation theory simply holds that humanity must look beyond immediate resource use and long-term use of the oceans as a receptacle for human organic and technological by-products.

At the opposite end of the continuum are similarly compelling arguments couched in social consciousness. To wit: in a world with an exponentially growing population, the immoral thing might be to not actively pursue use of the oceans and their resources for humanity's benefit. The overriding constraint from both a moral and a practical point of view, of course, is how to ensure sustainability.

The only long-term solution to this Malthusian problem in the oceans is to achieve control of the global human population explosion. Everything else is a variation on the famous shell game, moving the problem around, perhaps delaying things a decade or two, but without long-term solution. Ideally restoring the marine pelagic ecosystem to its state when the first fishing people baited hooks, to the time when human ancestors looked for wealth on land but were unable to recognize it in the oceans, might be dreamt of. This is undoubtedly impossible in the present global economy and political climate. But if humans are not aware of the breadth and depth of the problems, then society is not prepared to plan for effective and intelligent policy and action. The warning signs in the past have largely been ignored to date: evidence suggests that continued ignorance will lead to changes in the structure and function of the marine pelagic ecosystem that will overshadow every future human step taken, and that is likely to produce a new equilibrium that will be less anthroponomic.

21 · Polar and ice-edge marine systems

ANDREW CLARKE, ANDREW S. BRIERLEY, COLIN M. HARRIS, DAN LUBIN
AND RAYMOND C. SMITH

INTRODUCTION

Cold, seasonal and dominated by ice, the two polar regions are similar in many ways. Oceanographically, however, they are very different. The Arctic is a deep basin surrounded almost completely by extensive shallow continental shelves and continental land masses. Large river systems (Chapter 2) discharge significant volumes of fresh water and sediment into the Arctic basin, some of them carrying substantial pollution, and exchange with the Pacific and Atlantic Oceans is highly constrained. In direct contrast, the Antarctic is a single isolated land mass surrounded on all sides by a deep ocean contiguous with the three great ocean basins (Fig. 21.1).

Climate change has been identified as the key environmental trend in both polar regions, with the most likely effect in the marine environment being mediated through changes in sea-ice dynamics (Clarke & Harris 2003) (Table 21.1). The polar regions play a critical role in the

Aquatic Ecosystems, ed. N. V. C. Polunin. Published by Cambridge University Press. © Foundation for Environmental Conservation 2008.

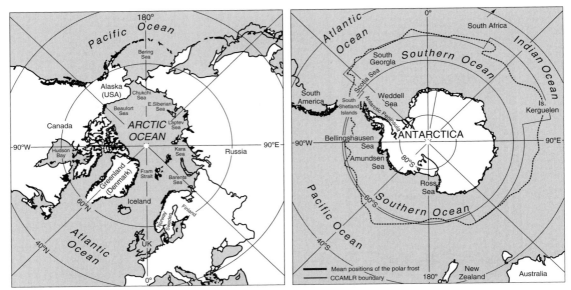

Fig. 21.1. The Arctic and Antarctic, showing the differences in topography and connections to lower latitude oceans. In the Antarctic, the dotted line shows the mean position of the Antarctic Polar Front. CCAMLR, Commission for the Conservation of Antarctic Marine Living Resources.

regulation of global climate, and it is expected that future climatic shifts will be particularly pronounced at high latitudes (IPCC [Intergovernmental Panel on Climate Change] 2001*a*). The precise patterns of change and the feedbacks that influence these are, however, poorly understood and our ability to predict the future is greatly limited by this uncertainty.

Major Arctic trends

Current expectations are that the Arctic region will be the first to experience substantial climate change, with a possible rise in average surface air temperature of 4–5 °C by the middle of this century (Stouffer *et al.* 1989; Manabe *et al.* 1991; IPCC 2001*a*). Increases in spring and summer air temperatures have already been detected (Martin *et al.* 1997), and both thinning and a decrease in spatial extent have been described for Arctic sea ice (Maslanik *et al.* 1996; Parkinson *et al.* 1999; Wadhams & Davis 2000; Rothrock *et al.* 2003). The IPCC (2001*a*) concluded that average Arctic sea-ice extents had declined by 15% in summer and 8% in spring, and there is some evidence that the rate of reduction may have been accelerating in recent years (Johannessen *et al.* 1995).

Reductions in sea-ice cover will have important oceanographic consequences, driven by changes in air–sea exchanges, irradiance and water-column stability. There are also probable impacts on Arctic marine biota at all levels in the food web (Tynan & DeMaster 1997). In addition, there has been a direct impact on the Arctic marine system through fishing and hunting. Many Arctic marine mammals have been overexploited in the past (Pagnan 2000) with many species still globally endangered. New developments in fishery technology have also increased the scale of fish catches in the Arctic, and overharvesting has led to the collapse of stocks of several species (Pagnan 2000).

Human activities have also had a major influence on the Arctic through the dissemination of pollution from both local and global sources (AMAP [Arctic Monitoring and Assessment Programme] 1998). These contaminants include persistent organic pollutants, heavy metals, acid rain, sulphur deposition, hydrocarbons from oil and gas extraction, and radioactivity. Industrial chlorinated hydrocarbons have also led to the depletion of stratospheric ozone over the Arctic (Pyle 2000). This pattern of pollution reflects both the long-distance dispersal from industry at lower latitudes and the topography of the Arctic as a basin almost completely surrounded by continental land masses.

Table 21.1. *Summary of major environmental trends and principal threats to polar marine ecosystems*

Category	Threats and trends
Arctic	
Environmental trends	(1) Some evidence for significant change in sea-ice extent and thickness in Arctic Ocean, though spatial patterns of change non-uniform
	(2) Moderate ozone depletion
	(3) Fishing has severely depleted stocks of fish and some marine mammals
Threats (long-term)	(1) Climate change, with consequent impacts on sea ice, hydrological cycle and UV-B flux
Threats (short-term)	(1) Fishing impact
	(2) Mineral extraction (especially pollution impacts)
Antarctic	
Environmental trends	(1) Significant atmosphere warming in Antarctic peninsula, but not over continent
	(2) Significant depletion of springtime ozone levels since *c.*1970, with consequent increase in UV-B flux. Believed to have peaked, with improvement expected over next 25 years
	(3) Significant decrease in extent of winter sea ice in Bellingshausen/Amundsen seas, but not in other sectors, since advent of satellite observations. Longer-term trends contentious
	(4) Southern Ocean fishery has severely depleted stocks of some fish, and some marine mammals. Recovery of fish and whale stocks slow; southern fur seal stock recovery rapid. Fishing outside Southern Ocean has by-catch impact on some seabirds that breed south of the Antarctic Polar Front, especially albatrosses and larger petrels
Threats (long-term)	(1) Climate change in Antarctic Peninsula area, with consequent impacts on sea ice
Threats (short-term)	(1) Illegal, unreported and unregulated (IUU) fishing activity, both in terms of impact on exploited stocks and on seabird by-catch

Source: From Clarke and Harris (2003).

Major Antarctic trends

Environmental trends in the Antarctic are similar in many ways to the Arctic, with climate change, stratospheric ozone depletion and fishing pressure being the most important. Pollution is currently much less of a threat to the Southern Ocean than in the Arctic (Clarke & Harris 2003). It is, however, difficult to detect any clear environmental trends for the Antarctic marine ecosystem as a whole, although there are localized trends in some areas. The Antarctic Peninsula is warming rapidly, but the Antarctic continent itself shows no spatially consistent trend (King 1994). The reason for this difference is not clear, though the climate of the Antarctic Peninsula area is highly sensitive to the complex feedbacks between atmosphere, oceans and sea ice, and is strongly influenced by climatic variations in the subtropical and tropical Pacific Ocean (King & Harangozo 1998; Yuan & Martinson 2000, 2001).

In recent years, a reduction in sea ice in some areas (Jacobs & Comiso 1993) has been balanced by an increase in others, and there is currently no evidence for a decline in overall sea-ice extent around Antarctica (Zwally *et al.* 2002). The Amundsen and Bellingshausen Seas have shown a significant decrease in sea ice, however, and this may be related to the regional climate warming of the Antarctic Peninsula. In the western Antarctic Peninsula region, sea-ice extent has decreased with consequent impact on all trophic levels of the marine ecosystem (Smith *et al.* 2003*a, b*). This may offer a valuable analogue for future changes elsewhere in polar regions.

Like the Arctic, the Southern Ocean living resources have a long history of human exploitation. Following the exploratory voyage of Captain James Cook in the eighteenth century, a fishery for southern fur seal (*Arctocephalus gazella*) developed rapidly. By 1822, a period of only 35 years, the industry collapsed (Bonner 1982), and exploitation shifted to the great whales. By the late 1960s, whale stocks had been severely depleted and attention switched to finfish, the exploitation of which continues today, together with a fishery for Antarctic krill (*Euphausia superba*).

In this chapter, the predominant environmental trends and threats identified by Clarke and Harris (2003) are used as a basis for predicting likely states of the two polar marine environments in 2025. Climate, ozone, sea ice and fishing are separately addressed, and the chapter concludes with a discussion of how human influences might be mitigated by regulatory actions.

CLIMATE

The future of both the Arctic and Antarctic marine eco-systems is linked very closely with climate change. During the past decade, the polar regions have exhibited many of the first unmistakable signs of climate warming, and the ecological manifestations of this warming are magnified by its dramatic impact on sea-ice cover. Observationally, climate warming throughout much of the Arctic, and in Antarctic locations particularly relevant for marine ecology, is well established (e.g. Stammerjohn & Smith 1996; Rigor *et al.* 2000). Satellite passive microwave observations have documented the significant retreat of Arctic sea-ice cover, along with a slight but statistically significant increase in Antarctic sea-ice cover when averaged over the entire Southern Ocean (Cavalieri *et al.* 1997; Johannessen *et al.* 1999; Zwally *et al.* 2002). These trends in sea-ice cover in both hemispheres are consistent with predictions from global climate model (GCM) simulations (Cavalieri *et al.* 1997; discussed below). The sea-ice retreat in the western Antarctic Peninsula region has not yet been reproduced in GCM simulations, although a plausible mechanism was proposed by Thompson and Solomon (2002). The mechanisms behind high-latitude climate warming are dynamically very complex, but are very likely to have an anthropogenic component.

Feedbacks in the polar climate system

For many years, high-latitude climate warming has been expected as a result of classical climate-feedback mechanisms involving surface albedo (the ratio of reflected to incident electromagnetic radiation), solar and terrestrial radiation, and cloud cover (Crane & Barry 1984; Somerville & Remer 1984; Ledley 1993; Curry & Webster 1999). In the ice–albedo feedback, which should be significant at temperatures close to the triple-point of water, a warming induces a localized melting of snow or ice, which thereby reduces the surface albedo. This reduced albedo leads to enhanced absorption of short-wave radiation by the surface, which thus accelerates the melting rate and further decreases the albedo. The result is a reduction in short-wave radiation backscattered to space by the Earth–atmosphere system, a concomitant increase in shortwave absorption by the surface and a decrease in snow/ice cover, which together constitute a strong positive feedback to climate warming. A warmer climate may also allow the atmosphere to hold more water vapour, and this water vapour will partially close infrared windows (the wavelength ranges over which the atmosphere is largely transparent to radiation) such that atmospheric long-wave emission warms the surface to a greater extent. This water-vapour feedback is another potentially important positive feedback at high latitudes (Curry *et al.* 1995). There is also a general cloud–radiation feedback (Curry & Webster 1999), through which changes in atmospheric precipitable water force changes in cloud amount, vertical distribution, optical depth, thermodynamic phase, effective droplet radius and ice particle size and temperature. The present consensus is that the net effect of these feedback mechanisms should be positive at high latitudes, although analysis of trends in satellite temperature data has challenged this conclusion (Wang & Key 2003). In contrast to these positive feedbacks, a negative feedback can arise if there is increasing precipitation on sea ice resulting from a warming atmosphere containing more moisture (Ledley 1993). In this scenario, the snow cover build-up increases the surface albedo, thus decreasing the absorbed short-wave radiation, and also decreases the turbulent energy flux from the ocean to the atmosphere.

The Arctic Oscillation

Recent research in atmospheric dynamics has introduced a new perspective on high-latitude climate change. Thompson and Wallace (1998) identified the Arctic Oscillation (AO), a persistent mode of variability in atmospheric circulation, as a major regulator of Arctic surface temperature. In its simplest conception, the AO can be thought of as a see-saw in sea-level pressure, alternately rising over the central Arctic Ocean and then in a sub-Arctic belt ranging from southern Alaska to Europe. The AO and the North Atlantic Oscillation (NAO) index which has been known for several decades are often considered part of the same dynamic phenomenon, the

Northern Annular Mode (NAM), although this view remains somewhat controversial. Physically, the NAM is manifest as an oscillation in the strength of counter-clockwise zonal atmospheric flow at temperate and high latitudes. Two important aspects of annular modes are relevant to climate change. Firstly, they do not necessarily vary in a periodic way; they can vary both monthly and annually. Secondly, annular modes are easily excited by external physical forcing such as atmospheric warming.

The interaction between the AO and climate is described by the AO index, defined as the leading principal component of the wintertime (November–April) monthly mean sea-level pressure anomaly field poleward of 20° N. In the positive phase of this oscillation (high AO index), stronger westerlies isolate colder air to the north, allowing Arctic temperatures at many locations to become warmer. In the negative phase (low AO index), there are relatively weak westerly winds, and colder air is allowed to spill out further south at most longitudes. During the past two decades, there has been a shift toward a positive phase in the NAM, and this may be the direct cause of much of the observed Arctic warming (Thompson & Wallace 2001).

In addition to this direct effect on Arctic surface air temperatures, the AO has been dynamically linked to Arctic sea-ice concentrations (Rigor *et al.* 2002). During low AO index conditions there is a substantial clockwise circulation in the Beaufort Gyre, which confines much sea ice into the colder central Arctic where it tends to thicken. During high AO index conditions this circulation weakens, ice residence time in the central Arctic is reduced, and there is a more rapid passage of ice through the Fram Strait into the North Atlantic. The AO thus has both thermodynamic and direct dynamic impacts on Arctic sea-ice cover.

Although the NAM mechanism would appear to lessen the role of direct greenhouse and similar forcings in high-latitude warming, these anthropogenic factors may still have a prominent indirect role. Current GCM simulations are beginning to reproduce the NAM (Moritz *et al.* 2002), and some models suggest that a strengthening of the AO may result from anthropogenic greenhouse forcing (Shindell *et al.* 1999).

The Southern Annular Mode and Antarctic ozone

The springtime decrease in stratospheric ozone over Antarctica (Fig. 21.2), popularly known as the ozone hole, is one of the first large-scale atmospheric changes to have a

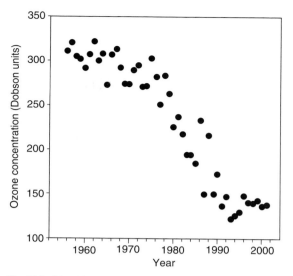

Fig. 21.2. Monthly mean total ozone in October, the month of minimum ozone levels above the British Antarctic Survey research station at Halley (British Antarctic Survey data).

proven human origin. Initial concern about the onset of the ozone hole during the 1980s involved dramatically enhanced solar ultraviolet (UV) radiation (Lubin *et al.* 1989) and possible damage to the Antarctic marine food web. Inhibition of photosynthesis in Southern Ocean phytoplankton, directly attributable to enhanced UV radiation in the upper water column, has been reported (Smith *et al.* 1992; Holm-Hansen *et al.* 1993). However, satellite remote sensing of ocean colour suggests that the Antarctic ozone hole has probably not caused irreparable damage to the Antarctic marine ecosystem through enhanced UV radiation (Arrigo *et al.* 1998), although it is difficult to distinguish long-term change against the background of intense year-to-year variability. Nevertheless, ozone-related increases in UV-B radiation are well documented and may interact with other environmental factors to influence productivity in marine systems (Hader *et al.* 2003). They may also affect biogeochemical cycles in ways that influence overall system productivity and dynamics (Zepp *et al.* 2003).

In terms of climate, one important aspect of the Antarctic ozone hole lies in its interaction with the Southern Annular Mode (SAM), the Antarctic equivalent of the NAM (Thompson & Wallace 2000). Recent observations indicate a trend towards stronger westerly circumpolar flow, related to a high SAM index, over the last few decades (Thompson *et al.* 2000b). Thompson and Solomon

(2002) suggested that this trend was dominated by the development of the Antarctic ozone hole, and associated stratospheric radiative cooling related to the low ozone abundances. This is because observed changes in southern hemisphere stratospheric circulation are strongly related to total column ozone. Furthermore, analysis of geopotential heights indicates similar and concomitant trends towards a high-index polarity of the SAM occurring in the tropospheric circulation, but with some timing differences (Thompson & Solomon 2002). This implies a coupling between the SAM in the troposphere and the circulation in the lower stratosphere. Theoretical predictions suggest that the stratosphere–troposphere coupling should be the strongest when the polar vortex is either building or decaying, and observations indicate that this is indeed the case and therefore the impacts of the springtime ozone losses on the lower stratosphere extend to the circulation of the troposphere. A significant portion of the trends in the surface temperature anomalies over the Antarctic continent (cooling over eastern Antarctica and the Antarctic plateau with a concurrent warming over the Antarctic Peninsula) could be explained by the SAM and related strengthening of the tropospheric westerlies in high SAM index conditions (Thompson & Solomon 2002).

High-latitude climate changes

The Arctic Ocean and the western Antarctic Peninsula are strongly influenced by changes in atmospheric dynamics that most likely have substantial anthropogenic origins, from both greenhouse warming in the troposphere and ozone depletion in the stratosphere. Anthropogenic ozone depletion is expected to persist for another three or four decades before significant recovery occurs under the Montreal Protocol. The most comprehensive assessments of climate change suggest a continued warming trend (IPCC 2001a). It is therefore reasonable to expect that these climate-related ecological changes, linked to high AO and SAM index conditions, will continue to the 2025 time horizon.

SEA ICE

The annual growth and decay of sea ice is one of the most prominent physical processes on Earth. Historically, Arctic sea-ice extent has varied between a summer minimum of about 7–9 million km^2 and a winter maximum of 15–16 million km^2 (Parkinson 2000). In the Antarctic, sea-ice extent has oscillated between a summer (February)

minimum extent of approximately 4 million km^2 and a winter (August) maximum of 20 million km^2 (Zwally *et al.* 2002). Some Arctic sea ice may persist for tens of years and form ridges more than 10 m thick (Maykut 1985), whereas most Antarctic sea ice lasts only one year and has a mean thickness of less than 3 m (Horner *et al.* 1992; Thomas & Dieckmann 2003).

The annual cycle of freezing and melting of sea ice has a profound oceanographic impact, leading alternately to the formation of cold dense high-salinity water during ice formation and to surface-stabilizing low-salinity water when the ice melts. This process in turn comprises a significant driver of global thermohaline circulation. The seasonal presence or absence of ice also has an important climate effect, influencing solar heat reflection or absorption (albedo), and mediating ocean–atmosphere interactions. In addition to influencing globally significant physical processes, sea ice provides habitats of major ecological importance. Numerous organisms live in or on, or are associated with, sea ice for some or all of their life cycle, and sea ice influences biological processes at all trophic levels. Sea ice is both an indicator of change and, via strong feedback systems, a mechanism affecting global climate. Further, sea ice is a major element of the polar environment and associated ecosystems. If the impact of climate variability is to be predicted, it is therefore essential that the variability and long-term trends in sea ice be understood.

Observations on ice thickness and extent come from historical sources such as whaling records or travel diaries, polar stations, ships, submarines, aircraft and satellites. Polar stations, while limited spatially, provide the longest time-series records. Conversely, passive-microwave satellite instruments can measure the global sea-ice cover every few days with a resolution of the order of 25–50 km, thus providing good spatial and extremely good temporal resolution. However, satellite data are available only for the past two decades and currently can provide no quantitative information on ice thickness (and hence volume), although the next generation of satellites will do so. All observations show very large seasonal variations in the thickness and extent of the sea-ice field. In spite of this variability, there are numerous reports of statistically significant changes in sea-ice extent and duration around both poles in recent decades.

Arctic sea ice

In the Arctic, field observations suggest that sea-ice reduction began in the 1960s (Fig. 21.3), and analyses of

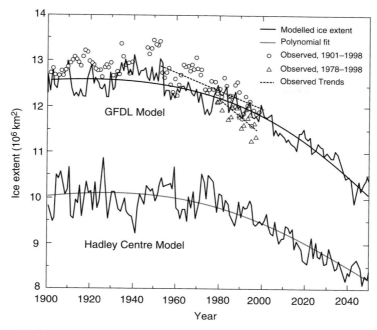

Fig. 21.3. Observed and modelled changes in sea-ice extent in the Arctic. Output is shown from two models, namely the Geophysical Fluid Dynamics Laboratory (GFDL) model and the Hadley Centre HadCM3 model. Although these models assume different ice extents at the start of the twentieth century, both show a decline in Arctic sea ice from the 1960s. The observed data (IPCC 2001a) also show a decrease in sea-ice extent from the 1960s.

satellite data suggest the rate of reduction has proceeded at roughly 3% per decade over the period 1979–96 (Serreze *et al.* 1995, 2000; Cavaliari *et al.* 1997; Johannessen *et al.* 1999; Parkinson *et al.* 1999; Deser *et al.* 2000). Decreases in Arctic sea-ice extent have occurred in all seasons although there is evidence for a stronger reduction in summer (4–6% per decade) compared with autumn and winter (0.6% per decade) (Chapman & Walsh 1993; Maslanik *et al.* 1996; Cavalieri *et al.* 1997; Parkinson *et al.* 1999; Vinnikov *et al.* 1999; Deser *et al.* 2000). Satellite observations permit the evaluation of trends by regions, and all regions show a statistically significant decrease in sea-ice cover over the past two decades (Parkinson *et al.* 1999).

Forty-year (1958–97) records of reanalysis products and corresponding sea-ice concentration data, combining conventional surface estimates with microwave satellite data, indicate that the recent and historically unprecedented trends in the wintertime NAO and AO circulation patterns over the past three decades have been imprinted upon the distribution of Arctic sea ice (Chapman & Walsh 1993; Deser *et al.* 2000). This and other work (Zhang *et al.*

2000; Hall & Visbek 2002) is consistent with the hypothesis that atmospheric circulation anomalies force the sea-ice variations. Over this 40-year record, Deser *et al.* (2000) showed a nearly monotonic decline (−4% per decade) in summer sea-ice extent for the Arctic as a whole.

The data for longer timescales are sparse, and there is a diversity of conclusions that Polyakov *et al.* (2003a) suggested was caused by problems in sampling spatially and temporally very variable ice thickness and extent. Making use of newly available historical Russian records from five polar stations spanning the Kara, Laptev, East Siberian and Chukchi Seas, Polyakov *et al.* (2003a, b) used a century-long time series to evaluate trends and long-term variability of August ice extent in Arctic marginal seas. Their analysis concentrated on fluctuations with a period of 50–80 years, and they argued that these low-frequency oscillations (LFOs) have played an important role in Arctic sea-ice variability. Time-series and wavelet analyses of these data from the Siberian marginal ice zone indicate two periods of maximum ice extent associated with positive LFO phases (and warming in the 1930s–1940s and late 1980s–1990s), and two periods of maximum ice extent associated with negative LFO phases (and cooling prior to the 1920s and in the 1960s–1970s). Analyses of these long-term, but regionally specific, data suggest that long-term trends are small and generally statistically insignificant (Polyakov *et al.* 2003*a*, *b*). This work shows the value of long-term records that permit analysis of low-frequency variability. It is also consistent with the more recent satellite-based analyses that show very large inter-annual variability among the various Arctic sea-ice regions, and emphasizes why a general conclusion for the Arctic as a whole cannot be generated from limited spatial observations.

Observations of sea-ice thickness are more restricted in both space and time. Holloway and Sou (2002) provided a recent review and analyses of ice-thickness observations. Ice draft data from submarine-based sonar profiling have led to a wide range (0–43%) of estimated trends in reduction of ice volume (Bourke & Garrett 1987; Bourke & McLaren 1992; McLaren *et al.* 1994; Shy & Walsh 1996; Rothrock *et al.* 1999; Wadhams & Davis 2000; Tucker *et al.* 2001; Winsor 2001). Holloway and Sou (2002) argued that these results are not necessarily contradictory, given the different time periods and locations of data. Indeed, all investigators have found large interannual and spatial variability. By making use of other data (atmosphere, rivers and ocean) in a dynamic ocean–ice–snow model, these authors attempted to constrain the inferences from the submarine-based data. Volume loss from 1987 to 1997 lies within the range of 16–25% and suggests that reports of more rapid loss are inconsistent with their more comprehensive data and model constraints (Holloway & Sou 2002). The best current estimates thus indicate a significant loss of Arctic sea-ice volume over recent decades.

An alternative perspective is provided by the use of models, making use of ice, ocean and atmospheric parameters, which permit longer time-series data to be used to examine sea-ice trends and explore the driving mechanisms associated with these trends. A thickness distribution sea-ice model coupled to an ocean model suggests that during the past two decades, the ice system has reacted to climate variability primarily through changes in ice advection, resulting in a change in distribution of ice mass rather than in thermal forcing (Zhang *et al.* 2000). Changes in Arctic ice cover may be responses to changes in atmospheric circulation and appear to be an integral part of the NAO and AO (Zhang *et al.* 2000). Linking ice dynamics to atmospheric

variability will be an important factor in distinguishing periodic behaviour from long-term trends.

Overall, there is general agreement that Arctic sea-ice extent and thickness have decreased during the past several decades. The data are consistent and in agreement with most models. Observations earlier than the late 1950s are sparse in both space and time so that establishing trends beyond the past four decades is difficult.

Antarctic sea ice

Satellite passive microwave observations have provided the most consistent sea-ice data for the Southern Ocean (Zwally *et al.* 1983, 2002; Gloersen & Campbell 1988; Gloersen *et al.* 1992; Parkinson 1992, 1994, 1998; Johannessen *et al.* 1995; Bjorgo *et al.* 1997; Watkins & Simmonds 2000). Studies prior to about 1998 suggested that there were no significant changes in Antarctic sea ice during the satellite era. However, a systematically calibrated and analysed data set for 1979–98 (Zwally *et al.* 2002) shows that the total extent of Antarctic sea ice (defined as a sea-ice concentration >15%) increased by $11\,180 \pm 4190$ km^2 per year (0.98 ± 0.37 % per decade) (Zwally *et al.* 2002). This is in contrast to the decreasing trend ($34\,300 \pm 3700$ km^2 per year) for the Arctic, making use of a similarly calibrated and analysed data set (Parkinson *et al.* 1999). Combination of the separate hemispheric sea-ice records into a global record indicates an overall decrease in global sea ice (Gloersen *et al.* 1999). The Antarctic record shows large regional oscillations within the hemisphere so that regional variability and trends are distinct from the hemispherical trends and variability. Regionally, the trends in sea-ice extent are positive in the Weddell Sea, Pacific Ocean and Ross Sea, slightly negative in the Indian Ocean, and strongly negative in the Bellingshausen and Amundsen Seas. Data from the period 1987–96 have also shown that total Antarctic sea-ice extent has increased significantly during this period (Watkins & Simmonds 2000). During the same period, the Antarctic circumpolar atmospheric pressure trough may have both deepened and moved further south (Simmonds *et al.* 1998). Such a shift would have a considerable impact upon the underlying sea-ice distribution through a shift in the net westerly wind regime and subsequent Ekman transport of sea ice (Stammerjohn & Smith 1997).

Thus, as found for the Arctic, shifts in Antarctic atmospheric patterns have an important influence on sea-ice distribution. Hall and Visbeck (2002) used a coupled ocean–atmosphere model to explore how the SAM generates ocean circulation and sea-ice variations, in the context of the Antarctic Circumpolar Wave (ACW) (White & Peterson 1996; White *et al.* 1998) and the Antarctic Dipole (Yuan & Martinson 2000). A 17-year passive-microwave data set demonstrated that coherent patterns of sea ice show opposite polarities during the two extremes of the Southern Oscillation Index (SOI), and that the climate anomalies in the Bellingshausen/Amundsen and Weddell Sea sectors of the Southern Ocean show the strongest link to the Southern Oscillation (Kwok & Comiso 2002). The composite patterns have been weighted by four strong El Niño–Southern Oscillation (ENSO) episodes over the last 17 years, and these warm events may have weighted the sea ice and climate anomalies towards patterns associated with the negative extremes of the SOI (Kwok & Comiso 2002).

Future sea-ice conditions

Global climate models are able to reproduce the observed rates of reduction in Arctic sea-ice extent, and forecasts from these models indicate that reduction in extent will continue to 2025 and beyond. Predictions for future sea-ice distribution vary, but it has been suggested that by 2100 there will be no permanent sea ice in the Arctic (Gregory *et al.* 2002). Aside from the natural physical and environmental impacts, large-scale reductions in Arctic sea-ice extent will enable increased shipping activities in the Arctic, opening trade routes, providing easier access for extraction of resources including oil, minerals and fish, and perhaps leading to increased political tensions between those nations that claim Arctic coastal waters as their own (Kerr 2002).

In the Antarctic, observations of the winter duration of fast sea ice (Murphy *et al.* 1995), inferences from whaling data (de la Mare 1997) and levels of methane-sulphonic acid in an ice core from Law Dome (Curran *et al.* 2003) have all suggested that a step reduction in sea-ice extent and duration occurred sometime between the 1950s and 1970s, before satellite data were available. However, later satellite observations and fast-ice duration records (Fig. 21.4) both indicated a 3–4-year periodicity in the sea-ice field during the 1980s and 1990s (later described as the ACW: White & Peterson 1996), demonstrating some congruence between the different

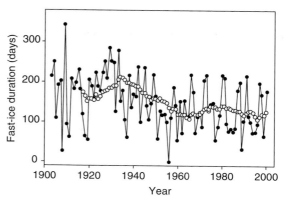

Fig. 21.4. Duration of winter fast-ice at South Orkney Islands. Data for individual years are shown in black, and the clear circles show a tapered 15-year running mean (British Antarctic Survey data, presentation modified from Murphy *et al.* 1995). Data to 1994 were collated from visual observations; data from 1995 were collected using an automatic camera which took four images every day through the winter. Note that the 15-year running mean indicates a period of secular change in mean winter fast-ice duration in the South Orkney Islands from the late 1930s to the late 1960s. This overlaps with, but does not precisely match, the periods of change in sea-ice extent reported from 20° to 30° E using whale catch position as a proxy for the ice-edge (de la Mare 1997), and for East Antarctica using methane-sulphonic acid in the Law Dome ice core to estimate sea-ice extent (Curran *et al.* 2003).

data streams and so providing an additional line of evidence that the inferred step change in Antarctic sea-ice extent was real. Satellite observations have revealed long-term trends in Antarctic sea ice, but the direction of change is not as clear-cut as in the Arctic. On balance, overall Southern Ocean sea-ice extent appears to have increased recently by about 1% per decade (Zwally *et al.* 2002), but this small overall change represents a balance between larger increases in the Weddell, Ross and Western Pacific sectors, and reductions in the Indian Ocean and Bellingshausen–Amundsen sectors. There are few data on Antarctic sea-ice thickness, and no trends have been detected (Murphy *et al.* 1995). Change in Antarctic sea-ice extent to 2025 and beyond is thus difficult to predict because, perhaps counter-intuitively, increased warming may lead to increased snow precipitation and hence to more sea ice.

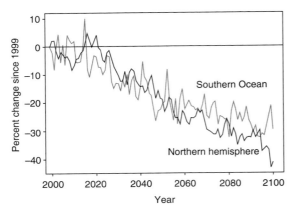

Fig. 21.5. Projected changes in total area of sea ice in the northern hemisphere and the Southern Ocean (Hadley Centre HadCM3 model http://www.met-office.gov.uk/research/hadleycentre/models/modeldata.htm).

For both the Arctic and the Antarctic, models suggest a 3% reduction in sea-ice extent by 2025 compared with 2000. For the Arctic, where longer-term observation data are available, this may equate to a reduction of perhaps 14% from the mid-twentieth-century state (see IPCC 2001*a*; Fig. 21.3). The UK Met Office's Hadley Centre HadCM3 model predicts that sea-ice reduction in both hemispheres will continue beyond 2025 (Fig. 21.5). However, because of complex climatic feedbacks that are not incorporated in some models, the reduction is unlikely to follow the sometimes-predicted quasi-linear trajectory over time.

ECOSYSTEM RESPONSES

Primary production

The open-water components of polar seas differ markedly in their rates of primary productivity from those that are ice covered. Changes by 2025 in the relative proportions of sea covered with ice will lead to changes in gross polar marine production, but at present it is difficult to predict by how much. For the Arctic, annual pelagic primary production is light-limited, and future reductions in sea-ice extent and duration are expected to reduce shading and increase total annual primary production (Rysgaard *et al.* 1999). This increase will be accentuated because the loss of sea ice will occur predominantly at the periphery of the Arctic Ocean, over the continental shelves (Chapter 19), and production in shelf waters is usually greater than in oceanic waters

(Longhurst 1998). By contrast, in the Antarctic, reduction in sea-ice extent will leave larger areas of less-productive deep ocean uncovered. The boundary between open water and solid pack ice is not abrupt but may span tens to hundreds of kilometres; this marginal ice zone (MIZ) is particularly important for primary production. The MIZ forms as sea ice melts and the pack is broken into smaller floes by wave and wind action. Fresh water released from melting sea ice stabilizes the upper water column, reducing downward mixing, and creates an environment favourable for the development of phytoplankton blooms. Monthly climatological maps of chlorophyll concentration and sea-ice extent derived from satellite observations suggest that mean annual primary productivity in the MIZ has been almost double that in the open ocean (0.66 versus 0.36 g C per m^2 per day). However, because of the relatively small size of the MIZ compared to that of the ice-free pelagic province, the MIZ contributed just 9.5% to the total Southern Ocean production (Arrigo *et al.* 1998).

It is difficult to evaluate by how much the reduction in sea-ice extent by 2025 might reduce MIZ production in the Southern Ocean, because the MIZ is irregularly shaped, highly dynamic and varies in size and position almost daily. Furthermore, recent estimates of annual primary productivity for the Southern Ocean vary by almost an order of magnitude (Smith *et al.* 1998). Complex interactions between position (latitude, underlying ocean depth) and time (irradiance levels vary through the year) will influence levels of production. However, modelling work suggests that as ice extent declines, the MIZ will become located further south earlier in the year and will be less extensive (IPCC 2001*a*). At the same time, there is no linear relationship between loss of sea-ice extent and the loss of MIZ length (length of ice-edge perimeter), and reduction in total production in percentage terms will be much less than the percentage reduction in sea-ice extent. A further complicating factor for the prediction of change in the Southern Ocean is the impact of ozone depletion on the spectral composition of solar radiation at the sea surface. Increased UV radiation (and especially the shorter-wavelength UV-B) may lead to a reduction in MIZ primary production (Smith *et al.* 1992).

Secondary production

Copepods and other grazers near and in the ice-edge zone in both hemispheres rely on ice-edge phytoplankton blooms to some extent, with numerous species timing their reproductive cycle to give juveniles optimal feeding conditions associated with the bloom (Conover & Huntley 1991; Kawall *et al.* 2001). The timing of melt, and distribution of ice, can have major impacts on production. Rysgaard *et al.* (1999) suggested that the increased primary production they expect in the Arctic following reduction in sea-ice coverage will lead to increased zooplankton production there. However, if the present situation in a polynya (a region of open water inside an otherwise ice-covered sector) is taken as a proxy for the likely situation in a future ice-free ocean, then the opposite may be inferred. Ashjian *et al.* (1995) found that significant proportions of primary production in the Arctic's North East Water polynya remained unconsumed.

In the Southern Ocean, euphausiids (krill) play a particularly important role in linking primary production to higher predators. Correlative evidence from the western Antarctic Peninsula suggests that reduction in sea-ice extent leads to a reduction in Antarctic krill (*Euphausia superba*) recruitment (see Loeb *et al.* 1997). This is because sea ice provides a favourable feeding habitat for adults, enabling increased reproductive output, and a nursery ground for juveniles (Brierley & Thomas 2002). Multiple seasons of reduced sea-ice extent may lead to reductions in total Antarctic krill population size. The area to the west of the Antarctic Peninsula is a major Antarctic krill breeding zone, and is also an area experiencing major regional warming (King 1994; Vaughan *et al.* 2001); these two factors combined could result in reduced krill biomass by 2025. Overall conclusions, however, are limited by the lack of detailed understanding of the links between ice dynamics and the population dynamics of krill.

In summer, krill appear to be concentrated along the ice edge, rather than being distributed evenly under ice (Brierley *et al.* 2002). As a result, loss of habitat for krill may not occur in proportion to loss of total sea-ice area but may be a function of the loss of ice-edge length. The 25% reduction in ice extent in the 1960s (de la Mare 1997) equates to only a 9% loss of ice-edge length. Changes between now and 2025 are unlikely to be on such a large scale. As with primary production, ozone depletion may impact some elements of secondary production in the Southern Ocean. Naganobu *et al.* (1999) found significant correlations between krill density in the Antarctic Peninsula area and ozone depletion parameters during the period 1977–97. Correlation, though, does not prove cause and effect, and more research is required to elucidate the full range of interactions between krill, sea ice and climate. In the absence of Antarctic krill, the salp *Salpa thompsoni* sometimes proliferates (Loeb *et al.* 1997). Because salps are not predated upon to a large extent, they may well represent a dead-end in the food chain, and the proliferation of salps has consequences for carbon cycling in the Southern Ocean in that it may divert energy from those areas of the food web leading to predators at higher trophic levels.

Higher predators

In the Arctic, copepods are predated upon predominantly by fish. The uncertainty surrounding the consequences of environmental change for these predators is illustrated by the conclusion that projected climate change could either halve or double average harvests of any given species (IPCC 2001*a*). This large range emphasizes the current inability to predict ecological consequences of climate change with any degree of certainty.

A key fish species in the Arctic sea-ice edge ecosystem is the polar cod (*Boreogadus saida*), which could either benefit or suffer from the changed copepod production, following a match or mismatch between spawning and larval food demand (see Pope *et al.* 1994). Changes in abundance of cod will, in turn, affect many seabird and mammal species, since the polar cod is the key link in the Arctic food chain (Legendre *et al.* 1992).

Although the consequences for fish are difficult to infer, the consequences of sea-ice reduction for polar bears seem more clear-cut. Polar bears depend on sea ice as a hunting ground, and reduction in sea ice will lead to a direct reduction of habitat, both in space and time. Bears leave the ice in spring to come ashore to breed. Every week earlier that the bears have to come ashore corresponds on average to a 10-kg reduction in pre-breeding body mass. Thus, as ice melts earlier, bears come ashore in poorer condition and are less likely to reproduce successfully (Stirling *et al.* 1999). As the reduction in sea ice proceeds, the link between the seasonal pack and land could eventually be broken, and it is possible to envisage a situation in the not-too-distant future when bears will be unable to make the transition between breeding ground and feeding ground. Impacts of climate change on polar bears will certainly be site-specific, but climate change may pose a very real threat to some local populations by 2025.

The poleward retreat of sea ice could also break a vital physical link for predators in the Antarctic. Many krill-dependent predators occupy breeding colonies well beyond the seasonal sea-ice maximum, but rely on ocean currents

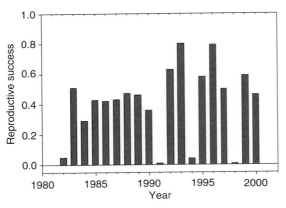

Fig. 21.6. Annual reproductive success of Gentoo penguin (*Pygoscelis papua*) at Bird Island (South Georgia), showing the poor breeding performance in years of low availability of Antarctic krill (*Euphausia superba*): 1982, 1991, 1994 and 1998. The supply of krill to South Georgia is affected by sea-ice conditions along the western Antarctic Peninsula and in the Scotia Sea, and hence is likely to change as the climate warms. (K. Reid, unpublished British Antarctic Survey data 2003.)

and the sea-ice edge to convey krill to them (Hofmann *et al.* 1998). Krill would probably be unable to survive the journey from the Antarctic Peninsula to South Georgia Island across an ice-free Scotia Sea without starving (Fach *et al.* 2002). Krill rely on favourable feeding conditions associated with sea ice, and if the sea ice were to retreat significantly then the essential food source for krill in transit would vanish, and krill would not survive to South Georgia. There is also a correlation between commercial krill catch at South Georgia and the ice-edge position (IPCC 2001*a*). Significant declines in many krill-dependent penguin, seal and albatross species at South Georgia have already occurred (Reid & Croxall 2001), and changes in the sea-ice environment may in part be responsible for these changes (Fig. 21.6).

Some species at higher trophic levels are able to forage in sectors of the ocean with sea ice ranging from zero to almost total coverage (for example minke whales [*Balaenoptera bonaerensis*]: Thiele & Gill 1999). These will probably be little affected by changes in the sea-ice environment *per se* by 2025. Others, such as emperor penguins (*Aptenodytes forsteri*) suffer if ice extent changes (Barbraud & Weimeskirch 2001), although paradoxically reductions are not always detrimental to all life-history stages. Seabirds are the best-studied vertebrate group in the Antarctic and many recent bird population-level

changes can be attributed to climate change (Croxall *et al.* 2002). A particularly marked example is the poleward shift of breeding distribution of pygoscelid penguins along the Antarctic Peninsula (Fraser *et al.* 1992; R. C. Smith *et al.* 1999; Convey *et al.* 2003). However, evidence of consistent patterns and causal links between ecological responses and climate variability along the western Antarctic Peninsula remains equivocal (Smith *et al.* 2003*a*, *b*), and a cautionary note should be adopted with regard to most predictions of the state of sea-ice systems in 2025. Although ecosystem meltdown is a metaphor used often in the popular press in descriptions of the state of the environment or environmental policy (see Bowles 2001; Lean 2003), for ice-edge ecosystems the metaphor may by 2025 transcend polar polemic and be well on the way to reality, with far-reaching consequences for the Earth system as a whole.

Food web

The discussion above emphasizes that the various parts of the polar marine food web can be expected to respond in different ways to the impact of climate change. The non-linear structure and intense complexity of oceanic food webs make it very difficult to predict how they will respond to environmental change (see review of Clarke & Harris 2003). Prediction is made even more difficult by the fact that both the Arctic and Antarctic oceanic food webs have been disrupted by intense fishing activity.

FISHING

The two polar regions have had very different histories of impact on their oceanic food webs, largely because of their very different geographies. The Arctic basin is surrounded by indigenous peoples who harvested the seas for millennia; the Southern Ocean is hostile and a long way from the populated part of the world, and consequently has been fished only relatively recently.

Harvesting of Arctic marine resources

Harvesting of marine living resources, including fish and marine mammals such as seals and whales, is vital to the sustenance and culture of Arctic peoples, and the Arctic is also one of the world's most important international fishing grounds. In the past, many Arctic marine mammals were overexploited (Pagnan 2000), and many are still considered globally endangered, for example bowhead (*Balaena mysticetus*), sei (*Balaenoptera borealis*), blue (*B. musculus*), fin (*B. physalus*) and northern right whales (*Eubalaena glacialis*) and Steller's sea lion (*Eumetopias jubatus*) (CAFF [Conservation of Arctic Flora and Fauna] 2001). Numerous other species present in the Arctic have been classified as globally vulnerable. Many of these species appear to be in decline in certain Arctic localities, although at others they appear stable or sometimes increasing; more generally however, there is a lack of data to determine overall trends (CAFF 2001).

New technologies and commercial fisheries have increased the scale of fish catches in Arctic regions. Historically, overharvesting has led to the localized collapse of numerous Arctic fish stocks, including herring (*Clupea harengus*), cod (*Gadus morhua*), whitefish (*Coregonus* sp.) and capelin (*Mallotus villosus*) (UNEP [United Nations Environment Programme] 1999*b*; Huntington 2000; Pagnan 2000; Northeast Fisheries Science Center 2002). While there is evidence of some recovery, not all stocks have recovered, and harvesting pressure remains high in some areas (Pagnan 2000). Moreover it has proved difficult to determine unequivocally the cause of decline, because for some species the effects of fishing pressure on stock size are confounded by climatically driven variability (OSPAR Commission 2000*b*). There is a high level of natural variability in Arctic fish stocks, suggesting that for the fishery to be sustainable, catch levels should be set more conservatively than has often previously been the case.

Harvesting of Antarctic marine living resources

Although harvesting of Antarctic marine living resources started long after that of the Arctic, fishing within the Southern Ocean also has a long history. In the eighteenth century, an active fishery for southern fur seal (*Arctocephalus gazella*) developed; the first sealers reached the Antarctic in 1778, and by 1791 over 10 million skins had been taken. Exploitation was initially at South Georgia, but soon moved to the South Shetland Islands as these stocks were depleted. By 1822, after only 35 years, the southern fur seal was almost extinct, and the industry collapsed (Bonner 1982).

The large numbers of great whales found in the Southern Ocean had also not escaped notice, and exploitation soon shifted to these. The first land-based whaling stations were established at South Georgia in 1906, but the critical innovations were the invention of the steam harpoon gun and the development of pelagic whaling ships (Headland 1984). The peak of whaling activity came between the two world wars, and by the mid 1960s the industry had collapsed economically, although significant unregulated whaling of protected stocks also took place subsequently, and some residual fishing for minke whales continues. In the first half of the twentieth century, taxes levied on whale products were used to fund the *Discovery* investigations based at South Georgia (Hardy 1967). Although concerned primarily with understanding the biology of the great whales being exploited, these pioneering studies provided the most comprehensive investigations of the biological oceanography of the Southern Ocean yet undertaken. In particular, they included a major study of the biology of *Euphausia superba*, a pivotal organism in the Southern Ocean ecosystem, and a major prey species for many fish, squid, seabirds and marine mammals.

In the late 1960s, attention switched to finfish, and in less than 5 years the stocks of marbled rockcod (*Notothenia rossii*) at South Georgia were reduced to uneconomic levels, from which they are recovering only slowly. The fishery then moved to new locations and species, notably icefish (*Champsocephalus gunnari*) (Kock 1992). The present fishery is directed at Patagonian toothfish (*Dissostichus eleginoides*), using bottom-set long-lines. In the 1970s, a trawl fishery for Antarctic krill was also developed. This was based principally around South Georgia in winter and the ice-free regions of the Scotia Sea in summer (Everson 2000). Experimental krill fishing was undertaken by eight nations in the 1970s (Agnew & Nicol 1996), prompted in part by the possibility that the decrease in the populations of krill-eating higher predators (notably baleen whales and fur seals) through fishing activities might have resulted in a krill surplus. The Southern Ocean food web is, however, complex and non-linear, and this makes the response of the system to fishing pressure very difficult to predict (see e.g. Clarke & Harris 2003).

The total catch of krill increased during the late 1970s to a maximum of over 500 000 tonnes in 1982; the current annual catch is of the order of 100 000 tonnes (Agnew & Nicol 1996). The original Antarctic Treaty, which applies south of 60° S latitude, made no attempt to regulate or manage fisheries. In the early 1970s, concern over possible environmental effects of overfishing led first to the most important series of international oceanographic studies since the *Discovery* investigations, the BIOMASS programme of the Scientific Committee for Antarctic

Research (SCAR). This was followed by the development and ratification of the Convention for the Conservation of Antarctic Marine Living Resources (CCAMLR). This Convention is now responsible for the regulation of all Southern Ocean fisheries (excluding exploitation of mammals such as whales and seals, which are covered by separate conventions). The CCAMLR area is different from that of the Antarctic Treaty, applying as it does to a region that is broadly coincident with the Southern Ocean as defined by the Polar Front, and including islands such as South Georgia and Kerguelen.

CCAMLR was the first such fisheries-management regime to take a holistic ecosystem-based view (Constable *et al.* 2000). Explicit attention is directed at the effects of a given fishery on the rest of the system, and specifically on the dependent predators. To monitor any such impact requires information on both harvested and dependent species, their interactions and the manner in which their populations vary naturally in size (Everson 2002). A key mechanism for achieving this is the CCAMLR Ecosystem Monitoring Programme (CEMP), which uses data from selected higher predators to monitor the upper trophic levels of the Southern Ocean marine ecosystem (Agnew & Nicol 1996; Constable *et al.* 2000; Everson 2002). This holistic approach may enable successful ecosystem management to be implemented even in the face of changes that may be manifest by 2025.

IUU and seabird bycatch

The major fisheries-related threats to the Southern Ocean ecosystem currently come from illegal, unreported and unregulated fishing (IUU) together with the impacts on seabirds feeding outside the Southern Ocean (Kock 2001). Although the fishery for Patagonian toothfish within the CCAMLR area is regulated, there is overwhelming evidence of significant IUU activity. For example, even minimum estimates of the IUU catch indicate that around one-third of Patagonian toothfish taken within the CCAMLR area in the late 1990s was from IUU fishing (Lack & Sant 2001). Whilst the absolute level of IUU catch is uncertain, what is clear is that it constitutes a significant issue for the Pantagonian toothfish fishery and for CCAMLR (Collins *et al.* 2003).

The other area of great conservation concern is the mortality of seabirds (specifically albatrosses and larger petrels), which are killed in long-line fisheries outside the Southern Ocean. This mortality is leading to a severe

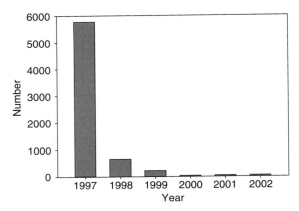

Fig. 21.7. Seabird by-catch in the licensed CCAMLR (Convention for the Conservation of Antarctic Marine Living Resources) fishery for Patagonian toothfish (*Dissostichus eleginoides*) at South Georgia. By-catch decreased significantly following the implementation of CCAMLR conservation measures. Note that seabird by-catch remains high but unquantified in the IUU (illegal, unreported and unregulated) fishery and thus continues to be a major conservation concern. (Data from Croxall & Nicol 2004.)

decline in all albatross populations (Croxall *et al.* 1997; Gales 1997), and is currently an area of very great scientific, conservation and public concern. A new international Agreement on the Conservation of Albatrosses and Petrels was concluded in 2001 under the Convention on the Conservation of Migratory Species of Wild Animals, although only seven nations were party to its agreement (New Zealand, Australia, Brazil, Peru, Chile, France and the UK). With numerous key fishing countries yet to sign up, it is too soon to assess this agreement's practical effect. However, it is clear that whilst implementation of guidelines for reducing seabird by-catch has significantly reduced incidental mortality within the CCAMLR-regulated fishery (Fig. 21.7), such mortality within the IUU fishery remains a major conservation concern.

THE POLAR MARINE ECOSYSTEMS IN 2025

Detailed prediction is impossible, but a general picture can be provided of how the two polar marine ecosystems might look in 2025.

Current climate warming is expected to continue, with consequences for the volume and dynamics of sea ice.

In the Arctic, less-extensive sea ice and a further extension of the summer open-water period are expected. For the Antarctic, a continued small overall increase in sea ice is likely, although with marked spatial variability. In particular, sea ice in the Bellingshausen and Amundsen Seas may continue to decline, associated with continued regional climate warming of the Antarctic Peninsula. Polar ecosystems are dominated by ice, and the narrow temperature threshold for an ice-to-water phase change may create a pronounced non-linear response to what is a relatively small temperature shift. Consequently, such non-linear amplifications of small climatic changes may increase the ecological response and amplify trends or change (Smith *et al.* 2003a; Welch *et al.* 2003).

It is likely that the change in extent and timing of ice cover will affect the intensity of primary production, although it is difficult to predict the overall quantitative impact, or the synergistic effect of enhanced UV flux from the springtime reduction in high-latitude ozone. Impacts on intermediate levels of the food web (zooplankton and nekton) remain largely unknown, although continued changes to sea-ice dynamics in the western Antarctic Peninsula region may affect the abundance of Antarctic krill, and there may be significant changes to the distribution of ice-associated higher predators.

The effects of climate change on the two polar marine ecosystems are likely to be substantial and far-reaching, and these could be exacerbated by the direct human effects through localized pollution or unregulated exploitation of living resources. Climate change is likely to lead to a very different polar ecosystem in 2025, although it is presently difficult to predict the magnitude, timing or distribution of many of the anticipated changes. For some iconic species, such as the polar bear, the consequences may be particularly severe.

There is, perhaps, some cause for cautious optimism when it is considered that the world's nations were capable of collectively agreeing to limit the production of ozone-destroying chemicals within just two years of the discovery of conclusive evidence for the damage they were doing to the Antarctic ozone layer (Farman *et al.* 1985). Balanced against this, however, should be remembered that even where concerted international action was taken, the 2003 ozone hole became the most extensive on record, and the world will suffer at least four more decades of ozone destruction before the system is expected to return to its original conditions. However, observed ozone depletion and climate change have also shown complex and incompletely understood linkages. Consequently, these assessments and predictions must be viewed within the context of continuing environmental change.

In the case of climate change, the situation is substantially more complex and difficult to predict, and there is an urgent need for more research to elucidate the mechanisms driving changes and to improve models as the basis for prediction. In the light of the uncertainties, which themselves are often used effectively to avoid taking corrective actions, the potential for unanticipated changes and possibly even more severe consequences should not be forgotten. Given the scale of the changes that the evidence seems to show are afoot, the case for more precautionary approaches to the management of human activities certainly can be made. There is now an urgent need to build actions into effective international agreements, such as those that address the more localized impacts of exploitation, for example through sustainable fisheries regimes, and those that address the more global changes, for example through the Kyoto Protocol. If the latter cannot be accepted by all key governments, then it is imperative that a viable alternative be found.

22 · The near future of the deep-sea floor ecosystems

CRAIG R. SMITH, LISA A. LEVIN, ANTHONY KOSLOW, PAUL A. TYLER
AND ADRIAN G. GLOVER

INTRODUCTION

The deep-sea floor lies between the shelf break (c.200 m depth) and the bottom of the Challenger Deep (c.11 000 m depth). It covers more than 300×10^6 km^2 and constitutes c.63% of the Earth's solid surface. The distinct habitats of the deep-sea floor are varied, and include sediment-covered slopes, abyssal plains and ocean trenches, the pillow basalts of mid-ocean ridges, rocky seamounts protruding above the sea floor and submarine canyons dissecting continental slopes. The sedimented plains of the slope and abyss are the largest in area, covering >90% of the deep-sea floor and often extending unbroken for over a thousand kilometres. Deep-sea trenches, where continental plates overrun oceanic crust, constitute 1–2% of the deep-ocean bottom. The rocky substrates of mid-ocean ridges (ribbons c.10 km wide, in total approximately 60 000 km long), seamounts (perhaps 50 000–100 000 in number: Epp & Smoot 1989; Smith 1991; Rogers 1994) and submarine

Aquatic Ecosystems, ed. N. V. C. Polunin. Published by Cambridge University Press. © Foundation for Environmental Conservation 2008.

canyons are relatively rare habitats in the enormous expanses of the deep sea, together estimated to occupy <4% of the sea floor.

Many deep-sea floor habitats share ecological characteristics that make them especially sensitive to environmental change and human impacts. Perhaps the most important characteristic is low biological productivity. Away from the occasional hydrothermal vent and cold seep, the energy for the deep-sea biota is ultimately derived from an attenuated 'rain' of organic matter from surface waters hundreds to thousands of metres above (typically 1–10 g C_{org} per m^2 per year). Detrital particles range in size from phytoplankton cells a few micrometres in diameter to dead whales 30 m long. The purely detrital base of most deep-sea food webs contrasts sharply with those of other marine and terrestrial ecosystems, which typically are sustained by local production (Polunin *et al.* 2001). The very low organic-energy flux, combined with low temperatures (−1–4 °C) in the deep sea, yields communities generally low in biomass and characterized by relatively low rates of growth, respiration, reproduction, recolonization and sediment mixing (Gage & Tyler 1991; Smith & Demopoulos 2003). In addition, many deep-sea animals, especially macrofaunal polychaetes, molluscs and crustaceans, are very small in body size and extremely delicate (Gage & Tyler 1991; Smith & Demopoulos 2003).

The deep-sea floor is also generally characterized by very low physical energy, including sluggish bottom currents (typically <0.25 knots), very low sediment accumulation rates (0.1–10 cm per thousand years) and an absence of sunlight (Gage & Tyler 1991; Smith & Demopoulos 2003). Nonetheless, the seemingly monotonous sediment plains of the deep sea often harbour communities with very high local species diversity, with one square metre of deep-sea mud containing hundreds of species of polychaetes, crustaceans and molluscs (Snelgrove & Smith 2002).

However, not all deep-sea habitats are low in energy and productivity. Hydrothermal vents, and to a lesser degree cold seeps, may sustain high productivity, animal growth rates and community biomass as a consequence of chemoautotrophic production by abundant microbial assemblages fuelled by reduced chemicals such as hydrogen sulphide (Van Dover 2000). Seamounts, canyons, whale falls and upwelling zones may also foster relatively high-biomass communities by enhancing bottom currents or by concentrating organic-matter flux (Wishner *et al.* 1990; Koslow 1997; Vetter & Dayton 1998; Smith & Baco 2003; Smith & Demopoulos 2003).

The sensitivity and recovery potential of deep-sea ecosystems in the face of human disturbance are not necessarily easy to predict. The relatively low levels of physical energy, productivity and biological rates, combined with small body size, suggest that most deep-sea habitats are especially sensitive to, and slow to recover from, deleterious human impacts compared to most other ecosystems (Table 22.1). The relatively high species diversity in the deep sea, in terms of the number of species per unit area, again may make the deep-sea habitats more sensitive to human impacts; there are more species to be extinguished by local disturbance, and more species to recolonize. Yet the size of the ecosystem is especially great and, for abyssal soft sediments, habitats are nearly continuous across ocean basins. The large habitats of the deep sea may make the fauna more resistant to extinctions caused by local perturbations, with a greater potential for recolonization from widespread source populations. However, the size of source populations depends to a large extent on the biogeographical ranges of deep-sea species, which are very poorly known (Glover *et al.* 2002). Large continuous habitats may also allow stressors such as disease agents, toxic chemicals or radioactive contaminants to disperse over vast distances and to become amplified over very large areas as they move up deep-sea food webs. Clearly, the unusual characteristics of deep-sea ecosystems present novel challenges to the prediction of anthropogenic impacts and requisite conservation actions.

While once considered remote and well buffered from the euphotic zone and from human impacts, deep-sea ecosystems are increasingly recognized to be linked to processes in the upper ocean. In fact, until the early 1980s, it was widely believed that the deep-sea floor was a remote and deliberate habitat, where low current speeds, low temperatures and a gentle rain of organic material drove biological processes at extremely slow and relatively constant rates (Smith 1994). However, more recent data from deep sediment-traps and time-series studies in the Atlantic and Pacific show marked temporal variability in particulate organic flux, accumulation of fresh phytodetritus, sediment-community respiration rates and reproductive patterns at the abyssal sea floor (Deuser & Ross 1980; Tyler *et al.* 1982; Tyler 1988; Thiel *et al.* 1989; Gage & Tyler 1991; Smith *et al.* 1996; Smith & Kaufmann 1999). In addition, the abundance and biomass of deep-sea benthos appear to be strongly correlated with large-scale spatial and temporal variations in export production from the euphotic zone (Smith *et al.* 1997; Smith *et al.* 2001; Smith &

Table 22.1. *A comparison of general ecological characteristics of the deep-sea relative to shallow-water (<100 m water depth) ecosystems, and the consequences of these differences for ecosystem sensitivity to, and recovery rates from, anthropogenic impacts*

Characteristic	Deep sea	Shallow water	Consequence for sensitivity to impacts	Consequence for recovery rates from impacts
Productivity	Usually low	High	Deep sea more sensitive	Deep sea slower
Growth, reproduction, recolonization rates	Usually low	High		Deep sea slower
Biomass	Usually low	High	Deep sea more sensitive	No effect
Physical energy	Usually low	High	Deep sea more sensitive	Deep sea slower
Size of habitat	Often large and continuous for 100–1000 km	Small, continuous only over 1–100 km	Deep sea more robust (extinction less likely?)	No effect
Species diversity	Usually high on local scales	Moderate, variable	Deep sea more sensitive	Deep sea slower (more species to recover)
Species distributions	Broad?, but poorly known	Narrow, variable	Deep sea less sensitive in some ways (extinctions less likely?)	Deep sea faster??

Demopoulos 2003). Clearly, changes in upper-ocean eco-systems resulting from climate change or direct human activities such as overfishing (Myers & Worm 2003; Chapters 19–21) are bound to alter the structure and functioning of deep-sea floor ecosystems.

Why should it be of concern whether the seemingly remote habitats of the ocean bottom are altered by human activities in the near future? There are a number of com-pelling reasons. (1) The deep-sea floor is one of Earth's largest ecosystems; the ecological 'health' of the planet can hardly be assessed without considering its vastest habitats. (2) The deep-sea floor is both an extreme environment and a substantial reservoir of biodiversity (Snelgrove & Smith 2002); its biota offers exciting insights into evolutionary novelty, as well as unusual biotechnological resources. (3) The floor of the ocean provides a number of ecosystem services fundamental to global geochemical cycles; for example, deep-sea sediments are major sites of nutrient recycling, and are responsible for much of the carbon burial occurring on the planet. (4) Because deep-sea

ecosystems are significantly buffered from the high-fre-quency variations in upper-ocean processes (Chapters 20 and 21), ecosystem parameters in the deep sea may act as 'low-pass filters', primarily responding to gradual, large-scale changes in production processes in the euphotic zone (see Hannides & Smith 2003). Thus, deep-sea ecosystems may be especially useful in elucidating basin-scale trends in ocean productivity driven by global climate change.

Existing anthropogenic alterations to deep-sea ecosys-tems and additional changes expected to occur to the 2025 time horizon are discussed in this chapter. The anthro-pogenic forcing factors considered include fishing, energy and mineral exploitation, waste disposal, climate change and general (i.e. 'non-point-source') pollution. The research needed to elucidate such changes and formulation of policies and regulations that might contribute to the mitigation of deleterious anthropogenic impacts are then considered. Conclusions about what the near future holds for deep-sea ecosystems, and what humankind might want to do about it, are addressed at the end of the chapter.

NATURAL AND ANTHROPOGENIC FORCING FACTORS

Effects of deepwater fisheries

ECOLOGY AND LIFE-HISTORY PATTERNS

Since the Second World War, global fishing activity has increasingly targeted deepwater species on the continental slope (Chapter 19), as well as on seamounts, banks and plateaus deeper than c.500 m (Chapter 1). In part, this fisheries expansion was made possible by technological developments, such as those in echo sounders, geographical positioning systems (GPS), net sondes, track plotters and rock-hopper trawls. This expansion was driven as well by the 'fishing-up' process, whereby fisheries increasingly exploit less accessible and more marginal fishing grounds, often containing less desirable species, following the depletion of traditional fisheries (Deimling & Liss 1994). Deepwater fishing activities, particularly those carried out by demersal trawling, have probably had a greater ecological impact than any other human activity in the deep sea.

Deepwater fisheries generally target different families of fishes, and often different orders, from conventional fisheries. These newly exploited groups of fishes generally exhibit markedly different life histories, and the ecological characteristics of the communities they inhabit differ as well. As a result, the direct and indirect impacts of many deepwater fisheries will be far more intense and persistent than those of traditional shallow-water fisheries, despite the relatively brief period for which deepwater stocks have been exploited. In fact, a marked 'boom-and-bust' cycle already characterizes deepwater fisheries; most have been massively depleted within 10 years of inception (Koslow et al. 2000).

The species exploited by deepwater fisheries vary considerably, depending upon habitat (for example deep banks and seamounts versus the continental slope) and also upon the ocean basin and biogeographical province. Seamount populations exhibit reproductive isolation across a range of spatial scales, such that different orders of fishes evolved to dominate seamounts in different ocean basins and climatic regions. For example, the pelagic armourhead *Pseudopentaceros wheeleri* (Pentacerotidae, Perciformes) dominates seamounts in the central North Pacific; several rockfish species (*Sebastes* spp., Scorpaenidae, Scorpaeniformes) prevail along the continental slope of the North Pacific and North Atlantic; orange roughy *Hoplostethus atlanticus* (Trachichthyidae, Beryciformes) and several

oreosomatid species (Zeiformes) dominate slopes and seamounts in the temperate South Pacific; and alfonsino (*Beryx* spp., Berycidae, Beryciformes) abounds in the tropics and subtropics (Koslow et al. 2000). Many, although not all, of these species are exceptionally long-lived. Orange roughy, some oreosomatids and species of *Sebastes* live more than 100 years (Smith & Stewart 1994; Tracey & Horn 1999; Caillet et al. 2001) and thus must experience extremely low natural mortality (0.05) and slow post-maturity growth. There are few time-series studies of recruitment for deepwater species, but these indicate life histories characterized by extreme iteroparity over a prolonged reproductive period punctuated by occasional episodes of recruitment (Murphy 1968; Stearns 1976). For example, populations of orange roughy and *Sebastes* may go for a decade or more with very low recruitment (Leaman & Beamish 1984; Clark 1995b).

Deepwater fisheries on the continental slope predominantly target several families of Gadiformes: grenadiers or rattails (Macrouridae), morid cods (Moridae), hakes (Merluciidae) and cusk eels (Brotulidae). Other key groups include flatfishes within the Pleuronectidae (e.g. Greenland halibut *Reinhardtius hippoglossoides*), one of the few families fished both on the continental shelf and in deep water, and various species of rockfish (*Sebastes* spp.). Species in these groups tend to be long-lived (50–100 years) but do not appear to exhibit the extreme ages of seamount-associated species (Bergstad 1990; Campana et al. 1990).

FISHERIES DEVELOPMENT AND DIRECT IMPACTS

The low productivity and episodic recruitment of deepwater species render them highly vulnerable to overexploitation. Since the 1960s, deepwater fisheries landings have fluctuated between 600 000 and 1 million tonnes per year, but underlying this apparent stability is a pattern of serial depletion (Koslow et al. 2000) (Fig. 22.1). Modern deepwater fisheries originated in the North Pacific and North Atlantic. Several species of *Sebastes* (for example *S. alutus* in the Pacific and the redfish complex *S. marinus*, *S. mentella* and *S. fasciatus* in the North Atlantic) were first fished along the shelf edge and upper slope, and are included among the deepwater fisheries because they share many life-history characteristics with deepwater species. These have been the largest and most stable deepwater fisheries, but Koslow et al. (2000) concluded 'uncertainties about the unit stocks, and in the North Atlantic even the taxonomic composition of the catch, appear to have

(a)

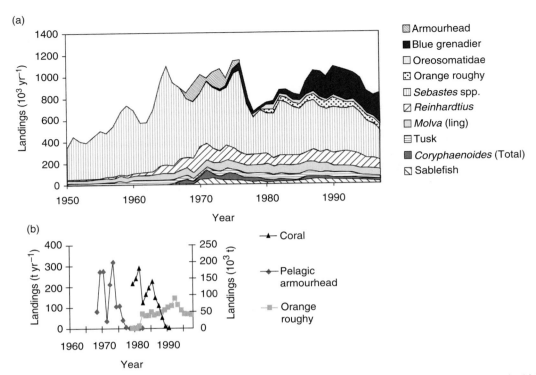

Fig. 22.1. (a) Overall annual landings for deepwater fisheries, 1950–99. (b) Landings for the pelagic armourhead, orange roughy (right axis) and *Corallium* sp. (left axis) fisheries.

masked the progressive fishing down of successive populations around the rim of both the North Pacific and the Atlantic'.

Coryphaenoides is a diverse genus of macrourids, dominating fish stocks over relatively flat portions of the deep sea in much of the world's oceans. There has been a substantial but diminishing fishery for *C. rupestris* (roundnose grenadier) in the North Atlantic and a small fishery for *C. acrolepis* (pacific grenadier) in the North Pacific (Merrett & Haedrich 1997). *Coryphaenoides rupestris* is most abundant in the 600–800-m depth range. The Soviet Union fleet developed the fishery in the north-west Atlantic, which peaked at 80 000 tonnes per year in 1971 and then rapidly declined; landings in 1997 were only a few hundred tonnes. These fisheries, which extend out into international waters along the mid-Atlantic ridge, are not actively managed.

The 'boom-and-bust' pattern is most marked for seamount-associated species, which aggregate on fixed topographical features and generally exhibit the most extreme life-history characteristics. Given the low productivity of these stocks, the fisheries tend to be unsustainable; a large accumulated biomass is fished down, usually within 10 years, and the fishery moves on to another area. The first seamount fishery targeted pelagic armourhead in the central North Pacific in the 1970s. Within about 10 years of their discovery by Soviet Union trawlers, the population was fished to commercial extinction (Boehlert 1986). A new species of *Corallium* (a precious red coral) was discovered and largely fished out on seamounts in the North Pacific in the 1980s (Grigg 1993).

Soviet trawlers discovered orange roughy on seamounts around New Zealand in the 1970s. In the 1980s and 1990s, orange roughy stocks were successively depleted around south-eastern Australia and New Zealand, followed by Namibia and most recently in the Indian Ocean (Clark 1999; Koslow *et al.* 2000). Australia and New Zealand intensively managed their larger orange roughy stocks, but with limited success owing to the low resilience of the stocks to overfishing (Clark 1999; Bax 2000); several key

New Zealand orange roughy fisheries have now been closed with depletion to only a few per cent of the virgin biomass. Smaller stocks, and stocks in international waters, are generally left unmanaged; fishing has ceased only when it has no longer been commercially viable.

The global depletion of deepwater fish species has been too recent to assess long-term impacts. The ability of these stocks to recover is not known, but, given their late maturity and sporadic recruitment, it will at best be slow compared to that of most shallow-water fish species.

INDIRECT FISHERIES IMPACTS

In addition to reducing target-species population sizes, fisheries may have pervasive secondary impacts on the target species, including density-dependent changes in growth, reproduction and recruitment. Fisheries may further dramatically impact other components of the ecosystem, altering the size structure, species composition and food webs of benthic communities due to by-catch and predator removal (Jennings & Kaiser 1998; Hall 1999). Trawl-based fisheries in particular may also cause substantial physical disruption of sea-floor habitats, by ploughing sediments and fragmenting the biogenic structure of corals, sponges and gorgonians (Jennings & Kaiser 1998; Hall 1999; Koslow et al. 2000). The large-scale removal of seabirds, top predators and fishes, and physical disruption of the sea floor, have raised the spectre of massive and possibly irreversible fishing down and removal of habitat complexity from marine ecosystems (Sainsbury 1988; Collie et al. 1997; Pauly et al. 1998; Thrush & Dayton 2002; Myers & Worm 2003).

Fishery impacts in the deep sea are generally poorly known. Fishing on the high seas is unregulated and poorly documented, and research on deepwater habitats and communities presents obvious challenges and great cost. Conversely, the recent development of many deepwater fisheries, coming at a time of greater awareness of fishery impacts, has enabled studies to be made of some deepwater fisheries at an early stage, which was not possible for many traditional fisheries that began decades or even centuries earlier. In general, those fishery-induced changes observed in populations and ecosystems on continental shelves (Chapter 19) also appear to be occurring in deepwater habitats.

Fishing appears to have induced density-dependent changes in some, though not all, target populations studied. The fecundity of the Tasmanian orange roughy population increased 20% following a 50% reduction in stock size owing to fishing (Koslow et al. 1995), but a similar effect was not observed for the Challenger Plateau (New Zealand) stock of orange roughy (Clark et al. 1994). Enhanced growth was observed in fished rockfish populations of S. alutus (Pacific ocean perch) and S. diploproa (splitnose rockfish) (Boehlert et al. 1989). The impact of reduced spawning-stock size on recruitment to deepwater populations has not been examined because of a lack of suitable recruitment time series.

Fishing has reduced the size structure of deepwater fish assemblages in the North Atlantic (Haedrich 1995). Perhaps surprisingly, trawl surveys in the North Atlantic (Lorance 1998) and New Zealand orange roughy fishing grounds (Clark & Tracey 1994; Clark 1995b) indicated no major shifts in species composition; however, this was because the relatively non-selective nature of the trawl fisheries reduced the abundances of most species simultaneously.

The impacts of deepwater fisheries on non-target populations, including demersal fish and hard-bottom benthos, are a major concern. For example, the take of non-targeted species (or by-catch) is considered responsible for 80–97% reductions over a 17-year period in the population densities of the demersal blue antimora Antimora rostrata, spinetail ray Bathyraja spinicauda and spinyback eel Notacanthus chemnitzii at 200–2500 m on the North Atlantic slope (Devine et al. 2006). In addition, deepwater trawl fisheries often target topographical features (such as seamounts, banks and canyons) with enhanced productivity; such features frequently also harbour distinct benthic communities dominated by hard and soft corals, sponges and other suspension feeders. Lophelia reefs in the North Atlantic and deepwater coral reefs in the South Pacific can be tens of metres high and represent millennia of growth (Rogers 1999). The biodiversity of these reefs rivals that of shallow-water tropical reefs (Chapter 16); seamounts also exhibit high levels of endemism, suggesting little genetic exchange between seamount chains (de Forges et al. 2000). Coral by-catch is reported from deepwater trawl fisheries as new fishing grounds are entered (Probert et al. 1997; Anderson & Clark 2003). Massive damage to deepwater coral habitats has been reported from fisheries off Tasmania, the European continental margin and North America (Koslow et al. 2001; Krieger 2001; Fossa et al. 2002; Mortensen et al. 2005). Recolonization processes of deepwater corals are unknown. Given the slow growth of deepwater corals (e.g. 5.5 mm per year for Lophelia pertusia: Rogers 1999) and uncertain rates of recruitment, the re-establishment of deepwater coral reefs will likely take centuries to millennia. As a result, deepwater marine protected areas (MPAs) have been declared off Tasmania, New Zealand, Norway and elsewhere (Koslow *et al.* 2000). Recognizing the grave environmental consequences of high-seas trawl fisheries, the United Nations General Assembly passed a resolution in 2002 to protect the biodiversity of the high seas. The extent of potentially irreversible damage already inflicted on seamount and deepwater coral habitats is unknown.

Energy and mineral exploitation

OIL AND GAS EXPLOITATION

In the 1990s, the offshore oil industry expanded exploratory drilling and production into the deep sea below 500 m, with the Brazilian oil company Petrobras now working at depths >2000 m. Very large oil reserves have been discovered in water depths >1000 m in the Gulf of Mexico and off West Africa (Douglas–Westwood 2002), suggesting that deep-sea oil production will develop substantially in the next decade.

The environmental effects of oil and gas drilling are reasonably well documented at shelf depths, and should be similar, albeit with higher disturbance sensitivities and slower recovery times, in the deep sea. On the shelf, principal benthic impacts result from release of drilling muds and drill cuttings. Drilling muds can include refined lubricant oils and other synthetic components, and are used to lubricate the drill bit and carry drill cuttings out of the well. Drill cuttings and drilling mud are separated on the drill platform and the cuttings discharged back into the sea, where they may accumulate on the seabed. The piles of cuttings are usually contaminated with drilling muds and pose a significant risk to marine life (Daan & Mulder 1996; Raimondi *et al.* 1997; Mauri *et al.* 1998; Grant & Briggs 2002).

Large quantities of drill cuttings have accumulated around drilling platforms; for example 30 years of drilling in the North Sea have left 1–1.5 million tonnes of drill cuttings on the sea floor (UKOOA [UK Offshore Operators Association] 2002). The principal impacts of drill cuttings on the seabed include physical smothering, organic enrichment and chemical contamination (by hydrocarbons, heavy metals, special chemicals and sulphides) of the benthos near the cuttings source (Daan & Mulder 1996; Mauri *et al.* 1998; Grant & Briggs 2002). Experimental studies indicate that drilling muds can also inhibit the settlement of marine invertebrate larvae (Raimondi *et al.* 1997). Deep-sea drilling also raises the risk of catastrophic failures in offshore operations (such as the explosion and sinking of the giant drilling rig P-36 in 2000 m of water off the Brazilian coast in 2001). The environmental impact of such disasters will depend on the frequency of their occurrence.

When drill cuttings are deposited in shelf areas with high current speeds, the fine rock chips are usually dispersed quickly, allowing biodegradation of drilling-mud contaminants. In the northern North Sea, where water is deeper and current speeds are lower, drill cuttings have accumulated, causing significant localized environmental impacts (UKOOA 2002). It is very likely that drill cuttings will cause even greater local environmental hazard in the deep sea, where current speeds, sediment accumulation rates and recolonization rates are often very low, yielding relatively low community resistance to burial and slow community recovery. In addition, the low background levels of productivity and biomass will make the impact of any organic enrichment more significant.

Forecasting the potential environmental impacts of deepwater drill cuttings is problematic given the uncertainties associated with future oil discoveries. On the Norwegian shelf, a single oil rig may impact a sea-floor area of roughly 100 km^2 (Olsgard & Gray 1995). Currently, the most active region of deepwater drilling in the world appears to be the Gulf of Mexico, where roughly 50 drilling rigs are in operation in waters deeper than 500 m; if each rig has an 'impact zone' of 100 km^2, this amounts to a total of 5000 km^2 of deep-sea impact. This area of impact is very small (0.4%) compared to the total area of the deep Gulf of Mexico (1.2 million km^2), but a 10-fold increase in drilling activities would provide cause for environmental concern. In addition, if drilling targeted petroleum-seep communities, which contain some very long-lived species (e.g. vestimentiferan tubeworms living for over 100 years: Fisher *et al.* 1997), this highly specialized seep ecosystem could be substantially impacted.

Current regulations in some countries require assessment of environmental impacts prior to drilling in deep water, and some high-profile baseline environmental surveys have been conducted in the deep sea (e.g. Gulf of Mexico and on the slope west of Scotland: Bett 2001). As a consequence, some vulnerable deepwater habitats, such as the species-rich deepwater coral (*Lophelia*) beds off Norway and Scotland, are likely to be protected from direct drilling impacts.

In the future, additional energy resources, in particular methane hydrates, are likely to be exploited in the deep sea. Methane hydrates are stable at ocean depths of 300–900 m, and constitute massive reservoirs of fossil fuel along many continental margins. Off the Chilean coast between 35° and 45° S, the methane gas potential (10^{13}–10^{14} m^3) exceeds by three orders of magnitude the annual consumption of gas in Chile (Esteban Morales 2003). Worldwide, gas hydrates may contain twice the amount of carbon found in all other fossil fuels combined. Although this resource probably will not be exploited for at least another decade, development of hydrate extraction technology may dramatically increase the environmental impacts of deep-sea fossil-fuel exploitation, by releasing methane (a greenhouse gas) into the atmosphere, by disturbing specialized deep-sea communities associated with methane seeps, and by causing slope instability, potentially yielding turbidity flows, earthquakes and even tsunamis.

MANGANESE NODULE MINING

The deep seabed contains a number of potentially extractable minerals, including manganese nodules, which abound on the abyssal sea floor beneath the oligotrophic ocean. Manganese nodules are potential sources of copper, nickel and cobalt. Manganese nodule mining is expected to occur by the year 2025 and could ultimately be the largest-scale human activity to directly impact the deep-sea floor. Twelve pioneer investor countries and consortia have conducted hundreds of prospecting cruises to investigate areas of high manganese nodule coverage in the Pacific and Indian Oceans, especially in the area between the Clarion and the Clipperton fracture zones, which extends over 6 million km^2 (Glasby 2000) and may contain 78 million tonnes of cobalt, 340 million tonnes of nickel and 265 million tonnes of copper (Ghosh & Mukhopadhyay 2000; Morgan 2000). Eight contractors are now licensed by the International Seabed Authority (ISA) to explore nodule resources and to test mining techniques within individual claim areas, each covering 75 000 km^2.

When mining begins, each mining operation could, through 'strip mining' of nodules, directly disrupt 300–800 km^2 of sea floor per year (Oebius et al. 2001; ISA, personal communication 2003) and disturb the sea-floor biota over a very poorly constrained area perhaps two to five times larger owing to redeposition of suspended sediments (Smith 1999; Thiel et al. 2001). In any given year, nodule mining by one to two contractors could disrupt sea-floor communities over areas of 600–8000 km^2, and 15 years of

such mining could conceivably impact 120 000 km^2 of sea floor. The most obvious direct impact of nodule mining will be removal of the nodules themselves, which will require millions of years to regrow (Ghosh & Mukhopadhyay 2000; McMurtry 2001). Essentially, nodule mining would thus permanently remove the only hard substrate present over much of the abyssal sea floor, yielding habitat loss and local extinction of the nodule fauna, which differs markedly from the fauna of surrounding sediments (Mullineaux 1987; Bussau et al. 1995).

Nodule-mining activities will also inevitably remove much of the top 5 cm of sediment, potentially distributing this material into the water column (Oebius et al. 2001; Thiel et al. 2001). Many sediment-dwelling animals in the path of the collector will be killed immediately, and communities in the general mining vicinity will be buried under varying depths of sediment (Jumars 1981; Smith 1999; Oebius et al. 2001; Sharma et al. 2001; Thiel et al. 2001). Because abyssal nodule habitats normally have high physical stability (these are possibly the most stable habitats on Earth) and are dominated by very small and/or fragile animals feeding on a thin veneer of organic matter near the sediment–water interface, the mechanical and burial disturbances resulting from commercial-scale nodule mining could be devastating (Jumars 1981; Glover & Smith 2003).

Because of the environmental risks posed by nodule mining, a number of in situ experiments have been conducted to evaluate the sensitivity and recovery times of these potentially very sensitive abyssal benthic communities to simulated mining disturbance. The experimental disturbances created were much lower in intensity and many orders of magnitude smaller in spatial scale than would result from commercial mining, but they provide important insights into the sensitivity and minimum recovery times of abyssal nodule communities subjected to mining (reviewed in Thiel et al. 2001; Glover & Smith 2003; Thiel 2003). Abyssal benthic communities will be substantially disturbed by even modest amounts (c.1 cm) of sediment redeposition resulting from mining activities, and full sediment-community recovery from major mining disturbance will take much longer than 7 years and possibly even centuries. These experiments do not allow prediction of the likelihood of species extinctions from nodule mining because the typical geographical ranges of species living within the nodule regions are unknown; the ranges may be large or small relative to the potential spatial scales of mining disturbance. Preliminary data based on morphological analyses suggest that some common benthic species may range well beyond the scales of impact for individual mining operations (Glover *et al.* 2002), but these results must be verified with molecular techniques due to the frequency of cryptic species amongst marine invertebrates (Knowlton 2000). To fully predict and manage commercial mining impacts, substantially more information is required concerning species ranges, sensitivity to sediment burial and the spatial-scale dependence of recolonization in abyssal benthic communities.

POLYMETALLIC SULPHIDE MINING

At present, there is commercial interest in polymetallic sulphides in the deep sea as sources of gold, silver, zinc, lead, copper and/or cobalt (Glover & Smith 2003). Polymetallic sulphides are generally associated with deep-sea spreading centres and occur as metalliferous muds or massive consolidated sulphides. Although metalliferous muds in the Red Sea sparked interest in the past (Degens & Ross 1969; Thiel 2003), current commercial interests focus on the massive sulphides around inactive hydrothermal vent sites at bathyal depths near New Guinea and New Zealand (Wiltshire 2001).

Hydrothermal-vent communities are characterized by low diversity and highly localized biomass (Van Dover 2000) when compared to both rocky and sedimentary substrata in other areas of the deep sea. In addition, vent communities appear to be highly resilient (i.e. they recover rapidly from disturbance). Although individual vent communities are often isolated and ephemeral, on a regional scale, vents are continually forming and decaying, and the adaptations of the dominant vent species ensure that new vents are colonized rapidly. Because of the ephemeral nature, high fragmentation and high energy flux of vent habitats, the primary vent species often grow rapidly, reproduce early and disperse widely (Lutz *et al.* 1994; Tyler & Young 1999). Nonetheless, as more vents are explored throughout the ocean basins, there is a growing realization that the global vent ecosystem is characterized by distinct biogeographical regions (Van Dover *et al.* 2002).

Mining of massive sulphides at active vents would be catastrophic to a local vent community, wiping out a large proportion of the fauna as well as modifying the stockwork (or 'plumbing') that underpins the vent site. At mined vents, biodiversity might increase initially as scavengers come in to exploit the dead biomass created by mining. The impacts of vent mining would differ from those of nodule mining in the rate of faunal community recovery.

At a vent site, even with the total destruction of the fauna and damage to the stockwork, new vents would form quickly. Colonization of vent sites proceeds rapidly (Tunnicliffe *et al.* 1997) and within 2 years the vents would be likely to be colonized by primary species. Analogous vent colonization and development have been analysed at the 9° N site on the East Pacific Rise, which showed a well-developed and evolving ecosystem 7 years after new vent formation (Shank *et al.* 1998).

Waste disposal

SOLID STRUCTURES

Between 1914 and 1990, more than 10 000 ships sank to the sea floor as a result of warfare and accidents (Thiel 2003). While the impacts of shipwrecks are poorly studied, they may generate reducing habitats (Dando *et al.* 1992), release petroleum hydrocarbons and other pollutants and/or provide a habitat for hard-substrate biota (Hall 2001). With the exception of oil-laden tankers, most shipwrecks are likely to have a localized impact. However, the scale and duration of the effects of sunken oil tankers on deep-sea floor communities clearly merit further study.

Proposals to scuttle oil- and gas-storage structures in the deep sea have received considerable attention (Rice & Owen 1998). Nevertheless, until both the negative and positive sea-floor impacts of such large structures are studied in more detail, it will be very difficult to rationally compare the relative merits of onshore versus deep-sea disposal of such structures (Dauterive 2000; Tyler 2003*a*).

MUNITIONS AND RADIOACTIVE WASTES

Several million tonnes of conventional and chemical weapons have been dumped onto continental slopes surrounding the UK and other European countries (Thiel *et al.* 1998). However, field surveys of major munitions dumpsites show little significant contamination of marine life, with many loose munitions and boxes colonized by sessile organisms (Thiel *et al.* 1998). Existing munitions dumpsites are unlikely to have significant widespread impacts on deep-sea ecosystems, but do pose a danger to deep-sea trawlers.

Low- and intermediate-level radioactive wastes have also been dumped into the deep sea (Smith *et al.* 1988; Thiel *et al.* 1998). From 1949 to the London Dumping Convention moratorium in 1983, about 220 000 drums of low-level waste were dropped in the north-east Atlantic by European countries, and over 75 000 drums were dumped by the USA in the North Atlantic and Pacific. The northeast Atlantic radioactive-waste dumpsites have revealed very little change in radionuclide levels, with a single anemone species exhibiting enhanced levels of ^{90}Sr and ^{137}Cs (Feldt et al. 1985). At the USA dumpsites, measurable levels of radioactive contamination have been recorded in sessile suspension feeders (anemones), deposit feeders (holothurians and asteroids) and mobile predators (grenadiers, bathypteroids and decapods) collected close to waste materials (Smith & Druffel 1998). Significant mechanisms of radionuclide transfer into the deep-sea ecosystem include bioturbation into local sediments, bioaccumulation in deposit and suspension feeders and uptake by mobile benthopelagic fish species (e.g. grenadier fishes), which may ultimately provide a pathway to humans through deep-sea fisheries.

SEWAGE SLUDGE AND DREDGE SPOILS

The disposal of sewage sludge at sea is permitted under the London Dumping Convention, and offshore dumping of sewage sludge is now under serious consideration by a number of countries (Thiel et al. 1998). The primary benthic impacts of sewage-sludge disposal are likely to be animal burial, clogging of feeding apparatus, dilution of natural food resources for deposit feeders, increases in turbidity, toxicity from sludge components, reductions in bottom-water oxygen concentrations and changes in community structure due to organic enrichment (Thiel et al. 1998). The only studies of sewage sludge impacts in the deep sea have been conducted at the Deepwater Dumpsite 106 (DWD 106), located in 2500 m depth 106 miles (170 km) offshore of New York (Van Dover et al. 1992; Takizawa et al. 1993; Bothner et al. 1994).

Approximately 36 million tonnes of wet sewage sludge were dumped in surface waters at DWD 106 from March 1986 to July 1992. The distribution of sewage sludge at the sea floor was traced from silver concentrations, linear alkylbenzenes (wetting agents in detergents), and coprostanol and spores of Clostridium perfringens, which originate from mammalian faeces (Bothner et al. 1994; Thiel 2003). During dumping, benthic macrofaunal abundance at the site increased significantly, although changes in total community structure were not substantial (Grassle 1991). Stable isotope ratios (δ^{15}N) indicated that sea urchins and other megafauna assimilated sewage-sludge material beneath the dump site (Van Dover et al. 1992). In addition, the levels of silver in sediments at DWD 106 were 20-fold higher than in background areas and contaminants

appeared to penetrate at least 5 cm into the sediment (Bothner et al. 1994). Environmental concerns halted the dumping at DWD 106 in 1992; the subsequent recovery of stable-isotope signatures to pre-dumping levels in sea urchins (indicating a return to natural phytoplankton-based food webs) took about 10 years (C. van Dover, personal communication 2004).

Sediments dredged from coastal waterways and harbours frequently contain high levels of contaminants such as hydrocarbons, heavy metals and synthetic organic substances (Thiel 2003). The disposal of dredge spoils in the deep sea on continental slopes has occurred in a number of countries and poses threats to sea-floor ecosystems similar to those from sewage-sludge disposal. However, there appear to be no published studies of the impacts of dredge-spoil dumping on deep-sea communities, with the exception of one study in the north-west Pacific offshore from San Francisco (J. Blake, personal communication 2007). Nonetheless, the often active nature of the slope biota and bioturbation rates on continental margins will cause many bioactive materials placed on the sediment surface to be rapidly incorporated into animal tissues and subducted into the sediment matrix (Levin et al. 1999; Miller et al. 2000; Smith et al. 2000).

Climate change

The Earth has warmed by approximately 0.6 °C during the past 100 years, and since 1976 the rate of warming has been greater than at any time during the last 1000 years (IPCC [Intergovernmental Panel on Climate Change] 2001e; Walther et al. 2002). A general warming trend is evident over large parts of the world ocean during the past 50 years (Levitus et al. 2000), and by 2020, the climate will be 0.3–0.6 °C warmer still (IPCC 2001e).

The influences of changing climate on regional patterns of circulation, upwelling and primary production in the surface ocean are very difficult to predict. However, many processes at the deep-sea floor are tightly linked to the surface ocean, allowing some speculation about how climate-driven changes in epipelagic ecosystems may propagate to the deep-sea floor. In particular, deep-sea benthic processes appear to be tightly coupled to the quantity and quality of food material sinking from the euphotic zone, and to the variations in this sinking flux over regional spatial scales and seasonal-to-decadal timescales. For example, benthic biomass and abundance, bioturbation rates, sediment mixed-layer depths and organic-carbon burial rates in sediments co-vary with particulate organic-carbon flux to the deep-sea floor and with primary production in overlying waters (Rowe 1971; Emerson 1985; Smith *et al.* 1997; Glover *et al.* 2001, 2002; Smith & Rabouille 2002; Smith & Demopoulos 2003). In addition, various studies suggest that smaller biota such as microbes and foraminiferans that reproduce more rapidly show rapid population responses to episodic (e.g. seasonal) food input to the deep-sea floor; larger longer-lived taxa generally integrate seasonal changes and respond primarily to interannual to decadal changes in primary production and organic flux (Gooday 2002). Thus, in very general terms, climate changes resulting in increased near-surface productivity and deep organic-carbon flux may be expected to enhance benthic standing crop, bioturbation rates and depths, and carbon sequestration in deep-sea sediments. Beyond this general prediction, the precise nature of ecosystem change (e.g. biodiversity levels) will depend on many factors, such as the temporal scale over which changes occur, the quality of particles reaching the sea floor (e.g. fresh diatom aggregates versus heavily reworked 'marine snow') and the original composition of the benthic fauna. Different types of deep-sea communities (e.g. oligotrophic abyss versus oxygen minimum zones) will respond in markedly different ways.

While it seems unlikely that drastic changes in deep-sea ecosystems will occur as a result of climate change to the 2025 time horizon, dramatic trends have been documented in deep-sea floor communities that ultimately may be ascribed to climate change. For example, in the north-east Pacific at 3800 m depth, the ratio between the sinking flux of particulate organic carbon (POC) and the consumption of such carbon by sea-floor metabolism (i.e. a ratio indicating surplus or deficit of food settling as small particles to the sea floor) progressively decreased over the period 1989–96 from *c*.0.99 to *c*.0.30 (Smith & Kaufmann 1999). This suggests that the proportion of food arriving as small particles from surface waters fell by two-thirds over this 7-year period, while the metabolic demands of the sediment community remained roughly the same. This decline in deep POC flux between 1989 and 1996 could have resulted from increasing sea-surface temperature in the eastern North Pacific (Roemmich 1992; Smith & Kaufmann 1999), which led to a reduction in the supply of nutrients to the euphotic zone and a decline in the export of primary production to the sea floor (McGowan *et al.* 1998).

In the North Atlantic abyss, a decade of trawl sampling has revealed a dramatic regional increase in the abundance of megafaunal holothurians, with one species (*Amperima rosea*) rising in density by up to three orders of magnitude (Billett *et al.* 2001). This 'regime shift' in the structure of an abyssal megafaunal community may be driven by systematic changes in the structure of the phytoplankton assemblages and the quality of export production from the euphotic zone far above (Billett *et al.* 2001; Wigham *et al.* 2003). Similar cause–effect scenarios, in which altered phytoplankton community structure modifies deep-sea detritivore assemblages, may be repeated in many parts of the ocean as a consequence of anthropogenic climate change.

Impacts of pollution

The deep sea is often considered one of the most pristine environments on Earth, relatively unaffected by anthropogenic pollutants because of its distance from pollution sources, the slow rates of physical exchange between near-surface and deep-water masses, and its vast diluent capacity. In fact, the deep sea and deep-sea organisms are an important sink and often the ultimate repository for many of the most persistent and toxic of human pollutants. Halogenated hydrocarbons such as polychlorinated biphenyls (PCBs), dichlorodiphenyltrichloroethane (DDT) and related pesticide compounds, the trace metals mercury and possibly cadmium, and certain radionuclides show particularly high levels of bioaccumulation in the deep sea.

The pollutants of most concern in the deep sea are predominantly transported into the oceans through the atmosphere, due to their volatility or atmospheric emission, and thus achieve a global distribution. About 80% of PCB and 98% of DDT and related compounds enter the ocean through the atmosphere (Clark *et al.* 1997). Approximately 25% of DDT enters the ocean within a year of production (Woodwell *et al.* 1971).

The substances of most concern are highly water insoluble, readily adsorbed onto particles and are generally lipophilic. They are therefore rapidly incorporated into marine food webs and transported into the deep sea in zooplankton and nekton faecal pellets and moults, and from feeding and vertical migration. The concentration of PCB in the faeces of the euphausiid *Meganyctiphanes norvegica* is 1.5 million times its concentration in the surrounding seawater (Clark *et al.* 1997), and methylmercury is concentrated by a similar magnitude in marine zooplankton (Mason *et al.* 1995). The residence time of organochlorides in near-surface waters is less than a year in the open ocean and as little as a few weeks in more productive waters, such as off Antarctica, before it is removed to deeper water (Tanabe & Tatsukawa 1983). As a result, despite the great diluting capacity of the deep ocean, PCB concentrations are virtually uniform throughout the oceanic water column (Harvey et al. 1974a; Tanabe & Tatsukawa 1983).

Reasonably high levels of many pollutants are now widespread in deepwater organisms. PCB levels were consistently >1 ppm and often >10 ppm (on a lipid basis) across a range of dominant midwater planktivorous and small predatory fishes in the North Atlantic and Gulf of Mexico in the 1970s (Harvey et al. 1974b; Baird et al. 1975). Thus, concentrations of pesticides and PCBs from midwater fishes are 'well within levels shown to be physiologically significant or even toxic to other fish species' (Baird et al. 1975). Methylmercury accumulates in species near the top end of the food chain and in larger older individuals. Mercury concentrations are several-fold higher in mesopelagic than in epipelagic planktivores (Monteiro et al. 1996), and concentrations in large long-lived deepwater fishes such as hake, orange roughy, sharks and grenadiers often approach or exceed maximum permissible levels for human consumption (0.5 ppm) (Barber et al. 1972; Cutshall et al. 1978; van den Broek & Tracey 1981; Hornung et al. 1993; Clark et al. 1997; Cronin et al. 1998). Significant behavioural impacts, such as inability to avoid predators, have been observed at mercury levels of 0.67 ppm (Kania & O'Hara 1974).

Much about the cycling of pollutants in the deep ocean is poorly understood. Although global production of PCBs was less than that of DDT compounds, the concentration of PCBs in midwater organisms is generally several-fold higher (Harvey et al. 1974a). It is also poorly understood why very high levels of particular pollutants are found in certain organisms. For example, cadmium levels in black scabbardfish *Aphanopus carbo* livers (6.98 ppm, wet weight basis) are 30-fold higher than in North Atlantic hake (0.22 ppm), another large midwater predator; both exceed permissible levels for human consumption of 0.2 ppm (Mormede & Davies 2001). Cadmium levels in the mesopelagic decapod shrimp *Systellaspis debilis* in the open north-east Atlantic Ocean (c.11–32 ppm) are >10-fold higher than in other midwater crustaceans in the region (Leatherland et al. 1973; Ridout et al. 1985). Radionuclides such as plutonium from nuclear testing (and naturally occurring ^{210}Po) are also scavenged from near-surface waters, taken up into the food chain and transported to deep water.

Public awareness has largely eliminated atmospheric nuclear testing and has dramatically reduced the production of PCBs, DDT and mercury, although DDT continues to be used in developing countries. Nonetheless, deep-sea pollution remains a potential problem because the organochlorides are highly persistent (especially in long-lived deep-sea animals), and the concentration of mercury in the atmosphere and upper ocean remains triple that of 100 years ago (Mason et al. 1994). A number of pollutants continue to be at critical or near-critical levels in specific groups of organisms. However, pollutant impacts on the deep-sea biota cannot be reasonably evaluated or controlled because the factors causing pollutant loading, and pollutant effects on behaviour, physiology, genetics and reproduction remain very poorly known.

RESEARCH NEEDS

In order to predict and manage the effects of direct human activities and climate change, much more must be learnt about the structure and functioning of deep-sea ecosystems. Three of the major knowledge gaps that severely hamper abilities to predict deep-sea environmental impacts are highlighted below.

Biogeography and habitat distributions

The biogeography of the deep sea and the distribution of its habitats are very poorly known for at least three reasons.

Firstly, deep-sea ecosystems are woefully undersampled. For example, in the entire North Pacific Basin, only 11 sites at abyssal depths (i.e. below 4000 m) are known from which macrofaunal species-level data have been collected (Smith & Demopoulos 2003). For the South Pacific and Indian Ocean respectively, apparently only one abyssal data set exists (Ingole et al. 2001; Thiel et al. 2001). Substantially more sites have been sampled on continental slopes, at least in the northern hemisphere, but the slopes are much more variable than the abyss in terms of habitat, with benthic community structure varying substantially with water depth, substrate type, current regime, overlying productivity, sea-floor efflux of reduced chemicals and depletion of bottom-water oxygen. All of these variables can change dramatically over distances of tens of kilometres on the continental slope (Levin & Gooday 2003; Smith & Demopoulos 2003; Tyler 2003b) and as a consequence, the distributions of habitats and species along the slope remain very poorly characterized on regional to basin scales.

Secondly, different sampling programmes in the deep sea have often used different sets of taxonomic specialists to identify the animals collected. Because many, if not most, species collected in the deep sea are new to science and have not been formally classified, it is often very difficult to relate the species list of one study to that of another, and thus to compare species and conduct biogeographical syntheses over basin scales.

Thirdly, most biodiversity and biogeographical studies in the deep sea have used traditional morphological methods for identifying species. However, recently developed molecular methods (e.g. DNA sequence data) suggest that morphological techniques typically underestimate the number of species and overestimate species ranges in marine habitats (Knowlton 1993, 2000; van Soosten *et al.* 1998; Creasey & Rogers 1999).

Major programmes designed to sample novel habitats and poorly studied regions of the deep sea (e.g. cold seeps, oxygen minimum zones, abyssal sediments in the South Pacific, hydrothermal vents and continental slopes in the Southern Ocean) must be considered a research priority. Within such programmes, it is critical to apply modern molecular techniques to resolve phylogenetic relationships, levels of biodiversity and species ranges. Perhaps most importantly, it is essential to integrate the taxonomic collections of many different expeditions and countries to allow biogeographical syntheses on regional and global scales.

The biogeography of seamounts merits particular attention because seamount communities are potentially very diverse, highly variable among locations, heavily impacted by fisheries and very poorly understood. For example, 1987 data suggested that only 15% of the invertebrates and 12% of the fishes collected on seamounts were endemic (Wilson & Kaufmann 1987). However, while over 100 seamounts had been sampled, few had been studied comprehensively, and 72% of the species recorded came from only five seamounts. A mere 27 species had been collected from seamounts in the entire south-west Pacific. The first intensive South Pacific study (de Forges *et al.* 2000) reported more than 850 species of benthic macro- and megafauna from seamounts in the Tasman and Coral Seas. This was 42% more than had previously been reported from all seamounts worldwide since the *Challenger* Expedition! Approximately 30% of these species were potential seamount endemics, and there was little overlap in species composition from one seamount chain to another (de Forges *et al.* 2000). Seamount chains may be effectively isolated from each other owing to topographical rectification of currents combined with the limited dispersal abilities of many deep-sea benthic organisms (de Forges *et al.* 2000).

This information might have profound implications for the conservation of threatened seamount faunas. However, it needs to be tested in other oceanic regions and over broader spatial scales on the many distinct seamount chains that remain virtually unexplored (e.g. in the south-west Pacific, Indian and Southern Oceans). As for other deep-sea habitats, species composition and population genetic structure (e.g. gene flow within and between seamount chains) must be evaluated to allow reasonable conservation measures (e.g. creation of MPAs) to be implemented.

Response of deep-sea ecosystems to human impacts

Numerous anthropogenic forcing factors will influence deep-sea ecosystems to the 2025 time horizon (Table 22.2), and in some regions, especially continental slopes, several impacts may occur simultaneously (e.g. fishing, energy exploitation and waste disposal) (Chapter 19). This means that prediction of environmental impacts will be especially difficult because deep-sea community responses even to single stressors are very poorly known. Research priorities must include experimental evaluation of the response of deep-sea benthos to (1) oil-drilling discharges, (2) burial, toxicant loading and organic enrichment associated with dredge-spoil dumping, and (3) the disturbance, predator removal and by-catch dumping effects resulting from deepwater trawling. Cleverly designed experimental studies will be necessary to resolve the response of deep-sea systems to varying levels of single and multiple stressors. It need also be recognized that community sensitivity and recovery processes may vary dramatically with habitat type across the deep sea (e.g. oxygen minimum zones versus well-oxygenated sediments), and are likely to be inversely related to the physical stability and flux of organic carbon within the habitat. In particular, the very stable food-poor habitats such as the abyssal plains are likely to be the most sensitive to, and slowest to recover from, human impacts, but many slope habitats are also likely to be very susceptible to anthropogenic impacts compared to shallow-water systems.

Research priorities should not be restricted to evaluation of specific environmental impacts, but must also be directed towards expanding the very limited knowledge of

Table 22.2. *Summary of anthropogenic forcing factors in the deep sea ordered by estimated severity level for 2025 in each category*

Forcing factor	Spatial scale[a]	Temporal scale (estimates)[b]	Estimated severity in 2025
Deepwater fisheries			
Direct effects of stock and biomass depletion	Regional–basin	1950s onwards	High
Indirect effects (trawling damage, by-catch, whale removal)	Regional–basin	1950s onwards	High
Energy and mineral extraction			
Polymetallic nodule mining	Basin	c.2010 onwards	High
Deepwater oil and gas extraction	Basin	1990s onwards	Moderate
Polymetallic sulphide mining	Local	Present day onwards	Moderate
Manganese crust mining	Local	Unknown	Low
Methane hydrate extraction	Regional–basin	Unknown	Low
Waste disposal			
Deepwater CO_2 sequestration	Local–regional	c.2015 onwards	High
Sewage and dredge spoil disposal	Local	c.2010 onwards	Moderate
Oil and gas structures disposal	Local	Isolated incidents	Low
Radioactive waste disposal	Local	1950s–1990s	Low
Munitions disposal	Local	1945–76 (now banned)	Low
Climate change			
Changes in oceanic production and phytoplankton community structure	Global	c.2050 onwards	Low
Other			
Research and bioprospecting at vents	Local	1970s onwards	Low
Shipwrecks	Local	Since shipping began, ongoing	Low
Underwater noise	Local	1960s onwards	Low

[a] Spatial scale of impact is indicated at the level of local (linear scale of 0–100 km), regional (100–1000 km) and basin (1000–10 000 km).

[b] In the deep sea, owing to low biological and chemical rates, the timescale of impact typically extends far beyond the timescale of activity (e.g. impacts of a large shipwreck or deep seabed mining are expected to last >100 years).

the general structure and dynamics of deep-sea ecosystems, especially those most likely to bear the brunt of human impacts. For a broad range of forcing factors and deep-sea habitats, there is a pressing need to identify (1) vulnerable species and ecological processes, (2) useful indicators of ecosystem health and (3) the linkages between terrestrial, coastal and atmospheric systems that may channel anthropogenic stressors to the deep-ocean floor. In addition, the remarkable biodiversity of many deep-sea habitats merits special attention. Knowledge of the roles played by biodiversity in ecosystem functions such as stability, nutrient cycling, biomass production and carbon burial, is fundamental to conservation and management decisions (e.g. location and sizing of MPAs, gear regulation, and fishery and oil-lease closures).

Long time-series studies

To evaluate and manage environmental impacts in the deep sea, ecosystem variability resulting from natural processes must be distinguished from anthropogenic change. Because many human-induced impacts such as global warming may occur gradually, relevant natural variability can only be elucidated through time-series studies covering many years to decades. Time-series studies of ecosystem structure and dynamics are extremely rare in the deep sea, and have been restricted to only a few sites in the ocean (i.e. the north-east Atlantic and the north-east Pacific). Nonetheless, these limited time-series data have documented surprising temporal variability in basic ecological parameters over multi-year timescales (Gage & Tyler 1991; Smith & Kaufmann 1999; Billett *et al.* 2001). Clearly, the few existing deep-sea time-series must be continued and new series initiated if there is to be a suitable ecological context in which to evaluate the deep-sea impacts of anthropogenic change in the coming decades. Since it will be impossible to inventory and monitor all the organisms and regions, it will be highly desirable to develop key indicators of ecosystem health (e.g. particular taxa, assemblages or biogeochemical processes) for specific types of deep-sea habitats, and to monitor these indicators at representative sites over long time periods to assess ecosystem change.

POLICY AND SOLUTIONS

Current and future anthropogenic forcing factors in the deep-sea ecosystem are many and varied (Table 22.2), targeting habitats ranging from slopes only a few kilometres offshore (e.g. dredge-spoil dumping off San Francisco) to the remote abyssal plains in the central Pacific (e.g. manganese nodule mining). The policy recommendations and management approaches to control these diverse impacts need to be carefully considered. However, environmental protection is urgent in the deep sea because some deep-sea ecosystems such as continental slopes and seamounts have already been substantially modified by human activities, and many others will be profoundly altered in the near future.

Many policies and regulations to protect the deep sea must come from national governments because substantial portions of the deep seabed, including those habitats most likely to sustain immediate human impacts, lie within the 200-mile limit of exclusive economic zones (EEZs). Nonetheless, most of the deep-ocean floor underlies the 'high seas' (i.e. waters outside of national jurisdictions) presenting unique policy challenges. The protection of seamount biota from the onslaught of deep-sea trawling in international waters provides an informative example of the challenges involved.

As a consequence of growing environmental awareness since 1999, Australia, New Zealand, Norway, the USA and other countries have declared deepwater MPAs designed to safeguard deep-sea coral and seamount habitats. In December 2002, the UN General Assembly responded by passing a resolution recognizing the need to protect biodiversity on the high-seas, including seamount biota. How can this be accomplished?

The United Nations Convention on the Law of the Sea (UNCLOS) specifically regulates mineral extraction on the high seas (the ISA set up by the United Nations is vested with that authority) but fails to address exploitation of biological-resources. This was because in the 1970s, when the UNCLOS was negotiated, the environmental implications of the global spread of deepwater fisheries were not yet appreciated. As a consequence, regulation of high-seas fishing and its environmental impacts has fallen into a legal limbo. However, several options are presently under consideration. There might be multinational or UN declaration of high-seas MPAs to protect sensitive and unique environments. The UN might broker a global moratorium on demersal trawling on the high seas (or on seamount and deepwater coral habitats, in particular). The ISA might be vested with the authority to regulate seabed biological-resource exploitation (or at least demersal fishing) on the high seas.

These options are not mutually exclusive. The declaration of MPAs on the high seas is seen as lacking recognition under international law and appears unlikely to be instituted, even on a trial basis, within the next 10 years. The advantage of a global moratorium on high-seas trawling (proposed by Greenpeace) is its immediate blanket impact if adopted by the UN General Assembly. In the longer term, vesting an internationally recognized body, such as the ISA, with the authority to regulate the seabed impacts of biological-resource exploitation could effectively close the gap within the international legal framework for the regulation of demersal fishing impacts on the high seas. Institution of a global and representative system of MPAs on the high seas might also be most effectively administered under the authority of the ISA.

CONCLUSIONS

The deep-sea floor (i.e. below 200 m depth) covers most of the Earth's solid surface and harbours a broad range of habitats and remarkable levels of biodiversity, yet it remains largely unexplored. Many deep-sea habitats and their resident biota are easily disturbed and slow to recover from anthropogenic disturbance due to low levels of productivity, low biological rates and delicate habitat structures. A surprising number of deep-sea habitats, especially seamounts and continental slopes, are currently threatened by deleterious human activities such as bottom fishing, waste disposal, fossil-fuel extraction and pollutant loading. The environmental impacts of these activities, and the requisite conservation measures to minimize these impacts, remain extremely poorly understood. Ecological processes at the deep-sea floor are often tightly coupled to the surface ocean, the atmosphere and the coastal zone. Climate change in the coming decades may thus alter deep-sea ecosystems on a global scale. Because deep-sea research has lagged far behind that in other habitats, there is an urgent need to better understand key attributes of deep-sea ecosystems, including species distributions, community responses to anthropogenic disturbance and natural temporal variability over years to decades. These needs can only be met by creative, well-funded research programmes which are essential to predicting and managing human impacts in the deep ocean. Anthropogenic impacts in the deep sea have substantially outstripped the predictive abilities of scientists, and this situation will worsen with accelerated human exploitation of the deep ocean. Environmental protection in the deep sea, perhaps more than in any other habitat, will require application of the 'precautionary principle', in which reasonable conservation measures are implemented prior to detailed scientific understanding of the ecosystem.

Part VIII
Synthesis

23 · Trends and global prospects of the Earth's aquatic ecosystems

NICHOLAS V. C. POLUNIN, BRIJ GOPAL, NICHOLAS A. J. GRAHAM, STEPHEN J. HALL, VENUGOPALAN ITTEKKOT AND ANNETTE MÜHLIG-HOFMANN

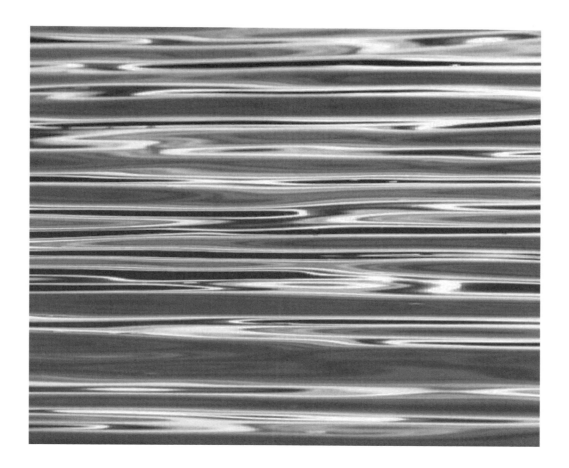

INTRODUCTION

The Earth's organisms have depended on water since they arose 3.5 billion years ago. In the last 10 000 years, human development of stone tools, learning of food cultivation, growth of civilizations and trade, and migration among others have increasingly impinged on aquatic ecosystems. During the last century, the human population has tripled, and water use for human purposes increased six-fold. Approximately half of all available fresh water is now used to meet human ends, twice what it was 40 years ago (World

Aquatic Ecosystems, ed. N. V. C. Polunin. Published by Cambridge University Press. © Foundation for Environmental Conservation 2008.

Water Council [WWC] 2000), and by 2025, the withdrawal of water is projected globally to increase by at least 50% (Martinez Austria & van Hofwegen 2006).

The availability of clean fresh water and sanitation is thus among the most important issues facing humanity today. The average annual per caput availability of renewable water resources is projected to fall by a quarter to 4 800 m^3 by 2025 and some people will suffer far more than others. By 2025, 3 billion women and men are expected to be living in arid and semi-arid regions with less than 1 700 m^3 of water per caput, the level below which people begin suffering from water stress (WWC 2000). In the 1990s, access to safe water greatly increased, yet 1.1 billion people still lack access to safe drinking water, and 2.4 billion lack basic sanitation. Most of the people affected in both cases are located in Asia; however, given the difference of population size between the two continents, the percentage of people affected is still greater in Africa (Martinez Austria & van Hofwegen 2006). In general, rich and poor countries vary greatly in water use; developing countries use 20 litres per person per day, while developed countries use on average 400–500 litres per person per day for all purposes (Martinez Austria & van Hofwegen 2006). Since the number of fatalities caused by natural disasters has tended to decrease over the last 30 years while 70% of all the deaths associated with disasters have been hydrometeorological in origin (Martinez Austria & van Hofwegen 2006), it appears that water and weather have become directly responsible for more and more human deaths. The costs of damages caused by water-related disasters reached approximately US$200 billion in 2005, borne mostly by the developed countries but with greater impact on the gross domestic product of developing countries (Martinez Austria & van Hofwegen 2006). Added to these disasters are the human health impacts of poor water quality and waterborne diseases, the future trajectory of which is unknown. At the same time, water specifically has huge potential to help meet energy demands. Eighty per cent of the global hydropower is untapped, the underutilization of this potential being 96% and 88% respectively for Africa and Asia (Martinez Austria & van Hofwegen 2006). As water becomes scarcer, conflicts amongst human users become more likely.

Sustaining the direct uses of water and its impacts such as flooding is the array of aquatic ecosystems; the array of goods and service these systems provide is even wider (WWC 2000). It is uncertain how much water ecosystems need if they are to be maintained. Providing six times

more water to humans now than 100 years ago alone has significant ecosystem implications (WWC 2000), added to which is global intensification of activities such as fishing, waste disposal, industry and land use, and processes such as climate warming. What are the key threats to the status of each of the aquatic ecosystems? To what extent might each of these ecosystems change between now and the 2025 time horizon? How do the patterns of threat and likely change vary among these ecosystems and what drives these variations? What information does this overview yield for the conservation science, management and governance of the world's aquatic ecosystems? This chapter summarizes major trends and possible states across all 21 ecosystems presented here (Chapters 2–22).

TRENDS WITHIN ECOSYSTEM TYPES

Flowing waters

In terms of wilderness areas, existence of human-unimpeded processes and structures, and delivery of vital natural services and goods to humanity, all three flowing-water ecosystems (Chapters 2–4) are expected to be significantly different by the year 2025, albeit for different reasons. Based on the expert group deliberations at the 5th International Conference on Environmental Future (hereafter '5th ICEF expert groups') (Chapter 1), major threats to rivers and streams include water pollution, habitat fragmentation, species additions and losses, gross catchment transfers of water and climate change (Chapter 2). For groundwaters, major issues include the lowering of water tables, biodiversity loss, salinization of soils, restriction of water functionality, land subsidence and changes in surface aquatic ecosystems (Chapter 3). Changes in water flow, physical modification of habitat, salinization and contamination, human displacements, invasive species and climate change are particular threats to flood plains (Chapter 4). Global concerns to the 2025 time horizon are changes in water flow through diversions, abstractions and other direct anthropogenic impacts, modifications of habitat including that in overall catchments and water-quality changes and pollution. However, the relative importance of these concerns varies between developed and developing countries. Habitat fragmentation and gross transfers of water are expected to intensify in both types of country, but water pollution and generally declines in water quality are likely to be more extensive in developing countries. How to reverse these threats has been under discussion since long before the Stockholm Conference on the Human

Environment (1972) and was well rehearsed at the United Nations Conference on Environment and Development, and specifically at the Dublin International Conference on Water and the Environment (both 1992). Following on from these and other deliberations, including the 5th ICEF expert groups, the maintenance of flowing-water ecosystems in a natural state to the 2025 time horizon will depend on many things. Above all, it will require wiser use of water, increased planning at basin scales (including transboundary agreements), more effective and extensive conservation measures (including restoration), and greater scientific and management capacity than exists at present. Since greater public awareness and involvement are considered key to these, it is clear that efforts in the direction of education, communication and participatory governance are crucial.

Still fresh waters

The extent and character of changes that have occurred and are expected to occur by the year 2025 to the world's lakes and ponds vary by ecosystem type and geographical location. Major concerns for small lakes include eutrophication, other forms of pollution (e.g. by heavy metals and organochlorines), acidification and acid precipitation, invasive species and climate change (Chapter 5). For large freshwater lakes, the 5th ICEF expert groups considered that eutrophication, pollution and invasive species are major threats, along with possible climate-change effects on lake dynamics and on these former three threats. Eutrophication continues as a major threat to the African Great Lakes, and invasive species and pollution to the St Lawrence Great Lakes (Chapter 6). Lake Baikal is threatened by accelerated land-use change, which, in the long term, is expected to increase land erosion and thus enhance siltation and eutrophication, even in other relatively undeveloped lake regions. The attractiveness of large lakes as a means to meet the growing human water demand will probably result in more schemes to abstract and divert water from them. In general, large and small lakes are threatened by growing human populations and increased demand for fresh water and other goods such as fish. Human activities have extensively also threatened or impacted saline lakes, especially through reduced availability of fresh water by diversions and/or climate change, mining and pollution, and introduction of invasive species is also of concern (Chapter 7). These activities have resulted in loss of biodiversity and salinization, as well as changes in the natural character of the lakes.

The 5th ICEF expert groups had a more optimistic view of the scope for dealing with environmental problems in developed countries, which have greater financial and personal resources to address them, than in developing countries, where the infrastructure for effectively reversing impacts such as overfishing and eutrophication scarcely exists. There has been some success in developed countries in reducing eutrophication, acidification and contaminant loading. In these countries, there are substantial prospects for fresh and saline water bodies on average remaining to the year 2025 in a state similar to the present, although even in the rich countries the future is far from bright. Although no lake anywhere can still be considered pristine, many of the most natural lake ecosystems exist in the developing regions of the world, yet the year 2025 is more likely to see accelerated losses of small lakes and ponds and greater degradation of lakes, both large and small.

There are several possible approaches to conserving lake ecosystems, including maintaining present structure and function, and where necessary reversing earlier declines. Thus, formation of effective protected areas can be used to exclude exploitation of all kinds, and approaches such as wastewater treatment and biomanipulation (e.g. species removal) can help rehabilitate once-degraded lakes (Chapter 5). The protected areas need to encompass wilderness areas as much as those close to human population centres, but remote sites require additional resources (e.g. through tourism) to be effective, and this is especially so in developing countries. Whatever form it takes, lake management consists largely of people management within the natural constraints on human use and abuse. Thus management actions need to include generation of salient and credible knowledge, raising public awareness of the environmental issues, developing economic incentives at regional levels to reduce unsustainable activities, and devising forms of co-management in consultation with major stakeholder groups. At the international level, which is particularly relevant to large lakes, transboundary instruments similar to the Great Lakes Water Quality Agreement between Canada and the USA need to be developed more extensively.

Freshwater wetlands

Understanding of freshwater wetlands in many temperate and all tropical regions is extremely poor, and the knowledge required for effective management will remain scarce for the foreseeable future. Peatlands hold about a quarter of the world's soil carbon, and peat mining as well as drainage are widespread threats to the future of the ecosystem, but the most important changes for cool temperate peatlands are expected to arise from global climate change (Chapter 8). In particular, the peatland capacity to accumulate atmospheric carbon could be greatly reduced but the extent of this reduction cannot be reliably estimated as the rate of carbon input to the catotelm remains unknown; the knowledge of the current contribution of this ecosystem to greenhouse gas emissions is also modest (e.g. Malmer *et al.* 2005). Climate change will be important for temperate wetlands (Chapter 9) and tropical wetlands (Chapter 10) because it will affect the hydrology of the landscape, but for both ecosystem types responses to this change are expected to differ between dry and wet regions. Growth of temperature and hydrological extremes in many cases, with greater dryness in dry regions and increased water availability in wet regions, will clearly impact the global extent, distribution and composition of freshwater wetlands.

However, at country level, social–economic and political circumstances will also influence future states of wetlands, through effects on human wetland use and recovery rates. At a regional within–country level, varying patterns of resource use will also moderate wetland state, intensifying or mitigating deterioration according to the circumstances. Human attitudes towards wetlands, and the nature and intensity of human impacts, have varied geographically depending on the social, cultural, economic and political characteristics of the region. Rapid degradation and loss of wetlands have brought about a change in the attitude of human societies in developed, mostly temperate, regions; a resource-oriented view of wetlands has been replaced by a landscape-level ecological perspective. The interactions between the ecological and societal factors over hundreds of years have shaped the present states of wetlands and will govern their future as well.

Various human impacts are directly related to gradients in human population density, economic status, and sociocultural and political milieu. In general, population densities are higher in the tropics, but high densities also occur in Europe and much of North America. Boreal regions where peatlands dominate are in general less populated. The economies of the tropical regions are less developed than those of most temperate regions. It is often accepted that the intensity of human impacts on wetlands is directly proportional to human densities, and pressure for development is greater at the lower economic status levels of the people in tropical countries. However, such impacts are related to the livelihood strategies and consumption patterns of the different user groups, and are driven both by local and global demand for goods and services (Chapter 10). Much of the large-scale degradation in developing countries occurs in response to the demands of people in developed countries, while the small-scale degradation often reflects livelihood needs of the people within the developing countries themselves. Also, the economic returns flowing to developing countries are insufficient to compensate for the environmental degradation or the associated social costs. An important dimension of economic gradients, linked through trade, is seen in the introduction of exotic species (e.g. water hyacinth *Eichhornia crassipes*), a major cause of degradation and loss of wetlands within the tropics.

The freshwater wetlands, through positive feedback effects on global warming, could play a pivotal role in the future of the Earth's ecosystems. The history of human impacts and their relation to sociocultural gradients provide significant pointers to the future of the freshwater wetlands. In the Asian tropics, human impacts have gradually increased over at least 5000 years. Sociocultural systems evidently helped to maintain natural wetland functions and diversity until recently, when traditional links with nature broke down and demand for goods and services accelerated. The transition is also recent in the African and South American tropics, driven by growth in demand for resources. In contrast, in temperate regions, especially North America, human impacts arose only recently but cultural attitudes and technological advances made it possible to decimate or alter wetlands very rapidly indeed. Such impacts still continue, for example in Canadian peatlands, and may do so elsewhere (e.g. Siberian peatlands with respect to oil and gas explorations).

Substantial commitments towards integrated water-resource management could help promote ecosystem protection, and increased awareness of ecological functions of wetlands in maintaining water quantity and quality might in some regions help revive traditional practices where these helped to conserve resources or whole ecosystems. Integrated water-resource management at the catchment scale will widely need to accommodate the smaller operational scales of livelihoods and different cultures' philosophies and practices that drive user behaviour at local level. There will be need for involving local communities;

and international cooperation will be crucial, especially in areas where rivers and wetlands transcend geopolitical boundaries. However, the 5th ICEF expert groups felt it is particularly difficult with freshwater wetlands to predict the course of human actions at local, national and global levels. Often, there is a large divide between the thoughts and actions in all countries where the compulsions of social, economic and political forces are the prime driving variables.

Coastal wetlands

These wetlands are best developed in coastal areas which coincide with high human population densities and are thus among the most anthropogenically impacted marine ecosystems of all. Destruction of saltmarsh habitat (Chapter 11) has declined in some countries (e.g. USA) but losses generally have continued with impacts including those of reclamation, relative sea-level rise and invasive species. Physical alteration and loss of habitat have occurred locally through dyking, ditching, leveeing, canal cutting, shore protection and dredging (e.g. related to salt production, aquaculture). Changes in freshwater flows and sediment supply related to upstream developments (e.g. dams, reservoirs and land-use modifications) have among other things altered the capacity of the system to accrete sediment which is important to its longevity. The economic importance of resources is a major driving force for severe mangrove habitat degradation and loss, and consequent spatial fragmentation of the constituent populations (Chapter 12). Inadequate ecosystem conservation in the face of low market values for mangrove goods and services in many countries has helped lead to overexploitation of mangrove timber and fishery resources (e.g. shrimp). Aquaculture, salt and rice production, and urban development have widely replaced and otherwise degraded mangrove habitat. Mangroves have proved particularly susceptible to eutrophication. While the changes to saltmarsh and mangrove may seem most visible, human impacts and ecosystem trends for the estuaries with which they are most often linked are similar (Chapter 13).

Up to the 2025 time horizon, the 5th ICEF expert groups identified habitat loss and alteration, eutrophication (including through land runoff and sewage inputs), fisheries overexploitation and relative sea-level rise as likely having far-reaching consequences, potentially modifying the structure (including biodiversity declines), function and controls of all three coastal wetland ecosystems. The most extensive changes may be the direct result of human actions at local and regional scales. Saltmarsh habitat loss and degradation will continue to the year 2025, but at variable rates in different countries, perhaps the most susceptible saltmarshes being those in developing countries. While many tropical countries (e.g. in South-East Asia) have established mangrove restoration projects, these are small compared to the spatial scale of major threats, such as logging and clearance for shrimp ponds, given the economic and social benefits potentially involved. Commercial exploitation of mangroves for timber, aquaculture and capture fisheries will continue in developing countries up to the 2025 time horizon. Estuaries are currently most heavily impacted on urbanized industrialized coastlines of the northern hemisphere; however, developing regions will probably emerge as foci of increasing anthropogenic impacts, as coastal watersheds become more developed and associated human activities escalate. Estuaries are major conduits of water and anthropogenic-nutrient flows to coastal seas and thus exemplify the trend towards greater inter-ecosystem impacts in future. These coastal wetland ecosystems are likely to suffer further declines and degradation, in both developed and developing regions, up to the 2025 time horizon.

Several means to depart from these trends are potentially available. Protected areas are one method of ring-fencing portions of these vulnerable ecosystems against local direct impacts such as resource extraction, but the information base for planning these as effective conservation tools is poor. This is especially true where hard choices have to be made such as between financing through tourism benefits and losses to traditional small-scale fisheries. In most cases, research is needed to understand the inherent processes and consequences of such area-based controls. In addition, those measures alone are rarely sufficient; for example, controls of fishing effort are essential if fishing impacts are not to be intensified, albeit over a smaller part of the overall area. Apart from possible restorative roles for depleted species, especially where these have limited home and migration ranges, widespread impacts dictate a role for protected areas as reference sites for better understanding of natural processes.

Alongside protected area measures, given sufficient incentive and resources, restoration of vegetation is feasible at modest scales in both mangroves and saltmarshes; in some cases this has the potential also to help locally address some biodiversity and fishery issues and regionally contribute to carbon-offsetting schemes. With their high biomass and in some cases also substantial reservoirs of detrital carbon, mangroves together with other forested ecosystems have potential as means of carbon offsetting. However, the wider environment also needs attention if such measures are to be secure. Thus, tighter controls on impacts across ecosystem boundaries (e.g. development in coastal watersheds and adjacent coastal waters) are essential if coastal wetlands are not to change substantially in structure and extent at many locations around the world. These controls have to address sources of nutrient, chemical-contaminant and sediment loading. In the case of nutrients, increasing nitrogen and decreasing silicate will alter the nutrient cocktail reaching estuaries and adjacent coastal ecosystems, changing their structure and function. Algal blooms and fish kills are well known, but less recognized is the potential especially for hypoxic coastal ecosystems to turn into enhanced sources of the greenhouse gas nitrous oxide derived from excess nitrate loading.

The meagre information base dictates the need for monitoring and research to improve overall knowledge, characterize ecosystem-level dynamics, identify impacts and improve the means of remedial action. A huge challenge is to understand how multiple stressors affect ecosystems at different ecological levels and spatial and temporal scales. This alone is a vast research agenda, and thus the approach has to be by means of detailed case studies which provide concepts that are then tried elsewhere. The 5th ICEF expert groups highlighted that the research challenge becomes further evident when the need for interdisciplinary work such as between natural and social sciences is considered.

Rocky shores

Depth, light and other limitations largely confine these ecosystems to narrow coastal bands rendering them accessible to both scientific study and resource exploitation, and susceptible to similar types of impact. Some threats are common to all three ecosystems. One of these is depletion of living resources, whether by small-scale or large-scale fisheries. In the rocky intertidal, principal targets including molluscs such as whelks, mussels and abalone have been widely depleted (Chapter 14). In the rocky subtidal ecosystem where kelps flourish, the original apex predators are little known (Chapter 15). On coral reefs, fishing is widely considered a major threat; it has depleted

a range of finfish and invertebrate species, and the areas affected have greatly expanded in the recent past (Chapter 16). In these ecosystems, extensive removal of fishery-target species is linked to trophic cascades which have widely altered ecosystem structure. Thus in kelp beds, predator removal has widely led to intensive sea-urchin grazing, leaving kelp-free patches where non-native herbivores and predators have become established. On coral reefs, the shift from large-bodied slow-growing fish and invertebrates to fast-growing, often generalist species, including sea urchins, has in many cases also changed community structure (coral–algal phase shift) and reduced system resilience.

Global and regional climate changes have also impacted all three ecosystems, for example through increased seawater temperatures and sea level. In kelp ecosystems, increasing El Niño frequencies, oceanographic regime shifts and storms may cause further kelp deforestation (Chapter 15). For tropical coral reefs, a general shift in benthic dominance from coral to algal taxa is expected, with higher organic production, and a consequent reduction in reef growth and substratum complexity, although some regions will be less susceptible to warming than others (Chapter 16). It is noticeable that these general drivers of change are generating common impacts. For example, coral bleaching and diseases, intense fishing, storms and nutrient enrichment are all tending to reduce coral cover, with immediate positive effects on macroalgae, potential long-term impacts on habitat complexity and thus profound consequences for the ecosystem as a whole. A common feature of kelp and coral-reef ecosystems is that the habitat is biogenic and the ecosystem-engineering organisms are proving highly vulnerable to a range of processes that are subject to human influence.

Yet, synthesis of the trends and future states of rocky shores is scarcely possible, because of substantial differences in biological composition and types and scales of impacts. Local impacts on these ecosystems are bound to vary due to their different geographical distributions, not so much for the rocky intertidal, which is ubiquitous, but certainly so in the case of kelp forests (cold waters) and coral reefs (tropics). In rocky intertidal ecosystems, threats such as oil spills and other forms of pollution including eutrophication, siltation and river discharge, are mostly localized, stemming from point sources and local human usage. Whilst these threats exist for the other two ecosystems alike, some threats such as sea-urchin herbivory are relatively specific to kelp ecosystems, whereas increases in seawater temperatures causing coral bleaching and coral diseases are substantial specific threats to tropical coral reefs. Another fundamental characteristic, namely rate of recovery, also varies between these ecosystems. Because of their non-biogenic foundations and generally greater exposure to physical stressors, rocky intertidal ecosystems are expected to recover more rapidly from major disturbances than kelp and coral reef ecosystems. To the 2025 time horizon, these ecosystems are unlikely to fail; however, they are already widely changed and may persist with entirely altered structures and food webs. Coral reefs in particular have shown a striking vulnerability, with important implications for biodiversity and adjacent ecosystems.

Climate change and the serial loss of biodiversity through overfishing appear to be the greatest threats to the structure and functioning of the three rocky shore systems over the 2025 time horizon. However, rocky shore ecosystems can not be understood, managed or conserved on their own, and even if they could be, their biodiversity and general evident complexity dictate that the scientific basis for rational prediction and decision-making is poor. The knowledge base exists for predicting some consequences for certain species of particular impacts such as fishing or sea-surface warming. However, especially at the ecosystem level, the uncertainty about such predictions becomes substantial, while the ability to forecast the combined interactive effects of several environmental factors is more modest still. The greatest ecological surprises are likely to occur where environmental change induces shifts between alternate states, an organism is particularly susceptible to a pollutant, or an exotic species has a much more prominent role in an invaded community than in its original environment.

The tropical and temperate countries with rocky shores have to develop appropriate policy and management priorities. The greatest prospects appear to be for co-management and ecosystem-based approaches involving protected areas, which together recognize the complexities of the ecosystem and realities of human impacts. Co-management can have huge advantages over previous top–down management and the principles are particularly relevant to rocky shores for reasons including: harvesting being a major factor, for the most part small scale and technically uncomplicated; rocky shores being difficult to monitor and police; it being possible to assign 'ownership' to local communities that live close by; the exploited resources being mostly site-attached; and local users being

able to benefit from the control measures to which they agree. Perhaps more so on rocky shores than anywhere else, biological, social and economic factors come together and could make co-management a good approach. In addition, regionally and globally, efforts are needed to identify and manage species that are prone to extinction. It is important to identify the environmental, ecological, and biological properties of rocky shore communities and organisms (including species recruitment and recovery) that will allow them to persist into the future. Consequently, one priority for the 2025 time horizon is to identify specific intertidal, kelp and reef areas and species that will require special protection in order to ensure their existence.

In many circumstances, management needs to focus on minimizing fishing impacts and restoring populations of functionally important species in these systems. Regionally and locally, there is a need to renew efforts for sustainable fisheries management through restrictions on areas, species, sizes, gears and fishing-effort levels. Restrictions on destructive fishing will challenge conservation as human numbers increase and competition for dwindling resources accelerates. In addition to legislation and enforcement to eliminate destructive fishing methods, there is a growing need to alleviate poverty by ensuring that fishing is undertaken sustainably in ways that contribute to human and economic development as part of a suite of livelihood options for the peoples of developing countries.

Management of watersheds and water quality are also important issues at the regional level. Both tropical and temperate nations need to take responsibility for balancing ecological and economic production, and the control of carbon, toxic (e.g. pesticides), inorganic and organic nutrient wastes and sediments to provide an improved future for rocky shores and their economic services. It is the ever-increasing demand for food, natural resources and economic growth that is the basis of anthropogenic global change, and not only in view of the conservation of these three ecosystems should efforts to control the emission of greenhouse gases be a global priority. The solution to this problem will require international cooperation at levels not seen before.

Soft shores

With respect to threats and impacts, the three soft-sediment marine ecosystems reveal three key differences

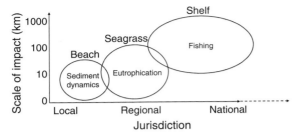

Fig. 23.1. Jurisdiction and impact scales of the principal threats to soft-sediment ecosystems.

and one point of similarity. The differences lie in the scales of impact, the identity of the primary threat and jurisdictional responsibility for management. The point of similarity is the likely divergence of trends for the developed and developing world, which is common for all systems. There is a continuum from local jurisdiction and relatively small-scale impacts such as changes in sediment budgets, the dominant threat to sandy beach ecosystems, to large-scale and national jurisdictions for fishing, the principal threat to the ecology of continental shelves (Fig. 23.1). Intermediate between these is eutrophication, the primary threat to seagrass systems. These scale relationships give an important indication of the level at which management action to mitigate threats needs to be focused.

Further coastal development and establishment of new coastal structures and activities will have the greatest impacts on sandy beaches in the coming years (Chapter 17). In these naturally physically dynamic ecosystems, coastal development will have profound effects on the resident biota of dunes, intertidal beaches and surf zones because the temporal and spatial patterns of physical disturbance and sand transport will change. Coastal development in the developing world is often intimately bound with tourism. In these areas the establishment of infrastructure to support beach access, in particular the continual hardening of surfaces in and above dune systems, is likely to constitute the greatest threat. In the developed world, further coastal development will result in an increase in beach nourishment programmes, which are likely to become the norm for many countries, especially in the face of relative sea-level rise. A key goal for management in these communities will be revegetation of dune systems to maintain their stability.

While coastal pollution is also a threat to sandy shores, this source of stress is most threatening to seagrass ecosystems (Chapter 18). Human population growth means increased sewage runoff and demand for water; combined with agricultural development and the associated increase in fertilizer application and land clearing, coastal eutrophication may be the key driver of change in developing countries. Seagrass losses are expected to accelerate with increased human population pressure, especially in South-East Asia and the Caribbean. Problems of eutrophication will also apply for developed countries, albeit to a lesser extent owing to the increased emphasis on, and resources for, pollution control. There may, however, be a parallel divergence within the developed world with institutional frameworks for controlling water quality becoming much more effective in Europe than in North America. For example, in contrast to the USA, emphasis appears to be growing in Europe on controlling inputs from agricultural as well as urban sources and this trend seems set to continue.

For soft-sediment continental shelves, fishing is the dominant threat, both in terms of geographical scale and severity of impact (Chapter 19). The 5th ICEF expert groups felt the institutional structures have yet to be established to ensure that continental-shelf fish stocks are restored to appropriate levels and fished sustainably. There are grounds, however, for cautious optimism. Firstly, continental shelves are largely under the control of individual nations since they form part of their exclusive economic zone (EEZ). Thus, in principle, the jurisdictional framework exists for sustainable exploitation. Secondly, it is broadly known where the problems lie and what needs to be done. Equally as important, some key drivers for social change are beginning to align, notably a growing interest in dealing with perverse incentives in the form of economic subsidies, the existence of high-level commitments to deal with the problem and the recognition that environmentally responsible fishing can lead to market advantage.

To secure a desirable future for the world's soft-shore ecosystems, all of the identified substantial threats need to be managed. Clearly, for any given system, some threats are more worrying than others and it is to these that particular attention must be paid. Nevertheless, while it is often convenient to treat aquatic systems as separate entities, they remain interconnected in many ways; treating each system individually is unrealistic and is in any case impossible. A good example of why this perspective matters can be seen in relationships between sand beaches and seagrass beds (Fig. 23.2). Choices about pollution control

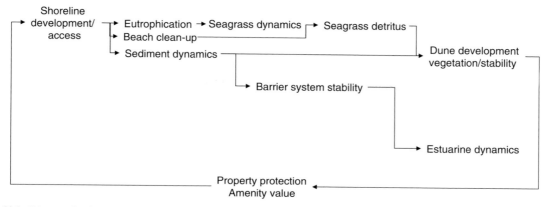

Fig. 23.2. Diagram showing how seagrass and sand beach community dynamics are interrelated and link to coastal development issues.

and beach clean-up have consequences for dune stability, property protection and amenity value. An appreciation of these linkages between systems is vital for sound management. An important role for scientists will be to understand and explain these linkages and illustrate the trade-offs that must be made with any management decision.

Vast marine ecosystems

The three vast marine ecosystems demonstrate that even the remotest, most pristine and most extensive ecosystems (e.g. the Antarctic, the deep-sea floor) bear human impacts. The magnitude and nature of threats vary between the ecosystems. Thus far, the most heavily altered of these ecosystems include the open ocean and the Arctic. For example, eutrophication and toxic algal blooms already affect the neritic zone (nearshore, over the continental shelf) (Chapter 20), while pollutant loading and global warming are major concerns for the Arctic (Chapter 21). The least impacted ecosystems thus far are the deep-sea floor (Chapter 22) and the Antarctic (Chapter 21).

The vast marine ecosystems also have a number of human impacts in common. Firstly, intensive fishing has fundamentally altered ecosystem structure and function on broad scales. In particular, overfishing has removed top-level predators throughout the marine pelagic and polar ecosystems, and intensive fishing pressure has extended far down the continental slope into the deep sea. The ecological consequences of this overfishing are very poorly understood for a variety of reasons, including management practices focused on single species rather than entire ecosystems, and lack of basic ecological knowledge of targeted species and their habitats (especially for the deep sea). The Antarctic and whole pelagic suffer from the after-effects of intensive whaling and the ongoing impacts of illegal unreported and unregulated (IUU) fishing activity. Human impacts at the deep-sea floor are the most poorly documented, yet past species extinctions resulting from whaling-induced loss of whale-fall habitats appear likely.

Secondly, most of these ecosystems now, or will soon, suffer from multiple anthropogenic stressors, including, besides overfishing, pollutant loading and climate change. The cumulative effects of multiple stressors are not well understood for any marine organism, yet impacts from multiple human activities are imminent for the vast marine ecosystems. Examples of non-linear responses and regime shifts in response to environmental changes include the enduring rarity of cod in the northwest Atlantic a full decade after the removal of major fishing pressure; the shift to a jellyfish food web in the Black Sea probably driven by eutrophication; and the interactions between global warming, sea-ice extent, and the abundance of foundation prey species (e.g. krill) and predators (e.g. polar bears, penguins, seals, whales) in polar regions. Few of the processes underlying these changes are well understood or realistically incorporated into mechanistic models, making their accurate prediction inherently unlikely.

Over the next decades, all the vast marine ecosystems have the potential to be markedly affected by climate change through changes in oceanic food webs and export production, and reductions in polar sea ice. However, ecological models are not up to the task of predicting ecosystem impacts; they lack the non-linear dynamics which appear to be part of system behaviour, and ecological time-series data to test outcomes are in any case almost non-existent.

Change is under way in all the vast marine ecosystems long before the scientific tools exist to predict or manage anthropogenic impacts. By their nature, these ecosystems are easily damaged and are slow to recover. An overarching urgent need is for better scientific understanding of the mechanisms driving ecosystem change (anthropogenic or otherwise), and for predictive models to utilize this improved understanding. The paucity of long time-series data severely hinders the task of placing anthropogenic trends within the context of natural variations. In addition, scientific and funding means need to be found to ascertain effects of multiple stressors on ecosystem structure and function. In some cases, particularly the abyssal deep sea, even the most basic ecosystem information, such as species distributions and community response to disturbance, is largely lacking. Creative well-funded research programmes are essential if the human impacts on all the vast marine ecosystems are to be adequately predicted and managed.

However, because of the complex nature of human impacts and marine ecosystem responses, not even the most productive research efforts will allow the complete prediction of all impending ecosystem changes in time to allow society to take corrective action. Because these changes can be large in scale and swift, and may be irreversible, precautionary environmental protection of the vast marine ecosystems must apply if these tracts of wilderness are to persist as they are. Reasonable protection measures will have to be implemented long before detailed understanding of ecosystem structure and dynamics is at hand. Application of the precautionary principle to international waters, which encompass much of the vast marine ecosystems, is particularly challenging because it requires cooperation from a broad range of countries with diverse interests and maritime cultures. Nonetheless, international precedents do already exist for environmental protection on the high seas (e.g. the International Whaling Commission, the Convention for the Conservation of Antarctic Marine Living Resources). A major goal for marine environmental science by the year 2025 must be to inform governments, international agencies and the general public of the changes occurring in the vast marine ecosystems and the measures required to mitigate them.

TRENDS AND PROSPECTS ACROSS ECOSYSTEMS

This assessment

This book has attempted to gain very broad expert overviews of the present and future states of all the major types of aquatic ecosystems, spanning both terrestrial and oceanic realms. To be comprehensive, this assessment relied on many decisions and it is inevitable that there are many weaknesses, real and potential. Thus the partitioning of the aquatic domain into the 21 ecosystems as designated brings with it some overlaps between systems (e.g. flood plains, rivers and streams). Also, some of the ecosystems are aggregate types which elsewhere have been subdivided (e.g. open ocean; Chapter 1) and the attempt at global overview subsumes much important geographical and other detail. Similar space has been allocated to each ecosystem in the book, yet the length and depth of these assessments might have varied according to considerations of productivity, area or biodiversity. To many scientists, it will be dissatisfying to see the lack of overarching data to verify the trends, but the only means to do this at a global level might have been from the outputs of modelling, whereas existing data and modelling approaches are clearly not yet up to the task, certainly not across all the ecosystems involved. By default, the tendency is to focus on particular experiences, for example case studies, always at the potential expense of not expressing the enormous gaps in knowledge. Although several of the 5th ICEF expert groups (Chapter 1) expressed the huge uncertainties inherent in predictions at a global level even to the 2025 time horizon, there has been a tendency to avoid the expression of uncertainty and ignorance, whereas the IPCC (Intergovernmental Panel on Climate Change) process has shown that to gain credibility, global environmental assessments if anything now need to express uncertainty and highlight what they do not know. In some cases and in some respects, in this assessment, uncertainty was more frequently expressed with respect to ecosystems about which rather a lot is known relative to their area (e.g. coral reefs, kelps), whereas experts on those ecosystems that have been less extensively researched (e.g. open ocean, deep sea) tended to be more focused in their predictions. Two circumstances among others may help explain any such differences. One is the likelihood that a relatively small amount of knowledge will raise far more questions than it can answer, whereas where knowledge is sparse and there are fewer broad experts it may appear easier to make generalizations; this will be especially so where the pattern of study tends to be inherently large-scale such as in oceanography. Another likely explanatory circumstance surrounding the divide between prediction and knowledge is a perceived need to highlight ecosystem jeopardy to garner interest.

The ultimate impacts of the present global environmental assessment of aquatic ecosystems are hard to predict. The salience, legitimacy and utility of the present assessment of environmental concern will surely be quickly superseded, only some elements of its content and structure persisting to future assessments. The book's target and most likely influence lie within the scientific community. However, the present assessment is one step towards the ultimate formulation of policy and appropriate action (Mitchell *et al.* 2006).

Patterns and prognoses for change

Across the 21 ecosystems, a number of patterns are evident in the assessments. For the freshwater and saline-lake systems (notably Chapters 2–7), there is predominant recognition of massive past and likely future changes, and the focus tends to be on restoration, widespread resource management, and conservation oriented largely to what is left that is deemed natural. At the other extreme, the nature and extent of impacts are nowhere as widespread and diverse, and more speculative, as for the vast marine ecosystems (Chapters 20–22), but the extent of these last remaining wilderness areas also drives a focus on strict large-scale conservation. This is not a simple marine/terrestrial divide, because in some of the larger saline and freshwater lakes an element of wilderness value, and sparsity of known substantial impacts, fortunately remains (e.g. Lake Baikal: Chapter 6). Clearly, the consequences of inter-ecosystem connectivity have been experienced to a greater extent on land, but at the ocean margins impacts of human development of wetlands, rivers and lakes are widely visible. Up to the year 2025, a dramatic increase in the amount of reactive nitrogen used by people in the Asian developing countries is expected to exacerbate the nitrogen problem along the river–sea continuum. Overall, the 5th ICEF expert groups felt that while climate change is a major concern in developed countries, due partly to prior modification of natural ecosystems but also capacity to regulate human threats, in the developing regions, human population represents a far greater direct threat to these systems (Fig. 23.3).

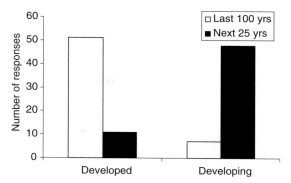

Fig. 23.3. Responses to the question 'Where have threats been worse in the last 100 years and where do you expect them to be worse to the 2025 time horizon: developed or developing regions?'

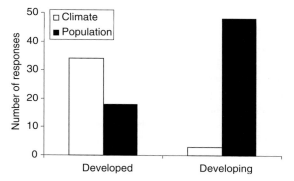

Fig. 23.4. Responses to the question 'Which of climate change or human population growth represents the greater environmental risk to the 2025 time horizon in developing and developed regions?'

Whereas the threats to natural ecosystems have been greatest in the developed countries in the last century, by far the greater concern in this regard in the 5th ICEF expert groups to the 2025 time horizon was with respect to the developing regions of the world (Fig. 23.4). In line with this, it appears that the management effort will need to increase far more in developing regions if aquatic ecosystems, their structure, functions, goods and services are to be adequately conserved over the 2025 time horizon (Fig. 23.5). Projections of 75% of developing countries having below-replacement fertility levels by 2050 (UN [United Nations] 2003) contrast with those for more developed regions where the human population will have been declining since 2030. For developed countries,

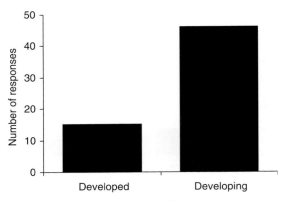

Fig. 23.5. Responses to the question 'For which does management effort need to increase more to the 2025 time horizon: developed or developing regions?'

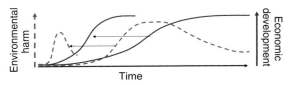

Fig. 23.6. A conceptual model of how economic development and environmental harm develop in relation to each other over time.

adequate resources for sound pollution control and environmental management are available, and there is an increasing social conscience and awareness among the populace concerning environmental issues. In contrast, the populace of the developing world has more immediate issues to contend with, resources for conservation at the disposal of governments are limited, and in broad terms poverty encourages environmental degradation. Experience in the developed countries suggests that initially, economic development is accompanied by environmental harm; however, as resources available for environmental management increase, impacts on ecosystems can be slowed, contained and ultimately reduced (Fig. 23.6). The challenge for regions developing at the present time is to compress this time-line and minimize the harm to ensure that economic growth and security deliver both human and environmental benefits.

The world is experiencing a demographic revolution at the same time as its natural ecosystems are almost universally feeling the effects of the rapid growth of human populations. This is a paradox critical to understanding human–environmental connections. There has been a tendency to view human population growth as an exogenous force that cannot be impeded (Cohen 2003). The experience of the last 40 years however, generates some hope for the future in terms of family planning and further decline in human fertility (Bongaarts *et al.* 1990; Heuveline 1999). In an era when the vast majority of people ever born reach the age of parenting, human fertility is the main determinant of population change. Human fertility has already fallen faster than any demographer had predicted, and although uncertain for the future, it is presently still falling.

It thus makes sense for those who seek the conservation of aquatic ecosystems to support policies that encourage the continued fall of human fertility, so long as this is the result of parents' conscious and free choices. At a 1994 United Nations conference (UN 1994), the governments of the world agreed to spending *c.* US$20 billion annually to make access to reproductive health services, including family planning, universal by the year 2020. However, actual spending is a fraction of that, and roughly stagnant. Ultimately, the future of aquatic ecosystems will be closely linked to the efforts of governments and others to make sure that all women and men have these opportunities, including the means to decide for themselves whether and when to have a child (Cohen 2003).

Anthropogenic environmental change is ultimately a consequence of the material interactions of the human population with its biosphere, of which the aquatic ecosystems are a very prominent part. How that consumption manifests itself, and the ways in which it causes change, are functions not only of the population size and structure but primarily the economic processes and governance regimes that drive and constrain them. It remains to be seen whether the environmental possibilities opened up by declining human fertility will or will not be compensated for by increased per caput demands for goods and services.

Across the range of ecosystems, the mode of proposed conservation action tends to differ, from co-management in some freshwater ecosystems (e.g. small lakes: Chapter 5) and marine-coastal ecosystems (e.g. rocky intertidal: Chapter 14) to more top–down patterns of governance at government and international levels for both large freshwater (e.g. large lakes: Chapter 6, and rivers: Chapter 2) and vast marine ecosystems (e.g. Antarctic: Chapter 21). Co-management, where user groups and government agencies contribute their respective strengths to monitoring and enforcing natural resource management, is especially appropriate where some element of resource ownership can be assigned to local user groups, which is clearly not feasible for uninhabited regions such as the Antarctic and open Pacific.

To be clear, in considering possible ways of reducing environmental harm (Fig. 23.6), the management action needed is not solely for developing countries themselves to take. Given globalized economic interests and impacts, actions such as in relation to anthropogenic climate change and international market demand are at least an equal challenge, and often are the greater responsibility, of the developed than of the developing countries. Co-management and collaborative governance of aquatic ecosystems thus need to build more ways of addressing both the regional and the global scales of threat to the world's aquatic ecosystems.

Because a major set of developed-country skills lies in science and technology, it is tempting to think that there are easy technical fixes to some of the underlying problems, for example through large-scale aquaculture or 'geoengineering' of climate. While massive measures such as sulphur injection into the upper atmosphere are utterly speculative (Crutzen 2006), aquaculture has already overtaken capture fisheries in many regions and is a major provider of nutrition and employment. However, many questions remain over the further expansion of aquaculture in the future, availability of fresh water and space being likely constraints, while intensification is limited by costs and commercial value of the product (FAO [Food and Agriculture Organization of the United Nations] 2006).

Another skills set of developed countries, namely in investment, trade and market management, could also benefit international collaboration over conservation of aquatic ecosystems in the developing countries where the resources are sorely needed. For example, there has been some success with environmental certification, and the large insurance businesses are knowledgeable about assessing future risks of 'natural disasters' and eager to know more. If the world's degraded aquatic ecosystems are to improve in state by the year 2025, it will be through a combination of many processes. Humanity and the natural world need to be brought into equilibrium through downward pressure on both population and resource consumption. To this end, appropriate environmental education and capacity building are needed in all spheres and across all geographic and economic zones. Efforts across the programmes of international and national agencies continue to contribute in both areas; however, there is still much to be done. One way to boost such efforts is through lobbying for research and development projects in developed countries that will allow their scientists to work in developing countries with the local scientists. Unless high-quality scientific personnel can be attracted to get involved in capacity building in developing countries and the relevant personnel have access to the necessary resources, the mid- and long-term impacts will miss the target that the international agenda demands. Realization that what happens in remote areas in developing countries of the world will have an impact on people's lives in developed countries would probably increase the will of institutions to conduct such projects.

Recognition of anthropogenic global warming and of potential positive feedbacks on this through processes in the open oceans and in wetlands have among other things focused attention on the world's aquatic ecosystems and what they deliver to humans at low to negligible cost. It has also encouraged forward thinking across a social and environmental front that is much broader than that of the economic domain, to which most comprehensive planning until recently was largely confined. Instruments such as the Montreal Protocol attest to the fact that strategy to address major anthropogenic environmental threats can be devised at an international level. Water encompasses all life and encapsulates most things that are good and those that are bad in human–environment relations. Management planning and action to address decades of decline, particularly in the freshwater, saline and coastal-marine ecosystems is a daunting task. Worry about 'natural disasters' and debate about the origins and trajectories of future climate change tend to deflect attention away from the main driver, namely the human species. Human behaviour is probably a greater barrier to sustaining the Earth's aquatic ecosystems than the large-scale ignorance about their intricate structures, functions and complex interactions.

References

Abbott, I. A. & Hollenberg, G. J. (1976) *Marine Algae of California*. Stanford, CA, USA: Stanford University Press.

Åberg, E. (1992) Tree colonization of three mires in southern Sweden. In: *Peatland Ecosystems and Man: An Impact Assessment*, ed. O. M. Bragg, P. D. Hulme, H. A. P. Ingram & R. A. Robertson, pp. 268–270. Dundee, UK: Department of Biological Sciences, University of Dundee.

Abernethy, R. K. & Turner, R. E. (1987) US forested wetlands: 1940–1980. *BioScience* **37**: 721–727.

Abramovitz, J. N. (1996) *Imperiled Waters, Impoverished Future: The Decline of Freshwater Ecosystems*, Worldwatch Paper No. 128. Washington, DC, USA: Worldwatch Institute.

Acharya, G. (2000) Approaches to valuing the hidden hydrological services of wetland ecosystems. *Ecological Economics* **35**: 63–74.

ACIA (Arctic Climate Impact Assessment) (2004) *Impacts of a Warming Arctic*. Cambridge, UK: Cambridge University Press.

Ackefors, H. & Enell, M. (1994) The release of nutrients and organic matter from aquaculture systems in Nordic countries. *Journal of Applied Ichthyology* **10**: 225–241.

Adam, P. (1978) Geographical variation in British saltmarsh vegetation. *Journal of Ecology* **66**: 339–366.

Adam, P. (1990) *Saltmarsh Ecology*. Cambridge, UK: Cambridge University Press.

Adam, P. (2000) Morecambe Bay saltmarshes: 25 years of change. In: *British Saltmarshes*, ed. B. R. Sherwood, B. G. Gardiner & T. Harris, pp. 81–107. Tresaith, UK: Forrest Text.

Adam, P. (2002) Saltmarshes in a time of change. *Environmental Conservation* **29**: 39–61.

Adams, N. L. (2001) UV radiation evokes negative phototaxis and covering behavior in the sea urchin *Strongylocentrotus droebachiensis*. *Marine Ecology Progress Series* **213**: 87–95.

Adams, W. M. (2001) *Green Development: Environment and Sustainability in the Third World*, 2nd edn. London, UK: Routledge.

Adey, W. H. & Steneck, R. S. (2001) Thermogeography over time creates biogeographic regions: a temperature/space/time-integrated model and an abundance-weighted test for benthic marine algae. *Journal of Phycology* **37**: 677–698.

Adey, W. H. (1964) The genus *Phymatolithon* in the Gulf of Maine. *Hydrobiologia* **24**: 377–420.

Adis, J. & Junk, W. J. (2002) Terrestrial invertebrates inhabiting lowland river floodplains of Central Amazonia and Central Europe: a review. *Freshwater Biology* **47**: 711–731.

Ærtebjerg, G., Andersen, J. H. & Hansen, O. S. (2003) Nutrients and eutrophication in Danish marine waters: a challenge for science and management. Unpublished report. Roskilde, Denmark: National Environmental Research Institute.

Aerts, R., Wallén, B. & Malmer, N. (1992) Growth-limiting nutrients in *Sphagnum*-dominated bogs subject to low and high atmospheric nitrogen supply. *Journal of Ecology* **80**: 131–140.

Agawin, N. S. R. & Duarte, C. M. (2002) Evidence of direct particle trapping by a tropical seagrass meadow. *Estuaries* **25**: 1205–1209.

Agnew, D. J. & Nicol, S. (1996) Marine disturbances: commercial fishing. In: *Foundations for Research West of the Antarctic Peninsula*, Antarctic Research Series vol. 70, ed. R. M. Ross, E. E. Hofmann & L. B. Quetin, pp. 417–35. Washington, DC, USA: American Geophysical Union.

Aioi, K. & Nakaoka, M. (2003) The seagrasses of Japan. In: *World Atlas of Seagrasses*, ed. E. P. Green & F. T. Short, pp. 201–210. Berkeley, CA, USA: University of California Press.

Airoldi, L. (1998) Roles of disturbance, sediment stress, and substratum retention on spatial dominance in algal turf. *Ecology* **79**: 2759–2770.

Aladin, N. V. & Potts, W. T. W. (1992) Changes in the Aral Sea ecosystems during the period 1960–1990. *Hydrobiologia* **237**(2): 67–79.

Aladin, N. V., Filippov, A. A., Plotnikov, I. S., Orlova, M. I. & Williams, W. D. (1998) Changes in the structure and function of biological communities in the northern part of the Aral Sea (Small Aral Sea), 1984–1994. *International Journal of Salt Lake Research* **7**: 301–343.

Albers, P. H. (2002) Sources, fate, and effects of PAHs in shallow water environments: a review with special reference to small watercraft. *Journal of Coastal Research* Special Issue **37**: 143–150.

Alcamo, J., Döll, P., Kaspar, F. & Siebert, S. (1997) *Global Change and Global Scenarios of Water Use and Availability: An Application of WaterGAP1.0*. Kassel, Germany: Center for Environmental Systems Research (CESR), University of Kassel.

Alcocer, J. & Escobar, E. (1990) The drying up of the Mexican Plateau axalapazcos. *Salinet* **4**: 44–46.

Alcocer, J. & Williams, W. D. (1996) Historical and recent changes in Lake Texcoco, a saline lake in Mexico. *International Journal of Salt Lake Research* **5**: 45–61.

Alcocer, J., Escobar, E. & Lugo, A. (2000) Water use (and abuse) and its effects on the crater-lakes of Valle de Santiago, Mexico. *Lakes and Reservoirs: Research and Management* **5**: 145–149.

Aldhous, P. (2003) The world's forgotten crisis. *Nature* **422**: 251–253.

Alewell, C., Manderscheid, B., Meesenburg, H. & Bittersohl, J. (2000) Is acidification still an ecological threat? *Nature* **407**: 856–857.

Al Hamed, M. I. (1966) Limnological studies on the inland waters of Iraq. *Bulletin of the Iraq Natural History Museum* **3**: 1–22.

Alin, S., Cohen, A., Bills, R., *et al.* (1999) Effects of landscaped disturbance on animal communities in Lake Tanganyika, East Africa. *Conservation Biology* **13**: 1017–1033.

Allan, J. D. & Flecker, A. S. (1993) Biodiversity conservation in running waters. *BioScience* **43**: 32–43.

Allan, J. D., Abell, R., Hogan, Z., *et al.* (2005) Overfishing of inland waters. *BioScience* **55**: 1041–1051.

Allen, J. R., Slinn, D. J., Shammon, T. M., Hartnoll, R. G. & Hawkins, S. J. (1998) Evidence for eutrophication of the Irish Sea over four decades. *Limnology and Oceanography* **43**: 1970–1974.

Allen, J. R. L. (2000) Morphodynamics of Holocene salt marshes: a review sketch from the Atlantic and Southern North Sea coasts of Europe. *Quaternary Science Reviews* **19**: 1155–1231.

Allen, M., Raper, S. & Mitchell, J. (2001) Uncertainty in the IPCC's *Third Assessment Report*. *Science* **293**: 430–433.

Allendorf, F. W. & Lundquist, L. L. (2003) Introduction: population biology, evolution, and control of invasive species. *Conservation Biology* **17**: 24–30.

Allison, G. W., Lubchenco, J. & Carr, M. H. (1998) Marine reserves are necessary but not sufficient for marine conservation. *Ecological Applications* **8**: 79–92.

Alm, J., Sculman, L., Walden, J., *et al.* (1999) Carbon balance of a boreal bog during a year with an exceptionally dry summer. *Ecology* **80**: 161–174.

Almendinger, J. E. & Leete, J. H. (1998a) Peat characteristics and groundwater chemistry of calcareous fens in the Minnesota River Basin, USA. *Biogeochemistry* **43**: 17–41.

Almendinger, J. E. & Leete, J. H. (1998b) Regional and local hydrogeology of calcareous fens in the Minnesota River Basin, USA. *Wetlands* **18**: 184–202.

Alongi, D. M. (1998) *Coastal Ecosystem Processes*. Boca Raton, FL, USA: CRC Press.

Alongi, D. M. (2002) Present state and future of the world's mangrove forests. *Environmental Conservation* **29**: 331–349.

Alroy, J. (2001) A multispecies overkill simulation of the end-Pleistocene megafaunal mass extinction. *Science* **292**: 1893–1896.

AMAP (1998) *AMAP Assessment Report: Arctic Pollution Issues*. Oslo, Norway: Arctic Monitoring and Assessment Programme.

Ambroggi, R. P. (1977) Underground reservoirs to control the water cycle. *Scientific American* **236**: 21–28.

Ambrose, R. F. (2000) Wetland mitigation in the United States: assessing the success of mitigation policies. *Wetlands* **19**: 1–27.

Amon, J. P., Thompson, C. A., Carpenter, Q. J. & Miner, J. (2002) Temperate zone fens of the glaciated midwestern USA. *Wetlands* **22**: 301–317.

Amoros, C. & Petts, G. E., eds. (1993) *Hydrosystèmes Fluviaux*. Paris, France: Masson.

Amsberry, L., Baker, M. A., Ewanchuk, P. J. & Bertness, M. D. (2000) Clonal integration and the expansion of *Phragmites australis*. *Ecological Applications* **10**: 1110–1118.

Amthor, J. S. (1995) Terrestrial higher-plant response to increasing atmospheric (CO_2) in relation to the global carbon cycle. *Global Change Biology* **1**: 243–274.

Andersen, U. V. (1995a) Resistance of Danish coastal vegetation types to human trampling. *Biological Conservation* **71**: 223–230.

Andersen, U. V. (1995b) Invasive aliens: a threat to the Danish coastal vegetation? In: *Directions in European Coastal Management*, ed. M. G. Healy & P. J. Doody, pp. 335–344. Cardigan, UK: Samara Publishing Ltd.

Anderson, D. M., Glibert, P. M. & Burkholder, J. M. (2002*a*) Harmful algal blooms and eutrophication: nutrient sources, composition, and consequences. *Estuaries* **25**: 704–726.

Anderson, D. M., Overpeck, J. T. & Gupta, A. K. (2002*b*) Increase in the Asian Southwest monsoon during the past four centuries. *Science* **297**: 596–599.

Anderson, J. A. R. (1983) The tropical peatswamps of western Malesia. In: *Mires: Swamp, Bog, Fen and Moor, Regional Studies, Ecosystems of the World 4B*, ed. A. J. P. Gore, pp. 181–199. Amsterdam, the Netherlands: Elsevier.

Anderson, O. F. & Clark, M. R. (2003) Analysis of bycatch in the fishery for orange roughy, *Hoplostethus atlanticus*, on the South Tasman Rise. *Marine Freshwater Research* **54**: 643–652.

Anderson, R. J., Carick, P., Levitt, G. J. & Share, A. (1997) Holdfasts of adult kelp *Ecklonia maxima* provide refuges from grazing for recruitment of juvenile kelps. *Marine Ecology Progress Series* **159**: 265–273.

Anderson, S. & Moss, B. (1993) How wetland habitats are perceived by children: consequences for children's education and wetland conservation. *International Journal for Science Education* **15**: 473–485.

Anderson, T. W. (1994) Role of macroalgal structure in the distribution and abundance of a temperate reef fish. *Marine Ecology Progress Series* **113**: 279–290.

Andersson, F. & Olsson, B., eds. (1985) Lake Gårdsjön: an acid forest lake and its catchment. *Ecological Bulletins (Stockholm)* Special Issue **37**.

Andersson, G., Berggren, H., Cronberg, G. & Gelin, C. (1978) Effects of planktivorous and benthivorous fish on organisms and water chemistry in eutrophic lakes. *Hydrobiologia* **59**: 9–15.

Andréasson-Gren, I.-M. & Groth, K.-H. (1995) Economic evaluation of Danube floodplains. Unpublished report. Gland, Switzerland: WWF International.

Andrew, N. L. (1993) Spatial heterogeneity, sea urchin grazing, and habitat structure on reefs in temperate Australia. *Ecology* **74**: 292–302.

Andrew, N. L. (1994) Survival of kelp adjacent to areas grazed by sea urchins in New South Wales, Australia. *Australian Journal of Ecology* **19**: 466–472.

Andrew, N. L. & Choat, J. H. (1982) The influence of predation and non-specific adults on the abundance of juvenile *Evechinus chloroticus* (Echinoidea: Echinodermata). *Oecologia* **54**: 80–87.

Andrew, N. L. & Jones, G. P. (1990) Patch formation by herbivorous fish in a temperate kelp forest. *Oecologia* **85**: 57–68.

Andrew, N. L. & O'Neill, A. L. (2000) Large-scale patterns in habitat structure on subtidal rocky reefs in New South Wales. *Marine and Freshwater Research* **51**: 255–263.

Andrew, N. L. & Underwood, A. J. (1993) Density-dependent foraging in the sea urchin *Centrostephanus rodgersii* on shallow subtidal reefs in New South Wales, Australia. *Marine Ecology Progress Series* **99**: 89–98.

Angel, A. & Ojeda, F. P. (2001) Structure and trophic organization of subtidal fish assemblages on the northern Chilean coast: the effect of habitat complexity. *Marine Ecology Progress Series* **217**: 81–91.

Angus, S. & Elliott, M. M. (1992) Erosion in Scottish machair with particular reference to the Outer Hebrides. In: *Coastal Dunes: Geomorphology, Ecology and Management for Conservation*, ed. R. W. G. Carter, T. G. F. Curtis & M. J. Sheehy-Skeffington, pp. 93–112. Rotterdam, the Netherlands: A. A. Balkema.

Annadotter, H., Cronberg, G., Aagren, R., *et al.* (1999) Multiple techniques for lake restoration. *Hydrobiologia* **395/396**: 77–85.

Anon. (1999) *Wetlands: A Source of Life, Conclusions of the 2nd International Conference on Wetlands and Development*, 10–14 November 1998, Dakar, Senegal. Wageningen, the Netherlands: Wetlands International; Gland, Switzerland: IUCN; and Dakar, Senegal: Ministry of Environment and Nature Protection.

Anon. (2003) Indien: Ende aller Sorgen. *Der Spiegel* **40**: 116–117 (in German).

Anon. (2004) A remedy for financial turbulence? *The Economist* **371**: 72.

Antila, C. K., Daehler, C. C., Rank, N. E. & Strong, D. R. (1998) Greater male fitness of a rare invader (*Spartina alterniflora* Poaceae) threatens a common native (*Spartina foliosa*) with hybridization. *American Journal of Botany* **85**: 1597–1601.

Antila, C. K., King, R. A., Ferris, C., Ayres, D. R. & Strong, D. R. (2000) Reciprocal hybrid formation of *Spartina* in San Francisco Bay. *Molecular Ecology* **9**: 765–770.

Applegate, V. & Moffett, J. (1955) The sea lamprey. *Scientific American* **192**: 36–41.

Archibold, O. W. (1995) *Ecology of World Vegetation*. London, UK: Chapman & Hall.

Armentano, T. V. & Menges, E. S. (1986) Patterns of change in the carbon balance of organic soil-wet lands of the temperate zone. *Journal of Ecology* **74**: 755–774.

Arntz, W. E., Valdivia, E. & Zeballos, J. (1988) Impact of El Niño 1982–83 on the commercially exploited invertebrates (Mariscos) of the Peruvian shore. *Meeresforschung* 32: 3–22.

Aronson, R. B. & Precht, W. F. (2001) Evolutionary paleoecology of Caribbean coral reefs. In: *Evolutionary Paleoecology: The Ecological Context of Macroevolutionary Change*, ed. W. D. Allmon & D. J. Bottjer, pp. 171–233. New York, USA: Columbia University Press.

Arp, W. J. & Drake, B. G. (1991) Increased photosynthetic capacity of *Scirpus olneyi* after 4 years of exposure to elevated CO_2. *Plant, Cell and Environment* 14: 1003–1006.

Arp, W. J., Drake, B. G., Pockman, W. T., Curtis, P. S. & Whigham, D. F. (1993) Interactions between C_3 and C_4 salt marsh plant species during four years of exposure to elevated atmospheric CO_2. *Vegetatio* 104/105: 133–143.

Arrigo, K. R., Worthen, D., Schnell, A. & Lizotte, M. P. (1998) Primary production in Southern Ocean waters. *Journal of Geophysical Research – Oceans* 103: 15587–15600.

Arrow, K., Bolin, B., Costanza, R., *et al.* (1995) Economic growth, carrying capacity and the environment. *Science* 268: 520–521.

Arthington, A. H. & Pusey, P. J. (2003) Flow restoration and protection in Australian rivers. *River Research and Applications* 19: 377–395.

Aselmann, I. & Crutzen, P. J. (1989) Global distribution of natural wetlands and rice paddies, their net primary productivity, seasonality and possible methane emissions. *Journal of Atmospheric Chemistry* 8: 307–358.

Ashjian, C. J., Smith, S. L. & Lane, P. V. Z. (1995) The northeast water polynya during summer 1992: distribution and aspects of secondary production of copepods. *Journal of Geophysical Research – Oceans* 100: 4371–4388.

Ashton, E. D., Macintosh, J. & Hogarth. P. (2003) A baseline study of the diversity and community ecology of crab and molluscan macrofauna in the Sematan mangrove forest, Sarawak, Malaysia. *Journal of Tropical Ecology* 19: 127–142.

Ashworth, W. (1987) *The Late, Great Lakes: An Environmental History*. Detroit, MI, USA: Wayne State University Press.

Asiaweek (2000) The bottom line. *Asiaweek* 24(45). Available online at www.asiaweek.com/asiaweek/magazine/2000/1117/bottomline.html

Aubrey, D. G. (1993) Coastal erosion's influencing factors include development, dams, wells, and climate change. *Oceanus* 36: 5–9.

Auster, P. J. & Langton, R. W. (1999) The effects of fishing on fish habitat. *American Fisheries Society Symposium* 22: 150–187.

Awosika, L. F., Ibe, A. C. & Ibe, C. E. (1993) Anthropogenic activities affecting sediment load balance along the west African coastline. In: *Coastlines of Western Africa*, ed. L. F. Awosika, A. C. Ibe & P. Shroader, pp. 26–39. New York, USA: American Society of Civil Engineers.

Ayres, D. R., Garcia-Rossi, D., Davis, H. G. & Strong, D. R. (1999) Extent and degree of hybridization between exotic (*Spartina alterniflora*) and native (*S. foliosa*) cordgrass (Poaceae) in California, USA determined by random amplified polymorphic DNA (RAPDs). *Molecular Ecology* 8: 1179–1186.

Ayres, D. R., Strong, D. R. & Baye, P. (2003) *Spartina foliosa* (Poaceae): a common species on the road to rarity? *Madroño* 50: 209–213.

Baan, P. J. A. & van Buuren, J. T. (2003) *Testing of Indicators for the Marine and Coastal Environment in Europe*. Copenhagen, Denmark: European Environment Agency.

Babcock, R.C, Kelly, S., Shears, N. T., Walker, J. W. & Willis, T. J. (1999) Changes in community structure in temperate marine reserves. *Marine Ecology Progress Series* 189: 125–134.

Bach, S. S., Borum, J., Fortes, M. D. & Duarte, C. M. (1998) Species composition and plant performance of mixed seagrass beds along a siltation gradient at Cape Bolinao, the Philippines. *Marine Ecology Progress Series* 174: 247–256.

Baden, S. P. & Pihl, L. (1990) Effects of oxygen depletion on the ecology, blood physiology and fishery of the Norway lobster, *Nephrops norvegicus*. *Marine Ecology Progress Series* 67: 151–155.

Baer, K. E. & Pringle C. M. (2000) Special problems of urban river conservation: the encroaching megapolis. In: *Global Perspectives in River Conservation*, ed. P. J. Boon, B. R. Davies & G. E. Petts, pp. 385–402. Chichester, UK: John Wiley.

Baeteman, C. (1994) Subsidence in coastal lowlands due to groundwater withdrawal: the geological approach. *Journal of Coastal Research* Special Issue 37: 61–75.

Bailey, R. G. (1998) *Ecoregions: The Ecosystem Geography of the Oceans and Continents*. New York, USA: Springer-Verlag.

Baird, R. C., Thompson, N. P., Hopkins, T. L. & Weiss, W. R. (1975) Chlorinated hydrocarbons in mesopelagic fishes of the eastern Gulf of Mexico. *Bulletin of Marine Science* 25: 473–481.

Baisre, J. (2000) *Chronicles of Cuban Marine Fisheries (1935–1995): Trend Analysis and Fisheries Potential*, FAO Fisheries Technical Paper 394. Rome, Italy: Food and Agriculture Organization of the United Nations.

Baker, A. C., Starger, C. J., McClanahan, T. R. & Glynn, P. W. (2004) Corals' adaptive response to climate change. *Nature* **430**: 741.

Bakker, J. B. (1989) *Nature Management by Grazing and Cutting: On the Ecological Significance of Grazing and Cutting – Regimes Applied to Restore Former Species-Rich Grassland Communities in the Netherlands*. Dordrecht, the Netherlands: Kluwer.

Bakker, S. A., Van den Berg, N. J. & Speleers, B. P. (1994) Vegetation transitions of floating wetlands in a complex of turbaries between 1937 and 1989 as determined from aerial photographs with GIS. *Vegetatio* **114**: 161–167.

Baldwin, N., Saalfeld, R., Ross, M. & Buettner, H. (1979) *Commercial Fish Production in the Great Lakes 1867–1977*, Technical Report No. 3. Ann Arbor, MI, USA: Great Lakes Fishery Commission.

Balgos, M. C., Ricci, N., Walker, L., *et al.* (2005) Draft: Compilation of summaries of national and regional ocean policies. The Nippon Foundation Research Task Force on National Ocean Policies, Gerard J. Mangone Center for Marine Policy, University of Delaware [www document]. URL http://www.globaloceans.org/tops2005/pdf/OceanPolicySummaries. pdf#search=%22%22national%20ocean%20policy%22%22

Ball, M. C., Cochrane, M. J. & Rawson, H. M. (1997) Growth and water use of the mangroves *Rhizophora apiculata* and *R. stylosa* in response to salinity and humidity under ambient and elevated concentrations of atmospheric CO_2. *Plant Cell and Environment* **20**: 1158–1166.

Balls, P. W., Macdonald, A., Pugh, K. & Edwards, A. C. (1995) Long-term nutrient enrichment of an estuarine system: Ythan, Scotland (1958–1993). *Environmental Pollution* **90**: 311–321.

Baltuck, M., Dickey, J., Dixon, T. & Harrison, G. A. (1996) New approaches raise questions about future sea-level change. *Eos* **77**: 385–388.

Bandeira, S. O. & Gell, F. (2003) The seagrasses of Mozambique and Southeastern Africa. In: *World Atlas of Seagrasses*, ed. E. P. Green & F. T. Short, pp. 105–112. Berkeley, CA, USA: University of California Press.

Barber, K. E. (1981) *Peat Stratigraphy and Climate Change*. Rotterdam, the Netherlands: A. A. Balkema.

Barber, R. T., Vijayakumar, A. & Cross, F. A. (1972) Mercury concentrations in recent and ninety-year-old benthopelagic fish. *Science* **178**: 636–639.

Barbier, E. B. (1997) The economic determinants of land degradation in developing countries. *Philosophical Transactions of the Royal Society of London B* **352**: 891–899.

Barbier, E. B. (2000) The economic linkages between rural poverty and land degradation: some evidence from Africa. *Agriculture, Ecosystems and Environment* **82**: 355–370.

Barbier, E. B. & Thompson, J. R. (1998) The value of water: floodplain versus large-scale irrigation benefits in Northern Nigeria. *Ambio* **27**: 434–440.

Barbraud, C. & Weimeskirch, H. (2001) Emperor penguins and climate change. *Nature* **411**: 183–186.

Barkai, A. & McQuaid, C. (1988) Predator–prey role reversal in a marine benthic ecosystem. *Science* **242**: 4875.

Barlow, M. & Clarke, T. (2002) *Blue Gold: The Battle Against Corporate Theft of the World's Water*. Toronto, Ontario Canada: Stoddart.

Barnes, R. S. & Hughes, R. N. (1988) *An Introduction to Marine Ecology*, 2nd edn. Oxford, UK: Blackwell, Scientific Publications.

Barnett, P. R. O. (1971) Some changes in intertidal sand communities due to thermal pollution. *Proceedings of the Royal Society of London B* **177**: 353–364.

Barnett, P. R. O. & Hardy, B. L. S. (1969) The effects of temperature on the benthos near the Hunterston generating station, Scotland. *Chesapeake Science* **16**: 255–256.

Barnett, T. P., Pierce, D. W. & Schnur, R. (2001) Detection of anthropogenic climate change in the world's oceans. *Science* **292**: 270–274.

Barraclough, S. & Finger-Stich, A. (1996) *Social Implications of Commercial Shrimp Farming in Asia*. Geneva, Switzerland: United Nations Research Institute for Social Development.

Barrales, H. L. & Lobban, C. (1975) The comparative ecology of *Macrocystis pyrifera* with emphasis on the forests of Chubut, Argentina. *Ecology* **63**: 657–677.

Bart, D. & Hartman, J. M. (2003) The role of large rhizome dispersal and low salinity windows in the establishment of common reed, *Phragmites australis*, in salt marshes: new links to human activities. *Estuaries* **26**: 433–436.

Bartley, D. M. (2000) International mechanisms for the conservation and sustainable use of wetlands. In: *Biodiversity in Wetlands: Assessment, Function and Conservation*, ed. B. Gopal, W. J. Junk & J. A. Davis, pp. 331–345. Leiden, the Netherlands: Backhuys.

Bascand, L. D. (1970) The roles of *Spartina* species in New Zealand. *Proceedings of the New Zealand Ecological Society* **17**: 33–40.

Baumgartner, T. R., Soutar, A. & Ferreira-Bartrina, V. (1992) Reconstruction of the history of Pacific sardine and northern anchovy populations over the past two millennia from sediments of the Santa Barbara basin, California. *CalCOFI Report* **33**: 24–40.

Bax, N. J. (2000) *Stock Assessment Report 2000: Orange Roughy* (Hoplostethus atlanticus). Hobart, Tasmania, Australia: South East Fishery Stock Assessment Group.

Baye, P. (1990) Ecological history of an artificial foredune ridge on a northeastern barrier spit. In: *Proceedings of the Symposium on Coastal Sand Dunes*, ed. R. G. D. Davidson-Arnott, pp. 389–403. Ottawa, Ontario, Canada: National Research Council Canada.

Bayley, P. B. (1995) Understanding large river–floodplain ecosystems. *BioScience* **45**: 143–158.

Bayliss, B., Brenman, K., Elliot, I., *et al.* (1997) *Vulnerability Assessment of Predicted Climate Change and Sea Level Rise in the Alligator Rivers Region, Northern Territory Australia*, Supervising Scientist Report No. 123. Canberra, ACT, Australia: Supervising Scientist.

Bayly, I. A. E. (1967) The general biological classification of aquatic environments with special reference to those in Australia. In: *Australian Inland Waters and their Fauna: Eleven Studies*, ed. A. H. Weatherley, pp. 78–104. Canberra, ACT, Australia: Australian National University Press.

Beamish, R. J. (1993) Climate and exceptional fish production off the west coast of North America. *Canadian Journal of Fisheries and Aquatic Sciences* **59**: 2270–2291.

Beardall, J., Beer, S. & Raven, J. A. (1998) Biodiversity of marine plants in an era of climate change: some predictions based on physiological performance. *Botanica Marina* **41**: 113–123.

Bebianno, M. J. & Langston, W. J. (1991) Metallothionein induction in *Mytilus edulis* exposed to cadmium. *Marine Biology* **108**: 91–96.

Bebianno, M. J. & Langston, W. J. (1992) Cadmium induction of metallothionein synthesis in *Mytilus galloprovincialis*. *Comparative Biochemistry and Physiology C* **103**: 79–85.

Bebianno, M. J. & Langston, W. J. (1995) Induction of metallothionein synthesis in the gill and kidney of *Littorina littorea* exposed to cadmium. *Journal of the Marine Biological Association of the United Kingdom* **75**: 173–186.

Bebianno, M. J., Nott, J. A. & Langston, W. J. (1993) Cadmium metabolism in the clam *Ruditapes decussata*: the role of metallothioneins. *Aquatic Toxicology* **27**: 315–333.

Beck, B. F. & Herring J. G. (2002) Geotechnical and environmental applications of karst geology and hydrology. In: *Proceedings of the 8th Interdisciplinary Conference on Sinkholes and the Engineering and Environmental Impacts of Karst*, ed. B. F. Beck & J. G. Herring, pp. 83–88. Rotterdam, the Netherlands: A. A. Balkema.

Beck, M. W., Heck Jr, K. L., Able, K. W., *et al.* (2003) The role of nearshore ecosystems as fish and shellfish nurseries. *Issues in Ecology* **11**: 1–12.

Beck, W. (1998) Warmer and wetter 6000 years ago? *Science* **279**: 1003–1004.

Becker, A. & Grünewald, U. (2003) Flood risk in Central Europe. *Science* **300**: 1099.

Becker, B. (1997) Sustainability assessment: a review of values, concepts and methodological approaches. *The World Bank – Issues in Agriculture* **10**: 1–63.

Bedford, B. L. (1996) The need to define hydrologic equivalence at the landscape scale for freshwater wetland mitigation. *Ecological Applications* **6**: 57–68.

Bedford, B. L. (1999) Cumulative effects on wetland landscapes: links to wetland restoration in the United States and southern Canada. *Wetlands* **19**: 775–788.

Bedford, B. L. & Godwin, K. S. (2003) Fens of the United States: distribution, characteristics, and scientific connection versus legal isolation. *Wetlands* **23**: 608–629.

Bedford, B. L. & Preston, E. M. (1988) Developing the scientific basis for assessing cumulative effects of wetland loss and degradation on landscape functions: status, perspectives, and prospects. *Environmental Management* **12**: 751–771.

Bedford, B. L., Leopold, D. J. & Gibbs, J. (2001) Wetland ecosystems. In: *Encyclopedia of Biodiversity*, vol. 5, ed. S. A. Levin, pp. 781–804. New York, USA: Academic Press.

Beeftink, W. G. (1966) Vegetation and habitat of the salt marshes and beach plains in the south-western part of the Netherlands. *Wentia* **15**: 83–108.

Beer, S. & Koch, E. (1996) Photosynthesis of marine macroalgae and seagrass in globally changing CO_2 environments. *Marine Ecology Progress Series* **141**: 199–204.

Beeton, A. (1965) Eutrophication of the St Lawrence Great Lakes. *Limnology and Oceanography* **10**: 240–254.

Beeton, A. (1969) Changes in the environment and biota of the Great Lakes. In: *Eutrophication: Causes, Consequences, Correctives Symposium Proceedings*, pp. 150–187. Washington, DC, USA: National Academy of Sciences.

Beeton, A. (1984) The world's Great Lakes. *Journal of Great Lakes Research* **10**: 106–13.

Beeton, A. (2002) Large freshwater lakes: present state, trends, and future. *Environmental Conservation* **29**: 21–38.

Beeton, A. & Edmondson, W. (1972) The eutrophication problem. *Journal of the Fisheries Research Board of Canada* **29**: 673–682.

Beeton, A. & Hageman, J. (1994) Impact of *Dreissena polymorpha* on the zooplankton community of western Lake Erie. *Verhandlungen Internationale Vereinigung für Limnologie* **25**: 2349 (abstract).

Beeton, A., Sellinger, C. & Reid, D. (1999) An introduction to the Laurentian Great Lakes ecosystem. In: *Great Lakes Fisheries Policy and Management*, ed. W. W. Taylor & C. P. Ferreri, pp. 3–54. East Lansing, MI, USA: Michigan State University Press.

Bellamy, D. J. & Rieley, J. (1967) Ecological statistics of a miniature bog. *Oikos* **18**: 33–40.

Bellan, G. & Bellan-Santini, D. (2001) A review of littoral tourism, sport and leisure activities: consequences on marine flora and fauna. *Aquatic Conservation: Marine and Freshwater Ecosystems* **11**: 325–333.

Beltman, B., Van den Broek, T. & Bloemen, S. (1995) Restoration of acidified rich-fen ecosystems in the Vechtplassen area: successes and failures. In: *Restoration of Temperate Wetlands*, ed. B. D. Wheeler, S. C. Shaw, W. J. Fojt & R. A. Robertson, pp. 273–286. Chichester, UK: John Wiley.

Beltman, B., Van den Broek, T., Van Maanen, K. & Vaneveld, K. (1996) Measures to develop a rich-fen wetland landscape with a full range of successional stages. *Ecological Engineering* **7**: 299–314.

Belyea, L. R. & Clymo, R. S. (2001) Feedback control of the rate of peat formation. *Proceedings of the Royal Society of London B* **268**: 1315–1321.

Belyea, L. R. & Malmer, N. (2004) Carbon sequestration in peatland: patterns and mechanisms of response to climate change. *Global Change Biology* **10**: 1043–1052.

Benayas, J. M. R., Colomer, M. G. S. & Levassor, C. (1999) Effects of area, environmental status and environmental variation on species richness per unit area in Mediterranean wetlands. *Journal of Vegetation Science* **10**: 275–80.

Benda, L. E., Poff, N. L., Tague, C., *et al.* (2002) How to avoid train wrecks when using science in environmental problem solving. *BioScience* **52**: 1127–1136.

Benke, A. C., Chaubey, I., Ward, M. & Dunn, E. L. (2000) Flood pulse dynamics of an unregulated river floodplain in the Southeastern US coastal plain. *Ecology* **81**: 2730–2741.

Bennett, E. M., Carpenter, S. R., Peterson, G. D., *et al.* (2003) Why global scenarios need ecology. *Frontiers in Ecology and Environment* **1**: 322–329.

Benson, L., Kashgarian, M., Rye, R., *et al.* (2002) Holocene multidecadal and multicentennial droughts affecting Northern California and Nevada. *Quaternary Science Reviews* **21**: 659–682.

Berdnikov, S. V., Selyutin, V. V., Vasilchenko, V. V. & Caddy, J. F. (1999) Trophodynamic model of the Black and Azov Sea pelagic ecosystem: consequences of the comb jelly, *Mnemiopsis leidyi*, invasion. *Fisheries Research* **42**: 261–289.

Berendonk, T. U. & Bonsall, M. B. (2002) The phantom midge and a comparison of metapopulation structures. *Ecology* **83**: 116–128.

Berendse, F., van Breemen, N., Rydin, H., *et al.* (2001) Raised atmospheric CO_2 levels and increased N deposition cause shifts in plant species composition and production in *Sphagnum* bogs. *Global Change Biology* **7**: 591–598.

Bergkamp, G., Pirot, J.-Y. & Hostetter, S., eds. (2000) *Integrated Management of Wetlands and Water Resources*. Proceedings of the 2nd International Wetlands Conference, Dakar, 1998. Gland, Switzerland: Ramsar.

Berglund, B. E., Malmer, N. & Persson, T. (1991) Landscape-ecological aspects of long-term changes in the Ystad area. *Ecological Bulletins* **41**: 405–424.

Bergstad, O. A. (1990) Distribution, population structure, growth and reproduction of the roundnose grenadier (*Coryphaenoides rupestris*) (Pisces: Macrouridae) in the deep waters of the Skagerrak. *Marine Biology* **107**: 25–39.

Berkes, F. (1987) The common property resource problem and the fisheries of Barbados and Jamaica. *Environmental Management* **11**: 225–235.

Berkes, F., Hughes, T. P., Steneck, R. S., *et al.* (2006) Globalization, roving bandits, and marine resources. *Science* **311**: 1557–1558.

Berkes, F., Mahon, R., McConney, P., Pollnac, R. & Pomeroy, R. (2001) *Managing Small-Scale Fisheries: Alternative Directions and Methods*. Ottawa, Ontario, Canada: International Development Research Centre.

Berkmüller, K., Mukherjee, S. & Mishra. B. (1990) Grazing and cutting pressures on Ranthambhore National Park, Rajasthan, India. *Environmental Conservation* **17**: 135–140.

Bernstein, B. B. & Jung, N. (1980) Selective pressures and coevolution in a kelp canopy community in Southern California. *Ecological Monographs* **49**: 335–355.

Bernstein, B. B. & Welsford, R. W. (1982) An assessment of feasibility of using high-calcium quicklime as an experimental tool for research into kelp bed/sea urchin ecosystems in Nova Scotia. *Canadian Technical Report of Fisheries and Aquatic Sciences* **968**: 1–51.

Bertness, M. D. (1984) Habitat and community modification by an introduced herbivorous snail. *Ecology* **65**: 370–381.

Bertness, M. D. & Ellison, A. M. (1987) Determinants of pattern in a New England marsh plant community. *Ecological Monographs* **57**: 129–147.

Bertness, M. D., Ewanchuk, P. & Silliman, B. R. (2002) Anthropogenic modification of New England salt marsh landscapes. *Proceedings of the National Academy of Sciences of the USA* **99**: 1395–1398.

Bethoux, J. P., Morin, P., Chaumery, C., *et al.* (1998) Nutrients in the Mediterranean Sea, mass balance and statistical analysis of concentrations with respect to environmental change. *Marine Chemistry* **63**: 155–169.

Bett, B. J. (2001) UK Atlantic Margin Environmental Survey: introduction and overview of bathyal benthic ecology. *Continental Shelf Research* **21**: 917–956.

Betts, R. A. (2000) Offset of the potential carbon sink from boreal forestation by decreases in surface albedo. *Nature* **408**: 187–190.

Beutel, M. W., Horne, A. J., Roth, J. C. & Barratt, N. J. (2001) Limnological effects of anthropogenic desiccation of a large, saline lake, Walker Lake, Nevada. *Hydrobiologia* **466**(1–3): 91–105.

Beyth, M. (2002) The Red Sea and the Mediterranean Dead Sea canals project [www document]. URL http://www.mfa.gov.il/mfa/go.asp?MFAH0mb90.

Biebl, R. & McRoy, C. P. (1971) Plasmatic resistance and rate of respiration and photosynthesis of *Zostera marina* at different salinities and temperatures. *Marine Biology* **8**: 48–56.

Bigot, J. (2002) Les stocks ne diminuent pas, ils se déplacent. *L'Industrie Nouvelle* **2851**: 10.

Bijlsma, L., Ehler, C. N., Klein, R. J. L., *et al.* (1996) Coastal zones and small islands. In: *Climate Change 1995: Impacts, Adaptations and Mitigation of Climate Change*, ed. Intergovernmental Panel on Climate Change, pp. 289–324. Cambridge, UK: Cambridge University Press.

Billett, D. S. M., Bett, B. J., Rice, A. L., *et al.* (2001) Long-term change in the megabenthos of the Porcupine Abyssal Plain (NE Atlantic). *Progress in Oceanography* **50**: 325–348.

Bird, E. C. F. (1985) *Coastline Changes: A Global Review.* Chichester, UK: John Wiley.

Bird, E. C. F. (1990) Artificial beach nourishment on the shores of Port Phillip Bay, Australia. *Journal of Coastal Research* Special Issue **6**: 55–68.

Bird, E. C. F. (1993) *Submerging Coasts: The Effects of a Rising Sea Level on Coastal Environments.* Chichester, UK: John Wiley.

Bird, E. C. F. & Cullen, P. W. (1996) Recreational uses and problems of Port Phillip Bay, Australia. In: *Recreational Uses of Coastal Areas*, ed. P. Fabbri, pp. 39–51. Dordrecht, the Netherlands: Kluwer.

Birkeland, C. (1982) Terrestrial runoff as a cause of outbreaks of *Acanthaster planci* (Echinodermata: Asteroidea). *Marine Biology* **69**: 175–185.

Birkeland, C., Dayton, P. K. & Engstrom, N. A. (1982) A stable system of predation on a holothurian by four asteroids and their top predator. *Australian Museum Memoirs* **16**: 175–189.

Biswas, A. K. (1998) Deafness to global water crisis: causes and risks. *Ambio* **27**: 492–493.

Bjorgo, E., Johannessen, O. M. & Miles, M. W. (1997) Analysis of merged SSMR–SSMI time series of Arctic and Antarctic sea ice parameters 1978–1995. *Geophysical Research Letters* **24**: 413–416.

Bjorndal, K. A. (1980) Nutrition and grazing behaviour of the green turtle *Chelonia mydas*. *Marine Biology* **56**: 147–154.

Blasco, F. (1983) *Mangroves of Gambia and Senegal.* Toulouse, France: University of Toulouse and UN Sahelian Office.

Blasco, F. (1984) Taxonomic considerations of the mangrove species. In: *The Mangrove Ecosystem*, ed. S. C. Snedaker & J. G. Snedaker, pp. 81–90. Paris, France: UNESCO.

Blasco, F., Aizpuru, M. & Gers, C. (2001) Depletion of the mangroves of Continental Asia. *Wetlands Ecology and Management* **9**: 255–266.

Blasco, F., Gauquelin, T., Rasolofoharinoro, M., Aizpuri, M. & Caldairou, V. (1998) Recent advances in mangrove studies using remote sensing data. *Marine and Freshwater Research* **49**: 287–296.

Blossey, B. (2003) A framework for evaluating potential ecological effects of implementing biological control of *Phragmites australis*. *Estuaries* **26**: 607–617.

Blowers, A. & Smith, S. (2003) Introducing environmental issues: the environment of an estuary. In: *Understanding Environmental Issues*, ed. S. Hinchliffe, A. Blowers & J. Freeland, pp. 3–48. Chichester, UK: John Wiley.

Bobbink, R., Ashmore, M., Braun, S., Flückiger, W. & Van den Wyngaert, I. J. J. (2003) Empirical nitrogen critical loads for natural and semi-natural ecosystems: 2002 update. In: *Empirical Critical Loads for Nitrogen*, Environmental Documentation No. 124, ed. B. Acherman & R. Bobbink, pp. 43–170. Berne, Switzerland: Swiss Agency for the Environment, Forests and Landscape.

Bodkin, J. L. (1988) Effects of kelp forest removal on associated fish assemblages in central California. *Journal of Experimental Marine Biology and Ecology* **117**: 227–238.

Boehlert, G. W. (1986) Productivity and population maintenance of seamount resources and future research directions. In: *Environment and Resources of Seamounts in the North Pacific*, NOAA Technical Report, ed. R. Uchida, S. Hayasi & G. W. Boehlert, pp. 95–101. Seattle, WA, USA: National Marine Fisheries Service.

Boehlert, G. W., Yoklavich, M. M. & Chelton, D. B. (1989) Time series growth in the genus *Sebastes* from the northeast Pacific Ocean. *Fishery Bulletin* **87**: 791–806.

Boesch, D. F., Brinsfield, R. B. & Magnien, R. E. (2001) Chesapeake Bay eutrophication: scientific understanding of ecosystem restoration and challenges for agriculture. *Journal of Environmental Quality* **30**: 303–320.

Boesch, D. F., Josselyn, M. N., Mehta, A. J., *et al.* (1994) Scientific assessment of coastal wetland loss, restoration, and management in Louisiana. *Journal of Coastal Research* Special Issue **20**: 1–103.

Bogan, A. E. (1993) Freshwater bivalve extinctions (Mollusca, Unionoida): a search for causes. *American Zoologist* **33**: 599–609.

Bohrer, B., Heidenreich, H., Schimmele, M. & Schultze, M. (1998) Numerical prognosis for salinity profiles of future lakes in the opencast mine of Merseburg-Ost. *International Journal of Salt Lake Research* **7**: 235–260.

Boje, R. & Tomczak, M. (1978) *Upwelling Ecosystems*. New York, USA: Springer-Verlag.

Bokn, T., Berge, J. A., Green, N. & Rygg, B. (1990) Invasion of the planktonic algae *Chrysochromulina polylepis* along south Norway in May–June 1988: acute effects on biological communities along the coast. In: *Eutrophication and Algal Blooms in North Sea Coastal Zones, the Baltic and Adjacent Areas: Prediction and Assessment of Preventative Actions*, Water Pollution Research Report No. 12, ed. C. Lancelot, G. Billen & H. Barth, pp. 183–193. Brussels, Belgium: Commission of the European Communities.

Bokuniewicz, H. C. (1990) Tailoring local responses to rising sea level: a suggestion for Long Island. *Shore and Beach* **58**(3): 22–25.

Bolen, E. G. & Guthery, F. S. (1982) Playas, irrigation, and wildlife in West Texas. *North American Wildlife Conference* **47**: 528–541.

Bologna, P. A. X. & Steneck, R. S. (1993) Kelp beds as habitat for American lobster *Homarus americanus*. *Marine Ecology Progress Series* **100**: 127–134.

Bolton, J. J. (1996) Patterns of species diversity and endemism in comparable temperate brown algal floras. *Hydrobiologia* **326/327**: 173–178.

Bolton, J. J. & Anderson, R. J. (1987) Temperature tolerances of two southern African *Ecklonia* species (Alariaceae: Laminariales) and of hybrids between them. *Marine Biology* **96**: 293–297.

Bonacci, O., Zdenko, T. & Trninic, D. (1992) Effects of dams and reservoirs on the hydrological characteristics of the Lower Drava River. *Regulated Rivers: Research and Management* **7**: 349–357.

Bondesan, M., Castiglioni, G. B., Elmi, C., *et al.* (1995) Coastal areas at risk from storm surges and sea-level rise in northeastern Italy. *Journal of Coastal Research* **11**: 1354–1379.

Bongaarts, J., Mauldin, W. P. & Phillips, J. F. (1990). The demographic impact of family planning programs. *Studies in Family Planning* **21**: 299–310.

Bonner, W. N. (1982) *Seals and Man*. Seattle, WA, USA: University of Washington Press.

Bonsdorff, E., Blomqvist, E. M., Mattila, J. & Norkko, A. (1997) Coastal eutrophication: causes, consequences and perspectives in the archipelago areas of the northern Baltic Sea. *Estuarine, Coastal and Shelf Science* **44**: 63–72.

Boorman, L. A. (1989) The grazing of British sand dune vegetation. *Proceedings of the Royal Society of Edinburgh B* **96**: 75–88.

Boorman, L. A. (2003) *Saltmarsh Review: An Overview of Coastal Saltmarshes, their Dynamic and Sensitivity Characteristics for Conservation and Management*, JNCC Report No. 334. Peterborough, UK: Joint Nature Conservation Council.

Bootsma, H. & Hecky, R. (1993) Conservation of the African Great Lakes: a limnological perspective. *Conservation Biology* **7**: 644–656.

Bootsma, H. & Hecky, R. (2003) A comparative introduction to the biology and limnology of the African Great Lakes. *Journal of Great Lakes Research* **29**(2): 3–19.

Borum, J. (1996) Shallow waters and land/sea boundaries. In: *Eutrophication in Coastal Marine Ecosystems*, ed. K. Richardson & B. B. Jorgensen, pp. 179–203. Washington, DC, USA: American Geophysical Union.

Boström, C., Baden, S. & Krause-Jensen, D. (2003) The seagrasses of Scandinavia and the Baltic Sea. In: *World Atlas of Seagrasses*, ed. E. P. Green & F. T. Short, pp. 35–54. Berkeley, CA, USA: University of California Press.

Botch, M. S. & Masing, V. V. (1983) Mire ecosystems in the USSR. In: *Mires: Swamp, Bog, Fen and Moor, Regional Studies, Ecosystems of the World 4B*, ed. A. J. P. Gore, pp. 95–152. Amsterdam, the Netherlands: Elsevier.

Bothner, M. H., Buchholtz ten Brink, M. & Manheim, F. T. (1998) Metal concentrations in surface sediments of Boston Harbor: changes with time. *Marine Environmental Research* **45**: 127–155.

Bothner, M. H., Takada, H., Knight, I. T., *et al.* (1994) Sewage contamination in sediments beneath a deep-ocean dump site off New York. *Marine Environmental Research* **38**: 43–59.

Botsford, L. W., Castilla, J. C. & Peterson, C. H. (1997) The management of fisheries and marine ecosystems. *Science* **277**: 509–515.

Botton, M. L., Loveland, R. E. & Jacobsen, T. R. (1988) Beach erosion and geochemical factors: influence on spawning success of horseshoe crabs (*Limulus polyphemus*) in Delaware Bay. *Marine Biology* **99**: 325–332.

Botton, M. L., Loveland, R. E. & Jacobsen, T. R. (1994) Site selection by migratory shorebirds in Delaware Bay, and its relationship to beach characteristics and abundance of Horseshoe Crab (*Limulus polyphemus*) eggs. *Auk* **111**: 605–616.

Boudouresque, C. F., Charbonel, E., Meinesz, A., *et al.* (2000) A monitoring network based on the seagrass *Posidonia oceanica* in the northwestern Mediterranean Sea. *Biologia Marina Mediterranea* **7**: 328–331.

Boulé, M. E. (1994) An early history of wetland ecology. In: *Global Wetlands: Old World and New*, ed. W. J. Mitsch, pp. 57–74. Amsterdam, the Netherlands: Elsevier.

Boulton A. J., Sheldon, F., Thoms, M. C. & Stanley, E. H. (2000) Problems and constraints in managing rivers with variable flow regimes. In: *Global Perspectives on River Conservation: Science, Policy and Practice*, ed. P. J. Boon, B. R. Davies & G. E. Petts, pp. 411–426. Chichester, UK: John Wiley.

Bourke, R. H. & Garrett, R. P. (1987) Sea ice thickness distribution in the Arctic Ocean. *Cold Regions Science and Technology* **13**: 259–280.

Bourke, R. H. & McLaren, A. S. (1992) Contour mapping of Arctic Basin ice draft and roughness parameters. *Journal of Geophysical Research* **97**: 17715–17728.

Bourque, B. J. (1995) *Diversity and Complexity in Prehistoric Maritime Societies: A Gulf of Maine Perspective*. New York, USA: Plenum Press.

Bowen-Jones, E. (1998) *A Review of the Commercial Bushmeat Trade, with Emphasis on Central/West Africa and the Great Apes*. Cambridge, UK: Ape Alliance/Flora International.

Bowes, G. (1993) Facing the inevitable: plants and increasing atmospheric CO_2. *Annual Review of Plant Physiology and Plant Molecular Biology* **44**: 309–332.

Bowles, I. (2001) Environmental meltdown in the White house. *Boston Globe* 22 April.

Bowman, M. J. (1995) The Ramsar Convention comes of age. *Netherlands International Law Review* **42**: 1–52.

Bowman, M. (2002) The Ramsar Convention on Wetlands: has it made a difference? In: *Yearbook of International Cooperation on Environment and Development 2002–03*, eds. O. S. Stokke & B. Øystein, pp. 61–68. London, UK: Earthscan Publications.

Bowman, R. S. & Lewis, J. R. (1977) Annual fluctuations in the recruitment of *Patella vulgata* L. *Journal of the Marine Biological Association of the United Kingdom* **57**: 793–815.

Boyd, P. W. & Newton, P. P. (1999) Does pelagic community structure determine downward particulate organic carbon flux in different oceanic provinces? *Deep-Sea Research I* **46**: 63–91.

Boyer, K. E. & Zedler, J. B. (1996) Damage to cordgrass by scale insects in a constructed salt marsh: effects of nitrogen additions. *Estuaries* **19**: 1–12.

Boyer, K. E. & Zedler, J. B. (1998) Effects of nitrogen additions on the vertical structure of a constructed cordgrass marsh. *Ecological Applications* **8**: 692–705.

Boyer, K. E. & Zedler, J. B. (1999) Nitrogen addition could shift species composition in a restored California salt marsh. *Restoration Ecology* **7**: 74–85.

Boyer, K. E., Callaway, J. C. & Zedler, J. B. (2000) Evaluating the progress of restored cordgrass (*Spartina foliosa*) marshes: belowground biomass and tissue N. *Estuaries* **23**: 711–721.

Boynton, W. R., Hagy, J. D., Murray, L., Stokes, C. & Kemp, W. M. (1996) A comparative analysis of eutrophication patterns in a temperate coastal lagoon. *Estuaries* **19**: 408–421.

Bradbury, R. & Seymour, R. (1997) Waiting for COTS. *Proceedings of the 8th International Coral Reef Symposium*, Panama, June 24–29 1996, vol. 1, ed. H. A. Lessios & I. E. Macintyre, pp. 1357–1362. Balboa, Panama: Smithsonian Tropical Research Institute.

Bradbury, R. H., Hammond, L. S., Moran, P. J. & Reichelt, R. E. (1985) Coral reef communities and the crown-of-thorns starfish: evidence for qualitatively stable cycles. *Journal of Theoretical Biology* **113**: 69–80.

Bradley, D. C. & Ormerod, S. J. (2001) Community persistence among stream invertebrates tracks the North Atlantic Oscillation. *Journal of Animal Ecology* **70**: 987–996.

Bradley, D. C. & Ormerod, S. J. (2002) Long-term effects of catchment liming on invertebrates in upland streams. *Freshwater Biology* **47**: 161–171.

Bradley, P. M. & Morris, J. T. (1991) The influence of salinity on the kinetics of NH_4+ uptake in *Spartina alterniflora*. *Oecologia* **85**: 375–380.

Bradley, P. M. & Morris, J. T. (1992) Effect of salinity on the critical nitrogen concentration of *Spartina alterniflora* Loisel. *Aquatic Botany* 43: 149–161.

Bradshaw, C. J. A., Thompson, C. M., Davis, L. S. & Lalas, C. (1999) Pup density related to terrestrial habitat use by New Zealand fur seals. *Canadian Journal of Zoology* 77: 1579–1586.

Bradshaw, M. J. (1977) *Earth, the Living Planet*. Sevenoaks, UK: Hodder & Stoughton.

Branch, G. M. & Moreno, C. A. (1994) Intertidal and subtidal grazers. In *Rocky Shores: Exploitation in Chile and South Africa*, ed. W. Siegfried, pp. 75–100. New York, USA: Springer-Verlag.

Branch, G. M. & Odendaal, F. (2003) Marine protected areas and wave action: impacts on a South African limpet, *Cymbula oculus*. *Biological Conservation* 114: 255–269.

Branch, G. M. & Steffani, C. N. (2004) Can we predict the effects of alien species? A case-history of the invasion of South Africa by *Mytilus galloprovincialis* (Lamarck). *Journal of Experimental Marine Biology and Ecology* 300: 189–215.

Branch, G. M., Eekhout, S. & Bosman, A. L. (1990) Short-term effects of the Orange River floods on the intertidal rocky-shore communities of the open coast. *Transactions of the Royal Society of South Africa* 47: 331–354.

Bravard, J. P. (1987) *Le Rhône du Léman à Lyon*. Lyon, France: La Manufacture.

Bray, R. N. (1981) Influence of water currents and zooplankton densities on daily foraging movements of blacksmith, *Chromis punctipinnis*, a planktivorous reef fish. *Fishery Bulletin of the US* 78: 829–841.

Breen, P. A. & Mann, K. H. (1976) Changing lobster abundance and the destruction of kelp beds by sea urchins. *Marine Biology* 34: 137–142.

Breitburg, D. L., Baxter, J. W., Hatfield, C. A., *et al.* (1998) Understanding effects of multiple stressors: ideas and challenges. In: *Successes, Limitations, and Frontiers in Ecosystem Science*, ed. M. L. Pace & P. M. Groffman, pp. 416–431. New York, USA: Springer-Verlag.

Breitburg, D. L., Sanders, J. G., Gilmour, C. C., *et al.* (1999) Variability in responses to nutrients and trace elements, and transmission of stressor effects through an estuarine food web. *Limnology and Oceanography* 44: 837–863.

Breitfuss, M. J., Connolly, R. M. & Dale, P. E. R. (2003) Mangrove distribution and mosquito control: transport of *Avicennia marina* propagules by mosquito-control runnels in southeast Queensland saltmarshes. *Estuarine, Coastal and Shelf Science* 56: 573–579.

Bren, L. J. (1988) Effects of river regulation on flooding of a riparian red gum forest on the River Murray, Australia. *Regulated Rivers: Research and Management* 2: 65–77.

Bricker, S. B. (1993) The history of Cu, Pb, and Zn inputs to Narragansett Bay, Rhode Island as recorded by saltmarsh sediments. *Estuaries* 16: 589–607.

Bricker, S. B., Clement, C. G., Pirhalla, D. E., Orlando, S. P. & Farrow, D. R. G. (1999) *National Estuarine Eutrophication Assessment: Effects of Nutrient Enrichment in the Nation's Estuaries*. Silver Spring, MD, USA: National Oceanic and Atmospheric Administration, National Ocean Service, Special Projects Office and the National Centers for Coastal Ocean Science.

Bricker-Urso, S., Nixon, S. W., Cochran, J. K., Hirschberg, D. J. & Hunt, C. (1989) Accretion rates and sediment accumulation in Rhode Island salt marshes. *Estuaries* 12: 300–317.

Brierley, A. S. & Thomas, D. N. (2002) Ecology of Southern Ocean pack ice. *Advances in Marine Biology* 43: 171–276.

Brierley, A. S., Fernandes, P. G., Brandon, M. A., *et al.* (2002) Antarctic krill under sea ice: elevated abundance in a narrow band just south of ice edge. *Science* 295: 1890–1892.

Briers, R. A., Cariss, H. M. & Gee, J. H. R. (2002) Dispersal of adult stoneflies (Plecoptera) from upland streams draining catchments with contrasting land-use. *Archiv für Hydrobiologie* 155: 627–644.

Briers, R. A., Gee, J. H. R., Cariss, H. H. & Geoghan, R. (2004) Interpopulation dispersal by adult stoneflies detected by stable isotope enrichment. *Freshwater Biology* 49: 425–431.

Brinson, M. M. (1990) Riverine forests. In: *Forested Wetlands*, ed. A. E. Lugo, M. M. Brinson & S. Brown, pp. 87–141. Amsterdam, the Netherlands: Elsevier.

Brinson, M. M. (1993) *A Hydrogeomorphic Classification for Wetlands*, Technical Report No. WRP-DE-4. Washington, DC, USA: US Army Corps of Engineers. Available online at http://el.erdc.usace.army.mil/wetlands/pdfs/wrpde4.pdf

Brinson, M. M. & Malvárez, A. I. (2002) Temperate fresh-water wetlands: types, status, and threats. *Environmental Conservation* 29: 115–133.

Brito, M. A., Sobrevila, C., Dalponte, J. C. & Borges, G. A. (undated) *Setting Conservation Priorities in the State of Mato Grosso, Brazil*, Relatório ao Centro de Dados para Conservação. Cuiabá, Brasil: Fundação Estadual de Meio Ambiente.

Brock, J. H. (1994) *Tamarix* spp. (salt cedar), and invasive exotic woody plant in arid and semi-arid riparian habitats of western USA. In: *Ecology and Management of Invasive Riverside Plants*, ed. L.C. de Waal, L. E. Child, P. M. Wade & J. H. Brock, pp. 27–44. New York, USA: John Wiley.

Brönmark, C. & Hansson, L.-A. (2002) Environmental issues in lakes and ponds: current state and perspectives. *Environmental Conservation* **29**: 290–306.

Brooks, T. M., Mittermeier, R. A., Mittermeier, C. G., *et al.* (2002) Habitat loss and extinction in the hotspots of biodiversity. *Conservation Biology* **16**: 909–923.

Brough, C. N. & White, K. N. (1990) Localization of metals in the gastropod *Littorina saxatilis* (Prosobranchia, Littorinoidea) from a polluted site. *Acta Zoologica* **71**: 77–88.

Brouns, J. J. (1994) Seagrasses and climate change. In: *Impact of Climate Change on Ecosystems and Species: Marine and Coastal Ecosystems*, ed. J. C. Pernetta, R. Leemnas, D. Elder & S. Humphrey, pp. 59–71. Gland, Switzerland: IUCN.

Broussard, J. (1983) National Audubon Society *v.* Superior Court, 33 Cal. 3d 419, California Supreme Court, CA, USA.

Browman, H. I., Rodriguez, C. A., Beland, F., *et al.* (2000) Impact of ultraviolet radiation on marine crustacean zooplankton and ichthyoplankton: a synthesis of results from the estuary and Gulf of St Lawrence, Canada. *Marine Ecology Progress Series* **199**: 281–311.

Brown, A. C. (1983) The ecophysiology of sandy-beach animals: a partial review. In: *Sandy Beaches as Ecosystems*, ed. A. McLachlan & T. Erasmus, pp. 575–605. The Hague, the Netherlands: Dr W. Junk.

Brown, A. C. (1985) *The Effects of Crude Oil Pollution on Marine Organisms*, Report No. 99. Pretoria, South Africa: South African National Scientific Programmes.

Brown, A. C. (1994) Plutonium and marine life. *Transactions of the Royal Society of South Africa* **49**: 213–224.

Brown, A. C. (1996) Behavioural plasticity as a key factor in the survival and evolution of the macrofauna on exposed sandy beaches. *Revista Chilena de Historia Natural* **69**: 469–474.

Brown, A. C. (2001) Biology of sandy beaches. In: *Encyclopedia of Ocean Sciences*, vol. 5, ed. J. H. Steele, S. A. Thorpe & K. K. Turekian, pp. 2496–2504. London, UK: Academic Press.

Brown, A. C. & McLachlan, A. (1990) *Ecology of Sandy Shores*. Amsterdam, the Netherlands: Elsevier.

Brown, A. C. & McLachlan, A. (2002) Sandy shores ecosystems and the threats facing them: some predictions for the year 2025. *Environmental Conservation* **29**: 62–77.

Brown, A. C., Davies, B. R., Day, J. A. & Gardiner, A. J. C. (1991) Chemical pollution loading of False Bay. *Transactions of the Royal Society of South Africa* **47**: 703–716.

Brown, A. C., McLachlan, A., Kerley, G. I. H. & Lubke, R. A. (2000) Functional ecosystems: sandy beaches and dunes. In: *Summary Marine Biodiversity Status Report for South Africa*, ed. B. D. Durham & J. C. Pauw, pp. 4–5. Pretoria, South Africa: National Research Foundation.

Brown Jr, E., Busiahn, T., Jones, M. & Argyle, R. (1999) Allocating Great Lakes forage bases in response to multiple demand. In: *Great Lakes Fisheries Policy and Management*, ed. W. W. Taylor & C. P. Ferreri, pp. 385–394. East Lansing, MI, USA: Michigan State University Press.

Brown, K. (2002) Water scarcity: forecasting the future with spotty data. *Science* **297**: 926–927.

Brown, L. R. (2003) *Plan B: Rescuing a Planet under Stress and a Civilization in Trouble*. London, UK: W.W. Norton.

Brown, R., Ebener, M. & Gorenflo, T. (1999) Great Lakes commercial fisheries: historical overview and prognosis for the future. In: *Great Lakes Fisheries Policy and Management*, ed. W. W. Taylor & C. Ferreri, pp. 307–354. East Lansing, MI, USA: Michigan State University Press.

Brünig, E. F. (1990) Oligotropic forested wetlands in Borneo. In: *Ecosystems of the World*, vol.15, *Forested Wetlands*, ed. A. E. Lugo, M. Brinson & S. Brown, pp. 299–334. Amsterdam, the Netherlands: Elsevier.

Brunke, M. & Gonser, T. (1997) The ecological significance of exchange processes between rivers and groundwater. *Freshwater Biology* **37**: 1–33.

Brunsden, D. & Moore, R. (1999) Engineering geomorphology on the coast: lessons from west Dorset. *Geomorphology* **31**: 391–409.

Bryan, G. W. & Langston, W. J. (1992) Bioavailability, accumulation, and effects of heavy metals in sediments with special reference to United Kingdom estuaries: a review. *Environmental Pollution* **76**: 89–131.

Bryan, G. W., Gibbs, P. E., Hummerstone, L. G. & Burt, G. R. (1986) The decline of the gastropod *Nucella lapillus* around southwest England: evidence for the effect of tributyl tin from antifouling paints. *Journal of the Marine Biological Association of the United Kingdom* **66**: 611–640.

Bryant, D., Burke, L., McManus, J. & Spalding, M. (1998) *Reefs at Risk*. Washington, DC, USA: World Resources Institute.

Bryant, J. C. & Chabreck, R. H. (1998) Effects of impoundment on vertical accretion of coastal marsh. *Estuaries* **21**: 416–422.

Budd, A. F. (2000) Diversity and extinction in the Cenozoic history of Caribbean reefs. *Coral Reefs* **19**: 25–35.

Budd, J., Beeton, A., Stumpf, R., Culver, D. & Kerfoot, C. (2001) Satellite observations of *Microcystis* blooms in western Lake Erie. *Verhandlungen Internationale Vereinigung für Limnologie* **27**: 3787–3793.

Buddemeier, R. W. & Smith, S. V. (1988) Coral reef growth in an era of rapidly rising sea level: predictions and suggestions for long-term research. *Coral Reefs* **7**: 51–56.

Buijse, A. D., Coops, H., Staras, M., *et al.* (2002) Restoration strategies for river floodplains along large lowland rivers in Europe. *Freshwater Biology* **47**: 889–907.

Bujang, J. S. & Zakeia, M. H. (2003) The seagrasses of Malaysia. In: *World Atlas of Seagrasses*, ed. E. P. Green & F. T. Short, pp. 166–176. Berkeley, CA, USA: University of California Press.

Bulgarian Ministry of Agriculture and Forests (2001) Strategy for the protection and restoration of floodplain forests on the Bulgarian Danube islands. Unpublished report. Vienna, Austria: WWF International Danube Carpathian Programme.

Bulkley, J., Wright, S. & Wright, D. (1984) Preliminary study of the diversion of 283 m^3 s^{-1} (10 000 cfs) from Lake Superior to the Missouri River basin. *Journal of Hydrology* **68**: 461–472.

Bull, D., Kemp, A. E. S. & Weedon, G. P. (2000) A 160-ky-old record of El Niño–Southern Oscillation in marine production and coastal run off from Santa Barbara Basin, California, USA. *Geology* **28**: 1007–1010.

Bumb, B. L. & Bannante, C. A. (1996) *The Role of Fertilizer in Sustaining Food Security and Protecting the Environment to 2020*, IFPRI Discussion Paper No. 17. Washington, DC, USA: World Resources Institute.

Bunn, S. E. & Arthington, A. H. (2002) Basic principles and ecological consequences of altered flow regimes for aquatic biodiversity. *Environmental Management* **30**: 455–467.

Bunn, S. E., Boon, P. I., Brock, M. A. & Schofield, N. J., eds. (1997) *National Wetlands R&D Program Scoping Review*, Special Publication No. 1/97. Canberra, ACT, Australia: Land and Water Resources R&D Corporation.

Bunt, J. S. (1992) Introduction. In: *Tropical Mangrove Ecosystems*, ed. A. I. Robertson & D. M. Alongi, pp. 1–6. Washington, DC, USA: American Geophysical Union.

Burby, R. J. (2002) Flood insurance and floodplain management: the US experience. *Environmental Hazards* **3**: 111–119.

Burd, F. (1992) *Erosion and Vegetation Change on the Salt Marshes of Essex and North Kent Between 1973 and 1978*, Research and Survey in Nature Conservation No. 42. Peterborough, UK: Nature Conservancy Council.

Burdick, D. M. & Konisky, R. A. (2003) Determinants of expansion of *Phragmites australis*, common reed, in natural and impacted coastal marshes. *Estuaries* **26**: 407–416.

Burdick, D. M. & Short, F. T. (1999) The effects of boat docks on eelgrass beds in coastal waters of Massachusetts. *Environmental Management* **23**: 231–240.

Burdick, D. M., Short, F. T. & Wolf, J. (1993) An index to assess and monitor the progression of the wasting disease in eelgrass, *Zostera marina*. *Marine Ecology Progress Series* **94**: 83–90.

Burger, J. (1991) Foraging behaviour and effects of human disturbance on the Piping Plover (*Charadrius melodus*). *Journal of Coastal Research* **7**: 39–52.

Burger, J. & Gochfeld, M. (1991) Human activity influence and diurnal and nocturnal foraging of Sanderlings (*Calidris alba*). *Condor* **93**: 259–265.

Burger, J., Gochfeld, M. & Niles, L. J. (1995) Ecotourism and birds in coastal New Jersey: contrasting responses of birds, tourists and managers. *Environmental Conservation* **22**: 56–65.

Burke, L., Kura, Y., Kassem, K., *et al.* (2001) *Pilot Analysis of Global Ecosystems (PAGE): Coastal Ecosystems*. Washington, DC, USA: World Resources Institute.

Burkett, V. & Kusler, J. (2000) Climate change: potential impacts and interactions in wetlands of the United States. *Journal of the American Water Resources Association* **36**: 313–320.

Burkholder, J. M. (1998) Implications of harmful microalgae and heterotrophic dinoflagellates in management of sustainable marine fisheries. *Ecological Applications* **8** (Suppl.): S37–S62.

Burns, K. A. & Codi, S. (1998) Contrasting impacts of localized versus catastrophic oil spills in mangrove sediments. *Mangroves and Salt Marshes* **2**: 63–74.

Burns, K. A. & Yelle-Simmons, L. (1994) The Galeta oil spill. IV. Relationship between sediment and organism hydrocarbon loads. *Estuarine, Coastal and Shelf Science* **38**: 397–412.

Burns, K. A., Levings, S. & Garrity, S. (1993) How many years before mangrove ecosystems recover from a catastrophic oil spill? *Marine Pollution Bulletin* **26**: 239–248.

Burrows, M. T., Hawkins, S. J. & Southward, A. J. (1992) A comparison of reproduction in co-occurring chthamalid barnacles, *Chthamalus stellatus* (Poli) and *Chthamalus montagui* (Southward). *Journal of Experimental Marine Biology and Ecology* **160**: 229–249.

Burrows, M., Kawai, K. & Hughes, R. (1999) Foraging by mobile predators on a rocky shore: underwater TV observations of movements of blennies *Lipophrys pholis* and crabs *Carcinus maenas*. *Marine Ecology Progress Series* **187**: 237–250.

Busnita, T. (1967) Die Ichthyofauna der Donau. In: *Limnologie der Donau*, ed. R. Liepolt, pp. 198–224. Stuttgart, Germany: Schweizerbartsche Verlagsbuchhandlung.

Bussau, C., Schriever, G. & Thiel, H. (1995) Evaluation of abyssal metazoan meiofauna from a manganese nodule area of the eastern South Pacific. *Vie Milieu* **45**: 39–48.

Bustamante, R. H. & Branch, G. M. (1996a) The dependence of intertidal consumers on kelp-derived organic matter on the west coast of South Africa. *Journal of Experimental Marine Biology Ecology* **196**: 1–28.

Bustamante, R. H. & Branch, G. M. (1996b) Large scale patterns and trophic structure of southern African rocky shores: the roles of geographic variation and wave exposure. *Journal of Biogeography* **23**: 339–351.

Bustamante, R. H., Branch, G. M. & Eekhout, S. (1995a) Maintenance of an exceptional intertidal grazer biomass in South Africa: subsidy by subtidal kelps. *Ecology* **76**: 2314–2329.

Bustamante, R. H., Branch, G. M., Eekhout, S., *et al.* (1995b) Gradients of intertidal productivity around the coast of South Africa and their relationship with consumer biomass. *Oecologia* **102**: 189–201.

Bustnes, J. O., Lonne, O. J., Skjoldal, H. R., *et al.* (1995) Sea ducks as predators on sea urchins in a northern kelp forest. In: *Ecology of Fjords and Coastal Waters*, ed. H. R. Skjoldal, C. Hopkins & K. E. Erikstad, pp. 599–608. Amsterdam, the Netherlands: Elsevier.

Busuioc, A., Chen, D. L. & Hellström, C. (2001) Performance of statistical downscaling models in GCM validation and regional climate change estimates: application for Swedish precipitation. *International Journal of Climatology* **21**: 557–578.

Cabello-Pasini, A., Lara-Turrent, C. & Zimmerman, R. C. (2002) Effects of storms on photosynthesis, carbohydrate content and survival of eelgrass populations from a coastal lagoon and the adjacent open ocean. *Aquatic Botany* **74**: 149–164.

Caddy, J., Csirke, J., Garcia, S. & Grainger, R. (1998) How pervasive is 'Fishing down marine food webs'? *Science* **282**: 183.

CAFF (Arctic Council Working Group on Conservation of Arctic Flora and Fauna). (2001) *Arctic Flora and Fauna: Status and Conservation*. Helsinki, Finland: Edita.

Cahill, T. A., Gill, T. E., Reid, J. S., Gearhart, E. A. & Gillette, D. A. (1996) Saltating particles, playa crusts and dust aerosols at Owens (dry) Lake, California. *Earth Surface Processes and Landforms* **21**: 621–639.

Caillet, G. M., Andrews, A. H., Burton, E. J., *et al.* (2001) Age determination and validation studies of marine fishes: do deep-dwellers live longer? *Experimental Gerontology* **36**: 739–764.

Cairns Jr, J. & Heckman, J. R. (1996) Restoration ecology: the state of an emerging field. *Annual Review of Energy and the Environment* **21**: 167–189.

Caldeira, K. & Wickett, M. E. (2003) Anthropogenic carbon and ocean pH. *Nature* **425**: 365.

Callaway, J. C. & Josselyn, M. N. (1992) The introduction and spread of smooth cordgrass *(Spartina alterniflora)* in South San Francisco Bay. *Estuaries* **15**: 218–226.

Cambell, G. W. & Lee, D. S. (1996) Atmospheric deposition of sulphur and nitrogen species in the UK. *Freshwater Biology* **36**: 151–167.

Cambridge, M. L., Chiffings, A. W., Brittan, C., Moore, L. & McComb, A. J. (1986) The loss of seagrass in Cockburn Sound, Western Australia. II. Possible causes of seagrass decline. *Aquatic Botany* **24**: 269–285.

Campana, S. E., Zwanenburg, K. C. T. & Smith, J. N. (1990) ^{210}Pb/^{226}Ra determination of longevity in red fish. *Canadian Journal of Fisheries and Aquatic Sciences* **47**: 163–165.

Campbell, I. D., Campbell, C., Yu, Z. C., Vitt, D. H. & Apps, M. J. (2000) Millennial-scale rhythms in peatlands in the western interior of Canada and in the global carbon cycle. *Quaternary Research* **54**: 155–158.

Campbell, L., Hecky, R. & Dixon, D. (2003) Review of mercury in Lake Victoria (East Africa): implications for human and ecosystem health. *Journal of Toxicology and Environmental Health B* **6**: 325–356.

Carbiener, R. & Schnitzler, A. (1990) Evolution of major pattern models and processes of alluvial forests of the Rhine in the rift valley (France/Germany). *Vegetatio* **88**: 115–129.

Carbonell, M., Nathai-Gyan, N. & Finlayson, C. M., eds. (2001) *Science and Local Communities: Strengthening Partnerships for Effective Wetland Management*. Memphis, TN, USA: Ducks Unlimited.

Cardinale, B. J., Palmer, M. A. & Collins, S. L. (2002) Species diversity enhances ecosystem functioning through interspecific facilitation. *Nature* **415**: 426–429.

Cardoso, P. G., Pardal, M. A., Raffaelli, D., Baeta, A. & Marques, J. C. (2004) Macroinvertebrate response to different species of macroalgal mats and the role of disturbance history. *Journal of Experimental Marine Biology and Ecology* **308**: 207–220.

Cargill, S. M. & Jefferies, R. L. (1984) Nutrient limitation of primary production in a sub-arctic salt marsh. *Journal of Applied Ecology* **21**: 657–668.

Carlson, D. B., O'Bryan, P. D. & Rey, J. R. (1994) The management of Florida's (USA) salt marsh impoundments for mosquito control and natural resource enhancement. In: *Global Wetlands: Old World and New*, ed. W. J. Mitsch, pp. 805–814. Amsterdam, the Netherlands: Elsevier.

Carlton, J. T. (1989) Man's role in changing the face of the oceans: biological conservation of near-shore environments. *Conservation Biology* **3**: 265–273.

Carlton, J. T. (1999) Molluscan invasions in marine and estuarine communities. *Malacologia* **41**: 439–454.

Carlton, J. T. (2000) Global change and biological invasions in the oceans. In: *Invasive Species in a Changing World*, ed. H. A. Mooney & R. J. Hobbs, pp. 31–53. Washington, DC, USA: Island Press.

Carlton, J. T. & Geller, J. B. (1993) Ecological roulette: the global transport of nonindigenous marine organisms. *Science* **261**: 78–82.

Carpenter, R. C. (1988) Mass-mortality of a Caribbean sea urchin: immediate effects on community metabolism and other herbivores. *Proceedings of the National Academy of Sciences of the USA* **85**: 511–514.

Carpenter, S. R., Caraco, N. F., Correll, D. L., *et al.* (1998) Nonpoint pollution of surface waters with phosphorus and nitrogen. *Ecological Applications* **8**: 559–568.

Carr, M. H. (1989) Effects of macroalgal assemblages on the recruitment of temperate zone reef fishes. *Journal of Experimental Marine Biology and Ecology* **126**: 59–76.

Carreiro-Silva, M., McClanahan, T. R. & Kiene, W. E. (2005) The role of inorganic nutrients and herbivory in controlling microbioerosion of carbonate substrate. *Coral Reefs* **24**: 214–221.

Carricart-Ganivet, J. P. & Beltran-Torres, A. U. (2000) Skeletal extension, density and calcification rate of the reef building coral *Montastraea annularis* (Ellis and Solander) in the Mexican Caribbean. *Bulletin of Marine Science* **66**: 215–224.

Carroll, P. & Crill, P. M. (1997) Carbon balance of a temperate poor fen. *Global Biogeochemical Cycles* **11**: 349–356.

Casagrande, R. A., Balme, G. & Blossey, B. (2003) *Rhizedra lutosa*, a natural enemy of *Phragmites australis* in North America. *Estuaries* **26**: 602–606.

Castellanos, E. M., Figueroa, M. E. & Davy, A. J. (1994) Nucleation and facilitation in saltmarsh succession: interactions between *Spartina maritima* and *Arthrocnemum perenne*. *Journal of Ecology* **82**: 239–248.

Castilla, J. C. (1996) Copper mine tailing disposal in northern Chile rocky shores: *Enteromorpha compressa* (Chlorophyta) as a sentinel species. *Environmental Monitoring and Assessment* **40**: 171–184.

Castilla, J. C. (1999) Coastal marine communities: trends and perspectives from human-exclusion experiments. *Trends in Ecology and Evolution* **14**: 280–283.

Castilla, J. C. (2000) Roles of experimental marine ecology in coastal management and conservation. *Journal of Experimental Marine Biology and Ecology* **250**: 3–21.

Castilla, J. C. (2001) Marine ecosystems, human impacts on. *Encyclopaedia of Biodiversity* **4**: 27–35.

Castilla, J.C & Camus, P. (1992) The Humboldt–El Niño scenario: coastal benthic resources and anthropogenic influences, with particular reference to the 1982/83 ENSO. *South African Journal of Marine Science* **12**: 703–712.

Castilla, J. C. & Duran, L. R. (1985) Human exclusion from the rocky intertidal zone of Central Chile: the effects on *Concholepas concholepas* (Gastropoda). *Oikos* **45**: 391–399.

Castilla, J. C. & Moreno, C. A. (1982) Sea urchins and *Macrocystis pyrifera*: experimental test of their ecological relations in southern Chile. In: *Echinoderms: Proceedings of the International Conference*, Tampa Bay, FL, USA, ed. J. M. Lawrence, pp. 257–263. Rotterdam, the Netherlands: A. A. Balkema.

Castillo, J. M., Fernández-Baco, L., Castellanos, E. M., *et al.* (2000) Lower limits of *Spartina densiflora* and *S. maritima* in a Mediterranean salt marsh determined by different ecophysiological tolerances. *Journal of Ecology* **88**: 801–812.

Castro, M. S., Driscoll, C. T., Jordan, T. E., Reay, W. G. & Boynton, W. R. (2003) Sources of nitrogen to estuaries in the United States. *Estuaries* **26**: 803–814.

Causey, B., Delany, J., Diaz, E., *et al.* (2002) Status of coral reefs in the US Caribbean and Gulf of Mexico: Florida, Texas, Puerto Rico, US Virgin Islands, Navassa. In: *Status of Coral Reefs of the World: 2002*, ed. C. Wilkinson, pp. 251–276. Townsville, Queensland, Australia: Australian Institute of Marine Science.

Cavalieri, D. J., Gloersen, P., Parkinson, C. L., Comiso, J. C. & Zwally, H. J. (1997) Observed hemispheric asymmetry in global sea ice changes. *Science* **278**: 1104–1106.

Cayan, D. R., Redmond, K. T. & Riddle, L. G. (1999) ENSO and hydrologic extremes in the western United States. *Journal of Climatology* **12**: 2881–2893.

Cearreta, A., Irabien, M. J., Leorri, E., *et al.* (2000) Recent anthropogenic impacts on the Bilbao estuary, northern Spain: geochemical and macrofaunal evidence. *Estuarine, Coastal and Shelf Science* **50**: 571–592.

Cerda, M. & Castilla, J. C. (2001) Diversity and biomass of macro-invertebrates in intertidal matrices of the tunicate *Pyura praeputialis* (Heller, 1878) in the Bay of Antofagasta, Chile. *Revista Chilena de Historia Natural* **74**: 841–853.

Ceulemans, R. & Mousseau, M. (1994) Effects of elevated atmospheric CO_2 on woody plants. *New Phytologist* **127**: 425–446.

Chalker-Scott, L. (1995) Survival and sex ratios of the intertidal copepod, *Tigriopus californicus*, following ultraviolet-B (290–320 nm) radiation exposure. *Marine Biology* **123**: 799–804.

Chambers, F. M. (1997) Bogs as treeless wastes: the myth and the implications for conservation. In: *Conserving Peatlands*, ed. L. Parkyn, R. E. Stoneman & H. A. P. Ingram, pp. 168–175. Wallingford, UK: CAB International.

Chambers, F. M., Mauquoy, D. & Todd, P. A. (1999*a*) Recent rise to dominance of *Molinia caerulea* in environmentally sensitive areas: new perspectives from palaeoecological data. *Journal of Applied Ecology* **36**: 719–733.

Chambers, R. M., Meyerson, L. A. & Saltonstall, K. (1999*b*) Expansion of *Phragmites australis* into tidal wetlands of North America. *Aquatic Botany* **64**: 261–273.

Chapelle, F. H. (2001) *Ground-Water Microbiology and Geochemistry*. New York, USA: John Wiley.

Chapman, A. R. O. (1986) Age versus stage: an analysis of age- and size-specific mortality and reproduction in a population of *Laminaria longicruris* Pyl. *Journal of Experimental Marine Biology and Ecology* **97**: 113–122.

Chapman, A. R. O. & Johnson, C. R. (1990) Disturbance and organization of macroalgal assemblages in the Northwestern Atlantic. *Hydrobiologia* **192**: 77–121.

Chapman, D. M. (1989) *Coastal Dunes of New South Wales: Status and Management*. Technical Report No. 89/3. Sydney, New South Wales, Australia: University of Sydney Coastal Studies Unit.

Chapman, J. L. & Reiss, M. (1998) *Ecology: Principles and Applications*, 2nd edn. Cambridge, UK: Cambridge University Press.

Chapman, L. J., Balirwa, J., Bugenyi, F. W. B., Chapman, C. & Crisman, T. L. (2001) Wetlands of East Africa: biodiversity, exploitation and policy perspectives. In: *Biodiversity in Wetlands: Assessment, Function and Conservation*, vol. 2, ed. B. Gopal, W. J. Junk & J. A. Davis, pp. 101–131. Leiden, the Netherlands: Backhuys.

Chapman, M. G. & Bulleri, F. (2003) Intertidal seawalls: new features of landscape in intertidal environments. *Landscape and Urban Planning* **62**: 159–172.

Chapman, P. M., Paine, M. D., Arthur, A. D. & Taylor, L. A. (1996) A triad study of sediment quality associated with a major, relatively untreated marine sewage discharge. *Marine Pollution Bulletin* **32**: 47–64.

Chapman, V. J. (1974) *Salt Marshes and Salt Deserts of the World*, 2nd edn. Lehre, Germany: Cramer.

Chapman, W. L. & Walsh, J. E. (1993) Recent variations of sea ice and air temperature in high latitudes. *Bulletin of the American Meteorological Society* **74**: 33–47.

Charles, C. D., Hunter, D. E. & Fairbanks, R. D. (1997) Interaction between the ENSO and the Asian Monsoon in a coral record of tropical climate. *Science* **277**: 925–928.

Charman, D. (2002) *Peatlands and Environmental Change*. Chichester, UK: John Wiley.

Chauhan, M. & Gopal, B. (2001) Biodiversity and management of Keoladeo National Park (India): a wetland of international importance. In: *Biodiversity in Wetlands: Assessment, Function and Conservation*, vol. 2, ed. B. Gopal, W. J. Junk & J. A. Davis, pp. 217–256. Leiden, the Netherlands: Backhuys.

Chavez, F. P., Ryan, J., Lluch-Cota, S. E. & Miguel Niquen, C. (2003) From anchovies to sardines and back: multidecadal change in the Pacific Ocean. *Science* **299**: 217–221.

Chazottes, V., Le Campion-Alsumard, T. & Peyrot-Clausade, M. (1995) Bioerosion rates on coral reefs: interactions between macroborers, microborers and grazers (Moorea, French Polynesia). *Palaeo* **113**: 189–198.

Chen, J., Carlson, B. E. & Del Genio, A. D. (2002) Evidence for strengthening of the tropical general circulation in the 1990s. *Science* **295**: 838–841.

Chesney, E. J., Baltz, D. M. & Thomas, R. G. (2000) Louisiana estuarine and coastal fisheries and habitats: perspectives from a fish's eye view. *Ecological Applications* **10**: 350–366.

Chesterfield, E. A. (1986) Changes in the vegetation of the river red gum forest at Barmah, Victoria. *Australian Forestry* **49**: 4–15.

Choat, J. H. & Ayling, A. M. (1987) The relationship between habitat and fish faunas of New Zealand reefs. *Journal of Experimental Marine Biology and Ecology* **110**: 228–284.

Choat, J. H. & Schiel, D. R. (1982) Patterns of distribution and abundance of large brown algae and invertebrate herbivores in subtidal regions of northern New Zealand. *Journal of Experimental Marine Biology and Ecology* **60**: 129–162.

Christensen, J. H. & Christensen, O. B. (2002) Severe summertime flooding in Europe. *Nature* **421**: 805.

Christensen, T. R. & Keller, M. (2004) Element interactions and trace gas exchange. In: *Interactions of the Major Biogeochemical Cycles: Global Change and Human Impacts*, ed. J. M. Melillo, C. B. Field & B. Moldan, pp. 247–258. Washington, DC, USA: Island Press.

Christensen, T. R., Lloyd, D., Svensson, B., *et al.* (2002) Biogenic controls on trace gas fluxes in northern wetlands. *Global Change News Letter* **51**: 9–15.

Christensen, V. (1998) Fishery-induced changes in the marine ecosystem: insights from models of the Gulf of Thailand. *Journal of Fish Biology* **53** (Suppl. A): 128–142.

Christensen, V. & Pauly, D., eds. (1993) *Trophic Models of Aquatic Ecosystems: ICLARM Conference Proceedings 26.* Manila, Philippines: International Center for Living Aquatic Resources.

Christensen, V., Guénette, S., Heymans, J. J., *et al.* (2003) Hundred-year decline of North Atlantic predatory fishes. *Fish and Fisheries* **4**: 1–24.

Christie, J., Becker, M., Cowden, J. & Vallentyne, J. (1986) Managing the Great Lakes Basin as a home. *Journal of Great Lakes Research* **12**: 2–17.

Chung, C.-H. (1985) The effects of introduced *Spartina* grass on coastal morphology in China. *Zeitschrift für Geomorphologie* (Suppl.) **57**: 169–174.

Chung, C.-H. (1994) Creation of *Spartina* plantations as an effective measure for reducing coastal erosion in China. In: *Global Wetlands: Old World and New*, ed. W. J. Mitsch, pp. 443–452. Amsterdam, the Netherlands: Elsevier.

Chung, C.-H., Zhuo, R. Z. & Xu, G. W. (2004) Creation of *Spartina* plantations for reclaiming Dongtai, China, tidal flats and offshore sands. *Ecological Engineering* **23**: 135–150.

Cincotta, R. P. & Engelman, R. (2000) *Nature's Place: Human Population and the Future of Biological Diversity.* Washington, DC, USA: Population Action International.

Cisneros-Mata, M. A., Botsford, L. & Quinn, J. F. (1997) Projecting the viability of *Totoba macdonaldi*, a population with unknown age dependent variability. *Ecological Applications* **7**: 968–980.

Ciupek, R. B. (1986) Protecting wetlands under Clean Water Act 404: EPA's conservation policy on mitigation. *Natural Wetlands Newsletter* **8**: 12–15.

Clark, J. R. (1983) *Coastal Ecosystem Management*. Malabar, FL, USA: R.E. Krieger.

Clark, J. R. (1995) *Coastal Zone Management Handbook*. Boca Raton, FL, USA: Lewis Publishers.

Clark, J. R. (1998) *Endangered and Threatened Species: Threatened Status for Johnson's Seagrass*, Department of Commerce, National Oceanic and Atmospheric Administration, 50 CFR Part 227. *Federal Register*, Vol. 63, No. 177. Available online at www.nmfs.noaa.gov/pr/pdfs/fr/fr63-49035.pdf

Clark, J. S., Carpenter, S. R., Barber, M., *et al.* (2001) Ecological forecasts: an emerging imperative. *Science* **293**: 657–660.

Clark, M. (1999) Fisheries for orange roughy (*Hoplostethus atlanticus*) on seamounts in New Zealand. *Oceanologica Acta* **22**: 593–602.

Clark, M. R. (1995) Experience with the management of orange roughy (*Hoplostethus atlanticus*) in New Zealand, and the effects of commercial fishing on stocks over the period 1980–1993. In: *Deep-Water Fisheries of the North Atlantic Oceanic Slope*, ed. A. G. Hopper, pp. 251–266. Dordrecht, the Netherlands: Kluwer.

Clark, M. R. & Tracey, D. M. (1994) Changes in a population of orange roughy, *Hoplostethus atlanticus*, with commercial exploitation on the Challenger Plateau, New Zealand. *Fishery Bulletin* **92**: 236–253.

Clark, M. R., Finncham, D. J. & Tracey, D. M. (1994) Fecundity of orange roughy (*Hoplostethus atlanticus*) in New Zealand waters. *New Zealand Journal of Marine and Freshwater Research* **28**: 193–200.

Clark, M. W. (1998) Management implications of metal transfer pathways from a refuse tip to mangrove sediments. *Science of the Total Environment* **222**: 17–34.

Clark, R., ed. (1999) *Global Environment Outlook 2000.* London, UK: Earthscan Publications for United Nations Environment Programme.

Clark, R. B. (1992) *Marine Pollution*, Oxford, UK: Clarendon Press.

Clark, R. B., Frid, C. L. J. & Atrill, M. (1997) *Marine Pollution*. Oxford, UK: Clarendon Press.

Clark, R. L. & Guppy, J. C. (1988) A transition from mangrove forest to freshwater wetland in the monsoon tropics of Australia. *Journal of Biogeography* **15**: 665–684.

Clark, W. C., Mitchell, R. B. & Cash, D. W. (2006) Evaluating the influence of global environmental assessments. In: *Global Environmental Assessments: Information and Influence,* ed. R. B. Mitchell, W. C. Clark, D. W. Cash, and N. M. Dickson, pp. 1–28. Cambridge, MA, USA: MIT Press.

Clarke, A. (1996) The influence of climate change on the distribution and evolution of organisms. In: *Animals and Temperature: Phenotypic and Evolutionary Adaptation*, ed. I. A. Johnston & A. F. Bennett, pp. 377–407. Cambridge, UK: Cambridge University Press.

Clarke, A. & Harris, C. M. (2003) Polar marine ecosystems: major threats and future change. *Environmental Conservation* 30: 1–25.

Clarke, F. W. (1920) *The Data of Geochemistry*, US Geological Survey Bulletin No. 695. Menlo Park, CA, USA: US Geological Survey.

Clarke, T. (2003) Lightning satellite survey. *Nature* 421: 228.

CLIMAP Project Members (1976) The surface of the ice-age Earth. *Science* 191: 1131–1137.

Cloern, J. E. (2001) Our evolving conceptual model of the coastal eutrophication problem. *Marine Ecology Progress Series* 210: 223–253.

Close, A. (1990) The impact of man on the natural flow regime. In: *The Murray*, ed. N. Mackay & D. Eastburn, pp. 61–74. Canberra, ACT, Australia: Murray–Darling Basin Commission.

Clough, B. F., Boto, K. G. & Attiwill, P. M. (1983) Mangroves and sewage : a re-evaluation. In: *Biology and Ecology of Mangroves*: vol. 8, *Tasks for Vegetation Science*, ed. H. J. Teas, pp. 151–161. The Hague, the Netherlands: Dr W. Junk.

Cluis, D. & Laberge, C. (2002) Analysis of El Niño effect on the discharge of selected rivers in the Asia Pacific region. *Water International* 27: 279–293.

Clymo, R. S. (1965) Experiments on breakdown of *Sphagnum* in two bogs. *Journal of Ecology* 53: 747–758.

Clymo, R. S. (1973) The growth of *Sphagnum*: some effects of environment. *Journal of Ecology* 61: 849–869.

Clymo, R. S. (1984) The limits to peat bog growth. *Philosophical Transactions of the Royal Society B* 303: 605–654.

Clymo, R. S. & Hayward, P. M. (1982) The ecology of *Sphagnum*. In: *Bryophyte Ecology*, ed. A. J. E. Smith, pp. 229–289. London, UK: Chapman & Hall.

Clymo, R. S. & Pearce, D. M. E. (1995) Methane and carbon dioxide production in, transport through, and efflux from a peatland. *Philosophical Transactions of the Royal Society of London B* 350: 249–259.

Clymo, R. S., Turunen, J. & Tolonen, K. (1998) Carbon accumulation in peatland. *Oikos* 81: 368–388.

Coale, K. H., Johnson, K. S., Fitzwater, S. E., *et al.* (1996) A massive phytoplankton bloom induced by an ecosystem-scale iron fertilization experiment in the equatorial Pacific Ocean. *Nature* 383: 495–501.

Cobb, K. M., Charles, C. D., Cheng, H. & Edwards, R. L. (2003) El Niño/Southern Oscillation and tropical Pacific climate during the last millennium. *Nature* 424: 271–276.

Cognetti, G. (2001) Marine eutrophication: the need for a new indicator system. *Marine Pollution Bulletin* 42: 163–164.

Cohen, A. N. & Carlton, J. T. (1998) Accelerating invasion rate in a highly invaded estuary. *Science* 279: 555–562.

Cohen, J. E. (1995) *How Many People Can the Earth Support?* New York, USA: W.W. Norton.

Cohen, J. E. (1997) Estimates of coastal population. *Science* 278: 1209.

Cohen, J. E. (2003) Human population @ the next half century. *Science* 302: 1172–1176.

Cohen, J. E., Small, C., Mellinger, A., Gallup, J. & Sachs, J. (1997) Estimates of coastal populations. *Science* 287: 1211–1212.

COHMAP Members (1988) Climatic changes over the last 18 000 years: observations and model simulations. *Science* 241: 1043–1052.

Colborn, T., Davidson, A., Green, S., *et al.* (1990) *Great Lakes, Great Legacy?* Washington, DC, USA and Ottawa, Ontario, Canada: The Conservation Foundation and the Institute for Research on Public Policy.

Cole, J., Dunbar, R., McClanahan, T. & Muthiga, N. (2000) Tropical Pacific forcing of decadal variability in SST in the western Indian Ocean. *Science* 287: 617–619.

Cole, J. J., Peierls, B. L., Caraco, N. F. & Pace, M. L. (1993) Nitrogen loading of rivers as a human-driven process. In: *Humans as Components of Ecosystems*, ed. M. J. McDonnell & S. T. A. Pickett, pp. 141–157. New York, USA: Springer-Verlag.

Cole, R. G. & Babcock, R. C. (1996) Mass mortality of a dominant kelp (Laminariales) at Goat Island, North-eastern New Zealand. *Marine Freshwater Resource* 47: 907–911.

Cole, R. G. & Syms, C. (1999) Using spatial pattern analysis to distinguish causes of mortality: an example from kelp in north-eastern New Zealand. *Journal of Ecology* 87: 963–972.

Coleman, R. A., Goss-Custard, J. D., LeV. dit Durell, S. E. A. & Hawkins, S. J. (1999) Limpet *Patella* spp. consumption by oystercatchers *Haematopus ostralegus*: a preference for solitary prey items. *Marine Ecology Progress Series* 183: 253–261.

Coles, R. G. (1996) *Coastal Management and Community Coastal Resource Planning in the Asia Pacific Region.* Canberra, ACT, Australia: Winston Churchill Memorial Trust.

Coles, R. G. & Fortes, M. (2001) Protecting seagrass: approach and methods. In: *Global Seagrass Research Methods*, ed. F. T. Short & R. G. Coles, pp. 445–464. Amsterdam, the Netherlands: Elsevier.

Coles, R. G., McKenzie, L. & Campbell, S. (2003a) The seagrasses of Eastern Australia. In: *World Atlas of Seagrasses*, ed. E. P. Green & F. T. Short, pp. 131–147. Berkeley, CA, USA: University of California Press.

Coles, R. G., McKenzie, L., Campbell, S., Fortes, M. & Short, F. (2003b) The seagrasses of the Western Pacific Islands. In: *World Atlas of Seagrasses*, ed. E. P. Green & F. T. Short, pp. 177–186. Berkeley, CA, USA: University of California Press.

Coles, S. L. & Al-Rayami, K. A. (1996) Beach tar concentrations on the Muscat coastline, Gulf of Oman, Indian Ocean, 1993–1995. *Marine Pollution Bulletin* **32**: 609–614.

Coles, S. L. & Brown, B. E. (2003) Coral bleaching: capacity for acclimatization and adaptation. *Advances in Marine Biology* **46**: 183–223.

Coles, S. M. (1979) Benthic microalgal populations on intertidal sediments and their role as precursors to salt marsh development. In: *Ecological Processes in Coastal Environments*, ed. R. L. Jefferies & A. J. Davy, pp. 25–42. Oxford, UK: Blackwell Scientific Publications.

Collie, J. S. & DeLong, A. K. (1999) Multispecies interactions in the Georges Bank fish community. In: *Alaska Sea Grant College Program, Ecosystem Approaches for Fisheries Management*, Publication No. AK-SG-99-01, pp. 187–210. Alaska, USA: Alaska Sea Grant.

Collie, J. S., Escanero, G. A. & Valentine, P. C. (1997) Effects of bottom fishing on the benthic megafauna of Georges Bank. *Marine Ecology Progress Series* **155**: 159–172.

Collie, J. S., Hall, S. J., Kaiser, M. J. & Poiner, I. R. (2000) A quantitative analysis of fishing impacts on shelf-sea benthos. *Journal of Animal Ecology* **69**: 785–798.

Collier, L. M. & Pinn, E. H. (1998) An assessment of the acute impact of sea lice treatment with ivermectin on a benthic community. *Journal of Experimental Marine Biology and Ecology* **23**: 131–147.

Collingham, Y. C. & Huntley, B. (2000) Impacts of habitat fragmentation and patch size upon migration rates. *Ecological Applications* **10**: 131–144.

Collins, M. A., Belchier, M. & Everson, I. (2003) Why the fuss about toothfish? *Biologist* **50**: 116–119.

Congressional Natural Hazards Caucus Work Group (2001) Discussion paper for the Congressional Natural Hazards Caucus [www document]. URL http://www.agiweb.org/workgroup/discussion_paper0101.html

Conlan, K., White, K. N. & Hawkins, S. J. (1992) The hydrography and ecology of a redeveloped brackish-water dock. *Estuarine, Coastal and Shelf Science* **35**: 435–452.

Conley, D. J., Stalnacke, P., Pitkanen, H. & Wilander, A. (2000) The transport and retention of dissolved silicate by rivers in Sweden and Finland. *Limnology and Oceanography* **45**: 1850–1853.

Connell, J. H. (1961) The influence of interspecific competition and other factors on the distribution of the barnacle *Chthamalus stellatus*. *Ecology* **42**: 710–723.

Connell, J. H. (1972) Community interactions on marine rocky intertidal shores. *Annual Review of Ecology and Systematics* **3**: 169–192.

Connell, J. H. (1997) Disturbance and recovery of coral assemblages. *Coral Reefs* **16**: S101–S113.

Connolly, R. M. (1999) Saltmarshes as habitat for fish and nektonic crustaceans: challenges in sampling designs and wetlands. *Australian Journal of Ecology* **24**: 422–430.

Connolly, R. M., Dalton, A. & Bass, D. A. (1997) Fish use of an inundated flat in a temperate Australian estuary. *Australian Journal of Ecology* **22**: 222–226.

Connolly, S. R. & Roughgarden, J. (1998) A latitudinal gradient in Northeast Pacific intertidal community structure: evidence for an oceanographically based synthesis of marine community theory. *American Naturalist* **151**: 311–326.

Conover, R. J. & Huntley, M. (1991) Copepods in ice covered seas: distribution, adaptations to seasonally limited food, metabolism, growth patterns and life cycle strategies in polar seas. *Journal of Marine Systems* **2**: 1–41.

Conserve Africa Foundation (2005) Poverty and environment in Africa: an overview [www document]. URL http://www.conserveafrica.org.uk/poverty.html

Constable, A. J., de La Mare, W. K., Agnew, D. J., Everson, I. & Miller, D. (2000) Managing fisheries to conserve the Antarctic marine ecosystem: practical implementation of the Convention on the Conservation of Antarctic Marine Living Resources. *ICES Journal of Marine Science* **57**: 778–791.

Contreras-Balderas, S. & Lozano-Vilano, M. L. (1994) Water, endangered fishes, and development perspectives in arid lands of Mexico. *Conservation Biology* **8**: 379–387.

Convey, P., Scott, D. & Fraser, W. R. (2003) Biophysical and habitat changes in response to climate alteration in the Arctic and Antarctic. *Advances in Applied Biodiversity Science* **4**: 79–84.

Cook, P., Robbins, J., Endicott, D., *et al.* (2003) Effects of aryl hydrocarbon receptor mediated early life stage toxicity on lake trout populations in Lake Ontario during the 20th century. *Environmental Science and Technology* **37**: 3864–3877.

Correa, J. A., Castilla, J. C., Ramirez, M., *et al.* (1999) Copper, copper mine tailings and their effect on marine algae in northern Chile. *Journal of Applied Phycology* **11**: 57–67.

Corredor, J. E., Howarth, R. W., Twilley, R. R. & Morell, J. M. (1999) Nitrogen cycling and anthropogenic impact in the tropical interamerican seas. *Biogeochemistry* **46**: 163–178.

Corredor, J. E., Morrell, J. M. & Del Castillo, C. E. (1990) Persistence of spilled crude oil in a tropical marine environment. *Marine Pollution Bulletin* **21**: 385–388.

Cortes, J. (1993) Comparison between Caribbean and eastern Pacific coral reefs. *Revistade Biologica Tropicale* **41** (Suppl. 1): 19–21.

Cortes, J., Macintyre, I. G. & Glynn, P. W. (1994) Holocene growth history of an eastern Pacific fringing reef, Punta Islotes, Costa Rica. *Coral Reefs* **13**: 65–73.

Cosgrove, W. J. & Rijsberman, F. R. (2000) *World Water Vision: Making Water Everybody's Business*. London, UK: Earthscan Publications.

Costa, L. T., Farinha, J. C., Hecker, N. & Tomàs Vives, P., eds. (1996) *Mediterranean Wetland Inventory*, vol. 1, *A Reference Manual* [www document]. URL http://www.wetlands.org/RSDB/Default.htm

Costanza, R., d'Arge, R., de Groot, R., *et al.* (1997) The value of the world's ecosystem services and natural capital. *Nature* **387**: 253–260.

Costello, M. J. & Read, P. (1994) Toxicity of sewage sludge to marine organisms: a review. *Marine Environmental Research* **37**: 23–42.

Coulson, J. C. & Butterfield, J. (1978) An investigation of the biotic factors determining the rates of plant decomposition on blanket bog. *Journal of Ecology* **66**: 631–650.

Coulter, G. & Mubamba, R. (1993) Conservation in Lake Tanganyika, with special reference to underwater parks. *Conservation Biology* **7**: 678–685.

Coutant, C. C. (1981) Foreseeable effects of CO_2-induced climate change. Fresh-water concerns. *Environmental Conservation* **8**: 285–297.

Covich, A. P. (1996) Stream biodiversity and ecosystem processes. *Bulletin of the North American Benthological Society* **13**: 294–303.

Covich, A. P., Austen, M., Bärlocher, F., *et al.* (2004) The role of biodiversity in the functioning of freshwater and marine benthic ecosystems. *BioScience* **54**: 767–775.

Covich, A. P., Fritz, S. C., Lamb, P. J., *et al.* (1997) Potential effects of climate change on aquatic ecosystems of the great plains of North America. *Hydrological Processes* **11**: 993–1021.

Covich, A. P., Palmer, M. A. & Crowl, T. A. (1999) Role of benthic invertebrate species in freshwater ecosystems. *BioScience* **49**: 119–127.

Cowardin, L. M., Carter, V., Golet, F.C & LaRoe, E. T. (1979) *Classification of Wetlands and Deepwater Habitats of the United States*, Biological Services Program No. FWS/OBS-79/31. Washington, DC, USA: US Fish and Wildlife Service, US Department of the Interior.

Cowen, R. K. (1983) The effect of sheephead (*Semicossyphus pulcher*) predation on red sea urchin populations: an experimental analysis. *Oecologia* **58**: 249–255.

Cowen, R. K., Lwiza, K. M. M., Sponaugle, S., Paris, C. B. & Olson, D. B. (2000) Connectivity of marine populations: open or closed? *Science* **287**: 857–859.

Craft, C. B. & Casey, W. P. (2000) Sediment and nutrient accumulation in floodplain and depressional freshwater wetlands of Georgia, USA. *Wetlands* **20**: 323–332.

Crain, A. D., Bolten, A. B. & Bjorndal, K. A. (1995) Effects of beach nourishment on sea turtles: review and research initiatives. *Restoration Ecology* **3**: 95–104.

Crane, R. G. & Barry, R. G. (1984) The influence of clouds on climate with a focus on high latitude interactions. *Journal of Climatology* **4**: 71–93.

Crawford, R. E. (2002) Secondary wake turbidity from small boat operation in a shallow sandy bay. *Journal of Coastal Research* Special Issue **37**: 49–64.

Creasey, S. S. & Rogers, A. D. (1999) Population genetics of bathyal and abyssal organisms. *Advances in Marine Biology* **35**: 1–151.

Creed, J. C. (2003) Seagrasses of South America: Brazil, Argentina and Chile. In: *World Atlas of Seagrasses*, ed. E. P. Green & F. T. Short, pp. 263–270. Berkeley, CA, USA: University of California Press.

Creed, J. C., Phillips, R. C. & Van Tussenbroek, B. I. (2003) The seagrasses of the Caribbean. In: *World Atlas of Seagrasses*, ed. E. P. Green & F. T. Short, pp. 254–262. Berkeley, CA, USA: University of California Press.

Cresser, M. S. (2000) The critical loads concept: milestone or millstone for the new millennium? *Science of the Total Environment* **249**: 51–62.

Cronin, M., Davies, I. M., Newton, A., *et al.* (1998) Trace metal concentrations in deep sea fish from the North Atlantic. *Marine Environmental Research* **45**: 225–238.

Crooks, S., Schutten, J., Sheern, G. D., Pye, K. & Davy, A. J. (2002) Drainage and elevation as factors in the restoration of salt marsh in Britain. *Restoration Ecology* **10**: 591–602.

Crow, G. E. (1993) Species diversity in aquatic angiosperms: latitudinal patterns. *Aquatic Botany* **44**: 229–258.

Crowe, T. P., Thompson, R. C., Bray, S. & Hawkins, S. J. (2000) Impacts of anthropogenic stress on rocky intertidal communities. *Journal of Aquatic Ecosystem Stress and Recovery* **7**: 273–297.

Croxall, J. P. & Nicol, S. (2004) Management of Southern Ocean fisheries global forces and future sustainability. *Antarctic Science* **16**: 568–584.

Croxall, J. P., Prince, P. A., Rothery, P. & Wood, A. G. (1997) Population changes in albatrosses at South Georgia. In: *Albatross Biology and Conservation*, ed. G. Robertson & R. Gales, pp. 69–83. Chipping Norton, NSW, Australia: Surrey Beatty.

Croxall, J. P., Trathan, P. N. & Murphy, E. J. (2002) Environmental change and Antarctic seabird populations. *Science* **297**: 1510–1514.

Crunkilton, R. (1985) Subterranean contamination of Meramec Spring by ammonium nitrate and urea fertilizer and its implication on rare biota. In: *Proceedings of the 1984 National Cave Management Symposium*, 25, pp. 151–158. Columbia, MO, USA: Missouri Speleological Survey.

Crutzen, P. (2006) Albedo enhancement by stratospheric sulphur injections: a contribution to resolve a policy dilemma? *Climatic Change* **77**: 211–220.

Cryer, M., Hartill, B. & Oshea, S. (2002) Modification of marine benthos by trawling: toward a generalization for the deep ocean. *Ecological Applications* **12**: 1824–1839.

Cullen, P. (2003) Salinity. In: *Ecology: An Australian Perspective*, ed. P. Attiwill & B. Wilson, pp. 474–488. Melbourne, Victoria, Australia: Oxford University Press.

Culver, D. C. & Holsinger, J. R. (1992) How many species of troglobites are there? *National Speleological Society Bulletin* **54**: 79–80.

Culver, D. C., Jones, W. K. & Holsinger, J. R. (1992) Biological and hydrological investigation of the Cedars, Lee County, Virginia: an ecologically significant and threatened karst area. In: *Proceedings of the 1st International Conference on Groundwater Ecology*, ed. J. A. Stanford & J. J. Simons, pp. 281–290. Bethesda, MD, USA: American Water Research Association.

Curran, M. A. J., van Ommen, T. D., Morgan, V. I., Phillips, K. L. & Palmer, A. S. (2003) Ice core evidence for Antarctic sea ice decline since the 1950s. *Science* **302**: 1203–1206.

Curry, J. A. & Webster, P. J. (1999) *Thermodynamics of Atmospheres and Oceans*. San Diego, CA, USA: Academic Press.

Curry, J. A., Schramm, J. L., Serreze, M. C. & Ebert, E. E. (1995) Water vapor feedback over the Arctic Ocean. *Journal of Geophysical Research* **100**: 14 223–14 229.

Curtis, P. S. & Wang, X. (1998) A meta-analysis of elevated CO_2 effects on woody plant mass, form, and physiology. *Oecologia* **113**: 299–313.

Curtis, P. S., Balduman, L. M., Drake, B. G. & Whigham, D. F. (1990) Elevated atmospheric CO_2 effects on belowground processes in C_3 and C_4 estuarine marsh communities. *Ecology* **71**: 2001–2006.

Curtis, P. S., Drake, B. G., Leadley, P. W., Arp, W. J. & Whigham, D. F. (1989a) Growth and senescence in plant communities exposed to elevated CO_2 concentrations on an estuarine marsh. *Oecologia* **78**: 20–26.

Curtis, P. S., Drake, B. G. & Whigham, D. F. (1989b) Nitrogen and carbon dynamics in C_3 and C_4 estuarine plants grown under elevated CO_2 *in situ*. *Oecologia* **78**: 297–301.

Cutshall, N. H., Naidu, J. R. & Pearcy, W. G. (1978) Mercury concentrations in Pacific hake, *Merluccius productus* (Ayres), as a function of length and latitude. *Science* **200**: 1489–1491.

Cwynar, L. C. & Spear, R. W. (1991) Reversion of forest to tundra in the central Yukon. *Ecology* **72**: 202–212.

Czaya, E. (1981) *Rivers of the World*. New York, USA: Van Nostrand Reinhold.

Daan, R. & Mulder, M. (1996) On the short-term and long-term impact of drilling activities in the Dutch sector of the North Sea. *ICES Journal of Marine Science* **53**: 1036–1044.

Dacey, J. W. H., Drake, B. G. & Klug, M. J. (1994) Stimulation of methane emission by carbon dioxide enrichment of marsh vegetation. *Nature* **370**: 47–49.

Dadnadji, K. K. & van Wetten, J. C. J. (1993) Traditional management systems and irrigation of small scale interventions in the Lagone floodplains of Chad. In: *Towards the Wise Use of Wetlands*, ed. T. J. Davis, pp. 74–81. Gland, Switzerland: Ramsar Convention Bureau.

Dadswell, M. (1974) *Distribution, Ecology, and Postglacial Dispersal of Certain Crustaceans and Fishes in Eastern North America*, Publications in Zoology No. 11. Ottawa, Ontario, Canada: National Museums of Canada.

Dahdouh-Guebas, F. (2002) The use of remote sensing and GIS in the sustainable management of tropical coastal ecosystems. *Environment, Development and Sustainability* **4**: 93–112.

Dahl, T. E. (1990) *Wetlands: Losses in the United States 1780s to 1980s*. Washington, DC, USA: US Fish and Wildlife Service.

Dahl, T. E. & Zoltai, S. C. (1997) Forested northern wetlands of North America. In: *Northern Forested Wetlands: Ecology and Management*, ed. C. C. Trettin, M. F. Jurgensen, D. F. Grigal, M. R. Gale & J. K. Jeglum, pp. 3–18. Boca Raton, FL, USA: CRC Press.

Dahm, C. N. & Molles Jr, M. C. (1991) Streams in semi-arid regions as sensitive indicators of global climate change. In: *Global Climate Change and Freshwater Ecosystems*, ed. P. Firth & S. G. Fisher, pp. 250–260. New York, USA: Springer-Verlag.

Dahm, C. N., Cleverly, J. R., Allred Coonroyd, J. E., *et al.* (2002) Evapotranspiration at the land/water interface in a semi-arid drainage basin. *Freshwater Biology* **47**: 831–843.

Daily, G. C. & Ehrlich, P. (1996) Socioeconomic equity, sustainability, and Earth's carrying capacity. *Ecological Applications* **6**: 991–1001.

Daka, E. R. & Hawkins, S. J. (2004) Tolerance to heavy metals in *Littorina saxatilis* from a metal-contaminated estuary in the Isle of Man. *Journal of the Marine Biological Association of the United Kingdom* **84**: 393–400.

Daka, E. R., Allen, J. R. & Hawkins, S. J. (2003) Heavy metal contamination in sediment and biomonitors from sites around the Isle of Man. *Marine Pollution Bulletin* **46**: 784–791.

Dale, P. E. R. (1994) An Australian perspective on coastal wetland management and vector control. In: *Global Wetlands: Old World and New*, ed. W. J. Mitsch, pp. 771–780. Amsterdam, the Netherlands: Elsevier.

Daly, H. E. (1991) Elements of environmental macroeconomics. In: *Ecological Economics: The Science and Management of Sustainability*, ed. R. Constanza, pp. 32–46. New York, USA: Columbia University Press.

Daly, M. A. & Mathieson, A. C. (1977) The effects of sand movement on intertidal seaweeds and selected invertebrates at Bound Rock, New Hampshire, USA. *Marine Biology* **43**: 45–55.

Dando, P. R., Southward, A. J., Southward, E. C., *et al.* (1992) Shipwrecked tube worms. *Nature* **356**: 667.

Dangles, O. & Malmqvist, B. (2004) Species richness–decomposition relationships depend on species dominance. *Ecology Letters* **7**: 395–402.

Danielopol, D. L., Griebler, C., Gunatilaka, A. & Notenboom, J. (2003) Present state and future prospects for groundwater ecosystems. *Environmental Conservation* **30**: 104–130.

Danielopol, D. L., Pospisil, P., Dreher, J., *et al.* (2000) A groundwater ecosystem in the wetlands of the Danube at Vienna (Austria). In: *Ecosystems of the World*, vol. 30, *Subterranean Ecosystems*, ed. H. Wilkens, D. C. Culver & W. F. Humphreys, pp. 581–602. Amsterdam, the Netherlands: Elsevier.

Darrigan, G. & Ezcurra de Drago, I. (2000) *Limnosperma fortunei* (Dunker, 1857) (Mytilidae) distribution in the Plata Basin. *Medio Ambiente* **13**: 75–79.

Dasgupta, P. (2000) Population and resources: an exploration of reproductive and environmental externalities. *Population and Development Review* **26**: 643–689.

Dasgupta, P. (2001) *Human Well-Being and the Natural Environment*. Oxford, UK: Oxford University Press.

da Silva, C. J., Wantzen, K. M., Nunes da Cunha, C. & Machado, F. de A. (2001) Biodiversity in the Pantanal Wetland, Brazil. In: *Biodiversity in Wetlands: Assessment, Function and Conservation*, vol. 2, ed. B. Gopal, W. J. Junk & J. A. Davis, pp. 187–215. Leiden, the Netherlands: Backhuys.

Daskalakis, K. D. & O'Connor, T. P. (1995) Distribution of chemical contaminants in US coastal and estuarine sediment. *Marine Environmental Research* **40**: 381–398.

Dauer, D. M., Rodi Jr, A. J. & Ranasinghe, J. A. (1992) Effects of low dissolved oxygen events on the macrobenthos of the lower Chesapeake Bay. *Estuaries* **15**: 384–391.

Daughton, C. G. & Ternes, T. A. (1999) Pharmaceuticals and personal care products in the environment: agents of subtle change? *Environmental Health Perspectives* **107**: 907–938.

Dauterive, L. (2000) *Rigs-to-Reefs Policy, Progress, and Perspective*, Gulf of Mexico OCS Region Report. New Orleans, LA, USA: US Department of the Interior, Minerals Management Service.

D'Avanzo, C. D. & Kremer, J. N. (1994) Diel oxygen dynamics and anoxic events in an eutrophic estuary of Waquoit Bay, Massachusetts. *Estuaries* **17**: 131–139.

Davidson, I., Vanderkam, R. & Padilla, M. (1999) Review of wetland inventory information in the Neotropics. In: *Global Review of Wetland Resources and Priorities for Wetland Inventory*, Supervising Scientist Report No. 144 / Wetlands International Publication No. 53, ed. C. M. Finlayson & A. G. Spiers, pp. 419–455. Canberra, ACT, Australia: Supervising Scientist.

Davis, J. (2000) Changes in a tidepool fish assemblage on two scales of environmental variation: seasonal and El Niño Southern Oscillation. *Limnology and Oceanography* **45**: 1368–1379.

Davis, R. & Short, F. T. (1997) Restoring eelgrass, *Zostera marina* L., habitat using a new transplanting technique: the horizontal rhizome method. *Aquatic Botany* **59**: 1–15.

Day, D. (1990) *Noah's Choice: True Stories of Extinction and Survival.* London, UK: Penguin Books.

Day, E. & Branch, G. M. (2002) Effects of sea urchins (*Parechinus angulosus*) on recruits and juveniles of abalone (*Haliotis midae*). *Ecological Monographs* **72**: 133–149.

Day Jr, J. W., Hall, C. A. S., Kemp, W. M. & Yáñez-Arancibia, A., eds. (1989) *Estuarine Ecology.* New York, USA: John Wiley.

Dayton, P. K. (1975) Experimental studies of algal–canopy interactions in a sea otter dominated kelp community at Amchitka Island, Alaska. *Fishery Bulletin, United States* **73**: 230–237.

Dayton, P. K. (1985*a*) Ecology of kelp communities. *Annual Review of Ecology and Systematics* **16**: 215–245.

Dayton, P. K. (1985*b*) The structure and regulation of some South American kelp communities. *Ecological Monographs* **55**: 447–468.

Dayton, P. K., Currie, V., Gerrodete, T., *et al.* (1984) Patch dynamics and stability of some California kelp communities. *Ecological Monographs* **54**: 253–289.

Dayton, P. K., Tegner, M. J., Edwards, P. B. & Riser, K. L. (1998) Sliding baselines, ghosts, and reduced expectations in kelp forest communities. *Ecological Applications* **8**: 309–322.

Dayton, P. K., Tegner, M. J., Edwards, P. B. & Riser, K. L. (1999) Temporal and spatial scales of kelp demography: the role of oceanography climate. *Ecological Monographs* **69**: 219–250.

Dayton, P. K., Tegner, M. J., Parnell, P. E. & Edwards, P. B. (1992) Temporal and spatial patterns of disturbance and recovery in a kelp forest community. *Ecological Monographs* **62**: 421–445.

Dayton, P. K., Thrush, S. F., Agardy, T. M. & Hofman, R. J. (1995) Environmental effects of fishing. *Aquatic Conservation: Marine and Freshwater Ecosystems* **5**: 205–232.

Dean, J. M. (1996) A crisis and opportunity in coastal oceans: coastal fisheries as a case study. In: *Sustainable Development in the Southeastern Coastal Zone*, ed. F. J. Vernberg, W. B. Vernberg & T. Siewicki, pp. 81–88. Columbia, SC, USA: University of South Carolina Press.

Dean, T. A. & Jewett, S. C. (2001) Habitat-specific recovery of shallow subtidal communities following the Exxon Valdez oil spill. *Ecological Applications* **11**: 1456–1471.

De Casabianca, M. L., Laugier, T. & Collart, D. (1997) Impact of shellfish farming eutrophication on benthic macrophyte communities in the Thau lagoon, France. *Aquaculture International* **5**: 301–314.

Dederen, L. H. T. (1992) Marine eutrophication in Europe: similarities and regional differences in appearance. In: *Marine Coastal Eutrophication*, ed. R. A. Vollenweider, R. Marchetti & R. Viviani, pp. 673–682. Amsterdam, the Netherlands: Elsevier.

de Forges, R., Koslow, J. A. & Poore, G. C. B. (2000) Diversity and endemism of the benthic seamount fauna in the southwest Pacific. *Nature* **405**: 944–947.

Degens, E. T. & Ross, D. A., eds. (1969) *Hot Brines and Recent Heavy Metal Deposits in the Red Sea: A Geochemical and Geophysical Account.* New York, USA: Springer-Verlag.

De Groot, A. J. (1995) Metals and sediments: a global perspective. In: *Metal Contaminated Aquatic Sediments*, ed. H. E. Allen, pp. 1–20. Ann Arbor, MI, USA: Ann Arbor Press.

de Groot, R. S., Wilson, M. A. & Boumans, R. M. J. (2002) A typology for the classification, description and valuation of ecosystem functions, goods and services. *Ecological Economics* **41**: 393–408.

Deichmann, U., Balk, D. & Yetman, G. (2001) Transforming population data for interdisciplinary usages: from census to grid [www document]. URL http://sedac.ciesin.columbia.edu/gpw/docs/gpw3_documentation_final.pdf

Deimling, E. A. & Liss, W. J. (1994) Fishery development in the eastern North Pacific: a natural–cultural system perspective, 1888–1976. *Fisheries Oceanography* **3**: 60–77.

De Jonge, V. N. & De Jong, D. J. (1992) Role of tide, light and fisheries in the decline of *Zostera marina* L. in the Dutch Wadden Sea. *Netherlands Institute for Sea Research Publication Series* **20**: 161–176.

De Jonge, V. N. (1990) Response of the Dutch Wadden Sea ecosystem to phosphorus discharges from the River Rhine. *Hydrobiologia* **195**: 49–62.

de la Mare, W. K. (1997) Abrupt mid-twentieth-century decline in Antarctic sea-ice extent from whaling records. *Nature* **389**: 57–59.

DeLaune, R. D. & Pezeshki, S. R. (1994) The influence of subsidence and saltwater intrusion on coastal marsh stability: Louisiana Gulf coast, USA. *Journal of Coastal Research* Special Issue **12**: 77–89.

Delgado, C. L., Wada, N., Rosegrant, M. W., Meijer, S. & Ahmed, M. (2003) *Fish to 2020: Supply and Demand in Changing Global Markets.* Manila, Philippines: International Food Policy Research Institute and World Fish Centre.

Delgado, O. & Lapointe, B. E. (1994) Nutrient-limited productivity of calcareous versus fleshy macroalgae in a eutrophic, carbonate-rich tropical marine environment. *Coral Reefs* **13**: 151–159.

Delgado, O., Ruiz, J., Perez, M., Romero, R. & Ballesteros, E. (1999) Effects of fish farming on seagrass in a Mediterranean Bay: seagrass decline after organic loading cessation. *Oceanologica Acta* **22**: 109–117.

De Marsily, G. (1992) Creation of 'hydrological nature reserves', a plea for the defence of ground water. *Ground Water* **30**: 658.

Denevan, W. M., ed. (1966) *The Aboriginal Cultural Geography of the Llanos de Mojos of Bolivia*. Berkeley, CA, USA: University of California Press.

Denevan, W. M. (1976) The aboriginal population of Amazonia. In: *The Native Population of the Americas*, ed. W. M. Denevan, pp. 205–234. Madison, WI, USA: University of Wisconsin Press.

den Hartog, C. (1970) *The Seagrasses of the World*. Amsterdam, the Netherlands: North Holland.

Denney, N. H., Jennings, S. & Reynolds, J. D. (2002) Life history correlates of maximum population growth rate in marine fishes. *Proceedings of the Royal Society of London B* **269**: 2229–2237.

Dennis, K. C., Niang-Diop, I. & Nicholls, R. J. (1995) Sea-level rise and Senegal: potential impacts and consequences. *Journal of Coastal Research* Special Issue **14**: 243–261.

Denny, M. & Paine, R. (1998) Celestial mechanics, sea-level changes, and intertidal ecology. *Biological Bulletin* **194**: 108–115.

Denny, P. (1992) Africa. In: *Wetlands*, ed. M. Finlayson & M. Moser, pp. 115–148. Oxford, UK: Facts on File.

Denny, P. (1993) Wetlands of Africa. In: *Wetlands of the World*, vol. 1, *Inventory, Ecology and Management*, ed. D. F. Whigham, D. Dykyjová & S. Hejný, pp. 1–128. Dordrecht, the Netherlands: Kluwer.

Department of Land and Water Conservation (2002) *Compensatory Wetlands*, a discussion paper under the NSW Wetlands Management Policy. Sydney, NSW, Australia: Department of Land and Water Conservation.

Depledge, M. H. & Billinghurst, Z. (1998) Ecological significance of endocrine disruption in marine invertebrates. *Marine Pollution Bulletin* **39**: 32–38.

Dermott, R. & Kerec, D. (1997) Changes in the deepwater benthos of eastern Lake Erie since the invasion of *Dreissena*: 1979–1993. *Canadian Journal of Fisheries and Aquatic Sciences* **54**: 922–930.

Derraik, J. G. B. (2002) The pollution of the marine environment by plastic debris: a review. *Marine Pollution Bulletin* **44**: 842–852.

Desai, U., ed. (1998) *Ecological Policy and Politics in Developing Countries: Economic Growth, Democracy, and Environment*. Albany, NY, USA: State University of New York Press.

Deser, C., Walsh, J. E. & Timlin, M. S. (2000) Arctic sea ice variability in the context of recent atmospheric circulation trends. *Journal of Climate* **13**: 617–633.

Desmond, J., Zedler, J. B. & Williams, G. D. (2000) Fish use of tidal creek habitats in two southern California salt marshes. *Ecological Engineering* **14**: 233–252.

Detenbeck, N. E., Galatowitsch, S. M., Atkinson, J. & Ball, H. (1999) Evaluating perturbations and developing restoration strategies for inland wetlands in the Great Lakes Basin. *Wetlands* **19**: 789–820.

Dette, H.-H. & Raudkivi, A. J. (1994) Beach nourishment and dune protection. In: *Coastal Engineering: Proceedings of the 24th Coastal Engineering Conference*, pp. 1934–1945. New York, USA: American Society of Civil Engineers.

Deuser, W. G. & Ross, E. H. (1980) Seasonal change in the flux of organic carbon to the deep Sargasso Sea. *Nature* **283**: 364–365.

Devine, J. A., Baker, K. D. & Haedrich, R. L. (2006) Deep-sea fishes qualify as endangered. *Nature* **439**: 29.

DFID/EC/UNDP/World Bank (2002) *Linking Poverty Reduction and Environmental Management: Policy Challenges and Opportunities*, World Bank Working Paper. Washington, DC, USA: The World Bank.

DHI Water and Environment (2001) Yellow River flood management sector project, China (2000). Unpublished report. Horsholm, Denmark: DHI Water and Environment.

Diaz, R. J. & Rosenberg, R. (1995) Marine benthic hypoxia: a review of its ecological effects and the behavioural responses of benthic macrofauna. *Oceanography and Marine Biology Annual Review* **33**: 245–303.

Dickhut, R. M. & Gustafson, K. E. (1995) Atmospheric inputs of selected polycyclic aromatic hydrocarbons and polychlorinated biphenyls to southern Chesapeake Bay. *Marine Pollution Bulletin* **30**: 385–396.

Dijkema, K. S. (1987) Geography of salt marshes in Europe. *Zeitschrift für Geomorphologie* **31**: 489–499.

Dionne, J.-C. (1989) An estimate of shore ice action in a *Spartina* tidal marsh, St Lawrence estuary, Québec, Canada. *Journal of Coastal Research* **5**: 281–293.

Diop, E. S. (1990) *La Côte Ouest-Africaine: du Saloum (Sénégal) à la Mellacorée (Rép. de Guinée)*, Collection Etudes et Thèses. Paris, France: ORSTOM.

Diop, E. S., Soumare, A., Diallo, N. & Guisse, A. (1997) Recent changes of the mangroves of the Saloum River estuary, Senegal. *Mangroves and Salt Marshes* 1: 163–172.

Dobson, A. P., Bradshaw, A. D. & Baker, A. J. M. (1997) Hopes for the future: restoration ecology and conservation biology. *Science* 277: 515–522.

Dobson, J. E., Bright, E. A., Ferguson, R. L., *et al.* (1995) *NOAA Coastal Change Analysis Program (C-CAP): Guidance for Regional Implementation*, NOAA Technical Report No. NFFS 123. Washington, DC, USA: US Department of Commerce.

Dodd, R. S. & Afzal-Rafii, Z. (2002) Evolutionary genetics of mangroves: continental drift to recent climate change. *Trees* 16: 80–86.

Dodd, R. S., Afzal-Rafii, Z., Kashani, N. & Budrick J. (2002) Land barriers and open oceans: effects on gene diversity and population structure in *Avicennia germinans* L. (Avicenniaceae). *Molecular Ecology* 11: 1327–1338.

Dodd, R. S., Blasco, F., Afzal-Rafii, Z. & Torquebiau, E. (1999) Mangroves of the United Arab Emirates: diversity at the bioclimatic extreme. *Aquatic Botany* 63: 291–304.

Donahue, D. L. (1999) *The Western Range Revisited: Removing Livestock from Public Lands to Conserve Native Biodiversity*. Norman, OK, USA: University of Oklahoma Press.

Done, T. (1992) Constancy and change in some Great Barrier Reef coral communities: 1980–1990. *American Zoologist* 32: 655–662.

Done, T. J. (1987) Simulation of the effects of *Acanthaster planci* on the population structure of massive corals in the genus *Porites*: evidence of population resilience? *Coral Reefs* 6: 75–90.

Donnelly, J. & Bertness, M. D. (2001) Rapid shoreward encroachment of salt marsh vegetation in response to sea-level rise. *Proceedings of the National Academy of Sciences of the USA* 98: 14 218–14 223.

Doody, P. (1989) Conservation and development of the coastal dunes in Great Britain. In: *Perspectives in Coastal Dune Management*, ed. F. van der Meulen, P. D. Jungerius & J. H. Visser, pp. 53–67, The Hague, the Netherlands: SPB Academic Publishing.

Dosdat, A., Gaumet, F. & Chartois, H. (1995) Marine aquaculture effluent monitoring: methodological approach to the evaluation of nitrogen and phosphorus excretion by fish. *Aquacultural Engineering* 14: 59–84.

Douglas, B. C. (1991) Global sea level rise. *Journal of Geophysical Research* 96: 6981–6992.

Douglas, B. C. & Peltier, W. C. (2002) The puzzle of global sea level rise. *Physics Today* March 2002: 35–40.

Douglas, M., Finlayson, C. M. & Storrs, M. J. (1998) Weed management in tropical wetlands of the Northern Territory, Australia. In: *Wetlands in a Dry Land: Understanding for Management*, ed. W. D. Williams, pp. 239–251. Canberra, ACT, Australia: Environment Australia.

Douglas–Westwood (2002) Into the deep [www document]. URL http://www.dw-1.com

Dove, S. (2004) Scleractinian coral with photoprotective host pigments are hypersensitive to thermal bleaching. *Marine Ecology Progress Series* 272: 99–116.

Dove, S., Hoegh-Guldberg, O. & Ranganathan, S. (2001) Major colour patterns in reef-building corals are due to a family of GFP-like proteins. *Coral Reefs* 19: 197–204.

Downing, J. A., McClain, M., Twilley, R., *et al.* (1999) The impact of accelerating land-use change on the N-cycle of tropical aquatic ecosystems: current conditions and projected changes. *Biogeochemistry* 46: 109–148.

Drake, B. G. (1992) A field study of the effects of elevated CO_2 on ecosystem processes in a Chesapeake Bay wetland. *Australian Journal of Botany* 40: 579–595.

Drake, B. G., Muehe, G., Peresta, M. A., González-Meler & Matamala, R. (1996a) Acclimation of photosynthesis, respiration and ecosystem carbon flux of a wetland on Chesapeake Bay, Maryland to elevated atmospheric CO_2 concentration. *Plant and Soil* 187: 111–118.

Drake, B. G., Peresta, G., Beugeling, F. & Matamala, R. (1996b) Long-term elevated CO_2 exposure in a Chesapeake Bay wetland: ecosystem gas exchange, primary production, and tissue nitrogen. In: *Carbon Dioxide and Terrestrial Ecosystems*, ed. G. W. Koch & H. A. Mooney, pp. 197–214. New York, USA: Academic Press.

Drake, P., Baldó, F., Sáenz, V. & Arias, A. M. (1999) Macrobenthic community structure in estuarine pollution assessment on the Gulf of Cádiz (SW Spain): is the phylum-level meta-analysis approach applicable? *Marine Pollution Bulletin* 38: 1038–1047.

Dreher, J. & Gunatilaka, A. (2001) Management of urban groundwater. I. Quantitative aspects. In: *Groundwater Ecology: A Tool For Management of Water Resources*, ed. C. Griebler, D. L. Danielopol, J. Gibert, H. P. Nachtnebel & J. Notenboom, pp. 197–212. Luxembourg: Office for Official Publications of the European Communities.

Dreyer, G. D. & Niering, W. A. (1995) Tidal marshes of Long Island Sound: ecology, history and restoration. *Connecticut College Arboretum Bulletin* **34**: 1–72.

Driscoll, C. T., Lawrence, G. B., Bulger, A. J., *et al.* (2001) Acidic deposition in the Northeastern United States: sources and inputs, ecosystem effects, and management strategies. *BioScience* **51**: 180–198.

Du Rietz, G. E. & Nannfeldt, J. A. (1925) *Ryggmossen und Stigsbo Rödmosse: Die letzten lebenden Hochmoore der Gegend von Uppsala*, Svenska Växtsociologiska Sällskapets Handlingar No. 3. Uppsala, Sweden: Amqvist & Wiksell.

Duarte, C. M. (1995) Submerged aquatic vegetation in relation to different nutrient regimes. *Ophelia* **41**: 87–112.

Duarte, C. M. (2001) Seagrass ecosystems. In: *Encyclopedia of Biodiversity*, vol. 5, ed. S. L. Levin, pp. 255–268. San Diego, CA, USA: Academic Press.

Duarte, C. M. (2002) The future of seagrass meadows. *Environmental Conservation* **29**: 192–206.

Duarte, C. M. & Chiscano, C. L. (1999) Seagrass biomass and production: a reassessment. *Aquatic Botany* **65**: 159–174.

Duarte, C. M., Martínez, R. & Barrón, C. (2002) Biomass, production and rhizome growth near the northern limit of seagrass (*Zostera marina*) distribution. *Aquatic Botany* **72**: 183–189.

Duarte, C. M., Terrados, J., Agawin, N. S. W., *et al.* (1997) Response of a mixed Philippine seagrass meadow to experimental burial. *Marine Ecology Progress Series* **147**: 285–294.

Dubinsky, Z. & Stambler, N. (1996) Marine pollution and coral reefs. *Global Change Biology* **2**: 511–526.

Duda, A. M. & El-Ashry, M. T. (2000) Addressing the global water and environment crises through integrated approaches to the management of land, water and ecological resources. *Water International* **25**: 115–126.

Dudgeon, D. (1999) *Tropical Asian Streams: Zoobenthos, Ecology and Conservation*. Hong Kong: Hong Kong University Press.

Dudgeon, D. (2000*a*) The ecology of tropical Asian rivers and streams in relation to biodiversity conservation. *Annual Review of Ecology and Systematics* **31**: 239–263.

Dudgeon, D. (2000*b*) Large-scale hydrological changes in tropical Asia: prospects for riverine biodiversity. *BioScience* **50**: 793–806.

Dudgeon, D. (2002) An inventory of riverine biodiversity in monsoonal Asia: present status and conservation challenges. *Water Science and Technology* **11**: 11–19.

Dudgeon, D., Arthington, A. H., Gessner, M. O., *et al.* (2006) Freshwater biodiversity: importance, threats, status and conservation challenges. *Biological Reviews* **81**: 163–182.

Dudley, N. (1990) *Nitrates: The Threat to Food and Water*. London, UK: Green Print.

Dugan, P. J., ed. (1990) *Wetland Conservation: A Review of Current Issues and Required Action*. Gland, Switzerland: IUCN.

Dugan, P. J. (1993) *Wetlands in Danger*. New York, USA: Oxford University Press.

Duggins, D. O. (1980) Kelp beds and sea otters: an experimental approach. *Ecology* **61**: 447–453.

Duggins, D. O. (1983) Starfish predation and the creation of mosaic patterns in a kelp dominated community. *Ecology* **64**: 1610–1619.

Duggins, D. O., Eckman, J. E. & Sewell, A. T. (1990) Ecology of understory kelp environments. II. Effects of kelps on recruitment of benthic invertebrates. *Journal of Experimental Marine Biology and Ecology* **143**: 27–45.

Duggins, D. O., Simenstad, C. A. & Estes, J. A. (1989) Magnification of secondary production by kelp detritus in coastal marine ecosystems. *Science* **245**: 170–173.

Duke, N. C. (1995) Genetic diversity, distributional barriers and rafting continents: more thoughts on the evolution of mangroves. *Hydrobiologia* **295**: 167–181.

Duke, N. C., Benzie, J. A. H., Goodall, J. A. & Ballment, E. R. (1998) Genetic structure and evolution of species in the mangrove genus *Avicennia* (Avicenniaceae) in the Indo-West Pacific. *Evolution* **52**: 1612–1626.

Duke, N. C., Pinzón, Z. S. & Prada, M. C. (1997) Large-scale damage to mangrove forests following two large oil spills in Panama. *Biotropica* **29**: 2–14.

Dulvy, N. K., Freckleton, R. P. & Polunin N. V. C. (2004) Coral reef cascades and the indirect effects of predator removal by exploitation. *Ecology Letters* **7**: 410–416.

Dulvy, N. K., Metcalfe, J. D., Glanville, J., Pawson, M. G. & Reynolds, J. D. (2000) Fishery stability, local extinctions, and shifts in community structure in skates. *Conservation Biology* **14**: 283–293.

Dulvy, N. K., Sadovy, Y. & Reynolds, J. D. (2003) Extinction vulnerability in marine populations. *Fish and Fisheries* **4**: 25–64.

Dumont, H. (1995) Ecocide in the Caspian Sea. *Nature* **377**: 673–674.

Dumortier, M., Verlinden, A., Beeckman, H. & van der Mijnsbrugge, K. (1996) Effects of harvesting dates and frequencies on above and below-ground dynamics in Belgian wet grasslands. *Ecoscience* **3**: 190–198.

Dunton, K. H. (1990) Growth and production in *Laminaria solidungula*: relation to continuous underwater light levels in the Alaskan High Arctic. *Marine Biology* **106**: 297–304.

Dunton, K. H. & Dayton, P. K. (1995) The biology of high latitude kelp. In: *Ecology of Fjords and Coastal Waters: Proceedings of the Mare Nor Symposium on the Ecology of Fjords and Coastal Waters*, Tromsø, Norway, 5–9 December 1994, ed. H. R. Skjoldal, C. Hopkins, K. E. Erikstad & H. P. Leinass, pp. 499–507. Amsterdam, the Netherlands: Elsevier.

Dunton, K. H. & Schell, D. M. (1987) Dependence of consumers on macroalgal (*Laminaria solidungula*) carbon in an Arctic kelp community: delta-carbon-13 evidence. *Marine Biology* **93**: 615–626.

Dunton, K. H. & Schonberg, S. V. (2002) Assessment of propeller scarring in seagrass beds of the South Texas Coast. *Journal of Coastal Research* Special Issue **37**: 99–109.

Duplisea, D. E., Jennings, S., Malcolm, S. J., Parker, R. & Sivyer, D. (2001) Modelling the potential impacts of bottom trawl fisheries on soft sediment biochemistry of the North Sea. *Geochemical Transactions* **14**: 1–6.

Dupont, L. M., Jahns, S., Marret, F. & Ning, S. (2000) Vegetation change in equatorial West Africa: time slices for the last 150 ka. *Paleogeography Paleoclimatology Paleoecology* **155**: 95–122.

Durako, M. J. & Moffler, M. D. (1987) Factors affecting the reproductive ecology of *Thalassia testudinum* Hydrocharitacea. *Aquatic Botany* **27**: 79–96.

Duran, L. R. & Castilla, J. C. (1989) Variation and persistence of the middle rocky intertidal community of central Chile, with and without human harvesting. *Marine Biology* **103**: 555–562.

Durgaprasad, M. K. & Anjaneyulu, Y., eds. (2003) *Lake Kolleru: Environmental Status (Past and Present)*. Hyderabad, Andhra Pradesh, India: B.S. Publications.

DWR (2003) Construction of the West Desert Pumping Project. Division of Water Resources, State of Utah [www document]. URL http://www.water.utah.gov/construction/gsl/construction.htm.

Dynesius, M. & Nilsson, C. (1994) Fragmentation and flow regulation of river systems in the northern third of the world. *Science* **266**: 753–762.

Eakin, C. M. (1996) Where have all the carbonates gone? A model comparison of calcium carbonate budgets before and after the 1982–1983 El Niño at Uva Island in the eastern Pacific. *Coral Reefs* **15**: 109–119.

Ebeling, A. W., Laur, D. R. & Rowley, R. J. (1985) Severe storm disturbance and reversal of community structure in a southern California kelp forest. *Marine Biology* **84**: 287–294.

Ebling, F. S., Hawkins, A. D., Kitching, J. A., Muntz, P. & Pratt, V. M. (1966) The ecology of Lough Ine. XVI. Predation and diurnal migrations in *Paracentrotus* community. *Journal of Animal Ecology* **35**: 559–566.

EC (2002) *Towards a Strategy to Protect and Conserve the Marine Environment*, Communication from the Commission to the Council and the European Parliament. Brussels, Belgium: European Commission.

EC (2005) *Directive of the European Parliament and of the Council Establishing a Framework for Community Action in the field of Marine Environmental Policy (Marine Strategy Directive)*, SEC (2005) 1290. Brussels, Belgium: European Commission.

Edelstein, T., Craigie, J. S. & McLachlan, J. (1969) Preliminary survey of the sublittoral flora of Halifax county. *Journal of the Fisheries Research Board of Canada* **26**: 2703–2713.

Edgar, G. J. & Barrett, N. S. (2002) Effects of catchment activities on macrofaunal assemblages in Tasmanian estuaries. *Estuarine, Coastal and Shelf Science* **50**: 639–654.

Edinger, E. N. (1998) Reef degradation and coral biodiversity in Indonesia: effects of land-based pollution, destructive fishing practices and changes over time. *Marine Pollution Bulletin* **36**: 617–630.

Edmondson, W. T. (1991) *The Uses of Ecology: Lake Washington and Beyond*. Seattle, WA, USA: University of Washington Press.

Edmunds, P. J. & Carpenter, R. C. (2001) Recovery of *Diadema antillarum* reduces macroalgal cover and increases abundance of juvenile corals on a Caribbean reef. *Proceedings of the National Academy of Sciences of the USA* **98**: 5067–5071.

Edwards, A. C. & Davis, D. E. (1974) Effects of herbicides on the *Spartina* salt marsh. In: *Ecology of Halophytes*, ed. R. J. Reimold & W. H. Queen, pp. 531–545. New York, USA: Academic Press.

Edyvane, K. (2003) *Conservation, Monitoring and Recovery of Threatened Giant Kelp* (Macrocystis pyrifera) *Beds in Tasmania*, Report for Environment Australia. Hobart, Tasmania, Australia: Department of Primary Industries, Water and Environment. Available online at www.dpiwe.tas.gov.au/inter.nsf/Publications/HMUY-5TT2P6?open

Eganhouse, R. P. & Sherblom, P. M. (2001) Anthropogenic organic contaminants in the effluent of a combined sewer overflow: impact on Boston Harbor. *Marine Environmental Research* **51**: 51–74.

Ehrenfeld, J. G. (1990) Dynamics and processes of barrier island vegetation. *Aquatic Science* **2**: 437–480.

Eiane, K., Aksnes, D. L., Bagoien, E. & Kaartvedt, S. (1999) Fish or jellies: a question of visibility? *Limnology and Oceanography* **44**: 1352–1357.

Eisenreich, S., Emmling, P. & Beeton, A. (1977) Atmospheric loading of phosphorus and other chemicals to Lake Michigan. *Journal of Great Lakes Research* **3**: 291–304.

Eisler, R. (1987) *Polycyclic Aromatic Hydrocarbon Hazards to Fish, Wildlife, and Invertebrates: A Synoptic Review*, Biological Report No. 85(1.11). Washington, DC, USA: US Fish and Wildlife Service.

Eisler, R. (2000) Polycyclic aromatic hydrocarbons. In: *Handbook of Chemical Risk Assessment*, vol. 2, pp. 1343–1411. Boca Raton, FL, USA: Lewis Publishers.

Eisma, D. (1998) *Intertidal Deposits: River Mouths, Tidal Flats, and Coastal Lagoons*. Boca Raton, FL, USA: CRC Press.

Eliot, I., Finlayson, C. M. & Waterman, P. (1999) Predicted climate change, sea-level rise and wetland management in the Australian wet–dry tropics. *Wetlands Ecology and Management* **7**: 63–81.

Ellenbroek, G. A., ed. (1987) *Ecology and Productivity of an African Wetland System: The Kafue Flats, Zambia*. The Hague, the Netherlands: Dr W. Junk.

Ellery, W. N. & McCarthy, T. S. (1998) Environmental change over two decades since dredging and excavation of the lower Boro River, Okavango Delta, Botswana. *Journal of Biogeography* **25**: 361–378.

Ellingsen, K. E. (2002) Soft-sediment benthic biodiversity on the continental shelf in relation to environmental variability. *Marine Ecology Progress Series* **232**: 15–27.

Elliott, W. R. (2000) Conservation of the North American cave and karst biota. In: *Ecosystems of the World*, vol. 30, Subterranean Ecosystems, ed. H. Wilkens, D. C. Culver & W. F. Humphreys, pp. 664–689. Amsterdam, the Netherlands: Elsevier.

Ellis, C. J. & Tallis, J. H. (2000) Climatic control of blanket mire development at Kentra Moss, north-west Scotland. *Journal of Ecology* **88**: 869–889.

Ellis, D. V. & Pattisina, L. A. (1990) Widespread neogastropod imposex: a biological indicator of global TBT pollution. *Marine Pollution Bulletin* **21**: 248–253.

Ellis, J., Cummings, V., Hewitt, J., Thrush, S. & Norkko, A. (2002) Determining effects of suspended sediment on condition of a suspension feeding bivalve (*Atrina zelandica*): results of a survey, a laboratory experiment and a field transplant experiment. *Journal of Experimental Marine Biology and Ecology* **267**: 147–174.

Ellis, L. M., Crawford, C. S. & Molles Jr, M. C. (2002) The role of the flood pulse in ecosystem-level process in southwestern riparian forests: a case study from the Middle Rio Grande. In: *Flood Pulsing and Wetlands: Restoring the Natural Balance*, ed. B. A. Middleton, pp. 51–108. New York, USA: John Wiley.

Ellis, R. (2003) *The Empty Ocean*. Washington, DC, USA: Island Press.

Ellison, A. M. & Farnsworth, E. J. (1996) Anthropogenic disturbance of Caribbean mangrove ecosystems: past impacts, present trends, and future predictions. *Biotropica* **28**: 549–565.

Ellison, A. M. & Farnsworth, E. J. (2001) Mangrove communities. In: *Marine Community Ecology*, ed. M. D. Bertness, S. D. Gaines & M. E. Hay, pp. 423–442. Sunderland, MA, USA: Sinauer Associates.

Ellison, A. M., Farnsworth, E. J. & Merkt, R. E. (1999) Origins of mangrove ecosystems and the mangrove biodiversity anomaly. *Global Ecology and Biogeography Letters* **8**: 95–115.

Ellison, J. C. & Stoddart, D. R. (1991) Mangrove ecosystem collapse during predicted sea-level rise: Holocene analogues and implications. *Journal of Coastal Research* **7**: 151–165.

El-Raey, M., Nasr, S., Frihy, O., Desouki, S. & Dewidar, K. (1995) Potential impacts of accelerated sea-level rise on Alexandria Governate, Egypt. *Journal of Coastal Research* Special Issue **14**: 190–204.

El-Sayed, S., Van Dijken, G. & Gonzalez-Rodas, G. (1996) Effects of ultraviolet radiation on marine ecosystems. *International Journal of Environmental Studies* **51**: 199–216.

Elster, C., Perdomo, L. & Schnetter, M.-L. (1999) Impact of ecological factors on the regeneration of mangroves in the Cíenaga Grande de Santa Marta, Colombia. *Hydrobiologia* **413**: 35–46.

EM-DAT (2004) The OFDA/CRED Emergency Disaster Database. Brussels, Belgium: Université Catholique de Louvain [www document]. URL http://www.em-dat.net/disasters/list.php.

Emerson, S. (1985) Organic carbon preservation in marine sediments. In: *The Carbon Cycle and Atmospheric CO_2: Natural Variations Archean to Present*, ed. E. T. Sundquist & W. Broecker, pp. 711–717. Washington, DC, USA: American Geophysical Union.

Emmerson, M. C., Solan, M., Emes, C., Paterson, D. M. & Raffaelli, D. (2001) Consistent patterns and the idiosyncratic effects of biodiversity in marine ecosystems. *Nature* **411**: 73–77.

Enfield, D. B. (1988) Is El Niño becoming more common? *Oceanography* **59**: 123–127.

Engelman, R. & LeRoy, P. (1993) *Sustaining Water: Population and the Future of Fresh Water Supplies.* Washington, DC, USA: Population Action International.

Engelman, R., Cincotta, R. P., Dye, B., Gardner-Outlaw, T. & Wisnewski, J. (2000) *People in the Balance: Population and Natural Resources at the Turn of the Millennium.* Washington, DC, USA: Population Action International.

Environment Canada (2002) Risk management strategy for road salts [www document]. URL http://www.ec.gc.ca/nopp/roadsalt/en/rms.htm

Epp, D. & Smoot, N. C. (1989) Distribution of seamounts in the North Atlantic. *Nature* **337**: 254–257.

Erlandson, J. M. (2002) Anatomically modern humans, maritime voyaging, and the Pleistocene colonization of the Americas. In: *The First Americans: The Pleistocene Colonization of the New World*, ed. N. G. Jablonski, pp. 1–19. San Francisco, CA, USA: California Academy of Sciences.

Erlandson, J. M., Kennett, D. J., Ingram, B. L., *et al.* (1996) An archaeological and paleontological chronology for Daisy Cave (CA-SMI-261), San Miguel Island, California. *Radiocarbon* **38**: 355–373.

Erlandson, J. M., Rick, T. C. & Vellanoweth, R. L. (2004) Human impacts on ancient environments: a case study from the Northern Channel Islands. In: *Voyages of Discovery: The Archaeology of Islands*, ed. S. M. Fitzpatrick, pp. 51–83. New York, USA: Praeger.

Esselink, J. (2000) *Nature Management of Coastal Salt Marshes: Interactions between Anthropogenic Influences and Natural Dynamics.* Haren, the Netherlands: Koeman en Bijkerk.

Esselink, P., Helder, G. J. F., Aerts, B. A. & Gerdes, K. (1997) The impact of grubbing by greylag geese (*Anser anser*) on the vegetation dynamics on a tidal marsh. *Aquatic Botany* **55**: 261–279.

Esteban Morales, G. (2003) Methane hydrates in the Chilean continental margin. *Journal of Biotechnology* **6**: 1–5.

Estes, J. A. & Duggins, D. O. (1995) Sea otters and kelp forests in Alaska: generality and variation in a community ecological paradigm. *Ecological Monographs* **65**: 75–100.

Estes, J. A. & Steinberg, P. D. (1988) Predation, herbivory, and kelp evolution. *Paleobiology* **14**: 19–36.

Estes, J. A., Duggins, D. O. & Rathbun, G. B. (1989) The ecology of extinctions in kelp forest communities. *Conservation Biology* **3**: 252–264.

Estes, J. A., Tinker, M. T., Williams, T. M. & Doak, D. F. (1998) Killer whale predation on sea otters linking oceanic and nearshore ecosystems. *Science* **282**: 473–476.

EU (2001) Common strategy on the implementation of the Water Framework Directive [www document]. URL http://www.europa.eu.int/comm/environment/water/water-framework

Evans, A. S., Webb, K. L. & Penhale, P. A. (1986) Photosynthetic temperature acclimation in two coexisting seagrasses *Zostera marina* and *Ruppia maritima*. *Aquatic Botany* **24**: 185–198.

Evans, C. D., Cullen, J. M., Alewell, C., *et al.* (2001) Recovery from acidification in European surface waters. *Hydrology and Earth System Sciences* **5**: 283–297.

Evans, N. T. & Short, F. T. (2005) Functional trajectory models for assessment of transplant development of seagrass, *Zostera marina* L., beds in the Great Bay Estuary, NH, USA. *Estuaries* **28**: 936–947.

Evans, R. S. (1989) Saline water disposal options in the Murray Basin. *BMR Journal of Australian Geology and Geophysics* **11**: 167–185.

Everard, M. (1997) Development of a British wetland strategy. *Aquatic Conservation: Marine and Freshwater Ecosystems* **7**: 223–238.

Everson, I. (2002) Consideration of major issues in ecosystem monitoring and management. *CCAMLR Science* **9**: 213–232.

Everson, I., ed. (2000) *Krill: Biology, Ecology and Fisheries.* Oxford, UK: Blackwell Science.

Ewel, K. C. & Odum, H. T., eds. (1984) *Cypress Swamps.* Gainesville, FL, USA: University of Florida Press.

Ewel, K. C., Twilley, R. R. & Ong, J. E. (1998) Different kinds of mangrove forests provide different goods and services. *Global Ecology and Biogeography Letters* **7**: 83–94.

Fach, B. A., Hofmann, E. E. & Murphy, E. J. (2002) Modeling studies of Antarctic krill *Euphausia superba* survival during transport across the Scotia Sea. *Marine Ecology Progress Series* **231**: 187–203.

Fagan, W. F., Lewis, M. A., Neubert, M. G. & van den Driessche, N. P. (2002) Invasion theory and biological control. *Ecology Letters* **5**: 148–157.

Fagbami, A. A., Udo, E. J. & Odu, C. T. I. (1988) Vegetation damage in an oil field in the Niger delta of Nigeria. *Journal of Tropical Ecology* **4**: 61–75.

Falkenmark, M. & Widstrand, C. (1992) Population and water resources: a delicate balance. *Population Bulletin* **47** (3): 1–36.

FAO (1990) *Fertilizer Yearbook.* New York, USA: United Nations.

FAO (1994) *FAO Yearbook: Fishery Statistics, Catches and Landings 1992*, vol. 74. Rome, Italy: Food and Agriculture Organization of the United Nations.

FAO (1996) *The State of World Fisheries and Aquaculture 1996*, Technical Report. Rome, Italy: Food and Agriculture Organization of the United Nations.

FAO (1998) *Wetland Characterization and Classification for Sustainable Agricultural Development*. Harare, Zimbabwe: Food and Agriculture Organization of the United Nations (FAO), Sub-Regional Office for East and Southern Africa (SAFR).

FAO (1999) *Review of the State of World Fishery Resources: Inland Fisheries*, FAO Fisheries Circular No. 942. Rome, Italy: Food and Agriculture Organization of the United Nations.

FAO (2001a) Asia-Pacific conference on early warning, prevention, preparedness, and management of disasters in food and agriculture. Unpublished report. Rome, Italy: Food and Agriculture Organization of the United Nations.

FAO (2001b) *What Is the Code of Conduct for Responsible Fisheries?* Rome, Italy: Food and Agriculture Organization of the United Nations.

FAO (2002) *World Agriculture: Towards 2015/2030 – An FAO Perspective*. Rome, Italy: Food and Agriculture Organization of the United Nations.

FAO (2003a) *International Year of Rice, 2004*. Rome, Italy: Food and Agriculture Organization of the United Nations. Available online at www.fao.org/rice2004/index_en.htm

FAO (2003b) *Status and Trends in Mangrove Area Extent Worldwide*, Forest Resources Assessment Working Paper No. 63, ed. M.L. Wilkie & S. Fortuna. Rome, Italy: Forest Resources Division, Food and Agriculture Organization of the United Nations.

FAO (2004) *The State of World Fisheries and Aquaculture 2004*. Rome, Italy: Food and Agriculture Organization of the United Nations. Available online at www.fao.org

FAO (2006) *The State of World Fisheries and Aquaculture 2006*. Rome, Italy: Food and Agriculture Organization of the United Nations. Available online at www.fao.org

Farina, J. M. & Castilla, J. C. (2001) Temporal variation in the diversity and cover of sessile species in rocky intertidal communities affected by copper mine tailings in northern Chile. *Marine Pollution Bulletin* 42: 554–568.

Farman, J. C., Gardiner, B. G. & Shanklin, J. D. (1985) Large losses of total ozone in Antarctica reveal seasonal ClO_X/NO_X interaction. *Nature* 315: 207–210.

Farnsworth, E. J., Ellison, A. M. & Gong, W. K. (1996) Elevated CO_2 alters anatomy, physiology, growth and reproduction of red mangrove (*Rhizophora mangle* L.). *Oecologia* 108: 599–609.

Farrier, D. & Tucker, L. (2000) Wise use of wetlands under the Ramsar Convention: a challenge for meaningful implementation of international law. *Journal of Environmental Law* 12: 21–32.

Feldt, W. G., Kanisch, G., Kanisch, M. & Vobach, M. (1985) Radioecological studies of sites in the northeast Atlantic used for dumping low-level radioactive wastes: results of the research cruises of FRV *Walter Herwig*. *Archiv für Fischereiwissenschaft* 35: 91–195.

Fellows, M. Q. & Zedler, J. B. (1999) Effects of the non-native grass, *Parapholis incurva* (Poaceae) on the rare and endangered hemiparasite, *Cordylanthus maritimus* subsp. *maritimus* (Scrophulariaceae). *Madroño* 52: 91–98.

Felzer, B. & Heard, P. (1999) Precipitation differences amongst GCMs used for the US National Assessment. *Journal of American Water Works Association* 35: 1327–1339.

Feminella, J. W. (1996) Comparison of benthic macroinvertebrate assemblages in small streams along a gradient of flow permanence. *Journal of the North American Benthological Society* 15: 651–669.

Fenchel, T. & Blackburn, T. H. (1979) *Bacteria and Mineral Cycling*. London, UK: Academic Press.

Ferguson, N. P. & Lee, J. A. (1979) Some effects of bisulphite and sulphate on the growth of *Sphagnum* species in the field. *Environmental Pollution Series A* 21: 59–71.

Fernandes, M. B., Sicre, M.-A., Boireau, A. & Tronczynski, J. (1997) Polycyclic hydrocarbon (PAH) distributions in the Seine River and its estuary. *Marine Pollution Bulletin* 34: 857–862.

Ferreira, L. V. & Stohlgren, T. J. (1999) Effects of river level fluctuation on plant species richness, diversity, and distribution in a floodplain forest in Central Amazonia. *Oecologia* 120: 582–587.

Ferrier-Pagès, C., Gattuso, J.-P., Dallot, S. & Jaubert, J. (2000) Effect of nutrient enrichment on growth and photosynthesis of the zooxanthellate coral *Stylophora pistillata*. *Coral Reefs* 19: 103–113.

Field, C. D. (1995) Impact of expected climate change on mangroves. *Hydrobiologia* 295: 75–81.

Field, C. D., ed. (1996) *Restoration of Mangrove Ecosystems*. Okinawa, Japan: International Society of Mangrove Ecosystems.

Field, C. D. (1998) Rehabilitation of mangrove ecosystems: an overview. *Marine Pollution Bulletin* 37: 383–392.

Field, C. D. (1999) Mangrove rehabilitation: choice and necessity. *Hydrobiologia* 413: 47–52.

Field, J. G., Griffiths, C. L., Griffiths, R. J., *et al.* (1980) Variation in structure and biomass of kelp communities along the south-west Cape coast. *Transactions of the Royal Society of South Africa* **44**: 145–203.

Field, J. G., Griffiths, C. L., Linley, E.A, Carter R. A. & Zoutendyk, P. (1977) Upwelling in the nearshore marine ecosystem and its biological implications. *Estuarine and Coastal Marine Science* **11**: 133–150.

Field, J. G., Hempel, G. & Summerhayes, C. P. (2002) *Oceans 2020: Science, Trends, and the Challenge of Sustainability.* Washington, DC, USA: Island Press.

Fielding, A. H. & Russell, G. (1976) The effect of copper on competition between marine algae. *Journal of Applied Ecology* **13**: 871–876.

Fields, P., Graham, J., Rosenblatt, R. & Somero, G. (1993) Effects of expected global climate change on marine faunas. *Trends in Ecology and Evolution* **8**: 361–367.

Figueras, M. J., Polo, F., Inza, I. & Guarro, J. (1997) Past, present and future perspectives of the EU bathing water directive. *Marine Pollution Bulletin* **34**: 148–156.

Figueroa, M. E., Castillo, J. M., Redondo, S., *et al.* (2003) Facilitated invasion by hybridization of *Sarcocornia* species in a salt-marsh succession. *Journal of Ecology* **91**: 616–626.

Findlay, R. H., Watling, L. & Mayer, L. M. (1995) Environmental impact of salmon net-pen culture on marine benthic communities in Maine: a case study. *Estuaries* **18**: 145–179.

Finlayson, C. M. (1996) Information required for wetland management in the South Pacific. In: *Wetland Conservation in the Pacific Islands Region: Proceedings of the Regional Workshop on Wetland Protection and Sustainable use in Oceania*, Port Moresby, Papua New Guinea, 6–10 June 1994, ed. R. Jaensch, pp. 185–201. Canberra, ACT, Australia: Wetlands International–Asia/Pacific.

Finlayson, C. M. & Davidson, N. C. (1999) Global review of wetland resources and priorities for wetland inventory: summary report. Unpublished report. Wageningen, the Netherlands: Wetlands International and Jabiru.

Finlayson, C. M. & Eliot, I. (2001) Ecological assessment and monitoring of coastal wetlands in Australia's wet-dry tropics: a paradigm for elsewhere? *Coastal Management* **29**: 105–115.

Finlayson, C. M. & Moser, M., eds. (1992) *Wetlands*. Oxford, UK: Facts on File.

Finlayson, C. M. & Rea, N. (1998) Reasons for the loss and degradation of Australian wetlands. *Wetlands Ecology and Management* **7**: 1–11.

Finlayson, C. M. & Spiers, A. G., eds. (1999) *Global Review of Wetland Resources and Priorities for Wetland Inventory*, Supervising Scientist Report No. 144/Wetlands International Publication No. 53. Canberra, ACT, Australia: Supervising Scientist.

Finlayson, C. M. & von Oertzen, I. (1993) Wetlands in Australia: Northern (tropical) Australia. In: *Wetlands of the World, vol.1, Inventory, Ecology and Management*, ed. D. F. Whigham, D. Dykyjová & S. Hejný, pp. 195–243. Dordrecht, the Netherlands: Kluwer.

Finlayson, C. M. & von Oertzen, I., eds. (1996) *Landscape and Vegetation Ecology of the Kakadu Region, Northern Australia*. Dordrecht, the Netherlands: Kluwer.

Finlayson, C. M., Begg, G. W., Howes, J., *et al.* (2002) *A Manual for an Inventory of Asian Wetlands: Version 1.0*. Kuala Lumpur, Malaysia: Wetlands International.

Finlayson, C. M., Davidson, N. C., Spiers, A. G. & Stevenson, N. J. (1999) Global wetland inventory: status and priorities. *Marine and Freshwater Research* **50**: 717–727.

Finlayson, C. M., D'Cruz, R. & Davidson, N., Synthesis Team Co-Chairs (2005) *Ecosystems and Human Well-Being: Wetlands and Water Synthesis*, a report of the Millennium Ecosystem Assessment. Washington, DC, USA: World Resources Institute.

Finlayson, C. M., Hall, R. & Bayliss, B. (1998) *Regional Review of Wetlands Management Issues: Wet–Dry Tropics of Northern Australia*, LWRRDC Occasional Paper No. 03/97. Canberra, ACT, Australia: Supervising Scientist.

Firth, P. & Fisher, S. G., eds. (1992) *Global Climate Change and Freshwater Ecosystems*. Berlin, Germany: Springer-Verlag.

Fischer, G. & Heilig, G. K. (1997) Population momentum and the demand on land and water resources. *Philosophical Transactions of the Royal Society of London B* **352**: 869–889.

Fish, C. (1960) *Limnological Survey of Eastern and Central Lake Erie, 1928–29* US Fish and Wildlife Service Special Science Report, Fisheries No. 334. Washington, DC, USA: US Fish and Wildlife Service.

Fishelson, L. (1983) Population ecology and biology of *Dotilla sulcata* (Crustacea, Ocypodidae) on sandy beaches of the Red Sea. In: *Sandy Beaches as Ecosystems*, ed. A. McLachlan & T. Erasmus, pp. 643–654. The Hague, the Netherlands: Dr W. Junk.

Fisher, C. R., Urcuyo, I. A., Simpkins, M. A. & Nix, E. (1997) Life in the slow lane: growth and longevity of cold-seep vestimentiferans. *Marine Ecology – Publicazioni della Stazione Zoologica di Napoli* **18**: 83–94.

Fitt, W. K., Brown, B. E., Warner, M. E. & Dunne, R. P. (2001) Coral bleaching: interpretation of thermal tolerance limits and thermal thresholds in tropical corals. *Coral Reefs* **20**: 51–65.

Fletcher, H. & Frid, C. (1996) Impact and management of visitor pressure on rocky intertidal algal communities. *Aquatic Conservation: Marine and Freshwater Ecosystems* **6**: 287–297.

Flower, R. (1994) A review of current biological and recent environmental research on Lake Baikal from a British perspective. *Freshwater Forum* **4**: 8–22.

Flower, R. (1998) Paleolimnology and recent environmental change in Lake Baikal: an introduction and overview of interrelated concurrent studies. *Journal of Paleolimnology* **20**: 107–117.

Fojt, W. J. (1994) Dehydration and the threat to East Anglian fens, England. *Biological Conservation* **69**: 163–175.

Foley, J. A., Kutzbach, J. E., Coe, M. T. & Levis, S. (1994) Feedbacks between climate and boreal forests during the Holocene epoch. *Nature* **371**: 52–54.

Folke, C. & Kaaberger, T., eds. (1991) *Ecology, Economy and Environment*, vol. 1, *Linking the Natural Environment and the Economy: Essays from the Eco-Eco Group*, Dordrecht, the Netherlands: Kluwer.

Folt, C. L., Chen, C. Y., Moore, M. V. & Burnaford, J. (1999) Synergism and antagonism among multiple stressors. *Limnology and Oceanography* **44**: 864–877.

Fong, P. & Glynn, P. W. (1998) A dynamic size-structured population model: does disturbance control size structure of a population of the massive coral *Gardineroseris panulata* in the Eastern Pacific? *Marine Biology* **130**: 663–674.

Fong, P. & Glynn, P. W. (2000) A regional model to predict coral population dynamics in response to El Niño–Southern Oscillation. *Ecological Applications* **10**: 842–854.

Fonseca, M. S., Kenworthy, W. J. & Thayer, G. W. (1998) *Guidelines for the Conservation and Restoration of Seagrasses in the United States and Adjacent Waters*, NOAA Coastal Ocean Program Decision Analysis Series No. 12. Silver Spring, MD, USA: NOAA Coastal Ocean Office.

Fonseca, M. S., Thayer, G. W. & Chester, A. J. (1984) Impact of scallop harvesting on eelgrass (*Zostera marina*) meadows: implications for management. *North American Journal of Fisheries Management* **4**: 286–293.

Fortes, M. D. (1988) Mangrove and seagrasses of East Asia: habitats under stress. *Ambio* **17**: 207–213.

Forum Biodiversität Schweiz (2004) *Biodiversität in der Schweiz: Zustand, Erhalt, Perspektiven.* Bern, Switzerland: Haupt-Verlag.

Fosberg, F. R. (1971) Mangroves versus tidal waves. *Biological Conservation* **4**: 38–39.

Fossa, J. H., Mortensen, P. B. & Furevik, D. M. (2002) The deep-water coral *Lophelia pertusa* in Norwegian waters: distribution and fishery impacts. *Hydrobiologia* **471**: 1–12.

Foster, D. R., King, G. A., Glaser, P. H. & Wright, H. E. (1983) Origin of string patterns in boreal peatlands. *Nature* **306**: 256–258.

Foster, G. & Carter, L. (1997) Mud sedimentation on the continental shelf at an accretionary margin: Poverty Bay, New Zealand. *New Zealand Journal of Geology and Geophysics* **40**: 157–173.

Foster, I. D. L. & Charlesworth, S. M. (1996) Heavy metals in the hydrological cycle: trends and explanation. *Hydrological Processes* **10**: 227–261.

Foster, M. & Schiel, D. R. (1985) *The Ecology of Giant Kelp Forests in California: A Community Profile*, Biological Report No. 85. Washington, DC, USA: US Fish and Wildlife Service.

Foster, M. S. (1990) Organization of macroalgal assemblages in the Northeast Pacific: the assumption of homogeneity and the illusion of generality. *Hydrobiologia* **192**: 21–33.

Foster, M. S., Tarpley, J. A. & Dern, S. L. (1990) To clean or not to clean: the rationale, methods, and consequences of removing oil from temperate shores. *Northwest Environmental Journal* **6**: 105–120.

Fowler, A. M. & Hennessey, K. J. (1995) Potential impacts of global warming on the frequency and magnitude of heavy precipitation. *Natural Hazards* **11**: 283–303.

Fox, J. J. (1998) Shoals and reefs in Australia–Indonesia relations: traditional Indonesian fishermen. In: *Australia in Asia: Episodes*, ed. A. Milner & M. Quilty, pp. 111–139. Melbourne, Victoria, Australia: Oxford University Press.

Francour, P., Ganteaume, A. & Poulain, M. (1999) Effects of boat anchoring in *Posidonia oceanica* seagrass beds in the Port-Cros National Park (north-western Mediterranean Sea). *Aquatic Conservation: Marine and Freshwater Ecosystems* **9**: 391–400.

Frankl, R. & Schmeidl, H. (2000) Vegetation change in a south German raised bog: ecosystem engineering by plant species, vegetation switch or ecosystem level feedback mechanisms? *Flora* **195**: 265–276.

Fraser, W. R., Trivelpiece, W. Z., Ainley, G. G. & Trivelpiece, S. G. (1992) Increases in Antarctic penguin populations: reduced competition with whales or loss of sea ice due to environmental warming. *Polar Biology* **11**: 525–531.

Freestone, A. L. & Nordstrom, K. F. (2001) Early development of vegetation in restored dune plant microhabitats on a nourished beach at Ocean City, New Jersey. *Journal of Coastal Conservation* 7: 105–116.

Frei, C., Schär, C., Lüthi, D. & Davies, H. C. (1998) Heavy precipitation processes in a warmer climate. *Geophysical Research Letters* 25: 1431–1434.

Fremlin, G., ed. (1974) *The National Atlas of Canada*, 4th edn. Toronto, Ontario, Canada: Macmillan.

French, G. T., Awisoka, L. F. & Ibe, C. E. (1995) Sea-level rise and Nigeria: potential impacts and consequences. *Journal of Coastal Research* Special Issue 14: 224–242.

Frenken, K., & Mharapara, I., eds. (2002) *Wetland Development and Management in SADC Countries, Proceedings of a sub-regional workshop*, 19–23 November 2001. Harare, Zimbabwe: Sub-regional Office for Southern and East Africa of the Food and Agriculture Organization of the United Nations.

Friday, A. & Ingram, D. S., eds. (1985) *The Cambridge Encyclopedia of Life Sciences*. Cambridge, UK: Cambridge University Press.

Friedlander, A. M. & Parrish, J. D. (1998) Habitat characteristics affecting fish assemblages on a Hawaiian coral reef. *Journal of Experimental Marine Biology and Ecology* 224: 1–30.

Friedman, J. M. & Auble, G. T. (2000) Floods, flood control, and bottomland vegetation. In: *Inland Flood Hazards: Human, Riparian, and Aquatic Communities*, ed. E. Wohl, pp. 219–237. Cambridge, UK: Cambridge University Press.

Friedrich, R. & Reis, S. (2000) *Tropospheric Ozone Abatement*. Berlin, Germany: Springer-Verlag.

Frissell, C. A., Liss, W. J., Warren, C. E. & Hurley, M. D. (1986) A hierarchical framework for stream habitat classification: viewing streams in a watershed context. *Environmental Management* 10: 199–214.

Froese, R. & Binohlan, C. (2000) Empirical relationships to estimate asymptotic length, length at first maturity, and length at maximum yield per recruit in fishes, with a simple method to evaluate length–frequency data. *Journal of Fish Biology* 56: 758–773.

Fryer, G. & Iles, T. D. (1972) *The Cichlid Fishes of the Great Lakes of Africa*. Edinburgh, UK: Oliver & Boyd.

Fujimoto, K., Miyagi, T., Kikuchi, T. & Kawana, T. (1996) Mangrove habitat formation and response to Holocene sea-level changes on Kosrae Island, Micronesia. *Mangroves and Salt Marshes* 1: 47–57.

Fujita, D. (1987) The report of interviews with fishermen on Isoyake in Taisei-cho, Hokkaido. *Suisanzoshoku* 35: 135–138.

Fujita, D. (1998) Strongylocentrotid sea urchin–dominated barren grounds on the Sea of Japan coast of northern Japan. In: *Echinoderms*, ed. R. Mooi & M. Telford, pp. 659–664. Rotterdam, the Netherlands: A.A. Balkema.

Fukami, T. & Morin, P. J. (2003) Productivity–biodiversity relationships depend on the history of community assembly. *Nature* 424: 423–426.

Funnell, B. F. & Boomer, I. (1998) Microbiofacies, tidal level and age deduction in Holocene saltmarsh on the North Norfolk Coast. *Bulletin of the Geological Society of Norfolk* 46: 31–55.

Funnell, B. F. & Pearson, I. (1989) Holocene sedimentation on the North Norfolk barrier coast in relation to relative sea-level. *Journal of Quaternary Science* 4: 25–36.

Gacia, E. & Duarte, C. M. (2001) Elucidating sediment retention by seagrasses: sediment deposition and resuspension in a Mediterranean (*Posidonia oceanica*) meadow. *Estuarine, Coastal and Shelf Science* 52: 505–514.

Gacia, E., Duarte, C. M., Marbà, N., *et al.* (2003) Sediment deposition and production in SE-Asia seagrass meadows. *Estuarine, Coastal and Shelf Science* 56: 909–919.

Gaden, M. (1998) Phosphorus targets achieved in Lake Erie [www document]. URL http://www.enviro-mich@great-lakes.net

Gage, J. D. & Tyler, P. A. (1991) *Deep-Sea Biology: A Natural History of Organisms at the Deep-Sea Floor*. Cambridge, UK: Cambridge University Press.

Gagne, J. A., Mann, K. H. & Chapman, A. R. O. (1982) Seasonal patterns of growth and storage in *Laminaria longicruris* in relation to differing patterns of availability of nitrogen in the water. *Marine Biology* 69: 91–101.

Gaines, S. D. & Roughgarden, J. (1985) Larval settlement rate, a leading determinant of structure in sessile marine species. *Nature* 360: 579–580.

Gaines, S. D. & Roughgarden, J. (1987) Fish in offshore kelp forests affect recruitment to intertidal barnacle populations. *Science* 235: 479–481.

Galatowitsch, S. M., Anderson, N. O. & Ascher, P. D. (1999) Invasiveness in wetland plants in temperate North America. *Wetlands* 19: 733–755.

Gales, R. (1997) Albatross populations: status and threats. In: *Albatross Biology and Conservation*, ed. G. Robertson & R. Gales, pp. 20–45. Chipping Norton, NSW, Australia: Surrey Beatty.

Galloway, J. (1995) Acid deposition: perspectives in time and space. *Water, Air and Soil Pollution* 85: 15–24.

Galloway, J. N., Levy II, H. & Kasibhatia, P. S. (1994) Year 2020: consequences of population growth and development on deposition of oxidized nitrogen. *Ambio* **23**: 123–132.

Gan, B. K. (1995) *A Working Plan for the Matang Mangrove Forest Reserve, Perak*, 4th revision. Ipoh, Malaysia: State Forestry Department of Perak Darul Ridzuan.

Garcia, S. & Grainger, R. (1996) Fisheries management and sustainability: a new perspective on an old problem. In: *Developing and Sustaining World Fisheries Resources: The State of Science and Management, 2nd World Fisheries Congress Proceedings*, ed. D. A. Hancock, D. C. Smith, A. Grant & J. P. Beumer, pp. 631–645. Collingwood, Victoria, Australia: CSIRO.

Garcia-Charton, J. A. & Pérez-Ruzafa, Á. (2001) Spatial pattern and habitat structure of a Mediterranean rocky reef fish local assemblage. *Marine Biology* **138**: 917–934.

Gardner, T. A., Cote, I. M., Gill, J. A., Grant, A. & Watkinson, A. R. (2003) Long-term region-wide declines in Caribbean corals. *Science* **301**: 958–960.

Garg, J. K., Singh, T. S. & Murthy, T. V. R., eds. (1998) *Wetlands of India*. Ahmedabad, Gujarat, India: Space Applications Centre (ISRO).

Gattuso, J. P., Frankignoulle, M. & Wollast, R. (1998) Carbon and carbonate metabolism in coastal aquatic ecosystems. *Annual Review of Ecology and Systematics* **29**, 405–434.

Gawler, M., ed. (2002) *Strategies for Wise Use of Wetlands: Best Practices in Participatory Management*, proceedings of a workshop held at the 2nd International Conference on Wetlands and Development, November 1998, Dakar, Senegal, WWF Publication No. 56. Wageningen, the Netherlands: Wetlands International and IUCN.

Geddes, M. C. & Williams, W. D. (1987) Comments on *Artemia* introductions and the need for conservation. In: *Artemia Research and its Applications*, vol. 3, ed. P. Sorgeloos, D. A. Bengston, W. Decleir & E. Jaspers, pp. 19–26. Wetteren, Belgium: Universa Press.

Geelen, L. H. W., Cousin, E. F. H. & Schoon, C. F. (1995) Regeneration of dune slacks in the Amsterdam Waterworks dunes. In: *Directions in European Coastal Management*, ed. M. G. Healy & P. J. Doody, pp. 525–532. Cardigan, UK: Samara Publishing.

George, D. G. (2002) Regional-scale influences on the long-term dynamics of lake plankton. In: *Phytoplankton Production*, ed. P. J. Le B. Williams, D. N. Thomas & C. S. Reynolds, pp. 265–290. Oxford, UK: Blackwell Scientific Publications.

George, D. G. & Taylor, A. H. (1995) UK lake plankton and the Gulf Stream. *Nature* **378**: 139.

George, D. G., Talling, J. F. & Rigg, E. (2000) Factors influencing the temporal coherence of five lakes in the English Lake District. *Freshwater Biology* **43**: 449–461.

Gerard, V. A. (1982) Growth and utilization of internal nitrogen reserves by the giant kelp *Macrocystis pyrifera* in a low nitrogen environment. *Marine Biology* **66**: 27–35.

Gerard, V. A. (1997) The role of nitrogen nutrition in high-temperature tolerance of kelp, *Laminaria saccharina*. *Journal of Phycology* **33**: 800–810.

GESAMP (1982) *Scientific Criteria for the Selection of Waste Disposal Sites at Sea*, Reports and Studies No. 16, Joint Group of Experts in the Scientific Aspects of Marine Environmental Protection. London, UK: Inter-Governmental Maritime Consultative Organization.

GESAMP (1990) *The State of the Marine Environment*, Reports and Studies No. 39, Joint Group of Experts on the Scientific Aspects of Marine Environmental Protection. London, UK: IMP.

GESAMP (2001) *Planning and Management for Sustainable Coastal Aquaculture Development*, Report and Studies No. 68, Joint Group of Experts in the Scientific Aspects of Marine Environmental Protection. Rome, Italy: Food and Agriculture Organization of the United Nations.

Getter, C. D., Ballou, T. G. & Koons, C. B. (1985) Effects of dispersed oil on mangroves, synthesis of a seven year study. *Marine Pollution Bulletin* **16**: 318–324.

Ghassemi, F., Jakeman, A. J. & Nix, H. A., eds. (1995) *Salinization of Land and Water Resources: Human Causes, Extent, Management and Case Studies*. Sydney, NSW, Australia: University of New South Wales Press.

Ghiorse, W. C. (1997) Subterranean life. *Science* **275**: 789–790.

Ghosh, A. K. & Mukhopadhyay, R. (2000) *Mineral Wealth of the Ocean*. Rotterdam, the Netherlands: A.A. Balkema.

Gibbs, P. E. & Bryan, G. W. (1987) TBT paints and the demise of the dogwhelk, *Nucella lapillus* (Gastropoda). In: *Oceans '87, Proceedings of the International Organotin Symposium*, vol. 4, pp. 1482–1487. New York, USA: Institute of Electrical and Electronic Engineers.

Gibert, J. (2001a) Basic attributes of groundwater ecosystems. In: *Groundwater Ecology: A Tool for Management of Water Resources*, ed. C. Griebler, D. L. Danielopol, J. Gibert, H. P. Nachtnebel & J. Notenboom, pp. 39–54. Luxembourg: Office for Official Publications of the European Communities.

Gibert, J. (2001*b*) Protocols for the assessment and conservation of aquatic life in the subsurface (PASCALIS): a European project. In: *Mapping Subterranean Biodiversity*, Special Publications No. 6, ed. D. C. Culver, L. Deharveng, J. Gibert & I. D. Sasowsky, pp. 19–21. Charles Town, WV, USA: Karst Waters Institute.

Gibert, J., Danielopol, D. L. & Stanford, J. A., eds. (1994) *Groundwater Ecology*. San Diego, CA, USA: Academic Press.

Gibert, J., Malard, F., Turquin, M.-J. & Laurent, R. (2000) Karst ecosystems in the Rhône River basin. In: *Ecosystems of the World*, vol. 30 *Subterranean Ecosystems*, ed. H. Wilkens, D. C. Culver & W. F. Humphreys, pp. 533–558. Amsterdam, the Netherlands: Elsevier.

Gibson, D. J. & Looney, P. B. (1994) Vegetation colonization of dredge spoil on Perdido Key, Florida. *Journal of Coastal Research* 10: 133–143.

Gibson, D. J., Ely, J. S. & Looney, P. B. (1997) A Markovian approach to modeling succession on a coastal barrier island following beach nourishment. *Journal of Coastal Research* 13: 831–841.

Gichuki, C. M. (2000) Community participation in the protection of Kenya's wetlands. *Ostrich* 71: 122–125.

Gido, K. B. & Brown, J. H. (1999) Invasion of North American drainages by alien fish species. *Freshwater Biology* 42: 387–399.

Giesen, W., Baltzer, M. & Baruadi, R., eds. (1991) *Integrating Conservation with Land-Use Development in Wetlands of South Sulawesi*. Bogor, Indonesia: PHPA and Asian Wetland Bureau.

Gilding, P., Hogarth, M. & Reed, D. (2002) Single bottom line sustainability: how a value centred approach to corporate sustainability can pay off for shareholders and society [www document]. ECOS Corporation, Sydney, Australia: 28 pp. URL http://www.ecoscorporation.com/think/essen/SingleBottomLine.htm

Gill, A. M., Ryan, P. G., Moore, P. H. R. & Gibson, M. (2000) Fire regimes of World Heritage National Park, Australia. *Austral Ecology* 25: 616–625.

Ginsburg, R. N. C. (1994) *Proceedings of the Colloquium on Global Aspects of Coral Reefs: Health, Hazards and History*. Miami, FL, USA: Rosenstiel School of Marine and Atmospheric Science, University of Miami.

Gitay, H., Brown, S., Easterling, W. & Jallow, B. (2001) Ecosystems and their goods and services. In: *Climate Change 2001: Impacts, Adaptations, and Vulnerability – Contribution of Working Group II to the Third Assessment Report of the Intergovernmental Panel on Climate Change*, ed.

J. J. McCarthy, O. F. Canziani, N. A. Leary, D. J. Dokken & K. S. White, pp. 235–342. Cambridge, UK: Cambridge University Press.

Gitay, H., Suárez, A., Watson, R. T. & Dokken, D. J. (2002) *Climate Change and Biodiversity*, IPCC Technical Paper No. 92-9169-104-7. Geneva, Switzerland: Intergovernmental Panel on Climate Change.

Glasby, G. P. (2000) Lessons learned from deep-sea mining. *Science* 289: 551–553.

Glasby, T. & Connell, S. (1999) Urban structures as marine habitats. *Ambio* 28: 595–598.

Glaser, P. H. & Janssens, J. A. (1986) Raised bogs in eastern North America: transitions in landforms and stratigraphy. *Canadian Journal of Botany* 64: 395–415.

Glaser, P. H., Hansen, B. C. S., Siegel, D. I., Reeve, A. S. & Morin, P. J. (2004*a*) Rates, pathways and drivers for peatland development in the Hudson Bay Lowlands, northern Ontario, Canada. *Journal of Ecology* 92: 1036–1053.

Glaser, P. H., Siegel, D. I., Reeve, A. S., Janssens, J. A. & Janecky, D. R. (2004*b*) Tectonic drivers for vegetation patterning and landscape evolution in the Albany River region of the Hudson Bay Lowlands. *Journal of Ecology* 92: 1054–1070.

Gleick, P. H., ed. (1993*a*) *Water in Crisis: A Guide to the World's Fresh Water Resources*. New York, USA: Oxford University Press.

Gleick, P. H. (1993*b*) Water in the 21st century. In: *Water in Crisis: A Guide to the World's Fresh Water Resources*, ed. P. H. Gleick, pp. 105–116. New York, USA: Oxford University Press.

Gleick, P. H. (1996) Basic water requirements for human activities: meeting basic needs. *Water International* 21: 83–122.

Gleick, P. H. (1999) The human right to water. *Water Policy* 1: 487–503.

Gleick, P., Cohen, M. & Mann, A. S. (2002) *The World's Water 2002–2003: The Biennial Report on Freshwater Resources*. Washington, DC, USA: Island Press.

Gleick, P. H., Singh, A. & Shi, H. (2001) *Threats to the World's Freshwater Resources*. Oakland, CA, USA: Pacific Institute; and Nairobi, Kenya: United Nations Environmental Programme.

Glenn, E. P., Brown, J. B. & O'Leary, J. W. (1998) Irrigating crops with seawater. *Scientific American* 279: 76–81.

Glenn, E. P., O'Leary, J. W., Watson, M. C., Thompson, T. L. & Kuehl, R. O. (1991) *Salicornia bigelovii* Torr.: an oilseed halophyte for seawater irrigation. *Science* 251: 1065–1067.

Gliwicz, Z. M. (2002) On the different nature of top-down and bottom-up effects in pelagic food webs. *Freshwater Biology* **47**: 2296–2312.

GLOBEC (2004) Global Ocean Ecosystem Dynamics [www document]. URL http://www.globec.org

Gloersen, P. & Campbell, W. J. (1988) Variations in the Arctic, Antarctic and global sea ice covers during 1978–1987 as observed with the Nimbus 7 scanning multichannel microwave radiometer. *Journal of Geophysical Research* **93**: 10666–10674.

Gloersen, P., Campbell, W. J., Cavalieri, D. J., *et al.* (1992) *Arctic and Antarctic Sea Ice, 1978–1987: Satellite Passive-Microwave Observations and analysis*, NASA Special Publication No. 511. Pasadena, CA: NASA.

Gloersen, P., Parkinson, C. L., Cavalieri, D. J., Comiso, J. C. & Zwally, H. J. (1999) Spatial distribution of trends and seasonality in the hemispheric sea ice covers: 1978–1996. *Journal of Geophysical Research* **104**: 20 827–20 835.

Glover, A. G. & Smith, C. R. (2003) The deep-sea floor ecosystem: current status and prospects for anthropogenic change by the year 2025. *Environmental Conservation* **30**: 219–242.

Glover, A. G., Paterson, G., Bett, B., *et al.* (2001) Patterns in polychaete abundance and diversity from the Madeira Abyssal Plain, northeast Atlantic. *Deep-Sea Research I* **48**: 217–236.

Glover, A. G., Smith, C. R., Paterson, G. L. J., *et al.* (2002) Polychaete species diversity in the central Pacific abyss: local and regional patterns, and relationships with productivity. *Marine Ecology Progress Series* **240**: 157–170.

Glynn, P. W. (1994) State of coral reefs in the Galapagos Islands: natural vs. anthropogenic impacts. *Marine Pollution Bulletin* **29**: 131–140.

Glynn, P. W. (2000) El Niño–Southern Oscillation mass mortalities of reef corals: a model of high temperature marine extinctions? In: *Carbonate Platform Systems: Components and Interactions*, ed. E. Insalaco, P. W. Skelton & T. J. Palmer, pp. 117–133. London, UK: Geological Society of London.

Glynn, P. W. & Ault, J. S. (2000) A biogeographic analysis and review of the far eastern Pacific coral reef region. *Coral Reefs* **19**: 1–23.

Glynn, P. W., Mate, J. L., Baker, A. C. & Calderon, M. O. (2001) Coral bleaching and mortality in Panama and Ecuador during the 1997–1998 El Niño–Southern Oscillation event: spatial/temporal patterns and comparisons with the 1982–1983 event. *Bulletin of Marine Science* **69**: 79–109.

Godcharles, M. F. (1971) *A Study of the Effects of a Commercial Hydraulic Clam Dredge on Benthic Communities in Estuarine Areas*, Technical Series No. 64. Tallahassee, FL, USA: Florida Department of Natural Resources.

Godwin, H. (1981) *The Archives of the Peat Bog.* Cambridge, UK: Cambridge University Press.

Godwin, K. S., Shallenberger, J., Leopold, D. J. & Bedford, B. L. (2002) Linking landscape properties to local hydrogeologic gradients and plant species occurrence in New York fens: a hydrogeologic setting (HGS) framework. *Wetlands* **22**: 722–737.

Gold, T. (1999) *The Deep Hot Biosphere.* New York, USA: Springer-Verlag.

Goldberg, E. D. (1994) Diamonds and plastics are forever? *Marine Pollution Bulletin* **25**: 1–4.

Goldberg, E. D. (1995) Emerging problems in the coastal zone for the twenty-first century. *Marine Pollution Bulletin* **31**: 152–158.

Goldberg, E. D. (1998) Marine pollution: an alternative view. *Marine Pollution Bulletin* **36**: 112–113.

Goldschmidt, T. (1996) *Darwin's Dreampond.* Cambridge, MA, USA: MIT Press.

Gong, W. K., Ong, J. E. & Wong, C. H. (1984) Demographic studies on *Rhizophora apiculata* Bl. in the Matang Mangrove Forest, Malaysia. In: *Ecological Effects of Increasing Human Activities on Tropical and Subtropical Forest Ecosystems*, pp. 64–67. Technical Report of the Malaysian National MAB Committee. Penang, Malaysia: Man and the Biosphere.

González Bernáldez, F., Pérez Pérez, C. & Sterling Carmona, A. (1985) Areas of evaporative discharge from aquifers: little known Spanish ecosystems deserving protection. *Journal of Environmental Management* **21**: 321–330.

González, E. (1991) El manglar de la Cíenaga Grande de Santa Marta, ecosistema en peligro de extinción: Colombia, sus gentes y regiones. *Instituto Geográfico 'Agustín Codazzi'* **21**: 2–21.

Goodall, D. W., ed. (1977–2000) *Ecosystems of the World*, vol. 1–30. Amsterdam, the Netherlands: Elsevier.

Gooday, A. J. (2002) Biological responses to seasonally varying fluxes of organic matter at the ocean floor: a review. *Journal of Oceanography* **58**: 305–332.

Goodland, R. (1991) The environment as capital. In: *Ecological Economics: The Science and Management of Sustainability*, ed. R. Costanza, pp. 168–175. New York, USA: Columbia University Press.

Goolsby, D. A., Battalin, W. A., Lawrence, G. B., *et al.* (1999) *Flux and Sources of Nutrients in the Mississippi–Atchafalaya*

River Basin, Topic 3, *Report for the Integrated Assessment on Hypoxia in the Gulf of Mexico*, NOAA Coastal Ocean Program Decision Analysis Series No. 17. Silver Spring, MD, USA: National Oceanic and Atmospheric Administration.

Gopal, B. (1987) *Water Hyacinth*. Amsterdam, the Netherlands: Elsevier.

Gopal, B. (1991) Wetland (mis)management by keeping people out: two examples from India. *Landscape and Urban Planning* 20: 53–59.

Gopal, B. (1999) Natural and constructed wetlands for wastewater treatment: potentials and problems. *Water Science and Technology* **40**(3): 27–35.

Gopal, B. (2002) Management issues of aquatic resources including wetlands. In: *Institutionalizing Common Pool Resources*, ed. D. K. Marothia, pp. 674–688. New Delhi, India: Concept Publishing.

Gopal, B. (2003) Aquatic biodiversity in arid and semi-arid zones of Asia and water management. In: *Conserving Biodiversity in Arid Regions: Best Practices in Developing Nations*, ed. J. Lemons, R. Victor and D. Schaeffer, pp. 199–215. Dordrecht, the Netherlands: Kluwer.

Gopal, B. (2005) Does inland aquatic biodiversity have a future in Asian developing countries? *Hydrobiologia* **542**: 69–75.

Gopal, B. & Chauhan, M. (2001) South Asian wetlands and their biodiversity: the role of monsoons. In: *Biodiversity in Wetlands: Assessment, Function and Conservation*, vol. 2, ed. B. Gopal, W. J. Junk & J. A. Davis, pp. 257–276. Leiden, the Netherlands: Backhuys.

Gopal, B. & Junk, W. J. (2001) Assessment, determinants, function and conservation of biodiversity in wetlands: present status and future needs. In: *Biodiversity in Wetlands: Assessment, Function and Conservation*, vol. 2, ed. B. Gopal, W. J. Junk & J. A. Davis, pp. 277–302. Leiden, the Netherlands: Backhuys.

Gopal, B. & Krishnamurthy, K. (1993) Wetlands of South Asia. In: *Wetlands of the World*, vol. 1, Inventory, Ecology and Management, ed. D. F. Whigham, D. Dykyjová & S. Hejný, pp. 345–414. Dordrecht, the Netherlands: Kluwer.

Gopal, B., Junk, W. J. & Davis, J. A., eds. (2000) *Biodiversity in Wetlands: Assessment, Function and Conservation*, vol. 1. Leiden, the Netherlands: Backhuys.

Gopal, B., Junk, W. J. & Davis, J. A., eds. (2001) *Biodiversity in Wetlands: Assessment, Function and Conservation*, vol. 2. Leiden, the Netherlands: Backhuys.

Gopal, B., Kvet, J., Löffler, H., Masing, V. & Patten, B. C. (1990) Definition and classification. In: *Wetlands and Shallow Continental Water Bodies*. vol. 1, Natural and Human Relationships, ed. B. C. Patten *et al.*, pp. 9–16. The Hague, the Netherlands: SPB Academic Publishing.

Gopal, B., Zutshi, D. P. & Van Duzer, C. (2003) Floating islands in India: control or conserve? *International Journal of Ecology and Environmental Sciences* 29: 157–169.

Gordon, C. (1992) Sacred groves and conservation in Ghana. *Newsletter of the IUCN SSC African Reptile and Amphibian Specialist Group* 1: 3–4.

Gordon, C. (2003) Aquatic resource management and freshwater ecosystems of West Africa. In: *Conservation, Ecology and Management of African Freshwaters*, ed. T. L. Crisman, L. J. Chapman, C. A. Chapman & L. S. Kaufman, pp. 62–78. Gainesville, FL, USA: University of Florida Press.

Gore, A. J. P., ed. (1983) *Ecosystems of the World*, vol. 4, *Mires, Swamp, Bog, Fen and Moor*. Amsterdam, the Netherlands: Elsevier.

Goreau, T., McClanahan, T., Hayes, R. & Strong, A. (2000) Conservation of coral reefs after the 1998 global bleaching event. *Conservation Biology* 14: 5–15.

Gorham, E. (1991) Northern peatlands: role in the carbon cycle and probable responses to climate warming. *Ecological Applications* 1: 182–195.

Gorke, M. (2003) *The Death of our Planet's Species*. Washington, DC, USA: Island Press.

Goss-Custard, J. D. & Moser, M. E. (1988) Rates of change in the numbers of dunlin *Calidris alpina* wintering in British estuaries in relation to the spread of *Spartina anglica*. *Journal of Applied Ecology* 25: 95–109.

Gosselink, J. G. & Maltby, E. (1990) Wetland gains and losses. In: *Wetlands: A Threatened Landscape*, ed. M. Williams, pp. 296–322. Oxford, UK: Blackwell Scientific Publications.

Gottgens, J. F., Perry, J. E., Fortney, R., *et al.* (2001) The Paraguay–Paraná Hidrovia: protecting the Pantanal with lessons from the past. *BioScience* 51: 301–308.

Goulet, T. L. (2006) Most corals may not change their symbionts. *Marine Ecology Progress Series* 321: 1–7.

Gowan, R. J., Mills, D. K., Trimmer, M. & Nedwell, D. B. (2000) Production and its fate in two coastal regions of the Irish Sea: the influence of anthropogenic nutrients. *Marine Ecology Progress Series* 208: 51–64.

Grachev, M. (1991) On the present state of the ecological system of Lake Baikal. Unpublished report. Irkutsk, USSR: Limnological Institute, Siberian Division, USSR Academy of Sciences.

Graham, M. H. (1996) Effect of high irradiance on recruitment of the giant kelp *Macrocystis* (Phaeophyta) in shallow water. *Journal of Phycology* **32**: 903–906.

Graham, M. H. (1997) Factors determining the upper limit of giant kelp, *Macrocystis pyrifera* Agardh, along the Monterey Peninsula, central California, USA. *Journal of Experimental Marine Biology and Ecology* **218**: 127–149.

Graham, M. H., Dayton, P. K. & Erlandson, J. M. (2003) Ice ages and ecological transitions on temperate coasts. *Trends in Ecology and Evolution* **18**: 33–40.

Graham, M. H., Harrold, C., Lisin, S., Light, K., Watanabe, J. M. & Foster, M. S. (1997) Population dynamics of giant kelp *Macrocystis pyrifera* along a wave exposure gradient. *Marine Ecology Progress Series* **148**: 269–279.

Graham, N. E. (1995) Simulation of recent global temperature trends. *Science* **267**: 666–671.

Graham, N. E. & Diaz, H. F. (2001) Evidence for intensification of North Pacific winter cyclones since 1948. *Bulletin of the American Meteorological Society* **82**: 1869–1893.

Granger, S. L., Traver, M. S. & Nixon, S. W. (2000) Propagation of *Zostera marina* L. from seed. In: *Seas at the Millennium: An Environmental Evaluation*, vol. 3, *Global Issues and Processes*, ed. C. R. C. Sheppard, pp. 4–5. Amsterdam, the Netherlands: Pergamon.

Granina, L. (1997) The chemical budget of Lake Baikal: a review. *Limnology and Oceanography* **42**: 373–378.

Grant, A. & Briggs, A. D. (2002) Toxicity of sediments from around a North Sea oil platform: are metals or hydrocarbons responsible for ecological impacts? *Marine Environmental Research* **53**: 95–116.

Grant, A., Hateley, J. G. & Jones, N. V. (1989) Mapping the ecological impact of heavy metals on the estuarine polychaete *Nereis diversicolor* using inherited metal tolerance. *Marine Pollution Bulletin* **20**: 235–238.

Grant, J., Hatcher, A., Scott, D. B., *et al.* (1995) A multidisciplinary approach to evaluating impacts of shellfish aquaculture on benthic communities. *Estuaries* **18**: 124–144.

Grassle, J. F. (1991) Effects of sewage sludge on deep-sea communities. *Eos* **72**: 84 (abstract).

Grassle, J. F. (2000) The ocean biogeographic system (OBIS): an on-line, worldwide atlas for accessing, modeling and mapping marine biological data in a multidimensional geographic context. *Oceanography* **13**: 5–7.

Gray, A. J. (1977) Reclaimed land. In: *The Coastline*, ed. R. S. K. Barnes, pp. 253–270. Chichester, UK: John Wiley.

Great Lakes Commission (2003) Toward a water resources management decision support system for the Great Lakes–St Lawrence River Basin. Unpublished report on CD-ROM. Ann Arbor, MI, USA: Great Lakes Commission.

Great Lakes Environmental Research Laboratory (2002) Great Lakes fish community impacted by *Diporeia* disappearance. Unpublished fact sheet. Ann Arbor, MI, USA: Great Lakes Environmental Research Laboratory.

Great Lakes Fishery Commission (2001*a*) TFM and sea lamprey control. Unpublished fact sheet no. 4. Ann Arbor, MI, USA: Great Lakes Fishery Commission.

Great Lakes Fishery Commission (2001*b*) Sea lamprey barriers. Unpublished fact sheet no. 5. Ann Arbor, MI, USA: Great Lakes Fishery Commission.

Green, E. P. & Short, F. T., eds. (2003) *World Atlas of Seagrasses*. Berkeley, CA, USA: University of California Press.

Green, M. O., Vincent, C. E., McCave, I. N., *et al.* (1995) Storm sediment transport: observations from the British North sea shelf. *Continental Shelf Research* **15**: 889–912.

Greene, K. (2002) *Beach Nourishment: A Review of the Biological and Physical Impacts*, ASMFC Habitat Management Series No.7. Washington, DC, USA: Atlantic States Marine Fisheries Commission.

Greenstein, B. J., Curran, H. A. & Pondolfi, J. M. (1998) Shifting ecological baselines and the demise of *Acropora cervicornis* in the western North Atlantic and Caribbean Province: a Pleistocene perspective. *Coral Reefs* **17**: 249–261.

Gregory, J. M., Stott, P. A., Cresswell, D. J., *et al.* (2002) Recent and future changes in Arctic sea ice simulated by the HadCM3 AOGCM. *Geophysical Research Letters* **29**: art. no. 2175.

Gregory, S. V., Swanson, F. J., McKee, W. A. & Cummins, K. W. (1991) An ecosystem perspective of riparian zones. *BioScience* **41**: 540–551.

Grevemeyer, I., Herber, R. & Essen, H. H. (2000) Microseismological evidence for changing wave climate in the northeast Atlantic Ocean. *Nature* **408**: 349–352.

Grevstad, F. S., Strong, D. R., Garcia-Rossi, M., Switzer, R. W. & Wecker, M. S. (2003) Biological control of *Spartina alterniflora* in Willapa Bay, Washington using the planthopper *Prokelisia marginata*: agent specificity and early results. *Biological Control* **27**: 32–42.

Griebler, C., Danielopol, D. L., Gibert, J., Nachtnebel, H. P. & Notenboom, J., eds. (2001) *Groundwater Ecology: A Tool for Management of Water Resources*. Luxembourg: Office for Official Publications of the European Communities.

Griebler, C., Mindl, B., Slezak, D. & Geiger-Kaiser, M. (2002) Distribution patterns of attached and suspended bacteria in pristine and contaminated shallow aquifers studied with an in situ sediment exposure microcosm. *Aquatic Microbial Ecology* **28**: 117–129.

Griffing, S. (2003) The wild world of West Nile. *Smithsonian Environmental Research Center Newsletter Spring Edition* **1**: 4–5.

Griffiths, C., Hockey, P., van Erkom Schurink, C. & le Roux, P. (1992) Marine invasive aliens on South African shores: implications for community structure and trophic functioning. *South African Journal of Marine Science* **12**: 713–722.

Grigal, D. F., Buttleman, C. G. & Kernick, L. K. (1985) Biomass and productivity of the woody strata of forested bogs in northern Minnesota. *Canadian Journal of Botany* **63**: 2416–2424.

Grigg, R. W. (1993) Precious coral fisheries of Hawaii and the US Pacific Islands. *Marine Fisheries Review* **55**: 50–60.

Grigg, R. W. (1997) Paleoceanography of coral reefs in the Hawaiian–Emperor Chain revisited. *Coral Reefs* **16**: S33–S38.

Grimm, N. B. & Fisher, S. G. (1992) Response of arid land streams to changing climate. In: *Global Climate Change and Freshwater Ecosystems*, ed. P. Firth & S. G. Fisher, pp. 211–233. Berlin, Germany: Springer-Verlag.

Grimm, N. B., Chacon, A., Dahm, C., *et al.* (1997) Sensitivity of aquatic ecosystems to climatic and anthropogenic changes: the basin and range, American southwest and Mexico. *Hydrological Processes* **11**: 1023–1041.

Grizzle, R. E., Short, F. T., Hoven, H., Kindbloom, L. & Newell, C. R. (1996) Hydrodynamically induced synchronous waving of seagrasses: 'monami' and its possible effects on larval mussel settlement. *Journal of Experimental Marine Biology and Ecology* **206**: 165–177.

Groombridge, B. (1992) *Global Biodiversity: Status of the Earth's Living Resources.* London, UK: World Conservation Monitoring Centre and Chapman & Hall.

Groombridge, B. & Jenkins, M. (1998) *Freshwater Biodiversity: A Preliminary Global Assessment*, WCMC Biodiversity Series No. 8. Cambridge, UK: World Conservation Monitoring Centre, World Conservation Press.

Grootjans, A. P., Everts, H., Bruin, K. & Fresco, L. (2001) Restoration of wet dune slacks on the Dutch Wadden Sea Islands: recolonization after large-scale sod cutting. *Restoration Ecology* **9**: 137–146.

Grosholz, E. (2002) Ecological and evolutionary consequences of coastal invasions. *Trends in Ecology and Evolution* **17**: 22–27.

Grosholz, E. D., Ruiz, G. M., Dean, C. A., *et al.* (2000) The impacts of a non-indigenous marine predator in a California bay. *Ecology* **81**: 1206–1224.

Grunawalt, R. J., King, J. E. & McClain, R. S., eds. (1996) *Protection of the Environment during Armed Conflict*, International Law Studies No. 69 Newport, RI, USA: Naval War College.

Guilford, S. & Hecky, R. (2000) Total nitrogen, total phosphorus and nutrient limitation in lakes and oceans: is there a common relationship? *Limnology and Oceanography* **45**: 1213–1223.

Guillén, J. E., Ramos, A. A., Martinez, L. & Sanchez Lizaso, J. L. (1994) Anti-trawling reefs and the protection of *Posidonia oceanica* (L.) Delile meadows in the western Mediterranean Sea: demand and aims. *Bulletin of Marine Science* **55**: 645–650.

Gunatilaka, A. & Dreher, J. (2001) Management of urban groundwater. II. Qualitative aspects. In: *Groundwater Ecology: A Tool for Management of Water Resources*, ed. C. Griebler, D. L. Danielopol, J. Gibert, H. P. Nachtnebel & J. Notenboom, pp. 213–230. Luxembourg: Office for Official Publications of the European Communities.

Gunn, J. & Keller, W. (1990) Biological recovery of an acidified lake after reductions in industrial emissions of sulphur. *Nature* **345**: 431–433.

Gunnarsson, U. & Rydin, H. (2000) Nitrogen fertilization reduces *Sphagnum* production in bog communities. *New Phytologist* **147**: 527–537.

Gunnarsson, U., Malmer, N. & Rydin, H. (2002) Dynamics or constancy in *Sphagnum* dominated mire ecosystems: a 40-year study. *Ecography* **25**: 685–704.

Gunnarsson, U., Rydin, H. & Sjörs, H. (2000) Diversity and pH changes after 50 years on the boreal mire Skattlösbergs Stormosse, central Sweden. *Journal of Vegetation Science* **11**: 277–286.

Gustavson, K., Garde, K., Wangberg, S.-A. & Selmer, J.-S. (2000) Influence of UV-B radiation on bacterial activity in coastal waters. *Journal of Plankton Research* **22**: 1501–1511.

Gutleb, A. C., Kranz., A, Nechay, G. & Toman, A. (1998) Heavy metal concentrations in livers and kidneys of the otter (*Lutra lutra*) from central Europe. *Bulletin of Environmental Contamination and Toxicology* **60**: 273–279.

Hacker, S. D., Heimer, C. E., Hellquist, C. E., *et al.* (2001) A marine plant (*Spartina anglica*) invades widely varying habitats: potential mechanisms of invasion and control. *Biological Invasions* **3**: 211–217.

Hader, D. P. (1997) Effects of UV radiation on phytoplankton. In: *Advances in Microbial Ecology*, ed. J. G. Jones, pp. 1–26. New York, USA: Plenum Press.

Hader, D. P., Kumar, H. D., Smith, R. C. & Worrest, R. C. (2003) Aquatic ecosystems: effects of solar ultraviolet radiation and interactions with other climatic change factors. *Photochemical and Photobiological Sciences* **2**: 39–50.

Haedrich, R. L. (1995) Structure over time of an exploited deep-water fish assemblage. In *Deep-Water Fisheries of the North Atlantic Oceanic Slope*, ed. A. G. Hopper, pp. 27–50. Dordrecht, the Netherlands: Kluwer.

Hagen, N. T. (1983) Destructive grazing of kelp beds by sea urchins in Vestfjorden, northern Norway. *Sarsia* **68**: 177–190.

Hahn, H. J. (2002) Distribution of the aquatic meiofauna of the Marbling Brook catchment (Western Australia) with reference to land-use and hydrological features. *Archiv für Hydrobiologie* **139** (Suppl.): 237–263.

Hakenkamp, C. C. & Morin, A. (2000) The importance of meiofauna to lotic ecosystem functioning. *Freshwater Biology* **44**: 165–176.

Hall, A. & Visbeck, M. (2002) Synchronous variability in the Southern Hemisphere atmosphere, sea ice, and ocean resulting from the Annular Mode. *Journal of Climate* **15**: 3043–3057.

Hall, C. M. (2001) Trends in ocean and coastal tourism: the end of the last frontier? *Ocean and Coastal Management* **44**: 601–618.

Hall, M. J. & Pilkey, O. H. (1991) Effects of hard stabilization on dry beach width for New Jersey. *Journal of Coastal Research* **7**: 771–785.

Hall, S. & Mills, E. (2000) Exotic species in large lakes of the world. *Aquatic Ecosystem Health and Management* **3**: 105–135.

Hall, S. J. (1999) *The Effects of Fishing on Marine Ecosystems and Communities*. Oxford, UK: Blackwell Science.

Hall, S. J. (2001) Is offshore exploration good for benthic conservation? *Trends in Ecology and Evolution* **16**: 601–618.

Hall, S. J. (2002) The soft-sediment continental shelf benthic ecosystem: current status, agents for change and future prospects. *Environmental Conservation* **29**: 350–374.

Hallegraeff, G. M. (1998) Transport of toxic dinoflagellates via ships' ballast water: bioeconomic risk assessment and efficacy of possible ballast water management strategies. *Marine Ecology Progress Series* **168**: 297–309.

Hallerman, E. M. & Kapuscinski, A. R. (1995) Incorporating risk assessment and risk management into public policies on genetically modified finfish and shellfish. *Aquaculture* **137**: 9–17.

Halling-Sørensen, B., Nors Nielsen, S., Lanzky, P. F., *et al.* (1998) Occurrence, fate and effects of pharmaceutical substances in the environment: a review. *Chemosphere* **36**: 357–393.

Halsey, L. A., Vitt, D. H. & Bauer, I. E. (1998) Peatland initiation during the Holocene in continental western Canada. *Climatic Change* **40**: 315–342.

Hameedi, M. J. (1997) Strategy for monitoring the environment in the coastal zone. In: *Coastal Zone Management Imperative for Maritime Developing Nations*, ed. B. U. Haq, pp. 111–142. Dordrecht, the Netherlands: Kluwer.

Hamelink, J. L., Landrum, P. F., Bergman, H. L. & Benson, W. H., eds. (1994) *Bioavailability: Physical, Chemical, and Biological Interactions*. Boca Raton, FL, USA: Lewis Publishers.

Hamilton, L. S. & Snedaker, S. C. (1984) *Handbook for Mangrove Area Management*. Honolulu, HI, USA: East–West Centre, IUCN and United Nations Environmental Programme.

Hamilton, S. K. (1999) Potential effects of a major navigation project (Paraguay–Paraná–Hidrovia) on inundation in the Pantanal floodplains. *Regulated Rivers: Research and Management* **15**: 289–299.

Hamilton, S. K. & Lewis Jr, W. M. (1990) Physical characteristics of the fringing floodplain of the Orinoco River, Venezuela. *Interciencia* **15**: 491–500.

Hamilton, S. K., Sippel, S. J. & Melack, J. M. (1996) Inundation patterns in the Pantanal wetland of South America determined by passive microwave sensing. *Archiv für Hydrobiologie* **137**: 1–23.

Hamm, L., Capobianco, M., Dette, H. H., *et al.* (2002) A summary of European experience with shore nourishment. *Coastal Engineering* **47**: 237–264.

Hammer, U. T. (1986) *Saline Lake Ecosystems of the World*. Dordrecht, the Netherlands: Kluwer.

Hammer, U. T. (1990) The effects of climate change on the salinity, water levels and biota of Canadian prairie saline lakes. *Verhandlungen Internationale Vereinigung für Limnologie* **24**: 321–326.

Hammerstrom, K. K., Kenworthy, W. J., Fonseca, M. S. & Whitfield, P. E. (2006) Seed bank, biomass, and productivity of *Halophila decipiens*, a deep water seagrass on the west Florida continental shelf. *Aquatic Botany* **84**: 110–120.

Hammerton, D. (1972) The Nile river: a case study. In: *River Ecology and Man*, ed. R. T. Oglesby, C. A. Carlson & M. J. McCann, pp. 171–214. New York, USA: Academic Press.

Handa, I. T., Harmsen, R. & Jefferies, R. L. (2002) Patterns of vegetation change and the recovery potential of degraded areas in a coastal marsh system of the Hudson Bay lowlands. *Journal of Ecology* **90**: 86–99.

Hannides, A. K. & Smith, C. R. (2003) The Northeastern Pacific abyssal plain. In: *Biogeochemistry of Marine Systems*, ed. K. D. Black & G. B. Shimmield, pp. 208–237. Sheffield, UK: Sheffield Academic Press.

Hansen, M. (1999) Lake trout in the Great Lakes: basinwide stock collapse and binational restoration. In: *Great Lakes Fisheries Policy and Management*, ed. W. W. Taylor & C. P. Ferreri, pp. 417–453. East Lansing, MI, USA: Michigan State University Press.

Hansen, P. J., Bjornsen, P. K. & Hansen, B. W. (1997) Zooplankton grazing and growth scaling within the 2–2000 μm body size range. *Limnology and Oceanography* **42**: 687–704.

Hanski, I. (1998) Metapopulation dynamics. *Nature* **396**: 41–49.

Hansson, L.-A., Annadotter, H., Bergman, E., *et al.* (1998) Biomanipulation as an application of food chain theory: constraints, synthesis and recommendations for temperate lakes. *Ecosystems* **1**: 558–574.

Haque, C. E. & Zaman, M. Q. (1993) Human responses to riverine hazards in Bangladesh: a proposal for sustainable development. *World Development* **21**: 93–107.

Haraszthy, L. (2001) The floodplain forests in Hungary. In: *The Floodplain Forests in Europe*, ed. E. Klimo & H. Hager, pp. 17–24. Leiden, the Netherlands: E. J. Brill.

Hardin, G. (1968) The tragedy of the commons. *Science* **162**: 1243–1248.

Hardy, A. (1967) *Great Waters*. London, UK: Collins.

Harley, C. D. G., Hughes, A. R., Hultgren, K. M., *et al.* (2006) The impacts of climate change in coastal marine systems. *Ecology Letters* **9**: 228–241.

Harmelin-Vivien, M., Francour, P. & Harmelin, J. G. (1999) Impact of *Caulerpa taxifolia* on Mediterranean fish assemblages: a six-year study. In: *Proceedings of the Workshop on Invasive* Caulerpa *in the Mediterranean*, MAP Technical Reports Series No. 125, pp. 127–138. Athens, Greece: United Nations Environmental Programme.

Harner, M. J. & Stanford, J. A. (2003) Differences in cottonwood growth between a losing and a gaining reach of an alluvial floodplain. *Ecology* **84**: 1453–1458.

Harper, D. M., Mavuti, K. M. & Muchiri, S. M. (1990) Ecology and management of Lake Naivasha, Kenya, in relation to climatic change, alien species introductions, and agricultural development. *Environmental Conservation* **17**: 328–336.

Harris, J. M., Branch, G. M., Sibiya, C. & Bill, C. (2003) The Sokhulu subsistence mussel-harvesting project: co-management in action. In: *Waves of Change: Co-Management of Fisheries and Coastal Resources in South Africa*, pp. 61–98. Cape Town, South Africa: University of Cape Town Press.

Harrison, I. J. & Stiassny, M. L. J. (1999) The quiet crisis: a preliminary listing of the freshwater fishes of the world that are extinct or 'missing in action'. In: *Extinctions in Near Time: Causes, Contexts and Consequences*, ed. R. D. E. MacPhee, pp. 271–331. New York, USA: Plenum Press.

Harrold, C. & Lisin, S. (1989) Radio-tracking rafts of giant kelp: local production and regional transport. *Journal of Experimental Marine Biology and Ecology* **30**: 237–252.

Harrold, C. & Reed, D. C. (1985) Food availability, sea urchin grazing, and kelp forest community structure. *Ecology* **66**: 1160–1169.

Harrold, C., Light, K. & Lisin, S. (1998) Organic enrichment of submarine-canyon and continental-shelf benthic communities by macroalgal drift imported from nearshore kelp forests. *Limnology and Oceanography* **43**: 669–678.

Hart, J. (1996) *Storm over Mono: The Mono Lake Battle and the California Water Future*. Berkeley, CA, USA: University of California Press.

Hartney, K. I. & Grorud, K. I. (2002) The effect of sea urchins as biogenic structures on the local abundance of a temperate reef fish. *Oecologia* **131**: 506–513.

Hartnoll, R. G. & Hawkins, S. J. (1985) Patchiness and fluctuations on moderately exposed rocky shores. *Ophelia* **24**: 53–63.

Harvell, C. D., Kim, K., Burkholder, J. M., *et al.* (1999) Emerging marine diseases: climate links and anthropogenic factors. *Science* **285**: 1505–1510.

Harvell, C. D., Mitchell, C. E., Ward, J. R., *et al.* (2002) Climate warming and disease risks for terrestrial and marine biota. *Science* **296**: 2158–2162.

Harvey, G. R., Miklas, H. P., Bowen, V. T. & Steinhauer, W. G. (1974a) Observations on the distribution of chlorinated hydrocarbons in Atlantic Ocean organisms. *Journal of Marine Research* **32**: 103–118.

Harvey, G. R., Steinhauer, W. G. & Miklas, H. P. (1974b) Decline of PCB concentrations in North Atlantic surface water. *Nature* **252**: 387–388.

Harwell, M. C. & Orth, R. J. (1999) Eelgrass (*Zostera marina* L.) seed protection for field experiments and implications for large-scale restoration. *Aquatic Botany* **64**: 54–61.

Hassan, F. A., Reuss, M., Trottier, J., *et al.* (2003) *History and Future of Shared Water Resources*, UNESCO-IHP Technical Documents in Hydrology, PCCP Series No. 6. Paris, France: UNESCO.

Hässelholms Vatten (2007) Home page [www document]. URL http://www.hasselholmsvatten.se (in Swedish).

Hatcher, B. G. (1983) Grazing in coral reef ecosystems. In: *Perspectives on Coral Reefs*, ed. D. J. Barnes, pp. 164–179. Canberra, ACT, Australia: Brian Clouster.

Hatcher, B. G., Kirkman, H. & Wood, W. F. (1987) Growth of the kelp *Ecklonia radiata* near the northern limit of its range in Western Australia. *Marine Biology* **95**: 63–73.

Hateley, J. G., Grant, A., Taylor, S. M. & Jones, N. V. (1992) Morphological and other evidence on the degree of genetic differentiation between populations of *Nereis diversicolor*. *Journal of the Marine Biological Association of the United Kingdom* **72**: 365–381.

Hatton, H. (1938) Essais de bionomie explicative sur quelques espèces intercotidals d'algues et d'animaux. *Annales de l'Institut Océanographique de Monaco* **17**: 241–348.

Hatton, T. & Salama, R. (1999) Is it feasible to restore the salinity affected rivers of the Western Australian wheatbelt? In: *Proceedings of the 2nd Australian Stream Management Conference*, vol. 1, ed. I. R. Rutherfurd & R. Bartley, pp. 313–317. Adelaide, SA, Australia: Cooperative Research Centre for Catchment Hydrology.

Hawkins, J. P. & Roberts, C. M. (1992) Effects of recreational SCUBA diving on fore-reef slope communities of coral reefs. *Biological Conservation* **62**: 171–178.

Hawkins, S. J. (1999) Experimental ecology and coastal conservation: conflicts on rocky shores. *Aquatic Conservation: Marine and Freshwater Ecosystems* **9**: 565–572.

Hawkins, S. J. & Hartnoll, R. G. (1985) Factors determining the upper limits of intertidal canopy forming algae. *Marine Ecology Progress Series* **20**: 265–271.

Hawkins, S. J. & Southward, A. J. (1992) The *Torrey Canyon* oil spill: recovery of rocky shore communities. In: *Restoring the Nation's Environment*, ed. G. W. Thayer, pp. 583–631. College Park, MD, USA: Maryland Sea Grant College.

Hawkins, S. J., Allen, J. R., Fielding, N. J., Wilkinson, S. B. & Wallace, I. D. (1999) Liverpool Bay and the estuaries: human impact, recent recovery and restoration. In: *Ecology and Landscape Development: A History of the Mersey Basin, Conference Proceedings*, ed. E. F. Greenwood, pp. 155–165. Liverpool, UK: Liverpool University Press.

Hawkins, S. J., Allen, J. R., Ross, P. M. & Genner, M. J. (2002*a*) Marine and coastal ecosystems. In *Handbook of Ecological Restoration*, vol. 2, *Restoration in Practice*, ed.

M. R. Perrow & A. J. Davy, pp. 121–148. Cambridge, UK: Cambridge University Press.

Hawkins, S. J., Corte-Real, H. B. S. M., Pannacciulli, F. G., Weber, L. C. & Bishop, J. D. D. (2000) Thoughts on the ecology and evolution of the intertidal biota of the Azores and other Atlantic Islands. *Hydrobiologia* **440**: 3–17.

Hawkins, S. J., Gibbs, P. E., Pope, N. D., *et al.* (2002*b*) Recovery of polluted coastal ecosystems: the case for long-term studies. *Marine Environmental Research* **54**: 215–222.

Hawkins, S. J., Proud, S. V., Spence, S. K. & Southward, A. J. (1994) From the individual to the community and beyond: water quality, stress indicators and key species in coastal ecosystems. In: *Water Quality and Stress Indicators in Marine and Freshwater Ecosystems: Linking Levels of Organisation (Individuals, Populations, Communities)*, ed. D. W. Sutcliffe, pp. 35–62. Ambleside, UK: Freshwater Biological Association.

Hawkins, S. J., Southward, A. J. & Barrett, R. L. (1983) Population structure of *Patella vulgata* during succession on rocky shores in southwest England. *Oceanologica Acta* Special Issue: 103–107.

Haworth, E. Y. (1969) The diatoms of a sediment core from Blea Tarn, Langdale. *Journal of Ecology* **57**: 429–439.

Haworth, E. Y. (1972) Diatom succession in a core from Pickerel Lake, north-eastern South Dakota. *Geological Society of America Bulletin* **83**: 157–172.

Hay, M. E. (1984) Patterns of fish and urchin grazing on Caribbean coral reefs: are previous results typical? *Ecology* **65**: 446–454.

Hay, M. E., Colburn, T. & Downing, D. (1983) Spatial and temporal patterns in herbivory on a Caribbean fringing reef: the effects on plant distribution. *Oecologia* **58**: 299–308.

Haycock, N. E., Burt, T., Goulding, K. & Pinay, G. (1997) *Buffer Zones: Their Processes and Potential in Water Protection*. Hertford, UK: Quest Environmental.

Hazen, T. C. (1997) Bioremediation. In: *The Microbiology of the Deep Subsurface*, ed. P. S. Amy & D. L. Haldeman, pp. 247–266. Boca Raton, FL, USA: Lewis Publishers.

Headland, R. (1984) *The Island of South Georgia*. Cambridge, UK: Cambridge University Press.

Hearn, C. J., Atkinson, M. J. & Falter, J. L. (2001) A physical derivation of nutrient-uptake rates in coral reefs: effects of roughness and waves. *Coral Reefs* **20**: 347–356.

Heathwaite, A. L., Johnes, P. J. & Peters, N. (1996) Trends in nutrients. *Hydrological Processes* **10**: 263–293.

Heck, K. L. & Valentine, J. F. (1995) Sea urchin herbivory: evidence for long-lasting effects in subtropical seagrass meadows. *Journal of Experimental Marine Biology and Ecology* **189**: 205–217.

Hecky, R. (1993) The eutrophication of Lake Victoria. *Verhandlungen Internationale Vereinigung für Limnologie* **25**: 39–48.

Hecky, R. & Bugenyi, F. (1992) Hydrology and chemistry of the African Great Lakes and water quality issues: problems and solutions. *Mitteilungen Internationale Vereinigung für Limnologie* **23**: 45–54.

Hecky, R., Botsma, H., Mugidde, R. & Bugenyi, F. (1996) Phosphorus pumps, nitrogen sinks, silicon drains: plumbing nutrients in the African Great Lakes. In: *The Limnology, Climatology, and Paleoclimatology of the East African Lakes*, ed. T. C. Johnson & E. Odada, pp. 205–224. Toronto, Ontario, Canada: Gordon & Breach.

Hecky, R., Bugenyi, F., Ochumba, P., *et al.* (1994) Deoxygenation of the deep water of Lake Victoria. *Limnology and Oceanography* **39**: 1476–1480.

Hecky, R., Smith, R., Barton, D., *et al.* (2004) The near shore phosphorus shunt: a consequence of ecosystem engineering by dreissenids in the Lawrentian Great Lakes. *Canadian Journal of Fisheries and Aquatic Sciences* **61**: 1285–1293.

Hector, A., Schmid, B., Beierkuhnlein, C., *et al.* (1999) Plant diversity and productivity experiments in European grasslands. *Science* **286**: 1123–1127.

Heeg, J. & Breen, C. M., eds. (1982) *Man and the Pongolo Floodplain*, South African National Scientific Programs Report No. 56. CSIR, Pretoria, South Africa:

Heemsbergen, D. A., Berg, M. P., Loreau, M., *et al.* (2004) Biodiversity effects on soil processes explained by interspecific functional dissimilarity. *Science* **306**: 1019–1020.

Heidelbaugh, W. S. & Nelson, W. G. (1996) A power analysis of methods for assessing change in seagrass cover. *Aquatic Botany* **53**: 227–233.

Hein, M. & Sand-Jensen, K. (1997) CO_2 increases oceanic primary production. *Nature* **388**: 526–527.

Heinz Center (2002) *Dam Removal: Science and Decision Making*. Washington, DC, USA: The Heinz Center.

Hellberg, M. E., Balch, D. P. & Roy, K. (2001) Climate-driven range expansion and morphological evolution in a marine gastropod. *Science* **292**: 1707–1710.

Hellström, C., Chen, D. L., Archberger, C. & Räisänen, J. (2001) Comparison of climate change scenarios for Sweden based on statistical and dynamical downscaling of monthly precipitation. *Climate Research* **19**: 45–55.

Helms, M., Buchele, B., Merkel, U. & Ihringer, J. (2002) Statistical analysis of the flood situation and assessment of the impact of diking measures along the Elbe (Labe) river. *Journal of Hydrology* **267**: 94–114.

Helmuth, B., Mieszkowska, N., Moore, P., Hawkins, S. J. (2006) Living on the edge of two changing worlds: forecasting responses of rocky intertidal ecosystems to climate change. *Annual Review of Ecology, Evolution and Systematics* **37**: 373–404.

Hemminga, M. A. & Duarte, C. M. (2000) *Seagrass Ecology*. Cambridge, UK: Cambridge University Press.

Hemond, H. F. & Benoit, J. (1988) Cumulative impacts on water quality functions of wetlands. *Environmental Management* **12**: 639–653.

Henley, W. J. & Dunton, K. H. (1997) Effects of nitrogen supply and continuous darkness on growth and photosynthesis of the arctic kelp *Laminaria solidungula*. *Limnology and Oceanography* **42**: 209–216.

Hensen, V. (1887) *Über die Bestimmung des Planktons oder des im Meere treibenden Materials an Pflanzen und Thieren*. Kommission zur wissenschaftlichen Untersuchungen der deutschen Meere in Kiel, 1882–1886. V. Bericht, Jahrgang 12–16.

Herbert, R. J. H., Hawkins, S. J., Sheader, M. & Southward, A. J. (2003) Range extension and reproduction of the barnacle *Balanus perforatus* in the Eastern English Channel. *Journal of the Marine Biological Association of the United Kingdom* **83**: 73–82.

Herdendorf, C. (1982) Large lakes of the world. *Journal of Great Lakes Research* **8**: 379–412.

Herdendorf, C. E. (1990) Distribution of the world's large lakes. In: *Large Lakes: Ecological Structure and Function*, ed. M. M. Tilzer & C. Serruya, pp. 3–38. New York, USA: Springer-Verlag.

Herman, J. S., Culver, D. C. & Salzman, J. (2001) Groundwater ecosystems and the service of water purification. *Stanford Environmental Law Journal* **20**: 479–498.

Hermsen, J. M., Collie, J. S. & Valentine, P. C. (2003) Mobile fishing gear reduces benthic megafaunal production on Georges Bank. *Marine Ecology Progress Series* **260**: 97–108

Hershey, A. (1985) Effects of predatory sculpin on the chironomid communities in an Arctic lake. *Ecology* **66**: 1131–1138.

Herut, B., Krom, M., Pan, G. & Mortimer, R. (1999) Atmospheric input of nitrogen and phosphorus to the Southeast Mediterranean: sources, fluxes, and possible impact. *Limnology and Oceanography* **44**: 1683–1692.

Hessen, D. O., Hindar, A. & Holtan, G. (1997) The significance of nitrogen runoff for eutrophication of freshwater and marine recipients. *Ambio* **26**: 312–320.

Heuveline, P. (1999) The global impact of mortality and fertility transitions, 1950–2000. *Population Development Review* **December**: 681–702.

Hewes, L. & Frandson, P. E. (1952) Occupying the wet prairie: the role of artificial drainage in Story County, Iowa. *Annals of the Association of American Geographers* **42**: 24–50.

Hiaasen, C. (1990) *Double Whammy*. London, UK: Pan Books.

Hickel, W., Mangelsdorf, P. & Berg, J. (1993) The human impact in the German Bight: eutrophication during 3 decades (1962–1991). *Helgoländer Meeresuntersuchungen* **47**: 243–263.

Highsmith, R. C. (1980) Geographic patterns of coral bioerosion: a productivity hypothesis. *Journal of Experimental Marine Biology and Ecology* **46**: 55–67.

Hildrew, A. G. & Giller, P. S. (1994) Patchiness, species interactions and disturbance in the stream benthos. In: *Aquatic Ecology: Scale, Pattern and Process*, ed. P. S. Giller, A. G. Hildrew & D. G. Raffaelli, pp. 21–62. Oxford, UK: Blackwell Scientific Publications.

Hildrew, A. G. & Ormerod, S. J. (1995) Acidification: causes, consequences and solutions. In: *The Ecological Basis for River Management*, ed. D. M. Harper & A. Ferguson, pp. 147–160. Chichester, UK: John Wiley.

Hill, B. J. (1977) The effect of heated effluent on egg production in the estuarine prawn *Upogebia africana* (Ortmann). *Journal of Experimental Marine Biology and Ecology* **29**: 291–302.

Hilton-Taylor, C., compiler (2000) *IUCN Red List of Threatened Species*. Gland, Switzerland and Cambridge, UK: IUCN.

Hily, C. (1991) Is the activity of benthic suspension feeders a factor controlling water quality in the Bay of Brest? *Marine Ecology Progress Series* **69**: 179–188.

Hily, C., van Katwijk, M. M. & den Hartog, C. (2003) The seagrasses of western Europe. In: *World Atlas of Seagrasses*, ed. E. P. Green & F. T. Short, pp. 46–55. Berkeley, CA, USA: University of California Press.

Himmelman, J. H. (1980) The role of the green sea urchin *Strongylocentrotus droebachiensis* in the rocky subtidal region of Newfoundland. In: *Proceedings of the Workshop on the Relationship between Sea Urchin Grazing and Commercial Plant/Animal Harvesting*, ed. J. D. Pringle, G. J. Sharp & J. F. Caddy, pp. 92–119. Halifax, NS, Canada: Resource Branch, Fisheries and Oceans Canada.

Hiscock, K., Southward, A., Tittley, I., Hawkins, S. (2004) Effects of changing temperature on benthic marine life in Britain and Ireland. *Aquatic Conservation: Marine and Freshwater Ecosystems*, **14**: 333–362.

Hjorleifsson, E., Kassa, O. & Gunnarsson, K. (1995) Grazing of kelp by green sea urchins in Eyyjafjordu, North Iceland. In: *Ecology of Fjords and Coastal Waters: Proceedings of the Mare Nor Symposium on the Ecology of Fjords and Coastal Waters*, Tromso, Norway, 5–9 December 1994, ed. H. R. Skjoldal, C. Hopkins, K. E. Erikstad & H. P. Leinass, pp. 593–597. Amsterdam, the Netherlands: Elsevier.

Hobday, A. J. (2000) Abundance and dispersal of drifting kelp *Macrocystis pyrifera* rafts in the Southern California Bight. *Marine Ecology Progress Series* **195**: 101–116.

Hockey, P. A. R. & Bosman, A. L. (1986) Man as an intertidal predator in Transkei: disturbance, community convergence and management of a natural food resource. *Oikos* **46**: 3–14.

Hockey, P. A. R. & van Erkom Schurink, C. (1992) The invasive biology of the mussel *Mytilus galloprovincialis* on the southern African Coast. *Transactions of the Royal Society of South Africa* **48**: 123–139.

Hodgkin, E. P. & Hamilton, B. (1998) Changing estuarine wetlands: a long-term perspective for management. In: *Wetlands for the Future*, ed. A. J. McComb & J. A. Davis, pp. 243–255. Adelaide, SA, Australia: Gleneagles Publishing.

Hodgson, G. & Dixon, J. A. (1992) Sedimentation damage to marine resources: environmental and economic analysis. In: *Resources and Environment in Asia Marine Sector*, ed. J. B. March, pp. 421–426. Washington, DC, USA: Taylor & Francis.

Hoegh-Guldberg, O. (1999) Climate change, coral bleaching and the future of the world's coral reefs. *Marine and Freshwater Research* **50**: 839–866.

Hoegh-Guldberg, O., Jones, R. J., Ward, S. & Loh, W. K. (2002) Is coral bleaching really adaptive? *Nature* **415**: 601–602.

Hoekstra, A. Y. & Hung, P. Q. (2002) *Virtual Water Trade: a Quantification of Virtual Water Flows between Nations in Relation to International Crop Trade*, Value of Water Research Report Series No. 11. Delft, the Netherlands: National Institute for Public Health and the Environment (IHE).

Hofmann, E. E., Klinck, J. M., Locarnini, R. A., Fach, B. & Murphy, E. J. (1998) Krill transport in the Scotia Sea and environs. *Antarctic Science* **10**: 406–415.

Hogarth, P. J. (1999) *The Biology of Mangroves*. Oxford, UK: Oxford University Press.

Holbrook, S. J., Schmitt, R. J. & Ambrose, R. F. (1990) Biogenic habitat structure and characteristics of temperate reef fish assemblages. *Australian Journal of Ecology* **15**: 489–503.

Holbrook, S. J., Swarbrick, S. L., Schmitt, R. J. & Ambrose, R. F. (1993) Reef architecture and reef fish: correlations of population densities with attributes of subtidal rocky environments. In: *The Ecology of Temperate Reefs: Proceedings of the 2nd International Temperate Reef Symposium*, Auckland, pp. 99–106. Wellington, New Zealand: National Institute of Water and Atmospheric Research.

Holdridge, L. R., Grenke, W. C., Hatheway, W. H., Liang, T. & Tosi Jr, J. A. (1971) *Forest Environments in Tropical Life Zones*. New York, USA: Pergamon Press.

Holland, R. (1993) Changes in planktonic diatoms and water transparency in Hatchery Bay, Bass Island area, western Lake Erie since the establishment of the zebra mussel. *Journal of Great Lakes Research* 19: 617–624.

Holland, R. & Beeton, A. (1972) Significance to eutrophication of spatial differences in nutrients and diatoms in Lake Michigan. *Limnology and Oceanography* 17: 88–96.

Holland, R., Johengen, T. & Beeton, A. (1995) Trends in nutrient concentrations in Hatchery Bay, Western Lake Erie, before and after *Dreissena polymorpha*. *Canadian Journal of Fisheries and Aquatic Sciences* 52: 1202–1209.

Hollis, G. E. (1998) Future wetlands in a world short of water. In: *Wetlands for the Future*, ed. A. J. McComb & J. A. Davis, pp. 5–18. Adelaide, SA, Australia: Gleneagles Publishing.

Holloway, G. & Sou, T. (2002) Has Arctic sea ice rapidly thinned? *Journal of the American Meteorological Society* 15: 1691–1701.

Holmer, M. (1999) The effect of oxygen depletion on anaerobic organic matter degradation in marine sediments. *Estuarine, Coastal and Shelf Science* 48: 383–390.

Holmer, M., Andersen, F. O., Nielsen, S. L. & Boschker, H. T. S. (2001) The importance of mineralization based on sulfate reduction for nutrient regeneration in tropical seagrass sediments. *Aquatic Botany* 71: 1–17.

Holmes, K. E., Edinger, E. N., Limmon, H. G. V. & Risk, M. J. (2000) Bioerosion of live massive corals and branching coral rubble on Indonesian coral reefs. *Marine Pollution Bulletin* 40: 606–617.

Holmes, N. T. H. (1999) Recovery of headwater stream flora following the 1989–1992 groundwater drought. *Hydrological Processes* 13: 341–354.

Holm-Hansen, O., Helbling, E. W. & Lubin, D. (1993) Ultraviolet radiation in Antarctica: inhibition of primary production. *Photochemistry and Photobiology* 58: 567–570.

Holmquist, C. (1959) Problems on marine–glacial relicts on account of investigations on the genus *Mysis*. Unpublished report. Lund, Sweden: Berlindska.

Hood, G. W. & Naiman, R. J. (2000) Vulnerability of riparian zones to invasion by exotic vascular plants. *Plant Ecology* 148: 105–114.

Hooper, R. (1980) Observations on algal–grazer interactions in Newfoundland and Labrador. In: *Proceedings of the Workshop on the Relationship between Sea Urchin Grazing and Commercial Plant/Animal Harvesting*, ed. J. D. Pringle, G. J. Sharp & J. F. Caddy, pp. 120–124. Halifax, NS, Canada: Resource Branch, Fisheries and Oceans Canada.

Hopkins, J. S., Sandifer, P. A., DeVoe, M. R., *et al.* (1995) Environmental impacts of shrimp farming with special reference to the situation in the continental United States. *Estuaries* 18: 25–42.

Hopkinson, C. (1992) A comparison of ecosystem dynamics in fresh-water wetlands. *Estuaries* 15: 549–562.

Horner, R., Ackley, S. F., Dieckmann, G. S., *et al.* (1992) Ecology of sea ice biota. I. Habitat, terminology and methodlogy. *Polar Biology* 12: 417–427.

Hornung, H., Krom, M. D., Cohen, Y. & Bernhard, M. (1993) Trace metal content in deep-water sharks from the eastern Mediterranean Sea. *Marine Biology* 115: 331–338.

Houde, E., Coleman, F., Dayton, P., *et al.* (2001) *Marine Protected Areas: Tools for Sustaining Ocean Ecosystems*. Washington, DC, USA: National Academy Press.

Houghton, J. (1999) *Global Warming: The Complete Briefing*, 2nd edn. Cambridge, UK: Cambridge University Press.

Houghton, J. T., Jenkins, G. J. & Ephrauns, J. J., eds. (1990) *Scientific Assessment of Climate Change: Report of Working Group 1*. Cambridge, UK: Cambridge University Press.

Houghton, J. T., Meira Filho, L. G., Bruce, J., *et al.* (1996) *Climate Change 1995: World Meteorological Organization/Intergovernmental Panel on Climate Change*. Cambridge, UK: Cambridge University Press.

Houlahan, J. E., Findlay, C. S., Schmidt, B. R., Meyer, A. H. & Kuzmin, S. L. (2000) Quantitative evidence for global amphibian population declines. *Nature* 404: 752–755.

Houston, J. R. (2002) The economic value of beaches: a 2002 update. *Shore and Beach* 70: 9–12.

Howard-Williams, C. & Thompson, K. (1985) The conservation and management of African wetlands. In: *The Ecology and Management of African Wetland Vegetation*, ed. P. Denny, pp. 203–230. The Hague, the Netherlands: Dr W. Junk.

Howarth, R. W., Anderson, D. M., Church, T. M., (2000a) *Clean Coastal Waters: Understanding and Reducing the Effects of Nutrient Pollution*. Washington, DC, USA: Ocean Studies Board and Water Science and Technology Board, National Academy Press.

Howarth, R. W., Anderson, D., Cloern, J., *et al.* (2000*b*) Nutrient pollution of coastal rivers, bays, and seas. *Issues in Ecology* 7: 1–15.

Howarth, R. W., Sharpley, A. & Walker, D. (2002) Sources of nutrient pollution to coastal waters in the United States: implications for achieving coastal water quality goals. *Estuaries* 25: 656–676.

Hu, S., Chapin, F. S., Firestone, M. K., Field, C. B. & Chlariello, N. R. (2001) Nitrogen limitation of microbial decomposition in a grassland under elevated CO_2. *Nature* 409: 188–191.

Hudson, P. F. & Colditz, R. R. (2003) Flood delineation in a large and complex alluvial valley, lower Pánuco basin, Mexico. *Journal of Hydrology* 280: 229–245.

Hughes, F. M. R. (1990) The influence of flooding regimes on forest distribution and composition in the Tana River floodplain, Kenya. *Journal of Applied Ecology* 27: 475–491.

Hughes, F. M. R., ed. (2003) *The Flooded Forest: Guidance for Policy Makers and River Managers in Europe on the Restoration of Floodplain Forests*, FLOBAR2. Cambridge, UK: European Union and Department of Geography, University of Cambridge.

Hughes, F. M. R. & Rood, S. B. (2003) The allocation of river flows for the restoration of floodplain forest ecosystems: a review of approaches and their applicability in Europe. *Environmental Management* 32: 12–33.

Hughes, P. D. M. & Barber, K. E. (2003) Mire development across the fen–bog transition on the Teifi floodplain at Tregaron Bog, Ceredigion, Wales, and a comparison with 13 other raised bogs. *Journal of Ecology* 91: 253–264.

Hughes, T. P. (1994) Catastrophes, phase shifts, and large-scale degradation of a Caribbean coral reef. *Science* 265: 1547–1551.

Hughes, T. P. & Tanner, J. E. (2000) Recruitment failure, life histories, and long-term decline of Caribbean corals. *Ecology* 81: 2250–2263.

Hughes, T. P., Baird, A. H., Bellwood, D. R., *et al.* (2003) Climate change, human impacts, and the resilience of coral reefs. *Science* 301: 929–933.

Hughes, T. P., Bellwood, D. R. & Connolly, S. R. (2002) Biodiversity hotspots, centres of endemicity, and the conservation of coral reefs. *Ecology Letters* 5: 775–784.

Hukkinen, J. (1998) Institutions, environmental management and long-term ecological sustenance. *Ambio* 27: 112–117.

Hulme, M., Jenkins, G. J., Lu, X., *et al.* (2002) *Climate Change Scenarios for the United Kingdom: The UKCIP02 Scientific Report*. Norwich, UK: Tyndall Centre for Climate

Change Research, School of Environmental Sciences, University of East Anglia.

Hultén, E. (1968) *Flora of Alaska and Neighboring Territories: A Manual of Vascular Plants*. Stanford, CA, USA: Stanford University Press.

Humborg, C., Ittekkot, V., Cociasu, A. & VonBodungen, B. (1997) Effect of Danube River dam on Black Sea biogeochemistry and ecosystem structure. *Nature* 386: 385–388.

Humphries, P. & Baldwin, D. S. (2003) Drought and aquatic ecosystems: an introduction. *Freshwater Biology* 48: 1141–1146.

Humphries, S. E., Groves, R. H. & Mitchell, D. S. (1991) Plant invasions of Australian ecosystems: a status review and management directions. Unpublished report, Kowari 2. Canberra, ACT, Australia: Australian National Parks and Wildlife Service.

Hunt, R. J. & Wilcox, D. A. (2003) Ecohydrology: why hydrologists should care. *Ground Water* 41: 289.

Huntington, H. P. (2000) Sustainable fisheries management. *WWF Arctic Bulletin* 1: 1–18.

Hurem, A. K. 1979. Rubble-mound structures as artificial reefs. In: *Coastal Structures 79*, pp. 1042–1051. New York, USA: American Society of Civil Engineers.

Hurrell, J. W. (1995) Decadal trends in the North Atlantic oscillation: regional temperatures and precipitation. *Science* 269: 676–679.

Huszar, P., Petermann, P., Leite, A., *et al.* (1999) *Fact or Fiction: A Review of the Hidrovia Paraguay–Paraná Official Studies*. Toronto, Ontario, Canada: WWF.

Hutchin, P. R., Press, M. C., Lee, J. A. & Ashenden, T. W. (1995) Elevated concentrations of CO_2 may double methane emissions from mires. *Global Change Biology* 1: 125–128.

Hutchings, J. A. (2000) Collapse and recovery of marine fishes. *Nature* 406: 882–885.

Hutchinson, G. E. (1957) *A Treatise on Limnology*, vol. 1. New York, USA: John Wiley.

Hwang, Y. H. & Morris, J. T. (1994) Whole plant gas exchange responses of *Spartina alterniflora* (Poaceae) to a range of constant and transient salinities. *American Journal of Botany* 81: 659–665.

Ingole, B., Ansari, Z. A., Rathod, V. & Rodrigues, N. (2001) Response of deep-sea macrobenthos to a small-scale environmental disturbance. *Deep-Sea Research II* 48: 3401–3410.

Ingram, H. A. P. (1978) Soil layers in mires: function and terminology. *Journal of Soil Science* 29: 224–227.

Ingrid, G., Andersen, T. & Vadstein, O. (1996) Pelagic food webs and eutrophication of coastal waters: impact of grazers on algal communities. *Marine Pollution Bulletin* **33**: 22–35.

Inman, D. J., Masters, P. M. & Stone, K. E. (1991) Induced subsidence: environmental and legal implications. In: *Coastal Zone 91*, pp. 16–27. New York, USA: American Society of Civil Engineers.

Innocenti, L. & Pranzini, E. (1993) Geomorphological evolution and sedimentology of the Ombrone River delta, Italy. *Journal of Coastal Research* **9**: 481–493.

International Board of Inquiry for the Great Lakes Fisheries (1943) Report and Supplement. Unpublished report. Washington, DC, USA: US Government Printing Office.

International Joint Commission (1989) Revised Great Lakes water quality agreement of 1978. Unpublished report. Washington, DC, USA and Ottawa, Ontario, Canada: International Joint Commission, United States and Canada.

International Joint Commission (2000) Tenth biennial report on Great Lakes water quality. Unpublished report. Washington, DC, USA and Ottawa, Ontario, Canada: International Joint Commission, United States and Canada.

International Organization for Migration (2000) *World Migration Report*. New York, USA: IOM.

International Tanker Owners Pollution Federation Limited (2000) Historical data [www document]. URL http://www.itopf.com/stats.html

Invers, O., Zimmerman, R. C., Alberte, R. S., Pérez, M. & Romero, J. (2001) Inorganic carbon sources for seagrass photosynthesis: an experimental evaluation of bicarbonate use in species inhabiting temperate waters. *Journal of Experimental Marine Biology and Ecology* **265**: 203–217.

IPCC (1995): *Climate Change 1995: The IPCC Second Assessment Synthesis of Scientific–Technical Information Relevant to Interpreting Article 2 of the UN Framework Convention on Climate Change*. Geneva, Switzerland: IPCC.

IPCC (1996) *Climate Change 1995: Impacts, Adaptations, and Mitigation of Climate Change – Contribution of Working Group II to the Second Assessment Report of the Intergovernmental Panel on Climate Change*, ed. R. T. Watson, M. C. Zinoyowera & R. H. Moss. Cambridge, UK: Cambridge University Press.

IPCC (2001a) *Climate Change 2001: The Scientific Basis – Contribution of Working Group I to the Third Assessment Report of the Intergovernmental Panel on Climate Change*, ed. J. T. Houghton, Y. Ding, D. J. Griggs, M. Noguer, P. J. van der Linden, X. Dai, K. Maskell & C. A. Johnson. Cambridge, UK: Cambridge University Press.

IPCC (2001b) *Climate Change 2001: Impacts, Adaptation, and Vulnerability – Contribution of Working Group II to the Third Assessment Report of the Intergovernmental Panel on Climate Change*, ed. J. J. McCarthy, O. F. Canziani, N. A. Leary, D. J. Dokken & K. S. White. Cambridge, UK: Cambridge University Press.

IPCC (2001c) *Climate Change 2001: Mitigation – Contribution of Working Group I to the Third Assessment Report of the Intergovernmental Panel on Climate Change*, ed. B. Metz, O. Davidson, R. Swart & J. Pan. Cambridge, UK: Cambridge University Press.

IPCC (2001d) *Climate Change 2001: Synthesis Report – A Contribution of Working Groups I, II, and III to the Third Assessment Report of the Intergovernmental Panel on Climate Change*, ed. R. T. Watson and the Core Writing Team. Cambridge, UK: Cambridge University Press.

IPCC (2001e) *IPCC Third Assessment Report: Climate Change 2001*. Cambridge, UK: Cambridge University Press.

IPCC (2002) *Climate Change and Biodiversity*, Intergovernmental Panel on Climate Change Technical Paper No. V, ed. H. Gitay, A. Suárez, R. T. Watson & D. J. Dokken. Geneva, Switzerland: IPCC. Available online at www.ipcc.ch/pub/tpbiodiv.pdf

ISSLR (2007) International Society for Salt Lake Research: home page [www document]. URL http://www.isslr.org.

Ittekkot, V., Jilan, S., Miles, E., *et al.* (1996) Oceans. In: *Climate Change 1995: Impacts, Adaptations and Mitigation of Climate Change – Contribution of Working Group II to the Second Assessment Report of the Intergovernmental Panel on Climate Change*, ed. R. T. Watson, M. C. Zinoyowera & R. H. Moss, pp. 267–288. Cambridge, UK: Cambridge University Press.

IUCN (1996) *IUCN Red List of Threatened Animals*. Gland, Switzerland: IUCN (World Conservation Union).

IUCN (2000) *Vision for Water and Nature: A World Strategy for Conservation and Sustainable Management of Water Resources in the Twenty-First Century*. Gland, Switzerland: IUCN (World Conservation Union).

Ivanov, K. E. (1981) *Water Movement in Mirelands*. London, UK: Academic Press.

Ivanov, V. P., Kamakin, A. M., Ushintzev, V. B., *et al.* (2000) Invasion of the Caspian Sea by the comb jellyfish *Mnemiopsis leidyi* (Ctenophora). *Biological Invasions* **2**: 255–258.

Ivanov, V. P., Vlasenko, A., Dkhodorevskaya, R. P. & Raspopov, V. M. (1999) Contemporary status of Caspian sturgeon (Acipenseridae) stock and its conservation. *Journal of Applied Ichthyology – Zeitschrift für Angewandte Ichthyologie* **15**: 103–105.

Iversen, T. M., Grant, R. & Nielsen, K. (1998) Nitrogen enrichment of European inland and marine waters with special attention to Danish policy measures. *Environmental Pollution* **102**: 771–780.

Iverson, R. L. (1990) Control of marine fish production. *Limnology and Oceanography* **35**: 1593–1604.

Iwata, H., Shinsuke, T., Ueda, K. & Tatsukawa, R. (1995) Persistent organochlorine residues in air, water, sediments, and soils from the Lake Baikal region, Russia. *Environmental Science and Technology* **29**: 792–801.

Jackson, G. (1998) Currents in the high drag environment of a coastal kelp stand off California. *Continental Shelf Research* **17**: 1913–1928.

Jackson, J. B. C. (1997) Reefs since Columbus. *Coral Reefs* **16**: S23–S32.

Jackson, J. B. C., Kirby, M. X., Berger, W. H., *et al.* (2001) Historical overfishing and the recent collapse of coastal ecosystems. *Science* **293**: 629–637.

Jackson, N. L. & Nordstrom, K. F. (1993) Depth of sediment activation by plunging waves on a steep sand beach. *Marine Geology* **115**: 143–151.

Jackson, N. L., Nordstrom, K. F. & Smith, D. R. (2002) Geomorphic–biotic interactions on sandy estuarine beaches. *Journal of Coastal Research* Special Issue **36**: 414–424.

Jacobs, S. S. & Comiso, J. C. (1993) A recent sea-ice retreat west of the Antarctic Peninsula. *Geophysical Research Letters* **20**: 1171–1174.

Jagtap, T. G., Komarpant, D. S. & Rodrigues, R. (2003) The seagrasses of India. In: *World Atlas of Seagrasses*, ed. E. P. Green & F. T. Short, pp. 113–120. Berkeley, CA, USA: University of California Press.

Jaksic, F. M. (1998) The multiple facets of El Niño Southern Oscillation in Chile. *Revista Chilena de Historia Natural* **71**: 121–131.

Jaramillo, C. & Bayona, G. (2000) Mangrove distribution during the Holocene in Tribugá Gulf, Colombia. *Biotropica* **32**: 14–22.

Jefferies, R. L. (1997) Long-term damage to sub-arctic coastal ecosystems by geese: ecological indicators and measures of ecosystem dysfunction. In: *Disturbance and Recovery in Arctic Lands: An Ecological Perspective*, ed. R. M. M. Crawford, pp. 151–166. Boston, MA, USA: Kluwer.

Jefferies, R. L. & Perkins, N. (1977) The effects on the vegetation of the additions of inorganic nutrients to salt marsh soils at Stiffkey, Norfolk. *Journal of Ecology* **65**: 847–865.

Jefferies, R. L., Henry, H. & Abraham, K. F. (2002) Agricultural nutrient subsidies to migratory geese and ecological change to Arctic coastal habitats. In: *Food Webs at the Landscape Level*, ed. G. A. Polis, M. E. Power & G. R. Huxel, pp. 268–283. Chicago, IL, USA: University of Chicago Press.

Jeglum, J. K. (1990) Peatland forestry in Canada: an overview. In: *Biomass Production and Element Fluxes in Forested Peatland Ecosystems*, ed. B. Hanell, pp. 19–28. Umeå, Sweden: Swedish University of Agricultural Science.

Jellison, R. & Melack, J. M. (1993) Algal photosynthetic activity and its response to meromixis in hypersaline Mono Lake, California. *Limnology and Oceanography* **38**: 818–837.

Jenkins, M., Green, R. E. & Madden, J. (2003) The challenge of measuring global change in wild nature: are things getting better or worse? *Conservation Biology* **17**: 20–23.

Jenkins, S. R., Arenas, F., Arrontes, J., *et al.* (2001) European-scale analysis of seasonal variability in limpet grazing activity and microalgal abundance. *Marine Ecology Progress Series* **211**: 193–203.

Jennings, S. & Kaiser, M. J. (1998) The effects of fishing on marine ecosystems. In *Advances in Marine Biology*, vol. 34, ed. J. Blaxter, A. J. Southward & P. A. Tyler, pp. 201–352. San Diego, CA, USA: Academic Press.

Jennings, S. & Lock, J. M. (1996) Population and ecosystem effects of fishing. In: *Reef Fisheries*, ed. N. V. C. Polunin & C. M. Roberts, pp. 193–218. London, UK: Chapman & Hall.

Jennings, S., Dinmore, T. A., Duplisea, D. E., Warr, K. J. & Lancaster, J. E. (2001) Trawling disturbance can modify benthic production processes. *Journal of Animal Ecology* **70**: 459–475.

Jennings, S., Greenstreet, S. P. R. & Reynolds, J. D. (1999) Structural change in an exploited fish community: a consequence of differential fishing effects on species with contrasting life histories. *Journal of Animal Ecology* **68**: 617–627.

Jennings, S., Reynolds, J. D. & Mills, S. C. (1998) Life history correlates of responses to fisheries exploitation. *Proceedings of the Royal Society of London B* **265**: 333–339.

Jeppesen, E. & Sammalkorpi, I. (2002) Lakes. In: *Handbook of Restoration Ecology*, vol. 2, ed. *Restoration in Practice*, ed. M. Perrow & T. Davy, pp. 297–324. Cambridge, UK: Cambridge University Press.

Jeppesen, E., Christoffersen, K., Landkildehus, F., Lauridsen, T. & Amsinck, S. (2001) Fish and crustaceans in northeast Greenland lakes with special emphasis on interactions between Arctic charr (*Salvelinus alpinus*), *Lepidurus arcticus* and benthic chydorids. *Hydrobiologia* **442**: 329–337.

Jeppesen, E., Jensen, J. P., Jensen, C., *et al.* (2003) The impact of nutrient state and lake depth on top-down control in the pelagic zone of lakes: study of 466 lakes from the temperate zone to the Arctic. *Ecosystems* **6**: 313–325.

Jeremiason, J., Hornbuckle, K. & Eisenreich, S. (1994) PCBs in Lake Superior, 1978–1992: decreases in water concentrations reflect loss by volatilization. *Environmental Science and Technology* **28**: 903–914.

JGOFS (2008) Joint Global Ocean Flux Study: home page [www document]. URL http://ijgofs.whoi.edu

Ji, C. Y., Liu, Q., Sun, D., *et al.* (2001) Monitoring urban expansion with remote sensing in China. *International Journal of Remote Sensing* **22**: 1441–1455.

Jickells, T. D. (1998) Nutrient biogeochemistry of the coastal zone. *Science* **281**: 217–222.

Joabsson, A. & Christensen, T. R. (2001) Methane emissions from wetlands and their relationship with vascular plants: an Arctic example. *Global Change Biology* **7**: 919–932.

Joabsson, A., Christensen, T. R. & Wallén, B. (1999) Vascular plant controls on methane emissions from northern peat forming wetlands. *Trends in Ecology and Evolution* **14**: 385–388.

Johannesburg Summit (2002) World summit on sustainable development, Johannesburg, 26 August–4 September 2002. UN News Service [www document]. URL http://www.un.org/events/wssd/

Johannessen, O. M., Miles, M. & Bjørgo, E. (1995) The Arctic's shrinking sea ice. *Nature* **376**: 126–127.

Johannessen, O. M., Shalina, E. & Miles, M. W. (1999) Satellite evidence for and Arctic sea ice cover in transition. *Science* **286**: 1937–1939.

Johannesson, K. & Warmoes, T. (1990) Rapid colonisation of Belgian breakwaters by the direct developer *Littorina saxatilis* (Olivi) (Prosobranchia, Mollusca). *Hydrobiologia* **193**: 99–108.

Johns Hopkins (1998) *Solutions for a Water-Short World*, Population Report Vol. XXVI, No. 1. Baltimore, MD, USA: Johns Hopkins Population Information Program. Available online at www.infoforhealth.org/pr/m14edsum.shtml

Johnson, L. C. & Damman, A. W. H. (1993) Decay and its regulation in *Sphagnum* peatlands. *Advances in Bryology* **5**: 249–296.

Johnson, L. C., Damman, A. W. H. & Malmer, N. (1990) *Sphagnum* macrostructure as an indicator of decay and compaction in peat cores from an ombrotrophic South Swedish peat bog. *Journal of Ecology* **78**: 633–647.

Johnson, M. P., Burrows, M. T. & Hawkins, S. J. (1998) Individual based simulations of the direct and indirect effects of limpets on a rocky shore *Fucus* mosaic. *Marine Ecology Progress Series* **169**: 179–188.

Johnson, N., Revenga, C. & Echeverria, J. (2001) Managing water for people and nature. *Science* **292**: 1071–1072.

Johnson, T., Scholz, C., Talbot, M., *et al.* (1996) Late Pleistocene desiccation of Lake Victoria and rapid evolution of cichlid fishes. *Science* **273**: 1091–1093.

Johnston, E. L. & Keough, M. J. (2003) Competition modifies the response of organisms to toxic disturbance. *Marine Ecology Progress Series* **215**: 15–26.

Jokiel, P. L., Hunter, C. L., Taguchi, S. & Watarai, L. (1993) Ecological impact of a fresh-water 'reef kill' in Kaneohe bay, Oahu, Hawaii. *Coral Reefs* **12**: 177–184.

Jolly, I. D. (1996) The effects of river management on the hydrology and hydroecology of arid and semi-arid flood-plains. In: *Floodplain Processes*, ed. M. G. Anderson, D. E. Walling & P. D. Bates, pp. 577–609. Chichester, UK: John Wiley.

Jonasson, S. & Chapin III, F. S. (1985) Significance of sequential leaf development for nutrient balance of the cotton sedge, *Eriophorum vaginatum* L. *Oecologia* **67**: 511–518.

Jones, C. G., Lawton, J. H. & Shachak, M. (1994) Organisms as ecosystem engineers. *Oikos* **69**: 373–386.

Jones, D., Cocklin, C. & Cutting, M. (1995) Institutional and landowner perspectives on wetland management in New Zealand. *Journal of Environmental Management* **45**: 143–161.

Jones, D. A., Watt, I., Woodhouse, T. D. & Richmond, M. D. (1994) Intertidal recovery in the Dawhat ad Dafi and Dawhat al Mussallamiya region (Saudi Arabia) after the Gulf War oil spill. *Courier Forschungsinstitut Senckenberg* **166**: 27–33.

Jones, D. K. C. (1993) Global warming and geomorphology. *Geographical Journal* **159**: 124–130.

Jones, G. (1994) Global warming, sea level change and the impact on estuaries. *Marine Pollution Bulletin* **28**: 7–14.

Jones, G. P. & Andrew, N. L. (1990) Herbivory and patch dynamics on rocky reefs in temperate Australasia: the roles of fish and sea urchins. *Australian Journal of Ecology* **15**: 505–520.

Jones, G. P. & Andrew, N. L. (1993) Temperate reefs and the scope of seascape ecology. In: *The Ecology of Temperate Reefs, Proceedings of the 2nd International Temperate Reef Symposium*, Auckland, pp. 63–76. Wellington, New Zealand: NIWA Publications.

Jones, G. P. (1988) Ecology of rocky reef fish of north-eastern New Zealand. *New Zealand Journal of Marine and Freshwater Research* **22**: 445–462.

Jones, N. S. (1946) Browsing of *Patella*. *Nature* **158**: 557.

Jones, N. S. & Kain, J. M. (1967) Subtidal algal colonization following the removal of *Echinus*. *Helgoländer Wissenschaftliche Meeresuntersuchungen* **15**: 160–166.

Jonsson, M. & Malmqvist, B. (2000) Ecosystem process rate increases with animal species richness: evidence from leaf-eating, aquatic insects. *Oikos* **89**: 519–523.

Jonsson, M. (2003a) Mechanisms behind diversity effects on ecosystem functioning: testing the facilitation and interference hypotheses. *Oecologia* **134**: 554–559.

Jonsson, M. (2003b) Importance of species identity and number for process rates within different functional feeding groups. *Journal of Animal Ecology* **72**: 453–459.

Jonsson, M., Dangles, O., Malmqvist, B. & Guérold, F. (2002) Simulating species loss following disturbance: assessing the effects on process rates. *Proceedings of the Royal Society of London B* **269**: 1047–1052.

Jonsson, M., Malmqvist, B. & Hoffsten, P. O. (2001) Leaf litter breakdown rates in boreal streams: does shredder species richness matter? *Freshwater Biology* **46**: 161–172.

Jorgensen, S., Riccardo, B., Ballatore, T. & Muhandiki, V. (2003) Lake Watch 2003: the changing state of the world's lakes. Unpublished report. Kusatsu, Japan: International Environment Committee Foundation.

Jumars, P. A. (1981) Limits in predicting and detecting benthic community responses to manganese nodule mining. *Marine Mining* **3**: 213–229.

Junk, W. J. & da Silva, V. M. F. (1997) Mammals, reptiles and amphibians. In: *The Central Amazon Floodplain: Ecology of a Pulsing System*, ed. W. J. Junk, pp. 409–417. Berlin, Germany: Springer-Verlag.

Junk, W. J. & Wantzen, K. M. (2004) The flood pulse concept: new aspects, approaches, and applications – an update. In: *Proceedings of the 2nd Large River Symposium (LARS)*, ed. R. Welcomme & C. Barow, pp. 117–149. Phnom Penh, Cambodia: Mekong River Commission and Food and Agriculture Organization of the United Nations.

Junk, W. J. (1993) Wetlands of tropical South America. In: *Wetlands of the World*, vol. 1, *Inventory, Ecology and Management*, ed. D. F. Whigham, D. Dykyjová & S. Hejný, pp. 679–739. Dordrecht, the Netherlands: Kluwer.

Junk, W. J. (1995) Human impact on neotropical wetlands: historical evidence, actual status and perspectives. *Scientia Guianae* **5**: 299–311.

Junk, W. J., ed. (1997) *The Central Amazon Floodplain: Ecology of a Pulsing System*. Berlin, Germany: Springer-Verlag.

Junk, W. J. (2000) Mechanisms for development and maintenance of biodiversity in neotropical floodplains. In: *Biodiversity in Wetlands: Assessment, Function and Conservation*, vol. 1, ed. B. Gopal, W. J. Junk & J. A. Davis, pp. 119–139. Leiden, the Netherlands: Backhuys.

Junk, W. J. (2002) Long-term environmental trends and the future of tropical wetlands. *Environmental Conservation* **29**: 414–435.

Junk, W. J. & Welcomme, R. L. (1990). Floodplains. In: *Wetlands and Shallow Continental Water Bodies*, ed. B. C. Patten *et al.* pp. 491–524. The Hague, the Netherlands: SPB Academic Publishing.

Junk, W. J., Bayley, P. B. & Sparks, R. E. (1989) The flood pulse concept in river–floodplain systems. *Canadian Special Publication of Fisheries and Aquatic Sciences* **106**: 110–127.

Junk, W. J., Brown, M., Campbell, I. S., *et al.* (2006) Comparative biodiversity of large wetlands: a synthesis. *Aquatic Sciences* **68**: 254–277.

Junk, W. J., Ohly, J. J., Piedade, M. T. F. & Soares, M. G. M., eds. (2000) *The Central Amazon Floodplain: Actual Use and Options for a Sustainable Management*. Leiden, the Netherlands: Backhuys.

Jutila, H. M. (1997) Vascular plant species richness in grazed and ungrazed coastal meadows, SW Finland. *Annales Botanici Fennici* **34**: 245–263.

Kabat, P., Schulze, R. E., Hellmuth, M. E. & Veraart, J. A., eds. (2002) Coping with impacts of climate variability and climate change in water management: a scoping paper. Unpublished report. Wageningen, the Netherlands: Dialogue on Water and Climate.

Kaczmarek, Z., Arnell, N. W., Stakhiv, E. Z., *et al.* (1996) Water resources management. In: *Climate Change 1995: Impacts, Adaptations and Mitigation of Climate Change – Contribution of Working Group II to the Second Assessment Report of the Intergovernmental Panel on Climate Change*, ed. R. T. Watson, M. C. Zinoyowera & R. H. Moss, pp. 469–486. Cambridge, UK: Cambridge University Press.

Kaczynski, V. M. & Fluharty, D. L. (2002) European policies in West Africa: who benefits from fisheries agreements? *Marine Policy* **26**: 75–93.

Kadioğlu, M., Şen, Z. & Batur, E. (1997) The greatest soda-water lake in the world and how it is influenced by climatic change. *Annales Geophysicae: Atmospheres, Hydrospheres and Space Sciences* **15**: 1489–1497.

Kain, J. M. (1975) Algal recolonization of some cleared subtidal areas. *Journal of Ecology* **63**: 739–765.

Kaiser, M. J. & de Groot, S. J., eds. (2000) *The Effects of Fishing on Non-Target Species and Habitats*. Oxford, UK: Blackwell Science.

Kaiser, M. J., Ramsay, K., Richardson, C. A., Spence, F. E. & Brand, A. R. (2000) Chronic fishing disturbance has changed shelf sea benthic community structure. *Journal of Animal Ecology* **69**: 494–503.

Kania, H. J. & O'Hara, J. (1974) Behavioral alterations in a simple predator–prey system due to sub-lethal exposure to mercury. *Transactions of the American Fisheries Society* **103**: 134–136.

Kannan, K., Smith Jr, R. G., Lee, R. F., *et al.* (1998) Distribution of total mercury and methylmercury in water, sediment, and fish from South Florida estuaries. *Archives of Environmental Contamination and Toxicology* **34**: 109–118.

Kapuscinski, A. M. & Hallerman, E. R. (1991) Implications of introduction of transgenic fish into natural ecosystems. *Canadian Journal of Fisheries and Aquatic Science* **48**: 99–107.

Karl, D. M. (1999) A sea of change: biogeochemical variability in the North Pacific Subtropical Gyre. *Ecosystems* **2**: 181–214.

Karl, D. M., Bidigare, R. R. & Letelier, R. M. (2001) Long-term changes in plankton community structure and productivity in the North Pacific Subtropical Gyre: the domain shift hypothesis. *Deep-Sea Research* II **48**: 1449–1470.

Karlson, K., Rosenberg, R. & Bonsdorff, E. (2002) Temporal and spatial large-scale effects of eutrophication and oxygen deficiency on benthic fauna in Scandinavian and Baltic waters: a review. *Oceanography and Marine Biology Annual Review* **40**: 427–489.

Karlsson, H., Muir, D., Teixiera, C., *et al.* (2000) Persistent chlorinated pesticides in air, water, and precipitation from the Lake Malawi area, Southern Africa. *Environmental Science and Technology* **34**: 4490–4495.

Karoly, D., Risbey, J. & Reynolds, A. (2003) Global warming contributes to Australia's worst drought. WWF Australia [www document]. URL http://www.wwf.org.au/News_and_information/Publications/PDF/Report/drought_report.pdf.

Karr, J. R. (1991) Biological integrity: a long-neglected aspect of water resource management. *Ecological Applications* **1**: 66–84.

Karr, J. R., Fausch, K. D., Angermeier, P. L., Yant, P. R. & Schlosser, I. J. (1986) *Assessing the Biological Integrity of Running Waters: A Method and Its Rationale*. Champaign, IL, USA: Illinois Natural History Survey.

Kathiresan, K. & Bingham, B. L. (2001) Biology of mangroves and mangrove ecosystems. *Advances in Marine Biology* **40**: 81–251.

Katunin, D. N. (2000) Trans-boundary diagnostic analysis of relevantly important commercial bioresources. Caspian Environment Programme [www document]. URL http://www.caspianenvironment.org/report_miscell3.htm.

Kaufman, L. (1992) Catastrophic change in species-rich freshwater ecosystems. *BioScience* **42**: 846–858.

Kautsky, N., Kautsky, H., Kautsky, U. & Waern, M. (1986) Decreased depth penetration of *Fucus vesiculosus* (L.) since the 1940s indicates eutrophication of the Baltic Sea. *Marine Ecology Progress Series* **28**: 1–8.

Kawall, H. G., Torres, J. J. & Geiger, S. P. (2001) Effects of the ice-edge bloom and season on the metabolism of copepods in the Weddell Sea, Antarctica. *Hydrobiologia* **453**: 67–77.

Keddy, P. A. (1981) Experimental demography of a dune annual: *Cakile edentula* growing along an environmental gradient in Nova Scotia. *Journal of Ecology* **69**: 615–630.

Keddy, P. A. (2000) *Wetland Ecology: Principles and Conservation*. Cambridge, UK: Cambridge University Press.

Keddy, P. A. & Wisheu, I. C. (1989) Ecology, biogeography, and conservation of coastal plain plants: some general principles from the study of Nova Scotian wetlands. *Rhodora* **91**: 72–94.

Keefer, D. K., Defrance, S. D., Moseley, M. E., *et al.* (1998) Early maritime economy and El Niño Events at Quebrada Tacahuay, Peru. *Science* **281**: 1833–1835.

Keenan, C. P. (1994) Recent evolution of population structure in Australian barramundi, *Lates calcarifer* (Bloch): an example of isolation by distance in one dimension. *Australian Journal of Marine and Freshwater Research* **45**: 1123–1148.

Kelleher, G., Bleakley, C. & Wells, S. (1995) *A Global Representative System of Marine Protected Areas*, vol. 1. Washington, DC, USA: World Bank.

Kelly, L. C., Rundle, S. D. & Bilton, D. T. (2002) Genetic population structure and dispersal in Atlantic Island caddisflies. *Freshwater Biology* **47**: 1642–1650.

Kelsey, K. A. & West, S. D. (1998) Riparian wildlife. In: *River Ecology and Management: Lessons from the Pacific Coastal Ecoregion*, ed. R. J. Naiman & R. E. Bilby, pp. 235–258. New York, USA: Springer-Verlag.

Kendall, M. A., Bowman, R. S., Williamson, P. & Lewis, J. R. (1985) Annual variation in recruitment of *Semibalanus balanoides* on the North Yorkshire coast 1969–1981. *Journal of the Marine Biological Association of the United Kingdom* **65**: 1009–1030.

Kendrick, G. A., Eckersley, J. & Walker, D. I. (1999) Landscape scale changes in seagrass distribution over time: a case study from Success Bank, Western Australia. *Aquatic Botany* **65**: 293–309.

Kennish, M. J. (1986) *Ecology of Estuaries: Physical and Chemical Aspects*. Boca Raton, FL, USA: CRC Press.

Kennish, M. J. (1992) *Ecology of Estuaries: Anthropogenic Effects*. Boca Raton, FL, USA: CRC Press.

Kennish, M. J. ed. (1997) *Practical Handbook of Estuarine and Marine Pollution*. Boca Raton, FL, USA: CRC Press.

Kennish, M. J. (1998a) *Pollution Impacts on Marine Biotic Communities*. Boca Raton, FL, USA: CRC Press.

Kennish, M. J. (1998b) Trace metal–sediment dynamics in estuaries: pollution assessment. *Reviews in Environmental Contamination and Toxicology* **155**: 69–110.

Kennish, M. J. ed. (2000) *Estuary Restoration and Maintenance: The National Estuary Programme*. Boca Raton, FL, USA: CRC Press.

Kennish, M. J. ed. (2001a) *Practical Handbook of Marine Science*, 3rd edn. Boca Raton, FL, USA: CRC Press.

Kennish, M. J. (2001b) Coastal salt marsh systems: a review of anthropogenic impacts. *Journal of Coastal Research* **17**: 731–748.

Kennish, M. J. ed. (2001c) Barnegat Bay–Little Egg Harbor, New Jersey: estuary and watershed assessment. *Journal of Coastal Research* Special Issue **32**.

Kennish, M. J. ed. (2002a) Impacts of motorized watercraft on shallow estuarine and coastal marine environments. *Journal of Coastal Research* Special Issue **37**.

Kennish, M. J. (2002b) Sediment contaminant concentrations in estuarine and coastal marine environments: potential for remobilization by boats and personal watercraft. *Journal of Coastal Research* Special Issue **37**: 150–177.

Kennish, M. J. (2002c) Environmental threats and environmental future of estuaries. *Environmental Conservation* **29**: 78–107.

Kennish, M. J. & Lutz, R. A., eds. (1984) *Ecology of Barnegat Bay, New Jersey*. New York, USA: Springer-Verlag.

Kentula, M. E. (2000) Perspectives on setting success criteria for wetland restoration. *Ecological Engineering* **15**: 199–209.

Kenworthy, W. J., Fonseca, M. S., Whitfield, P. E. & Hammerstrom, K. K. (2002) Analysis of seagrass recovery in experimental excavations and propeller-scar disturbances in the Florida Keys National Marine Sanctuary. *Journal of Coastal Research* Special Issue **37**: 150–177.

Kerr, R. A. (2002) A warmer Arctic means change for all. *Science* **297**: 1490–1492.

Kestner, F. J. T. (1962) The old coastline of the Wash: a contribution to the understanding of loose boundary processes. *Geographical Journal* **128**: 457–487.

Kickert, R. N., Tonella, G., Simonov, A. & Krupa, S. V. (1999) Predictive modeling of effects under global change. *Environmental Pollution* **100**: 87–132.

Kidd, K., Bootsma, H., Hesslein, R., Muir, D. & Hecky, R. (2001) Biomagnification of DDT through the benthic and pelagic food webs of Lake Malawi, East Africa: importance of trophic level and carbon source. *Environmental Science and Technology* **39**: 14–25.

Kiehl, K., Esselink, P. & Bakker, J. P. (1997) Nutrient limitation and plant species composition in temperate salt marshes. *Oecologia* **111**: 325–330.

Kimura, M. & Weiss, G. H. (1964) The stepping stone model of population structure and the decrease of genetic correlation with distance. *Genetics* **49**: 561–574.

King, J. C. (1994) Recent climate variability in the vicinity of the Antarctic Peninsula. *International Journal of Climatology* **14**: 357–369.

King, J. C. & Harangozo, S. A. (1998) Climate change in the western Antarctic Peninsula since 1945: observations and possible causes. *Annals of Glaciology* **27**: 571–575.

King, J. M. & Louw, D. (1998) Instream flow assessment for regulated rivers in South Africa using the building block methodology. *Aquatic Ecosystem Health and Management* **1**: 109–124.

Kingett, P. D. & Choat, J. H. (1981) Analysis of density and distribution patterns in *Chrysophrys auratus* (Pisces: Sparidae) within a reef environment: an experimental approach. *Marine Ecology Progress Series* **5**: 283–290.

Kingsford, M. J., Underwood, A. J. & Kennelly, S. J. (1991) Humans as predators on rocky reefs in New South Wales, Australia. *Marine Ecology Progress Series* **72**: 1–2.

Kingsford, R. T. (2000) Ecological impacts of dams, water diversions and river management on floodplain wetlands in Australia. *Austral Ecology* **25**: 109–127.

Kingsford, R. T. & Thomas, R. F. (1995) The Macquarie Marshes in arid Australia and their waterbirds: a 50-year history of decline. *Environmental Management* **19**: 867–878.

Kingsford, R. T., Boulton, A. J. & Puckridge, J. T. (1998) Challenges in managing dryland rivers crossing political boundaries: lessons from Cooper Creek and the Paroo River, central Australia. *Aquatic Conservation: Marine and Freshwater Ecosystems* **8**: 361–378.

Kingsley, C. (1890) *Glaucus, or the Wonders of the Shore.* London, UK: Macmillan.

Kinner, N. E., Harvey, R. W., Blakeslee, K., Novarino, G. & Meeker, L. D. (1998) Size-selective predation on groundwater bacteria by nanoflagellates in an organic-contaminated aquifer. *Applied and Environmental Microbiology* **64**: 618–625.

Kirby, A. (2002) Lake level fuels climate concern. BBC News World Edition, 28 October 2002 [www document]. URL http://news.bbc.co.uk/1/hi/sci/tech/2369333.stm

Kirkby, J. & O'Keefe, P. (1998) Conservation of inland deltas: a case study of the Gash Delta, Sudan. In: *The Sustainable Management of Tropical Catchments*, ed. D. Harper & A. G. Brown, pp. 209–223. Chichester, UK: John Wiley.

Kitching, J. A. & Ebling, F. J. (1961) The ecology of Lough Ine. XI. The control of algae by *Paracentrotus lividu*s (Echinoidea). *Journal of Animal Ecology* **30**: 373–383.

Kitching, J. A. & Thain, V. M. (1983) The ecological impact of the sea urchin *Paracentrotus lividus* (Lamarck) in Lough Ine, Ireland. *Philosophical Transactions of the Royal Society of London B* **300**: 513–552.

Kivinen, E. & Pakarinen, P. (1981) Geographical distribution of peat resources and major peatland complex types in the world. *Annales Academiae Scientiarum Fennicae, Series A, III, Geologica–Geographica* **132**: 1–28.

Kjerfve, B. (1989) Estuarine geomorphology and physical oceanography. In: *Estuarine Ecology*, ed. J. W. Day Jr, C. A. S. Hall, W. M. Kemp & A. Yañez-Arancibia, pp. 47–78. New York, USA: John Wiley.

Klarquist, M., Bohlin, E. & Nilsson, M. (2001) Long-term decline in apparent peat carbon accumulation in boreal mires in northern Sweden. *Acta Universitatis Agriculturae Suediae Silvestria* **203**(3): 1–22.

Klein, R. J. T., Smit, M. J., Goosen, H. & Hulsbergen, C. H. (1998) Resilience and vulnerability: coastal dynamics or Dutch dikes? *Geographical Journal* **164**: 259–268.

Kleypas, J. A., Buddemeier, R. W., Archer, D., *et al.* (1999*a*) Geochemical consequences of increased atmospheric carbon dioxide on coral reefs. *Science* **284**: 118–120.

Kleypas, J. A., McManus, J. W. & Menez, L. A. B. (1999*b*) Environmental limits to coral reef development: where do we draw the line? *American Zoologist* **39**: 146–159.

Klijn, F. & Witte, J.-P. M. (1999) Eco-hydrology: groundwater flow and site factors in plant ecology. *Hydrogeology Journal* **7**: 65–77.

Klimo, E. & Hager, H., eds. (2001) *The Floodplain Forests in Europe.* Leiden, the Netherlands: E. J. Brill.

Knapp, R., Matthews, K. R. & Sarnelle, O. (2001) Resistance and resilience of alpine lake fauna to fish introductions. *Ecological Monographs* **71**: 401–421.

Knopf, F. L. & Samson, F. B. (1994) Scale perspectives on avian diversity in western riparian ecosystems. *Conservation Biology* **8**: 669–676.

Knowlton, N. (1993) Sibling species in the sea. *Annual Review of Ecology and Systematics* **24**: 189–216.

Knowlton, N. (2000) Molecular genetic analyses of species boundaries in the sea. *Hydrobiologia* **420**: 73–90.

Knowlton, N., Lang, J. C. & Keller, B. D. (1988) Fates of staghorn coral fragments on hurricane-damaged reefs in Jamaica: the role of predators. In: *Proceedings of the 6th International Coral Reef Symposium*, Townsville, Australia, 8–12 August 1988, vol. 2, ed. J. H. Choat, D. Barnes, M. A. Borowitzka, *et al.*, pp. 83–88. Townsville, Queensland, Australia: 6th International Coral Reef Symposium Executive Committee.

Knutson, T. R., Tuleya, R. E. & Kurihara, Y. (1998) Simulated increase of hurricane intensities in a CO_2-warmed climate. *Science* **279**: 1018–1020.

Koch, E. W. & Orth, R. J. (2003) The seagrasses of the Mid-Atlantic Coast of the United States. In: *World Atlas of Seagrasses*, ed. E. P. Green & F. T. Short, pp. 234–243. Berkeley, CA, USA: University of California Press.

Kock, K.-H. (1992) *Antarctic Fish and Fisheries.* Cambridge, UK: Cambridge University Press.

Kock, K.-H. (2001) The direct influence of fishing and fishery-related activities on non-target species in the Southern Ocean with particular emphasis on longline fishing and its impact on albatrosses and petrels: a review. *Reviews in Fish Biology and Fisheries* **11**: 31–56.

Koike, K. (1993) The countermeasures against coastal hazards in Japan. *Geojournal* **38**: 301–312.

Kolar, C. S. & Lodge, D. M. (2001) Progress in invasion biology: predicting invaders. *Trends in Ecology and Evolution* **16**: 199–204.

Kolpin, D. W., Furlong, E. T., Meyer, M. T., *et al.* (2002) Pharmaceuticals, hormones, and other organic wastewater contaminants in US streams, 1999–2000: a national reconnaissance. *Environmental Science and Technology* **36**: 1202–1211.

Kooijman, A. M. & de Haan, M. W. A. (1995) Grazing as a measure against grass encroachment in Dutch dry dune grassland: effects on vegetation and soil. *Journal of Coastal Conservation* **1**: 127–134.

Koop, K., Booth, D., Broadbent, A. D., *et al.* (2001) ENCORE: the effect of nutrient enrichment on coral reefs – synthesis of results and conclusions. *Marine Pollution Bulletin* **42**: 91–120.

Kosarev, A. N. & Yablonskaya, E. A. (1994) *The Caspian Sea.* The Hague, the Netherlands: SPB Academic Publishing.

Koslow, J. A. (1997) Seamounts and the ecology of deep-sea fisheries. *American Scientist* **85**: 168–176.

Koslow, J. A., Bell, J., Virtue, P. & Smith, D. C. (1995) Fecundity and its variability in orange roughy: effects of population density, condition, egg size, and senescence. *Journal of Fish Biology* **47**: 1063–1080.

Koslow, J. A., Boehlert, G. W., Gordon, J. D. M., *et al.* (2000) Continental slope and deep-sea fisheries: implications for a fragile ecosystem. *ICES Journal of Marine Science* **57**: 458–557.

Koslow, J. A., Gowlett-Holmes, K., Lowry, J. K., *et al.* (2001) Seamount benthic macrofauna off southern Tasmania: community structure and impacts of trawling. *Marine Ecology Progress Series* **213**: 111–125.

Kota, S., Borden, R. C. & Barlaz, M. A. (1999) Influence of protozoan grazing on contaminant biodegradation. *FEMS Microbiology Ecology* **29**: 179–189.

Kothari, A., Pande, P., Singh, S. & Variava, D. (1989) *Management of National Parks and Sanctuaries in India: A Status Report.* New Delhi, India: Ministry of Environment and Forests.

Kotze, D. C. & Breen, C. M. (1994) *Agricultural Land-Use Impacts on Wetland Functional Values*, WRC Report No. 501/3/94. Pretoria, South Africa: Water Research Commission.

Kotze, D. C. & Breen, C. M. (2000) *WETLAND-USE: A Wetland Management Decision Support System for South African Freshwater Palustrine Wetlands.* Pretoria, South Africa: Department of Environmental Affairs and Tourism. Available online at www.ccwr.ac.za/wetlands/

Kozhov, M. (1963) *Lake Baikal and its Life.* The Hague, the Netherlands: Dr W. Junk.

Kozhova, O. & Silow, E. (1998) The current problems of Lake Baikal ecosystem. *Lakes and Reservoirs: Research and Management* **3**: 19–33.

Kraus, N. C. & Rankin, K. L., eds. (2004) Functioning and design of coastal groins: the interaction of groins and the beach-process and planning. *Journal of Coastal Research* Special Issue **33**.

Krieger, K. J. (2001) Coral (*Primnoa*) impacted by fishing gear in the Gulf of Alaska. In: *Proceedings of the 1st International Symposium on Deep-Sea Corals*, ed. J. Willison, J. Hall, S. Gass, *et al.*, pp. 106–116. Halifax, NS, Canada: Ecology Action Centre.

Kroes, D. & Brinson, M. M. (2004) Occurrence of riverine wetlands on floodplains along a climatic gradient. *Wetlands* **24**: 167–177.

Kroeze, C. & Seitzinger, S. P. (1998) Nitrogen inputs to rivers, estuaries and continental shelves and related nitrous oxide emissions in 1990 and 2050: a global model. *Nutrient Cycling in Agroecosystems* **52**: 195–212.

Kucklick, J. R. & Bidleman, T. F. (1994) Organic contaminants in Winyah Bay, South Carolina. I. Pesticides and polycyclic aromatic hydrocarbons in subsurface and microlayer waters. *Marine Environmental Research* **37**: 63–78.

Kucklick, J., Bidleman, T., McConnell, L., Walla, M. & Ivanov, G. (1994) Organochlorines in water and biota of Lake Baikal, Siberia. *Environmental Science and Technology* **28**: 31–37.

Kuffner, I. B. (2001) Effects of ultraviolet radiation and water motion on the reef coral *Porites compressa* Dana: a flume experiment. *Marine Biology* **138**: 467–476.

Kuhn, G. G. & Shepard, F. P. (1980) Coastal erosion in San Diego County, California. In: *Coastal Zone 80*, pp. 1899–1918. New York, USA: American Society of Civil Engineers.

Kuhnert, H., Patzold, J., Hatcher, B. G., *et al.* (1999) A 200-year coral stable oxygen isotope record from a high-latitude reef off Western Australia. *Coral Reef* **18**: 1–12.

Kumar, K. K., Rajagopalan, B. & Cane, M. A. (1999) On the weakening relationship between the Indian monsoon and ENSO. *Science* **284**: 2156–2159.

Kuriandewa, T. E., Kiswara, W., Hutomo, M., Soemodihardjo, S. & Hermana, I. (2003) The seagrasses of Indonesia. In: *World Atlas of Seagrasses*, ed. E. P. Green & F. T. Short, pp. 187–198. Berkeley, CA, USA: University of California Press.

Kurlansky, M. (1997) *Cod: A Biography of the Fish that Changed the World.* New York, USA: Walker & Co.

Kusler, J. & Larson, L. (1993) Beyond the ark: a new approach to US floodplain management. *Environment* **35**: 7–35.

Kwok, R. & Comiso, J. C. (2002) Southern ocean climate and sea ice anomalies associated with the Southern Oscillation. *Journal of Climate* **15**: 487–501.

Kyle, R., Pearson, B., Fielding, P. J., Robertson, W. D. & Birnie, S. L. (1997) Subsistence shellfish harvesting in the Maputaland Marine Reserve in northern KwaZulu-Natal, South Africa: rocky shore organisms. *Biological Conservation* 82: 183–192.

La Peyre, M. K., Mendelssohn, I. A., Reams, M. A., Templet, P. H. & Grace, J. B. (2001) Identifying determinants of nations' wetland management programs using structural equation modeling: an exploratory analysis. *Environmental Management* 27: 859–868.

Lacambra, C., Cutts, N., Allen, J., Burd, F. & Elliott, M. (2004) Spartina anglica: *A Review of its Status, Dynamics and Management.*, English Nature Research Report No.527. Peterborough, UK: English Nature.

Lack, M. & Sant, G. (2001) Patagonian toothfish: are conservation and trade measures working? *TRAFFIC Bulletin* 19: 1–18.

Lae, R. (1994) Effects of drought, dams, and fishing pressure on the fisheries of the central delta of the Niger River. *International Journal of Ecology and Environmental Sciences* 20: 119–128.

Lamb, M. & Zimmerman, M. H. (1964) Marine vegetation of Cape Ann, Essex County, Massachusetts. *Rhodora* 66: 217–254.

Lambert, G. & Steinke, T. D. (1986) Effects of destroying juxtaposed mussel-dominated and coralline algal communities at Umdoni Park, Natal South coast. *South African Journal of Marine Science* 4: 203–217.

Lambert, R. A., Hotchkiss, P. F., Roberts, N., *et al.* (1990) The use of wetlands (dambos) for micro-scale irrigation in Zimbabwe. *Irrigation and Drainage Systems* 4: 17–28.

Lambert, W. J., Levin, P. S. & Berman, J. (1992) Changes in the structure of a New England (USA) kelp bed: the effects of an introduced species? *Marine Ecology Progress Series* 88: 303–307.

Lambs, L. & Muller, E. (2002) Sap flow and water transfer in the Garonne River riparian woodland, France: first results on poplar and willow. *Annals of Forest Science* 59: 301–315.

Lamers, L. P. M., Smolders, A. J. P. & Roelofs, J. G. M. (2002) The restoration of fens in the Netherlands. *Hydrobiologia* 478: 107–130.

Lamers, L. P. M., Tomassen, H. B. M. & Roelofs, J. G. M. (1998) Sulfate-induced eutrophication and phytotoxicity in freshwater wetlands. *Environmental Science and Technology* 32: 199–205.

Lamparelli, C. C., Rodrigues, F. O. & de Moura, D. O. (1997) Long-term assessment of an oil spill in a mangrove forest in São Paulo, Brazil. In: *Mangrove Ecosystem Studies in Latin America and Africa*, ed. B. Kjerfve, L. D. Lacerda & E. H. S. Diop, pp. 191–203. Paris, France: UNESCO.

Lancaster, J., Real, M., Juggins, S., *et al.* (1996) Monitoring temporal changes in the biology of acid waters. *Freshwater Biology* 36: 179–201.

Landesman, L. (1994) Negative impacts of coastal tropical aquaculture developments. *World Aquaculture* 25(2): 12–17.

Lane, R. R., Day Jr, J. W. & Thibodeaux, B. (1999) Water quality analysis of a freshwater diversion at Caernarvon, Louisiana. *Estuaries* 22: 327–336.

Langdon, C. (2002) Overview of experimental evidence for effects of CO_2 on calcification of reef builders. In: *Proceedings of the 9th International Coral Reef Symposium*, 23–27 October 2000, vol. 2, ed. K. M. Moosa, S. Soemodihardjo, A. Soegiarto, *et al.*, pp. 1091–1098. Bali, Indonesia: Ministry of Environment, Indonesian Institute of Sciences, International Society for Reef Studies.

Langdon, C., Takahashi, T., Sweeney, C., Chapman, D. & Goddard, J. (2000) Effect of calcium carbonate saturation state on the calcification rate of an experimental coral reef. *Global Biogeochemical Cycle* 14: 639–654.

Langis, R., Zalejko, M. & Zedler, J. B. (1991) Nitrogen assessments in a constructed and a natural salt marsh of San Diego Bay, California. *Ecological Applications* 1: 40–51.

Lanier-Graham, S. D. (1993) *The Ecology of War: Environmental Impacts of Weaponry and Warfare*. New York, USA: Walker & Co.

Lant, C. L., Kraft, S. E. & Gillman, K. R. (1995) The 1990 farm bill and water quality in Corn Belt watersheds: conserving remaining wetlands and restoring farmed wetlands. *Journal of Soil and Water Conservation* 50: 201–205.

Lanyon, J. M., Limpus, C. J. & Marsh, H. (1989) Dugongs and turtles: grazers in seagrass systems. In: *Seagrasses: A Treatise on the Biology of Seagrasses with Special Reference to the Australian Region*, ed. A. W. D. Larkum, A. J. McComb & S. A. Shepherd, pp. 182–210. Amsterdam, the Netherlands: Elsevier/North Holland.

Lapointe, B. E. (1999) Simultaneous top-down and bottom-up forces control macroalgal blooms on coral reefs. *Limnology and Oceanography* 44: 1586–1592.

Larkum, A. W. D. & West, R. J. (1990) Long-term changes of seagrass meadows in Botany Bay, Australia. *Aquatic Botany* 37: 55–70.

Larned, S. T., Kinzie, R. A., Covich, A. P. & Chong, C. T. (2003) Detritus processing by endemic and non-native Hawaiian stream invertebrates: a microcosm study of species-specific effects. *Archiv für Hydrobiologie* 156: 241–254.

Larson, J. S. (1992) North America. In: *Wetlands*, ed. M. Finlayson & M. Moser, pp. 57–84. Oxford, UK: Facts on File.

Laurance, W. F., Cochrane, M. A., Bergen, S., *et al.* (2001) The future of the Brazilian Amazon. *Science* 291: 438–439.

Lawrence, G. B., David, M. B., Lovett, G. M., *et al.* (1999) Soil calcium status and the response of stream chemistry to changing acidic deposition rates. *Ecological Applications* 9: 1059–1072.

Lawrence, J. M. (1975) On the relationships between marine plants and sea urchins. *Oceanography and Marine Biology Annual Review* 13: 213–286.

Lawton, J. H. (2000) *Community Ecology in a Changing World.* Oldendorf/Kuhe, Germany: Ecology Institute.

Le Maitre, D. L., Scott, D. F. & Colvin, C. (1999) A review of information on interactions between vegetation and groundwater. *Water SA* 25: 137–152.

Leach, J., Mills, E. & Dochoda, M. (1999) Non-indigenous species in the Great Lakes: ecosystem impacts, binational policies, and management. In: *Great Lakes Fisheries Policy and Management*, ed. W. W. Taylor & C. P. Ferreri, pp. 185–207. East Lansing, MI, USA: Michigan State University Press.

Leach, S. J. (1988) Rediscovery of *Halimione pedunculata* (L.) Aellen in Britain. *Watsonia* 17: 170–171.

Leadbitter, D., Lee Long, W. & Dalmazzo, P. (1999) Seagrasses and their management implications for research. In: *Seagrass in Australia*, ed. A. Butler & P. Jernakoff, pp. 140–171. Canberra, ACT, Australia: CSIRO.

Leaman, B. M. & Beamish, R. J. (1984) Ecological and management implications of longevity in some Northeast Pacific groundfishes. *International North Pacific Fisheries Commission Bulletin* 42: 85–97.

Lean, G. (2003) 'Ecological meltdown': huge dust cloud threatens Asia. *The Independent* 26 January.

Lear, C. H., Elderfield, H. & Wilson, P. A. (2000) Cenozoic deep-sea temperatures and global ice volumes from Mg/Ca in benthic foraminiferal calcite. *Science* 287: 269–272.

Leatherland, T. M., Burton, J. D. C. F., McCartney, M. J. & Morris, R. J. (1973) Concentrations of some trace metals in pelagic organisms and of mercury in Northeast Atlantic Ocean water. *Deep-Sea Research* 20: 679–685.

Ledger, M. E. & Hildrew, A. G. (2001) Recolonization by the benthos of an acid stream following a drought. *Archiv für Hydrobiologie* 152: 1–17.

Ledger, M. E. & Hildrew, A. G. (2005) The ecology of acidification and recovery: changes in herbivore–algal food web linkages across a stream pH gradient. *Environmental Pollution* 137: 103–118.

Ledley, T. S. (1993) Variations in snow on sea ice: a mechanism for producing climate variations. *Journal of Geophysical Research* 98: 10401–10410.

Ledley, T. S., Sundquist, E. T., Schwartz, S. E., *et al.* (1999) Climate change and greenhouse gases. *Eos* 80: 453–458.

Lee, J. A. (1998) Unintentional experiments with terrestrial ecosystems: ecological effects of sulphur and nitrogen pollutants. *Journal of Ecology* 86: 1–12.

Lee, K.-S. & Lee, S. Y. (2003) The seagrasses of the Republic of Korea. In: *World Atlas of Seagrasses*, ed. E. P. Green & F. T. Short, pp. 211–216. Berkeley, CA, USA: University of California Press.

Lee, S. Y. & Kwok, P. W. (2002) The importance of mangrove species association to the population biology of the sesarmine crabs, *Perasesarma affinis* and *Parisesarma bidens*. *Wetland Ecology and Management* 10: 215–226.

Legendre, L., Ackley, S. F., Dieckmann, G. S., *et al.* (1992) Ecology of sea ice biota. II. Global significance. *Polar Biology* 12: 429–444.

Leggett, D., Bubb, J. C. & Lester, J. N. (1995) The role of pollutants and sedimentary processes in flood defence: a case study – salt marshes on the Essex coast, UK. *Environmental Toxicology* 16: 457–466.

Lehmusluoto, P., Machbub, B., Terangna, N., *et al.* (1999) Limnology in Indonesia. In: *Limnology in Developing Countries*, vol. 2, ed. R. G. Wetzel & B. Gopal, pp. 119–234. New Delhi, India: International Association of Theoretical and Applied Limnology.

Leighton, D. (1998) Control of sabellid infestation in green and pink abalone, *Haliotis fulgens* and *H. corrugata*, by exposure to elevated water temperatures. *Journal of Shellfish Research* 17: 701–705.

Leighton, D. L., Jones, L. G. & North, W. (1966) Ecological relationships between giant kelp and sea urchins in southern California. In: *Proceedings of the 5th International Seaweed Symposium*, ed. E. G. Young & J. L. McLachlan, pp. 141–153. Oxford, UK: Pergamon Press.

Leiva, G. E. & Castilla, J. C. (2001) A review of the world marine gastropod fishery: evolution of catches, management and the Chilean experience. *Reviews in Fish Biology and Fisheries* 11: 283–300.

Leland, A. (2002) A new apex predator in the Gulf of Maine: large, mobile Jonah crabs (*Cancer borealis*) control benthic community structure. M.Sc. thesis, University of Maine, Orono, ME, USA.

Lemmens, J. W. T. J, Clapin, G., Lavery, P. & Cary, J. (1996) Filtering capacity of seagrass meadows and other habitats of

Cockburn Sound, Western Australia. *Marine Ecology Progress Series* **143**: 187–200.

Lenihan, H. S., Micheli, F., Shelton, S. W. & Peterson, C. H. (1999) The influence of multiple environmental stressors on susceptibility to parasites: an experimental determination with oysters. *Limnology and Oceanography* **44**: 910–924.

Leonard, G. (2000) Latitudinal variation in species interactions: a test in the New England rocky intertidal zone. *Ecology* **81**: 1015–1030.

Leopold, A. (1924) Canada: journal entries dated June 11–25. Republished in Leopold, L. B., ed. (1953) *Round River: From the Journals of Aldo Leopold*. Oxford, UK: Oxford University Press.

Leopold, A. (1939) The farmer as conservationist. *American Forests* **45**: 294–299.

Leopold, A. (1941) Lakes in relation to terrestrial life patterns. In: *A Symposium on Hydrobiology*, ed. J. G. Needham, pp. 17–22. Madison, WI, USA: University of Wisconsin Press.

Leopold, L. B., Wolman, M. G. & Miller, J. P. (1964) *Fluvial Processes in Geomorphology*. San Francisco, CA, USA: W. H. Freeman.

Lessios, H. A. (1988) Mass mortality of *Diadema antillarum* in the Caribbean: what have we learned? *Annual Review of Ecology and Systematics* **19**: 371–393.

Lessios, H. A., Garrido, M. J. & Kessing, B. D. (2001) Demographic history of *Diadema antillarum*, a keystone herbivore on Caribbean reefs. *Proceedings of the Royal Society of London B* **268**: 1–7.

Letolle, R. & Mainguet, M. (1993) *L'Aral*. Paris, France: Springer-Verlag.

Levin, L. A. & Gooday, A. J. (2003) The deep Atlantic Ocean. In: *Ecosystems of the World*, vol. 28, *Ecosystems of the Deep Oceans*, ed. P. A. Tyler, pp. 111–178. Amsterdam, the Netherlands: Elsevier.

Levin, L. A., Blair, N. E., Martin, C. M., *et al.* (1999) Macrofaunal processing of phytodetritus at two sites on the Carolina margin: *in situ* experiments using C-13 labeled diatoms. *Marine Ecology Progress Series* **182**: 37–54.

Levin, P. S. (1994) Fine-scale temporal variation in recruitment of a temperate demersal fish: the importance of settlement versus post-settlement loss. *Oecologia* **97**: 124–133.

Levin, P. S. & Hay, M. E. (1996) Responses of temperate reef fishes to alterations in algal structure and species composition. *Marine Ecology Progress Series* **134**: 37–47.

Levin, P. S., Coyer, J. A., Petrik, R. & Good, T. P. (2002) Community-wide effects of non-indigenous species on temperate rocky reefs. *Ecology* **83**: 3182–3193.

Levin, R. B., Epstein, P. R., Ford, T. E., *et al.* (2002) US drinking water challenges in the twenty-first century. *Environmental Health Perspectives* **110** (Suppl. 1): 43–52.

Levitus, S., Antonov, J. I., Boyer, T. P. & Stephens, C. (2000) Warming of the world ocean. *Science* **287**: 2225–2229.

Levitus, S., Antonov, J. I., Wang, J., *et al.* (2001) Anthropogenic warming of Earth's climate systems. *Science* **292**: 267–270.

Lewis, J. R. (1964) *The Ecology of Rocky Shores*. London, UK: English Universities Press.

Lewis, J. R. (1976) Long-term ecological surveillance: practical realities in the rocky littoral. *Oceanography and Marine Biology Annual Review* **14**: 371–390.

Lewis, J. R. (1996) Coastal benthos and global warming: strategies and problems. *Marine Pollution Bulletin* **32**: 698–700.

Lewis, S. A. (1986) The role of herbivorous fishes in the organization of a Caribbean reef community. *Ecological Monographs* **56**: 183–200.

Li, M. S. & Lee, S. Y. (1997) Mangroves of China: a brief review. *Forest Ecology and Management* **96**: 241–259.

Liddle, M. J. & Greig-Smith, P. (1975) A survey of tracks and paths in a sand dune ecosystem. *Journal of Applied Ecology* **12**: 893–908.

Likens, G. E. & Bormann, F. H. (1974) Acid rain: a serious regional environmental problem. *Science* **184**: 1176–1179.

Lima, F. P., Queiroz, N., Ribeiro, P. A., Hawkins, S. J. & Santos, A. M. (2006) Recent changes in the distribution of a marine gastropod, *Patella rustica* Linnaeus, 1758, and their relationship to unusual climatic events. *Journal of Biogeography* **33**: 812–822.

Lindeman, K. C. & Snyder, D. B. (1999) Nearshore hardbottom fishes of southeast Florida and effects of habitat burial caused by dredging. *Fishery Bulletin* **97**: 508–525.

Lindig-Cisneros, R., Desmond, J., Boyer, K. & Zedler, J. B. (2003) Wetland restoration thresholds: can a degradation transition be reversed with increased effort? *Ecological Applications* **13**: 193–205.

Linley, E. A. S., Newell, R. C. & Bosma, S. A. (1981) Hetertrophic utilization of mucilage released during fragmentation of Kelp (*Ecklonia maxima* and *Laminaria pallida*) development of microbial communities associated with the degradation of kelp mucilage. *Marine Ecology Progress Series* **4**: 31–41.

Lipcius, R. N. & Stockhausen, W. T. (2002) Concurrent decline of spawning stock, recruitment, larval abundance and size of the blue crab *Callinectes sapidus* in Chesapeake Bay. *Marine Ecology Progress Series* **226**: 45–61.

Lipkin, Y. (1975) *Halophila stipulacea*, a review of a successful immigration. *Aquatic Botany* 1: 203–215.

Lipkin, Y., Beer S. & Zakai, D. (2003) The seagrasses of the Eastern Mediterranean and the Red Sea. In: *World Atlas of Seagrasses*, ed. E. P. Green & F. T. Short, pp. 75–83. Berkeley, CA, USA: University of California Press.

Little, C. & Kitching, J. A. (1996) *The Biology of Rocky Shores*. Oxford, UK: Oxford University Press.

Little, C. & Mettam, C. (1994) Rocky shore zonation in the Rance tidal power basin. *Biological Journal of the Linnean Society* 51: 169–182.

Littler, M. M. & Littler, D. S. (1980) The evolution of thallus form and survival strategies in benthic marine macroalgae: field and laboratory tests of a functional form model. *American Naturalist* 116: 25–44.

Littler, M. M., Littler, D. S. & Titlyanov, E. A. (1991) Comparisons of N- and P-limited productivity between high granitic islands versus low carbonate atolls in the Seychelles Archipelago: a test of the relative-dominance paradigm. *Coral Reefs* 10: 199–209.

Littler, M. M., Martz, D. R. & Littler, D. (1983) Effects of recurrent sand deposition on rocky intertidal organisms: the importance of substrate heterogeneity in a fluctuating environment. *Marine Ecology Progress Series* 11: 129–139.

Livingston, R. J. (1996) Eutrophication in estuaries and coastal systems: relationships of physical alterations, salinity stratification, and hypoxia. In: *Sustainable Development in the Southeastern Coastal Zone*, ed. F. J. Vernberg, W. B. Vernberg & T. Siewicki, pp. 285–318. Columbia, SC, USA: University of South Carolina Press.

Livingston, R. J. (1997) Trophic response of estuarine fishes to long-term changes of river runoff. *Bulletin of Marine Science* 60: 984–1004.

Livingston, R. J. (2000) *Eutrophication Processes in Coastal Systems: Origin and Succession of Plankton Blooms and Secondary Production in Gulf Coast Estuaries*. Boca Raton, FL, USA: CRC Press.

Livingston, R. J. (2001) *Management of the Apalachicola River Estuary: Relationships of River Flow and River-Bay Productivity*, Technical Report. Tallahassee, FL, USA: Florida State University.

Livingston, R. J. (2002) *Trophic Organization in Coastal Systems: Natural Food Web Processes and Effects of Natural Loading, Plankton Blooms, and Toxic Substances*. Boca Raton, FL, USA: CRC Press.

Livingston, R. J., Niu, X.-F., Lewis, F. G. & Woodsum, G. C. (1997) Freshwater input to a Gulf estuary: long-term control of trophic organization. *Ecological Applications* 7: 277–299.

Llewellyn, D. W., Shaffer, G. P., Craig, N. J., *et al.* (1996) A decision-support system for prioritizing restoration sites on the Mississippi River alluvial plain. *Conservation Biology* 10: 1446–1455.

Locarnini, S. J. P. & Presley, B. J. (1996) Mercury concentrations in benthic organisms from a contaminated estuary. *Marine Environmental Research* 41: 225–239.

Loeb, V., Siegel, V., Holm-Hansen, O., *et al.* (1997) Effects of sea-ice extent and krill or salp dominance on the Antarctic food web. *Nature* 387: 897–900.

Löffler, M. & Coosen, J. (1995) Ecological impact of sand replenishment. In: *Directions in European Coastal Management*, ed. M. G. Healy & P. J. Doody, pp. 291–299. Cardigan, UK: Samara Publishing.

Lofgren, B., Quinn, F., Clites, A., Assel, R. & Eberhardt, A. (2000) Impacts, challenges and opportunities: water resources, Great Lakes. In: *Preparing for a Changing Climate*, ed. P. J. Sousounis & J. M. Bisanz, pp. 29–37. Washington, DC, USA: US Environmental Protection Agency.

Lomborg, B. (2001) *The Skeptical Environmentalist: Measuring the Real State of the World*. Cambridge, UK: Cambridge University Press.

Long, E. R., Robertson, A., Wolfe, D. A., Hameedi, J. & Sloane, G. M. (1996) Estimates of the spatial extent of sediment toxicity in major US estuaries. *Environmental Science and Technology* 30: 3585–3592.

Longhurst, A. R. (1998) *Ecological Geography of the Sea*. San Diego, CA, USA: Academic Press.

Longhurst, A., Sathyendranath, S., Platt, T. & Caverhill, C. (1995) An estimate of global primary production in the ocean from satellite radiometer data. *Journal of Plankton Research* 17: 1245–1271.

Looney, P. B. & Gibson, D. J. (1993) Vegetation monitoring of beach nourishment. In: *Beach Nourishment: Engineering and Management Considerations*, ed. D. K. Stauble & N. C. Kraus, pp. 226–241. New York, USA: American Society of Civil Engineers.

Lorance, P. (1998) Structure du peuplement ichtyologique du talus continental à l'ouest des Iles Britanniques et impact de la pêche. *Cybium* 22: 209–231.

Lord, D. A., Paling, E. & Gordon, D. G. (1999) Review of seagrass rehabilitation and restoration programs in Australia. In: *Seagrass in Australia*, ed. A. Butler & P. Jernakoff, pp. 65–139. Canberra, ACT, Australia: CSIRO.

Loreau, M., Naeem, S., Inchausti, P., *et al.* (2001) Biodiversity and ecosystem functioning: current knowledge and future challenges. *Science* 294: 804–808.

Loreau, M., Naeem, S. & Inchausti, P., eds. (2002) *Biodiversity and Ecosystem Functioning: Synthesis and Perspectives*. Oxford, UK: Oxford University Press.

Lough, J. & Barnes, D. (1997) Several centuries of variation in skeletal extension, density and calcification in massive *Porites* colonies from the Great Barrier Reef: a proxy for seawater temperature and a background of natural variability against which to identify unnatural change. *Journal of Experimental Marine Biology and Ecology* 211: 29–67.

Løvas, S. M. & Tørum, A. (2001) Effect of the kelp *Laminaria hyperborea* upon sand dune erosion and water particle velocities. *Coastal Engineering* 44: 37–63.

Lowe-McConnell, R. H., ed. (1987) *Ecological Studies in Tropical Fish Communities*. New York, USA: Cambridge University Press.

Lowe-McConnell, R. (1994) The changing ecosystem of Lake Victoria, East Africa. *Freshwater Forum* 4: 76–89.

Loya, Y. & Rinkevich, B. (1980) Effects of oil pollution on coral reef communities. *Marine Ecology Progress Series* 3: 167–180.

Lu, J. (1995) Ecological significance and classification of Chinese wetlands. *Vegetatio* 118: 49–56.

Lubchenco, J., Olson, A. M., Brubaker, L. B., *et al.* (1991) The sustainable biosphere initiative: an ecological research agenda. *Ecology* 72: 371–412.

Lubin, D., Booth, C. R., Lucas, T., Neuschuler, D. & Frederick, J.E (1989) Measurements of enhanced springtime ultraviolet radiation at Palmer Station, Antarctica. *Geophysical Research Letters* 16: 783–785.

Ludgate, J. W. (1987) The economic and technical impact of TBT legislation on the USA marine industry. In: *Oceans '87 Proceedings*, vol. 4, *International Organotin Symposium*, pp. 1309–1313. New York, USA: Institute of Electrical and Electronic Engineers.

Ludwig, D. F., Iannuzzi, T. J. & Esposito, A. N. (2003) *Phragmites* and environmental management: a question of values. *Estuaries* 26: 624–630.

Lugo, A. E. & Snedaker, S. C. (1974) The ecology of mangroves. *Annual Reviews of Ecology and Systematics* 5: 39–64.

Lugo, A. E., Brinson, M. & Brown, S., eds. (1990) *Ecosystems of the World*, vol. 15, *Forested Wetlands*. Amsterdam, the Netherlands: Elsevier.

Lundquist, J. (1998) Avert looming hydrocide. *Ambio* 27: 428–433.

Lutz, R. A., Shank, T. M., Fornari, D. J., *et al.* (1994) Rapid growth at deep-sea vents. *Nature* 371: 663–664.

Lynam, C., Gibbons, M., Axelsen, B., *et al.* (2006) Jellyfish overtake fish in a heavily fished ecosystem. *Current Biology* 16: R492–R493.

MacIntyre, S., Flynn, K., Jellison, R. & Romero, J. (1999) Boundary mixing and nutrient fluxes in Mono Lake, California. *Limnology and Oceanography* 44: 512–529.

Mack, R. N., Simberloff, D., Lonsdale, W. M., *et al.* (2000) Biotic invasions: causes, epidemiology, global consequences, and control. *Ecological Applications* 10: 689–710.

Mackay, A. W. & Tallis, J. T. (1996) Summit-type blanket mire erosion in the Forest of Bowland, Lancashire, UK: predisposing factors and implications for conservation. *Biological Conservation* 76: 31–44.

MacKenzie, S. (1993) Ecosystem management in the Great Lakes: some observations from three RAP sites. *Journal of Great Lakes Research* 19: 136–144.

Mackereth, F. J.H. (1965) Chemical investigations of lake sediments and their interpretation. *Proceedings of the Royal Society of London B* 161: 293–375.

Madon, S. P., Williams, G. D., West, J. M. & Zedler, J. B. (2001) The importance of marsh access to growth of the California killifish, *Fundulus parvipinnis*, evaluated through bioenergetics modelling. *Ecological Modelling* 136: 149–165.

Madsen, T. V. & Sand-Jensen, K. (1991) Photosynthetic carbon assimilation in aquatic macrophytes. *Aquatic Botany* 41: 5–40.

Magnusson, J. J., Benson, B. J. & Kratz, T. K. (1990) Temporal coherence in the limnology of a suite of lakes in Wisconsin, USA. *Freshwater Biology* 23: 145–159.

Magnuson, J., Robertson, D., Benson, B., *et al.* (2000) Historical trends in lake and river ice cover in the Northern Hemisphere. *Science* 289: 1743–1746.

Magnuson, J. J., Webster, K. E., Assel, R. A., *et al.* (1997) Potential effects of climate changes on aquatic systems: Laurentian Great Lakes and Precambrian Shield Region. *Hydrological Processes* 11: 825–871.

Magoulick, D. D. (2000) Spatial and temporal variation in fish assemblages of drying stream pools: the role of abiotic and biotic factors. *Aquatic Ecology* 34: 29–41.

Maguire, T. L., Saenger, P., Baverstock, P. & Henry, R. (2000) Microsatellite analysis of genetic structure in the mangrove species *Avicennia marina* (Forsk.) Vierh. (Avicenniaceae). *Molecular Ecology* 9: 1853–1862.

Mäkilä, M., Saarnisto, M. & Kankainen, T. (2001) Aapa mires as a carbon sink and source during the Holocene. *Journal of Ecology* 89: 589–599.

Malakoff, D. (2002) Miscue raises doubt about survey data. *Science* 298: 515.

Malmer, N. (1962) Studies on mire vegetation in the archaean area of Southwestern Götaland (South Sweden). II. Distribution and seasonal variation in elementary constituents on some mire sites. *Opera Botanica* 7: 2–15.

Malmer, N. (1988) Patterns in the growth and the accumulation of inorganic constituents in the *Sphagnum*-cover on ombrotrophic bogs in Scandinavia. *Oikos* **53**: 105–120.

Malmer, N. & Wallén, B. (1993) Accumulation and release of organic matter in ombrotrophic bog hummocks: processes and regional variation. *Ecography* **16**: 193–211.

Malmer, N. & Wallén, B. (1996) Peat formation and mass balance in sub-arctic ombrotrophic peatlands around Abisko, northern Scandinavia. *Ecological Bulletins* **45**: 79–92.

Malmer, N. & Wallén, B. (1999) The dynamics of peat accumulation on bogs: mass balance of hummocks and hollows and its variation throughout a millennium. *Ecography* **22**: 736–750.

Malmer, N. & Wallén, B. (2004) Input rates, decay losses and accumulation rates of carbon in bogs during the last millennium: internal processes and environmental changes. *Holocene* **14**: 111–117.

Malmer, N. & Wallén, B. (2005) Nitrogen and phosphorus in mire plants: variation during 50 years in relation to supply rate and vegetation type. *Oikos* **109**: 539–554.

Malmer, N., Albinsson, C., Svensson, B. & Wallén, B. (2003) Interferences between *Sphagnum* and vascular plants: effects on plant community structure and peat formation. *Oikos* **100**: 469–482.

Malmer, N., Johansson, T., Olsrud M. & Christensen, T. R. (2005) Vegetation, climatic changes and net carbon sequestration in a North-Scandinavian sub-arctic mire over 30 years. *Global Change Biology* **11**: 1895–1909.

Malmer, N., Svensson, G. & Wallén, B. (1997) Mass balance and nitrogen accumulation in hummocks on a South Swedish bog during Late Holocene. *Ecography* **20**: 535–549.

Malmqvist, B. (2002) Aquatic invertebrates in riverine landscapes. *Freshwater Biology* **47**: 679–694.

Malmqvist, B. & Rundle, S. R. (2002) Threats to the running water ecosystems of the world. *Environmental Conservation* **29**: 134–153.

Malone, T. C., Màlej, A. & Smodlaka, N. (1996) Trends in land-use, water quality, and fisheries: a comparison of the northern Adriatic Sea and the Chesapeake Bay. *Periodicum Biologorum* **98**: 137–148.

Maltby, E. (1986) *Waterlogged Wealth: Why Waste the World's Wet Places?* London, UK: Earthscan Publications.

Maltby, E., Hogan, D. V., Immirzi, C. P., Tellam, J. H. & van der Peijl, M. J. (1994) Building a new approach to the investigation and assessment of wetland ecosystem functioning. In: *Global Wetlands: Old World and New*, ed. W. J. Mitsch, pp. 637–658. Amsterdam, the Netherlands: Elsevier.

Manabe, S., Stouffer, R. J., Spelman, M. J. & Bryan, K. (1991) Transient responses of a coupled ocean–atmosphere model to gradual changes in atmospheric CO2. I. Annual mean response. *Journal of Climate* **4**: 785–818.

Mann, K. H. (1973) Seaweeds: their productivity and strategy for growth. *Science* **182**: 975–981.

Mann, K. H. (1977) Destruction of kelp beds by sea urchins: a cyclical phenomenon or irreversible degradation? *Helgoländer Wissenschaftliche Meeresuntersuchungen* **30**: 455–467.

Mann, K. H. (2000) *Ecology of Coastal Waters, with Implications for Management*, vol. 2. Oxford, UK: Blackwell Science.

Mann, M. E., Bradley, R. S. & Hughes, M. K. (1998) Global-scale temperature patterns and climate forcing over the past six centuries. *Nature* **392**: 779–787.

Marbà, N. & Duarte, C. M. (1995) Coupling of seagrass (*Cymodocea nodosa*) patch dynamics to subaqueous dune migration. *Journal of Ecology* **83**: 381–389.

Marbà, N. & Duarte, C. M. (1997) Interannual changes in seagrass (*Posidonia oceanica*) growth and environmental change in the Spanish Mediterranean littoral. *Limnology and Oceanography* **42**: 800–810.

Marbà, N. & Duarte, C. M. (1998) Rhizome elongation and seagrass clonal growth. *Marine Ecology Progress Series* **174**: 269–280.

Marbà, N., Duarte, C. M., Cebrián, J., *et al.* (1996) Growth and population dynamics of *Posidonia oceanica* on the Spanish Mediterranean coast: elucidating seagrass decline. *Marine Ecology Progress Series* **137**: 203–213.

Marbà, N., Duarte, C. M., Holmer, M., *et al.* (2002) Assessing the effectiveness of protection on *Posidonia oceanica* populations in the Cabrera National Park (Spain). *Environmental Conservation* **29**: 509–518.

Marchant, C. J. (1968) Evolution in *Spartina* (Gramineae). II. Chromosomes, basic relationships and the problem of the *S. × townsendii* agg. *Botanical Journal of the Linnean Society* **60**: 381–409.

Marsh, P., Lesack, L. F. W. & Roberts, A. (1999) Lake sedimentation in the Mackenzie Delta, NWT. *Hydrological Processes* **13**: 2519–2536.

Martikainen, P. J., Nykanen, H., Crill, P. & Silvola, J. (1993) Effect of a lowered water table on nitrous oxide fluxes from northern peatlands. *Nature* **366**: 51–53.

Martin, S., Munoz, E. & Drucker, R. (1997) Recent observations of a spring–summer surface warming over the Arctic Ocean. *Geophysical Research Letters* **24**: 1259–1262.

Martinez Austria, P. & van Hofwegen, P., eds. (2006) *Synthesis of the 4th World Water Forum 2006*. Mexico DF,

Mexico: Comisión National de Agua [www document]. URL http://www.worldwatercouncil.org.

Maser, C. & Sedell, J. R. (1994) *From the Forest to the Sea: The Ecology of Wood in Streams, River, Estuaries, and Oceans.* Delray Beach, FL, USA: St Lucie Press.

Masing, V., Svirezhev, Y. M., Loffler, H. & Patten, B. C. (1990) Wetlands in the biosphere. In: *Wetlands and Shallow Continental Water Bodies*, vol. 1, ed. B. C. Patten, pp. 313–344. The Hague, the Netherlands: SPB Academic Publishing.

Maslanik, J. A., Serreze, M. C. & Barry, R. G. (1996) Recent decreases in summer Arctic ice cover and linkages to atmospheric circulation anomalies. *Geophysical Research Letters* 23: 1677–80.

Mason, C. F., Underwood, G. J.C., Baker, N. R., *et al.* (2003) The role of herbicides in the erosion of salt marshes in eastern England. *Environmental Pollution* 122: 41–49.

Mason, R. P., Fitzgerald, W. F. & Morel, F. M.M. (1994) The biogeochemical cycling of elementary mercury: anthropogenic influences. *Geochima et Cosmochimica Acta* 58: 3191–3198.

Mason, R. P., Reinfelder, J. R. & Morel, F. M.M. (1995) Bioaccumulation of mercury and methylmercury. *Water, Air, and Soil Pollution* 80: 915–921.

Mather, A. S. & Ritchie, W. (1977) *The Beaches of the Highlands and Islands of Scotland.* Perth, UK: Countryside Commission for Scotland.

Matsunaga, K., Kawaguchi, T., Suzuki Y. & Nigi, G. (1999) The role of terrestrial humic substances on the shift of kelp community to crustose coralline algae community of the southern Hokkaido Island in the Japan Sea. *Journal of Experimental Marine Biology and Ecology* 241: 193–205.

Matthiessen, P. & Gibbs, P. E. (1998) Critical appraisal of the evidence for tributyltin-mediated endocrine disruption. *Environmental Toxicology and Chemistry* 17: 37–43.

Mattice, A. (2003) Eliminating fishing subsidies as a way to promote conservation. *Economic Perspectives: An Electronic Journal of the US Department of State* 8 [www document]. URL http://usinfo.state.gov/journals/ites/0103/ijee/mattice.htm

Mauri, M., Polimeni, R., Modica, A. & Ferraro, M. (1998) Heavy metal bioaccumulation associated with drilling and production activities in middle Adriatic Sea. *Fresenius Environmental Bulletin* 7: 60–70.

Maykut, G. A. (1985) The ice environment. In: *Sea Ice Biota*, ed. R. A. Horner, pp. 21–82. Boca Raton, FL, USA: CRC Press.

Mazumder, A., Taylor, W. D., McQueen, D. J. & Lean, D. R.S. (1990) Effects of fish and plankton on lake temperature and mixing depth. *Science* 247: 312–315.

Mbewe, D. M.N. (1992) Agriculture as a component of wetlands conservation. In: *Managing the Wetlands of the Kafue Flats and Bangweulu Basin*, ed. R. C.V. Jeffery, H. N. Chabwela, G. Howard & P. J. Dugan, pp. 57–64. Gland, Switzerland: WWF and IUCN.

McAllister, D. E., Hamilton, A. L. & Harvey, B. (1997) Global freshwater biodiversity: striving for the integrity of freshwater ecosystems. *Sea Wind: Bulletin of Ocean Voice International* 11: 1–140.

McAllister, M., Craig, J. F., Davidson, N., Delany, S. & Seddon, M. (2000) Biodiversity impacts of large dams [www document]. URL http://www.dams.org/kbase/thematic/tr21.htm

McCabe, G. J. & Dettinger, M. D. (1999) Decadal variations in the strength of ENSO teleconnections with precipitation in the western United States. *International Journal of Climatology* 19: 1399–1410.

McCartney, M. P., Masiyandima, M. & Houghton-Carr, H. A. (2005) *Working Wetlands: Classifying Wetland Potential for Agriculture*, Research Report No. 90. Colombo, Sri Lanka: International Water Management Institute.

McClanahan, T. R. (1988) Seasonality in East Africa's coastal waters. *Marine Ecology Progress Series* 44: 191–199.

McClanahan, T. R. (1992) Resource utilization, competition and predation: a model and example from coral reef grazers. *Ecological Modelling* 61: 195–215.

McClanahan, T. R. (1995) A coral reef ecosystem–fisheries model: impacts of fishing intensity and catch selection on reef structure and processes. *Ecological Modelling* 80: 1–19.

McClanahan, T. R. (1999) Is there a future for coral reef parks in poor tropical countries? *Coral Reefs* 18: 321–325.

McClanahan, T. R. (2000a) Recovery of the coral reef keystone predator, *Balistapus undulatus*, in East African marine parks. *Biological Conservation* 94: 191–198.

McClanahan, T. R. (2000b) Bleaching damage and recovery potential of Maldivian coral reefs. *Marine Pollution Bulletin* 40: 587–597.

McClanahan, T. R. (2002a) The near future of coral reefs. *Environmental Conservation* 29: 460–483.

McClanahan, T. R. (2002b) The effects of time, habitat and fisheries management on Kenyan coral-reef associated gastropods. *Ecological Application* 12: 1484–1495.

McClanahan, T. R. & Kurtis, J. D. (1991) Population regulation of the rock-boring sea urchin *Echinometra mathaei* (de Blainville). *Journal of Experimental Marine Biology and Ecology* **147**: 121–146.

McClanahan, T. R. & Maina, J. (2003) Response of coral assemblages to the interaction between natural temperature variation and rare warm-water events. *Ecosystems* **6**: 551–563.

McClanahan, T. R. & Mangi, S. (2000) Spillover of exploitable fishes from a marine park and its effect on the adjacent fishery. *Ecological Application* **10**: 1792–1805.

McClanahan, T. R. & Mangi, S. (2001) The effect of closed area and beach seine exclusion on coral reef fish catches. *Fisheries Management and Ecology* **8**: 107–121.

McClanahan, T. R. & Obura, D. (1997) Sedimentation effects on shallow coral communities in Kenya. *Journal of Experimental Marine Biology and Ecology* **209**: 103–122.

McClanahan, T. R., Cokos, B. A. & Sala, E. (2002a) Algal growth and species composition under experimental control of herbivory, phosphorus and coral abundance in Glovers Reef, Belize. *Marine Pollution Bulletin* **44**: 441–451.

McClanahan, T. R., Graham, N. A.J., Calnan, J. M. & MacNeil, M. A. (2007) Towards pristine biomass: reef fish recovery in coral reef marine protected areas in Kenya. *Ecological Applications* **17**: 1055–1067.

McClanahan, T. R., Maina, J. & Pet-Soede, L. (2002b) Effects of the 1998 coral mortality event on Kenyan coral reefs and fisheries. *Ambio* **31**: 543–550.

McClanahan, T. R., Maina, J., Moothien Pillay, R. & Baker, A. C. (2005) Effects of geography, taxa, water flow, and temperature variation on coral bleaching intensity in Mauritius. *Marine Ecology Progress Series* **298**: 131–142.

McClanahan, T. R., Muthiga, N. A., Kamukuru, A. T., Machano, H. & Kiambo, R. (1999) The effects of marine parks and fishing on the coral reefs of northern Tanzania. *Biological Conservation* **89**: 161–182.

McClanahan, T. R., Muthiga, N. A. & Mangi, S. (2001) Coral and algal response to the 1998 coral bleaching and mortality: interaction with reef management and herbivores on Kenyan reefs. *Coral Reefs* **19**: 380–391.

McClanahan, T. R., Polunin, N. V.C. & Done, T. (2002c) Ecological states and the resilience of coral reefs. *Conservation Ecology* **6**: 18.

McClanahan, T. R., Sheppard, C. R. & Obura, D. (2000) *Coral Reefs of the Indian Ocean: Their Ecology and Conservation.* New York, USA: Oxford University Press.

McClanahan, T. R., Uku, J. N. & Machano, H. (2002d) Effect of macroalgal reduction on coral-reef fish in the Watamu Marine National park, Kenya. *Marine and Freshwater Research* **53**: 223–231.

McComb, A. J. & Lukatelich, R. J. (1995) The Peel–Harvey estuarine system, Western Australia. In: *Eutrophic Shallow Estuaries and Lagoons*, ed. A. J. McComb, pp. 5–17. Boca Raton, FL, USA: CRC Press.

McComb, A. J., ed. (1995) *Eutrophic Shallow Estuaries and Lagoons.* Boca Raton, FL, USA: CRC Press.

McConnaughey, R. A., Mier, K. L. & Dew, C. B. (2000) An examination of chronic trawling effects on soft-bottom benthos of the eastern Bering Sea. *ICES Journal of Marine Science* **57**: 1377–1388.

McCoy, M. B. & Rodriguez, J. M. (1994) Cattail (*Typha domingensis*) eradication methods in the restoration of a tropical, seasonal, freshwater marsh. In: *Global Wetlands: Old World and New*, ed. W. J. Mitsch, pp. 469–483. Amsterdam, the Netherlands: Elsevier.

McCulloch, M., Fallon, S., Wyndham, T., *et al.* (2003) Coral record of increased sediment flux to the inner Great Barrier Reef since European settlement. *Nature* **421**: 727–730.

McDowell, J. E. (1993) How marine animals respond to toxic chemicals in coastal ecosystems. *Oceanus* **36**: 56–63.

McGowan, J. A., Cayan, D. R. & Dorman, L. M. (1998) Climate–ocean variability and ecosystem response in the Northeast Pacific. *Science* **281**: 210–217.

McGrady-Steed, J., Harris, P. M. & Morin, P. J. (1997) Biodiversity regulates ecosystem predictability. *Nature* **390**: 162–165.

McGrath, D., Castro, F. de, Câmara, E. & Futemma, C. (1999) Community management of floodplain lakes and the sustainable development of Amazonian fisheries. In: *Várzea: Diversity, Development, and Conservation of Amazonia's Whitewater Floodplains, Advances in Economic Botany*, vol. 13, ed. C. Padoch, J. M. Ayres, M. Pinedo-Vasquez & A. Henderson, pp. 59–82. New York, USA: New York Botanical Garden Press.

McGuire, D., Melillo, J. M. & Joyce, L. A. (1995) The role of nitrogen in the response of forest net primary production to elevated atmospheric carbon dioxide. *Annual Review of Ecology and Systematics* **26**: 473–503.

McIntyre, A. D. (1992) The current state of the oceans. *Marine Pollution Bulletin* **25**: 1–4.

McIntyre, A. D. (1995) Human impact on the oceans: the 1990s and beyond. *Marine Pollution Bulletin* **31**: 147–151.

McKay, W. A., Johnson, C. E. & Branson, J. R. (1986) Environmental radioactivity in Caithness and Sutherland. III. Initial measurements and modeling of inshore waters. *Nuclear Energy* **27**: 321–335.

McKee, D., Atkinson, D., Collings, S. E., *et al.* (2003) Response of freshwater microcosm communities to nutrients, fish and elevated temperature during winter and summer. *Limnology and Oceanography* **48**: 707–722.

McKenzie, L. J., Lee Long, W. J., Coles, R. G. & Roder, C. A. (2000) Seagrass-Watch: community based monitoring of seagrass resources. *Biologia Marina Mediterranea* **7**: 393–396.

McLachlan, A. (1980) Exposed sandy beaches as semi-closed ecosystems. *Marine Environmental Research* **4**: 59–63.

McLachlan, A. (1996) Physical factors in benthic ecology: effects of changing sand particle size on beach fauna. *Marine Ecology Progress Series* **131**: 205–217.

McLachlan, A. & Harty, B. (1981) Effects of crude oil pollution on the supralittoral meiofauna of a sandy beach. *Marine Environmental Research* **7**: 71–80.

McLachlan, A. & Young, N. (1982) Effects of low temperatures on the burrowing rates of four sandy beach mollusks. *Journal of Experimental Marine Biology and Ecology* **65**: 275–284.

McLaren, A. S., Bourke, R. H., Walsh, J. E. & Weaver, R. L. (1994). Variability in sea-ice thickness over the North Pole from 1958 to 1992. In: *The Polar Oceans and Their Role in Shaping the Global Environment: The Nansen Centennial Volume*, ed. O. M. Johannessen, R. D. Muench & J. E. Overland, Geophysical Monograph No. 85, pp. 363–371. Washington, DC, USA: American Geophysical Union.

McMahon, K. & Walker, D. I. (1998) Fate of seasonal, terrestrial nutrient inputs to a shallow seagrass dominated embayment. *Estuarine Coastal and Shelf Science* **46**: 15–25.

McMahon, T. A. & Finlayson, B. L. (2003) Droughts and anti-droughts: the low flow hydrology of Australian rivers. *Freshwater Biology* **48**: 1147–1160.

McManus, J. W., Menez, L. A.B., Kesner-Reyes, K. N., Vergara, S. G. & Ablan, M. C. (2000) Coral reef fishing and coral–algal phase shifts: implications for global reef status. *Journal of Marine Science* **57**: 572–578.

McMillan, C. & Moseley, F. N. (1967) Salinity tolerances of five marine spermatophytes of Redfish Bay, Texas. *Ecology* **46**: 503–506.

McMurtry, G. (2001) Authigenic deposits. In: *Encyclopedia of Ocean Sciences*, ed. S. A. Thorpe & K. K. Turekian, pp. 201–220. London, UK: Academic Press.

McNeil, P. & Waddington, J. M. (2003) Moisture controls on *Sphagnum* growth and CO_2 exchange on a cutover bog. *Journal of Applied Ecology* **40**: 354–367.

McNeill, W. H. (1976) *Plagues and Peoples*. New York, USA: Anchor Books.

McQuaid, C. D. & Dower, K. M. (1990) Enhancement of habitat heterogeneity and species richness on rocky shores inundated by sand. *Oecologia* **84**: 142–143.

McRoy, C. P. & Helfferich, C. (1977) *Seagrass Ecosystems: A Scientific Perspective*. New York, USA: Marcel Dekker.

MDBMC (1995) *Floodplain Wetlands Management Strategy for the Murray–Darling Basin*. Canberra, ACT, Australia: Murray–Darling Basin Ministerial Council.

MDBMC (1999) *The Salinity Audit of the Murray–Darling Basin: A 100-Year Perspective*. Canberra, ACT, Australia: Murray–Darling Basin Commission.

Meade, R. H. (1969) Errors in using modern stream-loading data to estimate natural rates of denudation. *Bulletin of the Geological Society of America* **80**: 1265–1274.

Meakins, N. C., Bubb, J. C. & Lester, J. N. (1985) The mobility, partitioning and degradation of atrazine and simazine in the salt marsh environment. *Marine Pollution Bulletin* **30**: 812–819.

Medio, D., Sheppard, C. R.C. & Gasgoine, J. (2000) The Red Sea. In: *Coral Reefs of the Indian Ocean: Their Ecology and Conservation*, ed. T. R. McClanahan, C. R.C. Sheppard & D. O. Obura, pp. 231–255. New York, USA: Oxford University Press.

Mee, L. D. (2001) Eutrophication in the Black Sea and a basin-wide approach to its control. In: *Science and Integrated Coastal Management*, ed. B. von Bodungen & R. K. Turner, pp. 71–91. Berlin, Germany: Dahlem University Press.

Meehl, G. A. (1992) Effect of tropical topography on global climate. *Annual Review of Earth and Planet Planetary Science* **20**: 85–112.

Mees, C. C., Pilling, G. M. & Barry, C. J. (1999) Commercial inshore fishing activity in the British Indian Ocean Territory. In: *Ecology of the Chagos Archipelago*, ed. C. R.C. Sheppard & M. R.D. Seaward, pp. 327–345. London, UK: Linnean Society.

Meeuwig, J. J., Rasmussen, J. B. & Peters, R. H. (1998) Turbid waters and clarifying mussels: their moderation of empirical chl:nutrient relations in estuaries in Prince Edward Island, Canada. *Marine Ecology Progress Series* **171**: 139–150.

Meggers, B. J. (1987) The early history of man in Amazonia. In: *Biogeography and Quaternary History in Tropical America*, ed. T. C. Whitmore & G. T. Prance, pp. 151–174. Oxford, UK: Clarendon Press.

Mehner, T., Benndorf, J., Kasprzak, P. & Koschel, R. (2002) Biomanipulation of lake ecosystems: successful applications and expanding complexity in the underlying science. *Freshwater Biology* **47**: 2453–2465.

Meijer, M.-L., de Boois, I., Scheffer, M., Portielje, R. & Hosper, H. (1999) Biomanipulation in shallow lakes in the Netherlands: an evaluation of 18 case studies. *Hydrobiologia* **408/409**: 13–30.

Meinesz, A., de Vaugelas, J., Hesse, B. & Mari, X. (1993) Spread of the introduced tropical green alga *Caulerpa taxifolia* in northern Mediterranean waters. *Journal of Applied Phycology* **5**: 141–147.

Mekong River Commission (2002) International River Prize [www document]. URL http://www.mrcmekong.org/new_events/press_release/2002/

Melack, J. M. & Kilham, P. (1974) Photosynthetic rates of phytoplankton in East-African alkaline, saline lakes. *Limnology and Oceanography* **19**: 743–755.

Melillo, J. M., McGuire, A. D., Kicklighter, D. W., *et al.* (1993) Global climate change and terrestrial net primary production. *Nature* **363**: 234–240.

Mendelsohn, I. A. & Morris, J. T. (2000) Ecophysiological controls on the growth of *Spartina alterniflora*. In: *Concepts and Controversies in Tidal Marsh Ecology*, ed. M. P. Weinstein & D. P. Kreeger, pp. 59–80. Dordrecht, the Netherlands: Kluwer.

Mendelsohn, I. A., Hester, M. W., Monteferrante, F. J. & Talbot, F. (1991) Experimental dune building and vegetative stabilization in a sand-deficient barrier island setting on the Louisiana coast, USA. *Journal of Coastal Research* **7**: 137–149.

Menge, B. A. (1991) Relative importance of recruitment and other causes of variation in rocky intertidal community structure. *Journal of Experimental Marine Biology and Ecology* **145**: 69–100.

Menge, B. A. (1995) Indirect effects in marine rocky intertidal interaction webs: patterns and importance. *Ecological Monographs* **65**: 21–74.

Menge, B. A. (2000) Top-down and bottom-up community regulation in marine rocky intertidal habitats. *Journal of Experimental Marine Biology and Ecology* **250**: 257–289.

Menge, B. A., Daley, B. A., Wheeler, P. A., *et al.* (1997) Benthic–pelagic links and rocky intertidal communities: bottom-up effects on top-down control?. *Proceedings of the National Academy of Sciences of the USA* **94**: 14 530–14 535.

Menge, B. A., Lubchenco, J., Bracken, M. E. S., *et al.* (2003) Coastal oceanography sets the pace of rocky intertidal community dynamics. *Proceedings of the National Academy of Sciences of the USA* **100**: 12 229–12 234.

Merrett, N. R. & Haedrich, R. L. (1997) *Deep-Sea Demersal Fish and Fisheries*. London, UK: Chapman & Hall.

Merron, G. S., Bruton, M. N. & de Lalouviere, P. la H. (1993) Implications of water release from the Pongolapoort Dam for the fish and fishery of the Pongolo floodplain, Zululand. *South African Journal of Aquatic Sciences* **19**: 34–49.

Mertes, L. A. K. (2000) Inundation hydrology. In: *Inland Flood Hazards*, ed. E. E. Wohl, pp. 145–166. Cambridge, UK: Cambridge University Press.

Meyer, J. L., Sale, M. J., Mulholland, P. J. & Poff, N. L. (1999) Impacts of climate change on aquatic ecosystem functioning and health. *Journal of the American Water Resources Association* **35**: 1373–1386.

Meyerhoff, J. & Dehnhardt, A. (2002) Nachhaltige Entwicklung der Elbe. *Ökologisches Wirtschaften* **5**: 27–28.

Meyer-Reil, L. A. & Koster, M. (2000) Eutrophication of marine waters: effects on benthic microbial communities. *Marine Pollution Bulletin* **41**: 255–263.

Meyers, N., Mittermeier, R. A., Mittermeier, C. G., da Fonseca, G. A.B. & Kent, J. (2000) Biodiversity hotspots for conservation priorities. *Nature* **403**: 853–858.

Meyerson, L. A., Saltonstall, K., Windham, L., Kiviat, E. & Finlay, S. (2000) A comparison of *Phragmites australis* in freshwater and brackish marsh environments in North Australia. *Wetlands Ecology and Management* **8**: 89–103.

Michigan Department of Environmental Quality (2000) *Great Lakes trends: into the New Millennium.* Unpublished report. Lansing, MI, USA: Michigan Department of Environmental Quality.

Micklin, P. P. (1988) Desiccation of the Aral Sea: a water management disaster in the Soviet Union. *Science* **241**: 1170–1175.

Micklin, P. P. (1998) International and regional responses to the Aral crisis: an overview of efforts and accomplishments. *Post-Soviet Geography and Economics* **39**: 399–416.

Middleton, B. A. (1998) The water buffalo controversy in Keoladeo National Park, India. *Ecological Modelling* **106**: 93–98.

Middleton, B. A. (1999) *Wetland Restoration, Flood Pulsing and Disturbance Dynamics.* New York, USA: John Wiley.

Middleton, B. A. (2002*a*) Winter burning and the reduction of *Cornus sericea* in sedge meadows in southern Wisconsin. *Restoration Ecology* **10**: 1–8.

Middleton, B. A. (2002*b*) Nonequilibrium dynamics of sedge meadows grazed by cattle in southern Wisconsin. *Plant Ecology* **161**: 89–110.

Middleton, B. A. (2003) Ecology and objective based management: case study of the Keoladeo National Park, Bharatpur, Rajasthan. In: *Battles over Nature: Science and the Politics of Conservation*, ed. V. Saberwal & M. Rangarajan, pp. 86–116. New Delhi, India: Permanent Black Publishers.

Midun, Z. & Lee, S.-C. (1995) Implications of a greenhouse-induced sea-level rise: a national assessment for Malaysia. *Journal of Coastal Research* Special Issue 14: 96–115.

Mieszkowska, N., Leaper, R., Moore, P., *et al.* (2005) Marine biodiversity and climate change: assessing and predicting the influence of climatic change using intertidal rocky shore biota. *Occasional Publications, Marine Biological Association of the United Kingdom* 20: 1–53.

Miliman, J. D. & Meade, R. H. (1993) World-wide delivery of river sediment to the oceans. *Journal of Geology* 91: 1–21.

Millard, A. V. & Evans, P. R. (1984) Colonization of mudflats by *Spartina anglica*: some effects on invertebrate and shore bird populations at Lindisfarne. In: Spartina anglica *in Great Britain*, ed. P. Doody, pp. 41–48. Attingham Park, UK: Nature Conservancy Council.

Miller, M. W., Hay, M. E., Miller, S. L., *et al.* (1999) Effects of nutrients versus herbivores on reef algae: a new method for manipulating nutrients on coral reefs. *Limnology and Oceanography* 44: 1847–1861.

Miller, R. J., Smith, C. R., DeMaster, D. J. & Fornes, W. L. (2000) Feeding selectivity and rapid particle processing by deep-sea megafaunal deposit feeders: a Th-234 tracer approach. *Journal of Marine Research* 58: 653–673.

Miller, W. R. & Egler, F. E. (1950) Vegetation of the Wequetequock–Pawcatuck tidal marshes, Connecticut. *Ecological Monographs* 20: 143–172.

Milliman, J. D. & Meade, R. H. (1983) World-wide delivery of river sediment to the oceans. *Journal of Geology* 91: 1–21.

Mills, C. E. (2001) Jellyfish blooms: are populations increasing globally in response to changing ocean conditions? *Hydrobiologia* 451: 55–68.

Milner, A. M. (1994) System recovery. In: *The Rivers Handbook*, vol. 2, ed. P. Calow & G. E. Petts, pp. 76–97. Oxford, UK: Blackwell Scientific Publications.

Milner, J. (1874) Report on the fisheries of the Great Lakes: the results of inquiries prosecuted in 1871 and 1872. In: *Report of the Commissioner for 1872 and 1873*. Washington, DC, USA: US Commission of Fish and Fisheries.

Milton, K. (1998) Alternative futures to environmental disaster. *Global Ecology and Biogeography Letters* 7: 150–152.

Minchinton, T. & Bertness, M. D. (2003) Soil salinity, nitrogen availability and the expansion of *Phragmites australis* in New England salt marshes. *Ecological Applications* 13: 1400–1416.

Ministerio del Medio Ambiente (2001) *Politica nacional para humedales interiores de Colombia: estrategias para su conservación y uso racional*. Bogotá, Colombia: Ministerio del Medio Ambiente, Consejo Nacional Ambiental.

Minshall, G. W. (1992) Troubled waters of greenhouse earth: summary and synthesis. In: *Global Climate Change and Freshwater Ecosystems*, ed. P. Firth & S. G. Fisher, pp. 308–318. Berlin, Germany: Springer-Verlag.

Mirza, A. M. Q. (2002) Global warming and changes in the probability of occurrence of floods in Bangladesh and implications. *Global Environmental Change* 12: 127–138.

Mirza, M. M. (2003) Three recent extreme floods in Bangladesh: a hydro-meteorologic analysis. *Natural Hazards* 28: 35–64.

MIT (2003) Joint Program on the Science and Policy of Global Change, Massachusetts Institute of Technology [www document]. URL http://web.mit.edu/globalchange/www/

Mitchell, B. D. & Geddes, M. C. (1977) Distribution of the brine shrimps *Parartemia zietziana* (Sayce) and *Artemia salina* (L.) along a salinity and oxygen gradient in a South Australian salt-field. *Freshwater Biology* 7: 461–467.

Mitchell, J. K. (2003) European river floods in a changing world. *Risk Analysis* 23: 567–574.

Mitchell, R. B., Clark, W. C. & Cash, D. W. (2006) Information and influence. In: *Global Environmental Assessments: Information and Influence*, ed. R. B. Mitchell, W. C. Clark, D. W. Cash & N. M. Dickson, pp. 307–338. Cambridge, MA, USA: MIT Press.

Mitsch, W. J. & Gosselink, J. G. (2000) *Wetlands*, 3rd edn. New York, USA: John Wiley.

Mitsch, W. J. (1993) Ecological engineering: a cooperative role with the planetary life-support system. *Environmental Science and Technology* 27: 438–445.

Mitsch, W. J., Mitsch, R. H. & Turner, R. E. (1994) Wetlands of the Old and New Worlds: ecology and management. In: *Global Wetlands: Old World and New*, ed. W. J. Mitsch, pp. 3–56. Amsterdam, the Netherlands: Elsevier.

Moberg, F., Nystrom, M., Kautsky, N., Tedengren, M. & Jarayabhand, P. (1997) Effects of reduced salinity on the rates of photosynthesis and respiration in the hermatypic corals *Porites lutea* and *Pocillopora damicornis*. *Marine Ecology Progress Series* 157: 53–59.

Mono Lake Committee (2006) Home page [www document]. URL http://www.monolake.org

Monroe, M. W. & Kelly, J. (1992) *State of the Estuary*, Technical Report. Oakland, CA, USA: San Francisco Estuary Project.

Monson, D. H., Doak, D. F., Ballachey, B. E., Johnson, A. M. & Bodkin, J. L. (2000) Long-term impacts of the Exxon Valdez oil spill on sea otters, assessing through age-dependent mortality patterns. *Proceedings of the National Academy of Sciences of the USA* 97: 6562–6567.

Monteiro, L. R., Costa, V., Furness, R. W. & Santos, R. S. (1996) Mercury concentrations in prey fish indicate enhanced bioaccumulation in mesopelagic environments. *Marine Ecology Progress Series* **141**: 21–25.

Monteith, D. T., Hildrew, A. G., Flower, R. J., *et al.* (2005) Biological responses to the chemical recovery of acidified fresh waters in the UK. *Environmental Pollution* **137**: 83–101.

Mooney, H. A. & Hobbs, R. J., eds. (2000) *Invasive Species in a Changing World*. Washington, DC, USA: Island Press.

Moore, P. D. (1982) How to reproduce in bogs and fens. *New Scientist* 5 August: 369–371.

Moore, P. D. (1990) Soils and ecology: temperate wetlands. In: *Wetlands: A Threatened Landscape*, ed. M. Williams, pp. 95–114. Oxford, UK: Blackwell Scientific Publications.

Moore, P. D. (1993) The origin of blanket mire, revisited. In: *Climate Change and Human Impact on the Landscape*, ed. F. M. Chambers, pp. 217–224. London, UK: Chapman & Hall.

Moore, P. D. (2002) The future of cool temperate bogs. *Environmental Conservation* **29**: 3–20.

Moore, P. D. & Bellamy, D. J. (1973) *Peatlands*. London, UK: Paul Elek.

Moore, P. D., Chaloner, W. & Stott, P. (1996) *Global Environmental Change*. Oxford, UK: Blackwell Scientific Publications.

Moore, P. D., Merryfield, D. L. & Price, M. D. R. (1984) The vegetation and development of blanket mires. In: *European Mires*, ed. P. D. Moore, pp. 203–235. London, UK: Academic Press.

Moran, P. J. (1986) The *Acanthaster* phenomenon. *Oceanography and Marine Biology Annual Review* **24**: 379–480.

Moreno, C. A., Sutherland, J. P. & Jara, H. F. (1984) Man as a predator in the intertidal zone of southern Chile. *Oikos* **42**: 155–160.

Moreno-Casasola, P. (1986) Sand movement as a factor in the distribution of plant communities in a coastal dune system. *Vegetatio* **65**: 67–76.

Morgan, C. L. (2000) Resource estimates of the Clarion-Clipperton manganese nodule deposits. In: *Handbook of Marine Mineral Deposits*, ed. D. S. Cronan, pp. 145–170. Boca Raton, FL, USA: CRC Press.

Morgan, P. A. & Short, F. T. (2002) Using functional trajectories to track constructed salt marsh development in the Great Bay Estuary, Maine/New Hampshire, USA. *Restoration Ecology* **10**: 461–473.

Morita, K. & Yamamoto, S. (2002) Effects of habitat fragmentation by damming on the persistence of stream-dwelling charr populations. *Conservation Biology* **16**: 1318–1323.

Moritz, R. E., Bitz, C. M. & Steig, E. J. (2002) Dynamics of recent climate change in the Arctic. *Science* **297**: 1497–1502.

Mork, M. (1996) The effects of kelps on wave dumping. *Sarsia* **80**: 323–327.

Mormede, S. & Davies, I. M. (2001) Heavy metal concentrations in commercial deep-sea fish from the Rockall Trough. *Continental Shelf Research* **21**: 899–916.

Morris, B. L., Lawrence, A. R. L., Chilton, P. J. C., *et al.* (2003) *Groundwater and its Susceptibility to Degradation: A Global Assessment of the Problem and Options for Management*, Early Warning and Assessment Report Series Rs 03-3. Nairobi, Kenya: United Nations Environment Programme.

Morris, J. T. (1995) The salt and water balance of intertidal sediments: results from North Inlet, South Carolina. *Estuaries* **18**: 556–567.

Morris, J. T., Sundareshwar, P. V., Nietch, C. T., Kjerfve, B. & Cahoon, D. R. (2002) Responses of coastal wetlands to rising sea level. *Ecology* **83**: 2869–2877.

Morrison, J., Quick, M. C. & Foreman, M. G. G. (2002) Climate change in the Fraser River watershed: flow and temperature projections. *Journal of Hydrology* **263**: 230–244.

Mortensen, P. B., Buh-Mortensen, L., Gordon Jr, D. C., *et al.* (2005) Effects of fisheries on deep-water gorgonian corals in the Northeast Channel, Nova Scotia (Canada). In: *Benthic Habitats and the Effects of Fishing*, ed. P. Barnes & J. Thomas, pp. 369–382. Bethesda, MD, USA: American Fisheries Society.

Mortimer, C. (2004) *Lake Michigan in Motion: Responses Of An Inland Sea To Weather, Earth Spin, and Human Activities*. Madison, WI, USA: University of Wisconsin Press.

Morton, B. (1996) Protecting Hong Kong's marine biodiversity: present proposals, future challenges. *Environmental Conservation* **23**: 55–65.

Morton, S. R. & Brennan, K. G. (1991) Birds. In: *Monsoonal Australia: Landscape, Ecology and Man in Northern Australia*, ed. C. D. Haynes, M. G. Ridpath & M. A. J. Williams, pp. 133–149. Rotterdam, the Netherlands: A. A. Balkema.

Moser, F. C. & Bopp, R. F. (2001) Particle-associated contaminants in the Barnegat Bay–Little Egg Harbor Estuary. *Journal of Coastal Research* Special Issue **32**: 229–242.

Moser, M., Prentice, C. & Frazier, S. (1996) A global over-view of wetland loss and degradation. In: *Proceedings of the 6th Meeting of the Conference of the Contracting Parties*, Brisbane, Australia, 19–27 March 1996, vol. 10B/12: Technical Sessions B and D. Gland, Switzerland: Ramsar Convention Bureau. Available online of www.ramsar.org/about/about_wetland_loss.htm

Moser, M., Prentice, C. & Frazier, S. (2003) A global overview of wetland loss and degradation. Ramsar, Gland, Switzerland [www document]. URL http://www.ramsar.org/about/about_wetland_loss.htm

Moss, B. (1995) The emperor's clothes of knowledge and the seamless cloth of wisdom. In: *Science for the Earth*, 9th edn, ed. T. Wakeford & M. Walters, pp. 321–348. Chichester, UK: John Wiley.

Moss, B. (1998) *Ecology of Fresh Waters: Man and Medium, Past to Future*, 3rd edn. Oxford, UK: Blackwell Science.

Moss, B., McKee, D., Atkinson, D., *et al.* (2003) How important is climate? Effects of warming, nutrient addition and fish on phytoplankton in shallow lake microcosms. *Journal of Applied Ecology* **40**: 782–792.

Moss, R. H. & Schneider, S. H. (1997) Characterizing and communicating scientific uncertainty: building on the IPCC 2nd Assessment. In: *Elements of Change*, ed. S. J. Hassol & J. Katzenberger, pp. 90–135. Aspen, CO, USA: Aspen Global Change Institute.

Moy, L. D. & Levin, L. A. (1991) Are *Spartina* marshes a replaceable resource? A functional approach to evaluation of marsh creation efforts. *Estuaries* **14**: 1–16.

Moyle, P. B. (1999) Effects of invading species on freshwater and estuarine ecosystems. In: *Invasive Species and Biodiversity Management*, ed. O. T. Sandlund, P. J. Schei & Å. Viken, pp. 177–191. Dordrecht, the Netherlands: Kluwer.

Moyle, P. B. & Leidy, R. A. (1992) Loss of biodiversity in aquatic ecosystems: evidence from fish faunas. In: *Conservation Biology: The Theory and Practice of Nature Conservation, Preservation and Management*, ed. P. L. Fiedler & S. K. Jain, pp. 127–169. New York, USA: Chapman & Hall.

MSC (2004) Marine Stewardship Council: homepage [www document]. URL http://www.msc.org

Muehlstein, L. K., Porter, D. & Short, F. T. (1988) *Labyrinthula* sp., a marine slime mold producing the symptoms of wasting disease in eelgrass, *Zostera marina*. *Marine Biology* **99**: 465–472.

Muehlstein, L. K., Porter, D. & Short F.T, (1991) *Labyrinthula zosterae* sp. nov., the causative agent of wasting disease of eelgrass, *Zostera marina*. *Mycologia* **83**: 180–191.

Mugidde, R. (1993) The increase in phytoplankton primary productivity and biomass in Lake Victoria (Uganda). *Verhandlungen Internationale Vereinigung für Limnologie* **25**: 846–849.

Mulder, T. & Syvitski, J. P.M. (1995) Turbidity currents generated at river mouths during exceptional discharges to the world's oceans. *Journal of Geology* **103**: 285–299.

Muller, R. A. & MacDonald, G. J. (1997) Glacial cycles and astronomical forcing. *Science* **277**: 215–218.

Mullineaux, L. S. (1987) Organisms living on manganese nodules and crusts: distribution and abundance at three North Pacific sites. *Deep-Sea Research* **34**: 165–184.

Mumby, P. J., Chisholm, J. R. M., Edwards, A. J., *et al.* (2001) Unprecedented bleaching-induced mortality in *Porites* spp. at Rangiroa Atoll, French Polynesia. *Marine Biology* **139**: 183–189.

Munro, D. C. & Touron, H. (1997) The estimation of marshland degradation in southern Iraq using multitemporal Landsat TM images. *International Journal of Remote Sensing* **18**: 1597–1606.

Munro, J. L. (1989) Fisheries for giant clams (Tridacnidae: Bivalvia) and prospects for stock enhancement. In: *Marine Invertebrate Fisheries: Their Assessment and Management*, ed. J. F. Caddy, pp. 541–558. New York, USA: John Wiley.

Munro, J. L. (1996) The scope of tropical reef fisheries and their management. In: *Reef Fisheries*, vol. 20, ed. N. V. C. Polunin & C. M. Roberts, pp. 1–14. London, UK: Chapman & Hall.

Murawski, S. (2000) Definitions of overfishing from an ecosystem perspective. *ICES Journal of Marine Science* **57**: 649–658.

Murdoch, P. S., Baron, J. S. & Miller, T. L. (2000) Potential effects of climate change on surface-water quality in North America. *Journal of the American Water Resources Association* **36**: 347–366.

Murphy, E. J., Clarke, A., Symon, C. & Priddle, J. (1995) Temporal variation in Antarctic sea-ice: analysis of a long-term fast-ice record from the South Orkney Islands. *Deep-Sea Research I* **42**: 1045–1062.

Murphy, E. J., Watkins, J. L., Reid, K., *et al.* (1998) Interannual variability of the South Georgia marine ecosystem: biological and physical sources of variation in the abundance of krill. *Fisheries Oceanography* **7**: 381–390.

Murphy, G. I. (1968) Patterns in life history and the environment. *American Naturalist* **102**: 391–403.

Muthiga, N. A., Riedmiller, S., Carter, E., *et al.* (2000) Management status and case studies. In: *Coral Reefs of the Indian Ocean: Their Ecology and Conservation*, ed. T. R. McClanahan, C. S. Sheppard & D. Obura, pp. 473–505. New York, USA: Oxford University Press.

Muthuri, F. M., Jones, M. B. & Imbamba, S. K. (1989) Primary productivity of papyrus (*Cyperus papyrus*) in a tropical swamp: Lake Naivasha, Kenya. *Biomass* **18**: 1–14.

Myers, R. A. & Worm, B. (2003) Rapid worldwide depletion of predatory fish communities. *Nature* **423**: 280–283.

Myers, R. A., Hutchings, J. A. & Barrowman, N. J. (1997) Why do fish stocks collapse? The example of cod in Atlantic Canada. *Ecological Applications* **7**: 91–106.

Nachtnebel, H.-P. (2000) The Danube river basin environmental programme: plans and actions for a basin wide approach. *Water Policy* **2**: 113–129.

Naeem, S. (1998) Species redundancy in ecosystem reliability. *Conservation Biology* **12**: 39–45.

Naeem, S., Knops, J. M. H., Tilman, D., et al. (2000) Plant diversity increases resistance to invasion in experimental grassland plots. *Oikos* **91**: 97–108.

Naganobu, M., Kutsuwada, K., Sasai, Y., Taguchi, S. & Siegel, V. (1999) Relationships between Antarctic krill (*Euphausia superba*) variability and westerly fluctuations and ozone depletion in the Antarctic Peninsula area. *Journal of Geophysical Research: Oceans* **104**: 20651–20665.

Naiman, R. J., ed. (1992) *Watershed Management: Balancing Sustainability and Environmental Change*. New York, USA: Springer-Verlag.

Naiman, R. J. & Décamps, H. (1997) The ecology of interfaces: riparian zones. *Annual Review of Ecology and Systematics* **28**: 621–658.

Naiman, R. J., Bunn, S. E., Nilsson, C., et al. (2002) Legitimizing fluvial ecosystems as users of water: an overview. *Environmental Management* **30**: 455–467.

Naiman, R. J., Décamps, H. & McClain, M. E. (2005) *Riparia*. San Diego, CA, USA: Academic Press.

Naiman, R. J., Décamps, H. & Pollock, M. (1993) The role of riparian corridors in maintaining regional biodiversity. *Ecological Applications* **3**: 209–212.

Nakamura, T. & van Woesik, R. (2001) Differential survival of corals during the 1998 bleaching event is partially explained by water-flow rates and passive diffusion. *Marine Ecology Progress Series* **212**: 301–304.

Nalepa, T., Fahnenstiel, G. & Johengen, T. (1999) Impact of the zebra mussel *(Dreissena polymorpha)* on water quality: a case study in Saginaw Bay, Lake Huron. In: *Nonindigenous Freshwater Organisms, Vectors, Biology, and Impacts*, ed. R. Claudi & J. H. Leach, pp. 255–271. Boca Raton, FL, USA: Lewis Publishers.

Nalepa, T., Hartson, D., Buchanan, J., et al. (2000a) Spatial variation in density, mean size and physiological condition of the holarctic amphipod *Diporeia spp* in Lake Michigan. *Freshwater Biology* **43**: 107–119.

Nalepa, T., Lang, G. & Fanslow, D. (2000b) Trends in the macroinvertebrate populations in southern Lake Michigan. *Verhandlungen Internationale Vereinigung für Limnologie* **27**: 2540–2545.

Nasholm, T., Ekblad, A., Nordin, A., et al. (1998) Boreal forest plants take up organic nitrogen. *Nature* **392**: 914–916.

National Academy Press (1993) *Managing Wastewater in Coastal Urban Areas*. Washington, DC, USA: National Academy Press.

National Estuarine Research Reserve System (2001) *The National Estuarine Research Reserve's System-Wide Monitoring Program (SWMP): A Scientific Framework and Plan for Detection of Short-Term Variability and Long-Term Change in Estuaries and Coastal Habitats of the United States*, Technical Report, National Estuarine Research Reserve System. Silver Spring, MD, USA: National Oceanic and Atmospheric Administration.

National Estuary Program (1997a) *Nutrient Overloading*, Technical Report. Washington, DC, USA: US Environmental Protection Agency.

National Estuary Program (1997b) *Management Approaches Being Used to Address Critical Issues*, Technical Report. Washington, DC, USA: US Environmental Protection Agency.

National Estuary Program (1997c) *Declines in Fish and Wildlife Populations*, Technical Report. Washington, DC, USA: US Environmental Protection Agency.

Nauman, J. & Cory, R. L. (1969) Thermal additions and epifaunal organisms at Chalk Point, Maryland. *Chesapeake Science* **18**: 218–226.

Navodaru, I., Staras, M. & Cerniseucu, I. (2001) The challenge of sustainable use of the Danube Delta Fisheries, Romania. *Fisheries Management and Ecology* **8**: 323–332.

Naylor, E. (1965) Effects of heated effluents upon marine and estuarine organisms. *Advances in Marine Biology* **3**: 63–103.

Naylor, R. L., Goldberg, J., Primavera, J. H., et al. (2000) Effect of aquaculture on world fish supplies. *Nature* **405**: 1017–1024.

Naylor, R., Hindar, K., Fleming, I. A., et al. (2005) Fugitive salmon: assessing the risks of escaped fish from net-pen aquaculture. *BioScience* **55**: 427–437.

Neckles, H. A. & Neill, C. (1994) Hydrologic control of litter decomposition in seasonally flooded prairie marshes. *Hydrobiologia* **286**: 155–165.

Nedwell, D. B. (1975) Inorganic nitrogen metabolism in an eutrophicated tropical estuary. *Water Research* **9**: 221–231.

Neiff, J. J. (2001) Diversity in some tropical wetland systems of South America. In: *Biodiversity in Wetlands: Assessment, Function and Conservation*, vol. 2, ed. B. Gopal, W. J. Junk & J. A. Davis, pp. 157–186. Leiden, the Netherlands: Backhuys.

Nekola, J. C. (1994) The environment and vascular flora of northeastern Iowa fen communities. *Rhodora* 96: 121–169.

Nelson, W. G. (1993) Beach restoration in the southeastern US: environmental effects and biological monitoring. *Ocean and Coastal Management* 19: 157–182.

Neves, C. F. & Muehe, D. (1995) Potential impacts of sea-level rise on the metropolitan region of Recife, Brazil. *Journal of Coastal Research* Special Issue 14: 116–131.

New, M. (2002) Trends in freshwater and marine production systems. In: *Production Systems in Fishery Management*, Fisheries Centre Research Report No. 10(8), ed. D. Pauly & M. L. Palomares, pp. 21–27. Vancouver, BC, Canada: University of British Columbia.

Newell, R. C., Lucas, M. I., Velimirov, B. & Seiderer, L. J. (1980) The quantitative significance of dissolved organic losses following fragmentation of kelp (*Ecklonia maxima* and *Laminaria pallida*). *Marine Ecology Progress Series* 2: 45–59.

Newell, R. C., Moloney, C. L., Field, J. G., Lucas, M. I. & Probyn, T. A. (1988) Nitrogen models at the community level: plant–animal–microbe interactions. In: *Nitrogen Cycling in Coastal Marine Environments*, ed. T. H. Blackburn & J. Sorensen, pp. 379–414. New York, USA: John Wiley.

Newey, S. & Seed, R. (1995) The effects of the Braer oil spill on rocky intertidal communities in South Shetland, Scotland. *Marine Pollution Bulletin* 30: 274–280.

Newton, L. C., Parkes, E. V. H. & Thompson, R. C. (1993) The effects of shell collecting on the abundance of gastropods on Tanzanian shores. *Biological Conservation* 63: 241–245.

Nicholls, R. J. & Branson, J. (1998) Coastal resilience and planning for an uncertain future: an introduction. *Geographical Journal* 164: 255–258.

Nicholls, R. J. & Hoozemans, F. M. J. (1996) The Mediterranean: vulnerability to coastal implications of climate change. *Ocean and Coastal Management* 31: 105–132.

Nicholls, R. J. & Leatherman, S. P. (1996) Adapting to sea-level rise: relative sea-level trends to 2100 for the United States. *Coastal Management* 24: 301–324.

Nicholls, R. J. & Small, C. (2002) Improved estimates of coastal populations and exposure to hazards released. *Eos* 83: 301–307.

Nicholls, R. J., Hoozemans, F. M. J. & Marchand, M. (1999) Increasing flood risk and wetland losses due to global sea-level rise: regional and global analyses. *Global Environmental Change* 9: 69–87.

Nichols, F. H., Cloern, J. E., Luoma, S. N. & Peterson, D. H. (1986) The modification of an estuary. *Science* 231: 567–572.

Nickson, R., McArther, J., Burgees, W., *et al.* (1998) Arsenic poisoning of Bangladesh groundwater. *Nature* 395: 338.

Nilsson, C. & Berggren, K. (2000) Alterations of riparian ecosystems caused by river regulation. *BioScience* 50: 783–792.

Nilsson, C., Pizzuto, J. E., Moglen, G. E., *et al.* (2003) Ecological forecasting and the urbanization of stream ecosystems: challenges for economists, hydrologists, geomorphologists, and ecologists. *Ecosystems* 6: 659–674.

Nilsson, P. Å. (1999) Finjasjön, summer 1999: a summary of monitoring data. Hässleholm, Sweden (in Swedish) [www document]. URL http://www.hassleholmsvatten.se/

Ninio, R., Meekan, M., Done, T. & Sweatman, H. (2000) Temporal patterns in coral assemblages on the Great Barrier Reef from local to large spatial scales. *Marine Ecology Progress Series* 194: 65–74.

Nixon, S. W. (1980) Between coastal marshes and coastal waters: a review of twenty years of speculation and research on the role of salt marshes in estuarine productivity and water chemistry. In: *Estuarine and Wetlands Processes with Emphasis on Modeling*, ed. P. Hamilton & K. B. Macdonald, pp. 437–525. New York, USA: Plenum Press.

Nixon, S. W. (1995) Coastal marine eutrophication: a definition, social causes, and future concerns. *Ophelia* 41: 199–219.

Nixon, S. W. & Pilson, M. E. (1983) Nitrogen in estuarine and coastal marine ecosystems. In: *Nitrogen in the Marine Environment*, ed. E. J. Carpenter & D. G. Capone, pp. 565–648. New York, USA: Academic Press.

NOAA (1998) Draft Technical Guidance to NMFS for Implementing the Essential Fish Habitat Requirements for the Magnuson-Stevens Act [www document]. URL http://www.nmfs.noaa.gov/habitat/habitatprotection/essentialfishhabitat.htm

NOAA (2006) Florida Keys National Marine Sanctuary [www document]. URL http://www.fknms.nos.noaa.gov

Noaksson, E., Linderoth, M., Bosveld, A. T. C., *et al.* (2003) Endocrine disruption in brook trout (*Salvelinus fontinalis*) exposed to leachate from a public refuse dump. *Science of the Total Environment* 305: 87–103.

NODC (2003) World Ocean Circulation Experiment [www document]. URL http://woce.nodc.noaa.gov/wdice

Noe, G. B. & Zedler, J. B. (2001) Spatial and temporal variation of salt marsh seedling establishment in relation to the abiotic and biotic environment. *Journal of Vegetation Science* 12: 61–74.

Nogueira, F. & Junk, W. J. (2000) Mercury from goldmining in Amazon wetlands: contamination sites, intoxication levels and dispersion pathways. In: *The Central Amazon Floodplain: Actual Use and Options for a Sustainable Management*, ed. W. J. Junk, J. Ohly, M. T. F. Piedade & M. G. M. Soares, pp. 477–503. Leiden, the Netherlands: Backhuys.

Norby, R. J. & Cotrufo, M. F. (1998) A question of litter quality. *Nature* **396**: 17–18.

Nordstrom, K. F. & Mauriello, M. N. (2001) Restoring and maintaining naturally functioning landforms and biota on intensively developed barrier islands under a no-retreat scenario. *Shore and Beach* **69**(3): 19–28.

Nordstrom, K. F. (1992) *Estuarine Beaches*. London, UK: Elsevier.

Nordstrom, K. F. (2000) *Beaches and Dunes of Developed Coasts*. Cambridge, UK: Cambridge University Press.

Nordstrom, K. F. (2003) Restoring naturally functioning beaches and dunes on developed coasts using compromise management solutions: an agenda for action. In: *Values at Sea: Ethics for the Marine Environment*, ed. D. Dallmeyer, pp. 204–229. Athens, GA, USA: University of Georgia Press.

Nordstrom, K. F. & Roman, C. T., eds. (1996) *Estuarine Shores: Evolution, Environments, and Human Alterations*. New York, USA: John Wiley.

Nordstrom, K. F., Jackson, N. L., Bruno, M. S. & de Batts, H. A. (2002) Municipal initiative for managing dunes in coastal residential areas: a case study of Avalon, New Jersey, USA. *Geormorphology* **47**: 137–152.

Norkko, A., Thrush, S. F., Hewitt, J. E., *et al.* (2002) Smothering of estuarine sandflats by terrigenous clay: the role of wind–wave disturbance and bioturbation in site-dependent macrofaunal recovery. *Marine Ecology Progress Series* **234**: 23–41.

Norse, E. A. & Crowder, L. B. (2005) *Marine Conservation Biology*. Washington DC, USA: Island Press.

North, W. J. (1994) Review of *Macrocystis* biology. In: *Biology of Economic Algae*, ed. I. Akatsuka, pp. 447–527. The Hague, the Netherlands: SBP Academic Publishing.

Northeast Fisheries Science Center (2002) Large marine ecosystems of the world: regional summaries [www document]. URL http://www.edc.uri.edu/lme/clickable-map.htm

Norton, T. A. & Benson, M. R. (1983) Ecological interactions between the brown seaweed *Sargassum muticum* and its associated fauna. *Marine Biology* **75**: 169–177.

Notenboom, J. (2001) Managing ecological risks of groundwater pollution. In: *Groundwater Ecology: A Tool for Management of Water Resources*, ed. C. Griebler, D. L.

Danielopol, J. Gibert, H. P. Nachtnebel & J. Notenboom, pp. 248–262. Luxembourg: Office for Official Publications of the European Communities.

Notenboom, J., Plénet, S. & Turquin, M. J. (1994) Groundwater contamination and its impact on groundwater animals and ecosystems. In: *Groundwater Ecology*, ed. J. Gibert, D. L. Danielopol & J. A. Stanford, pp. 477–504. San Diego, CA, USA: Academic Press.

NRC (1993) *Managing Wastewater in Coastal Urban Areas*. Washington, DC, USA: National Academy Press.

NRC (1995) *Beach Nourishment and Protection*. Washington, DC, USA: National Academy Press.

NRC (1999) *Sustaining Marine Fisheries*. Washington, DC, USA: National Academy Press.

NRC (2000a) *Clean Coastal Waters: Understanding and Reducing the Effects of Nutrient Pollution*. Washington, DC, USA: National Academy Press.

NRC (2000b) *50 Years of Ocean Discovery*. Washington, DC, USA: National Academy Press.

NRC (2001) *Compensating for Wetland Losses under the Clean Water Act*. Washington, DC, USA: National Academy Press.

NRC (2002) *Riparian Areas: Functions and Strategies for Management*. Washington, DC, USA: National Academy Press.

Nuttle, W. K., Brinson, M. M., Cahoon, D., *et al.* (1997) Conserving coastal wetlands despite sea level rise. *Eos* **78**: 260–261.

Nyambe, N. & Breen, C. (2002) Environmental flows, power relations and the use of river system resources, ENVIRO FLOWS, Proceedings of the International Conference on Environmental Flows for River Systems, incorporating the 4th International Ecohydraulics Symposium. 3–8 March, unpublished proceedings, Cape Town, South Africa.

Nybakken, J. W. (1988) *Marine Biology: An Ecological Approach*, 2nd edn. New York, USA: Harper & Row.

Nyman, J. A., DeLaune, R. D., Roberts, H. H. & Patrick, W. H. (1993) Relationships between vegetation and soil formation in a rapidly submerging coastal marsh. *Marine Ecology Progress Series* **96**: 269–279.

Nystrom, M., Folke, C. & Moberg, F. (2000) Coral reef disturbance and resilience in a human-dominated environment. *Trends in Ecology and Evolution* **15**: 413–417.

Ochieng, C. A. & Erftemeijer, P. L. A. (2003) The seagrasses of Kenya and Tanzania. In: *World Atlas of Seagrasses*, ed. E. P. Green & F. T. Short, pp. 92–104. Berkeley, CA, USA: University of California Press.

O'Connor, T. P. & Beliaeff, B. (1995) *Recent Trends in Coastal Environmental Quality: Results from the Mussel Watch Project*, NOAA Technical Report. Rockville, MD, USA: National Oceanic and Atmospheric Administration.

O'Connor, T. P. & Paul, J. F. (2000) Misfit between sediment toxicity and chemistry. *Marine Pollution Bulletin* **40**: 59–64.

Odén, S. (1967) Nederbördens försurning. *Dagens Nyheter (Stockholm)* 24 October 1967 (in Swedish).

Odum, E. P. (1969) The strategy of ecosystem development. *Science* **164**: 262–270.

Odum, E. P., Finn, J. T. & Franz, E. H. (1979) Perturbation theory and the subsidy–stress gradient. *BioScience* **29**: 349–352.

Odum, W. E. & Heald, E. J. (1972) Trophic analysis of an estuarine mangrove community. *Bulletin of Marine Science* **22**: 671–738.

Oebius, H. U., Becker, H. J., Rolinski, S. & Jankowski, J. (2001) Parametrization and evaluation of marine environmental impacts produced by deep-sea manganese nodule mining. *Deep-Sea Research II* **48**: 3453–3467.

OECD Working Committees (2000) Analytical report on sustainable development: managing natural resources. Unpublished report. Paris, France: OECD General Secretariat.

Oechel, W. C., Cowles, S., Grulke, N., *et al.* (1994) Transient nature of CO_2 fertilization in Arctic tundra. *Nature* **371**: 500–503.

Oechel, W. C., Hastings, S. J., Vourlitis, G., *et al.* (1993) Recent change of Arctic tundra ecosystems from a net carbon dioxide sink to a source. *Nature* **361**: 520–523.

Oechel, W. C., Vourlitis, G., Hastings, S. J. & Bocharev, S. A. (1995) Change in arctic CO_2 flux over 2 decades: effects of climate-change at Barrow, Alaska. *Ecological Applications* **5**: 846–855.

Office of Technology Assessment (1986) *Pollutant Discharges to Surface Waters in Coastal Regions*. Washington, DC, USA: Office of Technology Assessment, USA Congress.

Ohly, J. J. & Hund, M. (2000) Floodplain animal husbandry in central Amazonia. In: *The Central Amazon Floodplain: Actual Use and Options for a Sustainable Management*, ed. W. J. Junk, J. J. Ohly & M. T. F. Piedade, pp. 313–343. Leiden, the Netherlands: Backhuys.

Ohmi, H. (1951) Studies on Isoyake or decrease of seaweeds along the coast of Northern Japan. *Bulletin of the Faculty of Fisheries Hakodate, Hokkaido University* **2**: 109–117.

Ojeda, F. & Santelices, B. (1984) Ecological dominance of *Lessonia nigrescens* (Phaeophyta) in central Chile. *Marine Ecology Progress Series* **19**: 83–91.

Oleksyn, J. & Reich, P. B. (1994) Pollution, habitat destruction, and biodiversity in Poland. *Conservation Biology* **8**: 943–960.

Olivero, A. M. (2001) *Classification and Mapping of New York's Calcareous Fen Communities*. Albany, NY, USA: New York Natural Heritage Program.

Olsgard, F. & Gray, J. S. (1995) A comprehensive analysis of the effects of offshore oil and gas exploration and production on the benthic communities of the Norwegian continental shelf. *Marine Ecology Progress Series* **122**: 277–306.

Olson, D. M. & Dinerstein, E. (1998) The global 200: a representation approach to conserving the earth's most biologically valuable ecosystems. *Conservation Biology* **12**: 502–515.

O'Neill, K. P. (2000) Role of bryophyte-dominated ecosystems in the global carbon budget. In: *Bryophyte Biology*, ed. A. J. Shaw & B. Goffinet, pp. 344–368. Cambridge, UK: Cambridge University Press.

Ong, J. E. (1982) Aquaculture, forestry and conservation of Malaysian mangroves. *Ambio* **11**: 252–257.

Ong, J. E. (1993) Mangroves: a carbon source and sink. *Chemosphere* **27**: 1097–1107.

Ong, J. E. (1995) The ecology of mangrove management and conservation. *Hydrobiologia* **295**: 343–351.

Ong, J. E., Gong, W. K. & Chan, H. C. (2001) Governments of developing countries grossly undervalue their mangroves? In: *Proceedings of the International Symposium on Protection and Management of Coastal Marine Ecosystems*, pp. 179–184. Bangkok, Thailand: United Nations Environment Programme.

Ong, J. E., Gong, W. K. & Wong, C. H. (1980) Ecological survey of the Sungai Merbok estuarine mangrove ecosystem. Unpublished report. Penang, Malaysia: Malaysian Fisheries Development Authority (MAJUIKAN).

Onuf, C. P., Phillips, R. C., Moncreiff, C. A., Raz-Guzman, A. & Herrera-Silveira, J. A. (2003) The seagrasses of the Gulf of Mexico. In: *World Atlas of Seagrasses*, ed. E. P. Green & F. T. Short, pp. 244–253. Berkeley, CA, USA: University of California Press.

Öquist, M. G., Svensson, B. H., Groffman, P. & Taylor, M. (1996) Non-tidal wetlands. In: *Climate Change 1995: Impacts, Adaptations and Mitigation of Climate Change – Contribution of Working Group II to the Second Assessment Report of the Intergovernmental Panel on Climate Change*, ed. R. T. Watson, M. C. Zinoyowera & R. H. Moss, pp. 215–239. Cambridge, UK: Cambridge University Press.

O'Reilly, C., Alin, S., Plisnier, P. -D., Cohen, A. & McKee, B. (2003) Climate change decreases aquatic ecosystem productivity of Lake Tanganyika, Africa. *Nature* **424**: 766–768.

Oren, A., ed. (2002) *Cellular Origin and Life in Extreme Habitats*, vol. 5, *Halophilic Microorganisms and Their Environments*. Boston, MA, USA: Kluwer.

O'Riordan, T. & Cameron, J., eds. (1994) *Interpreting the Precautionary Principle*. London, UK: Earthscan Publications.

Ormond, R., Bradbury, R., Bainbridge, S., *et al.* (1988) Test of a model of regulation of crown-of-thorns starfish by fish predators. In: Acanthaster *and the Coral Reef: a Theoretical Perspective*, ed. R. H. Bradbury, pp. 190–207. Townsville, Queensland, Australia: Springer-Verlag.

Orsi, J. J. & Mecum, W. (1986) Zooplankton distribution and abundance in the Sacramento–San Joaquin Delta in relation to certain environmental factors. *Estuaries* **9**: 326–335.

Orth, R. J. & Moore, K. A. (1983) Chesapeake Bay: an unprecedented decline in submerged aquatic vegetation. *Science* **222**: 53–57.

Orth, R. J., Fishman, J. R., Wilcox, D. J. & Moore, K. A. (2002) Identification and management of fishing gear impacts in a recovering seagrass system in the coastal bays of the Delmarva Peninsula, USA. *Journal of Coastal Research* Special Issue 37: 111–129.

OSPAR (2000*a*) *Quality Status Report 2000: Region II – Greater North Sea*. London, UK: OSPAR Commission for the Protection of the Marine Environment of the North-East Atlantic.

OSPAR (2000*b*) *Quality Status Report 2000: Region I – Arctic Waters*. London, UK: OSPAR Commission for the Protection of the Marine Environment of the North-East Atlantic.

OSPAR (2002) *OSPAR Convention for the Protection of the Marine Environment of the North-East Atlantic*, Report No. 02/21/1-E. London, UK: OSPAR Commission for the Protection of the Marine Environment of the North-East Atlantic.

OSPAR (2003) *Declaration of the Joint Ministerial Meeting of the Helsinki and OSPAR Commissions*, Report No. JMM 2003/3-E, London, UK: OSPAR Commission for the Protection of the Marine Environment of the North-East Atlantic.

Ostrander, G. K., Armstrong, K. M., Knobbe, E. T., Gerace, D. & Scully, E. P. (2000) Rapid transition in the structure of a coral reef community: the effects of coral bleaching and physical disturbance. *Ecology* **97**: 5297–5302.

Osvald, H. (1923) *Die Vegetation des Hochmoores Komosse, Svenska Växtsociologiska Sällskapets Handlingar 1*. Uppsala, Sweden: Almqvist & Wicksell.

Oudot, J. & Dutrieux, E. (1989) Hydrocarbon weathering and biodegradation in a tropical estuarine ecosystem. *Marine Environmental Research* **27**: 195–213.

Ozturk, M., Ozdemir, F. & Yucel, E. (1997) An overview of the environmental issues in the Black Sea region. In: *Scientific, Environmental, and Political Issues in the Circum-Caspian Region*, ed. M. H. Glantz & I. S. Zonn, pp. 213–226. Dordrecht, the Netherlands: Kluwer.

Packham, J. R. & Liddle, M. J. (1970) The Cefni saltmarsh and its recent development. *Field Studies* **3**: 331–356.

Paerl, H. W. (1985) Enrichment of marine primary production by nitrogen-enriched acid rain. *Nature* **316**: 747–749.

Paerl, H. W. (1995) Coastal eutrophication in relation to atmospheric nitrogen deposition: current perspectives. *Ophelia* **41**: 237–259.

Paerl, H. W. & Whitall, D. R. (1999) Anthropogenically-derived atmospheric nitrogen deposition, marine eutrophication and harmful algal bloom expansion: is there a link? *Ambio* **28**: 307–311.

Pagnan, J. L. (2000) Arctic marine protection. *Arctic* **53**: 469–473.

Pahl-Wostl, C. (2002) Towards sustainability in the water sector: the importance of human actors and processes of social learning. *Aquatic Sciences* **64**: 394–411.

Paijmans, K. (1990) Wooded swamps in New Guinea. In: *Ecosystems of the World*, vol. 15, *Forested Wetlands*, ed. A. E. Lugo, M. Brinson & S. Brown, pp. 335–355. Amsterdam, the Netherlands: Elsevier.

Paine, R. T. (1969) The *Pisaster–Tegula* interaction: prey patches, predator food preference, and intertidal community structure. *Ecology* **50**: 950–961.

Paine, R. T. (1980) Food webs: linkages, interaction strength and community infrastructure. *Journal of Animal Ecology* **49**: 667–685.

Paine, R. T. (1994) *Marine Rocky Shores and Community Ecology: An Experimentalist's Perspective*. Luhe, Germany: Ecology Institute.

Paine, R. T., Ruesink, J. L., Sun, A., *et al.* (1996) Trouble on oiled waters: lessons from the Exxon Valdez oil spill. *Annual Review of Ecology and Systematics* **27**: 197–235.

Paine, R. T., Tenger, M. J. & Johnson, E. A. (1998) Compounded perturbations yield ecological surprises. *Ecosystems* **1**: 535–545.

Painting, S. J., Moloney, C. L., Probyn, T. A. & Tibbles, B. (1992) Microhetrotrophic pathways in the southern

Benguela upwelling system. *South African Journal of Marine Sciences* **12**: 527–543.

Paling, E. I., van Keulen, M., Wheeler, K., Phillips, J. & Dryberg, R. (2001) Mechanical seagrass transplantation in Western Australia. *Ecological Engineering* **16**: 331–339.

PAME (2003) *Arctic Marine Strategic Plan.* Akureyri, Iceland: Protection of the Arctic Marine Environment (PAME) International Secretariat.

Pandolfi, J. M. (1996) Limited membership in Pleistocene reef coral assemblages from the Huon Peninsula, Papua New Guinea: constancy during global changes. *Paleobiology* **22**: 152–176.

Pandolfi, J. M. (1999) Response of Pleistocene coral reefs to environmental change over long temporal scales. *American Zoology* **39**: 113–130.

Pandolfi, J. M., Bradbury, R. H., Sala, E., *et al.* (2003) Global trajectories of the long-term decline of coral reef ecosystems. *Science* **301**: 955–958.

Pang, L. & Pauly, D. (2001) Part 1. Chinese marine capture fisheries from 1950 to the late 1990s: the hopes, the plans and the data. In: *The Marine Fisheries of China: Development and Reported Catches*, Fisheries Centre Research Report No. 9(2), ed. R. Watson, L. Pang & D. Pauly, pp. 1–27. Vancouver, BC, Canada: University of British Columbia.

Pantulu, V. R. (1986) The Mekong river system. In: *The Ecology of River Systems*, ed. B. R. Davies & K. F. Walker, pp. 695–719. The Hague, the Netherlands: Dr W. Junk.

Parker, D., Folland, C. K. & Jackson, M. (1995) Marine surface temperature: observed variations and data requirements. *Climatic Change* **31**: 559–600.

Parkinson, C. L. (1992) Spatial patterns of increases and decreases in the length of the sea ice season in the north polar region, 1979–1986. *Journal of Geophysical Research* **97**: 14 337–14 388.

Parkinson, C. L. (1994) Spatial patterns in the length of the sea ice season in the Southern Ocean. *Journal of Geophysical Research* **99**: 16 327–16 339.

Parkinson, C. L. (1998) Length of the sea ice season in the southern ocean, 1987–1994, in Antarctic sea ice: physical processes, interactions and variability. *Antarctic Research Series* **74**: 173–186.

Parkinson, C. L. (2000) Variability of Arctic sea ice: the view from space, an 18-year record. *Arctic* **53**: 341–358.

Parkinson, C. L., Cavalieri, D. J., Gloersen, P., Zwally, H. J. & Comiso, J. C. (1999) Arctic sea ice extents, areas and trends, 1978–96. *Journal of Geophysical Research* **104**: 20 837–20 856.

Parolin, P. (2000) Growth, productivity, and use of trees in white water floodplains. In: *The Central Amazon Floodplain: Actual Use and Options for a Sustainable Management*, ed. W. J. Junk, J. J. Ohly & M. T. F. Piedade, pp. 375–391. Leiden, the Netherlands: Backhuys.

Parrish, R. H. (1995) Lanternfish heaven: the future of world fisheries? *Naga* **18**(3): 7–9.

Parrish, R. H. (1998) Life history strategies for marine fishes in the late Holocene. In: *Global versus Local Change in Upwelling Areas*, ed. M. H. Durand, P. Cury, R. Mendelssohn, A. Bakun, C. Roy & D. Pauly, pp. 524–535. Séries Colloques et Séminaires ORSTOM. Paris, France: IRD.

Parsons, J. J. & Bowen, W. A. (1966) Ancient ridged fields of San Jorge River floodplain, Colombia. *Geographical Review* **56**: 317–343.

Pasqualini, V., Clabaut, P., Pergent, G., Benyoussef, L. & Pergent-Martini, C. (2000) Contribution of side scan sonar to the management of Mediterranean littoral ecosystems. *International Journal of Remote Sensing* **21**: 367–378.

Patrick, R. (1995) *Rivers of the United States*, vol. 2, *Chemical and Physical Characteristics*. New York, USA: John Wiley.

Patriquin, D. G. (1975) Migration of blowouts in seagrass beds at Barbados and Carriacou West Indies and its ecological and geological implications. *Aquatic Botany* **1**: 163–189.

Paul, A. J., Leavitt, P. R., Schindler, D. W. & Hardie, A. K. (1995) Direct and indirect effects of predation by a calanoid copepod (subgenus: *Hesperodiaptomus*) and of nutrients in a fishless alpine lake. *Canadian Journal of Fisheries and Aquatic Sciences* **52**: 2628–2638.

Paul, M. J. & Meyer, J. L. (2001) Streams in the urban landscape. *Annual Review of Ecology and Systematics* **32**: 333–365.

Pauly, D. (1980) On the interrelationships between natural mortality, growth parameters and mean environmental temperature in 175 fish stocks. *Journal du Conseil Internationale pour l'Exploration de la Mer* **39**: 175–192.

Pauly, D. (1997) Small scale fisheries in the tropics; marginality, marginalization, and some implications for fisheries management. In: *Global Trends in Fisheries Management*, ed. E. K. Pikitch, P. D. Huppert & M. P. Sissenweiwe, pp. 40–49. Bethesda, MD, USA: American Fisheries Society.

Pauly, D. (2000) Global change, fisheries and the integrity of marine ecosystems: the future has already begun. In: *Ecological Integrity: Integrating Environment, Conservation, and Health*, ed. D. Pimentel, L. Westra & R. F. Ross, pp. 227–239. Washington, DC, USA: Island Press.

Pauly, D. (2002) Review of *The Skeptical Environmentalist: Measuring the Real State of the World* by B. Lomborg. *Fish and Fisheries* **3**: 3–4.

Pauly, D. & Christensen, V. (1995) Primary production required to sustain global fisheries. *Nature* **374**: 255–257.

Pauly, D. & Chuenpagdee, R. (2003) Development of fisheries in the Gulf of Thailand Large Marine Ecosystem: analysis of an unplanned experiment. In: *Large Marine Ecosystems of the World: Trends in Exploitation, Protection, and Research*, ed. G. Hempel & K. Sherman, pp. 337–354. Amsterdam, the Netherlands: Elsevier.

Pauly, D. & Maclean, J. (2003) *In a Perfect Ocean: The State of Fisheries and Ecosystems in the North Atlantic Ocean.* Washington DC, USA: Island Press.

Pauly, D., Alder, J., Bennett, E., *et al.* (2003) The future for fisheries. *Science* **302**: 1359–1361.

Pauly, D., Alder, J., & Watson, R. (2005). Global trends in world fisheries: impacts on marine ecosystems and food security. *Philosophical Transactions of the Royal Society of London B* **360**: 5–12.

Pauly, D., Christensen, V., Dalsgaard, J., Froese, R. & Torres Jr, F. (1998) Fishing down marine food webs. *Science* **279**: 860–863.

Pauly, D., Christensen, V., Guénette, S., Pitcher, T. J., Sumaila, U. R., Walters, C. J., Watson, R. & Zeller, D. (2002) Toward sustainability in world fisheries. *Nature* **418**: 689–695.

Pauly, D., Palomares, M. L., Froese, R., *et al.* (2001) Fishing down Canadian aquatic food webs. *Canadian Journal of Fisheries and Aquatic Sciences* **58**: 51–62.

Payne, A. I. L., Brink, K. H., Mann, K. L. & Hilborn, R., eds. (1992) Benguela trophic functioning. *South African Journal of Marine Science* **12**.

Pearce, C. M. & Smith, D. G. (2003) Saltcedar: distribution, abundance, and dispersal mechanisms, northern Montana, USA. *Wetlands* **23**: 215–228.

Pearce, F. (2003) Mangrove plantations pose threat to coral reefs. *New Scientist* **178**: 11.

Peck, A. J. & Williamson, D. H. (1987) Effects of forest clearing on groundwater. *Journal of Hydrology* **94**: 47–65.

Peltier, W. R. & Tushingham, A. M. (1989) Global sea level rise and the greenhouse effect: might they be connected? *Science* **244**: 806–810.

Penang State Economic Planning Unit & DANCED (1999) *The Status of and Conservation Plans for the Mangroves of Penang*, Integrated Coastal Zone Management Project, Penang Component, Work Group 1: Mangroves, Final Report. Penang, Malaysia: State Economic Planning

Unit and Danish Cooperation for Environment and Development.

Penland, S. & Ramsey, K. E. (1990) Relative sea level rise in Louisiana and the Gulf of Mexico: 1908–1988. *Journal of Coastal Research* **6**: 323–342.

Percy, J. A. (1999) Whither the waters? Tidal and riverine restrictions in the Bay of Fundy. *Fundy Issues* **11**: 1–10.

Percy, J. A. (2000) Salt marsh saga: conserving Fundy's marine meadows. *Fundy Issues* **16**: 1–9.

Perelman, S. B., Leon, J. C. & Oesterheld, M. (2001) Cross-scale vegetation patterns of flooding pampa grasslands. *Journal of Ecology* **89**: 562–577.

Pergent, G., Mendez, S. & Pergent-Martini, C. (1999) Preliminary data on the impact of fish farming facilities on *Posidonia oceanica* meadows in the Mediterranean. *Oceanologia Acta* **22**: 95–107.

Persson, A. (1997) Phosphorus release by fish in relation to external and internal load in a eutrophic lake. *Limnology and Oceanography* **43**: 577–583.

Peters, A. F. & Breeman, A. M. (1993) Temperature tolerances and latitudinal range of brown algae from temperate Pacific South America. *Marine Biology* **115**: 143–150.

Peters, E. C. (1997) Diseases of coral-reef organisms. In: *Life and Death of Coral Reefs*, ed. C. Birkeland, pp. 114–139. New York, USA: Chapman & Hall.

Petersen, I., Masters, Z., Hildrew, A. G. & Ormerod, S. J. (2004) Dispersal of adult aquatic insects in catchments of differing land use. *Journal of Applied Ecology* **41**: 934–950.

Petersen, I., Winterbottom, J., Orton, S., *et al.* (1999) Emergence and lateral dispersal of adult Plecoptera and Trichoptera from Broadstone Stream, UK. *Freshwater Biology* **42**: 401–416.

Peterson, C. H. (1985) Patterns of lagoonal bivalve mortality after heavy sedimentation and their paleoecological significance. *Paleobiology* **11**: 139–153.

Peterson, C. H. & Estes, J. A. (2001) Conservation and management of marine communities. In: *Marine Community Ecology*, ed. M. D. Bertness, S. D. Gaines & M. E. Hay, pp. 469–508. Sunderland, MA, USA: Sinauer Associates.

Peterson, C. H., Hickerson, D. H. M. & Johnson, G. G. (2000*a*) Short-term consequences of nourishment and bulldozing on the dominant large invertebrates of a sandy beach. *Journal of Coastal Research* **16**: 368–378.

Peterson, C. H., McDonald, L. L., Green, R. H. & Erickson, W. P. (2000*b*) Sampling design begets conclusions: the statistical basis for detection of injury to and recovery of shoreline communities after the 'Exxon Valdez' oil spill. *Marine Ecology Progress Series* **210**: 255–283.

Peterson, C. H., Summerson, H. C. & Fegley, S. R. (1983) Relative efficiency of two clam rakes and their contrasting impacts on seagrass biomass. *Fisheries Bulletin* 81: 429–434.

Peterson, C. H., Summerson, H. C. & Fegley, S. R. (1987) Ecological consequences of mechanical harvesting of clams. *Fisheries Bulletin* 85: 281–298.

Pethick, J. (2002) Estuarine and tidal wetland restoration in the United Kingdom: policy versus practice. *Restoration Ecology* 10: 431–437.

Petts, G. E. (1984) *Impounded Rivers: Perspectives for Ecological Management.* Chichester, UK: John Wiley.

Petts, G. E. (1990) Forested river corridors: a lost resource. In: *Water, Engineering, and Landscapes: Water Control and Landscape Transformation in the Modern Period*, ed. D. Cosgrove & G. E. Petts, pp. 12–34. London, UK: Belhaven Press.

Pfadenhauer, J. & Kloetzli, F. (1996) Restoration experiments in middle European wet terrestrial ecosystems: an overview. *Vegetatio* 126: 101–115.

Pfeiffer, W. J. & Wiegert, R. G. (1981) Grazers on *Spartina* and their predators. In: *The Ecology of a Salt Marsh*, ed. L. R. Pomeroy & R. G. Wiegert, pp. 87–112. New York, USA: Springer-Verlag.

Phillips, R. C. (2003) The seagrasses of the Arabian Gulf and Arabian Region. In: *World Atlas of Seagrasses*, ed. E. P. Green & F. T. Short, pp. 84–91. Berkeley, CA, USA: University of California Press.

Phillips, V. D. (1998) Peatswamp ecology and sustainable development in Borneo. *Biodiversity and Conservation* 7: 651–671.

Phinn, S. R., Stow, D. A. & Zedler, J. B. (1996) Monitoring wetland habitat restoration in southern California using airborne digital multispectral video data. *Restoration Ecology* 4: 412–422.

Piazzi, L., Balestri, E. & Cinelli, F. (1994) Presence of *Caulerpa racemosa* in the north-western Mediterranean. *Cryptogamie et Algologie* 15: 183–189.

Piechota, T. C., Dracup, J. A. & Fovell, R. G. (1997) Western US streamflow and atmospheric circulation patterns during El Niño–Southern Oscillation. *Journal of Hydrology* 201: 249–271.

Pimbert, M. P. & Gujja, B. (1997) Village voice challenging wetland management policies: experiences in participatory rural appraisal from India and Pakistan. *Nature and Resources* 33: 34–42.

Pimentel, D., Lach, L., Zuniga, R. & Morrison, D. (2000) Environmental and economic costs of nonindigenous species in the United States. *BioScience* 50: 53–65.

Pinet, P. R. (2000) *Invitation to Oceanography*, 2 edn. Boston, MA, USA: Jones & Bartlett.

Pinnegar, J. K., Jennings, S., O'Brien, C. M. & Polunin, N. V. C. (2002) Long-term changes in the trophic level of the Celtic Sea fish community and fish market price distribution. *Journal of Applied Ecology* 39: 377–390.

Pinnegar, J. K., Polunin, N. V. C., Francour, P., *et al.* (2000) Trophic cascades in benthic marine ecosystems: lessons for fisheries and protected-area management. *Environmental Conservation* 27: 179–200.

Pinnegar, J. K., Viner, D., Hadley, D., *et al.* (2006) Alternative future scenarios for marine ecosystems: technical report. Unpublished report. Lowestoft, UK: Centre for Environment, Fisheries and Aquaculture Science.

Pitcher, T. (1995) Species changes and fisheries in African lakes: outline of the issues. In: *The Impact of Species Changes in African Lakes*, ed. T. J. Pitcher & P. J. B. Hart, pp. 1–16. London, UK: Chapman & Hall.

Pitcher, T. (2001) Fisheries managed to rebuild ecosystems? Reconstructing the past to salvage the future. *Ecological Applications* 11: 601–617.

Plaziat, J.-C., Cavagnetto, C., Koeniguer, J.-C. & Baltzer, F. (2001) History and biogeography of the mangrove ecosystem, based on a critical reassessment of the paleontological record. *Wetlands Ecology and Management* 9: 161–179.

Plotnikov, I. S., Aladin, N. V. & Philippov, A. A. (1991) The past and present of the Aral Sea Fauna. *Zoologichesky Zhurna* 70(4): 5–15 (in Russian).

PMSEIC (1998) *Dryland Salinity and Its Impact on Rural Industries and the Landscape*, Prime Minister's Science, Engineering and Innovation Council Occasional Paper No. 1. Canberra, ACT, Australia: Department of Industry, Science and Resources.

Poff, N. L. (1992) Regional hydrologic response to climate change: an ecological perspective. In: *Global Climate Change and Freshwater Ecosystems*, ed. P. Firth & S. G. Fisher, pp. 88–115. Berlin, Germany: Springer-Verlag.

Poff, N. L., Brinson, M. M. & Day Jr, J. W. (2002) *Aquatic Ecosystems and Global Climate Change: Potential Impacts on Inland Freshwater and Coastal Wetland Ecosystems in the United States*, Report No. 7. Arlington, VA, USA. Pew Center on Global Climate Change.

Poiani, K. A., Richter, B. D., Anderson, M. G. & Richter, H. E. (2000) Biodiversity conservation at multiple scales: functional sites, landscapes, and networks. *BioScience* 50: 133–146.

Poiner, I., Walker, D. I. & Coles, R. B. (1989) Regional studies: seagrasses of tropical Australia. In: *Biology of*

Seagrasses: A Treatise on the Biology of Seagrasses with Special Reference to the Australian Region, ed. A. W. D. Larkum, A. J. McComb & S. A. Shepherd, pp. 279–303. Amsterdam, the Netherlands: Elsevier.

Polis, G. A. & Hurd, S. D. (1996) Linking marine and terrestrial food webs: allochthonous inputs from the ocean supports high secondary productivity on small islands and coastal land communities. *American Naturalist* 147: 396–423.

Pollard, D. A. (1984) A review of ecological studies on seagrass–fish communities with particular reference to recent studies in Australia. *Aquatic Botany* 18: 3–42.

Polley, H. W., Wilsey, B. J. & Derner, J. D. (2003) Do species evenness and plant density influence the magnitude of selection and complementarity effects in annual plant species mixtures? *Ecology Letters* 6: 248–256.

Polunin, N., Morales-Nin, B., Pawsey, W., *et al.* (2001) Feeding relationships in Mediterranean bathyal assemblages elucidated by stable nitrogen and carbon isotope data. *Marine Ecology Progress Series* 220: 13–23.

Polyakov, I. V., Alekseev, G. V., Bekryaev, R. V., *et al.* (2003a) Long-term ice variability in Arctic marginal seas. *Journal of Climate* 16: 2078–2085.

Polyakov, I. V., Bekryaev, R. V., Alekseev, G. V., *et al.* (2003b) Variability and trends of air temperature and pressure in the maritime Arctic, 1875–2000. *Journal of Climate* 16: 2067–2077.

Ponce, V. M., ed. (1995) *Hydrological and Environmental Impact of the Paraná Paraguay Waterway on the Pantanal of Mato Grosso (Brazil)*. San Diego, CA, USA: San Diego State University.

Poole, G. C. (2002) Fluvial landscape ecology: addressing uniqueness within the river continuum. *Freshwater Biology* 47: 641–660.

Pope, J. G., Shepherd, J. G. & Webb, J. (1994) Successful surf-riding on size spectra: the secret of survival in the sea. *Philosophical Transactions of the Royal Society of London B* 343: 41–49.

Popp, M. (1995) Salt resistance in herbaceous halophytes and mangroves. *Progress in Botany* 56: 416–429.

Popper, K. R. (1972) *Objective Knowledge*. Oxford, UK: Oxford University Press.

Porter, J. W., Dustan, P., Jaap, W., *et al.* (2001) Patterns of spread of coral disease in the Florida Keys. *Hydrobiologia* 460: 1–24.

Postel, S. L. (2000) Entering an era of water scarcity: the challenges ahead. *Ecological Applications* 10: 941–948.

Postel, S. (2001) Growing more food with less water. *Scientific American* 284: 34–37.

Postel, S. L., Daily, G. C. & Ehrlich, P. R. (1996) Human appropriation of renewable fresh water. *Science* 271: 785–788.

Potts, M. (2000) The unmet need for family planning. *Scientific American* 282: 75–77.

Pounds, J. A. & Crump, M. L. (1994) Amphibian declines and climate disturbance: the case of the golden toad and the harlequin frog. *Conservation Biology* 8: 72–85.

Preen, A. R., Long, W. J. L. & Coles, R. G. (1995) Flood and cyclone related loss, and partial recovery, of more than 1000 km^2 of seagrass in Hervey Bay, Queensland, Australia. *Aquatic Botany* 52: 3–17.

Preston, C. D., Pearman, D. A. & Dines, T. D. (2002) *New Atlas of the British and Irish Flora*. Oxford, UK: Oxford University Press.

Pretty, J. N. & Guijt, I. (1992) Primary environmental care: an alternative paradigm for development assistance. *Environment and Urbanization* 4: 22–36.

Primavera, J. H. (1994) Shrimp farming in the Asia-Pacific: environmental and trade issues and regional cooperation [www document]. URL http://www.nautilus.org/papers/enviro.html/trade/shrimp.html

Pringle, C. M. (2001) Hydrologic connectivity and the management of biological reserves: a global perspective. *Ecological Applications* 11: 981–998.

Pringle, C. M. (2003) What is hydrologic connectivity and why is it ecologically important. *Hydrological Processes* 17: 2685–2689.

Pringle, J. D., Sharp, G. J. & Caddy, J. F. (1980) Proceedings of the workshop on the relationship between sea urchin grazing and commercial plant/animal harvesting. *Canadian Technical Report of Fisheries and Aquatic Sciences* 954.

Prinn, R. G. & Hartley, D. E. (1992) Atmosphere, ocean and land: critical gaps in earth system models. In: *Modeling the Earth System*, ed. D. Ojima, pp. 9–38. Boulder, CO, USA: University Corporation for Atmospheric Research.

Prinn, R. G., Jacoby, H., Sokolov, A., *et al.* (1999) Integrated global system model for climate policy assessment: feedbacks and sensitivity studies. *Climatic Change* 41: 469–546.

Priskin, J. (2003) Physical impact of four-wheel drive related tourism and recreation in a semi-arid, natural coastal environment. *Ocean and Coastal Management* 46: 127–155.

Probert, P. K., McKnight, D. G. & Grove, S. L. (1997) Benthic invertebrate bycatch from a deep-water trawl fishery, Chatham Rise, New Zealand. *Aquatic Conservation: Marine and Freshwater Ecosystems* 7: 27–40.

Procaccini, G., Buia, M. C., *et al.* (2003) The seagrasses of the Western Mediterranean. In: *World Atlas of Seagrasses*, ed. E. P. Green & F. T. Short, pp. 56–66. Berkeley, CA, USA: University of California Press.

Proctor, M. C. F. (1997) Aspects of water chemistry in relation to surface degradation on ombrotrophic mires. In: *Blanket Mire Degradation: Causes, Consequences and Challenges*, ed. J. H. Tallis, R. Meade & P. D. Hulme, pp. 140–152. Aberdeen, UK: Macaulay Land Use Research Institute.

Puckridge, J. T., Sheldon, F., Walker, K. F. & Boulton, A. J. (1998) Flow variability and the ecology of large rivers. *Marine and Freshwater Research* **49**: 55–72.

Pulfrich, A., Parkins, C. A. & Branch, G. M. (2003*a*) The effects of shore-based diamond-diving on intertidal and subtidal biological communities and rock lobsters in southern Namibia. *Aquatic Conservation: Marine and Freshwater Ecosystems* **13**: 233–255.

Pulfrich, A., Parkins, C. A., Branch, G. M., Bustamante, R. H. & Velasquez, C. R. (2003*b*) The effects of sediment deposits from Namibian diamond mines on intertidal and subtidal reefs and rock lobster populations. *Aquatic Conservation: Marine and Freshwater Ecosystems* **13**: 257–278.

Purcell, M. A., Malej, A. & Benovic, A. (1999) Potential links of jellyfish to eutrophication and fisheries. In: *Ecosystems at the Land–Sea Margin: Drainage Basin to Coastal Sea*, ed. T. C. Malone, A. Malej, L. W. Harding Jr, N. Smodlana & R. E. Turner, pp. 241–263. Washington, DC, USA: American Geophysical Union.

Pyle, J. A. (2000) Stratospheric ozone depletion: a discussion of our present understanding. In: *Causes and Environmental Implications of Increased UV-B Radiation*, ed. R. E. Hester & R. M. Harrison, pp. 1–16. Cambridge, UK: Royal Society of Chemistry.

Pysek, P. & Prach, K. (1993) Plant invasion and the role of riparian habitats: a comparison of four species alien to central Europe. *Journal of Biogeography* **20**: 413–420.

Quinn, F. & Croley, T. (1999) Potential climate change impacts on Lake Erie. In: *State of Lake Erie: Past, Present and Future*, ed. M. Munawar, T. Edsall & I. F. Munawar, pp. 23–30. Leiden, the Netherlands: Backhuys.

Rabalais, N. N. (1992) *An Updated Summary of Status and Trends in Indicators of Nutrient Enrichment in the Gulf of Mexico*, Report to the Gulf of Mexico Program, Nutrient Enrichment Subcommittee, Publication No. EPA/800-R-92-004. Stennis Space Center, MS, USA: US Environmental Protection Agency, Office of Water, Gulf of Mexico Program.

Rabalais, N. N. (2005) Eutrophication. In: *The Sea*, vol. 13, ed. A. R. Robinson & K. H. Brink, pp. 821–866. Cambridge, MA, USA: Harvard University Press.

Rabalais, N. N., Turner, R. E. & Wisemann, W. J. (2002) Hypoxia in the Gulf of Mexico, aka 'The Dead Zone'. *Annual Review of Ecology and Systematics* **33**: 235–263.

Raffaelli, D. & Hawkins, S. (1996) *Intertidal Ecology*. London, UK: Chapman & Hall.

Raimondi, P. T., Barnett, A. M. & Krause, P. R. (1997) The effects of drilling muds on marine invertebrate larvae and adults. *Environmental Toxicology and Chemistry* **16**: 1218–1228.

Rakocinski, C. F., Heard, R. W. & LeCroy, S. E. (1996) Responses by macrobenthic assemblages to extensive beach restoration at Perdido Key, Florida, USA. *Journal of Coastal Research* **12**: 326–353.

Ramanathan, V., Crutzen, P. J., Kiehl, J. T. & Rosenfeld, D. (2001) Aerosols, climate, and the hydrological cycle. *Science* **294**: 2119–2124.

Ramsar (2004) The Ramsar Convention on Wetlands [www document]. URL http://www.ramsar.org/

Ramsar (2006) Criteria for planning a wetland inventory [www document]. URL http://www.ramsar.org/res/lay_res_viii_06_e.htm

Ramsar & IUCN (1999) Wetlands and global change [www document]. URL http://www.ramsar.org/key_unfccc_bkgd.htm

Rangeley, R. W. & Kramer, D. L. (1995) Tidal effects on habitat selection and aggregation by juvenile pollock *Pollachius virens* in the rocky intertidal zone. *Marine Ecology Progress Series* **126**: 19–29.

Ranwell, D. S. (1972) *Ecology of Salt Marshes and Sand Dunes*. London, UK: Chapman & Hall.

Ranwell, D. S. (1981) Introduced coastal plants and rare species in Britain. In: *Biological Aspects of Rare Plant Conservation*, ed. H. Synge, pp. 413–419. Chichester, UK: John Wiley.

Rasmussen, E. (1977) The wasting disease of eelgrass (*Zostera marina*) and its effects on environmental factors and fauna. In: *Seagrass Ecosystems*, ed. C. P. McRoy & C. Helfferich, pp. 1–51. New York, USA: Marcel Dekker.

Raven, J. A. (1997) Inorganic carbon acquisition by marine autotrophs. *Advances in Botanical Research* **27**: 185–209.

Raven, J. A., Walker, D. I., Johnston, A. M., Handley, L. L. & Kubler, J. E. (1995) Implications of ^{13}C natural abundance measurements for photosynthetic performance by marine macrophytes in their natural environment. *Marine Ecology Progress Series* **123**: 193–205.

Rawson, D. S. (1946) Successful introduction of fish in a large saline lake. *Canadian Fish Culturalist* 1: 5–8.

Reaka-Kudla, M. L. (1997) The global biodiversity of coral reefs: a comparison with rain forests. In: *Biodiversity II: Understanding and Protecting our Biological Resources*, ed. M. L. Reaka-Kudla, D. E. Wilson & E. O. Wilson, pp. 83–108. Washington, DC, USA: Joseph Henry Press.

Reaka-Kudla, M. L., Feingold, J. S. & Glynn, P. W. (1996) Experimental studies of rapid bioerosion of coral reefs in the Galapagos Islands. *Coral Reefs* 15: 101–109.

Reati, G. J., Florin, M., Fernandez, G. J. & Montes, C. (1997) The Laguna de Mar Chiquita (Cordoba, Argentina): a little known, secularly fluctuating, salt lake. *International Journal of Salt Lake Research* 5: 187–219.

Redclift, M. (1992) *Sustainable Development: Exploring the Contradictions*. London, UK: Routledge.

Redfield, A. C. (1972) Development of a New England salt marsh. *Ecological Monographs* 42: 201–237.

Reed, D. (2001) *Economic Change, Governance and Natural Resource Wealth: The Political Economy of Change in Southern Africa*. London, UK: Earthscan Publications.

Reed, D. (2002) *Poverty Is Not a Number: The Environment Is Not a Butterfly*. Washington, DC, USA: WWF Macroeconomics Program Office.

Reed, D. C. & Foster, M. S. (1984) The effects of canopy shading on algal recruitment and growth in a giant kelp forest. *Ecology* 65: 937–948.

Reeves Jr, C. C. (1978) Economic significance of playa lake deposits. *Special Publications of the International Association for Sediments* 2: 279–290.

Rehfeldt, G. E., Tchebakova, N. M., Parfenova, Y. I., *et al.* (2002) Intraspecific responses to climate in *Pinus sylvestris*. *Global Change Biology* 8: 912–929.

Rehfeldt, G. E., Ying, C. C., Spittlehouse, D. L. & Hamilton, D. A. (1999) Genetic responses to climate in *Pinus contorta*: niche breadth, climate change, and reforestation. *Ecological Monographs* 69: 375–407.

Reid, D. & Beeton, A. (1992) Large lakes of the world: a global science opportunity. *Geojournal* 28: 67–72.

Reid, D. & Orlova, M. (2002) Geological and evolutionary underpinnings for the success of Ponto-Caspian species invasions in the Baltic Sea and North American Great Lakes. *Canadian Journal of Fisheries and Aquatic Sciences* 59: 1144–1158.

Reid, K. & Croxall, J. P. (2001) Environmental response of upper trophic-level predators reveals a system change in an Antarctic marine ecosystem. *Proceedings of the Royal Society of London B* 268: 377–384.

Reilly, F. J. & Bellis, V. J. (1983) *The Ecological Impact of Beach Nourishment with Dredged Materials on the Intertidal Zone at Bogue Banks, North Carolina*, Miscellaneous Report No. 83–3. Fort Belvoir, VA, USA: US Army Corps of Engineers, Coastal Engineering Research Center.

Reilly, J., Stone, P., Forest, C., *et al.* (2001) Uncertainty and climate change assessments. *Science* 291: 430–433.

Reise, K. (2002) More sand to the shorelines of the Wadden Sea: harmonizing coastal defense with habitat dynamics. In: *Marine Science Frontiers for Europe*, ed. G. Wefer, F. Lamy & F. Mantoura, pp. 203–216. Berlin, Germany: Springer-Verlag.

Rejmánková, E., Pope, K. O., Post, R. & Maltby, E. (1996) Herbaceous wetlands of the Yucatan Peninsula: communities at extreme ends of environmental gradients. *Internationale Revue der Gesamten Hydrobiologie* 81: 225–254.

Renn, C. E. (1936) The wasting disease of *Zostera marina*: a phytological investigation of the diseased plant. *Biological Bulletin* 70: 148–158.

Reschke, C. (1990) *Ecological Communities of New York State*. Latham, NY, USA: New York Natural Heritage Program, New York State Department of Environmental Conservation.

Retière, C. (1994) Tidal power and the aquatic environment of La Rance. *Biological Journal of the Linnean Society* 51: 25–36.

Revenga, C., Brunner, J., Henninger, N., Kassem, K. & Payne, R. (2000) *Pilot Analysis of Global Ecosystems. Freshwater Systems*. Washington, DC, USA: World Resources Institute.

Revenga, C., Murray, S., Abramovitz, J. & Hammond, A. (1998) *Watersheds of the World*. Washington, DC, USA: World Resources Institute.

Reyes, J., Sansón, M. & Afonso-Carrillo, J. (1995) Distribution and reproductive phenology of the seagrass *Cymodocea nodosa* (Ucria) Ascherson in the Canary Islands. *Aquatic Botany* 50: 171–180.

Reynolds, C. S. (1993) Scales of disturbance and their role in plankton ecology. *Hydrobiologia* 249: 157–171.

Ricciardi, A. (2001) Facilitative interactions among invaders: is an 'invasional meltdown' occurring in the Great Lakes? *Canadian Journal of Fisheries and Aquatic Sciences* 58: 2513–2525.

Ricciardi, A. (2003) Predicting the impacts of an introduced species from its invasion history: an empirical approach applied to zebra mussel invasions. *Freshwater Biology* 48: 972–981.

Ricciardi, A. (2006) Patterns of invasion in the Laurentian Great Lakes in relation to changes in vector activity. *Diversity and Distributions* 12: 425–433.

Ricciardi, A. & Bourget, E. (1999) Global patterns of macroinvertebrate biomass in marine intertidal communities. *Marine Ecology Progress Series* 185: 21–35.

Ricciardi, A. & MacIsaac, H. (2000) Recent mass invasion of the North American Great Lakes by Ponto-Caspian species. *Trends in Ecology and Evolution* 15: 62–65.

Ricciardi, A. & Rasmussen, J. B. (1999) Extinction rates of North American freshwater fauna. *Conservation Biology* 13: 1220–1222.

Ricciardi, A., Neves, R. J. & Rasmussen, J. B. (1998) Impending extinctions of North American freshwater mussels (Unionoida) following the zebra mussel (*Dreissena polymorpha*) invasion. *Journal of Animal Ecology* 67: 613–619.

Rice, A. L. & Owen, P. (1998) *Decommissioning the* Brent Spar. London, UK: E. and F.N. Spon.

Richardson, C. J., Reiss, P., Hussain, N. A., Alwash, A. J. & Pool, D. J. (2005) The restoration potential of the Mesopotamian marshes of Iraq. *Science* 307: 1307–1311.

Richardson, D. M., Pyšek, P., Rejmánek, M., *et al.* (2000) Naturalization and invasion of alien plants: concepts and definitions. *Diversity and Distributions* 6: 93–107.

Richardson, J. L. & Richardson, A. E. (1972) History of an African rift lake and its climatic implications. *Ecological Monographs* 42: 499–534.

Richardson, L. L. (1998) Coral diseases: what is really known? *Trends in Ecology and Evolution* 13: 438–443.

Ridd, P. V., Sandstrom, M. W. & Wolanski, E. (1988) Outwelling from tropical tidal salt flats. *Estuarine, Coastal and Shelf Science* 26: 243–253.

Ridout, P. S., Willcocks, A. D., Morris, R. J., White, S. L. & Rainbow, P. S. (1985) Concentrations of Mn, Fe, Cu, Zn and Cd in the mesopelagic decapod *Systellaspis cebilis* from the East Atlantic Ocean. *Marine Biology* 87: 285–288.

Riedel, R., Caskey, L. & Costa-Pierce, B. A. (2002) Fish biology and fisheries ecology of the Salton Sea, California. *Hydrobiologia* 473: 229–244.

Riemann, B. & Hoffmann, E. (1991) Ecological consequences of dredging and bottom trawling in the Limfjord, Denmark. *Marine Ecology Progress Series* 69: 171–178.

Rigor, I. G., Colony, R. L. & Martin, S. (2000) Variations in surface air temperature observations in the Arctic, 1979–97. *Journal of Climate* 13: 896–914.

Rigor, I. G., Wallace, J. M. & Colony, R. L. (2002) Response of the sea ice to the Arctic Oscillation. *Journal of Climate* 15: 2648–2668.

Risager, M. (1998) Impacts of nitrogen on *Sphagnum* dominated bogs with emphasis on critical load assessment. Unpublished Ph.D. thesis, University of Copenhagen, Copenhagen, Denmark.

Risk, M. J. & Sammarco, P. W. (1982) Bioerosion of corals and the influence of damselfish territoriality: a preliminary study. *Oecologia* 52: 376–380.

Risk, M. J. & Sammarco, P. W. (1991) Cross-continental shelf trends in skeletal density of the massive coral *Porites lobata* from the Great Barrier Reef. *Marine Ecology Progress Series* 69: 195–200.

Risk, M. J., Sammarco, P. W. & Edinger, E. N. (1995) Bioerosion in *Acropora* across the continental shelf of the Great Barrier Reef. *Coral Reefs* 14: 79–86.

Ritter, C. & Montagna, P. A. (1999) Seasonal hypoxia and models of benthic response in a Texas Bay. *Estuaries* 22: 7–20.

RIZA (2000) *Ecological Gradients in the Danube Delta Lakes*, Report No. 2000.015. Lelystad, the Netherlands: RIZA (Institute for Inland Water Management and Waste Water Treatment).

Roberts, C. & Polunin, N. (1993) Marine reserves: simple solutions to managing complex fisheries. *Ambio* 22: 363–368.

Roberts, C. M. & Hawkins, J. P. (1999) Extinction risk in the sea. *Trends in Ecology and Evolution* 14: 241–246.

Roberts, C. M., McClean, C. J., Veron, J. E. N., *et al.* (2002) Marine biodiversity hotspots and conservation priorities for tropical reefs. *Science* 295: 280–284.

Robertson, A. (1991) Effects of a toxic bloom of *Chrysomulina polylepis* on the common dogwhelk, *Nucella lapillus*, on the Swedish west coast. *Journal of the Marine Biological Association of the United Kingdom* 71: 569–578.

Robertson, A. I. & Mann, K. H. (1984) Disturbance by ice and life-history adaptations of the seagrass *Zostera marina*. *Marine Biology* 80: 131–141.

Robertson, A. L., Rundle, S. D. & Schmid-Araya, J. M. (2000) Putting the meio- into stream ecology: current findings and future directions for lotic meiofaunal research. *Freshwater Biology* 44: 177–183.

Robinson, C. T., Tockner, K. & Ward, J. V. (2002) The fauna of dynamic riverine landscapes. *Freshwater Biology* 47: 661–677.

Robinson, J. D. F. & Robb, G. A. (1995) Methods for the control and treatment of acid mine drainage. *Coal International* 243: 152–156.

Robinson, T. & Minton, C. (1990) The enigmatic banded stilt. *Birds International* 1990: 72–85.

Rodbell, D. T., Seltzer, G. O., Anderson, D. M., *et al.* (1999) An ~15 000-year record of El Niño-driven alluviation in Southwestern Ecuador. *Science* **283**: 516–520.

Rodgers, W. A. (1988) Domestic livestock and wildlife conservation: can they coexist? In: *International Rangeland Congress*, ed. P. Singh, V. Shankar & A. K. Srivastava, pp. 639–641. Jhansi, New Delhi, India: Range Management Society of India.

Rodhe, H., Grenfeldt, P., Wisniewski, J., *et al.* (1995) Acid reign '95? Conference summary statement. *Water, Air and Soil Pollution* **85**: 1–14.

Rodhe, W. & Herrera, R. (1988) *Acidification in Tropical Countries*. New York, USA: John Wiley.

Rodríguez, S. R. (2003) Consumption of drift kelp by intertidal populations of the sea urchin *Tetrapygus niger* on the central Chilean coast: possible consequences at different ecological levels. *Marine Ecology Progress Series* **251**: 141–151.

Rodríguez-Itrube, I. (2000) Ecohydrology: a hydrologic perspective of climate–soil–vegetation dynamics. *Water Resources Research* **36**: 3–9.

Rodríguez-Itrube, I. (2003) Hydrologic dynamics and ecosystem structure. *Water Science and Technology* **47**: 18–94.

Rodwell, M. J., Rowell, D. P. & Folland, C. K. (1999) Oceanic forcing of the wintertime North Atlantic Oscillation and European climate. *Nature* **398**: 320–323.

Roelse, P. (1990) Beach and dune nourishment in the Netherlands. *Coastal Engineering: Proceedings of the 22nd Coastal Engineering Conference*, pp. 1984–1997. New York, USA: American Society of Civil Engineers.

Roemmich, D. (1992) Ocean warming and sea level rise along the southwest US coast. *Science* **257**: 373–375.

Rogers, A. D. (1994) The biology of seamounts. *Advances in Marine Biology* **30**: 306–350.

Rogers, A. D. (1999) The biology of *Lophelia pertusa* (Linnaeus 1758) and other deep-water reef-forming corals and impacts from human activities. *International Review of Hydrobiology* **84**: 315–406.

Rogers, C. S. (1990) Responses of coral reefs and reef organisms to sedimentation. *Marine Ecology Progress Series* **62**: 185–202.

Roman, J. & Palumbi, S. R. (2003) Whales before whaling in the North Atlantic. *Science* **301**: 508–510.

Ronen, D., Magaritz, M. & Almon, E. (1988) Contaminated aquifers are a forgotten component of the global N$_2$O budget. *Nature* **335**: 57–59.

Roosevelt, A. C. (1999) Twelve thousand years of human–environment interaction in the Amazon Floodplain. In: *Várzea: Diversity, Development, and Conservation of Amazonia's Whitewater Floodplains, Advances in Economic Botany*, vol. 13, ed. C. Padoch, J. M. Ayres, M. Pinedo-Vasquez, A. Henderson, pp. 371–392. New York, USA: New York Botanical Garden Press.

Rose, C. S. & Risk, M. J. (1985) Increase in *Cliona delitrix* infestation of *Montastrea cavernosa* heads on an organically polluted portion of the Grand Cayman fringing reef. *Marine Ecology* **6**: 345–363.

Rose, K. A. (2000) Why are quantitative relationships between environmental quality of fish populations so elusive? *Ecological Applications* **10**: 367–385.

Rose, N., Appleby, P., Boyle, J., MacKay, A. & Flower, R. (1998) The spatial and temporal distribution of fossil-fuel derived pollutants in the sediment record of Lake Baikal. *Journal of Paleolimnology* **20**: 151–162.

Rosegrant, M., Cai, X. & Cline, S. (2002) *World Water and Food to 2025: Dealing with Scarcity*. Washington, DC, USA: International Food Policy Research Institute. Available online at http://www.ifpri.org/pubs

Rosenzweig, M. L. (2001) The four questions: what does the introduction of exotic species do to diversity? *Evolutionary Ecology Research* **3**: 361–367.

Rothgeb, J. M. (1996) *Foreign Investment and Political Conflict in Developing Countries*. Westport, CT, USA: Greenwood Press.

Rothrock, D. A., Yu, Y. & Maykut, G. A. (1999) Thinning of the Arctic sea ice cover. *Geophysical Research Letters* **26**: 3469–3472.

Rothrock, D. A., Zhang, J. & Yu, Y. (2003) The Arctic ice thickness anomaly of the 1990s: a consistent view from observations and models. *Journal of Geophysical Research* **108**: 3083.

Roughgarden, J., Gaines, S. & Possingham, H. (1988) Recruitment dynamics in complex life cycles. *Science* **241**: 1460–1466.

Rowan, D. & Rasmussen, J. (1992) Why don't Great Lakes fish reflect environmental concentrations of organic contaminants? An analysis of between lake variability in the ecological partitioning of PCBs and DDT. *Journal of Great Lakes Research* **18**: 724–741.

Rowe, G. T. (1971) Benthic biomass and surface productivity. In: *Fertility of the Sea*, ed. J. D. Costlow, pp. 441–454. New York, USA: Gordon & Breach.

RSPB (2003) *The RSPB's 'No Airport at Cliffe' Campaign*. Sandy, UK: Royal Society for the Protection of Birds.

Ruck, K. R. & Cook, P. A. (1998) Sabellid infestations in the shells of South African molluscs: implications for mariculture. *Journal of Shellfish Research* 17: 693–699.

Ruesink, J. L. & Srivastava, D. S. (2001) Numerical and per capita responses to species loss: mechanisms maintaining ecosystem function in a community of stream insect detritivores. *Oikos* 93: 221–234.

Ruiz, G. M. & Carlton, J. T. (2003) *Invasive Species: Vector and Management Strategies.* Washington, DC, USA: Island Press.

Ruiz, G. M., Fofonoff, P., Hines, A. H. & Grosholz, E. D. (1999) Non-indigenous species as stressors in estuarine and marine communities: assessing invasion impacts and interactions. *Limnology and Oceanography* 44: 950–972.

Ruiz-Luna, A. & Berlanga-Robles, C. A. (2003) Land use, land cover changes and coastal lagoon surface reduction associated with urban growth in northwest Mexico. *Landscape Ecology* 18: 159–171.

Rumbold, D. G., Davis, P. W. & Perretta, C. (2001) Estimating the effect of beach nourishment on *Caretta caretta* (loggerhead sea turtle) nesting. *Restoration Ecology* 9: 304–310.

Rundle, S. D., Bilton, D. T. & Shiozawa, D. K. (2000) Global and regional patterns in lotic meiofauna. *Freshwater Biology* 44: 123–134.

Russ, G. R. (2004) Marine reserves: long-term protection is required for full recovery of predatory fish populations. *Oecologia* 138: 622–627.

Russ, G. R. & Zeller, D. (2003) From *Mare Liberum* to *Mare Reservarum*. *Marine Policy* 27: 75–78.

Ryan, P. (1988) The characteristics and distribution of plastic particles on the sea surface off the southwestern Cape Province, South Africa. *Marine Environmental Research* 25: 249–273.

Rydin, H. & Clymo, R. S. (1989) Transport of carbon and phosphorus compounds about *Sphagnum. Proceedings of the Royal Society of London B* 237: 63–84.

Rysgaard, S., Nielsen, T. G. & Hansen, B. W. (1999) Seasonal variation in nutrients, pelagic primary production and grazing in a high-Arctic coastal marine ecosystem, Young Sound, Northeast Greenland. *Marine Ecology Progress Series* 179: 13–25.

Ryther, J. (1969) Photosynthesis and fish production in the sea. *Science* 166: 72–76.

Ryther, J. H. & Dunstan, W. N. (1971) Nitrogen, phosphorus and eutrophication in the coastal marine environment. *Science* 171: 1008–1013.

Sabo, J. L., Sponseller, R., Dixon, M., *et al.* (2005) Riparian zones increase regional species diversity by harboring different, not more, species. *Ecology* 86: 56–62.

Sadovy, Y. & Cheung, W. L. (2003) Near extinction of a highly fecund fish: the one that nearly got away. *Fish and Fisheries* 4: 86–99.

Saenger, P. (1998) Mangrove vegetation: an evolutionary perspective. *Marine and Freshwater Research* 49: 277–286.

Saenger, P. (2002) *Mangrove Ecology, Silviculture and Conservation.* Dordrecht, the Netherlands: Kluwer.

Saenger, P., Hegerl, E. J. & Davie, J. D. S. (1983) Global status of mangrove ecosystems. *Environmentalist* 3(Suppl.): 1–88.

Sainsbury, K. J. (1988) The ecological basis of multispecies fisheries and management of a demersal fishery in tropical Australia. In: *Fish Population Dynamics*, ed. J. Gulland, pp. 349–382. Chichester, UK: John Wiley.

Saintilan, N. & Williams, R. J. (1999) Mangrove transgression into saltmarsh environments in south-east Australia. *Global Ecology and Biogeography* 8: 117–124.

Saji, N. H., Goswami, B. N., Vinayachandran, P. N. & Yamagata, T. (1999) A dipole mode in the tropical Indian Ocean. *Nature* 401: 360–363.

Sakaguchi, Y. (1986) *Rivers in Japan.* Tokyo, Japan: Iwanami Shoten (in Japanese).

Sala, E., Boudouresque, C. F. & Harmelin-Viven, M. (1998) Fishing, trophic cascades, and the structure of algal assemblages: evaluation of an old but untested paradigm. *Oikos* 82: 425–439.

Sala, O. E., Chapin III, F. S., Armesto, J. J., *et al.* (2000) Global biodiversity scenarios for the year 2100. *Science* 287: 1770–1774.

Salama, R. B., Farrington, P., Bartle, G. A. & Watson, G. D. (1993) Salinity trends in the wheatbelt of Western Australia: results of water and salt balance studies from Cubaling catchment. *Journal of Hydrology* 145: 41–63.

Salih, A., Larkum, A., Cox, G., Kuhl, M. & Hoegh-Guldberg, O. (2000) Fluorescent pigments in corals are photoprotective. *Nature* 408: 850–853.

Salm, R. V. & Coles, S. L. (2001) *Coral Bleaching and Marine Protected Areas.* Honolulu, HI, USA: The Nature Conservancy, Asia Pacific Coastal Marine Program.

Saltonstall, K. (2002) Cryptic invasion by a non-native genotype of the common reed, *Phragmites australis*, into North America. *Proceedings of the National Academy of Sciences of the USA* 99: 2445–2449.

Saltonstall, K. (2003) Genetic variation among North American populations of *Phragmites australis*: implications for management. *Estuaries* 26: 444–451.

Sammarco, P. W. (1982) Echinoid grazing as a structuring force in coral communities: whole reef manipulations. *Journal of Experimental Marine Biology and Ecology* **61**: 31–35.

Sammarco, P. W. (1996) Comments on coral reef regeneration, bioerosion, biogeography, and chemical ecology: future directions. *Journal of Experimental Marine Biology and Ecology* **200**: 135–168.

Sammarco, P. W., Risk, M. J. & Rose, C. (1987) Effects of grazing and damselfish territoriality on internal bioerosion of dead corals: indirect effects. *Journal of Experimental Marine Biology and Ecology* **112**: 185–199.

Sampat, P. (2000) *Deep Trouble: The Hidden Threat of Groundwater Pollution*, World Watch Paper No. 154. Washington, DC, USA: World Watch Institute.

San Francisco Estuary Project (1998) *State of the Estuary: 1992–1997*, Technical Report. Oakland, CA, USA: San Francisco Estuary Project.

Sánchez, R. (2002) The role of science in an unsteady, market-driven fishery: the Patagonian case. In: *Oceans 2020: Science for Future Needs*, ed. J. G. Field, G. Hempel & C. P. Summerhayes, pp. 114–116. Washington, DC, USA: Island Press.

Sandalow, D. B. & Bowles, I. A. (2001) Fundamentals of treaty-making on climate change. *Science* **292**: 1839–1840.

Sanderson, E. W., Jaitheh, M., Levy, M. A., *et al.* (2002) The human footprint and the last of the wild. *BioScience* **52**: 891–904.

Santelices, B. & Ojeda, F. P. (1984*a*) Effects of canopy removal on the understory algal community structure of coastal forests of *Macrocystis pyrifera* from southern South America. *Marine Ecology Progress Series* **14**: 165–173.

Santelices, B. & Ojeda, F. P. (1984*b*) Population dynamics of coastal forests of *Macrocystis pyrifera* in Puerto Toro, Isla Navarino, Southern Chile. *Marine Ecology Progress Series* **14**: 175–183.

Sargeant, H. J. (2001) *Vegetation Fires in Sumatra, Indonesia: Oil Palm Agriculture in the Wetlands of Sumatra – Destruction or Development?*, European Union Forest Fire Prevention and Control Project with Dinas Kehutanan Propinsi Sumatera Selatan. Jakarta, Indonesia: European Union and Ministry of Forestry.

Sax, D. F. & Gaines, S. D. (2003) Species diversity: from general decrease to local increase. *Trends in Ecology and Evolution* **18**: 561–566.

Sayce, K. (1988) Introduced cordgrass *Spartina alterniflora* Loisel. In: *Salt Marshes and Tidelands of Willipa Bay, Washington*, Final Report, US Fish and Wildlife Service, USFWS FWSI-87058 TS. Ilwaco, WA: National Wildlife Réfuge.

Scatolini, S. R. & Zedler, J. R. (1996) Epibenthic invertebrates of natural and constructed marshes of San Diego Bay. *Wetlands* **16**: 24–37.

Scheffer, M., Carpenter, S., Foley, J. A., Folke, C. & Walter, B. (2001) Catastrophic shifts in ecosystems. *Nature* **413**: 591–596.

Scheibling, R. E., Hennigar, A. W. & Balch, T. (1999) Destructive grazing, epiphytism, and disease: the dynamics of sea urchin–kelp interactions in Nova Scotia. *Canadian Journal of Fisheries and Aquatic Sciences* **56**: 1–15.

Scherrer, P. & Mille, G. (1989) Biodegradation of crude oil in an experimentally polluted peaty mangrove soil. *Marine Pollution Bulletin* **20**: 430–432.

Schiel, D. R. (1990) Macroalgal assemblages in New Zealand: structure, interactions and demography. *Hydrobiologia* **192**: 59–76.

Schiel, D. R., Andrew, N. L. & Foster, M. S. (1995) The structure of subtidal algal and invertebrate assemblages at the Chatham Islands, New Zealand. *Marine Biology* **123**: 355–367.

Schimel, D. S. (1995) Terrestrial ecosystems and the carbon cycle. *Global Change Biology* **1**: 77–91.

Schimel, D. S. & Baker, D. (2002) The wildfire factor. *Nature* **420**: 29–30.

Schindler, D. W. (2001) The cumulative effects of climate warming and other human stresses on Canadian freshwaters in the new millennium. *Canadian Journal of Fisheries and Aquatic Sciences* **58**: 18–29.

Schindler, D. W., Curtis, P. J., Parker, B. R. & Stainton, M. P. (1996) Consequences of climate warming and lake acidification for UV-B penetration in North American boreal lakes. *Nature* **379**: 705–708.

Schlesinger, W. H. (1997) *Biogeochemistry: An Analysis of Global Change*, 2nd edn. New York, USA: Academic Press.

Schleyer, M. & Celliers, L. (2002) A consideration of biodiversity and future of southern African coral reefs. In: *Coral Reef Degradation in the Indian Ocean: Status Report 2002*, ed. O. Linden, D. Souter, D. Wilhelmsson & D. Obura. pp. 83–90. Kalmar, Sweden: CORDIO.

Schmidt, T. C., Morgenroth, E., Schirmer, M., Effenberger, M. & Haderlein, S. B. (2001) Use and occurrence of fuel oxygenates in Europe. In: *Oxygenates in Gasoline: Environmental Aspects*, ed. A. F. Diaz & D. L. Drogos, pp. 58–79. Washington, DC, USA: American Chemical Society.

Schmitz, D. C. & Simberloff, D. (1997) Biological invasions: a growing threat. *Issues in Science and Technology* **13**: 33–40.

Schneider, G. & Van Dijk, D. (1997) Environmental management measures for preventing further deterioration of African freshwater bodies: the case of the Lake Victoria Basin. In: *African Inland Fisheries, Aquaculture and the Environment*, ed. K. Remane, pp. 296–304. Oxford, UK: Fishing News Books.

Schneider, S. H. (1992) Introduction to climate modeling. In: *Climate System Modeling*, ed. K. Trenberth, pp. 3–26. Cambridge, UK: Cambridge University Press.

Schnitter, J. J. (1994) *A History of Dams: The Useful Pyramids*. Rotterdam, the Netherlands: A.A. Balkema.

Schofield, N. J., Stoneman, G. L. & Loh, I. C. (1989) Hydrology of the jarrah forest. In: *The Jarrah Forest: A Complex Mediterranean Ecosystem*, ed. B. Dell, J. J. Havel & N. Malajczuk, pp. 179–201. Dordrecht, the Netherlands: Kluwer.

Schofield, P. J. & Chapman, L. J. (1999) Interactions between Nile perch, *Lates niloticus*, and other fishes in Lake Nabugabo, Uganda. *Environmental Biology of Fishes* 55: 343–358.

Schonbeck, M. & Norton, T. (1978) Factors controlling the upper limits of fucoid algae. *Journal of Experimental Marine Biology and Ecology* 31: 303–313.

Schramm, W. (1996) The Baltic Sea and its transition zones. In: *Marine Benthic Vegetation: Recent Changes and the Effects of Eutrophication*, ed. W. Schramm & P. Nienhuis, pp. 131–164. Berlin, Germany: Springer-Verlag.

Schroeder, R. A., Orem, W. H. & Kharaka, Y. K. (2002) Chemical evolution of the Salton Sea, California: nutrient and selenium dynamics. *Hydrobiologia* 473: 23–45.

Schroeder, R. A., Setmire, J. G. & Wolfe, J. C. (1988) Trace elements and pesticides in the Salton Sea area, California. In: *Proceedings of Conference on Planning Now for Irrigation and Drainage*, Lincoln, NE, 19–21 July, pp. 700–707. New York, USA: American Society of Civil Engineers.

Schuijt, K. (2002) *Land and Water Use of Wetlands in Africa: Economic Value of African Wetlands*. Laxenburg, Austria: Institute for Applied System Analyses.

Schulte-Wülwer-Leidig, A. (1995) Ecological master plan for the Rhine catchment. In: *The Ecological Basis for River Management*, ed. D. M. Harper & A. J. D. Fergusen, pp. 505–514. Chichester, UK: John Wiley.

Schultze, E.-D. & Mooney, H. A., eds. (1993) *Biodiversity and Ecosystem Function*. Berlin, Germany: Springer-Verlag.

Schwartzlose, R. A., Alheit, J., Bakun, A., *et al.* (1999) Worldwide large-scale fluctuations of sardine and anchovy populations. *South African Journal of Marine Science* 21: 289–347.

Scott, D. A. (1992) Asia and the Middle East. In: *Wetlands*, ed. M. Finlayson & M. Moser, pp. 149–178. Oxford, UK: Facts on File.

Scott, D. A. (1993) Wetlands of West Asia: a regional overview. In: *Wetland and Waterfowl Conservation in South and West Asia, Proceedings of an International Symposium*, ed. M. Moser & J. van Vessem, pp. 9–22. Karachi, Pakistan: International Waterfowl and Wetlands Research Bureau.

Scott, D. A., ed. (1995) *A Directory of Wetlands in the Middle East*. Gland, Switzerland: IUCN, and Slimbridge, UK: IWRB.

Scott, D. A. & Jones, T. A. (1995) Classification and inventory of wetlands: a global overview. *Vegetatio* 118: 3–16.

Scott, M. L. & Auble, G. T. (2002) Conservation and restoration of semiarid riparian forests: a case study from the Upper Missouri River, Montana. In: *Flood Pulsing and Wetlands: Restoring the Natural Balance*, ed. B. A. Middleton, pp. 145–190. New York, USA: John Wiley.

Sculthorpe, C. D. (1968) *Biology of Aquatic Vascular Plants*. London, UK: Edward Arnold.

Seabloom, E. W. & Van der Valk, A. G. (2003) Plant diversity, composition, and invasion of restored and natural prairie pothole wetlands: implications for restoration. *Wetlands* 23: 1–12.

Seckler, D., Upali, A., Molden, D., de Silva, R. & Barker, R. (1998) *World Water Demand and Supply, 1990 to 2025: Scenarios and Issues*, Research Report No. 19. Colombo, Sri Lanka: International Water Management Institute.

Sedell, J. R., Steedman, R. J., Regier, H. A. & Gregory, S. V. (1991) Restoration of human impacted land–water ecotones. In: *Ecotones: The Role of Landscape Boundaries in the Management and Restoration of Changing Environments*, ed. M. M. Holland, P. G. Risser & R. J. Naiman, pp. 105–129. New York, USA: Chapman & Hall.

Seehausen, O., Van Aphen, J. & White, F. (1997) Cichlid fish diversity threatened by eutrophication that curbs sexual selection. *Science* 277: 1808–1811.

Seidenschwarz, I. A., ed. (1986) *Pioniervegetation von Flussufer und Kulturland im vorandinen Amazonasgebiet Perus: Ein pflanzensoziologischer Vergleich von vorandinem Flussufer und Kulturland*. Langen, Germany: TRIOPS Verlag.

Seigal, P. R. & Wenner, A. M. (1984) Abnormal reproduction of the sand crab *Emerita analoga* in the vicinity of a nuclear generating station in Southern California. *Marine Biology* 80: 341–345.

Seitzinger, S. P., Kroeze, C., Bouwman, A. F., *et al.* (2002) Global patterns of dissolved inorganic and particulate nitrogen inputs to coastal systems: recent conditions and future projections. *Estuaries* 25: 640–655.

Seki, M. (1994) *Rivers of the Earth*. Tokyo, Japan: Soshisya (in Japanese).

Serreze, M. C., Maslanik, J. A., Key, J. R., Kokaly, R. F. & Robinson, D. A. (1995) Diagnosis of the record minimum in Arctic sea ice area during 1990 and associated snow cover extremes. *Geophysical Research Letters* **22**: 2183–2186.

Serreze, M. C., Walsh, J. E., Chapin, F. S., *et al.* (2000) Observational evidence of recent change in the northern high-latitude environment. *Climatic Change* **46**: 159–207.

Serruya, C. & Pollingher, U. (1983) *Lakes of the Warm Belt*. Cambridge, UK: Cambridge University Press.

Servant-Vildary, S., Servant, M. & Jimenez, O. (2001) Holocene hydrological and climatic changes in the southern Bolivian Altiplano according to diatom assemblages in paleowetlands. *Hydrobiologia* **466**: 267–277.

Seymour, E. A. & Bergheim, A. (1991) Towards a reduction of pollution from intensive aquaculture with reference to the farming of salmonids in Norway. *Aquacultural Engineering* **10**: 73–88.

Seymour, R. J., Tegner, M. J., Dayton, P. K. & Parnell, P. E. (1989) Storm wave induced mortality of giant kelp, *Macrocystis pyrifera*, in Southern California. *Estuarine, Coastal and Shelf Science* **28**: 277–292.

Sægrov, H., Hobæk, A. & Lábe-Lund, H. H. (1996) Vulnerability of melanic *Daphnia* to brown trout predation. *Journal of Plankton Research* **18**: 2113–2118.

Shah, T., Molden, D., Sakthvadivel, R. & Seckler, D. (2000) *The Global Groundwater Situation: Overview of Opportunities and Challenges*. Colombo, Sri Lanka: International Water Management Institute. Available online at www.cgiar.org/iwmi/pubs/WWVisn/GrWater.htm

Shank, T. M., Fornari, D. J., Von Damm, K. L., *et al.* (1998) Temporal and spatial patterns of biological community development at nascent deep-sea hydrothermal vents (9° 50′ N, East Pacific Rise). *Deep-Sea Research II* **45**: 465–515.

Shannon, L. V., Lutjeharms, J. R. E. & Nelson, G. (1990) Causative mechanisms for intra- and interannual variability in the marine environment around southern Africa. *South African Journal of Science* **86**: 356–373.

Sharma, R., Nagender Nath, B., Parthiban, S. & Jai Sankar, S. (2001) Sediment redistribution during simulated benthic disturbance and its implications on deep seabed mining. *Deep-Sea Research I* **48**: 3363–3380.

Shatalov, V. (2001) The floodplain forests in the European part of the Russian federation. In: *The Floodplain Forests in Europe*, ed. E. Klimo & H. Hager, pp. 185–201. Leiden, the Netherlands: E. J. Brill.

Shaw, A. J. (2000) Phylogeny of the Sphagnopsida based on chloroplast and nuclear DNA sequences. *Bryologist* **103**: 277–306.

Shaw, D. G. (1992) The Exxon Valdez oil-spill: ecological and social consequences. *Environmental Conservation* **19**: 253–258.

Shaw, D. G. & Day, R. H. (1994) Color- and form-dependent loss of plastic micro-debris from the North Pacific Ocean. *Marine Pollution Bulletin* **28**: 39–45.

Sheaffer, J. R., Mullan, J. D. & Hinch, N. B. (2002) Encouraging wise use of floodplains with market-based incentives. *Environment* **44**: 33–43.

Shears, N. I. & Babcock, R. I. (2002) Marine reserves demonstrate top-down control of community structure on temperate reefs. *Oecologia* **132**: 131–142.

Sheppard, C. R. C. (2000) The Chagos Archipelago. In: *Coral Reefs of the Indian Ocean: Their Ecology and Conservation*, ed. T. R. McClanahan, C. R. C. Sheppard & D. O. Obura, pp. 445–470. New York, USA: Oxford University Press.

Sherman, D. J., Barron, K. M. & Ellis, J. T. (2002) Retention of beach sands by dams and debris basins in Southern California. *Journal of Coastal Research* Special Issue **36**: 662–674.

Sherman, D. J., Nordstrom, K. F., Jackson, N. L. & Allen, J. R. (1994) Sediment mixing depths on a low-energy reflective beach. *Journal of Coastal Research* **10**: 297–305.

Sherman, K. & Alexander, L. M., eds. (1986) *Variability and Management of Large Marine Ecosystems*. Boulder, CO, USA: Westview Press.

Sherman, K., Sissenwine, M., Christensen, V., *et al.* (2005) A global movement toward an ecosystem approach to management of marine resources. *Marine Ecology Progress Series* **300**: 275–279.

Shick, J. M., Lesser, M. P. & Jokiel, P. L. (1996) Effects of ultraviolet radiation on corals and other coral reef organisms. *Global Change Biology* **2**: 527–545.

Shick, J. M., Romaine-Lioud, S., Ferrier-Pages, C. & Gattuso, J. P. (1999) Ultraviolet-B radiation stimulates shikimate pathway-dependent accumulation of mycosporine-like amino acids in the coral *Stylophora pistillata* despite decreases in its population of symbiotic dinoflagellates. *Limnology and Oceanography* **44**: 1667–1682.

Shiklomanov, I. A. (1990) Global water resources. *Natural Resources* **26**: 34–43.

Shiklomanov, I. A. (1998) *World Water Resources: A New Appraisal and Assessment for the Twenty-First Century*. Paris, France: UNESCO.

Shiklomanov, I. A., ed. (1999) *World Water Resources: Modern Assessment and Outlook for the Twenty-First Century* (summary of the monograph *World Water Resources at the Beginning of the Twenty-First Century* (Shiklomanov & Rodda (2003), prepared in the framework of IHP-UNESCO). St Petersburg, Russia: Federal Service of Russia for Hydrometeorology and Environmental Monitoring, State Hydrological Institute.

Shiklomanov, I. A. (2000) World water resources and water use: present assessment and outlook for 2025. In: *Water Scenarios*, ed. F. R. Rijsberman, pp. 160–203. London, UK: Earthscan Publications.

Shiklomanov, I. A. & Rodda, J. C., eds. (2003) *World Water Resources at the Beginning of the Twenty-First Century*. Cambridge, UK: Cambridge University Press.

Shindell, D. T., Miller, R. L., Schmidt, G. A. & Pandolfo, L. (1999) Simulation of recent northern winter climate trends by greenhouse-gas forcing. *Nature* **399**: 452–455.

Shine, C. & de Klemm, C. (1999) *Wetlands, Water and the Law: Using Law to Advance Wetland Conservation and Wise Use*. Gland, Switzerland: IUCN.

Shinn, E. A., Reich, C. D., Hickey, T. D. & Lidz, B. H. (2003) Staghorn tempestites in the Florida Keys. *Coral Reefs* **22**: 91–97.

Shinn, E. A., Smith, G. W., Prospero, J. M., *et al.* (2000) African dust and the demise of Caribbean coral reefs. *Geophysical Research Letters* **27**: 3029–3032.

Shipley, F. S. & Kiesling, R. W., eds. (1994) *The State of the Bay: A Characterization of the Galveston Bay Ecosystem*, Publication No. GBNEP-44. Webster, TX, USA: Galveston Bay National Estuary Program.

Shireman, J. V. & Cichra, C. E. (1994) Evaluation of aquaculture effluents. *Aquaculture* **123**: 55–68.

Short, A. D. (1996) The role of wave height, period, slope, tide range and embaymentisation in beach classification: a review. *Revista Chilena de Historia Natural* **69**: 589–604.

Short, A. D. & Hesp, P. A. (1982) Wave, beach and dune interactions in southeastern Australia. *Marine Geology* **48**: 259–284.

Short, F. T. & Burdick, D. M. (1996) Quantifying eelgrass habitat loss in relation to housing development and nitrogen loading in Waquoit Bay, Massachusetts. *Estuaries* **19**: 730–739.

Short, F. T. & Neckles, H. (1999) The effects of global climate change on seagrasses. *Aquatic Botany* **63**: 169–196.

Short, F. T. & Short, C. A. (1984) The seagrass filter: purification of coastal water. In: *The Estuary as a Filter*, ed. V. S. Kennedy, pp. 395–413. New York, USA: Academic Press.

Short, F. T. & Short, C. A. (2003) Seagrasses of the western North Atlantic. In: *World Atlas of Seagrasses*, ed. E. P. Green & F. T. Short, pp. 225–233. Berkeley, CA, USA: University of California Press.

Short, F. T. & Wyllie-Echeverria, S. (1996) Natural and human-induced disturbances of seagrass. *Environmental Conservation* **23**: 17–27.

Short, F. T. & Wyllie-Echeverria, S. (2000) Global seagrass declines and effect of climate change. In: *Seas at the Millennium: An Environmental Evaluation*, vol. 3, ed. C. R. C. Sheppard, pp. 10–11. Amsterdam, the Netherlands: Elsevier.

Short, F. T., Burdick, D. M., Granger, S. & Nixon, S. W. (1996) Long-term decline in eelgrass, *Zostera marina* L., linked to increased housing development. In: *Seagrass Biology: Proceedings of an International Workshop*, Rottnest Island, Western Australia, 25–29 January 1996, ed. J. Kuo, R. C. Phillips, D. I. Walker & H. Kirkman, pp. 291–298. Nedlands, WA, Australia: Sciences UWA.

Short, F. T., Burdick, D. M., Short, C. A., Davis, R. C. & Morgan, P. A. (2000) Developing success criteria for restored eelgrass, salt marsh and mud flat habitats. *Ecological Engineering* **15**: 239–252.

Short, F. T., Davis, R. C., Kopp, B. S., Short, C. A. & Burdick, D. M. (2002*a*) Site selection model for optimal restoration of eelgrass, *Zostera marina* L. *Marine Ecology Progress Series* **227**: 253–267.

Short, F. T., Ibelings, B. W. & den Hartog, C. (1988) Comparison of a current eelgrass disease to the wasting disease of the 1930s. *Aquatic Botany* **30**: 295–304.

Short, F. T., Kopp, B. S., Gaeckle, J. & Tamaki, H. (2002*b*) Seagrass ecology and estuarine mitigation: a low-cost method for eelgrass restoration. *Japan Fisheries Science* **68**: 1759–1762.

Short, F. T., Mathieson, A. C. & Nelson, J. I. (1986) Recurrence of the eelgrass wasting disease at the border of New Hampshire and Maine, USA. *Marine Ecology Progress Series* **29**: 89–92.

Short, F. T., McKenzie, L. J., Coles, R. G. & Vidler, K. P. (2002*c*) *SeagrassNet Manual for Scientific Monitoring of Seagrass Habitat*. Cairns, Queensland, Australia: Queensland Department of Primary Industries. Available online at www.SeagrassNet.org

Shpigel, M. & Fishelson, L. (1991) Experimental removal of piscivorous groupers of the genus *Cephalopholis* (Serranidae) from coral habitats in the Gulf of Aqaba (Red Sea). *Environmental Biology of Fishes* **31**: 131–138.

Shuford, W. D., Warnock, N., Molina, K. C. & Sturm, K. K. (2002) The Salton Sea as critical habitat to migratory and resident waterbirds. *Hydrobiologia* **473**: 255–274.

Shuisky, Y. D. & Schwartz, M. L. (1988) Human impact and rates of shoreline retreat along the Black Sea coast. *Journal of Coastal Research* **4**: 405–416.

Shukla, J. B. & Dubey, B. (1996) Effect of changing habitat on species: application to Keoladeo National Park, India. *Ecological Modelling* **86**: 91–99.

Shy, T. L. & Walsh, J. E. (1996) North Pole ice thickness and association with ice motion history. *Geophysical Research Letters* **23**: 2975–2978.

Sidaway, R. (1991) *A Review of Marina Developments in Southern England.* Sandy, UK: Royal Society for the Protection of Birds.

Siebeck, O. (1988) Experimental investigations of UV tolerance in hermatypic corals (Scleractinia). *Marine Ecology Progress Series* **43**: 95–103.

Sienkiewicz, J., Kloss, M. & Grzyb, M. (2001) The floodplain forest ecosystem in Poland. In: *The Floodplain Forests in Europe*, ed. E. Klimo & H. Hager, pp. 249–297. Leiden, the Netherlands: E. J. Brill.

Silliman, B. R. & Bertness, M. D. (2002) A trophic cascade regulates salt marsh primary production. *Proceedings of the National Academy of Sciences of the USA* **99**: 10 500–10 505.

Simas, T., Nunes, J. P. & Ferreira, J. G. (2001) Effects of global climate change on coastal salt marshes. *Ecological Modelling* **139**: 1–15.

Simberloff, D. (2003) How much information on population biology is needed to manage introduced species? *Conservation Biology* **17**: 83–92.

Simberloff, D. & von Holle, B. (1999) Positive interactions of non-indigenous species: invasional meltdown? *Biological Invasions* **1**: 21–32.

Simenstad, C. A. & Fresh, K. L. (1995) Influence of intertidal aquaculture on benthic communities in Pacific northwest estuaries: scales of disturbance. *Estuaries* **18**: 43–70.

Simenstad, C. A., Estes, J. A. & Kenyon, K. W. (1978) Aleuts, sea otters, and alternate stable-state communities. *Science* **200**: 403–411.

Simmonds, I., Jones, D. A. & Walland, D. J. (1998) Multi-decadal climate variability in the Antarctic region and global change. *Annals of Glaciology* **27**: 617–622.

Simon, K. & Townsend, C. R. (2003) The impacts of freshwater invaders at different levels of ecological organization, with emphasis on ecosystem consequences. *Freshwater Biology* **48**: 982–994.

Simonovic, S. P. & Carson, R. W. (2003) Flooding in the Red River basin: lessons from post flood activities. *Natural Hazards* **28**: 345–365.

Sims, N. C. & Thoms, M. C. (2002) What happens when floodplains wet themselves: vegetation response to inundation on the Lower Balonne Floodplain. *International Association of Hydrological Sciences* **276**: 226–236.

Sippel, S. J., Hamilton, S. K., Melack, J. M. & Novo, E. M. M. (1998) Passive microwave observations of inundation area and the area/stage relation in the Amazon River floodplain. *International Journal of Remote Sensing* **19**: 3055–3074.

Sissenwine, M. P. & Rosenberg, A. A. (1996) Marine fisheries at a critical juncture. In: *Oceanography: Contemporary Readings in Ocean Sciences*, 3rd edn, ed. R. G. Pirie, pp. 293–302. New York, USA: Oxford University Press.

Sissenwine, M. P., Cohen, E. B. & Grosslein, M. D. (1984) Structure of the Georges Bank ecosystem, Rapport et procès-verbaux des réunions. *Journal du Conseil International pour l'Exploration de la Mer* **183**: 243–254.

Sival, F. P., Grootjans, A. P., Stuyfzand, P. J. & Verschoore de la Houssaye, T. (1997) Variation in groundwater composition and decalcification depth in a dune-slack: effects on basiphilous vegetation. *Journal of Coastal Conservation* **3**: 79–86.

Sivertsen, K. (1997) Geographic and environmental factors affecting the distribution of kelp beds and barren grounds and changes in biota associated with kelp reduction at sites along the Norwegian coast. *Canadian Journal of Fisheries and Aquatic Sciences* **54**: 2872–2887.

Sket, B. (1977) Gegenseitige Beeinflussung der Wasserpollution und des Hoehlenmilieus. *Proceedings of the 6th International Congress of Speleology*, Olomouc, Czechoslovakia, 1973, **5**: 253–262.

Sket, B. (1999) The nature of biodiversity in hypogean waters and how it is endangered. *Biodiversity and Conservation* **8**: 1319–1338.

Sloan, C. E. (1972) *Ground-Water Hydrology of Prairie Potholes in North Dakota*, US Geological Survey Professional Paper No. 585-C. Washington, DC, USA: US Government Printing Office.

Slootweg, R. & van Schooten, M. L. F. (1995) Partial restoration of floodplain functions at the village level: the experience of Gounougou, Benue Valley, Cameroon. In: *Tropical Freshwater Wetlands*, ed. H. Roggeri, pp. 159–166. Dordrecht, the Netherlands: Kluwer.

Small, M. P. & Gosling, E. M. (2001) Population genetics of a snail species complex in the British Isles: *Littorina saxatilis*

(Olivi), *L. neglecta* (Bean) and *L. tenebrosa* (Montagu), using SSCP analysis of cytochrome-B gene fragments. *Journal of Molluscan Studies* **67**: 69–80.

Smayda, T. J. (1997) Harmful algal blooms: their ecophysiology and general relevance to phytoplankton blooms in the sea. *Limnology and Oceanography* **42**: 1137–1153.

Smil, V. (1997) Global population and the nitrogen cycle. *Scientific American* **277**: 76–81.

Smith, A. B. (1992) *Pastoralism in Africa: Origins and Development Ecology*. London, UK: C. Hurst & Co.; and Athens, OH, USA: Ohio University Press.

Smith, A. H., Lopipero, P. A., Bates, M. N. & Steinmaus, C. M. (2002) Arsenic epidemiology and drinking water standards. *Science* **296**: 2145–2146.

Smith, C. R. (1994) Tempo and mode in deep-sea benthic ecology: punctuated equilibrium revisited. *Palaios* **9**: 3–13.

Smith, C. R. (1999) The biological environment in the nodule provinces of the deep sea. In: *Deep-Seabed Polymetallic Nodule Exploration: Development of Environmental Guidelines*, pp. 41–68. Kingston, Jamaica: International Seabed Authority.

Smith, C. R. & Baco, A. R. (2003) Ecology of whale falls at the deep-sea floor. *Oceanography and Marine Biology Annual Review* **41**: 311–354.

Smith, C. R. & Demopoulos, A. (2003) Ecology of the deep Pacific Ocean floor. In: *Ecosystems of the World*, vol. 28, *Ecosystems of the Deep Ocean*, ed. P. A. Tyler, pp. 179–218. Amsterdam, the Netherlands: Elsevier.

Smith, C. R. & Rabouille, C. (2002) What controls the mixed-layer depth in deep-sea sediments? The importance of POC flux. *Limnology and Oceanography* **47**: 418–426.

Smith, C. R., Berelson, W., Demaster, D. J., *et al.* (1997) Latitudinal variations in benthic processes in the abyssal equatorial Pacific: control by biogenic particle flux. *Deep-Sea Research II* **44**: 2295–2317.

Smith, C. R., Hoover, D. J., Doan, S. E., *et al.* (1996) Phytodetritus at the abyssal seafloor across 10° of latitude in the central equatorial Pacific. *Deep-Sea Research II* **43**: 1309–1338.

Smith, C. R., Levin, L. A., Hoover, D. J., McMurtry, G. & Gage, J. D. (2000) Variations in bioturbation across the oxygen minimum zone in the northwest Arabian Sea. *Deep-Sea Research II* **47**: 227–257.

Smith, C. R., Present, T. M. C. & Jumars, P. A. (1988) *Development of Benthic Biological Monitoring Criteria for Disposal of Low-Level Radioactive Waste in the Abyssal Deep Sea*, Final Report for EPA Contract

No. 68-02-4303. Washington, DC, USA: US Environmental Protection Agency.

Smith, D. C. & Stewart, B. D. (1994) *Development of Methods to Age Commercially Important Dories and Oreos*, Fisheries Research and Development Corporation Final Report. Queenscliff, Victoria, Australia: Victorian Fisheries Research Institute.

Smith, D. K. (1991) Seamount abundances and size distributions, and their geographic variations. *Reviews in Aquatic Sciences* **5**: 197–210.

Smith, J. E. (1968) Torrey Canyon *Pollution and Marine Life*. Cambridge, UK: Cambridge University Press.

Smith, K. L. & Druffel, E. R. M. (1998) Long time-series monitoring of an abyssal site in the NE Pacific: an introduction. *Deep-Sea Reseach II* **45**: 573.

Smith, K. L. & Kaufmann, R. S. (1999) Long-term discrepancy between food supply and demand in the deep eastern north Pacific. *Science* **284**: 1174–1177.

Smith, K. L., Kaufmann, R. S., Baldwin, R. J. & Carlucci, A. F. (2001) Pelagic–benthic coupling in the abyssal eastern North Pacific: an 8-year time-series study of food supply and demand. *Limnology and Oceanography* **46**: 543–556.

Smith, M. & Mettler, P. (2002) The role of the flood pulse in maintaining *Boltonia decurrens*, a fugitive plant species of the Illinois River floodplain: a case history of a threatened species. In: *Flood Pulsing in Wetlands: Restoring the Natural Balance*, ed. B. A. Middleton, pp. 109–144. New York, USA: John Wiley.

Smith, R. (1872) *Air and Rain*. London, UK: Longman, Green.

Smith, R. C., Aniley, D. G., Baker, K., *et al.* (1999) Marine ecosystem sensitivity to climate change. *BioScience* **49**: 393–404.

Smith, R. C., Baker, K. S., Byers, M. & Stammerjohn, S. E. (1998) Primary productivity of the Palmer Long Term Ecological Research area and the Southern Ocean. *Journal of Marine Systems* **17**: 245–259.

Smith, R. C., Fraser, W. R. & Stammerjohn, S. E. (2003*a*) Climate variability and ecological response of the marine ecosystem in the Western Antarctic Peninsula (WAP) region. In: *Climate Variability and Ecological Response*, ed. D. Greenland, D. Gooding & R. C. Smith, pp. 158–173. Oxford, UK: Oxford University Press.

Smith, R. C., Prezelin, B. B., Baker, K. S., *et al.* (1992) Ozone depletion: ultraviolet radiation and phytoplankton biology in Antarctic waters. *Science* **255**: 952–959.

Smith, R. C., Yuan, X., Liu, J., Martinson, D. G. & Stammerjohn, S. E. (2003*b*) The quasi-quintennial time scale climate variability and ecological response. In: *Climate Variability and Ecological Response*, ed. D. Greenland, D. Gooding & R. C. Smith, pp. 196–206. Oxford, UK: Oxford University Press.

Smith, R. D., Ammann, A., Bartoldus, C. & Brinson, M. M. (1995) *An Approach for Assessing Wetland Functions Using Hydrogeomorphic Classification, Reference Wetlands, and Functional Indices*, Technical Report No. TR-WRP-DE-9. Vicksburg, MS, USA: Waterways Experiment Station, Army Corps of Engineers. Available online at http://el.erdc.usace.army.mil/wetlands/pdfs/wrpde9.pdf

Smith, S. (1972) Factors of ecologic succession in oligotrophic fish communities of the Laurentian Great Lakes. *Journal of the Fisheries Research Board of Canada* 29: 717–730.

Smith, S. V. (1984) Phosphorus versus nitrogen limitation in the marine environment. *Limnology and Oceanography* 29: 1149–1160.

Smith, T. J. & Odum, W. E. (1981) The effects of grazing by snow geese on coastal salt marshes. *Ecology* 62: 98–106.

Smith III, T. J., Boto, K. G., Frusher, S. D. & Giddins, R. L. (1991) Keystone species and mangrove forest dynamics: the influence of burrowing by crabs on soil nutrient status and forest productivity. *Estuarine Coastal and Shelf Science* 33: 419–432.

Smith, V. H., Tilman, G. D. & Nekola, J. C. (1999) Eutrophication: impacts of excess nutrient inputs on freshwater, marine, and terrestrial ecosystems. *Environmental Pollution* 100: 179–196.

Smock, L. A., Smith, L. C., Jones, J. B. & Hooper, S. M. (1994) Effects of drought and a hurricane on a coastal headwater stream. *Archiv für Hydrobiologie* 131: 25–38.

Snedaker, S. C. & Araújo, R. J. (1998) Stomatal conductance and gas exchange in four species of Caribbean mangroves exposed to ambient and increased CO_2. *Marine and Freshwater Research* 49: 325–327.

Snedaker, S. C., Meeder, J. F., Ross, M. S. & Ford, R. G. (1994) Discussion of Ellison, J. C. and Stoddart, D. R. (1991) Mangrove ecosystem collapse during predicted sea-level rise: holocene analogues and implications. *Journal of Coastal Research* 10: 497–498.

Snelgrove, P. V. R. & Smith, C. R. (2002) A riot of species in an environmental calm: the paradox of the species-rich deep sea. *Oceanography and Marine Biology Annual Review* 40: 311–342.

Snow, A. A. & Vince, S. W. (1984) Plant zonation in an Alaskan salt marsh. II. An experimental study of the role of edaphic conditions. *Journal of Ecology* 72: 669–684.

Soares, A. G., Schlacher, T. A. & McLachlan, A. (1997) Carbon and nitrogen exchange between sandy beach clams (*Donax serra*) and kelp beds in the Benguela coastal upwelling region. *Marine Biology* 127: 657–664.

Somerville, R. C. J. & Remer, L. A. (1984). Cloud optical thickness feedbacks in the CO_2 climate problem. *Journal of Geophysical Research* 89: 9668–9672.

Sommaruga-Wögrath, S., Koinig, K. A., Schmidt, R., *et al.* (1997) Temperature effects on the acidity of remote alpine lakes. *Nature* 387: 64–67.

Song, M. & Dong, M. (2002) Clonal plants and plant species diversity in wetland ecosystems of China. *Journal of Vegetation Science* 13: 237–244.

Sophocleous, M. A., ed. (1998) Perspectives on sustainable development of water resources in Kansas. *Kansas Geological Survey Bulletin* 239.

Sophocleous, M. A. (2002) Interactions between groundwater and surface water: the state of the science. *Hydrogeology Journal* 10: 52–67.

Sophocleous, M. A. (2003) Environmental implications of intensive groundwater use with special regard to streams and wetlands. In: *Groundwater Intensive Use: Challenges and Opportunities*, ed. E. Custodio & R. Llamas, pp. 93–112. Lisse, the Netherlands: A.A. Balkema Publishers.

Soto, R. (1985) Efectos del fenómeno El Niño 1982–83. *Ecosistemas de la I Región: Investigación Pesquera (Chile)* 32: 199–206.

Soulsby, C., Turnbull, D., Langan, S. J., Owen, R. & Hirst, D. (1995) Long-term trends in stream chemistry and biology in Northeast Scotland: evidence for recovery. *Water, Air and Soil Pollution* 85: 689–694.

South Africa, Government of (1998) National Water Act 36. Available online at www.elaw.org/resources/text.asp?ID=1153

Southgate, T., Wilson, K., Myers, A. A. *et al.* (1984) Recolonization of a rocky shore in S. W. Ireland following a toxic bloom of the dinoflagellate *Gyrodinium aureolum*. *Journal of the Marine Biological Association of the United Kingdom* 64: 485–492.

Southward, A. J. (1967) Recent changes in abundance of intertidal barnacles in southwest England: a possible effect of climatic deterioration. *Journal of the Marine Biological Association of the United Kingdom* 47: 81–85.

Southward, A. J. (1980) Western English Channel: an inconstant ecosystem. *Nature* 285: 361–366.

Southward, A. J. (1982) An ecologist's view of the implications of the observed physiological and biochemical effects of petroleum compounds on marine organisms and ecosystems. *Philosophical Transactions of the Royal Society of London B* **297**: 241–255.

Southward, A. J. (1991) Forty years of changes in species composition and population density of barnacles on a rocky shore near Plymouth. *Journal of the Marine Biological Association of the United Kingdom* **71**: 495–513.

Southward, A. J. & Crisp, D. J. (1954) The distribution of certain intertidal animals around the Irish coast. *Proceedings of the Royal Irish Academy* **57B**: 1–29.

Southward, A. J. & Orton, J. H. (1954) The effects of wave-action on the distribution and numbers of the commoner plants and animals living on the Plymouth breakwater. *Journal of the Marine Biological Association of the United Kingdom* **33**: 1–19.

Southward, A. J. & Southward, E. C. (1978) Recolonization of rocky shores in Cornwall after use of toxic dispersants to clean up the *Torrey Canyon* spill. *Journal of the Fisheries Research Board of Canada* **35**: 682–706.

Southward, A. J., Hawkins, S. J. & Burrows, M. T. (1995) Seventy years of changes in distribution and abundance of zooplankton and intertidal organisms in the western English channel in relation to rising sea temperature. *Journal of Thermal Biology* **20**: 127–155.

Southward, A. J., Langmead, O., Hardman-Mountford, N. J., *et al.* (2005) Long-term oceanographic and ecological research in the western English Channel. *Advances in Marine Biology* **47**: 1–105.

Spalding, M., Taylor, M., Ravilious, C., Short, F. & Green, E. (2003) Global overview: the distribution and status of seagrasses. In: *World Atlas of Seagrasses*, ed. E. P. Green & F. T. Short, pp. 5–26. Berkeley, CA, USA: University of California Press.

Spalding, M. D., Blasco, F. & Field, C. D. (1997) *World Mangrove Atlas*. Okinawa, Japan: International Society for Mangrove Ecosystems.

Spalding, V. & Jackson, N. L. (2001) Effect of bulkheads on meiofaunal abundance in the foreshore of an estuarine sand beach. *Journal of Coastal Research* **17**: 363–370.

Spangler, D. P. (1984) Geologic variability among six cypress domes in North-Central Florida. In: *Cypress Swamps*, ed. K. C. Ewel & H. T. Odum, pp. 60–66. Gainesville, FL, USA: University of Florida Press.

Spanjol, Z., Tikvic, I. & Baricevic, D. (1999) Protected sites of floodplain forest ecosystems in the Republic of Croatia. *Ekológia (Bratislava)* **18**: 82–90.

Sparks, R. E., Nelson, J. C. & Yin, Y. (1998) Naturalization of the flood regime in regulated rivers. *BioScience* **48**: 706–720.

Spence, S. K., Bryan, G. W., Gibbs, P. E., *et al.* (1990) Effects of TBT contamination on *Nucella* populations. *Functional Ecology* **4**: 425–432.

Spencer, P. D. & Collie, J. S. (1997) Patterns of population variability in marine fish stocks. *Fisheries Oceanography* **6**: 188–204.

Squires, D., Grafton, R. Q., Alam, M. F. & Omar, M. H. (1998) *Where the Land Meets the Sea: Intergrated Sustainable Fisheries Development and Artisanal Fishing*, Department of Economics Discussion Paper No. 98–26. San Diego, CA, USA: University of California.

Stachowicz, J. J., Whitlatch, R. B. & Osman, R. W. (1999) Species diversity and invasion resistance in a marine ecosystem. *Science* **286**: 1577–1579.

Stammerjohn, S. E. & Smith, R. C. (1996) Spatial and temporal variability of western Antarctic Peninsula sea ice coverage. *Antarctic Research Series* **70**: 81–104.

Stammerjohn, S. E. & Smith, R. C. (1997) Opposing southern ocean climate patterns as revealed by trends in regional sea ice coverage. *Climatic Change* **37**: 617–639.

Stanford, J. A. & Ward, J. V. (1993) An ecosystem perspective of alluvial rivers: connectivity and the hyporheic corridor. *Journal of the North American Benthological Society* **12**: 48–60.

Stanley, E. H. & Doyle, M. W. (2003) Trading off: the ecological effects of dam removal. *Frontiers in Ecology and the Environment* **1**: 15–22.

Statistics Bureau of Japan (2003) State of the land [www document]. URL http://www.stat.go.jp/data/psi/3.htm (in Japanese).

Stearns, S. C. (1976) Life-history tactics: a review of the ideas. *Quarterly Review of Biology* **51**: 3–47.

Steele, J. H. (1974) *The Structure of Marine Ecosystems*. Oxford, UK: Blackwell Scientific Publications.

Steele, J. H. (1978) *Spatial Pattern in Plankton Communities*. New York, USA: Plenum Press.

Steele, J. H. (1991) Can ecological theory cross the land–sea boundary? *Journal of Theoretical Biology* **153**: 425–436.

Steele, J. H. (1995) Can ecological concepts span the land and ocean domains? In: *Ecological Time Series*, ed. T. M. Powell & J. H. Steele, pp. 5–19. New York, USA: Chapman & Hall.

Steele, J. H. (1998) Regime shifts in marine ecosystems. *Ecological Applications* **8**: S33–S36.

Steele, J. H. (2004) Regime shifts: reconciling observation and theory. *Progress in Oceanography* **60**: 135–141.

Steele, J. H. & Collie, J. S. (2005) Functional diversity and stability of coastal ecosystems. In: *The Sea*, vol. 13,

The Global Coastal Ocean: Interdisciplinary Regional Studies and Syntheses, ed. A.R. Robinson & K. Brink, pp. 785–819. Cambridge, MA, USA: Harvard University Press.

Steele, J.H. & Henderson, E.W. (1984) Modeling long-term fluctuations in fish stocks. *Science* **224**: 985–987.

Steele, J.H. & Hoagland, P. (2003) Are fisheries sustainable? *Fisheries Research* **64**: 1–3.

Steffani, C.N. & Branch, G.M. (2003*a*) Growth rate, condition, and shell shape of *Mytilus galloprovincialis*: responses to wave action. *Marine Ecology Progress Series* **246**: 197–209.

Steffani, C.N. & Branch, G.M. (2003*b*) Spatial comparisons of populations of an indigenous limpet *Scutellastra argenvillei* and the alien mussel *Mytilus galloprovincialis* along a gradient of wave energy. *South African Journal of Marine Science* **25**: 195–212.

Steffani, C.N. & Branch, G.M. (2003*c*) Temporal changes in an interaction between an indigenous limpet *Scutellastra argenvillei* and an alien mussel *Mytilus galloprovincialis*: effects of wave exposure. *South African Journal of Marine Science* **25**: 213–229.

Steidinger, K.A., Landsberg, J.H., Tomas, C.R. & Vargo, G.A., eds. (2004) *Harmful Algae 2002*. St Petersburg, FL, USA: Florida Fish and Wildlife Conservation Commission, Florida Institute of Oceanography, and Intergovernmental Oceanographic Commission of UNESCO.

Stein, B.A. (2001) A fragile cornucopia assessing the status of US biodiversity. *Environment* **43**: 11–22.

Steneck, R.S. (1997) Fisheries-induced biological changes to the structure and function of the Gulf of Maine ecosystem. In: *Proceedings of the Gulf of Maine Ecosystem Dynamics Scientific Symposium and Workshop*, RARGOM Report No. 91–1, pp. 151–165. Hanover, NH, USA: Regional Association for Research in the Gulf of Maine.

Steneck, R.S. (1998) Human influences on coastal ecosystems: does overfishing create trophic cascades? *Trends in Ecology and Evolution* **13**: 429–430.

Steneck, R.S. & Carlton, J.T. (2001) Human alterations of marine communities: students beware! In: *Marine Community Ecology*, ed. M. Bertness, S. Gaines & M. Hay, pp. 445–468. Sunderland, MA, USA: Sinauer Associates.

Steneck, R.S. & Dethier, M.N. (1994) A functional group approach to the structure of algal-dominated communities. *Oikos* **69**: 476–498.

Steneck, R.S. & Watling, L.E. (1982) Feeding capabilities and limitations of herbivorous molluscs: a functional group approach. *Marine Biology* **68**: 299–319.

Steneck, R.S., Graham, M.H., Bourque, B.J., *et al.* (2002) Kelp forest ecosystem: biodiversity, stability, resilience and their future. *Environmental Conservation* **29**: 436–459.

Steneck, R.S., Vavrinec, J. & Leland, A.V. (2004) Accelerating trophic level dysfunction in kelp forest ecosystems of the western North Atlantic. *Ecosystems* **7**: 323–331.

Stenson, J.A.E., Svensson, J.-E. & Cronberg, G. (1993) Changes and interactions in the pelagic community in acidified lakes in Sweden. *Ambio* **22**: 277–282.

Stephenson, T.A. & Stephenson, A. (1972) *Life between the Tidemarks on Rocky Shores*. San Francisco, CA, USA: W.H. Freeman.

Stevens, N.E. (1936) Environmental conditions and the wasting disease. *Science* **84**: 87–89.

Stevenson, J.C., Rooth, J.E., Kearney, M.S. & Sundberg, K.L. (2000) The health and long-term stability of natural and restored marshes in Chesapeake Bay. In: *Concepts and Controversies in Tidal Marsh Ecology*, ed. M.P. Weinstein & D.A. Kreeger, pp. 709–736. Dordrecht, the Netherlands: Kluwer.

Stevenson, J.C., Ward, L.G. & Kearney, M.S. (1986) Vertical accretion in marshes with varying rates of sea level rise. In: *Estuarine Variability*, ed. D. Wolfe, pp. 2412–2459. San Diego, CA, USA: Academic Press.

Stewart, K. & Diggins, T. (2002) Heavy metals in the sediments of Buffalo River at Buffalo, NY, USA. *Verhandlungen Internationale Vereinigung für Limnologie* **28**: 1262–1266.

Stiassny, M.L.J. (1996) An overview of freshwater biodiversity: with some lessons from African fishes. *Fisheries* **21**(9): 7–13.

Stirling, I., Lunn, N.J. & Iacozza, J. (1999) Long-term trends in the population ecology of polar bears in western Hudson Bay in relation to climatic change. *Arctic* **52**: 294–306.

Stoddard, J.L., Jeffries, D.S., Lükewille, A., *et al.* (1999) Regional trends in aquatic recovery from acidification in North America and Europe. *Nature* **401**: 575–578.

Stokstad, E. (2003) Can well-timed jolts keep out unwanted exotic fish? *Science* **301**: 157–158.

Stone, L., Huppert, A., Rajagopalan, B., Bhasin, H. & Loya, Y. (1999) Mass coral reef bleaching: a recent outcome of increased El Niño activity? *Ecological Letters* **2**: 325–330.

Storrs, M.J. & Finlayson, M. (1997) *Overview of the Conservation Status of Wetlands of the Northern Territory*, Report No. 116. Darwin, NT, Australia: Supervising Scientist.

Stouffer, R.J., Manabe, S. & Bryan, K. (1989) Interhemispheric asymmetry in climate response to a gradual increase in atmospheric CO_2. *Nature* **342**: 660–662.

Strand, J. A. & Weisner, S. B. (2001) Dynamics of submerged macrophyte populations in response to biomanipulation. *Freshwater Biology* **46**: 1397–1408.

Stromberg, J. C. & Chew, M. K. (2002) Flood pulses and the restoration of riparian vegetation in the American Southwest. In: *Flood Pulsing and Wetlands: Restoring the Natural Balance*, ed. B. A. Middleton, pp. 11–50. New York, USA: John Wiley.

Sturgess, P. (1992) Clear-felling dune plantations: studies in vegetation recovery. In: *Coastal Dunes: Geomorphology, Ecology and Management for Conservation*, ed. R. W. G. Carter, T. G. F. Curtis & M. J. Sheehy-Skeffington, pp. 339–349. Rotterdam, the Netherlands: A.A. Balkema.

Succow, M. & Lange, E. (1984) The mire types of the German Democratic Republic. In: *European Mires*, ed. P. D. Moore, pp. 149–175. London, UK: Academic Press.

Sugunan, V. V. & Gopal, B. eds. (2006) Wetlands, fisheries and livelihoods. *International Journal of Ecology and Environmental Sciences* Special Issue **32**.

Sullivan, T. (2000) *Aquatic Effects of Acid Deposition*. Boca Raton, FL, USA: Lewis Publishers.

Supanawanid, C. & Lewmanomont, K. (2003) The seagrasses of Thailand. In: *World Atlas of Seagrasses*, ed. E. P. Green & F. T. Short, pp. 158–165. Berkeley, CA, USA: University of California Press.

Supanawanid, C., Albertsen, J. O. & Mukai, H. (2001) Methods for assessing the grazing effects of large herbivores on seagrasses. In: *Global Seagrass Research Methods*, ed. F. T. Short & R. G. Coles, pp. 293–312. Amsterdam, the Netherlands: Elsevier.

Suzuki, T., Kuma, K., Kudo, I. & Matsunaga, K. (1995) Iron requirement of the brown macroalgae *Laminaria japonica, Undaria pinnatifida* and the crustose coralline algae, and their competition in the northern Japan Sea. *Phycologia* **34**: 201–205.

Svensson, B. H., Christensen, T. R., Johansson, E. & Öquist, M. (1999) Interdecadal changes in CO_2 and CH_4 fluxes of a subarctic mire: Stordalen revisited after 20 years. *Oikos* **85**: 22–30.

Svensson, G. (1988) Bog development and environmental conditions as shown by the stratigraphy of Store Mosse mire in southern Sweden. *Boreas* **17**: 89–111.

Swales, S., Storey, A. W., Roderick, I. D. & Figa, B. S. (1999) Fishes of floodplain habitats of the Fly River System, Papua New Guinea, and changes associated with El Niño droughts and algal blooms. *Environmental Biology of Fishes* **54**: 389–404.

Swanson, T. (1996) The underlying causes of biodiversity decline: and economic analysis [www document]. URL http://biodiversityeconomics.org/pdf/960401-08.pdf

Sweatman, H., Osborne, K., Smith, L., *et al.* (2002) Status of coral reefs of Australasia: Australia and Papua New Guinea. In: *Status of Coral Reefs of the World: 2002*, ed. C. Wilkinson, pp. 163–180. Townsville, Queensland, Australia: Australian Institute of Marine Science.

Sweeney, B. W., Bott, T. L., Jackson, J. K., *et al.* (2004) Riparian deforestation, stream narrowing, and loss of stream ecosystem services. *Proceedings of the National Academy of Sciences of the USA* **101**: 14132–14137.

Swiss Re (1998) *Floods: An Insurable Risk?* Zurich, Switzerland: Swiss Reinsurance Company.

Syms, C. & Jones, G. P. (1999) Scale of disturbance and the structure of a temperate fish guild. *Ecology* **80**: 921–940.

Symstad, A. J., Chapin III, F. S., Wall, D. H., *et al.* (2003) Long-term and large-scale perspectives on the relationship between biodiversity and ecosystem functioning. *BioScience* **53**: 89–98.

Szmant, A. M. (2001) Why are coral reefs world-wide becoming overgrown by algae? 'Algae, algae everywhere, and nowhere a bite to eat!'. *Coral Reefs* **19**: 299–302.

Taastrom, H. M. & Jacobsen, L. (1999) The diet of otters (*Lutra lutra* L.) in Danish freshwater habitats: comparisons of prey fish populations. *Journal of Zoology (London)* **248**: 1–13.

Tabacchi, E. & Planty-Tabacchi, A.-M. (2000) Riparian plant community composition and the surrounding landscape: functional significance of incomers. In: *Riparian Ecology and Management in Multi-Land Use Watersheds*, ed. P. J. Wigington Jr & R. L. Beschta, pp. 11–16. Corvallis, OR, USA: American Water Resources Association.

TAC/CGIAR (1989) *Sustainable Agricultural Production: Implications of an International Agricultural Research*, Research and Development Paper No. 4. Rome, Italy: Technical Advisory Committee/ Consultative Group on International Agricultural Research.

Takizawa, M., Straube, R. T. & Colwell, R. R. (1993) Near-bottom pelagic bacteria at a deep-water sewage sludge disposal site. *Applied and Environmental Microbiology* **59**: 3406–3410.

Talley, T. S., Crooks, J. A. & Levin, L. A. (2001) Habitat utilization and alteration by the invasive burrowing isopod, *Sphaeroma quoyanum* in California salt marshes. *Marine Biology* **138**: 561–573.

Tallis, J. H. (1985) Mass movement and erosion of a southern Pennine blanket peat. *Journal of Ecology* **73**: 283–315.

Tallis, J. H. (1987) Fire and flood at Holme Moss: erosion processes in an upland blanket mire. *Journal of Ecology* **75**: 1099–1129.

Tallis, J. H. (1998) Growth and degradation of British and Irish blanket mires. *Environmental Reviews* **6**: 81–122.

Tam, N. F. Y. & Wong, Y. S. (1997) Accumulation and distribution of heavy metals in a simulated mangrove system treated with sewage. *Hydrobiologia* **352**: 67–75.

Tamatamah, R., Duthie, H. & Hecky, R. (2004) The importance of atmospheric deposition to the phosphorus loading of Lake Victoria (East Africa). *Biogeochemistry* **73**: 1–20.

Tanabe, S. & Tatsukawa, R. (1983) Vertical transport and residence time of chlorinated hydrocarbons in the open ocean water column. *Journal of the Oceanographical Society of Japan* **39**: 53–62.

Tanner, J. E. (1995) Competition between scleractinian corals and macroalgae: an experimental investigation of coral growth, survival and reproduction. *Journal of Experimental Marine Biology and Ecology* **190**: 151–168.

Tao, F., Yokozava, M., Hayashi, Y. & Lin, E. (2003) Terrestrial water cycle and the impact of climate change. *Ambio* **32**: 295–301.

Tao, W. & Wei, W. (1997) Water resources and agricultural environment in arid regions of China. In: *Freshwater Resources in Arid Lands*, ed. J. Uitto & J. Schneider, pp. 44–60. New York, USA: United Nations Press.

Tapp, J. F., Shillabeer, N. & Ashman, C. M. (1993) Continued observations of the benthic fauna of the industrialized Tees estuary, 1979–1990. *Journal of Experimental Marine Biology and Ecology* **172**: 67–82.

Tarazona, J. & Parendes, C. (1992) Impacto de les eventos El-Niño sobre las comunidades bentonicas de playa arenosa durante 1976–1986. In: *Paleo ENSO Records International Symposium*, Lima, Extended Abstracts, pp. 299–303. Available online at www.crid.or.cr/crid/CD_El_Ni%F1o/pdf/spa/doc9263/doc9263.htm

Tarazona, J., Arntz, W., Canahuire, E., Ayala, Z. & Robles, A. (1985) Modificaciones producidas durante 'El-Niño' en la infauna bentonica de areas someras del ecosistema de afloriamento peruano. In: *El-Jernomeno El-Niño y su Impacto en la Fauna Marina*, ed. W. Arntz, A. Landa & J. Tarazona, pp. 55–63. Callao, Peru: Bolletín Instituto del Mardel, Special Issue 1985.

Tarazona, J., Salzwedel, H. & Arntz, W. (1988) Oscillations of marcobenthos in shallow waters of the Peruvian central coast induced by the El-Niño 1982–83. *Journal of Marine Research* **46**: 593–611.

Taylor, A. H. & Stephens, J. A. (1980) Latitudinal displacements of the Gulf Stream (1966 to 1977) and their relation to changes in temperature and zooplankton abundance in the NE Atlantic. *Oceanologica Acta* **3**: 145–149.

Taylor, P. (1993) The state of the marine environment: a critique of the work and role of the Joint Group of Experts on Scientific Aspects of Marine Pollution (GESAMP). *Marine Pollution Bulletin* **26**: 120–127.

Taylor, P. R. & Littler, M. M. (1982) The roles of compensatory mortality, disturbance and substrate retention in development and organization of a sand-influenced, rocky-intertidal community. *Ecology* **63**: 135–146.

Tegner, M. J. & Dayton, P. K. (1987) El Niño effects on southern California kelp forest communities. *Advances in Ecological Research* **17**: 243–279.

Tegner, M. J. & Dayton, P. K. (1991) Sea urchins, El Niños, and the long-term stability of Southern California kelp forest communities. *Marine Ecology Progress Series* **77**: 49–63.

Tegner, M. J. & Levin, L. A. (1983) Spiny lobsters and sea urchins: analysis of a predator–prey interaction. *Journal of Experimental Marine Biology and Ecology* **73**: 125–150.

Tegner, M. J., Basch, L. V. & Dayton, P. K. (1996a) Near extinction of an exploited marine invertebrate. *Trends in Ecology and Evolution* **11**: 278–279.

Tegner, M. J., Dayton, P. K., Edwards, P. B. & Riser, K. L. (1996b) Is there evidence for long-term climatic change in southern California kelp forests? *California Cooperative Oceanic Fisheries Investigations Reports* **37**: 111–126.

Tegner, M. J., Dayton, P. K., Edwards, P. B. & Riser, K. L. (1997) Large-scale, low-frequency oceanographic effects on kelp forest succession: a tale of two cohorts. *Marine Ecology Progress Series* **146**: 117–134.

Terrados, J., Duarte, C. M., Fortes, M. D., *et al.* (1998) Changes in community structure and biomass of seagrass communities along gradients of siltation in SE Asia. *Estuarine, Coastal and Shelf Science* **46**: 757–768.

Thayer, G. W., Adams, S. M. & LaCroix, M. W. (1975) Structural and functional aspects of a recently established *Zostera marina* community. In: *Estuarine Research*, vol. 1, *Chemistry, Biology, and the Estuarine System*, ed. L. E. Cronin, pp. 518–540. New York, USA: Academic Press.

The Nature Conservancy (2003) *Emiquon: Restoring the Illinois River.* Arlington, VA, USA: The Nature Conservancy. Available online at http://nature.org/wherewework/northamerica/states/illinois/preserves/art1112.html

The Times Atlas of the World (1999) London, UK: Collins Bartholomew.

Thesiger, W. (1964) *The Marsh Arabs*. London, UK: Penguin Books.

Thiel, H. (2003) Anthropogenic impacts on the deep sea. In: *Ecosystems of the World*, vol. 28, *Ecosystems of the Deep Oceans*, ed. P. A. Tyler, pp. 427–471. Amsterdam, the Netherlands: Elsevier.

Thiel, H., Angel, M. V., Foell, E. J., Rice, A. L. & Schriever, G. (1998) *Environmental Risks from Large-Scale Ecological Research in the Deep Sea*. Bremerhaven, Germany: Commission of the European Communities Directorate-General for Science, Research and Development.

Thiel, H., Pfannkuche, O., Schriever, G., *et al.* (1989) Phytodetritus on the deep-sea floor in a central oceanic region of the north-east Atlantic. *Biological Oceanography* **6**: 203–239.

Thiel, H., Schriever, G., Ahnert, A., *et al.* (2001) The large-scale environmental impact experiment DISCOL: reflection and foresight. *Deep-Sea Research II* **48**: 3869–3882.

Thiele, D. & Gill, P. C. (1999) Cetacean observations during a winter voyage into Antarctic sea ice south of Australia. *Antarctic Science* **11**: 48–53.

Thom, B. G. (1967) Mangrove ecology and deltaic geomorphology, Tabasco, Mexico. *Journal of Ecology* **55**: 301–343.

Thom, B. G. (1982) Mangrove ecology: a geomorphological perspective. In: *Mangrove Ecosystems in Australia*, ed. B. F. Clough, pp. 3–17. Canberra, ACT, Australia: Australian National University Press.

Thom, R. M., Shrefler, D. K. & MacDonald, K. B. (1994) *Shoreline Armoring Effects on Coastal Ecology and Biological Resources in Puget Sound, Washington*. Olympia, WA, USA: Shorelands and Coastal Zone Management Program, Washington Department of Ecology.

Thomas, C. A. & Bendell-Young, L. I. (1999) The significance of diagenesis versus riverine input in contributions to the sediment geochemical matrix of iron and manganese in an intertidal region. *Estuarine, Coastal and Shelf Science* **48**: 635–647.

Thomas, D. N. & Dieckmann, G. S. (2003) *Sea-Ice: An Introduction to its Physics, Chemistry, Biology and Geology*. Oxford, UK: Blackwell Science.

Thompson, B., Anderson, B., Hunt, J., Taberski, K. & Phillips, B. (1999) Relationships between sediment contamination and toxicity in San Francisco Bay. *Marine Environmental Research* **48**: 285–309.

Thompson, D. W. J. & Solomon, S. (2002) Interpretation of recent Southern Hemisphere climate change. *Science* **296**: 895–899.

Thompson, D. W. J. & Wallace, J. M. (1998) The Arctic Oscillation signature in the wintertime geopotential height and temperature fields. *Geophysical Research Letters* **25**: 1297–1300.

Thompson, D. W. J. & Wallace, J. M. (2000) Annular modes in the extratropical circulation. I. Month-to-month variability. *Journal of Climate* **13**: 1000–1016.

Thompson, D. W. J. & Wallace, J. M. (2001) Regional climate impacts of the Northern Hemisphere annular mode. *Science* **293**: 85–89.

Thompson, D. W. J., Wallace, J. M. & Hegerl, G. C. (2000) Annular modes in the extratropical circulation. II. Trends. *Journal of Climate* **13**: 1018–1036.

Thompson, G. B. & Drake, B. G. (1994) Insect and fungi on a C_3 sedge and a C_4 grass exposed to elevated atmospheric CO_2 concentrations in open-top chambers in the field. *Plant, Cell and Environment* **17**: 1161–1167.

Thompson, J. D. (1991) The biology of an invasive plant. *BioScience* **41**: 393–401.

Thompson, J. R. & Polet, G. (2000) Hydrology and land use in a Sahelian floodplain wetland. *Wetlands* **20**: 639–659.

Thompson, J. R. (1996) Africa's floodplains: a hydrologic overview. In: *Water Management and Wetlands in Sub-Saharan Africa*, ed. M. C. Acreman & G. E. Hollis, pp. 5–20. Gland, Switzerland: IUCN.

Thompson, R. C., Crowe, T. P. & Hawkins, S. J. (2002) Rocky intertidal communities: past environmental changes, present status and predictions for the next 25 years. *Environmental Conservation* **29**: 168–191.

Thompson, R. C., Norton, T. A. & Hawkins, S. J. (2004) Physical stress and biological control regulate the balance between producers and consumers in marine intertidal biofilms. *Ecology* **85**: 1372–1382.

Thompson, R. C., Roberts, M. F., Norton, T. A. & Hawkins, S. J. (2000) Feast or famine for intertidal grazing molluscs: a mis-match between seasonal variations in grazing intensity and the abundance of microbial resources. *Hydrobiologia* **440**: 357–367.

Thormann, M. N. & Bayley, S. E. (1997) Aboveground net primary production along a bog–fen–marsh gradient in southern boreal Alberta, Canada. *Ecoscience* **4**: 374–384.

Thresher, R., Proctor, C., Ruiz, G., *et al.* (2003) Invasion dynamics of the European shore crab, *Carcinus maenas*, in Australia. *Marine Biology* **142**: 867–876.

Thrush, S. F. & Dayton, P. K. (2002) Disturbance to marine benthic habitats by trawling and dredging: implications for marine biodiversity. *Annual Review of Ecology and Systematics* **33**: 449–473.

Thrush, S. F., Hewitt, J. E., Cummings, V. J., *et al.* (1998) Disturbance of the marine benthic habitat by commercial fishing: impacts at the scale of the fishery. *Ecological Applications* **8**: 866–879.

Thrush, S. F., Hewitt, J. E., Norkko, A., Cummings, V. J. & Funnell, G. A. (2003) Catastrophic sedimentation on estuarine sandflats: recovery of macrobenthic communities is influenced by a variety of environmental factors. *Ecological Applications* **13**: 1433–1455.

Tibbetts, J. (2002) Coastal cities: living on the edge. *Environmental Health Perspectives* **110**: 674–681.

Tiega, A. (2001) Priorities for wetland biodiversity conservation in Africa. In: *Wetland Inventory, Assessment and Monitoring: Practical Techniques and Identification of Major Issues*, ed. C. M. Finlayson & N. C. Davidson, pp. 112–120. Darwin, NT, Australia: Supervising Scientist.

Tilman, D., Fargione, J., Wolff, B., *et al.* (2001) Forecasting agriculturally driven global environmental change. *Science* **292**: 281–284.

Tilman, D., Wedin, D. & Knops, J. (1996) Productivity and sustainability influenced by biodiversity in grassland ecosystems. *Nature* **379**: 718–720.

Timchenko, V., Oksijuk, O. & Gore, J. (2000) A model for ecosystem state and water quality management in the Dnieper River delta. *Ecological Engineering* **16**: 119–125.

Timms, B. V. (1976) A comparative study of the limnology of three maar lakes in western Victoria. I. Physiography and physicochemical features. *Australian Journal of Marine and Freshwater Research* **27**: 35–60.

Tinker, P. B. (1997) The environmental implications of intensified land use in developing countries. *Philosophical Transactions of the Royal Society of London B* **352**: 1023–1033.

Titus, J. G. (1990) Greenhouse effect, sea level rise, and barrier islands: case study of Long Beach Island, New Jersey. *Coastal Management* **18**: 65–90.

Titus, J. G. & Narayanan, V. K. (1995) *The Probability of Sea Level Rise*, EPA Report No. 230-R-95-008. Washington, DC, USA: US Environmental Protection Agency.

Tockner, K. & Stanford, J. A. (2002) Riverine flood plains: present state and future trends. *Environmental Conservation* **29**: 308–330.

Tockner, K., Malard, F. & Ward, J. V. (2000) An extension of the flood pulse concept. *Hydrological Processes* **14**: 2861–2883.

Tockner, K., Pennetzdorfer, D., Reiner, N., Schiemer, F. & Ward, J. V. (1999) Hydrologic connectivity and the exchange of organic matter and nutrients in a dynamic river–floodplain system (Danube, Austria). *Freshwater Biology* **41**: 521–535.

Tockner, K., Ward, J. V., Arscott, B. A., *et al.* (2003) The Tagliamento River: a model ecosystem of European importance. *Aquatic Sciences* **65**: 239–253.

Todd, C. D. (1998) Larval supply and recruitment of benthic invertebrates: do larvae always disperse as much as we believe? *Hydrobiologia* **376**: 1–21.

Todd, C. D. & Lewis, J. R. (1984) Effects of low air temperature on *Laminaria digitata* L. in southwestern Scotland. *Marine Ecology Progress Series* **16**: 199–201.

Todd, C. D., Lambert, W. J. & Thorpe, J. P. (1998) The genetic structure of intertidal populations of two species of nudibranch molluscs with planktotrophic and pelagic lecithotrophic larval stages: are pelagic larvae 'for' dispersal? *Journal of Experimental Marine Biology and Ecology* **228**: 1–28.

Todhunter, P. E. & Rundquist, B. C. (2004) Terminal lake flooding and wetland expansion in Nelson County, North Dakota. *Physical Geography* **25**: 68–85.

Tolba, M. K. & El-Kholy, O. A., eds. (1992) *The World Environment 1972–1992: Two Decades of Challenge*. London, UK: Chapman & Hall and United Nations Environment Programme.

Tolonen, K. (1967) Über die Entwicklung der Moore im finnischen Nordkarelien. *Annales Botanici Fennici* **4**: 219–416.

Tomalin, B. J. & Kyle, R. (1998) Subsistence and recreational mussel (*Perna perna*) collecting in KwaZulu–Natal, South Africa: fishing mortality and precautionary management. *South African Journal of Science* **33**: 12–22.

Tomascik, T., Mah, A. J., Nontji, A. & Moosa, M. K. (1997) *The Ecology of the Indonesian Seas*, Part 2. Hong Kong: Periplus.

Tomasko, D. A., Dawes, C. J. & Hall, M. O. (1996) The effects of anthropogenic nutrient enrichment on turtle grass (*Thalassia testudinum*) in Sarasota Bay, Florida. *Estuaries* **19**: 448–456.

Tomicic, J. (1985) Efectos del fenómeno El Niño 1982–83 en las comunidades litorales de la península de Mejillones. *Investigación Pesquera (Chile)* **32**: 209–213.

Tóth, J. (1963) A theoretical analysis of groundwater flow in small drainage basins. *Journal of Geophysical Research* **68**: 4795–4812.

Tóth, J. (1999) Groundwater as a geologic agent: an overview of the causes, processes, and manifestations. *Hydrogeology Journal* **7**: 1–14.

Towns, D. R. & Ballantine, W. J. (1993) Conservation and restoration of New Zealand island ecosystems. *Trends in Ecology and Evolution* **8**: 452–457.

Townsend, C. R. (1996) Concepts in river ecology: pattern and process in the catchment hierarchy. *Archiv für Hydrobiologie* 113 (Suppl.): 3–21.

Townsend, C. R. (2003) Individual, population, community, and ecosystem consequences of a fish invader in New Zealand streams. *Conservation Biology* 17: 38–47.

Townsend, C. R., Dolédec, S., Norris, R., Peacock, K. & Arbuckle, C. J. (2003) The influence of scale and geography on relationships between stream community composition and landscape variables: description and prediction. *Freshwater Biology* 48: 768–785.

Tracey, D. M. & Horn, P. L. (1999) Background and review of ageing orange roughy (*Hoplostethus atlanticus*, Trachichthyidae) from New Zealand and elsewhere. *New Zealand Journal of Marine and Freshwater Research* 33: 67–86.

Travis, J. M. J. (2003) Climate change and habitat destruction: a deadly anthropogenic cocktail. *Proceedings of the Royal Society of London B* 270: 467–473.

Trupin, A. & Wahr, J. (1990) Spectroscopic analysis of global tide gauge sea level data. *Geophysical Journal International* 100: 441–453.

Truscott, A. (1984) Control of *Spartina anglica* on the amenity beaches of Southport. In: Spartina anglica *in Great Britain*, ed. P. Doody, pp. 64–69. Attingham Park, UK: Nature Conservation Council.

Tucker, W. B., Weatherly, J. W., Eppler, D. T., Farmer, L. D. & Bentley, D. L. (2001) Evidence for rapid thinning of sea ice in the western Arctic Ocean at the end of the 1980s. *Geophysical Research Letters* 28: 2851–2854.

Tudhope, A. W., Chilcott, C. P., McColluch, M. T., *et al.* (2001) Variability in the El Niño–Southern Oscillation through a glacial–interglacial cycle. *Science* 291: 1511–1517.

Tunnicliffe, V., Embley, R. W., Holden, J. F., *et al.* (1997) Biological colonization of new hydrothermal vents following an eruption on Juan de Fuca Ridge. *Deep-Sea Research I* 44: 1627–1644.

Turetsky, M., Wieder, K., Halsey, L. & Vitt, D. H. (2002) Current disturbance and the diminishing peatland carbon sink. *Geophysical Research Letters* 29: 1400.

Turner, A. (2000) Trace metal contamination in sediments from UK estuaries: an empirical evaluation of the role of hydrous iron and manganese oxides. *Estuarine, Coastal and Shelf Science* 50: 355–371.

Turner, G. (1994) Fishing and the conservation of the endemic fishes of Lake Malawi. *Archiv für Hydrobiologie* 44: 481–494.

Turner, R. E. & Lewis, R. R. (1997) Hydrologic restoration of coastal wetlands. *Wetlands Ecology and Management* 4: 65–72.

Turner, R. E. & Warren, R. S. (2003) Valuation of continuous and intermittent *Phragmites* control. *Estuaries* 26: 618–623.

Turner, R. K. (2000) Integrating natural and socio-economic science in coastal management. *Journal of Marine Systems* 25: 447–460.

Turner, R. K., Lorenzoni, I., Beaumont, N., *et al.* (1998) Coastal management for sustainable development: analysing environmental and socio-economic changes on the UK coast. *Geographical Journal* 164: 269–281.

Tweddle, D. (1992) Conservation and threats to the resources of Lake Malawi. *Mitteilungen Internationale Vereinigung für Limnologie* 24: 17–24.

Twenhöven, F. L. (1992a) Untersuchungen zur Wirkung stickstoffhaltiger Niederschläge auf die Vegetation von Hochmooren. *Mitteilungen der Arbeitsgemeinschaft Geobotanik in Schleswig-Holstein und Hamburg* 44: 1–172.

Twenhöven, F. L. (1992b) Competition between two *Sphagnum* species under different deposition levels. *Journal of Bryology* 17: 71–80.

Twilley, R. R., Chen, R. H. & Hargis, T. (1992) Carbon sinks in mangroves and their implications to carbon budget of tropical coastal ecosystems. *Water, Air, and Soil Pollution* 64: 265–288.

Twilley, R. R., Gottfried, R. R., Rivera-Monroy, V. H., *et al.* (1998a) An approach and preliminary model of integrating ecological and economic constraints of environmental quality in the Guayas River estuary, Ecuador. *Environmental Science and Policy* 1: 271–288.

Twilley, R. R., Rivera-Monroy, V. H., Chen, R. & Botero, L. (1998b) Adapting an ecological mangrove model to simulate trajectories in restoration ecology. *Marine Pollution Bulletin* 37: 404–419.

Tyler, P. A. (1988) Seasonality in the deep sea. *Oceanography and Marine Biology Annual Review* 26: 227–258.

Tyler, P. A. (2003a) Disposal in the deep sea: analogue of nature or *faux ami*? *Environmental Conservation* 30: 26–39.

Tyler, P. A. (2003b) The peripheral deep seas. In: *Ecosystems of the World*, vol. 28, *Ecosystems of the Deep Ocean*, ed. P. A. Tyler, pp. 261–293. Amsterdam, the Netherlands: Elsevier.

Tyler, P. A. & Young, C. M. (1999) Reproduction and dispersal at vents and cold seeps. *Journal of the Marine Biological Association of the United Kingdom* 79: 193–208.

Tyler, P. A., Grant, A., Pain, S. L. & Gage, J. D. (1982) Is annual reproduction in deep-sea echinoderms a response to variability in their environment? *Nature* 300: 747–750.

Tynan, C. T. & DeMaster, D. (1997) Observations and predictions of Arctic climate change: potential effects on marine mammals. *Arctic* **50**: 308–322.

Uchiyama, Y., Nadaoka, K., Rolke, P., Adachi, K. & Yagi, H. (2000) Submarine groundwater discharge into the sea and associated nutrient transport in a sandy beach. *Water Resources Research* **36**: 1467–1479.

UKOOA (2002) UKOOA Drill Cuttings Initiative Final Report. United Kingdom Offshore Operators Association [www document]. URL http://www.oilandgas.org.uk/issues/drillcuttings/pdfs/finalreport.pdf

UN (1994) *Report of the International Conference on Population and Development*, UN Document No. A/CONF.171/13, New York, USA: UN Population Division.

UN (1997) *World Urbanization Prospects: The 1996 Revision*. New York, USA: UN Population Division.

UN (1998) Report of the conference of the parties on its third session, held at Kyoto from 1 to 11 December 1997. United Nations Framework Convention on Climate Change [www document]. URL http://unfccc.int/resource/docs/cop3/07aol.pdf

UN (1999) *World Population Prospects: The 1998 Revision*. New York, USA: UN Population Division.

UN (2002) United Nations human rights [www document]. URL http://www.un.org/rights/

UN (2003) *World Population Prospects: The 2002 Revision*. New York, USA: UN Population Division.

UNCED (1992) *Chapter 17. Agenda 21: Earth Summit – The United Nations Programme of Action from Rio*, United Nations Conference on Environment and Development. New York, USA: UN Department of Public Information.

UN Country Team (2004) Kazakhstan: achievements, issues, and prospects [www document]. URL http://www.undp.kz/library_of_publications/files/1871-31465.pdf

Underwood, A. J. (1993) Exploitation of species on the rocky coast of New South Wales (Australia) and options for its management. *Ocean and Coastal Management* **20**: 41–62.

Underwood, A. J. (1999) History and recruitment in structure of intertidal assemblages on rocky shores: an introduction to problems for interpretation of natural change. In: *Aquatic Life Cycle Strategies Survival in a Variable Environment*, ed. M. Whitfield, J. Matthews & C. Reynolds, pp. 79–96. Plymouth, UK: Marine Biological Association.

Underwood, A. J. & Fairweather, P. G. (1989) Supply-side ecology and benthic marine assemblages. *Trends in Ecology and Evolution* **4**: 16–20.

Underwood, A. J., Denley, E. J. & Moran, M. J. (1983) Experimental analyses of the structure and dynamics of mid-shore rocky intertidal communities in New South Wales. *Oecologia* **56**: 202–219.

UNDP (1995) *Human Development Report*. New York, USA: Oxford University Press.

UNEP (1999a) Strategic Action Plan for the South China Sea [www document]. URL http://www.unep.org/dewa/giwa/areas/area54.asp

UNEP (1999b) *Global Environment Outlook 2000: UNEP's Millennium Report on the Environment*. London, UK: Earthscan Publications.

UNEP (2000a) Regional-Integrated Sustainable Management of Trans-Boundary Environmental Resources in South-western Djibouti and Northeastern Ethiopia, UNEP Division of GEF Coordination, Nairobi [www document]. URL http://www.iwlearn.net/docs/smter/smter10e.pdf

UNEP (2000b) Action Plan for the Protection of the Marine Environment and the Sustainable Development of the Coastal Areas of the Mediterranean [www document]. URL http://www.unep.org/regionalseas/Programmes/UN-EP_Administered_ Programmes/Mediterranean_Region/default2.asp

UNEP (2000c) *Regional Seas: A Survival Strategy for Our Oceans and Coasts*. Geneva, Switzerland: United Nations Environment Programme.

UNEP (2001a) *GLOBIO: Global methodology for mapping human impacts on the biosphere*, UNEP/DEWA/TR No. 01-3, Nairobi, Kenya: United Nations Environment Programme.

UNEP (2001b) The Mesopotamian marshlands: demise of an ecosystem. Unpublished report. Nairobi, Kenya: United Nations Environment Programme.

UNEP (2002) *GEO-3: Global Environment Outlook 3 – Past, Present and Future Perspectives*. Nairobi, Kenya: United Nations Environment Programme.

UNEP (2003) *Water for People, Water for Life*, UN World Water Development Report. Paris, France: UNESCO.

UNEP (2004) *Desk Study on the Environment in Liberia*. Nairobi, Kenya: United Nations Environment Programme.

UNEP–WCMC (2004) Biodiversity and Climate Change [www document]. URL http://www.unep-wcmc.org/climate/impacts.htm

UNESCO (2000) *Vision for the Aral Sea Basin*. Paris, France: UNESCO.

UNESCO–IHE (2003) Partner in Capacity Building [www document]. URL: http://www.unesco-ihe.org/education/intro.htm

UNPD (2002) *International Migration Report 2002*. New York, USA: UN Population Division.

UNPD (2006) *World Population Prospects: The 2006 Revision*. New York, USA: UN Population Division.

UN World Water Action Programme (2003) *Water for People, Water for Life*. Paris, France: UNESCO.

Urban, F. E., Cole, J. E. & Overpeck, J. T. (2000) Influence of mean climate change on climate variability from a 155-year tropical Pacific coral record. *Nature* 407: 989–993.

US Army Corps of Engineers (1989) *Benefits Reevaluation Study: Brigantine Island, New Jersey*. Philadelphia, PA, USA: US Army Corps of Engineers.

US Army Corps of Engineers (2001) *The New York District's Biological Monitoring Program for the Atlantic Coast of New Jersey, Asbury Park to Manasquan Section Beach Erosion Control Project*. Vicksburg, MS, USA: Engineer Research and Development Center, Waterways Experiment Station.

US Army Corps of Engineers (2003) *1988 National List of Plants that Occur in Wetlands*. Washington, DC, USA: US Army Corps of Engineers Available online at http://www. fws.gov/nwi/bha/list88.html

US Census Bureau (2003) International Data Base (IDB) [www document]. URL http://www.census.gov/ipc/www/idbnew.html

US EPA (1992) Framework for ecological risk assessment forum. Unpublished report EPA/630/R92/001. Washington, DC, USA: US Environmental Protection Agency.

US EPA (2000) *Report to Congress Groundwater Chapters*, Report No. EPA 816-R-00-013, EPA Office of Water, Washington, DC, USA: US Environmental Protection Agency.

US EPA (2001) *National Coastal Condition Report*, Report No. EPA-620/R-01/005. Washington, DC, USA: US Environmental Protection Agency.

US Fish and Wildlife Service (2003a) *Wildlife Institute of India*. Washington, DC, USA: US Fish and Wildlife Service, Office of International Affairs. Available online at http://international.fws.gov/laws/wii.html (accessed August 12, 2003)

US Fish and Wildlife Service (2003b) *US/South Africa Binational Commission*. Washington, DC, USA: US Fish and Wildlife Service, Office of International Affairs. Available online at http://international.fws.gov/laws/ussabc.html

US Geological Survey (1996) *National Water Summary of Wetland Resources*, US Geological Survey Water-Supply Paper No. 2425. Washington, DC, USA: US Government Printing Office.

Uthicke, S. & Benzie, J. A. H. (2001) Effect of beche-de-mer fishing on densities and size structure of *Holothuria nobilis* (Echinodermata: Holothuroidea) populations on the Great Barrier Reef. *Coral Reefs* 3: 271–276.

Vadas, R. L. & Steneck, R. S. (1988) Zonation of deep water benthic algae in the Gulf of Maine. *Journal of Phycology* 24: 338–346.

Valentine, J. P. & Johnson, C. R. (2003) Establishment of the introduced kelp *Undaria pinnatifida* in Tasmania depends on disturbance to native algal assemblages. *Journal of Experimental Marine Biology and Ecology* 295: 63–90.

Valette-Silver, N. J. (1993) The use of sediment cores to reconstruct historical trends in contamination of estuarine and coastal sediments. *Estuaries* 16: 577–608.

Valiela, I. (1995) *Marine Ecological Processes*, 2nd edn. New York, USA: Springer-Verlag.

Valiela, I., Cole, M. L., McClelland, J., *et al.* (2000) Role of salt marshes as part of coastal landscapes. In: *Concepts and Controversies in Tidal Marsh Ecology*, ed. M. P. Weinstein & D. P. Kreeger, pp. 23–38. Dordrecht, the Netherlands: Kluwer.

Valiela, I., Foreman, K., LaMontagne, M., *et al.* (1992) Couplings of watersheds and coastal waters: sources and consequences of nutrient enrichment in Waquoit Bay, Massachusetts. *Estuaries* 15: 443–457.

Valiela, I., Kremer, J., Lajtha, K., *et al.* (1997) Nitrogen loading from coastal watersheds to receiving estuaries: new method and application. *Ecological Applications* 7: 358–380.

Vallentyne, J. & Beeton, A. (1988) The ecosystem approach to managing human uses and abuses of natural resources in the Great Lakes Basin. *Environmental Conservation* 15: 59–62.

Valtýsson, H. (2001) The sea around Icelanders: catch history and discards in Icelandic waters. In: *Fisheries Impacts on North Atlantic Ecosystems: Catch, Effort and National/Regional Data Sets*, Fisheries Centre Research Report No. 9 (3), ed. D. Zeller, R. Watson & D. Pauly, pp. 52–87. Vancouver, BC, Canada: University of British Columbia.

Valtýsson, H. & Pauly, D. (2003) Fishing down the food web: an Icelandic case study. In: *Competitiveness within the Global Fisheries, Proceedings of a Conference*, Akureyri, Iceland, 6–7 April 2000, ed. E. Guðmundsson & H. Valtýsson. Akureyri, Iceland: University of Akureyri.

Van Boxel, J. H., Jungerius, P. D., Kieffer, N. & Hampele, N. (1997) Ecological effects of reactivation of artificially stabilized blowouts in coastal dunes. *Journal of Coastal Conservation* 3: 57–62.

Van Buynder, P., Sam, G., Russell, R., *et al.* (1995) Barmah Forest Virus epidemic on the south coast of New South Wales. *Communicable Diseases Intelligence* **19**: 188–191.

Van Dam, R., Gitay, H., Finlayson, C. M., Davidson, N. & Orlando, B. (2002) Climate Change and Wetlands: Impacts, Adaptation and Mitigation. Information Paper. Ramsar COP8, Doc.11 + annex [www document]. URL http://www.ramsar.org/cop8/cop8_doc_11_e.htm.

Van Dam, R. A., Finlayson, C. M. & Watkins, D., eds. (1999) *Vulnerability Assessment of Major Wetlands in the Asia-Pacific Region to Climate Change and Sea Level Rise*, Supervising Scientist Report No. 149, Canberra, ACT, Australia: Supervising Scientist.

Vandelannoote, A., Robberecht, H., Deelstra, H., Vyumvohore, F., Bitetera, L. & Ollevier, F. (1996) The impact of the River Ntahangwa, the most polluted Burundian affluent of Lake Tanganyika, on the water quality of the lake. *Hydrobiologia* **328**: 161–171.

Van den Broek, W. L. F. & Tracey, D. M. (1981) Concentration and distribution of mercury in flesh of orange roughy (*Hoplostethus atlanticus*). *New Zealand Journal of Marine and Freshwater Research* **15**: 255–260.

Van der Nat, D., Schmidt, A., Edwards, P. J., Tockner, K. & Ward, J. V. (2002) Inundation dynamics in braided flood-plains (Fiume Tagliamento). *Ecosystems* **5**: 636–647.

Vanderploeg, H., Johengen, T., Strickler, J., Liebig, J. & Nalepa, T. (2001) Zebra mussels (*Dreissena polymorpha*) selective filtration promoted *Microcystis* blooms in Saginaw Bay (Lake Huron) and Lake Erie. *Canadian Journal of Fisheries and Aquatic Sciences* **58**: 1208–1221.

Vanderploeg, H., Nalepa, T., Jude, D., *et al.* (2002) Dispersal and emerging ecological impacts of Ponto-Caspian species in the Laurentian Great Lakes. *Canadian Journal of Fisheries and Aquatic Sciences* **59**: 1209–1228.

Van der Wal, D. (1998) The impact of the grain-size distribution of nourishment sand on aeolian sand transport. *Journal of Coastal Research* **14**: 620–631.

Van der Wal, D. (2000) Grain-size-selective aeolian sand transport on a nourished beach. *Journal of Coastal Research* **16**: 896–908.

Van Dijk, G. M., Van Liere, L., Bannik, B. A. & Cappon, J. J. (1994) Present state of the water quality of European rivers and implications for management. *Science of the Total Environment* **145**: 187–195.

Van Dover, C. L. (2000) *The Ecology of Deep-Sea Hydrothermal Vents*. Princeton, NJ, USA: Princeton University Press.

Van Dover, C. L., German, C. R., Speer, K. G., Parson, L. M. & Vrijenhoek, R. C. (2002) Evolution and biogeography of deep-sea vent and seep invertebrates. *Science* **295**: 1253–1257.

Van Dover, C. L., Grassle, J. F., Fry, B., Garritt, R. H. & Staczak, V. R. (1992) Stable isotopic evidence for entry of sewage-derived organic material into a deep-sea food web. *Nature* **360**: 153–156.

Van Duzer, C. (2004) *Floating Islands: A Global Bibliography, with an Edition and Translation of G. C. Munz' Exercitatio academico de insulis natantibus (1711)*. Los Altos Hills, CA, USA: Cantor Press.

Van Lent, F., Nienhuis, P. H. & Verschuure, J. M. (1991) Production and biomass of the seagrass *Zostera noltii* Hornem. and *Cymodocea nodosa* (Ucria) Ascherson at the Banc d'Arguin (Mauritania, NW Africa): a preliminary approach. *Aquatic Botany* **41**: 353–367.

Van Noordwijk, J. M. & Peerbolte, E. B. (2000) Optimal sand nourishment decisions. *Journal of Waterway, Port, Coastal and Ocean Engineering* **126**: 30–38.

Van Soosten, C., Schmidt, H. & Westheide, W. (1998) Genetic variability and relationships among geographically widely separated populations of *Petitia amphophthalma* (Polychaeta: Syllidae): results from RAPD-PCR investigations. *Marine Biology* **131**: 659–669.

Van Wilgen, B. W., Richardson, D. M., Le Maitre, D. C., Marais, C. & Magadlela, D. (2001) The economic consequences of alien plant invasions: examples of impacts and approaches to sustainable management in South Africa. *Environment, Development and Sustainability* **3**: 145–168.

Vasquez, J. A. (1993) Abundance, distributional patterns and diets of main herbivorous and carnivorous species associated with *Lessonia trabeculata* kelp beds in northern Chile. *Serie Ocasional de Universidad Católica del Norte* **2**: 213–229.

Vasquez, J. A. & Buschmann, A. H. (1997) Herbivore–kelp interactions in Chilean subtidal communities: a review. *Revista Chilena de Historia Natural* **70**: 41–52.

Vasquez, J. A., Castilla, J. C. & Santelices, B. (1984) Distributional patterns and diets of four species of sea urchins in giant kelp forest (*Macrocystis pyrifera*) of Puerto Toro, Navarino Island, Chile. *Marine Ecology Progress Series* **19**: 55–63.

Vaughan, D. G., Marshall, G. J., Connolley, W. M., King, J. C. & Mulvaney, R. (2001) Devil in the detail. *Science* **293**: 1777–1779.

Velimirov, B., Field, F. G., Griffiths, C. L. & Zoutendyk, P. (1977) The ecology of kelp bed communities in the Benguela upwelling system. *Helgoländer Wissenschaftliche Meeresuntersuchungen* **30**: 495–518.

Venema, H. D., Schiller, E. J., Adamowski, K. & Thizy, J.-M. (1997) A water resources planning response to climate change in the Senegal River basin. *Journal of Environmental Management* **49**: 125–155.

Verberg, P., Hecky, R. & Kling, H. (2003) Ecological consequences of a century of warming in Lake Tanganyika. *Science* **301**: 505–507.

Verhoeven, J. T. A. & Bobbink, R. (2001) Plant diversity of fen landscapes in the Netherlands. In: *Biodiversity in Wetlands: Assessment, Function and Conservation*, ed. B. Gopal, pp. 65–87. Leiden, the Netherlands: Backhuys.

Verissimo, J. (1895) *A Pesca na Amazonia*. Rio de Janeiro, Brazil: Livraria Alves.

Verity, P. G. & Smetacek, V. (1996) Organism life cycles, predation, and the structure of marine pelagic ecosystems. *Marine Ecology Progress Series* **130**: 277–293.

Verity, P. G., Smetacek, V. & Smayda, T. J. (2002) Status, trends, and the future of the marine pelagic ecosystem. *Environmental Conservation* **29**: 207–237.

Vermaat, J. E., Beijer, J. A., Gijlstra, J., *et al.* (1993) Leaf dynamics and standing stocks of intertidal *Zostera noltii* Hornem. and *Cymodocea nodosa* (Ucria) Ascherson on the Banc d'Arguin (Mauritania). *Hydrobiologia* **258**: 59–72.

Vermeer, J. G. & Joosten, J. H. J. (1992) Conservation and management of bog and fen reserves in the Netherlands. In: *Fens and Bogs in the Netherlands: Vegetation, History, Nutrient Dynamics and Conservation*, ed. J. T. A. Verhoeven, pp. 433–478. Dordrecht, the Netherlands: Kluwer.

Vermeij, G. J. (1986) Survival during biotic crises: the properties and evolutionary significance of refuges. In: *Dynamics of Extinction*, ed. D. K. Elliott, pp. 231–246. New York, USA: John Wiley.

Veron, J. E. N., ed. (1995) *Corals in Space and Time: The Biogeography and Evolution of the Scleractinia*. Sydney, NSW, Australia: University of New South Wales Press.

Veron, J. E. N. & Minchin, P. R. (1992) Correlations between sea surface temperature, circulation patterns and the distribution of hermatypic corals of Japan. *Continental Shelf Research* **12**: 835–857.

Vetter, E. W. (1995) Detritus-based patches of high secondary production in the nearshore benthos. *Marine Ecology Progress Series* **120**: 251–262.

Vetter, E. W. & Dayton, P. K. (1998) Macrofaunal communities within and adjacent to a detritus-rich submarine canyon system. *Deep-Sea Research II* **45**: 25–54.

Vezina, A. R. & Platt, T. (1988) Food web dynamics in the ocean. I. Best estimates of flow networks using inverse methods. *Marine Ecology Progress Series* **42**: 269–287.

Viles, H. & Spencer, T. (1995) *Coastal Problems: Geomorphology, Ecology and Society of the Coast*. London, UK: Edward Arnold.

Vinnikov, K. Y., Robock, A., Stouffer, R. J., *et al.* (1999) Global warming and northern hemisphere sea ice extent. *Science* **286**: 1934–1937.

Vinueza, L., Branch, G. M., Branch, M. L. & Bustamante, R. H. (2006) Top-down herbivory and bottom-up El Niño effects on Galapagos rocky-shore communities. *Ecological Monographs* **76**: 111–131.

Vitousek, P. M., Aber, J. D., Howarth, R. W., *et al.* (1997*a*) Human alteration of the global nitrogen cycle: sources and consequences. *Ecological Applications* **7**: 737–750.

Vitousek, P. M., D'Antonio, C. M., Loope, L. L. & Westbrooks, R. (1996) Biological invasions as global environmental change. *American Scientist* **84**: 468–478.

Vitousek, P. M., Mooney, H. A., Lubchenco, J. & Melillo, J. M. (1997*b*) Human domination of earth's ecosystems. *Science* **277**: 494–499.

Vitt, D. H. & Chee, W. (1990) The relationships of vegetation to surface water chemistry and peat chemistry in fens of Alberta, Canada. *Vegetatio* **89**: 87–106.

Vitt, D. H., Halsey, L. A., Bauer, I. E. & Campbell, C. (2000) Spatial and temporal trends in carbon storage of peatlands of continental western Canada through the Holocene. *Canadian Journal of Earth Sciences* **37**: 683–693.

Vitt, D. H., Halsey, L. A., Wieder, K. & Turetsky, M. (2003) Response of *Sphagnum fuscum* to nitrogen deposition: a case study of ombrogenous peatlands in Alberta, Canada. *Bryologist* **106**: 235–245.

Vitt, D. H., Yenhung, L. & Belland, R. J. (1995) Patterns of bryophyte diversity in peatlands of continental western Canada. *Bryologist* **98**: 218–227.

Vollmer, M., Weiss, R. & Bootsma, H. (2002) Ventilation of Lake Malawi/Nyasa. In: *The East African Great Lakes: Limnology, Palaeolimnology, and Biodiversity*, ed. E. Odada & D. Olago, pp. 209–234. Dordrecht, the Netherlands: Kluwer.

Volonte, C. R. & Nicholls, R. J. (1995) Uruguay and sea-level rise: potential impacts and responses. *Journal of Coastal Research* Special Issue **14**: 262–285.

Vörösmarty, C. J., Askew, A., Barry, R., *et al.* (2001) Global water data: a newly endangered species. *Eos* **82**(5): 54, 56, 58.

Vörösmarty, C. J., Green, P., Salisbury, J. & Lammers, R. B. (2000) Global water resources: vulnerability from climate change and population growth. *Science* **289**: 284–288.

Wadhams, P. & Davis, N. R. (2000) Further evidence of ice thinning in the Arctic Ocean. *Geophysical Research Letters* 27: 3973–3975.

Wagner, C. S., Brahmakulam, I., Jackson, B., Wong, A. & Yoda, T. (2001) *Science and Technology Collaboration: Building Capacity in Developing Countries?*, Report prepared for the World Bank. Santa Monica, CA, USA: RAND Science & Technology. Available online at: www.rand.org/pubs/monograph_reports/MR1357.0/MR1357.0.pdf

Wagner, T. (1994) Remedial Action Plans: looking back and moving forward. *Focus on International Joint Commission Activities* 19: 16–17.

Wagner, W., Gawel, J., Furumai, H., *et al.* (2002) Sustainable watershed management: an international multi-watershed case study. *Ambio* 31: 2–13.

Wake, D. B. (1991) Declining amphibian populations. *Science* 253: 860.

Walden, D., van Dam, R., Finlayson, M., *et al.* (2004) A risk assessment of the tropical wetland weed *Mimosa pigra* in northern Australia. In: *Research and Management of Mimosa pigra*, ed. M. Julien, G. Flanagan, T. Heard, *et al.*, pp. 11–21. Canberra, ACT, Australia: CSIRO.

Walker, D. I. (1989) Regional studies: seagrass in Shark Bay, the foundations of an ecosystem. In: *Biology of Seagrasses: A Treatise on the Biology of Seagrasses with Special Reference to the Australian Region*, ed. A. W. D. Larkum, A. J. McComb & S. A. Shepherd, pp. 182–210. Amsterdam, the Netherlands: Elsevier.

Walker, D. I. (1991) The effect of sea temperature on seagrasses and algae on the Western Australian coastline. *Journal of the Royal Society of Western Australia* 74: 71–77.

Walker, D. I. (2003) The seagrasses of Western Australia. In: *World Atlas of Seagrasses*, ed. E. P. Green & F. T. Short, pp. 121–130. Berkeley, CA, USA: University of California Press.

Walker, D. I. & McComb, A. J. (1992) Seagrass degradation in Australian coastal waters. *Marine Pollution Bulletin* 25: 191–195.

Walker, D. I. & Woelkerling, W. J. (1988) A quantitative study of sediment contribution by epiphytic coralline red algae in seagrass meadows in Shark Bay, Western Australia. *Marine Ecology Progress Series* 43: 71–77.

Walker, D. I., Hillman, K. A., Kendrick, G. A. & Lavery, P. (2001) Ecological significance of seagrasses: assessment for management of environmental impact in Western Australia. *Ecological Engineering* 16: 323–330.

Walker, D. I., Kendrick, G. A. & McComb, A. J. (1988) The distribution of seagrasses in Shark Bay, Western Australia, with notes on their ecology. *Aquatic Botany* 30: 305–317.

Walker, D. I., Lukatelich, R. J., Bastyan, G. & McComb, A. J. (1989) Effect of boat moorings on seagrass beds around Perth, Western Australia. *Aquatic Botany* 36: 69–77.

Walker, J. & Peet, R. K. (1984) Composition and species-diversity of pine–wiregrass savannas of the Green Swamp, North Carolina. *Vegetatio* 55: 163–179.

Walker, K. F. & Thoms, M. C. (1993) Environmental effects of flow regulation on the lower River Murray, Australia. *Regulated Rivers: Research and Management* 8: 103–119.

Wallace, J. B. & Webster, J. R. (1996) The role of macroinvertebrates in stream ecosystem function. *Annual Review of Entomology* 41: 115–139.

Wallén, B. (1986) Above and below ground dry mass of the three main vascular plants on hummocks on a sub-arctic peat bog. *Oikos* 46: 51–56.

Wallén, B. (1992) Methods for studying below-ground production in mire ecosystems. *Suo* 43: 155–162.

Walters, B. B. (2000) Local mangrove planting in the Philippines: are fisherfolk and fishpond owners effective restorationists? *Restoration Ecology* 8: 237–246.

Walters, M. J. (1992) *A Shadow and a Song: The Struggle to Save an Endangered Species*. Post Mills, VT, USA: Chelsea Green Publishing.

Walther, G.-R., Post, E., Convey, P., *et al.* (2002) Ecological responses to recent climate change. *Nature* 416: 391–395.

Wang, C., Prinn, R. & Sokolov, A. (1998) A global interactive chemistry and climate model: formulation and testing. *Journal of Geophysical Research* 103: 3399–3417.

Wang, X. & Key, J. R. (2003) Recent trends in Arctic surface, cloud, and radiation properties from space. *Science* 299: 1725–1728.

Wania, F., Axelman, J. & Broman, D. (1998) A review of processes involved in the exchange of persistent organic pollutants across the air–sea interface. *Environmental Pollution* 102: 3–23.

Wantzen, K. M. (1998) Siltation effects on benthic communities in first order streams in Mato Grosso, Brazil. *Verhandlungen Internationale Vereinigung für Limnologie* 26: 1155–1159.

Ward, J. V., Tockner, K. & Schiemer, F. (1999) Biodiversity of floodplain river ecosystems: ecotones and connectivity. *Regulated Rivers: Research and Management* 15: 125–139.

Ward, J. V., Tockner, K., Arscott, D. B. & Claret, C. (2002) Riverine landscape diversity. *Freshwater Biology* 47: 517–539.

Wardle, D. (2002) *Communities and Ecosystems: Linking the Aboveground and Belowground Components*. Princeton, NJ, USA: Princeton University Press.

Wares, J. P., Goldwater, D. S., Kong, B. Y. & Cunningham, C. W. (2002) Refuting a controversial case of a human-mediated marine species introduction. *Ecology Letters* **5**: 577–584.

Warne, A. G., Toth, L. A. & White, W. A. (2000) Drainage-basin-scale geomorphic analysis to determine reference conditions for ecologic restoration: Kissimmee River, Florida. *Geographical Society of America Bulletin* **112**: 884–899.

Warren, R. S. & Niering, W. A. (1993) Vegetation change on a north east tidal marsh: interaction of sea-level rise and marsh accretion. *Ecology* **74**: 96–103.

Warrick, R. A., Le Provost, C., Meier, M. F., Oerlemans, J. & Woodworth, P. L. (1996) Changes in sea level. In: *Climate Change 1996: The Science of Climate Change*, ed. J. T. Houghton, L. G. Meira Filho, B. A. Callander, *et al.*, pp. 359–405. Cambridge, UK: Cambridge University Press.

Wassmann, P. (1998) Retention versus export food chains: processes controlling sinking loss from marine pelagic systems. *Hydrobiologia* **63**: 29–57.

Watkins, A. B. & Simmonds, J. (2000) Current trends in Antarctic sea ice: the 1990s impact on a short climatology. *Journal of Climate* **13**: 4441–4451.

Watling, L. & Norse, E. A. (1998) Disturbance of the seabed by mobile fishing gear: a comparison to forest clearcutting. *Conservation Biology* **12**: 1180–1197.

Watson, R. & Pauly, D. (2001) Systematic distortions in world fisheries catch trends. *Nature* **414**: 534–536.

Watson, J. G. (1928) *Mangrove Forests of the Malay Peninsula*, Malayan Forest Records No. 6. Singapore, Malaysia: Fraser & Neave.

WCD (2000) *Dams and Development: A New Framework for Decision-Making*. London, UK: Earthscan Publications for World Commission on Dams.

WCMC (1998) *Freshwater Biodiversity: a preliminary global assessment*. Biodiversity Series No. 8. By B. Groombridge & M. Jenkin. World Conservation Monitoring Centre, Cambridge, UK: World Conservation Press.

Weber, P. (1994) It comes down to the coasts. *World Watch* – March/April: 20–29.

Webster, M. D., Babiker, M., Mayer, M., *et al.* (2002) Uncertainty in emissions projections for climate models. *Atmospheric Environment* **36**: 3659–3670.

Webster, M. D., Forest, C., Reilly, J., *et al.* (2003) Uncertainty analysis of climate change and policy response. *Climatic Change* **61**: 295–320.

Wehrmeyer, W. & Mulugetta, Y., eds. (1999) *Growing Pains: Environmental Management in Developing Countries*. Sheffield, UK: Greenleaf Publishing.

Weier, J. (2000) Bright lights, big city [www document]. URL http://earthobservatory.nasa.gov/Study/Lights/

Weinstein, M. & Kreeger, D. A., eds. (2000) *Concepts and Controversies in Tidal Marsh Ecology*. Dordrecht, the Netherlands: Kluwer.

Weis, P., Weis, J. S. & Proctor, T. (1993) Copper, chromium, and arsenic in estuarine sediments adjacent to wood treated with chromated-copper-arsenate (CCA). *Estuarine, Coastal and Shelf Science* **36**: 71–79.

Welch, K. A., Lyons, W. B., McKnight, D. M., *et al.* (2003) Climate and hydrologic variations and implications for lake and stream ecological response in the McMurdo Dry Valleys, Antarctica. In: *Climate Variability and Ecological Response*, ed. D. Greenland, D. Goodin & R. C. Smith, pp. 174–195. Oxford, UK: Oxford University Press.

Welcomme, R. L. (1975) *The Fisheries Ecology of African Floodplains*, CIFA Technical Paper No. 3. Rome, Italy: Food and Agriculture Organization of the United Nations.

Welcomme, R. L. (1979) *Fisheries Ecology of Floodplain Rivers*. Harlow, UK: Longman.

Welcomme, R. L., ed. (1985) *River Fisheries*, Fisheries Technical Paper No. 262. Rome, Italy: Food and Agriculture Organization of the United Nations.

Welcomme, R. L. (2006) Providing for the water requirements of river fish and fisheries. *International Journal of Ecology and Environmental Sciences* **32**: 85–97.

Wellington, G. M. & Glynn, P. W. (2007) Coral reef responses to El Niño–Southern Oscillation sea warming events. In: *Geological Approaches to Coral Reef Ecology: Placing the Current Crisis in the Historical Context*, ed. R. Aronson, pp. 342–385. New York, USA: Springer-Verlag.

Wells, L. & McLain, A. (1973) *Lake Michigan: Man's Effects on Native Fish Stocks and Other Biota*, Technical Report No. 20. Ann Arbor, MI, USA: Great Lakes Fishery Commission.

Wenger, E. L., Zinke, A. & Gutzweiler, K.-A. (1990) Present situation of European floodplain forests. *Forest Ecology and Management* **33/34**: 5–12.

West, J. M. & Salm, R. V. (2003) Resistance and resilience to coral bleaching: implications for coral reef conservation and management. *Conservation Biology* **17**: 956–967.

West, J. M. & Zedler, J. B. (2000) Marsh–creek connectivity: fish use of a tidal salt marsh in southern California. *Estuaries* **23**: 699–710.

Western Australia, Government of (2002) *Policy for the Implementation of Ecologically Sustainable Development for Fisheries and Aquaculture within Western Australia*, Report No. 157. Perth, WA, Australia: Department of Fisheries.

Weston, D. P. (1990) Quantitative examination of macrobenthic community changes along an organic and organic-enrichment gradient. *Marine Ecology Progress Series* **61**: 233–247.

Wetlands International (2002) A Directory of Wetlands of International Importance, 7th edn [www document]. URL http://www.wetlands.org/RDB/Directory.html

Wheatcroft, R. A. (2000) Oceanic flood sedimentation: a new perspective. *Continental Shelf Research* **20**: 2059–2066.

Wheatcroft, R. A., Borgeld, J. C., Born, R. S., *et al.* (1996) The anatomy of an oceanic flood deposit. *Oceanography* **9**: 158–162.

Wheeler, B. D. (1984) British fens: a review. In: *European Mires*, ed. P. Moore, pp. 237–282. New York, USA: Academic Press.

Wheeler, B. D. & Proctor, M. C. F. (2000) Ecological gradients, subdivisions and terminology of north-west European mires. *Journal of Ecology* **88**: 187–203.

Wheeler, B. D., Shaw, S. C., Fojt, W. J. & Robertson, R. A. (1995) *Restoration of Temperate Wetlands*. New York, USA: John Wiley.

White, W. B. & Peterson, R. G. (1996) An Antarctic circumpolar wave in surface pressure, wind, temperature and sea-ice extent. *Nature* **380**: 699–702.

White, W. B., Chen, S. C. & Peterson, R. G. (1998) The Antarctic circumpolar wave: a beta effect in ocean–atmosphere coupling over the Southern Ocean. *Journal of Physical Oceanography* **28**: 2345–2361.

Whitehead, P. J. & Blomquist, G. O. (1991) Measuring contingent values for wetlands: effects of information about related environmental goods. *Water Resources Research* **27**: 2523–2531.

Whitfield, P. E., Kenworthy, W. J., Hammerstrom, K. K. & Fonseca, M. S. (2002) The role of a hurricane in the expansion of disturbances initiated by motor vessels on seagrass banks. *Journal of Coastal Research* Special Issue **37**: 85–98.

Whitlock, J. E., Jones, D. T. & Harwood, V. J. (2002) Identification of the sources of fecal coliforms in an urban watershed using antibiotic resistance analysis. *Water Research* **36**: 4273–4282.

Whittaker, R. H. (1975) *Communities and Ecosystems*, 2nd edn. New York, USA: Macmillan.

WHO (1996) Water and Sanitation, Fact Sheet No. 112, World Health Organization [www document]. URL http://www.who.int/inffs/en/fact112html

Wickens, P. A. & Field, J. G. (1986) The effects of water transport on nitrogen flow through a kelp-bed community. *South African Journal of Marine Sciences* **4**: 79–92.

Wiegers, J. (1990) Forested wetlands in western Europe. In: *Ecosystems of the World*, vol. 15, *Forested Wetlands*, ed. A. E. Lugo, M. M. Brinson & S. Brown, pp. 407–436. Amsterdam, the Netherlands: Elsevier.

Wiggins, B. A., Andrews, R. W., Conway, R. A., *et al.* (1999) Use of antibiotic resistance analysis to identify nonpoint sources of fecal pollution. *Applied and Environmental Microbiology* **65**: 3483–3486.

Wigham, B. D., Tyler, P. A. & Billett, D. S. M. (2003) Reproductive biology of the abyssal holothurian *Amperima rosea*: an opportunistic response to variable flux of surface derived organic matter? *Journal of the Marine Biological Association of the United Kingdom* **83**: 175–188.

Wigley, T. M. L. (1989) Measurement and prediction of global warming. In: *Ozone Depletion: Health and Environmental Consequences*, ed. R. R. Jones & T. Wigley, pp. 85–97. Chichester, UK: John Wiley.

Wigley, T. M. L. (1995) Global-mean temperature and sea level consequences of greenhouse gas stabilization. *Geophysical Research Letters* **22**: 45–48.

Wilcock, H. R., Hildrew, A. G. & Nichols, R. A. (2001) Genetic differentiation of a European caddisfly: past and present gene flow among fragmented larval habitats. *Molecular Ecology* **10**: 1821–1834.

Wilcock, H. R., Nichols, R. A. & Hildrew, A. G. (2003) Genetic population structure and neighbourhood population size estimates of the caddisfly *Plectrocnemia conspersa*. *Freshwater Biology* **48**: 1813–1824.

Wilcove, D. S., Rothstein, D., Dubow, J., Phillips, A. & Losos, E. (1998) Quantifying threats to imperiled species in the United States. *BioScience* **48**: 607–615.

Wild, S. R. & Jones, K. C. (1995) Polynuclear aromatic hydrocarbons in the United Kingdom environment: a preliminary source inventory and budget. *Environmental Pollution* **88**: 91–108.

Wildlife Institute of India (2003) Homepage [www document]. URL http://www.wii.gov.in/

Wilkens, H., Culver, D. C. & Humphreys, W. F., eds. (2000) *Ecosystems of the World*, vol. 30, *Subterranean Ecosystems*. Amsterdam, the Netherlands: Elsevier.

Wilkie, M. J. & Fortuna, S. (2003) *Status and Trends in Mangrove Area Extent Worldwide*, Forest Resources Assessment Working Paper No. 63. Rome, Italy: Food and Agriculture Organization of the United Nations.

Wilkinson, C. (1996) Global change and coral reefs: impacts on reefs, economies and human cultures. *Global Change Biology* **2**: 547–558.

Wilkinson, C. R. & Sammarco, P. W. (1983) Effects of fish grazing and damselfish territoriality on coral reef algae. II. Nitrogen fixation. *Marine Ecology Progress Series* 13: 15–19.

William, P. B. & Orr, M. K. (2002) Physical evolution of restored breached levee salt marshes in the San Francisco Bay estuary. *Restoration Ecology* 10: 527–542.

Williams, D. M. (2001) Impacts of terrestrial run-off on the Great Barrier Reef World Heritage Area. Unpublished report. Townsville, Queensland, Australia: CRC Reef Research Centre.

Williams, G. D. & Zedler, J. B. (1999) Fish assemblage composition in constructed and natural tidal marshes of San Diego Bay: relative influence of channel morphology and restoration history. *Estuaries* 22: 702–716.

Williams, I. D. & Polunin, N. V. C. (2001) Large-scale associations between macroalgal cover and grazer biomass on mid-depth reefs in the Caribbean. *Coral Reefs* 19: 358–366.

Williams, I. D., Polunin, V. C. & Hendrick, V. J. (2001) Limits to grazing by herbivorous fishes and the impact of low coral cover on macroalgal abundance on a coral reef in Belize. *Marine Ecology Progress Series* 222: 187–196.

Williams, M., ed. (1990) *Wetlands: A Threatened Landscape.* Oxford, UK: Blackwell Scientific Publications.

Williams, W. D. (1981) Problems in the management of inland saline lakes. *Verhandlungen Internationale Vereinigung für Limnologie* 21: 688–692.

Williams, W. D. (1993a) Conservation of salt lakes. *Hydrobiologia* 267: 291–306.

Williams, W. D. (1993b) The worldwide occurrence and limnological significance of falling water-levels in large, permanent saline lakes. *Verhandlungen Internationale Vereinigung für Limnologie* 25: 980–983.

Williams, W. D. (1995) Lake Corangamite, Australia, a permanent saline lake: conservation and management issues. *Lakes and Reservoirs: Research and Management* 1: 55–64.

Williams, W. D. (1996) The largest, highest and lowest lakes of the world: saline lakes. *Verhandlungen Internationale Vereinigung für Limnologie* 26: 61–79.

Williams, W. D. (1998a) *Management of Inland Saline Waters,* United Nations Environment Programme/International Lake Environment Committee Guidelines of Lake Management Series No. 6. Kusatsu, Japan: ILEC/UNEP.

Williams, W. D. (1998b) Salinity as a determinant of the structure of biological communities in salt lakes. *Hydrobiologia* 381: 191–201.

Williams, W. D. (1998c) Dryland wetlands. In: *Wetlands for the Future,* ed. A. J. McComb & J. A. Davis, pp. 33–47. Adelaide, SA, Australia: Gleneagles Publishing.

Williams, W. D. (1999) Salinization: a major threat to water resources in the arid and semi-arid regions of the world. *Lakes and Reservoirs: Research and Management* 4: 85–91.

Williams, W. D. (2000) Biodiversity in temporary wetlands of dryland regions. *Verhandlungen Internationale Vereinigung für Limnologie* 27: 141–144.

Williams, W. D. (2001) Salinization: unplumbed salt in a parched landscape. *Water Science and Technology* 43(4): 85–91.

Williams, W. D. (2002a) Environmental threats to salt lakes and the likely status of inland saline ecosystems in 2025. *Environmental Conservation* 29: 154–167.

Williams, W. D. (2002b) Anthropogenic salinization of inland waters. *Hydrobiologia* 466: 329–337.

Williams, W. D. & Aladin, N. V. (1991) The Aral Sea: recent limnological changes and their conservation significance. *Aquatic Conservation: Marine and Freshwater Ecosystems* 1: 3–23.

Williamson, M. (1999) Invasions. *Ecography* 22: 5–12.

Williamson, M. & Fitter, A. (1996) The varying success of invaders. *Ecology* 77: 1661–1666.

Willott, E. (2004) Restoring Nature, without mosquitoes? *Restoration Ecology* 12: 147–153.

Willrich, T. L. & Smith, G. E., ed. (1970) *Agricultural Practices and Water Quality.* Ames, IA, USA: Iowa State University Press.

Wilsey, B. J. & Potvin, C. (2000) Biodiversity and ecosystem functioning: the importance of species evenness in an old field. *Ecology* 81: 887–892.

Wilson, R. R. & Kaufmann, R. S. (1987) Seamount biota and biogeography. In: *Seamounts, Islands and Atolls,* ed. B. H. Keating, P. Fryer, R. Batiza & G. W. Boehlert, pp. 355–377. Washington, DC, USA: American Geophysical Union.

Wilson, S. K., Graham, N. A. J., Pratchett, M. S., Jones, G. P. & Polunin, N. V. C. (2006) Multiple disturbances and the global degradation of coral reefs: are reef fishes at risk or resilient? *Global Change Biology* 12: 1–15.

Wilson, W. D., Fatemi, S. M. R., Shokri, M. R. & Claereboudt, M. R. (2002) Status of coral reefs of Persian/Arabian Gulf and Arabian Sea Region. In: *Status of Coral Reefs of the World: 2002,* ed. C. Wilkinson, pp. 53–62. Townsville, Queensland, Australia: Australian Institute of Marine Science.

Wiltshire, J. (2001) Future prospects for the marine minerals industry. *Underwater* 13: 40–44.

Windom, H. L. (1992) Contamination of the marine environment from land-based sources. *Marine Pollution Bulletin* 25: 32–36.

Winger, P. V. (2002) Toxicological assessment of aquatic ecosystems: application to watercraft contaminants in shallow water environments. *Journal of Coastal Research* Special Issue **37**: 178–190.

Winn, P. J. S., Young, R. M. & Edwards, A. M. C. (2003) Planning for the rising tides: the Humber Estuary Shoreline Management Plan. *Science of the Total Environment* **314–316**: 13–30.

Winsor, P. (2001) Arctic sea ice thickness remained constant during the 1990s. *Geophysical Research Letters* **28**: 1039–1041.

Winter, T. C. (1988) A conceptual framework for assessing cumulative impacts on hydrology of nontidal wetlands. *Environmental Management* **12**: 605–620.

Winter, T. C. (1992) A physiographic and climatic framework for hydrologic studies of wetlands. In: *Aquatic Ecosystems in Semi-Arid Regions: Implications for Resource Management*, ed. R. D. Robarts & M. L. Bothwell, pp. 127–148. Saskatoon, Alberta, Canada: Environment Canada.

Winter, T. C. (1999) Relation of streams, lakes, and wetlands to groundwater flow systems. *Hydrogeology Journal* **7**: 28–45.

Winter, T. C. (2001) The concept of hydrological landscapes. *Journal of the American Water Resources Association* **37**: 335–349.

Wiseman, R., Taylor, D. & Zingstra, H., ed. (2003) Wetlands and agriculture, Proceedings of the Workshop on Agriculture, Wetlands and Water Resources, 17th Global Biodiversity Forum, Valencia, Spain, November 2002. *International Journal of Ecology and Environmental Sciences* Special Issue **29**: 1–122.

Wishner, K., Levin, L., Gowing, M. & Mullineaux, L. (1990) Involvement of the oxygen minimum in benthic zonation on a deep seamount. *Nature* **346**: 57–59.

Wittmann, F. (2002) Species distribution, community structure, and adaptations of Amazonian Várzea forests depending on annual flood-stress, using remote sensing methods. Unpublished Ph.D. thesis, University of Mannheim, Germany.

Woinarski, J. C. Z., Tidemann, S. C. & Kerin, S. (1988) Birds in a tropical mosaic: the distribution of bird species in relation to vegetation patterns. *Australian Wildlife Research* **15**: 171–196.

Wolanski, E. (1989) Measurements and modelling of the water circulation in mangrove swamps. Unpublished report, RAF/87/038, Série Documentaire No. 3, pp. 1–43, COMARF Regional Project for Research and Training in Coastal Marine Systems in Africa. Paris, France: UNESCO and Coastal Marine Project in Africa.

Wolanski, E., Mazda, Y. & Ridd, P. (1992) Mangrove hydrodynamics. In: *Tropical Mangrove Ecosystems*, ed. A. I. Robertson & D. M. Alongi, pp. 43–62. Washington, DC, USA: American Geophysical Union.

Wolff, W. J. (1997) Case studies and reviews: development of the conservation of Dutch coastal waters. *Aquatic Conservation: Marine and Freshwater Ecosystems* **7**: 165–177.

Wolff, W. J. (2000) The south-eastern North Sea: losses of vertebrate fauna during the past 2000 years. *Biological Conservation* **95**: 209–217.

Wong, Y. S., Lan, C. Y., Chen, G. Z., et al. (1995) Effect of wastewater discharge on nutrient contamination of mangrove soils and plants. *Hydrobiologia* **295**: 243–354.

Wong, Y. S., Tam, N. F. Y. & Lan, C. Y. (1997) Mangrove wetlands as wastewater treatment facility: a field trial. *Hydrobiologia* **352**: 49–59.

Wood, C. M. & McDonald, D. G., eds. (1997) *Global Warming: Implications for Freshwater and Marine Fish.* Cambridge, UK: Cambridge University Press.

Wood, P. J. & Petts, G. E. (1999) The influence of drought on chalk stream macroinvertebrates. *Hydrological Processes* **13**: 387–399.

Woodroffe, C. D. (1987) Pacific island mangroves: distribution and environmental settings. *Pacific Science* **41**: 166–185.

Woodroffe, C. D. (1990) The impact of sea-level rise on mangrove shorelines. *Progress in Physical Geography* **14**: 483–520.

Woodroffe, C. D. (1992) Mangrove sediments and geomorphology. In: *Tropical Mangrove Ecosystems*, ed. A. I. Robertson & D. M. Alongi, pp. 7–41. Washington, DC, USA: American Geophysical Union.

Woodroffe, C. D., Thom, B. G. & Chappell, J. (1985) Development of widespread mangrove swamps in mid-Holocene times in northern Australia. *Nature* **317**: 711–713.

Woodwell, G. M., Craig, P. P. & Johnson, H. A. (1971) DDT in the biosphere: where does it go? *Science* **174**: 1101–1107.

Wooster, M. (2000) Niagara River progress (?). *Great Lakes News* **15**: 17–18.

World Bank (2002) *Bangladesh: Climate Change and Sustainable Development*, Report No. 21104-BD. Dhaka, Bangladesh: The World Bank.

World Commission on Environment and Development (1987) *Our Common Future: The Brundtland Report.* Oxford, UK: Oxford University Press.

Worm, B. & Myers, R. A. (2003) Meta-analysis of cod–shrimp interactions reveals top-down control in oceanic food webs. *Ecology* **84**: 162–173.

Worm, B., Lotze, H. K., Bostrom, C., *et al.* (1999) Marine diversity shift linked to interactions among grazers, nutrients and propagule banks. *Marine Ecology Progress Series* **185**: 309–314.

Worthington, S. & Worthington, E. B. (1933) *Inland Waters of Africa*. London, UK: Macmillan.

Wren, C. D., Harris, S. & Harttrup, N. (1995) Ecotoxicology of mercury and cadmium. In: *Handbook of Ecotoxicology*, ed. D. J. Hoffman, B. A. Rattner, G. A. Burton Jr & J. Cairns Jr, pp. 392–423. Boca Raton, FL, USA: Lewis Publishers.

WRI (1998) *World Resources 1998–1999*. New York, USA: Oxford University Press.

Wright, L. D. & Short, A. D. (1984) Morphodynamic variability of surf zones and beaches: a synthesis. *Marine Geology* **58**: 93–118.

Wright, S. (1955) *Limnological Survey of Western Lake Erie*, Special Scientific Report: Fisheries No. 139. Washington, DC, USA: US Fish and Wildlife Service.

WTO (2003) *World Trade Organization: Doha Declarations*. Geneva, Switzerland: World Trade Organization.

Wu, J. F., Sunda, W., Boyle, E. A. & Karl, D. M. (2000) Phosphate depletion in the western North Atlantic Ocean. *Science* **289**: 759–762.

Wu, M. Y., Hacker, S., Ayres, D. & Strong, D. A. (1999) Potential of *Prokelesia* spp. as biological control agents of English cordgrass, *Spartina anglica*. *Biological Control* **16**: 267–273.

Wu, R. S. S. (1999) Eutrophication, water borne pathogens and xenobiotic compounds: environmental risks and challenges. *Marine Pollution Bulletin* **39**: 11–22.

Wubben, D. L. (2000) UV-induced mortality of zoea I larvae of the brown shrimp *Crangon crangon* (Linnaeus 1758). *Journal of Plankton Research* **22**: 2095–2104.

Wurtsbaugh, W. A. & Gliwicz, Z. M. (2001) Limnological control of brine shrimp population dynamics and cyst production in the Great Salt Lake, Utah. *Hydrobiologia* **466**: 119–132.

WWC (2000) World Water Vision: Making Water Everybody's Business. Commission report, World Water Council [www document]. URL http//:www.watervision.org

WWC (2002) UN consecrates water as public good, human right [www document]. URL http://www.worldwatercouncil.org/download/UM_water_publicgood.pdf

WWC (2003*a*) Water management [www document]. URL http://www.worldwatercouncil.org/wwa_contents_fr.shtml

WWC (2003*b*) *World Water Council 3rd World Water Forum*. Marseille, France: World Water Council.

WWF (1993) Living rivers. Unpublished report. Zeist, the Netherlands: World Wide Fund for Nature.

WWF (1999) *Living Planet Report*, ed. J. Loh, J. Randers, A. MacGillivray, *et al.* Gland, Switzerland: World Wide Fund for Nature.

WWF (2006) *Living Planet Report 2006*, ed. J. Loh & S. Goldfinger. Gland, Switzerland: World Wide Fund for Nature.

WWF & EU (2001) Implementing the EU Water Freshwater Directive: Seminar 2 – The role of wetlands in river basin management. Unpublished report, 9–10 November 2000. Brussels, Belgium: European Union.

Wyatt, G. (2003) Proposed introduction of *Atriplex pedunculata* (annual sea purslane) to a tidally-influenced site on the Essex coast. *BSBI News* **92**: 19–22.

Wyllie-Echeverria, S. & Ackerman, J. D. (2003) The seagrasses of the Pacific Coast of North America. In: *World Atlas of Seagrasses*, ed. E. P. Green & F. T. Short, pp. 217–224. Berkeley, CA, USA: University of California Press.

Wyllie-Echeverria, S., Olson, A. M., Hershman, M. J., eds. (1994) *Seagrass Science and Policy in the Pacific NorthWest: Proceedings of a Seminar Series*, (SMA 94-1) EPA Report No. 910/R-94-004. Washington, DC, USA: US Environmental Protection Agency.

Wyman, R. L. (1990) What's happening to the amphibians? *Conservation Biology* **4**: 350–353.

Wynberg, R. P. & Branch, G. M. (1997) Trampling associated with bait-collection for sandprawns *Callianassa kraussi* Stebbing: effects on the biota of an intertidal sandflat. *Environmental Conservation* **24**: 139–148.

Xiao, X., Melillo, J. M., Kicklighter, D. W., *et al.* (1998) Transient climate change and net ecosystem production of the terrestrial biosphere. *Global Biogeochemical Cycles* **12**: 345–360.

Xu, C. Y. (1999) From GCMs to river flow: a review of downscaling methods and hydrological modelling approaches. *Progress in Physical Geography* **23**: 229–249.

Yang, H. & Zehnder, A. J. B. (2001) China's regional water scarcity and implications for grain supply and trade. *Environment and Planning A* **33**: 79–95.

Yang, H. & Zehnder, A. J. B. (2002) Water scarcity and food import: a case study for southern Mediterranean countries. *World Development* **30**: 1413–1430.

Yap, H. T. (1992) Marine environmental problems: experiences of developing regions. *Marine Pollution Bulletin* **25**: 37–40.

Yarnell, B. (1998) Integrated regional assessment and climate change impacts in river basins. *Climate Research* **11**: 65–74.

Yendo, K. (1902) Kaiso Isoyake Chosa Hokoku. *Suisan Chosa Hokoku* **12**: 1–33 (in Japanese).

Yendo, K. (1903) Investigations on 'Isoyake' (decrease of seaweed). *Journal of the Imperial Fisheries Bureau* **12**: 1–33 (in Japanese).

Yendo, K. (1914) On the cultivation of seaweeds, with special accounts of their ecology. *Economic Proceedings of the Royal Dublin Society* **2**: 105–122.

Yentsch, C. S., Yentsch, C. M., Cullen, J. J., *et al.* (2002) Sunlight and water transparency: cornerstones in coral research. *Journal of Experimental Marine Biology and Ecology* **268**: 171–183.

Yin, K. D., Harrison, P. J., Chen, J., Huang, W. & Qian, P. Y. (1999) Red tides during spring 1998 in Hong Kong: is El Niño responsible? *Marine Ecology Progress Series* **187**: 289–294.

Yon, D. & Tendron, G. (1981) *Alluvial Forests of Europe*. Brussels, Belgium: European Committee for the Conservation of Nature and National Resources.

Yuan, X. & Martinson, D. G. (2000) Antarctic sea ice extent variability and its global connectivity. *Journal of Climate* **13**: 1697–1717.

Yuan, X. & Martinson, D. G. (2001) The Antarctic dipole and its predictability. *Geophysical Research Letters* **28**: 3609–3612.

Zachos, J., Pagani, M., Sloan, L., Thomas, E. & Billups, K. (2001) Trends, rhythms, and aberrations in global climate 65 Ma to present. *Science* **292**: 686–693.

Zahir, H. (2002) Status of coral reefs of Maldives. In: *Coral Reef Degradation in the Indian Ocean: Status Report 2002*, ed. O. Linden, D. Souter, D. Wilhelmsson & D. Obura, pp. 119–124. Kalmar, Sweden: CORDIO.

Zedler, J. B. (1993) Canopy architecture of natural and planted cordgrass marshes: selecting habitat evaluation criteria. *Ecological Applications* **3**: 123–138.

Zedler, J. B. (1996) Costal mitigation in southern California: the need for a regional restoration strategy. *Ecological Applications* **6**: 84–93.

Zedler, J. B. & Callaway, J. C. (1999) Tracking wetland restoration: do mitigation sites follow desired trajectories? *Restoration Ecology* **7**: 69–73.

Zedler, J. B., Callaway, J. C., Desmond, J., *et al.* (1999) Californian salt marsh vegetation: an improved model of spatial pattern. *Ecosystems* **2**: 19–35.

Zedler, J. B., Callaway, J. C. & Sullivan, G. (2001) Declining biodiversity: why species matter and how their functions might be restored. *BioScience* **51**: 1005–1017.

Zedler, J. B., Nordby, C. S. & Kus, B. E. (1992) *The Ecology of Tijuana Estuary: A National Estuarine Research Reserve*. Washington, DC, USA: NOAA Office of Coastal Resource Management, Sanctuaries and Reserves Division.

Zedler, J. B., Paling, E. & McComb, A. (1990) Differential salinity responses to help explain the replacement of native *Juncus kraussii* by *Typha orientalis* in Western Australian salt marshes. *Australian Journal of Ecology* **15**: 57–72.

Zedler, P. H. (1987) *The Ecology of Southern California Vernal Pools: A Community Profile*, Biological Report No. 85 (7.11). Washington, DC, USA: US Fish and Wildlife Service.

Zedler, P. H. (2003) Vernal pools and the concept of 'isolated wetlands'. *Wetlands* **23**: 597–607.

Zehnder, A. J. B., Yang, H. & Schertenleib, R. (2003) Water issues: the need for action at different levels. *Aquatic Sciences: Research across Boundaries* **65**: 1–20.

Zenkevitch, L. (1963) *Biology of the Seas of the USSR*. London, UK: Allen & Unwin.

Zepp, R. G., Callaghan, T. V. & Erickson, D. J. (2003) Interactive effects of ozone depletion and climate change on biogeochemical cycles. *Photochemical and Photobiological Sciences* **2**: 51–61.

Zhang, J. D., Rothrock, D. A. & Steele, M. (2000) Recent changes in Arctic sea ice: the interplay between ice dynamics and thermodynamics. *Journal of the American Meteorological Society* **13**: 3099–3114.

Zieman, J. C. (1976) The ecological effects of physical damage from motor boats on turtle beds in Southern Florida. *Aquatic Botany* **2**: 127–139.

Zimmerman, R. C., Kohrs, D. G., Steller, D. L. & Alberte, R. S. (1997) Impacts of CO_2 enrichment on productivity and light requirements of eelgrass. *Plant Physiology* **115**: 599–607.

Zinke, A. & Gutzweiler, K.-A. (1990) Possibilities for regeneration of floodplain forests within the framework on the flood-protection measures on the Upper Rhine, West Germany. *Forest Ecology and Management* **33/34**: 13–20.

Zobell, C. E. (1971) Drift seaweeds on San Diego county beaches. *Nova Hedwigia* **32**: 269–314.

Zoltai, S. C. (1988) Wetland environment and classification. In: *Wetlands of Canada*, ed. National Wetlands

Working Group, pp. 3–26. Montreal, Quebec, Canada: Polyscience.

Zong, Y. & Chen, X. (2000) The 1998 flood on the Yangtze, China. *Natural Hazards* **22**: 165–184.

Zwally, H. J., Comiso, J. C., Parkinson, C. L., Cavalieri, D. J. & Gloersen, P. (2002) Variability of Antarctic sea ice 1979–1998. *Journal of Geophysical Research* **107**(C5): 10, doi: 1029/2000JC000733.

Zwally, H. J., Parkinson, C. L. & Comiso, J. C. (1983) Variability of Antarctic sea ice and changes in carbon dioxide. *Science* **220**: 1005–1012.

Zwiers, F. W. & Kharin, V. V. (1998) Changes in the extremes of the climate simulated by CCC GCM2 under CO_2-doubling. *Journal of Climate* **11**: 2200–2222.

Zwiers, F. W. (2002) The 20-year forecast. *Nature* **416**: 690–691.

Index